Eukaryotic Microbes

Eukaryotic Microbes

Edited by

Moselio Schaechter

AMSTERDAM • BOSTON • HEIDELBERG • LONDON • NEW YORK • OXFORD
PARIS • SAN DIEGO • SAN FRANCISCO • SINGAPORE • SYDNEY • TOKYO
Academic Press is an imprint of Elsevier

Academic Press is an imprint of Elsevier
525 B Street, Suite 1900, San Diego, California 92101–4495, USA
225 Wyman Street, Waltham, MA 02451, USA
Radarweg 29, PO Box 211, 1000 AE Amsterdam, The Netherlands
The Boulevard, Langford Lane, Kidlington, Oxford, OX51GB, UK

Material in the work originally appeared in *Encyclopedia of Microbiology* (Elsevier Inc, 2009), *Current Opinion in Microbiology, Volume 13, Issue 4* (Elsevier Ltd 2010), *Trends in Microbiology Volume 18, Issue 5* (Elsevier Ltd 2010), *Trends in Biotechnology Volume 27, Issue 2* (Elsevier Ltd 2008)

Front cover image: An Early Oligocene coccosphere of *Reticulofenestra umbilica* from the Falkland Plateau. Reprinted from Jordan, R.W. (2009) Coccolithophores. In: *Encyclopedia of Microbiology*. (Schaechter, M., ed.) Third Edition. Vol. 5: 593-605. Oxford, Elsevier. Academic Press., with permission from Elsevier.

Library of Congress Cataloging-in-Publication Data

Eukaryotic microbes / edited by Moselio Schaechter.
p. ; cm.
Abridgement of: Encyclopedia of microbiology. 3rd ed. / editor-in-chief, Moselio Schaechter. c2009.
Includes bibliographical references and index.
ISBN 978-0-12-810366-1 (alk. paper)
1. Microbiology. I. Schaechter, Moselio. II. Encyclopedia of microbiology.
[DNLM: 1. Eukaryota. QX 50]
QR9.E85 2012
579–dc23 2011014726

British Library Cataloguing-in-Publication Data

A catalogue record for this book is available from the British Library.

ISBN: 978-0-12-810366-1

For information on all Academic Press publications
visit our Web site at www.books.elsevier.com

Printed and bound in U.S.A

11 12 13 9 8 7 6 5 4 3 2 1

Contents

Contributors xiii
Preface xv

Part I
Fungi

1. Yeasts

G.M. Walker

Defining Statement 3
Definition and Classification of Yeasts 3
Definition and Characterization of Yeasts 3
Yeast Taxonomy 3
Yeast Biodiversity 4
Yeast Ecology 4
Natural Habitats of Yeast Communities 4
Yeasts in the Food Chain 4
Microbial Ecology of Yeasts 5
Yeast Cell Structure 5
General Cellular Characteristics 5
Methods in Yeast Cytology 5
Subcellular Yeast Architecture and Function 6
Nutrition, Metabolism, and Growth of Yeasts 6
Nutritional and Physical Requirements for Yeast Growth 6
Carbon Metabolism by Yeasts 8
Nitrogen Metabolism by Yeasts 9
Yeast Growth 10
Yeast Genetics 12
Life Cycle of Yeasts 12
Genetic Manipulation of Yeasts 12
Yeast Genome and Proteome Projects 13
Industrial, Agricultural, and Medical Importance of Yeasts 14
Industrial Significance of Yeasts 14
Yeasts of Environmental and Agricultural Significance 16
Medical Significance of Yeasts 16
Overview of Recent Progress in Yeast Science and Technology 16
Further Reading 17

2. Aspergillus

C. Scazzocchio

Defining Statement 19
Note 19
What is *Aspergillus*? 19
Food Contamination by Aspergilli 21
***Aspergillus* as Pathogens** 21
Aspergillus as Human Pathogens 21
Aspergillus in Veterinary Medicine 24
A. sydowii, a Specific Pathogen for Gorgonian Corals 24
Useful Aspergilli 25
Oriental Food Uses of *Aspergillus* 25
Extracellular Enzymes Produced by Aspergilli: Aspergilli as Hosts for Recombinant Proteins 25
Aspergillus and Production of Organic Acids 26
Medically Useful Secondary Metabolites 26
***A. (Emericella) nidulans* as a Model Organism** 26
The *A. nidulans* Genetic System 26
The Mitochondrial DNA of *A. nidulans* 27
***A. nidulans* as a Model for Genetic Metabolic Diseases** 28
Control of Gene Expression 29
A. nidulans as a Model for Cell Biology 33
A. nidulans Developmental Pathways 35
The Genus *Aspergillus* in the Genomic Era 38
Further Reading 39

3. Clavicipitaceae

M.S. Torres and J.F. White

Defining Statement 41
Introduction 41
Free-Living Saprotrophs and Insect Parasites 41
Species on Soft-Bodied Scale Insects 43
Plant Biotrophs 43
Chemical Diversity in the Clavicipitaceae 45
Biological Activities of Secondary Metabolites 46
Economic, Agricultural, and Ecological Importance of Clavicipitacious Endophytes 47
Origin and Evolution of Ergot Alkaloids 47

The Proposal of Functionality for Ergot Alkaloids 47
The Evidence for Defensive Mutualism in the Clavicipitaceae Endophyte/Grass Association 47
The Defensive Mutualism Controversy: A Debate Regarding Function 48
Evolution of Defensive Mutualism from Self-Defense to Host Defense 48
Conclusion 48
Acknowledgment 49
Further Reading 49

4. Endophytic Microbes

M. Tadych and J.F. White

Defining Statement 51
Introduction 51
Fungal Endophytes 52
Some Common Endophyte–Grass Associations 55
Tall Fescue–*Neotyphodium coenophialum* Association 55
Perennial Ryegrass–*Neotyphodium lolii* Association 55
Darnel Ryegrass–*Neotyphodium occultans* Association 56
The Sleepygrass–*Neotyphodium chisosum* Association 56
The ‘Drunken Horse Grass’–*Neotyphodium gansuense* Association 56
The ‘Dronkgras’–*Neotyphodium melicicola* Association 56
Forest Hedgehog Grass–*Neotyphodium* Association 57
‘Hueću’ Grasses–*Neotyphodium tembladerae* Association 57
Ecological Impacts of Endophytes 57
Production of Nontoxic Endophytes 58
Endophytes as Sources of Bioactive Metabolites 58
Ergot Alkaloids 59
Indole-Diterpene Metabolites 59
Peramine 60
Lolines 60
Recent Developments 61
Further Classification of Fungal Endophytes 62
New Endophytic Associations 62
Novel Mechanism of Hyphal Growth 63
Defensive Mutualism and Mechanisms of Stress Tolerance in Hosts 63
Bioactive Compounds 63
New Characterization Methods in the Study of Endophytes 63
Conclusion 64
Acknowledgment 64
Further Reading 64

5. Microsporidia: A Model for Minimal Parasite–Host Interactions

Catherine Texier, Cyril Vidau, Bernard Viguès, Hicham El Alaoui and Frédéric Delbac

Introduction 65
The Polar Tube and the Spore Wall of Microsporidian Spores are Involved in Host Cell Invasion 66
The Reduced and Compact Microsporidian Genomes Reflect Host Dependency 66
Host Immune Response: From Protective to Overwhelmed Defences 68
Conclusion and Future Directions 70
References and Recommended Reading 70

6. Mycorrhizae

J. Dighton

Defining Statement 73
Introduction 73
Types of Mycorrhizae 74
Endomycorrhizae 74
Ectomycorrhizae 75
Orchid Mycorrhiza 75
Other Mycorrhizae 76
Mycorrhizal Function 76
Nutrient Acquisition 76
Water Acquisition 77
Plant Defense 77
Ecology of Mycorrhizae 78
Global Distribution of Mycorrhizal Types and Soil Nutrients 78
Influence of Mycorrhizae on Plant Communities 78
Mycorrhizae as Food 79
Mycorrhizae and Plant Production 79
Agriculture 79
Commercial Forestry 79
Restoration 80
Mycorrhizae and Pollution 80
Acidifying Pollutants 80
Heavy Metal Pollutants 80
Radionuclide Pollutants 81
Organic Pollutants 81

Climate Change 81
Fungal Conservation 81
Recent Developments 82
Further Reading 82

7. Lichens

S. Hammer

Defining Statement 85
Introduction 85
Historical Note 89
Lichen Evolution 90
The Symbiosis 92
Discussion 94
Further Reading 94

8. Plant Pathogens and Disease

J.J. Burdon, P.H. Thrall and L. Ericson

Defining Statement 97
Introduction 97
Where Disease Emergence Stems from Simple Ecological Changes in Distribution 98
Pathogens 'Catching Up' with Anthropogenically Generated Changes in the Distribution of Their Hosts 98
Jumps to New, Previously Unexposed Hosts 98
Where Disease Emergence Stems from Genetic Change in Pathogens 99
Where Disease Emergence Stems from Environmental Change 102
Can We Predict the Identity of Future Emerging Diseases? 102
Countering Invasive Plant Diseases 103
Conclusions 103
Further Reading 104

9. Fungal and Protist Plant Pathogens

A.B. Gould

Defining Statement 105
Introduction 105
Fungal Characteristics 106
Hyphae and Fungal Cells 106
Nutrition 106
Reproduction 108
Dispersal and Survival 109
Fungi and the Environment 110
Symptoms of Plant Disease Caused by Fungi 111
Abnormal Growth 111
Abscission 111
Host Tissue Replacement 111
Necrosis 111
Permanent Wilt 112
Pathogen Groups 113
Kingdom Protozoa 113
Kingdom Stramenopila 113
Kingdom Eumycota (Fungi) 116
Diagnosis 123
Control 124
Further Reading 125

10. Entomogenous Fungi

R.A. Humber

Defining Statement 127
Introduction 127
Infection Processes and Pathobiology 128
Fungus–Host Interactions During Cuticular Penetration 128
Fungal Development in the Hemocoel 132
Practical Uses of Entomogenous Fungi 134
Development of Fungal Biological Control Agents 134
Is a Fungus a Pathogen or Saprobe? 136
Safety of Fungi as Biological Control Agents 136
A Look Forward for Entomogenous Fungi 137
Modernization of Systematics and Taxonomy 137
Nontraditional and Nonorganismal Uses of Entomogenous Fungi 137
Integration of Multidisciplinary Inputs to Insect Mycology 138
Recent Developments 139
Further Reading 140

11. Fungal Infections, Systemic

J.F. Staab and B. Wong

Defining Statement 143
Introduction 143
Classification of Pathogenic Fungi 143
Host Defenses against Fungal Infections 145
Molecular Approaches for Studying Fungal Pathogenesis 147
Common Fungal Diseases 150
Diseases Caused by True Pathogens 150
Diseases Caused by Opportunistic Fungi 157
Recent Developments 165
Summary and Conclusion 166
Further Reading 166

12. Fungal Infections, Cutaneous

C.J. Watts, D.K. Wagner and P.G. Sohnle

Defining Statement 167
Cutaneous Host Defenses 167
Structure of the Skin 167

Keratinization and Epidermal Proliferation 167
Antifungal Substances 167
The Innate Immune System 168
The Inflammatory Response 168
The Cutaneous Immune System 168
Description of the Diseases 169
Superficial Fungal Infections 169
Cutaneous and Subcutaneous Mycoses 171
Systemic Mycoses with Cutaneous Manifestations 173
Recent Developments 173
Further Reading 173

Part II
Protists

13. Amitochondriate Protists (Diplomonads, Parabasalids, and Oxymonads)

A.G.B. Simpson and I. Čepička

Defining Statement 177
Introduction 177
Amitochondriate Protists and Eukaryote Evolution 178
Systematics 179
Habitats 182
Cell Organization 182
Diplomonads 182
Parabasalids 183
Oxymonads 185
Hydrogenosomes and Mitosomes 185
Genetics and Genomics 187
Important Pathogenic Species 188
Further Reading 189

14. Amoebas, Lobose

A. Smirnov

Defining Statement 191
Introduction 191
Some Noteworthy Dates 192
Systematics and Phylogeny 192
Morphology 193
Morphotypes of Gymnamoebae 200
Diversity 200
Amoeboid Movement 207
Biology and Ecology 209
Importance 210
Further Reading 211

15. Ciliates

D. Lynn

Defining Statement 213
Introduction 213
Morphological Features 215
Origin of Ciliates 215
Evolution of the Ciliate Cortex 216
Evolution of Nuclear Dimorphism 217
Major Clades 218
Ultrastructure and the Structural Conservatism of the Ciliate Cortex 219
Molecular Systematics and the Major Clades 219
Diversity at the Class Level 219
The Two Subphyla 219
The 11 Classes 219
Probing Diversity at the Species Level 224
Morphological Diversity at the Species Level 224
Molecular Techniques at the Species Level 224
Further Reading 226

16. Secretive Ciliates and Putative Asexuality in Microbial Eukaryotes

Micah Dunthorn and Laura A. Katz

Putative Asexual Microbial Eukaryotes 227
The Powerlessness of Observation 228
An Improved Phylogenetic Framework 229
Reversing the Loss of Sex? 229
Ancient Asexuality? 231
Concluding Remarks and Outstanding Questions 231
Acknowledgements 231
References 232

17. Coccolithophores

R.W. Jordan

Defining Statement 235
Morphology 235
General Features of the Coccolithophore Cell 235
Coccolith Morphology 237
Coccolith Production 238
Functions of Coccoliths 238
Taxonomy 238
From the Chrysophyceae to the Haptophyceae 238
Haptophyte Taxonomy 239

Taxonomic Concepts Based on Morphology 239
Taxonomic Concepts Based on Molecular Genetics 240
Collection Methods 240
Biogeography and Ecology 241
General Distribution 241
Seasonality and Depth Preferences 241
Blooms 242
Detection of Coccolithophore Blooms 242
Coccolithophore Viruses 244
Impact on the Regional Climate and Environment 244
Carbon Release and Sinking of Blooms 245
Evolution 245
First Appearance of the Coccolithophores 245
Coccolithophore Diversity and Extinctions 246
Past Coccolithophore Blooms 246
Biostratigraphic Use 247
Further Reading 247

18. The Glass Menagerie: Diatoms for Novel Applications in Nanotechnology

Richard Gordon, Dusan Losic, Mary Ann Tiffany, Stephen S. Nagy and Frithjof A.S. Sterrenburg

Why Diatoms? 249
Diatom Silica Structure 251
Silica Biomineralization and Diatom Genomics 251
Diatom Biophotonics 255
Microfluidics Within Diatoms 256
Diatoms for Drug Delivery 257
Selective Breeding of Diatoms Using a Compustat 258
Computing With Diatoms 258
Conclusions 258
Acknowledgements 258
References 259

19. Dinoflagellates

M.-O. Soyer-Gobillard

Defining Statement 263
Introduction 263
Dinoflagellate Evolution 263
Dinoflagellate Diversity 265
Crypthecodinium cohnii Biecheler 265
Noctiluca scintillans McCartney 268
Prorocentrum micans 269
Mixotrophic Dinoflagellates 270
Evolution of the Mitotic Apparatus 271
Conclusions 274
Further Reading 277

20. Dictyostelium

C. Scott and P. Schaap

Defining Statement 279
Taxonomy, Evolution, and Ecology 279
Genomics and Tractability 281
The *D. discoideum* Genome 281
Experimental Tractability 282
The Developmental Program of *D. discoideum* 283
Morphogenesis 283
Gene Expression and Cell Differentiation 286
Recent Developments in the *Dictyostelium* Field 288
The Sporulation Cascade 288
Phylogeny Wide Genome Sequencing 288
Evolutionary History of cAMP Signaling 288
Conclusions 289
Further Reading 289

21. Foraminifera

J. Pawlowski

Defining Statement 291
Introduction 291
Cytological and Morphological Characteristics 292
Granuloreticulopodia 292
Nuclei and Other Organelles 292
Test Morphology 293
Life Cycle and Reproduction 294
Ecology 295
Distribution and Abundance 295
Feeding Strategies 295
Symbiosis 295
Collection and Maintenance 297
Morphology-Based Classification 297
Molecular Phylogeny and Diversity 298
Phylogenetic Position 298
Macroevolutionary Relationships 300
Molecular Diversity 308
Evolutionary History and Geological Importance 308
Further Reading 309

22. Euglenozoa

M.A. Farmer

Defining Statement 311
Taxonomy 311
Cell Structure 312
Flagellar Structure 312
Mitochondria 314
Nuclei 314
Cytoskeleton 314
Euglenids 314
Pellicle 314
Flagella 315
Chloroplasts 315
Feeding Apparatus 316
Kinetoplastids 317
The Kinetoplast 317
RNA Editing in Kinetoplastids 318
The Glycosome 318
Diplonemids and Other Euglenozoa 319
Evolutionary Relationships 320
Further Reading 321

23. Protozoan, Intestinal

K. Pierce and C.D. Huston

Defining Statement 323
Introduction 323
The Apicomplexa: *Cryptosporidium* Species 324
The Microsporidia 326
The Flagellates: *Giardia intestinalis* 327
The Amoebae: *E. histolytica* 329
Recent Developments 331
Cryptosporidium species 331
G. intestinalis 332
E. histolytica 332
Conclusion 332
Further Reading 332

24. Leishmania

G.B. Ogden and P.C. Melby

Defining Statement 335
Classification and Morphology 335
Life Cycle and Ecology 336
Cellular Biology 337
Plasma Membrane and Surface Molecules 337
Flagellum 338
Other Membrane-Bound Organelles 338
Molecular Biology and Control of Gene Expression 339
Genomic Organization 339
mRNA Processing 339
Control of Gene Expression 340
DNA Transfection and Gene Targeting 340
Pathogenesis and Host Response 340
Epidemiology and Disease 341
Diagnosis, Treatment, and Control 343
Recent Developments 344
Further Reading 344

25. Oomycetes (Water Mold, Plant Pathogenic)

S. Kamoun

Defining Statement 347
Impact on Agriculture and Environment 347
Biology 348
Evolutionary History 348
General Biological Features 349
Unique Biological Features 350
Genome Structure 350
Pathology 350
Infection Cycle 350
Adhesion, Penetration, and Colonization of Host Tissue 350
Induction of Defense Responses and Disease-Like Symptoms 351
Inhibition of Host Enzymes 351
Effectors 352
Conclusions 352
Further Reading 353

26. Picoeukaryotes

R. Massana

Defining Statement 355
Introduction 355
What Are Marine Picoeukaryotes? 355
Method-Driven History of Marine Picoeukaryotes 356
Biology of Cultured Marine Picoeukaryotes 358
Cultured Strains 358
Cellular Organization 359
Physiological Parameters 360
The Implications of Being Small 360
Picoeukaryotes in the Marine Environment 361
Bulk Abundance and Distribution 361
Ecological Role of PP 362
Ecological Role of HP 362
Molecular Tools to Study Picoeukaryote Ecology 363

Elusive View of In Situ Diversity by Nonmolecular Tools 363
Cloning and Sequencing Environmental Genes 363
Beyond Clone Libraries: FISH and Fingerprinting Techniques 364
In Situ Phylogenetic Diversity 365
Overview of 18S rDNA Libraries 365
Relatively Well-Known Groups 365
Marine Alveolates and Marine Stramenopiles 367
Novel High-Rank Phylogenetic Groups 367
Biogeography 367
The Genomic Era 368
Genome Projects on Cultured Picoeukaryotes 368
Environmental Genomics or Metagenomics 368
Recent Developments 368
Concluding Remarks 370
Further Reading 370

27. Stramenopiles

H.S. Yoon, R.A. Andersen, S.M. Boo and D. Bhattacharya

Defining Statement 373
Evolutionary History of the Stramenopiles 373
Origin of the Stramenopiles 374
Fossil Record and Divergence Times for Stramenopiles 374
Diversity of the Stramenopiles 376
The Stramenopile Plastid 377
Cell Covering 379
Flagella 379
Phylogeny and Classification of the Stramenopiles 380
Colorless Stramenopiles 380
Photosynthetic Stramenopiles 381
Phaeophyceae 383
Conclusion 383
Further Reading 383

28. Toxoplasmosis

J.C. Boothroyd

Defining Statement 385
Introduction 385
Classification 386
Life Cycle 386
Clinical Aspects and Public Health 387
Symptoms 387
Diagnosis 387
Treatment 387
Public Health 388
Population Biology 388
Major Genotypes 388
Strain-Specific Virulence 389
Molecular Biology and Genetics 389
Genome and Gene Expression 389
Genetics 389
Molecular Genetic Tools Available for the Study of *Toxoplasma* 390
Cell Biology 390
Organelles 390
The Lytic Cycle 391
Host Immune Response 393
Nature of the Host Response 393
Genetics of Host Susceptibility 394
Immunization Studies 394
Effect on Behavior 394
Prospects for Future Improvements in Controlling Toxoplasmosis 394
Public Health 394
Vaccination 394
Chemotherapy 395
Further Reading 395

29. Trypanosomes

L.V. Kirchhoff, C.J. Bacchi, F.S. Machado, H.D. Weiss, H. Huang, S. Mukherjee, L.M. Weiss and H.B. Tanowitz

Defining Statement 397
Trypanosoma Cruzi 397
Transmission 398
Organism and Life Cycle 398
Epizootiology and Epidemiology 400
Clinical Manifestations 401
Pathogenesis 402
Diagnosis 403
Treatment 403
Prevention 404
Trypanosoma Brucei 404
Organism and Life Cycle 404
Epidemiology 405
Pathology and Pathogenesis 406
Clinical Manifestations 407
Diagnosis 407
Treatment 407
Prophylaxis and Prevention 409
Further Reading 409

30. Sleeping Sickness

S.C. Welburn, K. Picozzi, I. Maudlin and P.P. Simarro

Defining Statement 411
Background 411
Epidemiology of Sleeping Sickness 412
Distribution 412
Reservoirs of Disease 413
Transmission Cycles 414
Causes of Epidemics 415
Diagnosis 416
Clinical Signs 416
Diagnostic Tests 416
Treatment 417
Disease Burden 418
Economic Impact 418
Sleeping Sickness Control 419
Controlling Gambian Sleeping Sickness 419
Controlling Rhodesian Sleeping Sickness 419
Future Prospects for Controlling Sleeping Sickness 420
A Neglected Disease 420
Treatment 422
Diagnostics 422
Vector Control 423
The Future: Control or Eliminate? 423
Gambian Sleeping Sickness 423
Rhodesian Sleeping Sickness 423
Further Reading 425

31. Secondary Endosymbiosis

J.M. Archibald

Defining Statement 427
The Origin of Eukaryotic Photosynthesis 427
Primary Endosymbiosis 427
Endosymbiotic Gene Transfer and Plastid Protein Import 429
Secondary Endosymbiosis 430
Diversity of Secondary Plastid-Containing Algae 430
Gene Transfer and Protein Import in Secondary Plastids 431
Number of Secondary Endosymbiotic Events 431
Nucleomorphs and Their Genomes 433
Further Reading 434

32. Algal Blooms

P. Assmy and V. Smetacek

Defining Statement 435
Introduction 435
Physical Environment of Blooms 436
Chemical Environment of Blooms 437
Major Contributors to Algal Blooms 438
Cyanobacteria 438
Haptophytes (Prymnesiophytes) 440
Dinoflagellates 441
Other Groups 442
Pathogens and Grazers of Algal Blooms 442
Recurrent and Unusual Algal Blooms 443
Spring Blooms 443
Autumn Blooms 445
Blooms in Upwelling Regions of Low Latitudes 445
Miscellaneous Algal Blooms 446
Harmful Algal Blooms 446
Iron-Fertilized Blooms 447
Future Research Avenues 448
Recent Developments 448
Marine Genomics 448
Abandoning Sverdrup's Critical Depth Hypothesis 449
The Role of Grazing in Suppressing Nondiatom Blooms and Recycling Iron in Pelagic Ecosystems 449
Acknowledgment 450
Further Reading 450

33. Food Webs, Microbial

E.B. Sherr and B.F. Sherr

Defining Statement 451
Introduction 451
Understanding Microbial Food Webs 452
Components and Pathways 452
Microbes in Aquatic Food Webs 454
Heterotrophic Prokaryotes 454
Autotrophic Prokaryotes 454
Autotrophic Eukaryotes 455
Heterotrophic Eukaryotes 455
Marine Pelagic Habitats 456
Benthic Habitats 457
Role of Microbial Food Webs in Biogeochemical Cycling 458
Food Resource for Metazoans 460
Modeling Microbial Food Webs 460
Chemical Interactions Between Microbes 462
Spatial Structure of Microbial Food Webs 463
Recent Developments Biogeography of Form and Function in Microbial Food Webs 464
Further Reading 465

Index 467

Contributors

Numbers in the paraentheses indicate the pages on which the authors' contributions begin.

I. Čepička (175), Charles University in Prague, Prague, Czech Republic

R.A. Andersen (373), Bigelow Laboratory for Ocean Sciences, West Boothbay Harbor, ME, USA

J.M. Archibald (427), Dalhousie University, Halifax, NS, Canada

P. Assmy (435), Alfred Wegener Institute for Polar and Marine Research, Bremerhaven, Germany

C.J. Bacchi (397), Pace University, New York, NY, USA

D. Bhattacharya (373), University of Iowa, Iowa City, IA, USA

S.M. Boo (373), Chungnam National University, Daejeon, Republic of Korea

J.C. Boothroyd (385), Stanford University School of Medicine, Stanford, CA, USA

J.J. Burdon (97), CSIRO, Canberra, Australia

Frédéric Delbac (65), Clermont Université, Université Blaise Pascal, Laboratoire Microorganismes: Génome et Environnement, BP 10448, F-63000 Clermont-Ferrand, France
CNRS, UMR 6023, LMGE, F-63177 Aubiere, France

J. Dighton (73), Rutgers University Pinelands Field Station, New Lisbon, NJ, USA

Micah Dunthorn (227), Organismic and Evolutionary Biology, University of Massachusetts, Amherst, MA 01003, USA
Corresponding author: Dunthorn, M. (dunthorn@rhrk.uni-kl.de) Present address: Department of Ecology, University of Kaiserslautern, 67653 Kaiserslautern, Germany.

Hicham El Alaoui (65), Clermont Université, Université Blaise Pascal, Laboratoire Microorganismes: Génome et Environnement, BP 10448, F-63000 Clermont-Ferrand, France
CNRS, UMR 6023, LMGE, F-63177 Aubiere, France

L. Ericson (97), Umeå University, Umeå, Sweden

M.A. Farmer (311), University of Georgia, Athens, GA, USA

Richard Gordon (249), Department of Radiology, University of Manitoba, Winnipeg MB R3A 1R9, Canada

A.B. Gould (105), Rutgers University, New Brunswick, NJ, USA

S. Hammer (85), Boston University, Boston, MA, USA

H. Huang (397), Albert Einstein College of Medicine, Bronx, NY, USA

R.A. Humber (127), USDA-ARS Biological Integrated Pest Management Research Unit, Ithaca, NY, USA

C.D. Huston (323), University of Vermont College of Medicine, Burlington, VT, USA

R.W. Jordan (235), Yamagata University, Yamagata, Japan

S. Kamoun (347), The Sainsbury Laboratory, Norwich, UK

Laura A. Katz (227), Organismic and Evolutionary Biology, University of Massachusetts, Amherst, MA 01003, USA
Department of Biological Sciences, Smith College, Northampton, MA 01063, USA

L.V. Kirchhoff (397), University of Iowa, Iowa City, IA, USA

Dusan Losic (249), Ian Wark Research Institute, University of South Australia, Mawson Lakes, Adelaide SA 5095, Australia

D. Lynn (213), University of Guelph, Guelph, ON, Canada

F.S. Machado (397), Federal University of Minas Gerais, Belo Horizonte, Minas Gerais, Brazil

R. Massana (355), Institut de Ciències del Mar, Barcelona, Catalonia, Spain

I. Maudlin (411), Centre for Infectious Diseases, The University of Edinburgh, 1 Summerhall Square, Edinburgh, UK

P.C. Melby (335), South Texas Veterans Health Care System and The University of Texas Health Science Center at San Antonio, San Antonio, TX, USA

S. Mukherjee (397), Albert Einstein College of Medicine, Bronx, NY, USA

Stephen S. Nagy (249), Montana Diatoms, PO Box 5714, Helena MT 59604, USA

G.B. Ogden (335), St. Mary's University and The University of Texas Health Sciences Center at San Antonio, San Antonio, TX, USA

J. Pawlowski (291), University of Geneva, Geneva, Switzerland

K. Picozzi (411), Centre for Infectious Diseases, The University of Edinburgh, 1 Summerhall Square, Edinburgh, UK

K. Pierce (323), University of Vermont College of Medicine, Burlington, VT, USA

C. Scazzocchio (19), Institut de Génétique et Microbiologie, Université Paris-Sud, Orsay, France; Department of Microbiology, Imperial College London, London, UK

P. Schaap (279), University of Dundee, Dundee, UK

C. Scott (279), University of Dundee, Dundee, UK

B.F. Sherr (451), Oregon State University, Corvallis, OR, USA

E.B. Sherr (451), Oregon State University, Corvallis, OR, USA

P.P. Simarro (411), World Health Organization, Control of Neglected Tropical Diseases, Innovative and Intensified Disease Management, Geneva, Switzerland

A.G.B. Simpson (175), Dalhousie University, Halifax, Nova Scotia, Canada

V. Smetacek (435), Alfred Wegener Institute for Polar and Marine Research, Bremerhaven, Germany

A. Smirnov (191), St. Petersburg State University, St. Petersburg, Russia

P.G. Sohnle (167), Division of Infectious Diseases, Department of Medicine, Medical College of Wisconsin, and the Section of Infectious Diseases, VA Medical Center, Milwaukee, WI, USA

M.-O. Soyer-Gobillard (263), Université Pierre et Marie Curie Paris 6, Laboratoire Arago, Banyuls-sur-Mer, France

J.F. Staab (143), Oregon Health Science University, Portland, OR, USA

Frithjof A.S. Sterrenburg (249), Stationsweg 158, Heiloo 1852LN, The Netherlands

M. Tadych (51), Rutgers University, New Brunswick, NJ, USA

H.B. Tanowitz (397), Albert Einstein College of Medicine, Bronx, NY, USA

Catherine Texier (65), Clermont Université, Université Blaise Pascal, Laboratoire Microorganismes: Génome et Environnement, BP 10448, F-63000 Clermont-Ferrand, France
CNRS, UMR 6023, LMGE, F-63177 Aubiere, France

P.H. Thrall (97), CSIRO, Canberra, Australia

Mary Ann Tiffany (249), Center for Inland Waters, San Diego State University, 5500 Campanile Drive, San Diego CA 92182, USA

M.S. Torres (41), Rutgers University, New Brunswick, NJ, USA

Cyril Vidau (65), Clermont Université, Université Blaise Pascal, Laboratoire Microorganismes: Génome et Environnement, BP 10448, F-63000 Clermont-Ferrand, France
CNRS, UMR 6023, LMGE, F-63177 Aubiere, France

Bernard Viguès (65), Clermont Université, Université Blaise Pascal, Laboratoire Microorganismes: Génome et Environnement, BP 10448, F-63000 Clermont-Ferrand, France
CNRS, UMR 6023, LMGE, F-63177 Aubiere, France

D.K. Wagner (167), Division of Infectious Diseases, Department of Medicine, Medical College of Wisconsin, and the Section of Infectious Diseases, VA Medical Center, Milwaukee, WI, USA

G.M. Walker (1), University of Abertay Dundee, Dundee, Scotland, UK

C.J. Watts (167), Division of Infectious Diseases, Department of Medicine, Medical College of Wisconsin, and the Section of Infectious Diseases, VA Medical Center, Milwaukee, WI, USA

H.D. Weiss (397), Albert Einstein College of Medicine, Bronx, NY, USA

L.M. Weiss (397), Albert Einstein College of Medicine, Bronx, NY, USA

S.C. Welburn (411), Centre for Infectious Diseases, The University of Edinburgh, 1 Summerhall Square, Edinburgh, UK

J.F. White (41), (51), Rutgers University, New Brunswick, NJ, USA
Rutgers University, New Brunswick, NJ, USA

B. Wong (143), Oregon Health Science University, Portland, OR, USA

H.S. Yoon (373), Bigelow Laboratory for Ocean Sciences, West Boothbay Harbor, ME, USA

Preface

The eukaryotic microbes, fungi and protists, vie with the prokaryotic ones for importance in the affairs of this planet. Fungi are the great recyclers, protists the great photosynthesizers (or, if you prefer, the great predators), to name but two of their life-driving attributes. If these organisms were lacking, the biochemical cycles of matter in nature would have a different quality, one that would not sustain our biosphere as we know it.

Given their importance, it pleases us to present a compendium of chapters on these organisms derived from the *Encyclopedia of Microbiology*, 3rd edition. A few chapters from other sources have also been included; all were written by investigators with authority in their fields. Where appropriate, these contributions have been updated with current references and a section on future developments. The organization of this book does not follow a proscribed taxonomy, which would be an uncertain endeavor. Rather, the chapters are presented in an order we believe is convenient to the readers.

The terms *fungi* and *protists* are used here *sensu latissimo*. The housing of these two groups between the same covers is unusual; most treatises deal with one or the other. We find this union to be justified. Although these two huge groups do not share a common phylogeny, unless, of course, one traces them back to the early branches, they do share vast habitats where they surely play collaborative roles. Those mainly interested in but one of the two should find ample reasons herein to also explore the other.

Part I

Fungi

1. Yeasts 3
2. Aspergillus: A Multifaceted Genus 19
3. Clavicipitaceae: Free-Living and Saprotrophs to Plant Endophytes 41
4. Endophytic Microbes 51
5. Microsporidia: A Model for Minimal Parasite–Host Interactions 65
6. Mycorrhizae 73
7. Lichens 85
8. Plant Pathogens and Disease: Newly Emerging Diseases 97
9. Fungal and Protist Plant Pathogens 105
10. Entomogenous Fungi 127
11. Fungal Infections, Systemic 143
12. Fungal Infections, Cutaneous 167

Chapter 1

Yeasts

G.M. Walker

University of Abertay Dundee, Dundee, Scotland, UK

Chapter Outline

Abbreviations	3	Nutrition, Metabolism, and Growth of Yeasts	6
Defining Statement	3	Yeast Genetics	12
Definition and Classification of Yeasts	3	Industrial, Agricultural, and Medical Importance of Yeasts	14
Yeast Ecology	4	Overview of Recent Progress in Yeast Science and Technology	16
Yeast Cell Structure	5	Further Reading	17

ABBREVIATIONS

AFLP Amplified fragment length polymorphism
AFM Atomic force microscopy
CDI Cyclin-dependent kinase inhibitor
DEAE Diethylaminoethyl
ER Endoplasmic reticulum
FACS Fluorescence-activated cell sorting
GAP General amino acid permease
NAD Nicotinamide adenine dinucleotide
RAPD Random amplified polymorphic DNA
YEPG Yeast extract peptone glucose
YNB Yeast nitrogen base

DEFINING STATEMENT

Yeasts are eukaryotic unicellular microfungi that play important roles in industry, the environment, and medical science. This chapter describes the classification, ecology, cytology, metabolism, and genetics of yeast, with particular reference to *Saccharomyces cerevisiae* – baker's yeast. The biotechnological potential of yeasts is also discussed, including their exploitation in food, fermentation, and pharmaceutical industries.

DEFINITION AND CLASSIFICATION OF YEASTS

Definition and Characterization of Yeasts

Yeasts are recognized as unicellular fungi that reproduce primarily by budding, and occasionally by fission, and that do not form their sexual states (spores) in or on a fruiting body. Yeast species may be identified and characterized according to various criteria based on cell morphology (e.g., mode of cell division and spore shape), physiology (e.g., sugar fermentation tests), immunology (e.g., immunofluorescence), and molecular biology (e.g., ribosomal DNA phylogeny, DNA reassociation, DNA base composition and hybridization, karyotyping, random amplified polymorphic DNA (RAPD), and amplified fragment length polymorphism (AFLP) of D1/D2 domain sequences of 26 S rDNA). Molecular sequence analyses are being increasingly used by yeast taxonomists to categorize new species and to understand yeast biodiversity and interrelationships among the yeasts. Due to advances in phylogenetic analysis of gene sequences, the number of characterized yeast species has increased dramatically in recent years.

Yeast Taxonomy

The most commercially exploited yeast species, *S. cerevisiae* (baker's yeast), belongs to the fungal kingdom subdivision Ascomycotina. Table 1 summarizes the taxonomic hierarchy of yeasts, with *S. cerevisiae* as an example.

Other yeast genera are categorized under Basidiomycotina (e.g., *Cryptococcus* spp. and *Rhodotorula* spp.) and Deuteromycotina (e.g., *Candida* spp. and *Brettanomyces* spp.). There are around 100 recognized yeast genera and the reader is directed to Kurtzman and Fell (1998) and Kurtzman and Piskur (2006) for additional information on yeast taxonomy.

TABLE 1 Taxonomic hierarchy of yeast

Taxonomic category	Example (*Saccharomyces cerevisiae*)
Kingdom	Fungi
Division	Ascomycota
Subdivision	Ascomycotina
Class	Hemiascomycete
Order	Endomycetales
Family	Saccharomycetacae
Subfamily	Saccharomyetoideae
Genus	*Saccharomyces*
Species	*cerevisiae*

Yeast Biodiversity

Around 1500 species of yeast have been described, but new species are being characterized on a regular basis and there is considerable untapped yeast biodiversity on Earth. For example, it has been estimated (in 1996) that only 0.065% of yeast genera (total 62 000) and 0.22% of yeast species (total 669 000) have been isolated and characterized. This means that there is an immense gap in our knowledge regarding biodiversity and the available 'gene pool' of wild natural isolates of yeast. Several molecular biological techniques are used to assist in the detection of new yeast species in the natural environment, and together with input from cell physiologists, they provide ways to conserve and exploit yeast biodiversity. *S. cerevisiae* is the most studied and exploited of all the yeasts, but the biotechnological potential of non-*Saccharomyces* yeasts is gradually being realized, particularly with regard to recombinant DNA technology (see Table 8).

YEAST ECOLOGY

Natural Habitats of Yeast Communities

Yeasts are not as ubiquitous as bacteria in the natural environment. Nevertheless, yeasts can be isolated from soil, water, plants, animals, and insects. Preferred yeast habitats are plant tissues (leaves, flowers, and fruits), but a few species are found in commensal or parasitic relationships with animals. Many new yeast species are being isolated from insect intestinal tracts. Some yeasts, most notably *Candida albicans*, are opportunistic human pathogens. Several species of yeast may be isolated from specialized or extreme environments, such as those with low water potential (i.e., high sugar or salt concentrations), low temperature (e.g., some psychrophilic yeasts have been isolated from polar regions), and low oxygen availability (e.g., intestinal tracts of animals). Table 2 summarizes the main yeast habitats.

Yeasts in the Food Chain

Yeasts play important roles in the food chain. Numerous insect species, notably *Drosophila* spp., feed on yeasts that colonize plant material. As insect foods, ascomycetous

TABLE 2 Natural yeast habitats

Habitat	Comments
Soil	Soil may only be a reservoir for the long-term survival of many yeasts, rather than a habitat for growth. However, yeasts are ubiquitous in cultivated soils (about 10 000 yeast cells per gram of soil) and are found only in the upper, aerobic soil layers (10–15 cm). Some genera are isolated exclusively from soil (e.g., *Lipomyces* and *Schwanniomyces*)
Water	Yeasts predominate in surface layers of fresh and salt waters, but are not present in great numbers (about 1000 cells per liter). Many aquatic yeast isolates belong to red pigmented genera (*Rhodotorula*). *Debaryomyces hansenii* is a halotolerant yeast that can grow in nearly saturated brine solutions
Atmosphere	A few viable yeast cells may be expected per cubic meter of air. From layers above soil surfaces, *Cryptococcus*, *Rhodotorula*, *Sporobolomyces*, and *Debaryomyces* spp. are dispersed by air currents
Plants	The interface between soluble nutrients of plants (sugars) and the septic world are common niches for yeasts (e.g., the surface of grapes); the spread of yeasts on the phyllosphere is aided by insects (e.g., *Drosophila* spp.); a few yeasts are plant pathogens. The presence of many organic compounds on the surface and decomposing areas (exudates, flowers, fruits, phyllosphere, rhizosphere, and necrotic zones) creates conditions favorable for growth of many yeasts
Animals	Several nonpathogenic yeasts are associated with the intestinal tract and skin of warm-blooded animals; several yeasts (e.g., *Candida albicans*) are opportunistically pathogenic toward humans and animals; numerous yeasts are commensally associated with insects, which act as important vectors in the natural distribution of yeasts
Built environment	Yeasts are fairly ubiquitous in buildings, for example, *Aureobasidium pullulans* (black yeast) is common on damp household wallpaper and *Saccharomyces cerevisiae* is readily isolated from surfaces (pipework and vessels) in wineries

yeasts convert low-molecular-weight nitrogenous compounds into proteins beneficial to insect nutrition. In addition to providing a food source, yeasts may also affect the physiology and sexual reproduction of drosophilids. In marine environments, yeasts may serve as food for filter feeders.

Microbial Ecology of Yeasts

In microbial ecology, yeasts are not involved in biogeochemical cycling to the same extent as bacteria or filamentous fungi. Nevertheless, yeasts can use a wide range of carbon sources and thus play an important role as saprophytes in the carbon cycle, degrading plant detritus to carbon dioxide. In the cycling of nitrogen, some yeasts can reduce nitrate or ammonify nitrite, although most yeasts assimilate ammonium ions or amino acids into organic nitrogen. Most yeasts can reduce sulfate, although some are sulfur auxotrophs.

YEAST CELL STRUCTURE

General Cellular Characteristics

Yeasts are unicellular eukaryotes that have ultrastructural features similar to that of higher eukaryotic cells. This, together with their ease of growth, and amenability to biochemical, genetic, and molecular biological analyses, makes yeasts excellent model organisms in studies of eukaryotic cell biology. Yeast cell size can vary widely, depending on the species and conditions of growth. Some yeasts may be only 2–3 μm in length, whereas others may attain lengths of 20–50 μm. Cell width appears less variable, between 1 and 10 μm. *S. cerevisiae* is generally ellipsoid in shape with a large diameter of 5–10 μm and a small diameter of 1–7 μm. Table 3 summarizes the diversity of yeast cell shapes.

Several yeast species are pigmented and various colors may be visualized in surface-grown colonies, for example, cream (e.g., *S. cerevisiae*), white (e.g., *Geotrichum* spp.), black (e.g., *Aureobasidium pullulans*), pink (e.g., *Phaffia rhodozyma*), red (e.g., *Rhodotorula* spp.), orange (e.g., *Rhodosporidium* spp.), and yellow (e.g., *Bullera* spp.). Some pigmented yeasts have applications in biotechnology. For example, the astaxanthin pigments of *P. rhodozyma* have applications as fish feed colorants for farmed salmonids, which have no means of synthesizing these red compounds.

Methods in Yeast Cytology

By using various cytochemical and cytofluorescent dyes and phase contrast microscopy, it is possible to visualize several subcellular structures in yeasts (e.g., cell walls,

TABLE 3 Diversity of yeast cell shapes

Cell shape	Description	Examples of yeast genera
Ellipsoid	Ovoid-shaped cells	*Saccharomyces*
Cylindrical	Elongated cells with hemispherical ends	*Schizosaccharomyces*
Apiculate	Lemon shaped	*Hanseniaspora, Saccharomycodes*
Ogival	Elongated cell rounded at one end and pointed at other	*Dekkera, Brettanomyces*
Flask shaped	Cells dividing by bud fission	*Pityrosporum*
Pseudohyphal	Chains of budding yeast cells, which have elongated without detachment. Pseudohyphal morphology is intermediate between a chain of yeast cells and a hypha	Occasionally found in starved cells of *Saccharomyces cerevisiae* and frequently in *Candida albicans* (filamentous cells form from 'germ tubes', and hyphae may give rise to buds called blastospores)
Hyphal	Basidiomycetous yeast cells grow lengthwise to form branched or unbranched threads or true hyphae, occasionally with septa (cross walls) to make up mycelia. Septa may be laid down by the continuously extending hyphal tip	*Saccharomycopsis* spp.
Dimorphic	Yeasts that grow vegetatively in either yeast or filamentous forms	*Candida albicans, Saccharomycopsis fibuligera, Kluyveromyces marxianus, Malassezia furfur, Yarrowia lipolytica, Ophiostoma novo-ulmi, Sporothrix schenkii, Histoplasma capsulatum*
Miscellaneous	Triangular	*Trigonopsis*
	Curved	*Cryptococcus*
	Stalked	*Sterigmatomyces*
	Spherical	*Debaryomyces*

capsules, nuclei, vacuoles, mitochondria, and several cytoplasmic inclusion bodies). The *GFP* gene from the jellyfish (*Aequorea victoria*) encodes the green fluorescent protein (GFP, which fluoresces in blue light) and can be used to follow the subcellular destiny of certain expressed proteins when GFP is fused with the genes of interest. Immunofluorescence can also be used to visualize yeast cellular features when dyes such as fluorescein isothiocyanate and rhodamine B are conjugated with monospecific antibodies raised against yeast structural proteins. Confocal scanning laser immunofluorescence microscopy can also be used to detect the intracellular localization of proteins within yeast cells and to give three-dimensional ultrastructural information. Fluorescence-activated cell sorting (FACS) has proven very useful in studies of the yeast cell cycle and in monitoring changes in organelle (e.g., mitochondrial) biogenesis. Scanning electron microscopy is useful in revealing the cell surface topology of yeasts, as is atomic force microscopy, which has achieved high-contrast nanometer resolution for yeast cell surfaces (Figure 1). Transmission electron microscopy, however, is essential for visualizing the intracellular fine structure of ultrathin yeast cell sections (Figure 2).

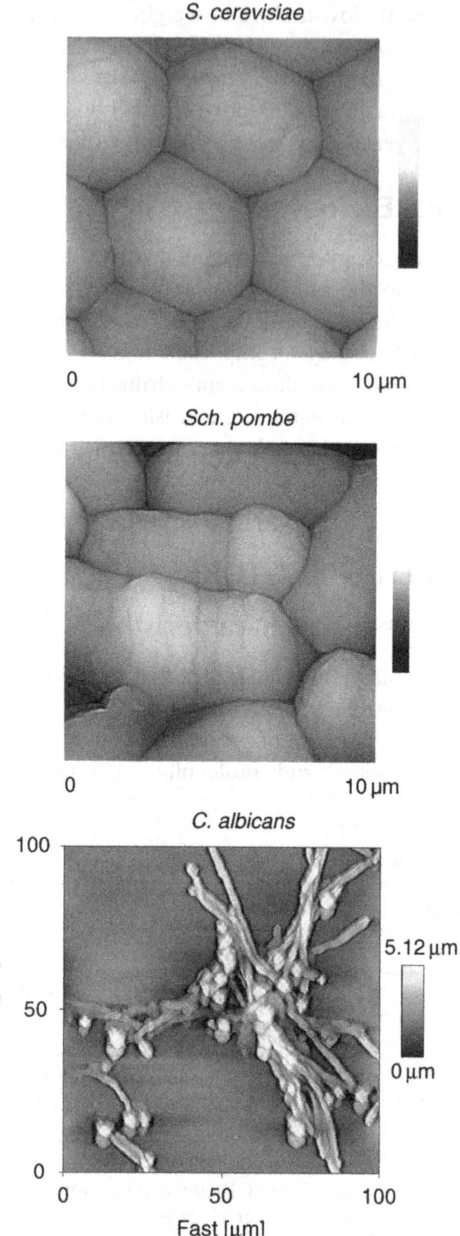

FIGURE 1 Atomic force microscopy (AFM) of yeast cell surfaces. *Courtesy of Dr. A. Adya and Dr. E. Canetta, University of Abertay Dundee.*

Subcellular Yeast Architecture and Function

Transmission electron microscopy of a yeast cell typically reveals the cell wall, nucleus, mitochondria, endoplasmic reticulum (ER), Golgi apparatus, vacuoles, microbodies, and secretory vesicles. Figure 2 shows an electron micrograph of a typical yeast cell.

Several of these organelles are not completely independent of each other and derive from an extended intramembranous system. For example, the movement and positioning of organelles depends on the cytoskeleton, and the trafficking of proteins in and out of cells relies on vesicular communication between the ER, Golgi apparatus, vacuole, and plasma membrane. Yeast organelles can be readily isolated for further studies by physical, chemical, or enzymatic disruption of the cell wall, and the purity of organelle preparations can be evaluated using specific marker enzyme assays.

In the yeast cytoplasm, ribosomes and occasionally plasmids (e.g., 2 µm circles) are found, and the structural organization of the cell is maintained by a cytoskeleton of microtubules and actin microfilaments. The yeast cell envelope, which encases the cytoplasm, comprises (from the inside looking out) the plasma membrane, periplasm, cell wall, and, in certain yeasts, a capsule and a fibrillar layer. Spores encased in an ascus may be revealed in those yeasts that undergo differentiation following sexual conjugation and meiosis. Table 4 provides a summary of the physiological functions of the various structural components found in yeast cells.

NUTRITION, METABOLISM, AND GROWTH OF YEASTS

Nutritional and Physical Requirements for Yeast Growth

Yeast nutritional requirements

Yeast cells require macronutrients (sources of carbon, nitrogen, oxygen, sulfur, phosphorus, potassium, and magnesium) at the millimolar level in growth media, and they require

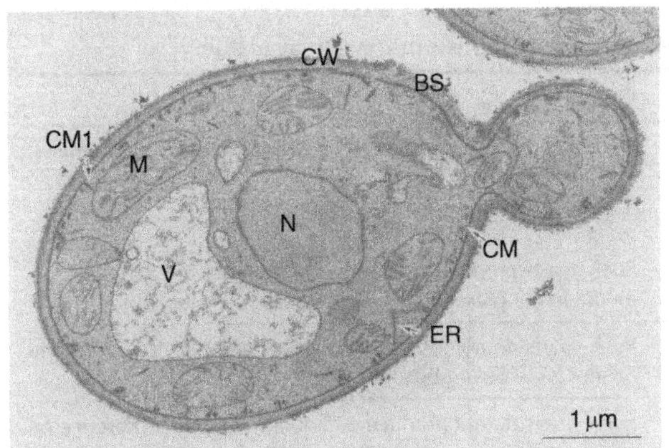

FIGURE 2 Ultrastructural features of a yeast cell. The transmission electron micrograph is of a *Candida albicans* cell. BS, bud scar; CM, cell membrane; CMI, cell membrane invagination; CW, cell wall; ER, endoplasmic reticulum; M, mitochondrion; N, nucleus; and V, vacuole. *Courtesy of M. Osumi, Japan Women's University, Tokyo.*

trace elements (e.g., Ca, Cu, Fe, Mn, and Zn) at the micromolar level. Most yeasts grow quite well in simple nutritional media, which supply carbon–nitrogen backbone compounds together with inorganic ions and a few growth factors. Growth factors are organic compounds required in very low concentrations for specific catalytic or structural roles in yeast, but are not used as energy sources. Yeast growth factors include vitamins, which serve vital functions as components of coenzymes, purines and pyrimidines, nucleosides and nucleotides, amino acids, fatty acids, sterols, and other miscellaneous compounds (e.g., polyamines and choline). Growth factor requirements vary among yeasts, but when a yeast species is said to have a growth factor requirement, it indicates that the species cannot synthesize the particular factor, resulting in the curtailment of growth without its addition to the culture medium.

Yeast culture media

It is quite easy to grow yeasts in the laboratory on a variety of complex and synthetic media. Malt extract or yeast extract supplemented with peptone and glucose (as in YEPG) is commonly employed for the maintenance and growth of most yeasts. Yeast nitrogen base (YNB) is a commercially available chemically defined medium that contains ammonium sulfate and asparagine as nitrogen sources, together with mineral salts, vitamins, and trace elements. The carbon source of choice (e.g., glucose) is usually added to a final concentration of 1% (w/v). For the continuous cultivation of yeasts in chemostats, media that ensure that all the nutrients for growth are present in excess except one (the growth-limiting nutrient) are usually designed. Chemostats can therefore facilitate studies on the influence of a single nutrient (e.g., glucose, in carbon-limited chemostats) on yeast cell physiology, with all other factors

TABLE 4 Functional components of an ideal yeast cell

Organelle or cellular structure	Function
Cell envelope	Comprises the plasma membrane that acts as a selectively permeable barrier for transport of hydrophilic molecules in and out of fungal cells; the periplasm containing proteins and enzymes unable to permeate the cell wall; the cell wall that provides protection and shape and is involved in cell–cell interactions, signal reception, and specialized enzyme activities; fimbriae involved in sexual conjugation; and capsules to protect cells from dehydration and immune cell attack
Nucleus	Contains chromosomes (DNA–protein complexes) that pass genetic information to daughter cells during cell division and the nucleolus, which is the site of ribosomal RNA transcription and processing
Mitochondria	Responsible, under aerobic conditions, for respiratory metabolism and, under anaerobic conditions, for fatty acid, sterol, and amino acid metabolism
Endoplasmic reticulum	Ribosomes on the rough endoplasmic reticulum are the sites of protein biosyntheses (translation of mRNA nucleotide sequences into amino acid sequences in a polypeptide chain)
Proteasome	Multi-subunit protease complexes involved in regulating protein turnover
Golgi apparatus and vesicles	Secretory system for import (endocytosis) and export (exocytosis) of proteins
Vacuole	Intracellular reservoir (amino acids, polyphosphate, and metal ions), proteolysis, protein trafficking, and control of intracellular pH
Peroxisome	Present in some methylotrophic (methanol-utilizing) yeasts for oxidative utilization of specific carbon and nitrogen sources (contain catalase and oxidases). Glyoxysomes contain enzymes of the glyoxylate cycle

Reproduced from Walker, G. M. and White, N. A. (2005). Introduction to fungal physiology. In: K. Kavanagh (Ed.), *Fungi: Biology and applications* (pp. 1–34, Chapter 2). Chichester, UK: Wiley.

TABLE 5 Classification of yeasts based on fermentative property/growth response to oxygen availability

Class	Examples	Comments
Obligately fermentative	*Candida pintolopesii* (*Saccharomyces telluris*)	Naturally occurring respiratory-deficient yeasts. Only ferment, even in the presence of oxygen
Facultatively fermentative		
Crabtree-positive	*Saccharomyces cerevisiae*	Such yeasts predominantly ferment high-sugar-containing media in the presence of oxygen (respirofermentation)
Crabtree-negative	*Candida utilis*	Such yeasts do not form ethanol under aerobic conditions and cannot grow anaerobically
Nonfermentative	*Rhodotorula rubra*	Such yeasts do not produce ethanol, in either the presence or absence of oxygen

being kept constant. In industry, yeasts are grown in a variety of fermentation feedstocks, including malt wort, molasses, grape juice, cheese whey, glucose syrups, sulfite liquor, and lignocellulosic hydrolysates.

Physical requirements for yeast growth

Most yeast species thrive in warm, dilute, sugary, acidic, and aerobic environments. Most laboratory and industrial yeasts (e.g., *S. cerevisiae* strains) grow best from 20 to 30 °C. The lowest maximum temperature for growth of yeasts is around 20 °C, whereas the highest is around 50 °C. Yeasts cannot really be described as thermophiles when compared to certain thermophilic bacteria and fungi that can grow at temperatures far in excess of 50 °C.

Yeasts need water in high concentration for growth and metabolism. Several food spoilage yeasts (e.g., *Zygosaccharomyces* spp.) are able to withstand conditions of low water potential (i.e., high sugar or salt concentrations), and such yeasts are referred to as osmotolerant or xerotolerant.

Most yeasts are acidophilic and grow very well between pH 4.5 and 6.5. Media acidified with organic acids (e.g., acetic and lactic) are more inhibitory to yeast growth than are media acidified with mineral acids (e.g., hydrochloric). This is because undissociated organic acids can lower intracellular pH following their translocation across the yeast cell membrane. This forms the basis of the action of weak acid preservatives in inhibiting food spoilage yeast growth. Actively growing yeasts acidify their growth environment through a combination of differential ion uptake, proton secretion during nutrient transport (see later), direct secretion of organic acids (e.g., succinate and acetate), and carbon dioxide evolution and dissolution. Intracellular pH is regulated within relatively narrow ranges in growing yeast cells (e.g., around pH 5 in *S. cerevisiae*), mainly through the action of the plasma membrane proton-pumping ATPase.

Most yeasts are aerobes. Yeasts are generally unable to grow well under completely anaerobic conditions because, in addition to providing the terminal electron acceptor in respiration, oxygen is needed as a growth factor for membrane fatty acid (e.g., oleic acid) and sterol (e.g., ergosterol) biosynthesis. In fact, *S. cerevisiae* is auxotrophic for oleic acid and ergosterol under anaerobic conditions and this yeast is not, strictly speaking, a facultative anaerobe. Table 5 categorizes yeasts based on their fermentative properties and growth responses to oxygen availability.

Carbon Metabolism by Yeasts

Carbon sources for yeast growth

As chemorganotrophic organisms, yeasts obtain carbon and energy in the form of organic compounds. Sugars are widely used by yeasts. *S. cerevisiae* can grow well on glucose, fructose, mannose, galactose, sucrose, and maltose. These sugars are also readily fermented into ethanol and carbon dioxide by *S. cerevisiae*, but other carbon substrates such as ethanol, glycerol, and acetate can be respired by *S. cerevisiae* only in the presence of oxygen. Some yeasts (e.g., *Pichia stipitis* and *Candida shehatae*) can use five-carbon pentose sugars such as D-xylose and L-arabinose as growth and fermentation substrates. A few amylolytic yeasts (e.g., *Saccharomyces diastaticus* and *Schwanniomyces occidentalis*) that can use starch exist, and several oleaginous yeasts (e.g., *Candida tropicalis* and *Yarrowia lipolytica*) can grow on hydrocarbons, such as straight-chain alkanes in the C_{10}–C_{20} range. Several methylotrophic yeasts (e.g., *Hansenula polymorpha* and *Pichia pastoris*) can grow very well on methanol as the sole carbon and energy source, and these yeasts have industrial potential in the production of recombinant proteins using methanol-utilizing genes as promoters.

Yeast sugar transport

Sugars are transported into yeast cells across the plasma membrane by various mechanisms such as simple net diffusion (a passive or free mechanism), facilitated (catalyzed) diffusion, and active (energy-dependent) transport. The precise mode of sugar translocation will depend on the sugar, yeast species, and growth conditions. For example, *S. cerevisiae* takes up glucose by facilitated diffusion and maltose by active transport. Active transport means that the plasma membrane ATPases act as directional proton pumps in accordance with chemiosmotic principles. The pH gradients thus drive nutrient transport either via proton symporters (as is the case with certain sugars and amino acids) or via proton antiporters (as is the case with potassium ions).

Yeast sugar metabolism

The principal metabolic fates of sugars in yeasts are the dissimilatory pathways of fermentation and respiration (shown in Figure 3) and the assimilatory pathways of gluconeogenesis and carbohydrate biosynthesis. Yeasts described as fermentative are able to use organic substrates (sugars) anaerobically as electron donors, electron acceptors, and carbon sources. During alcoholic fermentation of sugars, *S. cerevisiae* and other fermentative yeasts reoxidize the reduced coenzyme NADH to NAD (nicotinamide adenine dinucleotide) in terminal step reactions from pyruvate. In the first of these terminal reactions, catalyzed by pyruvate decarboxylase, pyruvate is decarboxylated to acetaldehyde, which is finally reduced by alcohol dehydrogenase to ethanol. The regeneration of NAD is necessary to maintain the redox balance and prevent the stalling of glycolysis. In alcoholic beverage fermentations (e.g., of beer, wine, and distilled spirits), other fermentation metabolites, in addition to ethanol and carbon dioxide, that are very important in the development of flavor are produced by yeast. These metabolites include fusel alcohols (e.g., isoamyl alcohol), polyols (e.g., glycerol), esters (e.g., ethyl acetate), organic acids (e.g., succinate), vicinyl diketones (e.g., diacetyl), and aldehydes (e.g., acetaldehyde). The production of glycerol (an important industrial commodity) can be enhanced in yeast fermentations by the addition of sulfite, which chemically traps acetaldehyde.

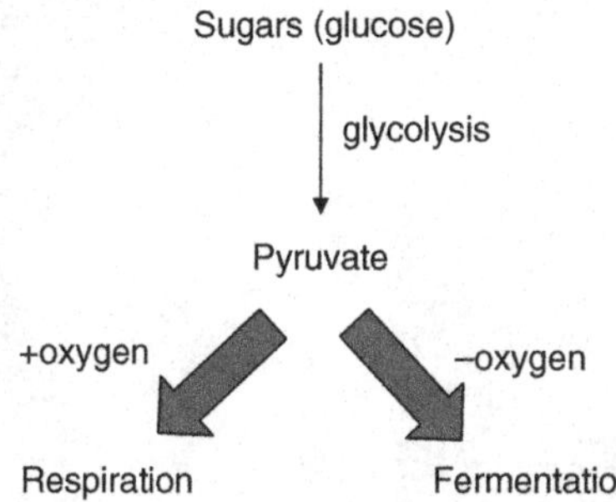

FIGURE 3 Overview of sugar catabolic pathways in yeast cells. *Reproduced from Walker, G. M. (1998)* Yeast physiology and biotechnology. *Chichester, UK: Wiley.*

$$\text{Glucose} + HSO_3^- \rightarrow \text{Glycerol} + \text{Acetaldehyde-}HSO_3^- + CO_2$$

Aerobic respiration of glucose by yeasts is a major energy-yielding metabolic route and involves glycolysis, the citric acid cycle, the electron transport chain, and oxidative phosphorylation. The citric acid cycle (or Krebs cycle) represents the common pathway for the oxidation of sugars and other carbon sources in yeasts and filamentous fungi and results in the complete oxidation of one pyruvate molecule to $2CO_2$, 3NADH, 1FADH2, $4H^+$, and 1GTP.

Of the environmental factors that regulate respiration and fermentation in yeast cells, the availability of glucose and oxygen is best understood and is linked to the expression of regulatory phenomena, referred to as the Pasteur effect and the Crabtree effect. A summary of these phenomena is provided in Table 6.

Nitrogen Metabolism by Yeasts

Nitrogen sources for yeast growth

Although yeasts cannot fix molecular nitrogen, simple inorganic nitrogen sources such as ammonium salts are widely used. Ammonium sulfate is a commonly used nutrient in

TABLE 6 Summary of regulatory phenomena in yeast sugar metabolism

Phenomenon	Description	Examples of yeasts
Pasteur effect	Activation of sugar metabolism by anaerobiosis	*Saccharomyces cerevisiae* (resting or starved cells)
Crabtree effect (short-term)	Rapid ethanol production in aerobic conditions due to sudden excess of glucose (that acts to inactivate respiratory enzymes)	*S. cerevisiae* and *Schizosaccharomyces pombe*
Crabtree effect (long-term)	Ethanol production in aerobic conditions when excess glucose acts to repress respiratory genes	*S. cerevisiae* and *Sch. pombe*
Custers effect	Stimulation of ethanol fermentation by oxygen	*Dekkera* and *Brettanomyces* spp.
Kluyver effect	Anaerobic fermentation of glucose, but not of certain other sugars (disaccharides)	*Candida utilis*

yeast growth media because it provides a source of both assimilable nitrogen and sulfur. Some yeasts can also grow on nitrate as a source of nitrogen, and, if able to do so, may also use subtoxic concentrations of nitrite. A variety of organic nitrogen compounds (amino acids, peptides, purines, pyrimidines, and amines) can also provide the nitrogenous requirements of the yeast cell. Glutamine and aspartic acids are readily deaminated by yeasts and therefore act as good nitrogen sources.

Yeast transport of nitrogenous compounds

Ammonium ions are transported in *S. cerevisiae* by both high-affinity and low-affinity carrier-mediated transport systems. Two classes of amino acid uptake systems operate in yeast cells. One is broadly specific, the general amino acid permease (GAP), and effects the uptake of all naturally occurring amino acids. The other system includes a variety of transporters that display specificity for one or a small number of related amino acids. Both the general and the specific transport systems are energy dependent.

Yeast metabolism of nitrogenous compounds

Yeasts can incorporate either ammonium ions or amino acids into cellular protein, or these nitrogen sources can be intracellularly catabolized to serve as nitrogen sources. Yeasts also store relatively large pools of endogenous amino acids in the vacuole, most notably arginine. Ammonium ions can be directly assimilated into glutamate and glutamine, which serve as precursors for the biosynthesis of other amino acids. The precise mode of ammonium assimilation adopted by yeasts will depend mainly on the concentration of available ammonium ions and the intracellular amino acid pools. Amino acids may be dissimilated (by decarboxylation, transamination, or fermentation) to yield ammonium and glutamate, or they may be directly assimilated into proteins.

Yeast Growth

The growth of yeasts is concerned with how cells transport and assimilate nutrients and then integrate numerous component functions in the cell in order to increase in mass and eventually divide. Yeasts have proven invaluable in unraveling the major control elements of the eukaryotic cell cycle, and research with the budding yeast, *S. cerevisiae*, and the fission yeast, *Schizosaccharomyces pombe*, has significantly advanced our understanding of cell cycle regulation, which is particularly important in the field of human cancer. For example, two scientists, Leland Hartwell and Paul Nurse, were awarded the Nobel Prize for Medicine in 2002 for their pioneering studies on the control of cell division in budding and fission yeasts, respectively.

Vegetative reproduction in yeasts

Budding is the most common mode of vegetative reproduction in yeasts and is typical in ascomycetous yeasts such as *S. cerevisiae*. Figure 4 shows a scanning electron micrograph of budding cells of *S. cerevisiae*. Yeast buds are initiated when mother cells attain a critical cell size at a time that coincides with the onset of DNA synthesis. This is followed by localized weakening of the cell wall and this, together with tension exerted by turgor pressure, allows the extrusion of the cytoplasm in an area bounded by the new cell wall material. The mother and daughter bud cell walls are contiguous during bud development. Multilateral budding is common in

(a)

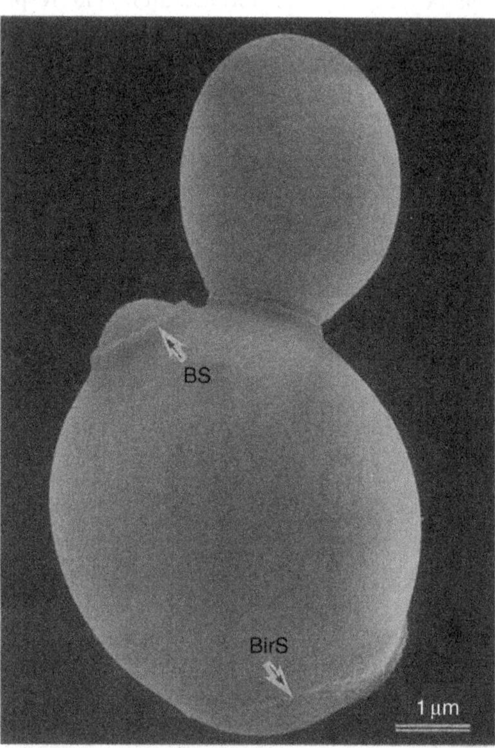

(b)

FIGURE 4 Scanning electron micrographs of budding yeast. (a) Individual cell. BS, bud scar; and BirS, birth scar. *Courtesy of M. Osumi, Japan Women's University: Tokyo. (b) Cluster of cells.*

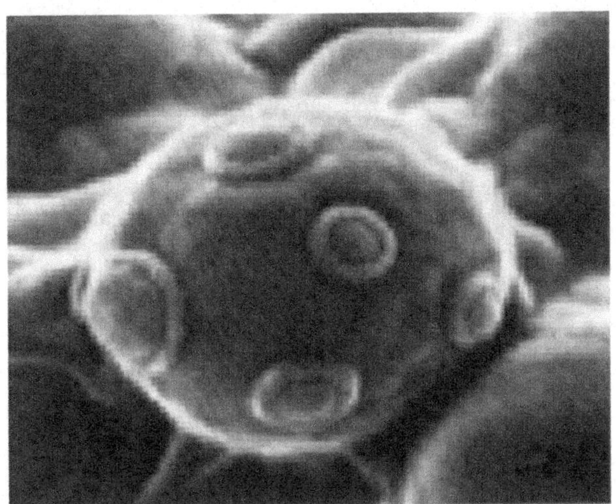

FIGURE 5 Bud scars in a single cell of *Saccharomyces cerevisiae*. The micrograph shows multilateral budding on the surface of an aged cell of *S. cerevisiae*. *Courtesy of Prof. A. Martini, University of Perugia, Italy.*

which daughter buds emanate from different locations on the mother cell surface. Figure 5 shows multilateral budding in *S. cerevisiae*. In *S. cerevisiae*, cell size at division is asymmetrical, with buds being smaller than mother cells when they separate (Figure 6). Some yeast genera (e.g., *Hanseniaspora* and *Saccharomycodes*) undergo bipolar budding, where buds are restricted to the tips of lemon-shaped cells. Scar tissue on the yeast cell wall, known as the bud and birth scars, remain on the daughter bud and mother cells, respectively. These scars are rich in chitin (a polymer of *N*-acetyl glucosamine) and can be stained with fluorescent dyes (e.g., calcoflour white) to provide useful information regarding cellular age in *S. cerevisiae*, since the number of scars represents the number of completed cell division cycles.

Fission is a mode of vegetative reproduction typified by species of *Schizosaccharomyces*, which divide exclusively by forming a cell septum that constricts the cell into two equal-size daughters. In *Sch. pombe*, which has been used extensively in eukaryotic cell cycle studies, newly divided daughter cells grow lengthways in a monopolar fashion for about one-third of their new cell cycle. Cells then switch to bipolar growth for about three-quarters of the cell cycle until mitosis is initiated at a constant cell length stage.

Filamentous growth occurs in numerous yeast species and may be regarded as a mode of vegetative growth alternative to budding or fission. Some yeasts exhibit a propensity to grow with true hyphae initiated from germ tubes (e.g., *C. albicans*, Figure 7), but others (including *S. cerevisiae*) may grow in a pseudohyphal fashion when induced to do so by unfavorable conditions. Hyphal and pseudohyphal growth represent different developmental pathways in yeasts, but cells can revert to unicellular growth upon return to more conducive growth conditions. Filamentation may therefore represent an adaptation to foraging by yeasts when nutrients are scarce.

Population growth of yeasts

As in most microorganisms, when yeast cells are inoculated into a liquid nutrient medium and incubated under optimal physical growth conditions, a typical batch growth curve will result when the viable cell population is plotted against time. This growth curve is made up of a lag phase (period of no growth, but physiological adaptation of cells to their new environment), an exponential phase (limited period of logarithmic cell doublings), and a stationary phase (resting period with zero growth rate).

Diauxic growth is characterized by two exponential phases and occurs when yeasts are exposed to two carbon growth substrates that are used sequentially. This occurs during aerobic growth of *S. cerevisiae* on glucose (the second substrate being ethanol formed from glucose fermentation).

In addition to batch cultivation of yeasts, cells can also be propagated in continuous culture in which exponential

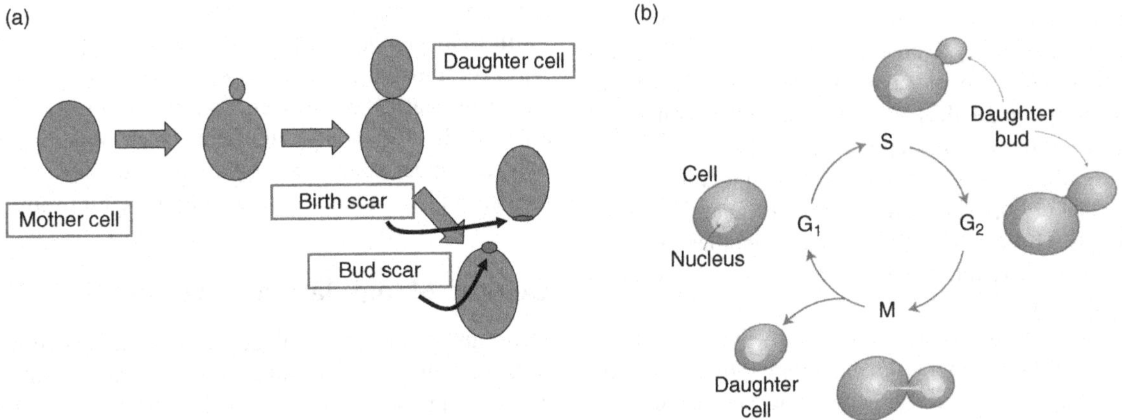

FIGURE 6 Budding processes in yeast. (a) Schematic diagram of budding. (b) Budding cell cycle, as typified by *Saccharomyces cerevisiae*. S, DNA synthesis period; G_1, pre-DNA synthesis gap period; G_2, post-DNA synthesis gap period; and M, mitosis. *Reproduced from Madhani, H. (2007).* From a to α. Yeast as a model for cellular differentiation. *New York: Cold Spring Harbor Laboratory Press.*

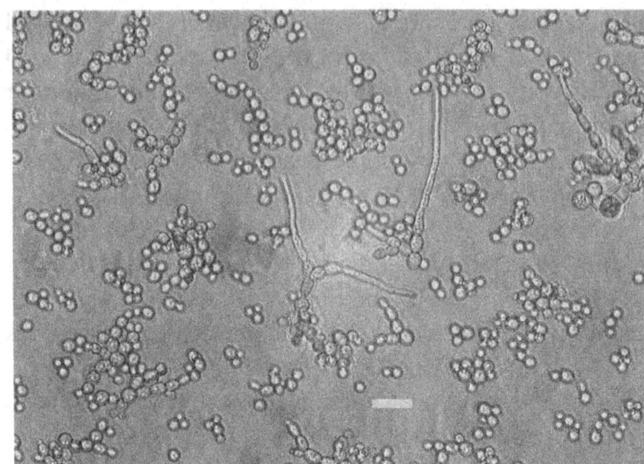

FIGURE 7 Dimorphism in *Candida albicans*. The micrograph shows a mixture of budding cells and hyphal forms of the yeast, which is an important human pathogen.

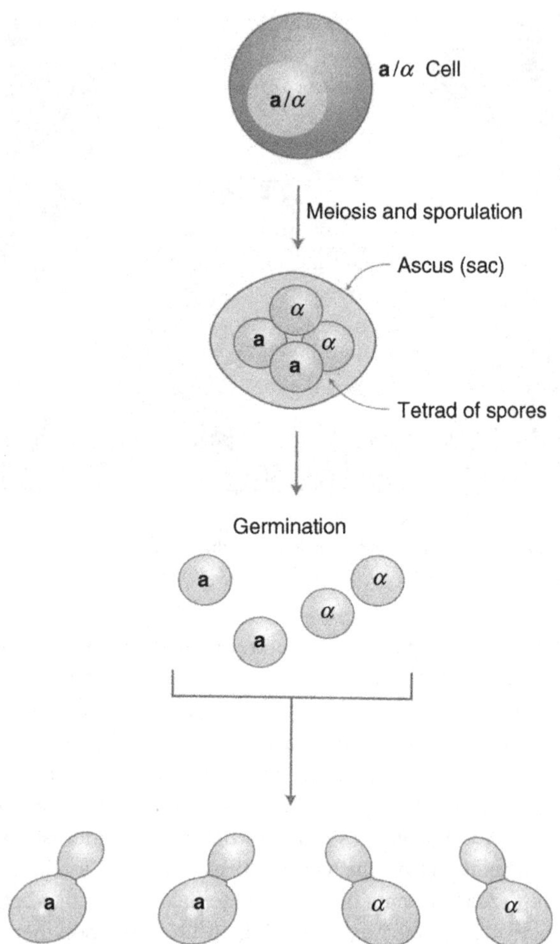

FIGURE 8 Sexual life cycle of *Saccharomyces cerevisiae. Reproduced from Madhani, H. (2007).* From a to α. Yeast as a model for cellular differentiation. *New York: Cold Spring Harbor Laboratory Press.*

growth is prolonged without lag or stationary phases. Chemostats are continuous cultures that are based on the controlled feeding of a sole growth-limiting nutrient into an open culture vessel, which permits the outflow of cells and spent medium. The feeding rate is referred to as the dilution rate, which is employed to govern the yeast growth rate under the steady-state conditions that prevail in a chemostat.

Specialized yeast culture systems include immobilized bioreactors. Yeast cells can be readily immobilized or entrapped in a variety of natural and synthetic materials (e.g., calcium alginate gel, wood chips, hydroxyapatite ceramics, diethylaminoethyl (DEAE) cellulose, or microporous glass beads), and such materials have applications in the food and fermentation industries.

YEAST GENETICS

Life Cycle of Yeasts

Many yeasts have the ability to reproduce sexually, but the processes involved are best understood in the budding yeast, *S. cerevisiae*, and the fission yeast, *Sch. pombe*. Both species have the ability to mate, undergo meiosis, and sporulate. The development of spores by yeasts represents a process of morphological, physiological, and biochemical differentiation of sexually reproductive cells.

Mating in *S. cerevisiae* involves the conjugation of two haploid cells of opposite mating types, designated as **a** and α (Figure 8). These cells synchronize one another's cell cycles in response to peptide mating pheromones, known as **a** factor and α factor.

The conjugation of mating cells occurs by cell wall surface contact followed by plasma membrane fusion to form a common cytoplasm. Karyogamy (nuclear fusion) then follows, resulting in a diploid nucleus. The stable diploid zygote continues the mitotic cell cycles in rich growth media, but if starved of nitrogen, the diploid cells sporulate to yield four haploid spores. These germinate in rich media to form haploid budding cells that can mate with each other to restore the diploid state. Figure 9 shows mating and sporulation in *S. cerevisiae*.

In *Sch. pombe*, haploid cells of the opposite mating types (designated h^+ and h^-) secrete mating pheromones and, when starved of nitrogen, undergo conjugation to form diploids. In *Sch. pombe*, however, such diploidization is transient under starvation conditions and cells soon enter meiosis and sporulate to produce four haploid spores.

Genetic Manipulation of Yeasts

There are several ways of genetically manipulating yeast cells, including hybridization, mutation, rare mating, cytoduction, spheroplast fusion, single chromosome transfer, and transformation using recombinant DNA technology. Classic genetic approaches in *S. cerevisiae* involve mating of haploids of opposite mating types. Subsequent meiosis

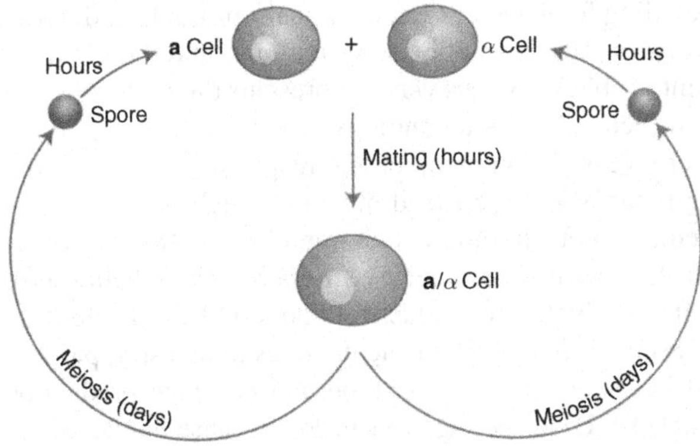

FIGURE 9 Meiosis and sporulation in *Saccharomyces cerevisiae*. Diploid (**a**/α) cells can undergo meiosis and sporulation to form spores that can germinate into **a** and α haploid cells. *Reproduced from Madhani, H, (2007).* From a to α. Yeast as a model for cellular differentiation. *New York: Cold Spring Harbor Laboratory Press.*

and sporulation result in the production of a *tetrad ascus* with four spores, which can be isolated, propagated, and genetically analyzed (i.e., tetrad analysis). This process forms the basis of genetic breeding programs for laboratory reference strains of *S. cerevisiae*. However, industrial (e.g., brewing) strains of this yeast are polyploid, are reticent to mate, and exhibit poor sporulation with low spore viability. It is, therefore, generally fruitless to perform tetrad analysis and breeding with brewer's yeasts. Genetic manipulation strategies for preventing the sexual reproductive deficiencies associated with brewer's yeast include spheroplast fusion and recombinant DNA technology.

Intergeneric and intrageneric yeast hybrids may be obtained using the technique of spheroplast fusion. This involves the removal of yeast cell walls using lytic enzymes (e.g., glucanases from snail gut juice or microbial sources), followed by the fusion of the resulting spheroplasts in the presence of polyethylene glycol and calcium ions.

Recombinant DNA technology (genetic engineering) of yeast is summarized in Figure 10 and transformation strategies in Figure 11. Yeast cells possess particular attributes for expressing foreign genes and have now become the preferred hosts, over bacteria, for producing certain human proteins for pharmaceutical use (e.g., insulin, human serum albumin, and hepatitis vaccine). Although the majority of research and development in recombinant protein synthesis in yeasts has been conducted using *S. cerevisiae*, several non-*Saccharomyces* species are being studied and exploited in biotechnology. For example, *H. polymorpha* and *P. pastoris* (both methylotrophic yeasts) exhibit particular advantages over *S. cerevisiae* in cloning technology (see Table 8).

Yeast Genome and Proteome Projects

A landmark in biotechnology was reached in 1996 with completion of the sequencing of the entire genome of a eukaryotic cell – the yeast, *S. cerevisiae*. The *Sch. pombe* genome was sequenced in 2002, and many other yeast species have now had their genomes sequenced (e.g., *C. albicans*, *Kluyveromyces lactis*, *P. stipitis*). The functional analysis of the many orphan genes of *S. cerevisiae*, for which no function has yet been assigned, is under way through international research collaborations. Elucidation by cell physiologists of the biological function of all *S. cerevisiae* genes, that is, the complete analysis of the yeast proteome, will not only lead to an understanding of how a simple eukaryotic cell works but also provide an insight into molecular biological aspects of heritable human disorders.

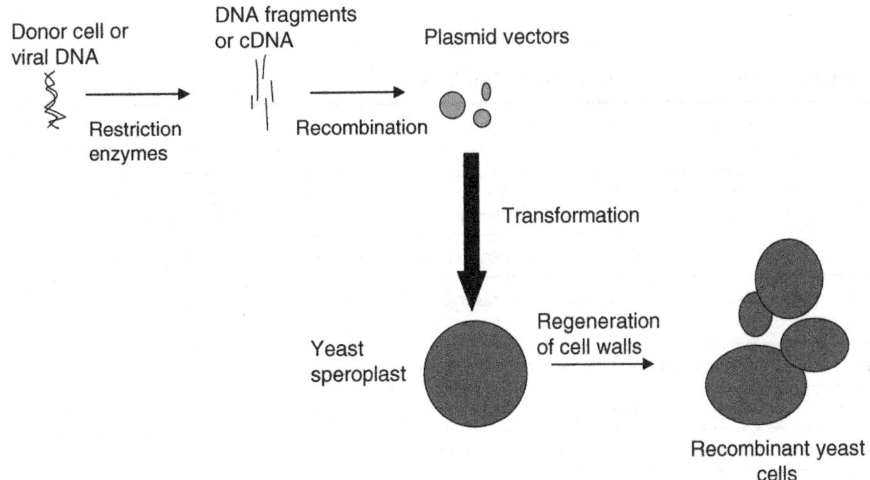

FIGURE 10 Basic procedures in yeast genetic engineering. *Reproduced from Walker, G. M. (1998).* Yeast physiology and biotechnology. *Chichester, UK: Wiley.*

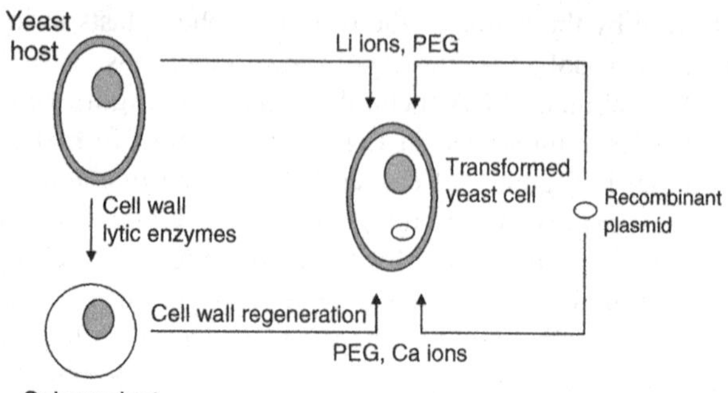

FIGURE 11 Yeast transformation strategies. PEG, polyethylene glycol.

INDUSTRIAL, AGRICULTURAL, AND MEDICAL IMPORTANCE OF YEASTS

Industrial Significance of Yeasts

Yeasts have been exploited for thousands of years in traditional fermentation processes to produce beer, wine, and bread. The products of modern yeast biotechnologies impinge on many commercially important sectors, including food, beverages, chemicals, biofuels, industrial enzymes, pharmaceuticals, agriculture, and the environment (Table 7). *S. cerevisiae* represents the primary yeast 'cell factory' in biotechnology and is the most exploited microorganism known, being responsible for producing potable and industrial ethanol, which is the world's premier biotechnological commodity. However, other non-*Saccharomyces* species are increasingly being used in the production of industrial commodities (Table 8).

Some yeasts play detrimental roles in industry, particularly as spoilage yeasts in food and beverage production (Table 9). Food spoilage yeasts do not cause human infections or intoxications, but do deleteriously affect food nutritive quality and are of economic importance for food producers.

In addition to their traditional roles in food and fermentation industries, yeasts are finding increasingly important roles in the environment and in the health care sector of biotechnology. Yeasts are also invaluable as model eukaryotic cells in fundamental biological and biomedical research (Figure 12).

TABLE 7 Industrial commodities produced by yeasts

Commodity	Examples
Beverages	Potable alcoholic beverages: beer, wine, cider, sake, and distilled spirits (whisky, rum, gin, vodka, and cognac)
Food and animal feed	Baker's yeast, yeast extracts, fodder yeast, livestock growth factor, and feed pigments
Chemicals	Fuel ethanol (bioethanol) carbon dioxide, glycerol, and citric acid vitamins; yeasts are also used as bioreductive catalysts in organic chemistry
Enzymes	Invertase, inulinase, pectinase, lactase, and lipase
Recombinant proteins	Hormones (e.g., insulin), viral vaccines (e.g., hepatitis B vaccine), antibodies (e.g., IgE receptor), growth factors (e.g., tumor necrosis factor), interferons (e.g., leukocyte interferon-α), blood proteins (e.g., human serum albumin), and enzymes (e.g., gastric lipase and chymosin)

TABLE 8 Uses of non-*Saccharomyces* yeasts in biotechnology

Yeast	Uses
Candida spp.	Many uses in foods, chemicals, pharmaceuticals, and xylose fermentation (*C. shehatae*)
Kluyveromyces spp.	Lactose, inulin-fermented, rich sources of enzymes (lactase, lipase, pectinase, and recombinant chymosin)
Hansenula and *Pichia*	Cloning technology. Methylotophic yeasts (*H. polymorpha* and *P. pastoris*)
Saccharomycopsis and *Schwanniomyces*	Amylolytic yeasts (starch-degrading)
Schizosaccharomyces	Cloning technology, fuel alcohol, some beverages (rum), and biomass protein
Starmerella	Wine flavor during fermentation
Yarrowia	Protein from hydrocarbons (*Y. lipolytica*)
Zygosaccharomyces	High salt/sugar fermentations (soy sauce)

TABLE 9 Some yeasts important in food production and food spoilage

Yeast genus	Importance in foods
Candida spp.	Some species (e.g., *C. utilis, C. guilliermondii*) are used in the production of microbial biomass protein, vitamins, and citric acid. Some species (e.g., *C. zeylanoides*) are food spoilers in frozen poultry
Cryptococcus spp.	Some strains are used as biocontrol agents to combat fungal spoilage of postharvest fruits. *C. laurentii* is a food spoilage yeast (poultry)
Debaryomyces spp.	*D. hansenii* is a salt-tolerant food spoiler (e.g., meats and fish). Also used in biocontrol of fungal fruit diseases
Kluyveromyces spp.	Lactose-fermenting yeasts are used to produce potable alcohol from cheese whey (*K. marxianus*). Source of food enzymes (pectinase, microbial rennet, and lipase) and found in cocoa fermentations. Spoilage yeast in dairy products (fermented milks and yoghurt)
Metschnikowia spp.	*M. pulcherrimia* is used in biocontrol of fungal fruit diseases (post-harvest). Osmotolerant yeasts
Phaffia spp.	*P. rhodozyma* is a source of astaxanthin food colorant used in aquaculture (feed for salmonids)
Pichia spp.	Production of microbial biomass protein, riboflavin (*P. pastoris*). *P. membranefaciens* is an important surface film spoiler of wine and beer
Rhodotorula spp.	*R. glutinis* is used as a source of food enzymes such as lipases. Some species are food spoilers of dairy products
Saccahromyces spp.	*S. cerevisiae* is used in traditional food and beverage fermentations (baking, brewing, winemaking, etc.), source of savory food extracts, and food enzymes (e.g., invertase). Also used as fodder yeast (livestock growth factor). *S. bayanus* is used in sparkling wine fermentations, *S. diastaticus* is a wild yeast spoiler of beer, and *S. boulardii* is used as a probiotic yeast
Schizosaccharomyces spp.	*Sch. pombe* is found in traditional African beverages (sorghum beer), rum fermentations from molasses, and may be used for wine deacidification. Regarded as an osmotolerant yeast
Schwanniomyces spp.	Starch-utilizing yeasts. *Sch. castellii* may be used for production of microbial biomass protein from starch
Yarrowia spp.	*Y. lipolytica* is used in production of microbial biomass protein, citric acid, and lipases
Zygosaccharomyces spp.	*Z. rouxii* and *Z. bailii*, being osmotolerant, are important food and beverage (e.g., wine) spoilage yeasts. *Z. rouxii* is also used in soy sauce production

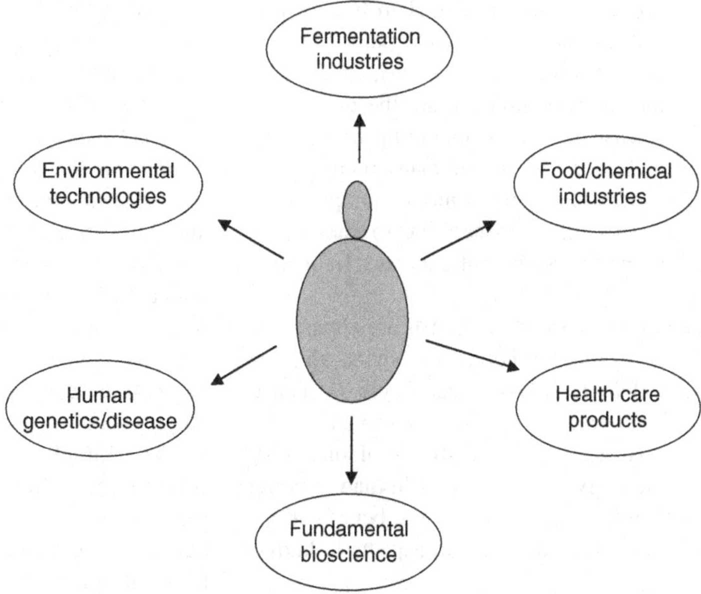

FIGURE 12 Uses of yeasts in biotechnology.

Yeasts of Environmental and Agricultural Significance

A few yeast species are known to be plant pathogens. For example, *Ophiostoma novo-ulmi* is the causative agent of Dutch Elm disease, and members of the genus *Eremothecium* cause diseases such as cotton ball in plants. On the contrary, several yeasts have been shown to be beneficial to plants in preventing fungal disease. For example, *S. cerevisiae* has potential as a phytoallexin elicitor in stimulating cereal plant defenses against fungal pathogens, and several yeasts (e.g., *Cryptococcus laurentii*, *Metschnikowia pulcherrima*, *Pichia anomala*, and *Pichia guilliermondii*) may be used in the biocontrol of fungal fruit and grain spoilage, especially in preventing postharvest fungal deterioration. Other environmental benefits of yeasts are to be found in aspects of pollution control. For example, yeasts can effectively biosorb heavy metals and detoxify chemical pollutants from industrial effluents. Some yeasts (e.g., *Candida utilis*) can effectively remove carbon and nitrogen from organic wastewaters, and others (e.g., *P. anomala*) can bioremediate toxic polluting chemicals such as polyaromatic hydrocarbons.

In agriculture, live cultures of *S. cerevisiae* have been shown to stabilize the rumen environment of ruminant animals (e.g., cattle) and improve nutrient availability to increase animal growth or milk yields. The yeasts may be acting to scavenge oxygen and prevent oxidative stress to rumen bacteria, or they may provide malic and other dicarboxylic acids to stimulate rumen bacterial growth.

Medical Significance of Yeasts

The vast majority of yeasts are beneficial to human life. However, some yeasts are opportunistically pathogenic toward humans. Mycoses caused by *C. albicans*, collectively referred to as candidosis (candidiasis), are the most common opportunistic yeast infections, including nosocomial (hospital-derived) infections. There are many predisposing factors to yeast infections, but immunocompromised individuals appear particularly susceptible to candidosis. *C. albicans* infections in AIDS patients are frequently life-threatening.

The beneficial medical aspects of yeasts are apparent in the provision of novel human therapeutic agents through yeast recombinant DNA technology (see Table 7). Yeasts are also extremely valuable as experimental models in biomedical research, particularly in the fields of oncology, pharmacology, toxicology, virology, and human genetics (Table 10). It is therefore apparent that the beneficial aspects of yeasts far outweigh detrimental aspects in human health and wellbeing.

TABLE 10 Value of yeasts in biomedical research

Biomedical field	Examples
Oncology	Basis of cell cycle control, human oncogene (e.g., Ras) regulation; telomere function, tumor suppressor function, and design of (cyclin-dependent kinase inhibitors) CDIs/anticancer drugs
Aging	Mechanisms of cell aging, longevity genes, and apoptosis
Pharmacology	Multidrug resistance, drug action/metabolism, and drug screening assays
Virology	Viral gene expression, antiviral vaccines, and prion structure/function
Human genetics	Basis of human hereditary disorders and genome/proteome projects

OVERVIEW OF RECENT PROGRESS IN YEAST SCIENCE AND TECHNOLOGY

Since publication of *Yeasts* (Walker, GM: *Enclyclopedia of Microbiology*), there have been several advances in both fundamental and applied aspects of yeast science and technology, and the current chapter has been updated to reflect some of these developments. Additional new references have been added to the Further Reading list.

In relation to yeast biodiversity, there have been many more yeast species isolated from the environment that have now been characterized and identified using advanced phylogenetic analysis of gene sequences. Currently described yeast species now stands at over 1500 with new yeasts being named on a regular basis. Some natural habitats, such as insects, are proving rich sources of yeast biodiversity. Major advances are being made in the fields of yeast 'omics'. For example, many yeasts in addition to *S. cerevisiae* and *Sch. pombe* (two very well studied model yeasts) have now had their genomes fully sequenced and these include *C. albicans* (an opportunistic human pathogen), *K. lactis* (a lactose-fermenting yeast), and *P. stipitis* (a xylose-fermenting yeast). In terms of yeast genomic and proteomic analysis, several recent advances have been made – not least transcriptional profiling and the identification of differential gene expression during industrial yeast fermentations and following exposure of yeast cells to environmental stress. Improvement of existing yeast strains by metabolic engineering has also progressed to the point where complete metabolic pathways can be manipulated for optimizing production of industrial commodities. One of these commodities which has seen a very dramatic rise in recent years is bioethanol. The production of fuel alcohol by yeast fermentation is now contributing in a major way to reducing dependence on imported fossil fuels and reducing greenhouse gas emissions. The two countries that are dominating world bioethanol production are the United States and Brazil, exploiting maize and sugarcane as their respective feedstocks. Emerging on the horizon is bioethanol produced from second-generation substrates such as lignocellulosic wastes and the so-called energy crops. These processes are near to commercial reality and involve a combination of yeast and enzyme technology. Other biofuels from yeast are also being considered for exploitation, for example, butanol (which has several key advantages over ethanol as a transportation fuel). In the biopharmaceutical sector, several recombinant yeast strains are being exploited for the biosynthesis of valuable novel drugs, vaccines, growth factors, hormones, etc. This sector is likely to see further developments as new yeasts are being constructed through recombinant DNA technology and metabolic engineering. Many of these yeasts are non-*Saccharomyces* species that possess several advantages over *S. cerevisiae* in terms of efficiency and yields of recombinant protein production. In addition to biotechnological exploitation of yeasts, it should not be forgotten that species such as *S. cerevisiae* and *Sch. pombe* represent the most valuable eukaryotic model microbes. Fundamental molecular biological knowledge gained from studies with yeast cells is continuing to provide us with fascinating insight into the basis of human disease and genetic disorders.

Finally, although yeasts have been inextricably linked with human activities for millennia, they remain at the forefront of biotechnology and basic science and will continue to occupy a position as the world's premier industrial microbes and model eukaryotes for many years to come.

FURTHER READING

Barnett, J. A., Payne, R. W., & Yarrow, D. (2000). *Yeasts: Characteristics and identification* (3rd edn.). Cambridge University Press: Cambridge.

Boekhout, T., Robert, V., Smith, M. T., *et al.* (2002). Yeasts of the world. Morphology, physiology, sequences and identification. World Biodiversity Database CD-ROM Series. In ETI: University of Amsterdam.

Boulton, C., & Quain, D. (2006). *Brewing yeast and fermentation*. Blackwell Science: Oxford.

Deak, T. (2008). *Handbook of food spoilage yeasts* (2nd edn.). Taylor and Francis: Boca Raton, FL.

De Winde, J. H. (2003). *Functional genetics of industrial yeasts*. Springer: Berlin & Heidelberg.

Fantes, P., & Beggs, J. (2000). *The yeast nucleus*. Oxford University Press: Oxford.

Gerngross, T. U. (2004). Advances in the production of human therapeutic proteins in yeast and filamentous fungi. *Nature Biotechnology, 22*, 1409.

Guthrie, C., & Fink, G. R. (2002). Guide to yeast genetics and molecular biology. In *Methods in enzymology: Vol. 351*. Academic Press: Amsterdam.

Kurtzman, C. P., & Fell, J. W. (1998). *The yeasts: A taxonomic study* (4th edn. Elsevier Science: Amsterdam.

Kurtzman, C. P., & Piskur, J. (2006). Taxonomy and phylogenetic diversity among the yeasts. In: P. Sunnerhagen & J. Piskur (Eds.), *Comparative genomics: using fungi as models*: *Vol. 15* (pp. 29–46). Springer: Berlin.

Linder, P., Shore, D., & Hall, M. N. (2006). *Landmark papers in yeast biology*. Cold Spring Harbor Laboratory Press: New York.

Madhani, H. (2007). *From a to α: Yeast as a model for cellular differentiation*. Cold Spring Harbor Laboratory Press: New York.

Martini, A. (2005). *Biological diversity of yeasts DVD*. Insight Media: New York.

Nevoigt, E. (2008). Progress in metabolic engineering of *Saccharomyces cerevisiae*. *Microbiology and Molecular Biology Reviews, 72*, 379–412.

Querol, A., & Fleet, G. H. (2006). *Yeasts in food and beverages*. Springer: Berlin.

Rosa, C. A., & Peter, G. (2006). *The yeast handbook*. Springer-Verlag: Berlin & Heidelberg.

Strathern, J. N. (2002). *The molecular biology of the yeast Saccharomyces*. Cold Spring Harbor Laboratory Press: New York.

Walker, G. M., & White, N. A. (2005). Introduction to fungal physiology. In: K. Kavanagh (Ed.), *Fungi: Biology and applications* (pp. 1–34). Wiley: Chichester, UK ch. 2.

Wolf, K., Breunig, K., & Barth, G. (Eds.), (2003). *Non-conventional yeasts in genetics, biochemistry and Biotechnology*. Springer: Berlin.

Chapter 2

Aspergillus

A Multifaceted Genus

C. Scazzocchio
Institut de Génétique et Microbiologie, Université Paris-Sud, Orsay, France; Department of Microbiology, Imperial College London, London, UK

Chapter Outline

Abbreviations **19**
Defining Statement **19**
Note 19
What is *Aspergillus*? **19**
Food Contamination by Aspergilli **21**
***Aspergillus* as Pathogens** **21**
Aspergillus as Human Pathogens 21
Aspergillus in Veterinary Medicine 24
A. sydowii, a Specific Pathogen for Gorgonian Corals 24
Useful Aspergilli **25**
Oriental Food Uses of *Aspergillus* 25
Extracellular Enzymes Produced by Aspergilli: Aspergilli as Hosts for Recombinant Proteins 25
Aspergillus and Production of Organic Acids 26
Medically Useful Secondary Metabolites 26
A. *(Emericella) nidulans* as a Model Organism **26**
The *A. nidulans* Genetic System 26
The Mitochondrial DNA of *A. nidulans* 27
***A. nidulans* as a Model for Genetic Metabolic Diseases** **28**
Control of Gene Expression 29
A. nidulans as a Model for Cell Biology 33
A. nidulans Developmental Pathways 35
The Genus *Aspergillus* in the Genomic Era **38**
Further Reading **39**
Relevant Websites **40**

ABBREVIATIONS

APC Anaphase-promoting complex
HMG-CoA 3-Hydroxy-3-methylglutaryl-coenzyme A
ORF Open reading frame
ROS Reactive oxygen species

DEFINING STATEMENT

In this article, different aspects of ongoing work in the genus *Aspergillus* are discussed, ranging from toxin production, pathogenicity to humans and animals, traditional and modern biotechnological uses, genomics, and the use of *Aspergillus nidulans* as a model organism to study fundamental problems of cell and molecular biology.

Note

For *Aspergillus* genes and proteins, the standard nomenclature is followed, for example, the *benA* gene encodes the BenA protein, which is a β-tubulin. When genes or proteins of other species are mentioned, the standard nomenclature for each species is used.

WHAT IS *ASPERGILLUS*?

An *Aspergillum* is an instrument used in the Roman Catholic mass to sprinkle holy water over the heads of the faithful. *Aspergillus* is a genus of the ascomycete fungi (see below). In 1729, Pietro Antonio Micheli, priest and botanist, described the asexual spore heads (conidiophores; see below) of a number of common molds. The heads of some of these molds showed rows of spores radiating from a globular central structure, which he thought resembled the *Aspergillum* he was familiar with. The morphology of the conidiophore is still an essential taxonomic marker. Figure 1 compares the original Micheli's drawing with modern observations of conidiophores.

The classification of the kingdom fungi into major groups (Phyla, such as Ascomycota and Basidiomycota) is based on the morphology of the sexual reproductive structures. The Aspergilli should be placed among the

Eukaryotic Microbes.

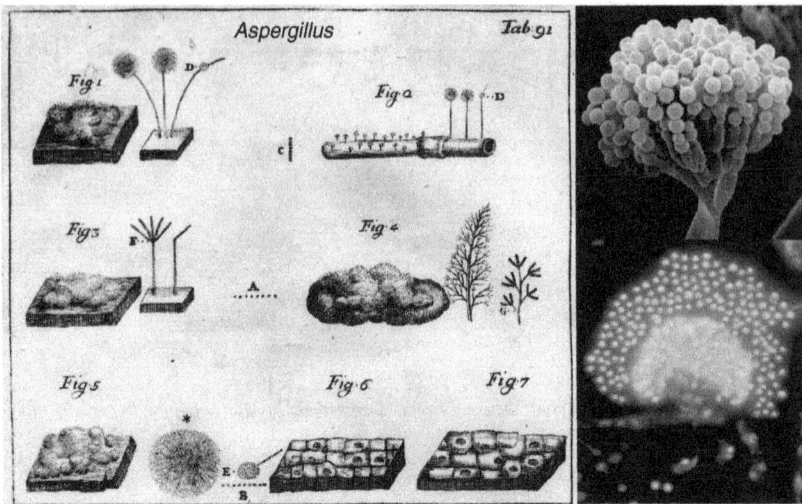

FIGURE 1 The left panel shows a scan of copper-engraving 91 from Micheli's *Nova plantarum genera*, showing his drawings of *Aspergillus* conidiophores. The description in Micheli's text suggest that Figure 1 of the engraving, called *Aspergillus capitatus* (*muffa turchina*, blue mold) by Micheli, may correspond to *Aspergillus fumigatus* or a close relative. The right panel shows on the top a scanning electron micrograph of the conidiophore of *Aspergillus nidulans* and at the bottom an epifluorescence micrograph. The preparation is stained with DAPI which reveals the nuclei of the conidia and of the subjacent structures of the conidiophore. Both pictures on the right panel have kindly provided by Reinhard Fischer. Reproduced with permission from Kues, U., & Fischer, R. (Eds.) (2006) *The micota I: Growth differentiation and sexuality*. Berlin: Springer. Micheli describes in his text the condiophore as been formed by a stalk, and a head, which he called 'placenta' carrying the conidia. See legend to Figure 13 for the correspondence with the modern terminology.

ascomycetes (see below), those fungi that have the products of meiosis placed in a sac or ascus. However, most of the fungi we call Aspergilli (see below) do not have sexual reproduction, and thus no asci. To solve this problem, mycologists created the group fungi imperfecti or otherwise called deuteromycetes in which they placed all fungi without known sexual reproduction. This is a mixed bag without any phylogenetic significance. This provokes ridiculous situations, by which a fungus would change genus, and in fact phylum, every time sexual reproduction is detected. Thus, the 'imperfect' fungus *Aspergillus nidulans* (see below) becomes the 'perfect' fungus *Emericella nidulans*, and it is placed in a different phylum from *Aspergillus sydowii*, in spite of the fact that morphological and molecular data show the two organisms to be close relatives. Names like *Emericella*, *Eurotium*, and *Neosartorya* design Aspergilli with a sexual cycle (also called teleomorphs). Thus, *E. nidulans* is the teleomorph (perfect form) of *A. nidulans* (anamorph, imperfect form). No one but a mycologist would know that we are talking about one and same organism.

This situation exists for many other genera. The only solution to this conundrum is to completely abandon the division 'fungi imperfecti' and choose in each case one and only one name for a given genus. As early as in 1926, Thom and Church, Thom and Raper (1945), and Raper and Fenell (1973) proposed that "the generic name *Aspergillus* should be applied to all these fungi whether or not an ascosporic (sexual) stage was produced." The main morphological characteristics of the genus, drastically abbreviated from Raper and Fenell (1973), are "vegetative mycelium consisting of septate branching hyphae ... Conidial apparatus developed as condiophores ... conidiophores ... broadening into turbinate elliptical, hemispherical, or globose fertile vesicles ... bearing fertile cell or sterigmata ... conidia (asexual spores) ... produced successively from the sterigmata. Ascocarps (asci, containing sexual spores) found in certain groups only, unknown in most species." Recent work shows that the Aspergilli are as a whole a monophyletic group, where loss of sexual reproduction has occurred many times independently.

Not everyone agrees with the reasonable proposal of Raper and Fenell. A recent classification (2005) of the ascomycetes places the *Aspergillus*-related teleomorph genus names in the kingdom Fungi, subphylum Pezizomycotina, class Pezizomycetes, family Trichocomaceae. The genera *Aspergillus* and the related *Penicillium* do not appear in this list as they are considered imperfect forms! Throughout this article, the generic name *Aspergillus* is used, as a genus comprising all the related 'teleomorphs' together with the forms where sexual reproduction is absent. The teleomorph name is also indicated when appropriate, as this is used in some important databases (as NCBI). The kiss of death to the concept of 'fungi imperfecti' was delivered by recent molecular data that show that the genomes of 'imperfect' Aspergilli, include the genes that determine mating types, these genes being clearly homologous to, and even placed in the same place in the chromosome as, the ones found in the 'perfect forms' (see section 'The Genus *Aspergillus* in the Genomic Era').

Fungi of the genus *Aspergillus*, which includes about 200 species, are important in public health as toxin-producing food contaminants, as human and animal pathogens, as useful fungi in traditional and modern biotechnological processes, and finally one species has been used as a model to study a number of cellular processes. A recent development is the availability of eight complete genomes within the genus, a matter of obvious practical, taxonomical, and evolutionary importance.

FOOD CONTAMINATION BY ASPERGILLI

Many organisms, including bacteria, fungi, and plants, produce secondary metabolites. These are molecules that can be very complex and are not obviously necessary for the viability of the organism. In fungi, they are produced during the stationary phase and their synthesis is usually coordinated with asexual sporulation (see below). Some secondary metabolites are extremely toxic, and when fungi grow on stored foods, they secrete them, provoking food spoilage and eventually intoxications that may be fatal. Among the Aspergilli, the two main culprits are strains of *Aspergillus flavus* and *Aspergillus parasiticus*, which secrete aflatoxins, a group of highly substituted coumarins. Strains that are closely related may vary drastically in their ability to produce the toxin. These saprophytic fungi can grow on a variety of foodstuffs, or even on plants before harvesting. In fact, *A. flavus* can be considered a weak opportunistic, nonspecific plant pathogen. The aflatoxins were discovered in 1960 when thousands of turkeys died in an English hatchery. The contaminated food was a ground peanut meal. The most serious contamination is that of maize. While this contamination results in loss of hundreds of million dollars every year to farmers in developed countries, the impact on human health is extremely serious in developing countries. Of the related compounds called aflatoxins, Aflatoxin B1 is one of the most toxic and carcinogenic compounds known, as judged by tests on laboratory animals. Maize stored under warm and humid conditions becomes contaminated with aflatoxigenic Aspergilli, and when consumed by humans or animals, this can lead to liver failure and death. Periodic outbreaks of acute aflatoxin poisoning occurred in East Africa, the latest in 2004, leading to 125 deaths. It is more difficult to assess the damage caused by chronic aflatoxin poisoning and the correlation of the toxin in food with the frequency of liver cancer. Controls on aflatoxin levels are tight in developed countries; they are, however, impossible to be enforced in developing countries, where people would store grains in their homes and the stored grain may be the only available food. Human aflatoxicosis is a disease of poverty.

The Aspergilli can contaminate food with other toxic molecules. Only the ochratoxins, produced by a number of Aspergilli and Penicillia will be discussed below. The ochratoxins comprise an isocumarin moiety and a phenylalanine ring joined by an amide bond. Ochratoxin contamination has been reported in many foodstuffs, including grapes, nuts, cacao, coffee beans, and spices. In poultry, and laboratory animals, ochratoxins provoke serious kidney lesions. It is difficult to assess damage to human health caused by chronic exposure to ochratoxins. They have been implicated as a cause of testicular cancer. The similarity of symptoms of porcine mycotoxin nephropathy with that of Balkan endemic nephropathy, a disease localized to regions of Bulgaria, Romania, and former Yugoslavia, has implicated ochratoxins as causal agents of the disease. A similar case can be made for chronic interstitial nephropathy of northern Africa. A case of acute renal failure, almost certainly due to the exposure to an ochratoxin, has revived the hypothesis that exposure to this mycotoxin is the cause of the 'mummy curse,' which is alleged to have killed archeologists who have braved the prohibition to open royal tombs.

ASPERGILLUS AS PATHOGENS

The common fungus diseases are mild and superficial, while those that are deep-seated and endanger life are so rare that one man can seldom see enough cases to make any extensive study of them. (Henrici, Presidential Address to the Society of American Bacteriologists, 1939)

Aspergillus as Human Pathogens

The emergence of species of the genus *Aspergillus* as, in many cases, intractable human pathogens, has gone hand in hand with the progress of medicine. All Aspergilli encountered as causal agents of human or animal diseases are opportunistic pathogens. The Aspergilli are all saprophytes, usually growing in decomposing vegetal material. The main pathogen, *Aspergillus fumigatus*, thrives on compost.

Before the transplant era, *Aspergillus* infections were only encountered sporadically. Farmer's lung is a general name for an allergic disease that could be due to different causal agents, bacteria or fungi, of which the Aspergilli are the main culprits. It is an occupational disease associated with high exposure of spores, in environments such as grain silos. In the nineteenth century, two exotic occupational diseases associated with *A. fumigatus* were described: the *maladie de gaveurs de pigeons* and the *maladie de peigneurs de cheveux*. These pulmonary diseases were associated with people who force-fed pigeons and with people who sorted hair for wigs, respectively. A perusal of the Pathogenesis chapter by Austwick, included in Raper and Fennell's monograph of 1973, leaves the impression that a large number of *Aspergillus* species could be opportunistic pathogens, that pulmonary disease was basically an occupational hazard, that virtually every organ could be colonized by one or other *Aspergillus* species, and that once the fungus was established the prognosis was bleak. Henrici, compared invasive fungal diseases to autocatalytic processes, sluggish to start, but eventually becoming unstoppable. The comparison still holds today, except that immunodepression gives the fungus a head start. *A. fumigatus*, was then as now, the prevalent species, followed by *A. flavus*.

Three types of respiratory pathologies are associated with the Aspergilli. Exposure to the fungus can result in allergic diseases, such as farmer's lung and allergic broncopulmonary aspergillosis, encountered mainly in asthmatic and cystic fibrosis patients. *Aspergillus* spores can germinate in preexisting cavities such as the sinuses or those present in the lung as a result of tuberculosis. This leads to localized Aspergillomas in immunocompetent subjects, which can be treated surgically and/or with appropriate drugs. Finally, the most threatening form is the invasive Aspergillosis, associated, in most but not in every case, with a depression of the immune system.

The ability to perform grafts of bone marrow cells in leukemic patients, of solid organs such as kidney, liver, and lung, has been accompanied by the emergence of invasive Aspergillosis. AIDS patients are also at risk, but *Aspergillus* spp. are encountered less frequently in these patients than *Pneumocystis carinii*, *Candida* spp., or *Cryptococcus neoformans*. Susceptible patients include those affected by neutropenia. Neutropenia can result from leukemia or from the chemotherapy used to control it, or be subsequent to treatment with immunodepressants used in bone marrow, stem cell, or organ transplants. Patients of systemic diseases treated with immunodepressing drugs, mainly corticosteroids, are also at danger. In all these patients, germination of *Aspergillus* spores leads to invasive aspergillosis, usually of the lung, which breaking through the blood vessels can infect other organs. There seems to be no organ in which the fungus cannot grow in the absence of an appropriate immune response. In almost all cases, spores enter through the respiratory tract and germinate in the parenchyma of the lung, leading to invasion of the bronchiolar walls and the adjacent blood vessels. Invasive Aspergillosis has been classified into angioinvasive and bronchioinvasive forms, but this classification is somewhat artificial, as invasion of both bronchioles and arterioles can be seen in the same patient. This leads eventually to respiratory failure and death. Figure 2 shows an *Aspergillus* mycelium grown in lung tissue.

A very recent review estimates an eightfold increase in *Aspergillus*-disseminated infections, from the 1970s to the present day. Between 9% and 17% of all deaths in transplant recipients are due to *Aspergillus* infections according to recent data. The prognosis of invasive aspergillosis is grim; mortality in transplant patients infected with *Aspergillus* sp. is never lower than 60% for patients treated with antifungals and 100% in nontreated patients.

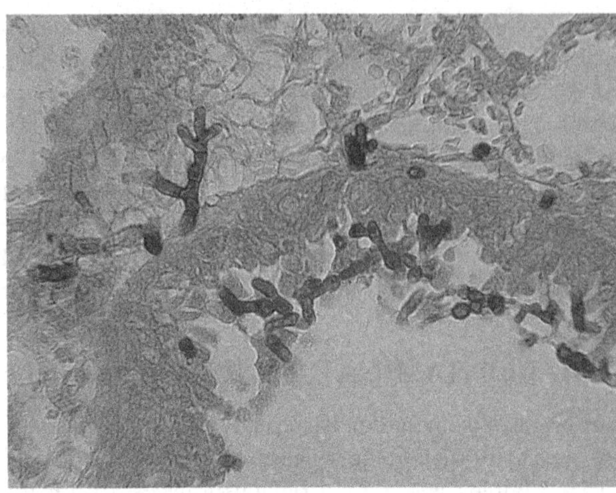

FIGURE 2 A neutropenic mouse lung tissue, experimentally infected with *Aspergillus nidulans* is shown at 20 h postinfection. The fixed sections were stained with Grocotts Methanamine Silver. The pictures show hyphae (stained brown) actively growing in the lung tissue (stained green). Photograph kindly provided by Elaine Bignell.

The most common encountered species in all *Aspergillus*-related pathologies is *A. fumigatus*; *A. flavus*, *Aspergillus niger*, *A. nidulans*, and *Aspergillus ustus* have also been recorded. One recent study of nosocomial infection reports that of 458 patients 154 were infected with *A. fumigatus* and 101 with *A. flavus*. The same and other studies establish a link between construction or renovation work in the vicinity of the hospital and frequency of invasive aspergillosis and conclude that even very low spore counts (1 spore per cubic meter) are dangerous to immunocompromised patients. Genotyping has shown that there are no specific pathogenic strains and suggests that every *A. fumigatus* strain present in the environment is a potential risk for immunodepressed patients. Recently, an upsurge of *Aspergillus terreus* infections has been observed. This is particularly worrying, as the organism is resistant to Amphotericin B, the drug most widely used to treat invasive Aspergillosis. The prevalence of *A. fumigatus* infection has not been explained. We are all continuously exposed to fungal spores and two obvious factors can be considered to explain the prevalence of one or other species. The first is the spore density in specific environments. Unfortunately, many early studies simply report the density of '*Aspergillus*' without any further species discrimination, let alone genotyping. It is generally accepted that the high frequency of *A. fumigatus* infections cannot be explained by a prevalence of the organism in the environment. The second parameter to be considered is spore size. The smaller the spores, the most likely they are to reach the alveolar tissue of the lung, as they will be less susceptible to removal by the mucociliary tissue of the respiratory tract.

A. fumigatus spores are usually about 2–3 μm in diameter, at the lower end of the genus. Specific gravity of spores has, to my knowledge, never been measured. Another obvious parameter is thermotolerance, especially in relation to spore germination. However, it is unlikely that the combination of small spores and ability to germinate rapidly at 37 °C be sufficient to explain the prevalence of *A. fumigatus*. Both characteristics are shared by *A. fumigatus* and *A. nidulans*, the latter being rarely encountered as an opportunistic pathogen. Another possibly interesting parameter is spore hydrophobicity. This is determined by a family of proteins called hydrophobins. Strains of *A. fumigatus* lacking a specific hydrophobin become more sensitive to macrophage killing. Sensitivity of different species to neutrophil and macrophage killing has been sporadically, but not systematically, assessed.

It is important to distinguish putative-specific virulence determinants from essential metabolic processes, even if the latter can be potential drug targets. Only those processes, that when blocked, by mutation or otherwise result in reduced virulence but do not affect the growth of the fungus outside infected tissues, can be considered proper virulence determinants. This is of course conditional to the media in which the fungus is tested, my feeling is that the more we know about the metabolism of the fungus in the wild, the less we will be inclined to call a specific metabolic step a 'virulence determinant.' It is not surprising that engineered strains, deficient in essential biosynthetic pathways, or cell wall biosynthesis show reduced or no virulence. As an example, strains blocked in lysine biosynthesis show reduced virulence, but this tells nothing about virulence, it reveals that lysine is limiting in the alveolar environment. However, as some of these processes are fungal-specific, they are potential targets for antifungal drugs.

Secondary metabolites and nonribosomal peptides vary considerably from one fungal species to the other, and thus they represent an interesting avenue of research bearing on virulence. These metabolites may have evolved in saprophytic organisms in response to the presence of competing organisms in a common environment. As such, they may be cytoxic and eventually involved in pathogenicity. One of these metabolites, gliotoxin, a substituted diketopiperazine, has received considerable attention. It been implicated in the suppression of the innate immune response, including inducing apoptosis of neutrophils. However, specific inability to synthesize gliotoxin does not affect the virulence of A. *fumigatus* in the neutropenic mouse model. However, deletion of *laeA*, a gene necessary for the transcription of a large number of genes encoding enzymes of secondary metabolite synthesis (see sections 'Medically Useful Secondary Metabolites' and 'Regulation of secondary metabolism'), including gliotoxin, does affect the virulence of *A. fumigatus*. Absence of LaeA leads to a pleiotropic phenotype, and the decreased virulence may result from a combination of factors. It seems that at present we simply do not know why *A. fumigatus* is the prevalent pathogen and why other Aspergilli are occasional pathogens. It is likely that a complex combination of characters is responsible for triggering the autocatalytic process proposed by Henrici. Opportunistic pathogens have not evolved as such, in a coevolutionary relationship with a host organism, it thus may be completely fortuitous that one or other of them be able to thrive in the tissues of immunocompromised patients.

It is important to determine which are the barriers that prevent fungal infections in immunocompetent subjects. Alveolar macrophage would get rid of ungerminated conidia, while polymorphonuclear neutrophils destroy hyphae mainly through the action of reactive oxygen species (ROS). One proposed mechanism involves the recognition of fungal cell wall constituents, such as β-1,3-glucans, by macrophage membrane receptors, leading to phagocytosis. Recent studies, however, imply a less clear-cut distribution of labor, with neutrophils also having an important role in preventing conidial germination, which correlates with the susceptibility of neutropenic patients. Dendritic cells are able to ingest *Aspergillus* spores, thus being able to present specific antigens to T cells, a role shared with macrophage. Both $CD4^+$ T and $CD8^+$ T cells respond to fugal antigens, $CD4^+$ T cells produce cytokines, which further recruit neutrophils. The protective role of specific antibodies is subject to discussion, as they are found in infected patients, which they fail to protect. Recently, protective roles have been postulated for surfactants secreted by epithelial cells that interact with conidia and may facilitate phagocytosis. A crucial role in innate immunity to opportunistic fungi is carried out by PTX3, a protein belonging to the pentraxin family of secreted, soluble proteins. PTX3 is essential in conidial recognition by macrophage and dendritic cells, and homozygous knocked-out mice genes are highly susceptible to experimental infection. The study of the immune response to infection by *Aspergillus* spores has made considerable progress in recent years and may lead to treatments, which promote the recovery of the immune response of the patient as an alternative or in association with antifungal drugs.

Fungi are eukaryotes, more closely related to metazoans than to plants, that is why ascomycetes such as *Saccharomyces cerevisiae* and *A. nidulans* are useful models in molecular and cell biology. Many cell processes are common to the fungal and the animal cell, and that, in order to find effective antifungal agents, is necessary to identify those processes that will inhibit growth of the fungal cell without damaging the host. Flucytosine (5-fluorocytosine) has been used as an antimycotic since 1968. In clinical practice, it is used mainly in candidiasis. It affects nucleic acid synthesis and thus can hardly be considered a specific antifungal agent. Four other classes of compounds are currently used to treat fungal infections in clinical practice. The polyenes, such as Amphotericin B, interact directly with ergosterol in the fungal cell membranes leading to leakage of potassium ions and cell death. Amphotericin B, one of the most used antifungals, interacts with animal cell membranes also, which can lead to acute kidney failure. The azoles, such as fluconazole or the newly developed voriconazole, inhibit specifically lanosterol demethylase, blocking the synthesis of ergosterol. They are less toxic than Amphotericin B, and a case has been made to use voriconazole as a first line, rather than a second line, drug for the treatment of invasive Aspergillosis. However, they are not free of secondary effects. The allylamines such as Terbinafine also result in ergosterol depletion by inhibiting squalene epoxidase. Finally, the echinocandins are really specific antifungal drugs, as they affect the fungal-specific process, the synthesis of the glucans of the fungal cell wall by inhibiting noncompetitively β-1,3-glucan synthase. Better knowledge of fungal development and metabolism, the search for genes essential for the pathogen, but absent in, or not essential for the host should lead to the development of new-specific antifungal drugs. A different and complementary approach is to reinforce the immunological response of the host. This includes the possible development of an antifungal vaccine. Besides the uncertainty as to whether protective antibodies can be produced, the large variety of fungi that can affect immunodepressed patients posits an additional difficulty. Recently, a whole roaster of new fungi appeared as opportunistic pathogens, such as *Fusarium*, black molds, and zygomycetes. Success against *Candida* has been followed by an increase of *Aspergillus* infections. Preventive treatment with voriconazole, effective against *Aspergillus*, has been followed by infections by a whole variety of zygomycetes. It has been proposed that a vaccine using β-1,3-glucan as antigen, which is a universal component of fungal cell walls, may be worth exploring. An early diagnosis is essential in the successful treatment of invasive fungal infections. Immunological detection of cell wall components such as galactomannan and 1,3-β-D-glucan and detection of fungal DNA by PCR are being developed and evaluated.

Aspergillus in Veterinary Medicine

Aspergilli are encountered, even if uncommonly, in veterinary practice. Here, as in the human disease, *A. fumigatus* is the most frequently encountered pathogen, followed by *A. flavus*. In mammals, canine sinonasal Aspergillosis, guttural pouch mycosis of horses, and bovine mycotic abortion are the most common diseases, but infection of other species and pulmonary and generalized aspergillosis has also been described. The horse disease is correlated with the presence of an extension of the Eustachian tube, the guttural pouch, an organ of uncertain physiological significance exclusive of horses, other Equidæ, and rhinos and tapirs. This organ could provide temperature and humidity conditions suitable for the growth of *Aspergillus*. Bovine mycotic abortion is correlated with confinement to sheds, which leads to exposure to high concentrations of spores. More surprising is the finding of pulmonary Aspergillosis in free-range dolphins. If the finding of *Aspergillus* infections in mammals is sporadic and infrequent, birds are at a much higher risk. The main pathogen is *A. fumigatus* and the route of entry is the respiratory tract. In a large postmortem study, 4% of more than 10000 birds showed fungal infection of the respiratory tract, probably in most cases due to *A. fumigatus*. Aspergillosis affects both free-ranging and domestic birds. Turkeys, poultry, and waterfowl are commonly affected but fatal infections of penguins, ostriches, and rheas have also been reported. The susceptibility of birds to Aspergilli has been explained by both anatomical characteristics of the respiratory system and cellular differences related to innate immunity such as the absence of alveolar macrophage.

A. sydowii, a Specific Pathogen for Gorgonian Corals

An ecologically menacing new Aspergillosis affects Gorgonian (fan) corals, mainly but not exclusively *Gorgonia ventalina* (infections of *Gorgonia flabellum* and one outbreak affecting *Pseudopterogorgia americana* have been reported). Up to the present time, it has been recorded only in the Caribbean Sea, first identified in Saba in 1996 and studied intensively in the Florida Keys. In an epizootic starting in 1997, more than 50% of the sea fan corals were lost. A subsidence of the epizootic has been since reported. The organism responsible is exclusively *A. sydowii*. The restricted host–pathogen specificity contrasts with the situation described above for mammals and birds. This species is a common saprophyte, which can be isolated from a number of environments. Cultures isolated from diseased *G. ventalina* are infectious, while strains isolated from nonmarine environments are not. As only three and two strains respectively were analyzed, this experiment is not definitive. The pathogenic and nonpathogenic strains do not form separate clades when molecular markers are analyzed. As *A. sydowii* does not sporulate in seawater, aerial dissemination has been suspected. One hypothesis is that the spores are carried by dust storms, originating in the North Africa. While fungal spores are surely carried by dust storms, no genotyping work confirming this hypothesis has been reported. Warming and nutrient effluents, including nitrates, have also been blamed for the outbreak. Obviously, these possible causes are nonexclusive. It is possible that the decrease of the epizootic is due to selection for resistant strains of *G. ventalina*. Thus, sea fan infection by *A. sydowii*, besides being an ecological menace, provides an interesting opportunity to study a specific host–parasite interaction involving an *Aspergillus*, and

FIGURE 3 A specimen of the fan coral *Gorgonia ventalina* infected with *Aspergillus sydowii*. The infected areas are deep purple. The purple gall-like growths may be a result of the infection. A necrotic area surrounded by a deep-purpled ring can be seen at the bottom left of the colony. Bar 5 cm. The photograph has been kindly provided by Kiho Kim.

the elucidation of the mechanism of resistance could lead to the discovery of new antifungal compounds, for which there is a crying need. An infected sea fan coral is shown in Figure 3.

USEFUL ASPERGILLI

Aspergillus biotechnology ranges from the first steps of sake fermentation to the production of recombinant mammalian proteins. These processes are briefly summarized below.

Oriental Food Uses of *Aspergillus*

The use of Aspergilli in the food industry in the far East relies on the extracellular enzymes secreted by the fungus when grown on solid or semisolid substrates. These technologies originated in China more than 2000 years ago. An old review cites more than a hundred such different fungal fermentations. The main products are soyu (soy sauce), miso (fermented soybean paste), and sake (rice wine). The production of soy sauce involves the fermentation of a mixture of cooked soybeans and wheat. The mixture is inoculated in traditional production by 'koji,' which derives from a previous fermentation, or in more modern procedures by a spore suspension of specific strains of *Aspergillus oryzae* or *Aspergillus sojae*. A second fermentation is carried out by lactic acid bacteria and yeasts. In the production of sake, the Japanese wine derived from rice, the steamed rice is inoculated with spores of *A. oryzae*, and the hydrolyzed product is used as the substrate for alcoholic fermentation by *Saccharomyces sake*. The *Aspergillus* strains used in the soy sauce production differ from those used for sake production, the former are selected for high protease, the latter for high amylase titers. Both *A. oryzae* and *A. sojae* belong to the *A. flavus* groups, and genomic analysis has confirmed the very close relationship between *A. oryzae* and *A. flavus*, while *A. sojae* is considered a domesticated strain of *A. parasiticus*. Through the centuries, the organisms have been in use, they have been selected for both high extracellular enzyme titers and nil toxin production, at least under fermentation conditions. *Aspergillus* fermented food products represent, according to a recent source, 2% of the gross national product of Japan.

Extracellular Enzymes Produced by Aspergilli: Aspergilli as Hosts for Recombinant Proteins

The Aspergilli are major producers of enzymes such as carbohydrate hydrolases, lipases, and proteases, used in a variety of industries such as food, beverages, detergent, and animal food additives industries. The first microbial enzyme to be marketed (1894) was an amylase, 'takadiastase,' produced from *A. oryzae*. At least 27 different enzymes are produced industrially by the Aspergilli. Different species, mainly but not exclusively, of the *A. niger*, *A. oryzae*, and *A. sojae* groups have been optimized for the production of specific enzymes. In some cases, increased production has been achieved through proprietary recombinant procedures, which allows an increase in the copy number of homologous and in a few cases heterologous enzyme genes. Chymosin (rennin) is an enzyme essential for cheese production, which prior to its heterologous production by *A. niger* var. *awamori* (and other microorganisms), had to be extracted from calf's stomach. The stunning efficiency of some of the Aspergilli in the process of enzyme secretion ($>20\ g\ l^{-1}$), the considerable experience of the fermentation industry, and the fact that many procedures involving Aspergilli are generally regarded as safe (GRAS) had suggested that the Aspergilli could be used as 'cell factories' for the production of heterologous proteins. This has been successful for some recombinant enzymes (chymosin, lipase, and phytase), but not for high valued, medically important mammalian proteins. Lactoferrin is produced in commercial quantities by recombinant strains of *A. awamori*. More research is needed to understand why some filamentous fungi are so efficient at secreting many fungal proteins but are inefficient as heterologous hosts. Tissue plasminogen activator and interleukin have been experimentally produced at a rate of 12–25 mg l^{-1} in a protease-less mutant of *A. niger*. A number of bottlenecks, such as specificity of glycolsylation and the onset of the unfolded protein response by the translation of foreign proteins, are under active investigation.

Aspergillus and Production of Organic Acids

Depending on culture conditions, strains of *A. niger* are able to excrete a number of organic acids such as oxalic (used in metal leaching), citric, and itaconic acids and are thus used in their industrial production. Citric acid, a tricarboxylic acid, is an intermediate of the Krebs cycle. It is used in the food, beverage, and pharmaceutical industries. The annual production of citric acid, quoted for 2001, was 1 million tons. The main producing organisms are strains of *A. niger*. Since the ability of the organism to divert its metabolism to the production of citric acid was detected, industrial strains, that can convert over 90% of the carbon source in the culture media (carbohydrates) into citric acid were selected. Industrial carbon sources are low-grade molasses (typically sugar beet), but in principle many other residues of industrial process could be used. Specific culture conditions such as high concentrations of carbon source, low pH, and limitation of ions such as manganese are essential. It is not clear how the metabolism of the organism is diverted to citric acid overproduction. The production of citric acid implies that there is a bottleneck in the Krebs cycle, so that much more citric acid is produced than that is utilized in the cycle. Citrate itself inhibits phosphofructokinase I, the enzyme that catalyzes the conversion of fructose-6-phosphate to fructose-1,6-bisphosphate, a crucial step in glycolysis. It has been proposed that under the culture conditions used, this inhibition is counteracted by other metabolites, thus removing this feedback inhibition of citrate production. Alternatively or additionally, tricarboxylic acid mitochondrial transporters leak out citrate from the mitochondrion, thus depleting the cycle.

Itaconic acid is a dicarboxylic acid, which is used in industry as a precursor of polymers used in plastics, adhesives, and coatings. New uses of itaconic acid-derived polymers are under active investigation. The production of itaconic acid for 2001 was quoted as 15 000 tons. There is a renewed interest in this chemical as industry searches for substitutes of petroleum-derived chemicals. Virtually all itaconic acid produced is by fermentation by specific strains of *A. terreus*. Itaconic acid production is a further perversion of the Krebs cycle, citrate is converted as normally into *cis*-aconitate, which for reasons unknown is, in some organisms, decarboxylated into itaconitate, which has no known metabolic role in the cell.

The fact that different strains of *Aspergillus* and more generally of fungi can divert metabolic pathways to the overproduction and secretion of useful chemicals, coupled with the fact that these organisms can grow on residues of processes such as sugar and ethanol production, open the possibility of engineering pathways to produce high value chemicals through 'green,' low polluting, waste-eliminating procedures.

Medically Useful Secondary Metabolites

Of the useful fungal secondary metabolites, the most well-known are the β-lactam antibiotics, penicillin and cephalosporin, and their derivatives. Some Aspergilli, such as *A. nidulans*, produce low titers of isopenicillin-N. This has been useful in the elucidation of the genomic organization and regulation of the pathway. It has already been mentioned above (see section '*Aspergillus* as Human Pathogens') that the echinocandins are specific antifungal drugs. Anidulafungin is a semisynthetic derivative of echinocandin B_0, produced by *A. nidulans* var. *echinulatus*. Anidulafungin has been recently introduced in clinical practice and it is specifically indicated to treat *Candida* infections of the digestive tract. The most widely used secondary metabolite produced by an *Aspergillus* is lovastatin, produced by *A. terreus*. This metabolite, as other statins, is used in medical practice to reduce cholesterol levels. The market for statins has been estimated at more than US$12 billion annually. Statins are specific inhibitors of the 3-hydroxy-3-methylglutaryl-coenzyme A (HMG-CoA) reductase, which reduces HMG-CoA to mevalonate. Statins are built around a common polyketide skeleton, have a structure similar to HMG, and act as competitive inhibitors of the HMG-CoA reductase. As other secondary metabolites, statins are synthesized by complex sequential steps, involving polyketide synthases, the genes coding for the cognate enzymes map in a 64 kb gene cluster. Gene clustering and its possible role in the synthesis of secondary metabolites is discussed in section 'Regulation of secondary metabolism.'

A. (EMERICELLA) NIDULANS AS A MODEL ORGANISM

The *A. nidulans* Genetic System

In 1953, Guido Pontecorvo published a 97-page review on the 'Genetics of *Aspergillus nidulans*.' Why did Pontecorvo and coworkers spend a considerable amount of time and energy to develop a genetic system for what was then an exotic organism? One of the key problems of biology was, at the time, that of the nature of the gene. The classical image of the gene was that of a discrete element, and mutations were considered to be alternative states of this element. The gene was an abstract concept whose molecular nature was elusive. The image of the gene as an indivisible, discrete unit was based on the experimental fact that mutations that did not complement were not separable in recombination experiments. That is, crosses of individuals carrying different alleles of the same gene, wild-type progeny were never obtained. 'Never' was a few dozen progeny in mice, a few thousands in *Drosophila melanogaster*. The modern reader may not grasp how fundamental the problem was at the time. In the 1940s, there were already exceptions to the 'nonrecombination' rule. In *D. melanogaster*, a few noncomplementing mutations could recombine in crosses to yield rare wild-type individuals. These mutations were called 'pseudo-alleles.' Pontecorvo was looking for a system where hundreds of thousands progeny could be scored. *A. nidulans* happened to be such an organism. By early 1950s, it became clear in Pontecorvo's laboratory, through the work of Alan Roper, followed by Bob Pritchard, that the gene was divisible. The paradigmatic work on the divisibility of the gene was carried out by Seymour Benzer using the bacteriophage T4. "That is, that the gene as a working unit in physiological action is based on a chromosome segment larger than the unit of mutation or recombination" Roper (1953) and "The classical 'gene' which served at once as the unit of recombination, of mutation, and of function, is no longer adequate. These units require separate definition. A lucid discussion of this problem has been given by Pontecorvo" Benzer (1957). The phage system of Benzer was so powerful and elegant that *A. nidulans*, as a system to study the fine structure of the gene, seemed redundant. Nevertheless, a beautiful genetic system was there, ready for the taking. I dare say that in 1953 no system afforded the same degree of sophistication.

This system allows conventional meiotic genetics, carried out by analyzing the progeny contained in a fruiting body (cleistothecium). The cleistothecium may contain as many as 100 000 thousand asci, each containing eight ascospores, the product of a single meiosis and an additional mitosis. In the standard genetic analysis, the products are analyzed 'in bulk,' without isolating single asci, as those are small and difficult to dissect. However, tetrad analysis is possible and was employed in early work. The power of resolution of the 'in bulk' genetic analysis has permitted fine structure mapping to the extent that mutations separated by 11 nucleotides have been resolved by recombination. *A. nidulans* strains carrying different markers can form heterokaryons. Nuclei in heterokaryons can rarely fuse, giving origin to stable diploids, which allow another layer of genetic analysis, developed by Pontecorvo and Etta Käffer, the parasexual cycle. Diploids can revert to haploids in which all markers in one chromosome segregate as a unit, allowing the rapid assignment to any new mutation to one of the eight chromosomes of the organism. Mitotic recombination also occurs and can be selected for in diploids, permitting the mapping of markers in relation to the centromere. The discovery of the parasexual cycle led to two developments. The first one was the possibility to carry out genetic analysis in the Aspergilli and the related Penicillia where a sexual cycle was not described. It is not known why stable diploids, different from the transient diploids that occur during the sexual cycle, can be obtained in these organisms and not in other filamentous ascomycetes. The second was the analogous development of systems, initiated by Pontecorvo, and based on cell fusion to carry out somatic genetics in mammalian cells. *A. nidulans* entered the molecular era when relatively efficient transformation techniques were worked out in 1983 followed by the development of replicating plasmids. We are witnessing a second methodological revolution, with the development of techniques and modified strains that allow to inactivate genes, introduce point mutations, change promoters, or add tags in a very simple and rapid way, opening the possibility of a high throughput reverse genetics. This, together with the availability of a complete genome and microarrays is changing again the prospects for this model organism. Usually a technique is first worked out in *A. nidulans*, and then that is applied to the other Aspergilli, such as the pathogenic or industrially important organisms mentioned in previous sections. The life cycle of *A. nidulans* is shown in Figure 4.

Work carried out with *A. nidulans* has served as a model mainly in three aspects of cell and molecular biology (but a few others will be briefly discussed below). Historically, each of these three aspects can be related to a specific scientific school. David Cove and John Pateman initiated the investigation on control of gene expression in *A. nidulans*. The cell biology work derives from an early article by Ron Morris (1976) where he isolated mutations blocked in the cell cycle and in nuclear migration. Bill Timberlake initiated an analysis of the development of the conidiophore, work which profited from the early genetic work of John Clutterbuck, himself a product of the Glasgow school of genetics. Work on secondary metabolism stems from a confluence of this work with work carried out in other species of Aspergilli producing noxious chemicals.

The Mitochondrial DNA of *A. nidulans*

The possibility to construct heterokaryons allows the genetic study of mutations that occur in the mitochondrial genome, as these are cytoplasmically inherited. A few markers were characterized and a circular genetic map was established. In the late 1970s, two groups, led respectively by Hans Künzel and Wayne Davies, attempted to sequence the whole 33 000 bp mitochondrial DNA of *A. nidulans*. This was almost accomplished, except for a gap of around 200 bp. At the time, where the longest DNA sequenced was the 16 000 bp mitochondrial DNA of *Homo sapiens*, this was a less than trivial enterprise.

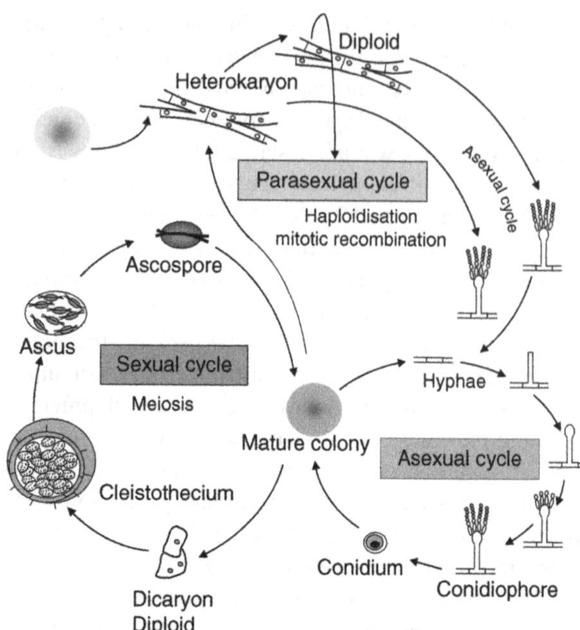

FIGURE 4 Schematized life cycle of *Aspergillus nidulans*. Three cycles are shown, the asexual cycle, which has been described in the text (see sections '*A. nidulans* as a Model for Cell Biology' and '*A. nidulans* Developmental Pathways'). In the sexual cycle, two nuclei divide synchronously as a dikaryon in a specialized structure. Eventually the nuclei fuse to give a diploid, which does not divide as such but undergo meiosis. No differentiated mating types exist; any given mycelium can generate fertile cleisthotecia. To cross two strains, nutritional and color markers are used. Classical genetics procedures are facilitated by the fact that one cleisthotecium derives from only one fertilization event. In the parasexual cycle nuclei fuse in mycelia, outside the specialized developmental structure to yield a diploid, which does not undergo meiosis (at variance with diploids produced during the sexual cycle) but divides as such. Breaking up of the diploid (haplodization) and mitotic recombination are additional genetic tools. For clarity we have supposed that the heterokaryon and diploid formation are carried out between green (yA^{+}) yellow strains (yA^{-}) and nuclei are colored accordingly. Both diploids and heterokaryons can undergo the asexual cycle. The different structures are not shown to scale. Scheme kindly provided by Stéphane Demais and modified by the author.

A number of important results derived from this sequencing effort. It was possible to compare the whole DNA organization with that of the completely sequenced human mitochondrial DNA, and the ongoing sequence of the *S. cerevisiae* mDNA, an organism where sophisticated genetic and molecular studies were actively carried out. These comparisons were extended to other mitochondrial DNA sequencing projects such as those of *Neurospora crassa*, *Podospora anserina*, and *Schizosaccharomyces pombe*. A number of unidentified open reading frames (ORFs) were found in the mitochondrial DNA of *A. nidulans* and the human mitochondrial DNA, but not on that of *S. cerevisiae* or of *S. pombe*. These reading frames correspond to genes coding for complex I, the NADH dehydrogenase complex, which is absent in both model yeasts. The highlight of this work was the elucidation of the structure of class I introns. Metazoan mitochondrial genes have no introns, while introns of two different classes are present in the mitochondria of fungi and plants. Introns on the nuclear genome of the fungi are small, typically of >100 bp. Mitochondrial fungal introns are large, usually >1000 bp, and can contain ORFs. A comparison of the sequences of the mitochondrial introns of *A. nidulans* with those of wild-type and mutant strain of *S. cerevisiae* led to a model of the secondary structure of class I introns, including the proposal of a mechanism of intron splicing. Around the same time, it was published that the intron of the precursor of the ribosomal 26s gene of *Tetrahymena termophila* was self-splicing. It was noticed that pairings that were postulated by us and by Bernard Dujon and François Michel for mitochondrial group I introns were conserved in this self-splicing intron; thus the model for splicing of mitochondrial class I introns became a model of self-splicing, which was confirmed experimentally. Self-splicing of introns was essential to the concept of ribozyme and eventually to that of a primeval RNA world. Thus, the sequence of the mitochondrial DNA of *A. nidulans* contributed, albeit somewhat indirectly, to present ideas on the origin of life.

A. NIDULANS AS A MODEL FOR GENETIC METABOLIC DISEASES

The metabolic versatility of the Aspergilli led a group of Spanish scientists to use mutants blocked in amino acid degradation to identify the enzymes and the genes of human metabolic diseases, including those of aromatic and branched amino acid catabolism. As stated in a review article "The metabolic capacity of *A. nidulans* for amino acid degradation largely resembles that of human liver." Figure 5 shows the breakdown of phenylalanine and the cognate blocks in human diseases affecting this metabolism. The gene of *A. nidulans* coding for the fumaryl-acetoacetate hydrolase was cloned as a cDNA highly expressed in the presence of phenylacetic acid (*fahA*, Figure 5). Mutations in the human homologue result in the serious disease tyrosinemia I, and the *A. nidulans* ORF shows 47% identity with the human gene. Satisfactorily, the growth of *A. nidulans* is strongly inhibited by the accumulation of this metabolite in *fahA*-deleted strains. In a second step, suppressor mutations of this inhibited phenotype were isolated. These pinpointed the gene (*hgmA*) coding for homogentisate dioxygenase. Not only these mutations suppressed the toxicity of phenylalanine seen in *fahA* nulls, they also resulted in the accumulation of a purple pigment (see Figure 5). This is exactly the pigment that is accumulated in the urine of patients affected by a milder disease, alkaptonuria. The identification of the gene was straightforward and its sequence served to identify the hitherto unknown human gene and to identify the loss-of-function mutations

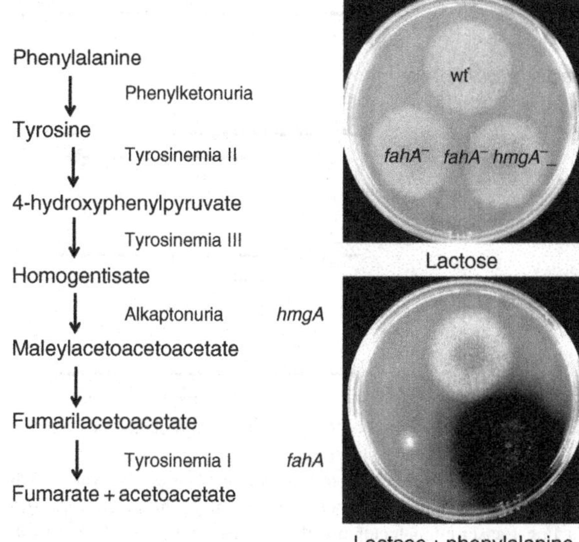

FIGURE 5 *Aspergillus nidulans* as a model for human metabolic diseases. To the left the degradation of phenylalanine is shown, to the right of the pathway, the steps blocked in the cognate human metabolic diseases. In italics the relevant corresponding genes of *A. nidulans* are shown. To the right the toxicity of fumaryl-acetoacetate and the suppression of the toxicity by the upstream *hmgA* null is shown together with the secretion of the purple oxidation product of homogentisic acid. Lactose is used as a poor, nonrepressing carbon source, as phenylalanine catabolism is subject to carbon catabolite repression (see section 'Nitrogen and carbon utilization'). See text for details. Photographs of plates were kindly provided by Miguel Peñalva.

present in a number of patients. Alkaptonuria was identified by Garrod in 1902 as an 'inborn error of metabolism' and shown to be inherited as single Mendelian gene. The work of Beadle and Tatum in *N. crassa*, and the one gene-one enzyme proposal arising from it, can be seen as a completion of Garrod early proposals. The identification of the human gene through the cloning of the *A. nidulans* gene is of more than historical importance and underlines the necessity of choosing the correct model system for a particular problem.

Control of Gene Expression

Nitrogen and carbon utilization

The Aspergilli can utilize a large number of metabolites as nitrogen and/or carbon sources. Early work with *A. nidulans* has established some fundamental concepts pertaining to the control of gene expression and metabolic regulation. Some of the first eukaryotic pathway-specific regulatory genes (*nirA*, *uaY*; see below) were characterized in the 1960s. The genes encoding the key regulators for nitrogen (*areA*) and carbon catabolite repression (*creA*) were the first such genes to be described in any eukaryotic organism. Their mode of action was established by formal genetic analysis long before recombinant DNA technology came into existence.

In general, the genes coding for the enzymes involved in the utilization of a specific metabolite are only transcribed in the presence of a specific inducer. Inducers act by activating specific transcription factors, which in turn elicit the transcription of specific catabolic genes. Thus, nitrate activates the NirA protein, necessary for the transcription of the genes encoding nitrate and nitrite reductase and the nitrate transporters, acetaldehyde activates AlcR, regulating ethanol utilization, uric acid activates UaY, regulating at least eight scattered genes encoding enzymes and transporters involved in purine utilization, proline activates PrnA, regulating all the other genes of the proline utilization gene cluster, while β-alanine activates AmdR/IntA a protein that positively regulates the *amdS* gene (encoding acetamidase) and the *gabA* gene (encoding the γ-aminobutyrate transporter).

Almost all the pathway-specific transcription factors belong to a group of proteins that bind DNA through a specific fungal motif, the Zn binuclear cluster (Cys_6Zn_2). Most bind DNA as dimers, including the paradigmatic *S. cerevisiae* protein GAL4. AlcR is an exception, which uniquely binds as a monomer. NirA and UaY are localized in the nucleus as a result of induction. Nitrate induces by breaking the association of NirA with KapK (the orthologue of the mammalian and *S. cerevisiae* exportins Crm1P and CRM1). PrnA and AlcR are always nuclear, PrnA necessitating induction to bind its cognate sequences in the promoter.

The induction of genes involved in the utilization of nitrogen sources does not occur in the presence of preferred sources such as ammonium and glutamine, while the induction of genes involved in the utilization of carbon sources is strongly diminished in the presence of glucose. These processes, nitrogen metabolite repression and carbon catabolite repression, involve two additional regulators, AreA and CreA. AreA is a GATA factor, acting positively in synergy with the specific regulators (such as NirA or UaY). Ammonium and glutamine negate AreA function at a number of levels, including the stability of its cognate mRNA. The dependence on AreA is absolute for the *niiA-niaD* bidirectional promoter, driving the genes encoding nitrate and nitrite reductases, less marked for some of the genes of the purine utilization pathway.

CreA acts as a genuine repressor in the presence of favored carbon sources, negating the activation by or competing with the binding of the pathway-specific factors such as AlcR.

CreA is a Zn finger protein, with a Zn finger sequence extremely similar to Mig1p, the repressor mediating carbon catabolite repression in *S. cerevisiae* and related organisms. However, the similarity between CreA and Mig1p stops there. Little sequence conservation can be seen outside the DNA-binding domain. Neither the glucose signaling mechanism nor the downstream mechanism of transcriptional repression seems to be shared by Mig1p and CreA. Mig1 represses transcription by recruiting the Tup1/Ssn6p corepressor complex, which is not the case in *A. nidulans* and most likely in the fungi where a CreA, rather than a Mig1p orthologue, is present.

The Aspergilli can use a number of metabolites as both carbon and nitrogen sources, the mechanism of regulation having been elucidated for the *prn* gene cluster (comprising five genes involved in the utilization of proline) and the *amdS* gene. Repression occurs only when both repressing carbon (glucose) and nitrogen sources (ammonium or glutamine) are present. This can be rationalized by thinking that if a repressing nitrogen source is present, it will be advantageous for the organism to use proline or acetamide as a carbon source, while if only a favored carbon source is present, it will still be advantageous to use proline or acetamide as a nitrogen source. Carbon metabolite repression requires the CreA repressor, while nitrogen metabolite repression operates through the inactivation of the AreA GATA factor. While, for example, in the nitrate assimilation pathway AreA is always essential for transcription to occur, for *prn* and *amdS*, it is only necessary when the CreA repressor is activated by a repressing carbon source. These regulatory patterns are conserved in the Aspergilli and more generally in the filamentous ascomycetes and are schematized in Figures 6–8, while the nuclear-cytoplasmic shuffling of NirA is illustrated in Figure 9. The *gabA* gene, encoding the γ-aminobutyrate transporter, is subject to an even more complex pattern of regulation. It is induced by ω-amino acids and subject to concomitant repression by nitrogen, carbon, and alkaline pH.

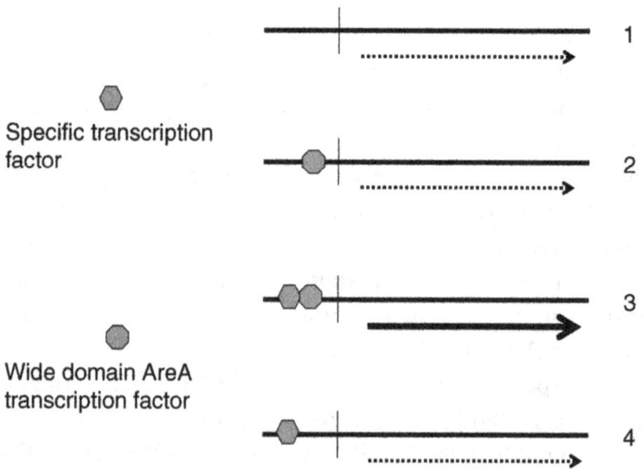

FIGURE 6 General scheme of the transcriptional regulation of genes involved in the utilization of nitrogen sources.
(1) In the absence of a specific inducer (such as nitrate) and in the presence of a preferred, repressing nitrogen source (ammonium, glutamine), neither the specific transcription factor nor the broad-domain GATA factor AreA is activated. No or only basal transcription is seen.
(2) In the absence of a specific inducer in the presence of a nonrepressive nitrogen source, only AreA is active. No or only basal transcription is seen.
(3) In the presence of a specific inducer and in the absence of a repressing nitrogen source, both transcription factors are active, full transcription is seen.
(4) In the presence of both a specific inducer and a repressing nitrogen source, the specific transcription factor is active, but the AreA factor is inactive and no transcription is seen. In the nitrate utilization pathway a further mechanism is in act, as AreA is necessary both indirectly through its regulation of transporters for the uptake of the specific inducer (nitrate) and for the binding of the specific transcription factor (NirA) to DNA. Thus, the situation will be identical to that seen in scheme 1.

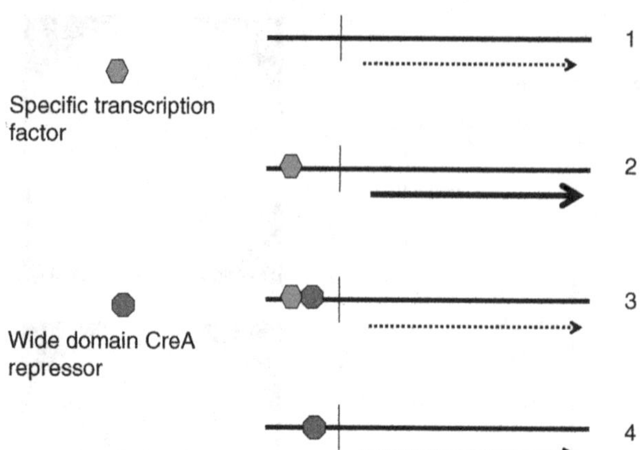

FIGURE 7 General scheme of the transcriptional regulation of genes involved in the utilization of carbon sources.
(1) In the presence of 'neutral' carbon source (such as glycerol) and the absence of an inducing carbon source, neither the specific positive-acting transcription factor nor the CreA repressor are bound to the promoter. No or only basal transcription is seen.
(2) In the presence of an inducer carbon source, the specific transcription factor (such as AlcR in the ethanol utilization pathway) is bound to DNA and active, full transcription is seen.
(3) In the presence of both inducing and repressing carbon sources, the specific transcription factor is active but the CreA repression partially or totally negates its effect. No or only basal transcription is seen.
(4) In the presence of only a repressing carbon source, only the CreA repressor is bound to DNA, no or only basal transcription is seen.

Regulation of gene expression by external pH

Soil organisms, such as the Aspergilli, respond to a variety of environments and it is not surprising that a system that regulates gene expression as a function of external pH has evolved. External pH regulates genes coding for extracelullar enzymes or transporters or those encoding steps in the synthesis of exported metabolites. In neutropenic mice experimentally infected with *A. nidulans*, this process is necessary for virulence. Penicillin is synthesized by some Aspergilli but only at alkaline pH. The synthesis and uptake of siderophores is also regulated by pH.

The elucidation of the mechanism of pH regulation is a superb scientific achievement of the groups of Herb Arst and Miguel Angel Peñalva. The signal transduction pathway described below is conserved throughout the ascomycetes. The key actor is PacC, a transcription factor of the classical Zn finger type. In its active form, PacC acts as a positive transcription factor of alkaline-expressed genes

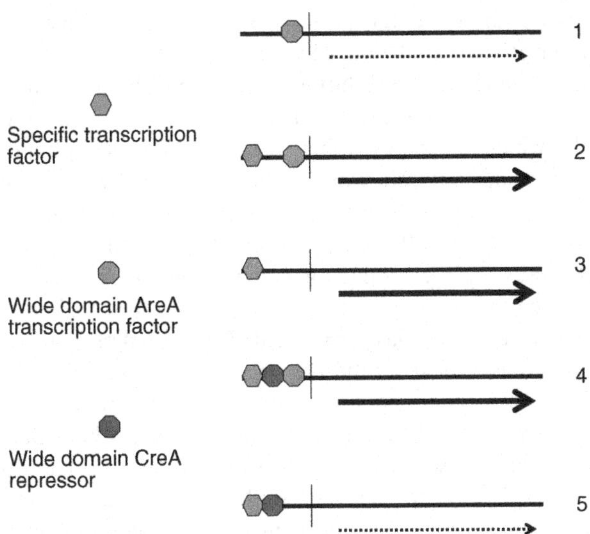

FIGURE 8 General scheme of the transcriptional regulation of genes involved in the utilization of metabolites that can serve as both nitrogen and carbon sources.

(1) In conditions where the inducer is not present the specific transcription factor (PrnA in the proline utilization gene cluster) is not bound to the promoter. In the scheme shown, under neutral conditions (e.g., urea as nitrogen source, lactose as carbon source) AreA would be bound, and CreA would not be bound. No or only basal transcription is seen. This applies to every other combination (not shown) where the specific inducer is absent.

(2) Same conditions but in the presence of the inducer (proline in the example given in the text), both the specific transcription factor (such as PrnA) and AreA are bound, full transcription is seen.

(3) In the presence of the inducer and a repressing nitrogen source (ammonium or glutamine) but no repressing carbon source. Only the specific transcription factor is bound. Full or almost full transcription.

(4) In the presence of the inducer and a repressing carbon source (glucose), but no repressing nitrogen source. The three regulatory proteins are bound; AreA negates the repressing action of CreA. Full or almost full transcription.

(5) In the presence of inducer (such as proline) and both carbon and nitrogen repressing metabolites (ammonium or glutamine and glucose). The specific transcription factor (such as PrnA) is bound and CreA negates its action. Efficient repression. No or only basal transcription is seen.

(such as alkaline phosphatase or isopenicillin synthase) and as a repressor of acid-expressed genes (such as acid phosphatase or the γ-aminobutyrate transporter). At acidic pH (pH usually tested 4.0), there is no activation signal, and the protein is in an inactive form. In the full-length PacC ($PacC^{72}$), intramolecular interactions hold the protein in a folded inactive form, which is largely excluded from the nucleus. At alkaline pH values (usually 8.0), the protein is activated by two proteolytic steps. The *palA*, *palB*, *palC*, *palF*, *palH*, and *palI* genes encode proteins involved in pH sensing and in signal transduction. Mutations in all these *pal* genes have an acidity-mimicking phenotype, while mutation in the transcription factor *pacC* can lead to acidity-mimicking (loss-of-function mutations), alkalinity-mimicking, or neutrality-mimicking phenotypes, where both 'alkaline' and 'acidic' genes are expressed. The pH sensor is probably PalH assisted by PalI, both of which are plasma membrane proteins. The C-terminus cytoplasmic tail of PalH interacts directly with PalF, a member of the arrestin family, which is, similarly to the mammalian arrestins, phosphorylated and ubiquinated. These modifications occur at alkaline pH and are dependent on the PalH and PalI proteins. Under alkaline pH conditions, PalA binds to motifs flanking a specific protease-sensitive sequence in the C-terminus of the full-length PacC ($PacC^{72}$). PalA interacts directly with the *A. nidulans* orthologue of Vps32, a protein involved in the formation of multivesicular endosomes. PacC/PalA interaction renders PacC sensitive to a specific cleavage, catalyzed by PalB, a protease of the calpain family. $PacC^{72}$ is cleaved to $PacC^{53}$. The cleaved form of PacC becomes susceptible to further processing by the proteasome, yielding $PacC^{27}$. This is the active form of PacC, strictly localized in the nucleus, where it activates genes expressed at alkaline pH and represses genes expressed at acid pH. This account leaves open the mechanism of pH sensing and the connection between the arrestin-like PalF and PalA-PalB. The interactions of PalA with components of the mature endosome, and recent work

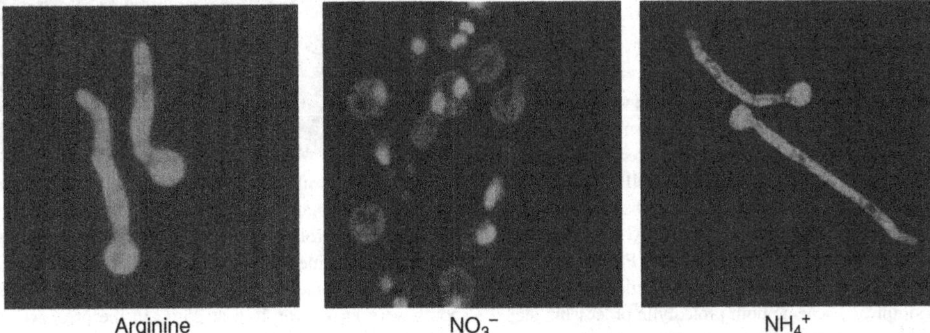

FIGURE 9 Cytoplasmic and nuclear localization of the NirA transcription factor. A construction where the whole NirA transcription factor is fused to green fluorescent protein (GFP) substitutes the NirA wild-type gene. This construction is competent to mediate induction by nitrate. NirA is localized in the cytoplasm in the presence of a noninducing, nonrepressing nitrogen source (arginine), of a repressing nitrogen source (ammonium) and localizes in the nucleus only when an inducing nitrogen source is present. See text for details. This figure illustrates also the technology of gene fusions and epiflourescence microscopy, which is been extended to all other Aspergilli and many filamentous fungi, including animal and plant pathogens. The original pictures were kindly provided by Joseph Strauss.

on the related signal transduction pathway in *S. cerevisiae*, strongly suggests that endocytosis provides this connection. PalC, the unplaced actor of the process, has a functionally important Bro1 domain (also present in PalA), a domain of possible interaction with Vps32, strengthening the endosomal connection of the pH signaling pathway. The YPXL/I motif recognized by PalA is also recognized by its putative mammalian orthologue, AIP1/Alix, a protein involved in a variety of functions, including the budding of the human HIV virus from infected cells. The whole process has tantalizing similarities with the Hedgehog signaling pathway in metazoans, leading to the proteolytic activation of the Zn finger transcription factor *cubitus interruptus/Gli*, posing the question of whether these pathways are evolutionarily related. A simplified version of the pH signaling process is shown in Figure 10.

Specific regulatory mechanisms acting at the level of transporters

The control of transporter synthesis and activity is a key step in metabolic regulation, as the activity of specific transporters modulates the entry of metabolites that serve as inducers or repressors of specific pathways. Work with *A. nidulans* has led to the identification of two new control processes affecting transporters, besides their tight specific transcriptional regulation. The transcription of a number of transporters is activated during the isotropic phase of conidial germination (see below). This is a developmental control, which bypasses other specific control systems. Recent transcriptomic work suggests that this mechanism occurs for many transporters and is general for the filamentous ascomycetes. It can be proposed that germinating fungal spores explore an unknown environment by expressing a whole range of transporters, to progress to specific induction once the spore has germinated. The second mechanism is posttranslational. In the presence of a favored nitrogen source such as ammonium, both purine and amino acid transporters are internalized to the vacuole, where they are possibly destroyed. This posttranslational mechanism is synergistic with but independent from the nitrogen metabolite repression mechanism (described above). Figure 11 illustrates this process.

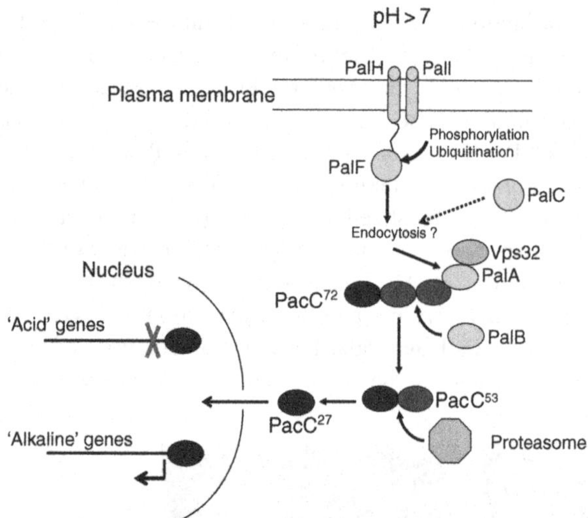

FIGURE 10 Simplified scheme of the regulation of gene expression by external pH, redrawn from a number of articles of the groups of Arst and Peñalva. Alkaline pH is sensed by the PalH and PalI proteins, the signal is transduced to the PalF arrestin via the C-terminus of PalH. PalF is phosphorylated and ubiquinated and signals PalA, which has been shown to interact with the endosomal protein Vps32. The role of PalC is hypothetical. PalA leads to the opening of PacC72, which is cleaved by PalB at a specific site. PacC53 is further processed by the proteasome, leading to the active form PacC27. In this simplified scheme both proteolytic processing steps are shown in the cytoplasm, in fact the second step may occur in either or both the cytoplasm and the nucleus. In *light* grey proteins of the pH signal transducing pathway, in *middle* grey cellular proteins or complexes nonspecific for the pathway. The 'active' portion of PacC is shown in *dark grey* the inhibitory, cleaved portions are shown in black.

Regulation of secondary metabolism

Fungi produce an astonishing variety of secondary metabolites. Fungal toxins, the β-lactam antibiotics and lovastatin, have already been mentioned. In the pregenomic days, conventional genetic analysis led to the identification, cloning, and sequencing of a number of genes encoding

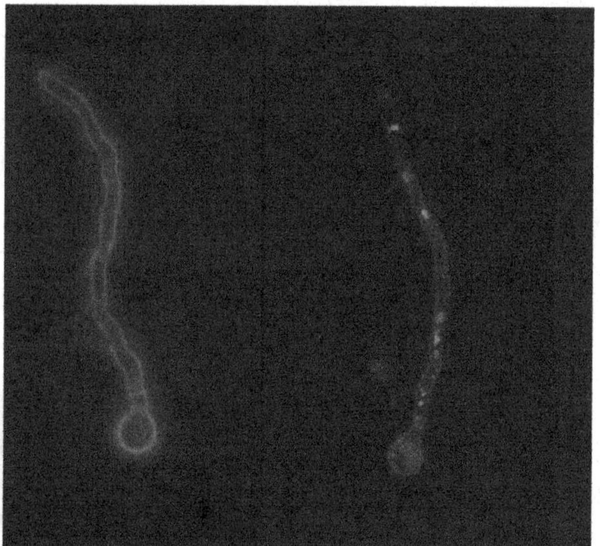

FIGURE 11 Posttranslational regulation of transporters. The left panel shows the membrane localization of a fusion with the green fluorescent protein of the proline transporter (PrnB-GFP). The cognate gene *prnB* maps in the proline gene cluster, its transcriptional regulation is schematized in Figure 8 (see section 'Nitrogen and carbon utilization'). Conidiospores were grown for 16 h at 25 °C in the presence of urea as nitrogen source, and induced with L-proline after 10 h. In the right-hand panel ammonium was added for the last 2 h. In the left panel, the PrnB-GFP fusion strongly stains also the basal septum. Localization in septa is a characteristic of all membrane proteins studied up to now. Confocal microscope images were kindly provided by Vicky Sophianopoulou.

biosynthetic steps for a number of secondary metabolites, while many more metabolites were identified as secreted by a variety of Aspergilli. As many secondary metabolites involve nonribosomal peptide or polyketide synthases, putative fungal metabolite gene clusters have been identified in a number of fungal genomes. In *A. fumigatus*, the estimate is that of 22 secondary metabolite gene clusters. The best studied pathways are those leading to the biosynthesis of isopenicillin in *A. nidulans*, aflatoxin in *A. flavus*, and the aflatoxin precursor sterigmatocystin in *A. nidulans*. Secondary metabolism synthesis occurs late during mycelial growth and is generally correlated with conidiation and shares with this process some of its signaling pathway. Pathway-specific transcription factors have been characterized for the aflatoxin, sterigmatocystin, and gliotoxin pathways. The clusters of aflatoxin and sterigmatocystin biosynthesis include the regulatory gene *aflR*, necessary for the expression of the rest of the genes of the cluster. AflR belongs to the Cys_6Zn_2 family of specific fungal activators. It is not known whether the activation of AflR involves a specific metabolite or if it is only activated by the 'fluffy' signaling pathway to be described below (see section '*A. nidulans* Developmental Pathways'). No pathway-specific activator has been described for isopenicillin biosynthesis, which is regulated by a number of environmental parameters, including extracellular pH.

Clustering of genes is variable for genes involved in primary metabolism. In contrast, the genes of secondary metabolism biosynthesis are as a rule organized in large clusters. The 70 kb aflatoxin cluster comprises 25 coregulated genes. The gliotoxin gene cluster comprises 12 genes. Does the clustering of secondary metabolism genes have an evolutionary and/or functional significance? Possibly the two divergently transcribed genes responsible for isopenicillin-N synthesis have been horizontally transferred from a *Streptomyces* to an ancestor of the Aspergilli and Penicillia. There is no evidence for horizontal transfer for any other secondary metabolite gene cluster. Comparative genomics is providing some clues, even if not yet an answer, to the significance of secondary metabolite gene clustering. There is a significant bias toward the location of secondary metabolite clusters in subtelomeric regions. The fact that species of Aspergilli differ widely in the secondary metabolites they produce correlates with the mapping of the cognate genes in genomic regions where syntheny between species is broken.

A fundamental advance in the understanding of the regulation of secondary metabolism arises from the discovery of the global regulator LaeA in the laboratory of Nancy Keller. LaeA is conserved in filamentous fungi, but not in yeasts. LaeA shows a domain typical of histone methyltransferases, the SAM domain, while lacking a second domain found in these enzymes, the SET domain. LaeA regulates positively the synthesis of isopenicillin, sterigmatocystin, gliotoxin, and lovastatin. The global role of LaeA has recently been investigated by transcriptomic studies with *A. fumigatus*. Of the 22 gene clusters, a deletion of *laeA* diminishes clearly the transcription of 13. Thus, LaeA is a broad, but not a universal regulator of secondary metabolism. Recent work points to a role of LaeA in remodeling chromatin structure. In *A. nidulans*, deletion of a number of genes universally involved in gene silencing in heterochromatin result in premature secondary metabolite production. More strikingly, these deletions act as partial suppressors of a *laeA* deletion. Thus, the exciting possibility arises that LaeA acts by reversing a heterochromatic state of the secondary metabolite gene clusters. Thus, the study of the regulation of secondary metabolism may lead to an understanding of the role and genomic distribution of heterochromatin in filamentous ascomycetes.

A. nidulans as a Model for Cell Biology

The life cycle of the Aspergilli includes a number of tightly regulated developmental pathways, from the germination of conidia or ascospores to the formation of complex structures involved in sexual (cleistothecia) or asexual (conidiophore) spore formation. The germination of conidiospores, but not that of ascospores, has been well studied. Conidiospores can stay dormant and viable for many years and contain (in *A. nidulans*) one nucleus arrested in the G1 phase. When plated on suitable media, they go through a phase of isotropic growth, where the conidium swells. The first mitosis may occur in this phase or after the emergence of the germ tube (Figure 12). Mitosis occurs

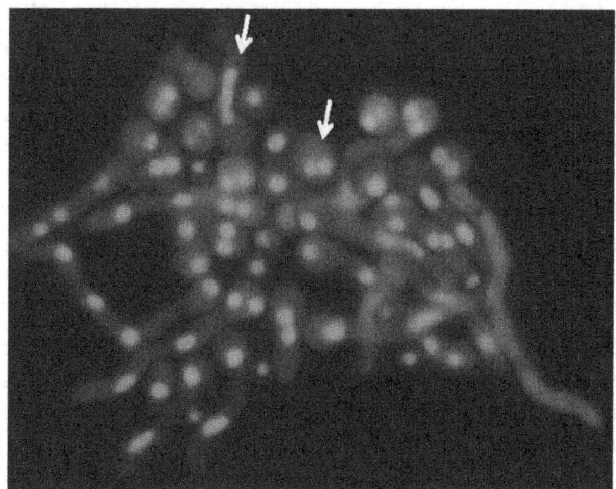

FIGURE 12 Conidial germination. A group of germinating conidia from *A. nidulans* are shown. They are stained with the green fluorescent protein (GFP) fused to a strong nuclear localization signal, driven by a strong constitutive promoter. One white arrow indicates a conidia where the first mitosis has occurred before the production of the germinal tube, another mitosis occurring concomitantly with germination. Note that the signal is not lost during mitosis, which as in other fungi is closed. Photograph by Ana Pokorska in the laboratory of the author.

synchronically, up to the eight nuclei stage when a perforated septum appears basally (Figure 11). Other septa are laid during hyphal growth out every three to four nuclei. Only the nuclei comprised between the septum and hyphal tip are competent to divide and they do so synchronously. A second germ tube can arise from the conidiospore at 180° from the first one. Nuclei in nonapical compartments became again competent to divide when the conidiophore is developed (see below) and when branches arise from subapical compartments. Thus, a highly coordinated process occurs, involving the regulation of mitosis, the establishment of a primary polar axis, the establishment of secondary polar axes in branches, nuclear migration, the laying down of septa and finally the appearance of another highly polarized structure, the conidiophore. Some processes, such as hyphal polar growth and the deposition of septa, are specific of fungi, while others are common to all eukaryotes, and the *A. nidulans* work matches and has added considerable information to the work carried out in *S. pombe* and *S. cerevisiae*. In both yeasts and most cells in higher eukaryotes, mitosis is followed by cytokinesis, where the two daughter cells separate. This is not exactly the case in filamentous fungi, where the whole mycelium is one syncytium, subdivided by perforated septa. It must be stressed that during condiogenesis the situation resembles budding, with proper cytokinesis, as metullae, phialides, and conidia are uninucleate cells, while ascospores are binucleate (see section '*A. nidulans* Developmental Pathways'). Thus, an understanding of *Aspergillus* cytokinesis involves understanding the generation of these different patterns. The determination of hyphal polarity and the related problem of the relationship of mitosis with septum formation are active fields of research at present, and recent work has shown that while some of the determinants of polarity and cytokinesis are common with the yeasts, some are entirely novel. In particular, a specific ceramide synthase is essential for polarity, probably by generating specific lipid rafts at the growing tip, which in turn would be involved in the localization of other polarity determinants such as formin, an actin nucleating protein. Particular to filamentous fungal growth is the Spitzenkörper, a subapical organelle that acts as a vesicle supply center. The challenge for future research is to understand the coordination of signaling pathways, the polarization of the actin cytoskeleton, the formation of lipid rafts, and the activity of the Spitzenkörper to reach a complete understanding of polarity determination.

Some of the highlights of the work relating to the cell cycle are indicated below, where *A. nidulans* has served as an eukaryotic model, while some specific aspects of *Aspergillus* development are summarized in '*A. nidulans* Developmental Pathways.'

The judicious use of mutants resistant to the tubulin inhibitor benomyl led to the identification of the first α- and β-tubulin-encoding genes in any organism. It was then shown that the tubulins are involved in nuclear and chromosomal movement. The crowning of this work was the discovery of γ-tubulin by Berl and Liz Oakley. A *benA* (encoding one of the isoforms of β-tubulin) temperature-sensitive mutant, *benA33*, results in microtubules that are hyperstable (rather than nonfunctional) at the nonpermissive temperature. Three suppressors of *benA33* mapped in a gene that when cloned and sequenced was shown to code for a new tubulin. This tubulin is critical for the nucleation of microtubules in all eukaryotes where it has been studied. The establishment of the function of γ-tubulin illustrates the use of *A. nidulans* as a model organism. While the inactivation of the cognate gene is lethal, the mutation could be maintained in a heterokaryon (see section 'The *A. nidulans* Genetic System'). As conidia are uninucleate, heterokaryons will produce two types of conidia, one of which carries the disrupted allele, where the phenotype caused by the mutation during conidial germination can be assessed microscopically. The disruption does not affect germination, but blocks nuclear division and to some extent nuclear migration. DNA is replicated, chromosomes condense, but spindles are not assembled. Thus, work that started with the isolation of tubulin inhibitor-resistant mutants led to the discovery of a new tubulin, which in all organisms is crucial for microtubule nucleation in centrosomes and in fungi (which have a closed mitosis) in spindle polar bodies.

In the seminal Morris article of 1976, a large number of conditional mutants were characterized. These were temperature-sensitive mutants, which either failed to enter mitosis at the nonpermissive temperature (*nim*, never in mitosis), were blocked at different stages (*bim*, blocked in mitosis), or where the nuclei failed to migrate (*nud*, nuclear distribution), while *sep* mutants are defective in septum formation. Eventually, the cognate genes were cloned and sequenced, suppressors were isolated and identified, to give a growing picture of the genes involved in basic processes of cell biology.

Cellular motors of the myosin class associate with actin filaments, while kinesis and dyneins move cargo (vesicles and organelles) along microtubules. *nudA* encodes the dynein heavy chain, *nudG* the dynein light chain, while other *nud* mutants defined hitherto undescribed regulatory proteins of the dynein complex. In particular, *nudF* encodes a close homologue of the human protein LIS1, which is mutated in Miller–Dicker lysencephaly, a human hereditary disease of the nervous system where neurons fail to migrate in the hemizygote. NudC, a protein that interacts with NudF is also conserved from fungi to mammals. It is likely that the primary effects of NudC/NudF in organisms with an open mitosis are in cytokinesis, a role obviously that can only be partially conserved in a syncytial organism with a closed mitosis such as *A. nidulans*. This pioneering work, which exploited both the *A. nidulans* genetic system and its specific morphology, has guided the work leading to the understanding of the function of the dynein complex in the nervous system.

At variance with dyneins, kinesin genes are highly redundant and only one was identified through mutant screens. This is *bimC*, which defines a specific class of plus-end conserved kinesins. Mutants in this gene are defective in spindle pole separation and are thus blocked in nuclear division and provided the first direct evidence the kinesins are involved in mitosis.

In the genetic screen, no mutants blocked in G1 were found, mutants blocked in the S-phase map at five loci, others blocked in the transition of G2 to mitosis map at six loci. Among the genes so defined, some are orthologues of genes previously known from *S. pombe*. *nimX* (not identified in the screen) encodes the orthologue of the cyclin-dependent *S. pombe* cdc2 kinase. The homologue of the cdc13 cyclin B is encoded by *nimE*, while the phosphatase activity necessary for the activation of NimX^{cdc2} is encoded by *nimT*. NimA, on the contrary, is a newly discovered serine/threonine kinase, which defines a whole class of proteins conserved throughout the eukaryotes. NimA functions downstream of NimX^{cdc2}/cyclin B, which would then have two independent functions, one to promote spindle formation, through the activation of other kinases, the second to activate NimA, which in turn is necessary for chromosome condensation. NimA is necessary for entry into mitosis, mutants showing duplicated spindle polar bodies, while its destruction by proteolysis is necessary for exit from mitosis. There is a considerable evidence for similar roles in mitosis for NimA homologues in higher eukaryotes. A human protein, Pin1, interacting with NimA was identified in a two-hybrid screen. Pin1 mutants have a phenotype reciprocal to that of NimA mutants, suggesting that Pin1 (PinA in *A. nidulans*) is involved in the inactivation of NimA. Pin1 is a universally (in eukaryotes) conserved peptidyl-prolyl isomerase that catalyzes specifically the isomerization of prolyl bonds in a P-Ser/Thr-Pro dipeptide, increasing its rate by about 1000 times, thus allowing a drastic change in the peptide backbone conformation, NimA is only one of its substrates, another one being cdc2/cyclin B. It has recently been shown in HeLa cells that Pin1 is necessary for entry into mitosis, associates with mitotic chromosomes, and it strongly stimulates cdc2 phosphorylation. The discovery of Pin1 and its involvement in mitosis has led to flurry of activity, concerning its possible role in cancer, but more cogently in the onset of Alzheimer's disease. Mice homozygously deleted for the *Pin1* gene develop a neuronal degeneration with many of the histological characteristics of Alzheimer's. Both *tau*, a microtubule-associated protein, and APP (amyloid precursor protein) are phosphorylated at Ser/Thr-Pro motifs. These proteins are hyperphosphorylated and insoluble in Alzheimer's. It had been proposed that the key regulator of the state of these proteins is actually Pin1, which would displace the equilibrium toward the nonphosphorylated, soluble forms.

The anaphase-promoting complex (APC) is an ubiquitin ligase that targets key mitotic proteins such as cyclins and directs them to the proteosome. Mutants in its components will be expected to be blocked in metaphase and to show a *bim* phenotype. Two such components were first identified among the *bim* mutants. BimE was identified first as a negative regulator of mitosis. Biochemical work in *Xenopus* oocytes showed that a protein that copurified with APC (APC1) is the orthologue of BimE. *bimA* encodes the APC3 component. Once all chromosomes are attached to microtubules, APC activation results in degradation of securin. This releases and activates separase, a protease that cleaves cohesin. As cohesin keeps sister chromosomes together, this cleavage is the prerequisite for anaphase. *bimB* encodes separase. A component of cohesin, *sudA*, was identified as a suppressor of a *bimD* allele, which itself results in an anaphase block characterized for defective chromosome separation. Finally, mutations in *bimG* result in large, polyploid nuclei that fail to complete anaphase. Nuclei are clumped and conidia fail to germinate highlighting the link between the regulation of mitosis and the establishment of polarity. BimG is a phosphatase, showing striking identity with mammalian phosphatases of the PP1 class. BimG is localized to the spindle polar bodies, to the nucleolus, to the tip of the hypha, and transiently in the septum. There is no hint as to what are the substrates of BimG in the mitosis, septum formation, and polarity establishment.

A. nidulans Developmental Pathways

In the sexually reproducing Aspergilli, the mycelial mat can follow two different developmental pathways. Meiosis and the formation of asci that contain ascospores, occurs in specialized structures, the cleistothecia. In *A. nidulans*, mature cleistothecia are globose, darkly pigmented structures of 100–200 μm in diameter. Ascospores have a characteristic bivalve morphology (about 4 μm × 3.5 μm) showing two equatorial crests. Ascospore ornamentation is a valuable taxonomical character in the sexually reproducing Aspergilli. The protocleistothecium is generated from vegetative hyphae, which coil in a spherical structure developing into a cleistothecium, surrounded by specialized, modified hyphal cells called Hülle cells. *A. nidulans* is homothallic; two genetically identical nuclei can fuse to give diploids which, as in all other filamentous ascomycetes, are immediately committed to meiosis. In heterothallic Aspergilli, the sexual cycle occurs only when nuclei of opposite mating types meet in heterokaryons. Some nonsexual Aspergilli (*A. flavus* and *A. parasiticus*) form structures, sclerotia, which may be developmentally related to the cleistothecium. Conceptually, we can distinguish two processes in the development of

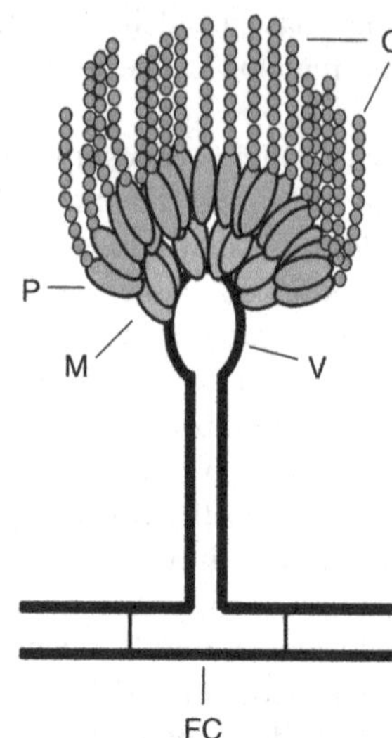

FIGURE 13 The condiophore of *Aspergillus nidulans*. In the left panel scanning electron microscopy images of condiophore of the wild-type and two mutant strains are shown, these carry loss-of-function mutations in the *brlA* (bristle) and *abaA* (abacus) gene respectively. The center panel illustrates the developmental process by showing the expression of a membrane protein (the UapA transporter fused to the green fluorescent protein (GFP)), which is specifically expressed in the metula stage, which is then diluted on in the phialides and conidia, the same transporter is then expressed again during conidial germination (see text). The right panel show a schematic representation of the conidiophore of *A. nidulans*, metulae and phialides are arbitrarily colored to facilitate identification. FC, foot compartment; S, conidiophore stalk; V, Vesicle; M, metula; P, phialide; C, conidia. Notice that for some metulae in the left and center panels the two cognate phialides can be clearly seen. The pictures in the left hand panel has been kindly provided by Reinhrad Fischer. Reproduced with permission from Kues, U., & Fischer, R. (Eds.) (2006) *The micota I: Growth differentiation and sexuality*. Berlin: Springer. One of the center panel by George Diallinas and Areti Pantazopoulou.

the mature cleistothecium. One is the morphological process that leads to cleisthotecia, surrounded by h¨lle cells. The second is the behavior of nuclei, which in the primordium of the cleisthotecium form dikaryons, in which two nuclei divide synchronously. Dicaryotic nuclei fuse into transient diploids, which undergo immediate meiosis, followed by two mitoses leading to eight binucleate haploid ascospores per ascus. These processes can be experimentally separated, as it is possible to obtain morphologically perfect cleisthothecia that do not contain asci. Many genes, including transcription factors and G-coupled receptors, have been implicated in either or both processes. The availability of the genome and possibility of following tagged proteins through the developmental processes should lead to an understanding of the sexual maturation process, including the roles of mating types in homothallic and also heterothallic species. Very recent work has established that both α and HMR mating type genes (see section 'The Genus *Aspergillus* in the Genomic Era') are necessary for fertility but not for cleistothecial formation. At present, we cannot yet draw a scheme of the developmental pathway leading to the formation of sexually mature, fertile cleistothecia.

The second developmental pathway is the formation of asexual conidia, which is present in all Aspergilli. These are formed from a specific structure, the conidiophore, which is the taxonomic marker of the genus. Conidiophores sizes range from 50 to 70 μm long in *A. nidulans* as much as 5 cm in *Aspergillus giganteus*. The structure of the conidiophore is shown in Figure 13. From the flat mycelial mat, a stalk grows from a foot compartment at a right angle from the mat. The stalk then swells into a multinucleate vesicle. From the vesicle a first series of cells arise, the metulae or primary sterigmata. About 60 metulae are formed in each vesicle. Each metula buds at its tip to give two or three uninuclear phialides, also called secondary sterigmata. From the phialide, uninuclear conidia bud, only one nucleus enters each conidium. The process is repeated, in such a way that clonal rows of conidia are formed, the last conidium to be formed is adjacent to the metula, the first and oldest being the most distal one. This process is not identical in all Aspergilli; some species, called uniseriate (such as *A. fumigatus*), have only one series of sterigmata from where conidia arise directly, while in some Aspergilli, conidia contain more than one nucleus (such as *A. oryzae*).

Two approaches were used to study this process. In the first one, mutants were isolated and blocked in different steps of conidiophore development; in the second, mycelia were synchronized, and by the technique of 'cascade hybridization,' an early methodology to define a transcriptome, it was determined which genes were expressed at different stages of conidiophore development. John Clutterbuck published in 1969 the seminal article of the study of the conidiophore developmental pathway, while the first cascade experiment in any organism was published by the Timberlake Laboratory in 1980. Clutterbuck described a number of mutants blocked in different steps of

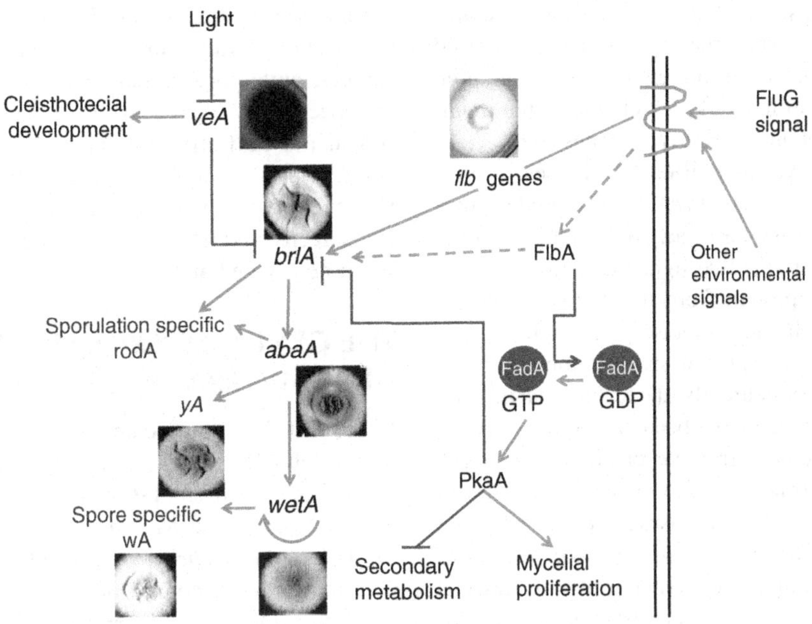

FIGURE 14 Simplified scheme of the development of the conidiophore. The central developmental pathway, involving the products of the BrlA, AbaA, and WetA transcription factors is shown. In blue, target genes. Three are specifically shown, *rodA* encoding a conidial-specific hydrophobin (see section '*Aspergillus* as Human Pathogens'), *yA* encoding the last step of the synthesis of the wild-type green pigment and *wA*, encoding a polyketyde synthase which is the first step in the synthesis of the conidial pigment. The insets show the morphological phenotypes of *veA1*, *brlA*, *abaA*, *wetA*, *yA*, and *wA* mutations. To the right a simplified version of the signalization pathway defined by dominant and recessive fluffy mutations. The inset shows the morphological phenotype of a fluffy mutant. The transmembrane receptor of the FluG signal has not been identified. The insets correspond approximately to the size of colonies grown on complete medium at 37 °C for 48 h. Pictures from the laboratory of the author or kindly provided by Reinhard Fischer. All signals are shown as affecting the expression of *brlA*, but this is for simplicity's sake, as we do not know if these are direct or indirect. PkaA may affect the BrlA protein posttranslationally. To simplify the scheme the β- and γ-subunits interacting with FadA are not shown.

conidiophore development. In *bristle* mutants (*brlA*), conidiophore stalks that fail to complete the developmental pathway originate from the mycelial mat. In *abacus* mutants (*abaA*), sterigmata continue to give row after row of additional sterigmata, without ever terminally differentiate phialidae or conidia (Figure 14). *Stunted* mutants (*stuA*) result in short conidiophores with conidia being made directly from the vesicle, while in *medusa* (*medA*), metulae do not immediately differentiate and produce series on metulae before giving origin to phialides. Finally, *wet* mutants (*wetA*) do not affect the development of the conidiophore, but result in defective conidia that autolyze. Once these genes were cloned, their function was analyzed by inactivating them, following their expression pattern and overexpressing them conditionally using the tightly regulated *alcA* promoter (see above). BrlA is a Zn finger transcription factor, which directly regulates the expression of *abaA*. AbaA is also a transcription factor, which regulates *brlA* in a feedback loop and *wetA*. WetA regulates late-expressed and conidial-specific genes such as cell wall genes, and it is supposed to be a transcription factor. StuA has the characteristics of a transcription factor and limits in some unknown way the spatial distribution of the BrlA and AbaA proteins as well as being involved in conidiophore elongation and wall thickening, while MedA regulates the temporal expression of *brlA* along the developing conidiophore. Downstream of the three core regulators, BrlA, AbaA, and WetA, there are target genes, which are activated at different stages of the conidial developmental process. Some of these, as genes involved in spore pigmentation, were known from the earlier days of *A. nidulans* genetics, others were identified by the cascade hybridization methodology. These target genes show a variety of regulation patterns, some like *wA* (see Figure 14) being under the control of WetA, some like *yA* under the control of AbaA, some others requiring, in order to be expressed, various combinations of the three transcription factors. The number of downstream genes has been estimated >100 by genetical procedures and at about 1200 by cascade hybridization. This huge difference could be partly explained by the fact that the inactivation of at least some sporulation-specific genes does not lead to any visible phenotype and that overexpression of many metabolic genes occurs during sporulation.

This scheme does not account for the signaling pathway that leads to the onset of conidiation, that is, the formation from the mycelial mat of the conidiophore stalk. All the early work was carried out in a standard Glasgow strain, which is really a hyperconidiating constitutive mutant carrying the *veA1* mutation. This partial loss-of-function mutation leads to profuse conidiation in the dark. In strains carrying this mutation, when the mycelia are transferred from submerged culture to an air interphase, conidiation occurs synchronously in the dark, which allows the monitoring of the expression of relevant genes. The *veA*$^+$ (wild-type) strains behave quite differently. In these strains, conidiation is light inducible, and *veA*$^+$ strains produce profusely cleisthothecia in the dark. *veA* null mutants do not produce cleisthotecia at all. Blue light irradiation result in exclusion of the VeA protein from the nucleus, whereas the VeA1 mutant protein lacks precisely the nuclear localization signal and is always present in the cytoplasm. VeA mediates also, directly or indirectly, the response to polyunsaturated fatty acids, which have been shown to provide a sporogenic signal that is alternative or additive to light, the ratio of different unsaturated fatty acids driving development toward the sexual or the asexual cycle. The *veA* gene product behaves formally in the absence of light as a repressor of conidiation and an activator of cleistothecial development. Thus, VeA must directly or indirectly repress the expression of *brlA*. Figures 13 and 14 illustrate and summarize the process of conidiophore development.

In order to study the signalization of *brlA*, another set of mutants was isolated. These are strains where a conidiophore does not develop, 'fluffy' mutants, which make a fast growing, colorless, undifferentiated mycelium. The study of fluffy mutations has revealed a complex signaling pathway, which determines both the conidiation pathway and the production of secondary metabolites. Recessive mutations of the *fluG* gene lead to a fluffy phenotype. FluG codes for an enzyme that is responsible for the synthesis of a diffusible product, which activates, through an unknown receptor, a signaling cascade specified by several *flb* genes. Dominant mutations in the *fadA* gene also lead to a fluffy phenotype. FadA is a subunit of a trimeric G-protein, which in its GTP-bound form activates a protein kinase (PkaA) that through phosphorylation of target proteins represses secondary metabolism, promotes growth, and represses conidiation. The FlbA protein, also characterized by fluffy loss-of-function mutations, acts downstream of the FluG signal to shift FadA to the GDP-bound inactive form. Figure 14 summarizes the 'fluffy' signaling pathway and its possible relationship with the developmental pathway leading to conidiation. Recent work has shown that the 'fluffy' class of mutations has not been saturated and that there are at least three concurrent signaling pathways. This complex signaling pathway (which has been drastically simplified here and in Figure 14) conceals, however, a conceptual problem. Fluffy mutants are not only aconidial but they are also foremost, fast proliferating mutants. By studying fluffy mutants, we are studying the signals that control and limit mycelial proliferation, and the complete or partial loss of conidiation could be a necessary result of the enhanced proliferative activity. In fact, if we slow down the growth of fluffy mutants by using very poor carbon sources, quite good conidiation can be seen. Mutants that are specifically altered in the signals upstream of bristle should not be fluffy, but 'bald', where condiophores do not arise from a normally proliferating mycelial flat mat. However, 'bald' mutants were not reported in the early mutant screens, and it will be most interesting to know if they could be isolated at all.

THE GENUS *ASPERGILLUS* IN THE GENOMIC ERA

In the last few years, complete genomic sequences have been established for *A. nidulans*, *A. oryzae*, *A. fumigatus*, *A. terreus*, two different strains of *A. niger*, *A. flavus*, *Aspergillus clavatus*, and *Neosartorya fischeri* ('perfect' name for *Aspergillus fischerianus*). Articles have been published describing the genomes of *A. nidulans*, *A. fumigatus*, *A. oryzae*, and *A. niger*. Dedicated Web sites exist for all the species except for the ongoing projects of *A. clavatus* and *N. fisheri*; the predicted genes and proteins of the latter can be found in the NCBI database. For *A. nidulans*, where a detailed genetic map exists, a correlation of the genetic and physical maps is available, even if the *in silico* identification of many of the classically mapped genes with autocalled genes is far from complete. Genome annotation is in different stages of completion in different species and so is the availability of microarrays. The availability of complete genomes has stimulated technological developments that allow high throughput gene inactivation and substitution. The genomes reflect the high metabolic versatility of the genus and have highlighted the evolutionary divergence of the different species. Telomeres have been assigned to each *A. nidulans* chromosome and this can be carried out for all the other member of the genus. Putatively active transposons of different families have been identified, including the interesting finding that helitrons, a newly described family of rolling-circle replicating eukaryotic transposons, are present and active in *A. nidulans* but not in the other species. The last release of the *A. nidulans* genome predicts 10 701 proteins as coded by the genome of this species, while for *A. niger*, the predicted number is 14 165. Interestingly, all the genomes sequenced contain mating type genes and other genes known or presumed to be involved in the sexual cycle. The homothallic *A. nidulans* and *N. fischeri* contains unlinked genes for both the α and HGM mating types, but their location in the genome suggest that homothallism has evolved independently in these species. Interestingly, for *A. fumigatus*, *A. oryzae*, and *A. flavus*, strains of opposite mating types have been found in nature, opening the possibility that in fact all (or at least many) Aspergilli can undergo the sexual cycle.

Another interesting finding is the variability and phylogenetic scatter of the genes putatively involved in the production of secondary metabolites. As an example, the genome of *A. niger* contains a gene cluster similar to that involved in the synthesis of fumonisin in *Giberella (Fusarium) moniliformis*, while these genes are absent in other Aspergilli, thus positing the question of whether convergent evolution or horizontal transfer is involved. The considerable knowledge accumulated in *A. nidulans* may permit to extrapolate to other Aspergilli, for example, by asking which are the genes regulated by transcription factors such as NirA, AreA, or PacC (see section 'Control of Gene Expression'), investigate promoter structure *in silico*, or more subtly to investigate the factors that make the conidiophore of *A. nidulans* biseriate, that of *A. fumigatus* uniseriate and that of *A. clavatus* characteristically elongated and uniseriate. At present, genome data have grown faster than the capacity of the research community to make use of them, a general problem of the genomic era. It is hoped that the comparison of genomes will lead to a better understanding of problems such as the evolution of metabolism, the evolution of the sexual and parasexual cycles, the evolution of silencing mechanisms, the isolation of useful new secondary metabolites, the possible engineering of pathways to produce new metabolites, and the identification of specific fungal essential genes that could lead the development of highly specific antifungal agents.

FURTHER READING

Abad, A., Victoria Fernández-Molina, J., Bikandi, J., Ramírez, A., Margareto, J., Sendino, J., Luis Hernando, F., Pontón, J., Garaizar, J., & Rementeria, A. (2010). What makes *Aspergillus fumigatus* a successful pathogen? Genes and molecules involved in invasive aspergillosis. *Revista Iberoamericana de Micología, 27*(4), 155–182.

Bernreiter, A., Ramon, A., Fernández-Martínez, J., *et al.* (2007). Nuclear export of the transcriptional factor NirA is a regulatory checkpoint for nitrate induction in Aspergillus nidulans. *Molecular and Cellular Biology, 27*, 791–802.

Brakhage, A. A. (2005). Systemic fungal infections caused by *Aspergillus* species: Epidemiology, infection processes and virulence determinants. *Current Drug Targets, 6*, 875–876.

Brakhage, A. A., Bruns, S., Thywissen, A., Zipfel, P. F., & Behnsen, J. (2010). Interaction of phagocytes with filamentous fungi. *Current Opinion in Microbiology, 13*(4), 409–415.

Brakhage, A. A., & Schroeckh, V. (2011). Fungal secondary metabolites – Strategies to activate silent gene clusters. *Fungal Genetics and Biology, 48*(1), 15–22.

Davies, R. W., Waring, R. B., Ray, J. A., Brown, T. A., & Scazzocchio, C. (1982). Making ends meet: A model for RNA splicing in fungal mitochondria. *Nature, 300*, 719–724.

Dyer, P. S. (2007). Sexual reproduction and significance of MAT in the Aspergilli. In J. Heitman, J. W. Kronstad, J. W. Taylor & L. A. Casselton (Eds.), *Sex in fungi: Molecular determination and evolutionary principles* (pp. 123–142). ASM Press: Washington, DC.

Felenbok, B., Flipphi, M., & Nikolaev, I. (2001). Ethanol catabolism in *Aspergillus nidulans*: A model system for studying gene regulation. *Progress in Nucleic Acid Research and Molecular Biology, 69*, 149–204.

Galagan, J. E., Calvo, S. E., Cuomo, C., *et al.* (2005). Sequencing of Aspergillus nidulans and comparative analysis with A. fumigatus and A. oryzae. *Nature, 438*, 1105–1115.

Gastebois, A., Clavaud, C., Aimanianda, V., & Latgé, J. P. (2009). *Aspergillus fumigatus*: Cell wall polysaccharides, their biosynthesis and organization. *Future Microbiology, 4*(5), 583–595.

Goldman, G. H., Osmani, S. A., *et al.* (2008). *The Aspergilli: Genomics, medical applications, biotechnology, and research methods* (Vol. 19). CRC Press/Taylor and Francis: Boca Raton, FL; pp. 321–342.

Keller, N., Turner, G., & Bennett, J. W. (2005). Fungal secondary metabolism – From biochemistry to genetics. *Nature Reviews. Microbiology, 3*, 937–947.

Kiho, K., & Harvell, C. D. (2004). The rise and fall of a six-year oral-Fungal Epizootic. *The American Naturalist, 164*(Suppl.), 52–63.

Kwon-Chung, K. J., & Sugui, J. A. (2009). Sexual reproduction in *Aspergillus* species of medical or economical importance: Why so fastidious? *Trends in Microbiology, 17*(11), 481–487.

Li, S., Du, L., Yuen, G., & Harris, S. D. (2006). Distinct ceramide synthases regulate polarized growth in the filamentous fungus *Aspergillus nidulans*. *Molecular Biology of the Cell, 17*, 1218–1227.

Machida, M., Asai, K., Sano, M., *et al.* (2005). Genome sequencing and analysis of Aspergillus oryzae. *Nature, 438*, 1157–1161.

Martinelli, S., & Kinghorn, J. R. (Eds.), (1994). *Aspergillus: 50 years on.* In: *Progress in industrial microbiology* Vol. 29. Elsevier Scientific: Amsterdam.

McCormick, A., Loeffler, J., & Ebel, F. (2010). *Aspergillus fumigatus*: Contours of an opportunistic human pathogen. *Cellular Microbiology, 12*(11), 1535–1543.

Meyer, V., Wu, B., & Ram, A. F. (2011). *Aspergillus* as a multi-purpose cell factory: Current status and perspectives. *Biotechnology Letters, 33*(3), 469–476.

Nierman, W. C., Pain, A., Anderson, M. J., *et al.* (2005). Genomic sequence of the pathogenic and allergenic filamentous fungus Aspergillus fumigatus. *Nature, 438*, 1151–1156.

Park, S. J., & Mehrad, B. (2009). Innate immunity to *Aspergillus* species. *Clinical Microbiology Reviews, 22*(4), 535–551.

Pei, H. J., de Winde, J. H., Archer, D. B., *et al.* (2007). Genome sequencing and analysis of the versatile cell factory Aspergillus niger CBS 513.88. *Nature Biotechnology, 25*, 221–231.

Peñalva, M. A., & Arst, H. N., Jr. (2004). Recent advances in the characterization of ambient pH regulation of gene expression in filamentous fungi and yeasts. *Annual Review of Microbiology, 58*, 425–451.

Pontecorvo, G., Roper, J. A., Hemmons, L. M., MacDonald, K. D., & Bufton, A. W. J. (1953). The genetics of *Aspergillus nidulans*. *Advances in Genetics, 5*, 141–238.

Raper, K. B., & Fennell, D. I. (1973). *The genus* Aspergillus. Robert Krieger: Huntington, New York.

Rhome, R., & Del Poeta, M. (2009). Lipid signaling in pathogenic fungi. *Annual Review of Microbiology, 63*, 119–131.

Singh, A., & Del Poeta, M. (2011). Lipid signalling in pathogenic fungi. *Cellular Microbiology, 13*(2), 177–185.

Smith, J. E., Pateman, J. A., *et al.* (1978). *Genetics and physiology of Aspergillus nidulans*. Academic Press: London.

Tell, L. A. (2005). Aspergillosis in mammals and birds: Impact on veterinary medicine. *Medical Mycology: Official Publication of the International Society for Human and Animal Mycology*, *43*(Suppl.), 571–573.

Yu, J. H., & Keller, N. (2005). Regulation of secondary metabolism in filamentous fungi. *Annual Review of Phytopathology*, *43*, 437–458.

RELEVANT WEBSITES

http://www.fgsc.net/. – Fungal Genetics Stock Center.
http://www.aspergillus.org. – Fungal Research Trust.
http://docterfungus.org. – Mycoses Study Group.
http://www.fieldmuseum.org. – The Field Museum.

Chapter 3

Clavicipitaceae

Free-Living and Saprotrophs to Plant Endophytes

M.S. Torres and J.F. White
Rutgers University, New Brunswick, NJ, USA

Chapter Outline

Defining Statement	41
Introduction	41
Free-Living Saprotrophs and Insect Parasites	41
Species on Soft-Bodied Scale Insects	43
Plant Biotrophs	43
Chemical Diversity in the Clavicipitaceae	45
Biological Activities of Secondary Metabolites	46
Economic, Agricultural, and Ecological Importance of Clavicipitacious Endophytes	47
Origin and Evolution of Ergot Alkaloids	47
The Proposal of Functionality for Ergot Alkaloids	47
The Evidence for Defensive Mutualism in the Clavicipitaceae Endophyte/Grass Association	47
The Defensive Mutualism Controversy: A Debate Regarding Function	48
Evolution of Defensive Mutualism from Self-Defense to Host Defense	48
Conclusion	48
Acknowledgment	49
Further Reading	49

DEFINING STATEMENT

Fungal endophytes are found in most of the plant species examined to date. Fungal endophytes on grass plants commonly referred as grass endophytes have received more attention due to their economic and ecological relevance. Members of the fungal family Clavicipitaceae show a wide range of ecologies that offer an opportunity to analyze the plant fungal interactions on different biological associations.

INTRODUCTION

The family Clavicipitaceae (Hypocreales and Ascomycota) includes saprotrophic and symbiotic species associated with insects and fungi (*Cordyceps* spp.) or grasses, rushes, and sedges (*Balansia* spp., *Epichloë* spp., and *Claviceps* spp.). In a phylogenetic context, the Clavicipitaceae has been considered to be a monophyletic group derived from within the Hypocreales. The most primitive and widespread members of the family are soil saprophytes and insect pathogens. The soil-inhabiting genera are deeply rooted in phylogenetic trees based on gene sequences (Figure 1), suggesting that they are the most primitive members of the family. These are often identified as species of the teleomorphic genus *Cordyceps*. Frequently, Clavicipitaceae found in soil are known only in their anamorphic states. These include several hyphomycetous soil genera (e.g., *Acremonium*, *Chaunopycnis*, *Paecilomyces*, *Metarhizium*, *Lecanicillium*, *Beauveria*, and *Hirsutella*).

The family Clavicipitaceae is distinguished by its wide host range and diverse ecologies. To explain the evolutionary history of the family, it has been proposed that interkingdom host jumps occurred among the animal, fungal, and plant host species. Recent studies based on multigene phylogenetic analysis and ancestral character state reconstructions have further shown that the grass endophytes were derived from insect parasitic species through a process of successive host jumping.

FREE-LIVING SAPROTROPHS AND INSECT PARASITES

Chaunopycnis alba has a high saprobic capacity, being commonly isolated from soils, plant litter, and lichens. Species of this genus are widespread in all continents; and they have been found in extreme habitats, including Antarctic

Eukaryotic Microbes.

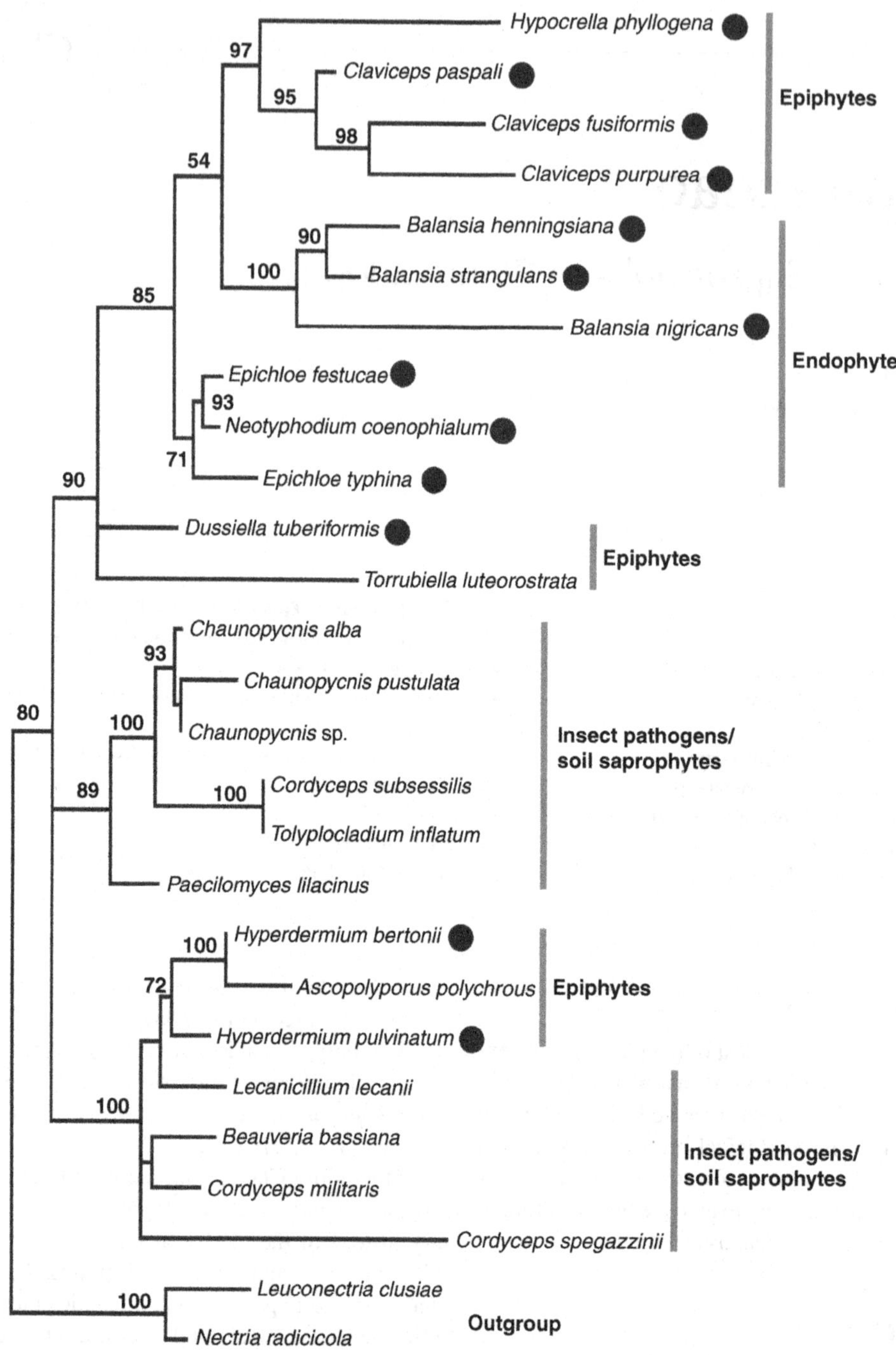

FIGURE 1 The most likely tree (– ln 3006.913) as determined by Paup using the GTR + I + G model of evolution. The numbers on the branches indicate the posterior probability as a percentage for the node they proceed (only ≥50% shown). •, ergot alkaloids reported.

soils. They are phylogenetically related to the insect parasitic genera *Cordyceps* spp.

Species of *Beauveria* are widespread in all continents as insect pathogens. *Beauveria bassiana* is widely used as a biocontrol agent in different parts of the world. These fungi infect insects and internally degrade them, but they are also able to survive as saprophytes in the absence of the host. They have been found to grow into plant leaves and become dormant in leaves until consumption by insects after which they reinitiate growth and consume the insect.

Lecanicillium (anamorph of *Torrubiella*) is another genus containing common insect pathogens. Species of *Simplicillium* have been isolated from soil-borne nematodes. *Paecilomyces* is another genus that is frequently isolated from insects, nematodes, and soil. Many soil inhabitants and insect parasites of the family Clavicipitaceae have potential to be used as biological control agents.

The enzymatic capabilities of species of Clavicipitaceae that grow in the soil environment are comparable to those of many other groups of soil fungi. Numerous soil fungi have been shown to degrade a wide range of substrates including pectins, xylans, cellulose, and a range of amino acids and inorganic nitrogen sources, making them capable of gaining nutrients from many organic substances present in soils. Soil inhabitants in the Clavicipitaceae tend to be generalists in their enzymatic capabilities.

SPECIES ON SOFT-BODIED SCALE INSECTS

Several species of Clavicipitaceae grow on soft-bodied scale insects consuming the entire body of the insect where they first develop the anamorphic stage and less commonly the teleomorph. *Hypocrella* spp. (Figures 2(a)–2(d) and 3(c)–3(d)), *Dussiella* spp. (Figure 3(a)), *Hyperdermium* spp. (Figure 3(b)) are examples of Clavicipitaceae that infect scale insects.

Scale insects are particularly distributed and diverse in the tropics, developing on both scale insect subfamilies Aleyrodiidae and Coccidae. In this environment, Clavicipitaceae with the entomogenous habit maintain a balance of energy allocation to the production of the anamorphic stage that provides large amounts of conidia for short-distance dispersal and the teleomorphic stage often with dry spores for long-distance dispersal. The anamorph stages of these species have been shown to be important biological control agents. *Aschersonia aleyrodes* has been used for control of the whitefly of greenhouse (*Trialeurodes vaporariorum*).

The large stromata that some species develop compared to the scale insect body suggests that the fungus is able to access plant nutrients through the stylet hole remaining from the scale insect. This mechanism has been suggested for *Hypocrella* and *Dussiella* species.

PLANT BIOTROPHS

The most derived members in the Clavicipitaceae are the plant biotrophs. *Claviceps* spp. are exclusively plant biotrophic. Some species have a very narrow host range while others infect numerous grass species, with a worldwide distribution along with grass hosts. These species infect developing ovules of grasses, eventually replacing the seeds by a fungal fruiting body (sclerotium) (Figure 4(b)).

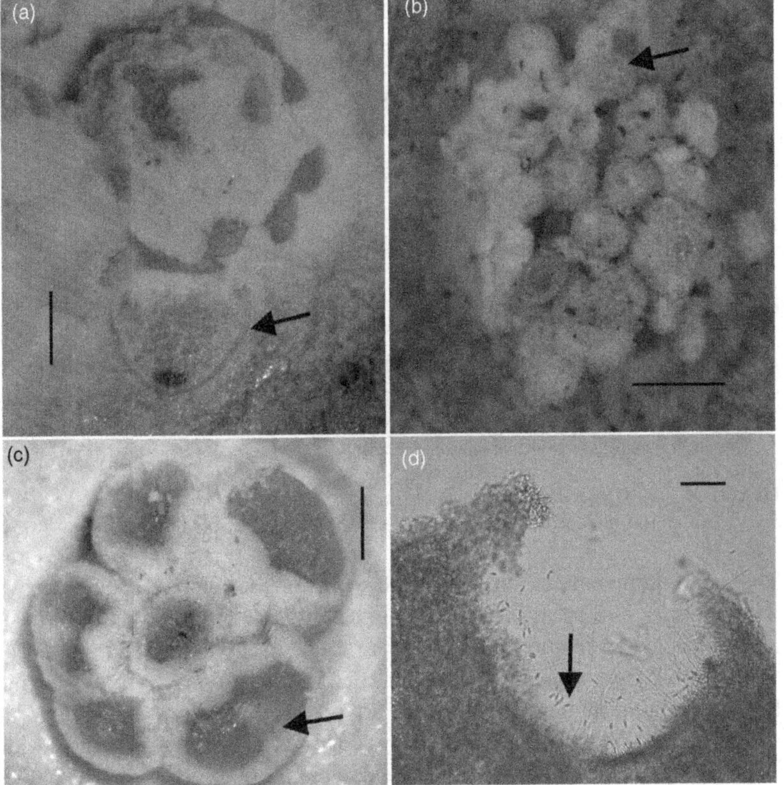

FIGURE 2 *Hypocrella* spp. (a) Growing on scale insect (arrow, bar = 150 μm). (b) Perithecia (arrow, bar = 0.2 mm). (c) Pouch-like conidiomata of anamorphic stage (arrow, bar = 200 μm). (d) Longitudinal section on pouch-like conidiomata showing conidia (arrow, bar = 20 μm).

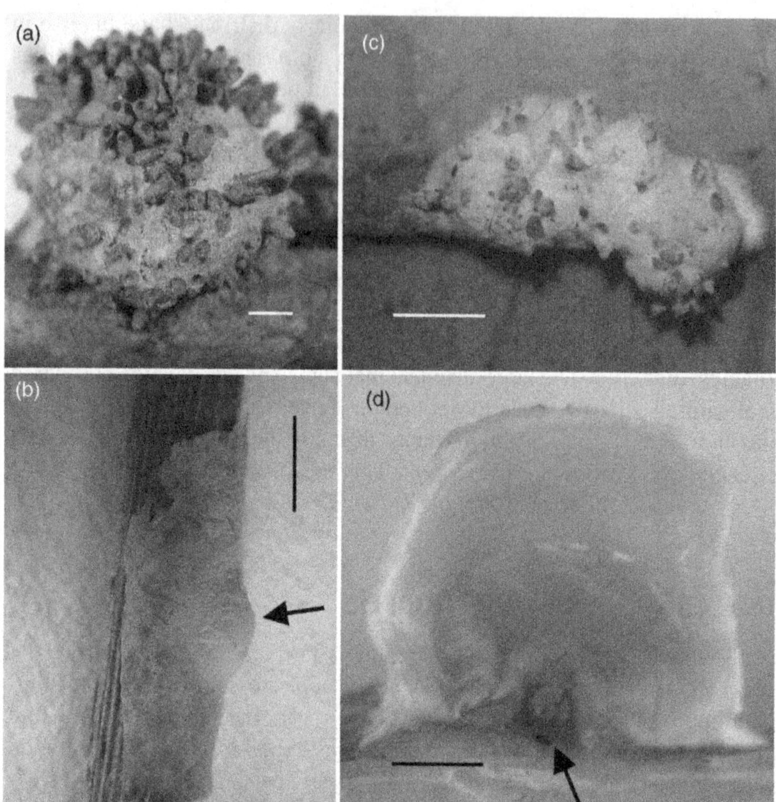

FIGURE 3 (a) *Dussiella tuberiformis* (bar = 0.2 mm). (b) *Hyperdermium bertonii* (bar = 0.4 mm). (c–d) *Hypocrella* spp. (c) Developing stromata on leaf (bar = 0.5 mm). (d) Longitudinal section of stromata showing stylet wound (arrow, bar = 500 μm).

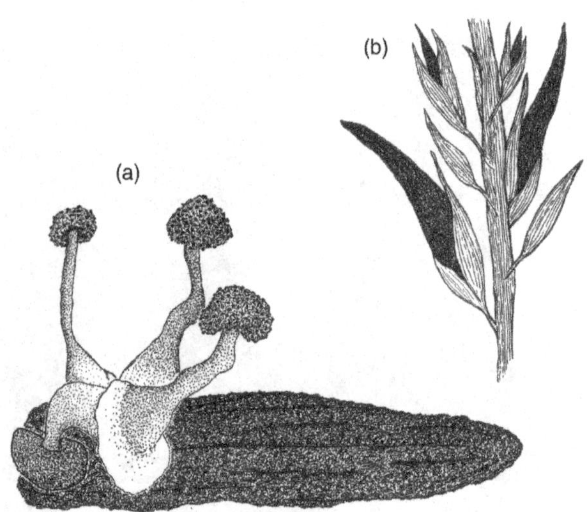

FIGURE 4 (a) Germinating sclerotium of *Claviceps purpurea* from *Spartina* sp., × 100. (b) Florets of *Spartina* sp. infected by *Claviceps* sp., × 4.

The sclerotium remains limited to the ovary of the plant without extensive invasion of plant tissues by the fungal mycelium. When mature, the sclerotium drops to the soil and eventually germinates to produce a stipe with a globose head containing numerous perithecia (Figure 4(a)). Ascospores ejected from perithecia infect grass ovules and reinitiate the cycle. Fungi of this genus are commonly referred to as ergots. The ergot sclerotia of *Claviceps purpurea* were the initial source of the ergot alkaloid derivative lysergic acid diethylamide (LSD). The 'psychedelic era' of the 1960s was largely a product of this fungus.

Myriogenospora atramentosa is epibiotic on grasses, infecting developing leaves and forming its fruiting body directly on the surface of the leaf blade. *Myriogenospora* forms a fine epibiotic mycelium on meristematic leaves, gaining nutrients that diffuse to the mycelium on the leaf surface. Through the production of auxins or other compounds that modify plant cell development, the fungus prevents the formation of the waxy cuticle on the epidermis of leaves. This enables the fungus to maintain a flow of nutrients and water to the fungal stroma that matures on leaf surfaces. Perithecia are produced that eject ascospores that may land on host grasses in the vicinity and reinitiate new infections.

In the genus *Balansia*, parasitic on warm-season grasses, some species show the endophytic habit. These species grow in the intercellular spaces of healthy grass plants and may reproduce by forming a fungal stroma on living leaves or inflorescences. The mycelium at the base of the stromata of the fungus generally penetrates intercellularly into the vascular tissues of the host and directly absorbs nutrients for fungal development. Generally in *Balansia*, populations of the endophytes consist of two

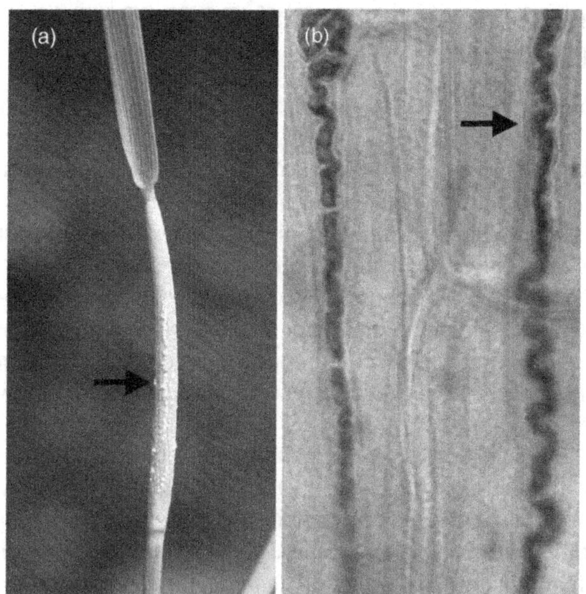

FIGURE 5 (a) *Epichloë* spp. on *Agropyron* spp. (b) Endophytic mycelia of *Neotyphodium coenophialum* on tall fescue.

mating types. *Ephelis* conidia (spermatia) of one mating type must be transferred to the stromata of the opposite mating type in order for perithecia to develop. Grasshoppers and other grass feeding insects have been observed to visit stromata to feed and are hypothesized to vector spermatia between mating types in the process of feeding.

In genus *Epichloë*, fungi are endophytic of cool-season grasses. Mycelium occurs in the intercellular spaces of leaves, culms, and rhizomes of grasses (Figure 5(b)). When the grass flowers, the fungus grows over the developing inflorescence and forms the fungal stroma (Figure 5(a)). The inflorescence primordium remains in an arrested stage of development within the fungal mycelium that prevents the development of the seed head, a condition known as 'choke disease.' Some *Epichloë* species exhibit choke disease only in a proportion of the tillers, allowing partial seed production and thus vertical transmission. *Epichloë* has been also shown to produce auxins that may play a role in altering the development of plant tissues. As in *Balansia*, the stroma bears spermatia and fungal populations contain two mating types that must be transferred between stromata before perithecia and ascospores develop. Symbiotic flies of genus *Botanophila* (Diptera: Anthomyiidae) act as 'pollinators' of stromata.

The female fly visits stromata consuming spermatia that pass through its gut undigested. When the female deposits her eggs on the stroma, she defecates depositing a mix of spermatia from stromata previously visited. Fertilized stromata produce perithecia containing ascospores that are ejected and may infect uninfected plants. Developing larvae feed on the fertilized stromata, consuming some of the perithecia, but the benefit of fertilization of stromata outweighs perithecial consumption by the fly larvae. In cultivated fine fescues (*Festuca* spp.) in Oregon, the *Botanophila–Epichloë* relationship seems to reflect a simple fungivory. Fine fescues infected with *Epichloë festucae* manifest a low frequency of fertilized stromata in the presence of the *Botanophila* fly while the fly is still able to complete its life cycle. In commercial cultivars, other factors like asymmetrical distribution of mating types could play a role in the association. It is still unknown precisely how ascospores infect plants.

The sexual cycle of *Epichloë* occurs only on grasses in the northern hemisphere while common in the southern hemisphere, all representatives of this genus of endophytes exhibit only asexual reproduction. The explanation for this biogeographical phenomenon is presently unknown. Some species of *Epichloë* have lost capacity for development of the sexual stage and instead are transmitted vertically through seed or may rely on conidial transmission. The asexual stages of *Epichloë* are generally recognized as species in the imperfect genus *Neotyphodium*. Asexual *Neotyphodium* endophytes have evolved numerous times and are distributed worldwide. These species are endophytic of leaves, culms, rhizomes, and colonize inflorescence primordia. As inflorescences develop, the endophyte mycelium grows in the ovules, developing seeds, scutelum, and embryo axis prior to germination of the seed. No symptoms of *Neotyphodium* infections are observable at any stage of plant development. The asexual endophytes are vertically transmitted and they rely on the host plant for survival and dissemination through the seeds. Asexual endophytes are not entirely clonal since phylogenetic analysis has demonstrated interspecific hybridization. This interspecies hybridization has been proposed as a mechanism to maintain genetic variability among asexual endophytes and may represent a means whereby asexual endophytes may evolve and adapt to hosts.

CHEMICAL DIVERSITY IN THE CLAVICIPITACEAE

Secondary metabolites are bioactive compounds nonessential for growth or reproduction produced in large amounts in diverse groups of organisms. Species in the Clavicipitaceae are rich sources of secondary metabolites. Insect pathogenic species of genus *Cordyceps* and *Hypocrella* produce numerous secondary compounds with biological activity with possible applications in medicine and insect control. These include ergot alkaloids (ergopeptides, lysergic acid amides, and clavines), pyrrolopyrazines, indole diterpenes, lolines, sterols, and cyclopeptides.

Cyclosporine A, a cyclic peptide produced by species of genus *Tolypocladium*, is used as an immune suppressant drug and commonly used to prevent organ rejection after transplant surgery. This compound inhibits T cell activation and is also used in the treatment of nephritic syndrome, biliary cirrhosis, aplastic anemia, and rheumatoid arthritis.

The sterols, ergokonins, have been shown to be active inhibitors of fungi such as *Candida albicans*.

Hypocrellins A and B are pigments isolated from the insect parasite *Hypocrella bambusae*. These compounds have a long history of use in traditional medicine in China and they also have antimicrobial activity.

Indole diterpenoids are produced broadly in the family. Indole diterpenoids from *C. alba* have been shown to inhibit maxi-K channels in mammalian cells and to have applications in the treatment of glaucoma without tremorgenic effects. The indole diterpenoid lolitrem B is produced by the ryegrass endophyte *Neotyphodium lolii* and has been shown to be the cause of the cattle toxicosis 'ryegrass staggers' common in endophyte-infected pastures of perennial ryegrass.

The ergot alkaloids have a long history of use in medicine. Midwives in Europe as early as in the sixteenth century used ergot alkaloids in sclerotia of ergot (*C. purpurea*) to induce and increase uterine contractions to facilitate the process of childbirth. Women in labor were given several sclerotia to chew, thereby extracting the ergot alkaloids. The early physicians who used sclerotia reported that with the use of sclerotia the birthing process seldom lasted longer than 3 h. The 1836 Dispensatory of the United States recommended the use of 15–20 sclerotia to induce uterine contractions. The practice of using sclerotia in childbirth was later abandoned due to the resulting high mortality of babies and mothers. Ergot alkaloids have also been shown to increase and decrease blood pressure depending on the particular ergot alkaloid and dose. Ergotamine tartrate is currently used in the treatment of migraine. This compound stimulates the contraction of capillaries and other vessels in the brain, thus alleviating the pressure contributing to migraine headaches. In the Middle Ages, ergot alkaloids in sclerotia that were incorporated in grain products and consumed by humans resulted in mass poisoning of populations in Europe. The resulting ergotism was caused by the vasoconstrictive properties of the numerous alkaloids present in sclerotia and later in food products.

A cattle toxicosis called 'fescue foot' is commonly seen in cattle grazing on pastures of endophyte (*Neotyphodium coenophialum*)-infected tall fescue (*Festuca arundinacea*). Here the ergot alkaloid ergovaline causes vasoconstriction, effectively reducing the flow of blood to the animal's body extremities; hooves and tails may show gangrenous necrosis as a result.

Loline alkaloids are saturated 1-amino pyrrolizidines that are produced in a number of grass associations with endophytes of the genera *Epichloë* and *Neotyphodium*. These compounds are known to be potent feeding deterrents to insects.

BIOLOGICAL ACTIVITIES OF SECONDARY METABOLITES

To date, there is a lack of literature addressing the individual effects of many secondary metabolites. Because most Clavicipitaceae produce an array of secondary metabolites, it is difficult to separate their individual effects.

In one study, small mammals (rabbits) were used to test the effect of ergot alkaloids on food preference and satiety. Rabbits preferred perennial ryegrass plants that were endophyte infected but free of ergot alkaloids to endophyte-free plants. In satiety tests, consumption of ergovaline during an initial meal had a negative effect on subsequent rabbit chow consumption.

Loline alkaloids are compounds with a defensive capacity for the grass host due to their insecticidal activities. Loline alkaloids have insecticidal activity against aphids (*Rhopalosiphum padi*), Japanese beetle (*Popillia japonica*), fall armyworm (*Spodoptera frugiperda*), and a broad range of other insects. They tend to accumulate to high levels in plant tissues with concentrations up to 20 mg g^{-1} of leaf material. The highest concentrations of loline alkaloids are associated with developing inflorescences and seeds, but they are distributed in all plant parts due to their high solubility. It is estimated that in tall fescue 10–15% of the total loline and smaller amounts of ergot alkaloids may occur in the roots. Moreover, the highest concentrations of loline alkaloids are associated with plant parts related to survival such as developing inflorescences and seeds.

The sclerotia of ergots contain a diversity of ergot alkaloids. The ergot alkaloids are believed to be defensive in nature, functioning to repel consumers of the fungal fruiting body. Infections of rye crops were the cause of numerous mass poisonings of humans throughout history. Hallucinogenic and other toxic effects of the ergot alkaloids are believed to have been the trigger for at least some of the witch trial episodes of the middle ages. *Claviceps* has adapted biotrophically to its host through the use of only a narrow range of substances that are available in the phloem nutrient stream of host grasses.

Among the secondary metabolites produced by Clavicipitaceae are pigments, generally polycyclic aromatics. Functions of the pigments produced by the Clavicipitaceae may relate to the avoidance of predation. For example, in *Epichloë*, white stromata-bearing spermatia are produced on grass culms. These attract flies in genus *Botanophila* that feed on spermatia and mycelium of the stromata and in the process vector spermatia between mating types of the fungus, effectively fertilizing stromata. This process is comparable to pollination in plants. Once fertilization has occurred and perithecia begin to form, the stroma produces a yellow pigment. The yellow stromata blend in with the yellowed leaf tips of the aging grass leaves, making it difficult to locate the stromata. This yellow coloration of fertilized stroma may serve to camouflage stromata and reduce the consumption of stromata by fungivorous insects such as beetles and weevils.

In experimental studies using models of stromata that were either white or yellow and placed randomly among natural populations of *Epichloë* stromata on grasses in the field, white models attracted significantly more insects than yellow stromata. Coleopterans including beetles and weevils were seen to be especially attracted to white stromata with white models showing approximately 2.8 times the numbers of coleopterans that visited yellow stroma models. Anthraquinones, such as skyrin, impart an orange to red pigmentation to mature stromata of *Hyperdermium bertonii*. This pigment isolated from *Hypocrella* has been demonstrated to be toxic to insects in laboratory studies.

ECONOMIC, AGRICULTURAL, AND ECOLOGICAL IMPORTANCE OF CLAVICIPITACIOUS ENDOPHYTES

Because of the alkaloids and other secondary metabolites produced by *Epichloë* and *Neotyphodium*, infected grasses are often toxic to animals. Perennial ryegrass infected by *N. lolii* contains fungus-produced lolitrems that cause cattle to lose muscular control. In tall fescue grass infected by *N. coenophialum*, the alkaloid ergovaline acts as a vasoconstrictor and results in a condition similar to ergotism, resulting in restriction of blood flow to body extremities and eventual gangrene and crippling or death.

A *Neotyphodium* species in 'sleepygrass' (*Achnatherum robustum*), common in the western United States, produces lysergic acid amide that causes animals (especially horses) that consume the grass to sleep for up to 3 days. In general, these endophytes produce an array of secondary metabolites that have multiple impacts on herbivores and perhaps host plants. The biological role of the majority of the secondary metabolites and their impacts on herbivores is an area of active investigation.

Plant endophytes (*Epichloë* spp. and *Neotyphodium* spp.) in the Clavicipitaceae may also confer benefits to grasses that include insect resistance, increased drought tolerance, and disease suppression. This has made the use of endophytes in turfgrass cultivars an attractive and ecologically sound alternative to enhance turfgrass performance, to reduce pesticide applications, and to reduce water usage for maintenance of high-quality turf.

ORIGIN AND EVOLUTION OF ERGOT ALKALOIDS

In the grass endophytes, infection of the grass host is associated with the production of an array of compounds, ergot alkaloids, and nonalkaloid secondary metabolites. In the past three decades, the focus of research has been centered on the associations between *Neotyphodium* endophytes and cool-season forage grasses due to their economic impacts related to the detrimental effects of endophyte-infected grasses on livestock.

Ergot alkaloid production is not unique to grass endophytes. Vascular plants in the family Convulvulaceae (e.g., *Ipomea violaceae* and *Turbina corymbosa*) also have been long known to contain ergot alkaloids. Recent research has demonstrated that epiphytic members of the Clavicipitaceae are also responsible for the ergot alkaloids in these plants. Removal of the fungi from plants using fungicides resulted in the elimination of ergot alkaloids from plants. Several other genera in the families Hypocreaceae (Hypocreales) and Clavicipitaceae (Hypocreales), such as *Hypomyces*, *Dussiella*, *Hyperdermium*, and *Hypocrella*, have been reported to produce ergot alkaloids in vitro. Apparently independently derived in the phylogenetically distant order Eurotiales, several *Penicillium* species and *Aspergillus fumigatus* also produce ergot alkaloids.

THE PROPOSAL OF FUNCTIONALITY FOR ERGOT ALKALOIDS

Typically, ergot alkaloids are produced in a pathway in which the end product and intermediary products accumulate in large quantities. This suggests that ergot alkaloid diversity has been selected for due to functional reasons. The functions of many secondary metabolites are uncertain and frequently a matter of speculation. In *Aspergillus* and *Penicillium*, the fungi concentrate ergot alkaloids in sclerotia in a comparable way to *Claviceps* spp. Since the sclerotium is important in terms of survival, a metabolic effort is expended to defend and preserve it. Sclerotia of *Aspergillus flavus* are avoided by the dried-fruit beetle (*Carpophilus hemipterus*). This beetle will consume other fungal parts but avoid the sclerotia.

THE EVIDENCE FOR DEFENSIVE MUTUALISM IN THE CLAVICIPITACEAE ENDOPHYTE/GRASS ASSOCIATION

The defensive mutualism concept for the role of symbiotic endophytism in the Clavicipitaceae was proposed by Dr. Keith Clay of Indiana University and has been generally accepted in the scientific community. This hypothesis holds that plants are defended from herbivory by animals and insect herbivores through the production of secondary metabolites. In this view, grass endophytes are considered as plant mutualists primarily because they deter herbivore feeding. Thus, the primary driving selective force behind the endophyte–plant mutualism is considered to be defense against herbivores.

Endophyte-infected grasses contain a variety of secondary metabolites that are not found in noninfected plants. These compounds, mainly alkaloids, are considered the primary mechanism for antiherbivore and antimicrobial activity. Several factors influence the types of alkaloids produced and concentrations. Some factors include the strain of endophyte, plant part, age, growing season, and fertilization status. In general terms, alkaloid production and concentration correlate positively with hyphal density and the survival value of the plant part, with meristems, leaf sheaths, and seeds exhibiting high concentrations of alkaloids.

Lolines, ergot alkaloids, and lolitrems tend to be limited to leaf sheaths, meristematic zones, inflorescences, and seeds, while peramine and lolines are also present in the blade. This broader distribution of loline and peramine is likely due to their greater solubility in the aqueous apoplast. Peramine has been shown to be mobilized from within the fungus, taken up into the xylem, and secreted by hydathodes at the leaf tip. The ergot alkaloids and lolitrems are less soluble and are not likely translocated to great distances through plant tissues.

Approximately 65% of the endophytes are present in asexual form and they produce some degree of toxicity in herbivores through the production of several kinds of alkaloids. The persistence of the alkaloids in a highly evolved relationship suggests that they are likely advantageous for both the host plant and the endophyte. Alkaloids with known specific biological activities are lolines, peramines, lolitrems, and the ergot alkaloids. Loline alkaloids are more active against insects than mammals. Peramine is not toxic to insects but is a feeding deterrent. Ergot alkaloids and lolitrems are active against mammals and both of them have received far more attention for being involved in animal toxicosis such as 'fescue foot' and 'ryegrass staggers.'

THE DEFENSIVE MUTUALISM CONTROVERSY: A DEBATE REGARDING FUNCTION

The defensive mutualism explanation for endophyte distribution came under scrutiny recently due to observations that some endophytes do not appear to impart obvious defensive benefits to host plants. Dr. Stanley Faeth at Arizona State University has argued that a strong anti-insect effect in endophyte-infected grasses appears to be the exception rather than the rule. The case of Arizona fescue (*Festuca arizonica*) is notable where plants are often infected by endophytes that do not appear to confer insect or animal herbivory deterrence benefits to hosts. Dr. Faeth has also argued that herbivory is a weak selective force on grasses and unlikely to provide significant selective pressure to drive the evolution and spread of endophyte associations. Grasses are adapted to animal herbivory by developing basal meristems that are rarely consumed by animals. The endophytes grow in these protected parts of plants and are therefore protected in grass rhizomes and meristems and are not susceptible to extinction by herbivory. It would therefore follow that antiherbivore selection pressure, at least from mammals, is negligible.

The production of defensive compounds is the most accepted explanation for antiherbivore properties in infected grasses. Several hypotheses have been proposed to explain endophyte-mediated disease resistance and tolerance to abiotic stresses but the underlying mechanisms involved are still unknown.

EVOLUTION OF DEFENSIVE MUTUALISM FROM SELF-DEFENSE TO HOST DEFENSE

Ergot alkaloids in endophytes have been associated with host defense. However, it is not clear that this is the case in insect necrotrophs/plant biotrophs (e.g., *Hypocrella* spp.) or the epibiotic species, such as *Claviceps*. In these species, mycelium is restricted on plants either on the scale insect carcass or in the plant ovary. There is no evidence either ecologically or chemically that the alkaloids diffuse in plant tissues or that plants are defended in any way by the presence of the epibiotic fungi. In the epibiotic species (*Hyperdermium* spp., *Hypocrella* sp., *Claviceps* sp., etc.) where the fungal organism spends the entire life cycle as an external symbiont, it is unlikely that there is a transfer of the compounds from stromata to the plant. In this case, alkaloids could act as a 'self-defense mechanism' to avoid herbivores of the stromata and prevent colonization by other organisms, without any apparent benefit for the plant host that is supplying nutrients. When endophytes acquired the endophytic habit as in genera *Balansia* and *Epichloë*, fungal-defensive metabolites apparently developed the 'host-defensive' function. This host-defensive feature seems to stem from at least two characteristics of the endosymbionts including the internal systemic location of the endophytes in leaves, stems, and seeds and the water-soluble and diffusible nature of their defensive metabolites. Once endophytism was acquired, the host-defensive strategy followed.

CONCLUSION

The Clavicipitaceae constitute a group of fungi that contain forms that demonstrate the evolutionary transition from free-living generalist soil and insect-infecting species to plant mutualistic endophyte species. Within this family, all stages of evolution to plant biotrophy are evident. Because many species of this group of fungi are accessible and culturable, the family is an ideal model for understanding the evolution of plant biotrophy and the dynamics and ecology of endophytism and defensive mutualisms.

ACKNOWLEDGMENT

This research was in part supported by The Fogarty International Center (Nih) under U01tw006674 For International Cooperative Biodiversity Groups. We are also grateful to Rachna Patel for ink drawings of *Claviceps purpurea*.

FURTHER READING

Belesky, D. P., & Bacon, C. W. (2009). Tall fescue and associated mutualistic toxic fungal endophytes in agroecosystems. *Toxin Reviews*, *28*(2–3), 102–117.

Cheplick, G. P., & Faeth, S. (2009). *Ecology and evolution of the grass–endophyte symbiosis*. Oxford University Press: New York.

Clay, K., & Schardl, C. L. (2002). Evolutionary origins and ecological consequences of endophyte symbiosis with grasses. *The American Naturalist*, *160*, S99–S127.

Faeth, S. H., Hayes, C. J., & Gardner, D. R. (2010). Asexual endophytes in a native grass: Tradeoffs in mortality, growth, reproduction, and alkaloid production. *Microbial Ecology*, *60*(3), 496–504, 10.1007/s00248-010-9643-4.

Hoveland, C. S. (1993). Economic importance of *Acremonium* endophytes. *Agricultural Ecosystem Environment*, *44*, 3.

Rodriguez, R., White, J. F., Jr., Arnold, A. E., & Redman, R. (2009). Fungal endophytes: Diversity and ecological roles. *New Phytologist*, *182*, 314–330.

Schardl, C. L., Grossman, R. B., Nagabhyru, R. B., Faulkner, J. R., & Malik, U. P. (2007). Loline alkaloids: Currencies of mutualism. *Phytochemistry*, *68*, 980–996.

Schardl, C. L., Leuchtmann, A., & Spiering, M. J. (2004). Symbioses of grasses with seedborne fungal endophytes. *Annual Review of Plant Biology*, *55*, 315–340.

Schardl, C. L., Panaccione, D. G., & Tudzynski, P. (2006). Ergot alkaloids biology and molecular biology. *The Alkaloids: Chemistry and Biology*, *63*, 45–86.

White, J. F., Jr., & Torres, M. S. (2009). Symbiosis, defensive mutualism and variation on the theme. In J. F. White & M. S. Torres (Eds.), *Defensive mutualism in microbial symbiosis* (pp. 3–8). CRC Press: Boca Raton, FL.

White, J. F., Jr., & Torres, M. S. (2010). Is plant endophyte-mediated defensive mutualism the result of oxidative stress protection? *Physiologia Plantarum*, *138*, 440–446.

Chapter 4

Endophytic Microbes

M. Tadych and J.F. White
Rutgers University, New Brunswick, NJ, USA

Chapter Outline

Abbreviations	51	Endophytes as Sources of Bioactive Metabolites	58
Defining Statement	51	Recent Developments	61
Introduction	51	Conclusion	64
Fungal Endophytes	52	Acknowledgment	64
Some Common Endophyte–Grass Associations	55	Further Reading	64
Ecological Impacts of Endophytes	57		

ABBREVIATIONS

VOC Volatile organic compound

DEFINING STATEMENT

Endophytic associations between microbial and autotrophic organisms in nature are ubiquitous. Because of the complexity of the endophytic associations with their hosts we are only starting to understand these interactions. In this article, we compare the biology and ecology of different groups of endophytes associated with hosts. The functional role of endophytic microbes is also discussed. We also evaluate the potential of endophytic microorganisms as sources of bioactive metabolites with potential use in agriculture and medicine.

INTRODUCTION

Microorganisms are found virtually in every biotic and abiotic niche on earth. This includes extremophiles living in deserts, rocks, thermal springs, freshwater, marine, and arctic environments, and associated with terrestrial and aquatic animals. A wide diversity of microorganisms may be isolated from most terrestrial and marine plants. Microorganisms are present on the surface of and within tissues of most parts of plants, especially the leaves. When microbial organisms colonize a plant and the plant tissue is healthy, the relationship between the microorganism and its host plant may range from latent pathogenesis to mutalistic symbiosis. Therefore, the microbial organisms may be latent pathogens, epiphytes (epibionts), or endophytes (endobionts).

Endophytes are generally any organisms that under normal circumstances are contained within tissues of living plants (usually autotrophs) without causing noticeable symptoms of disease, and the host tissues remain intact and functional. However, the same organisms may also be described as saprobic or pathogenic at another time; for example, when the host is stressed, some endophytes may become pathogenic. The delicate balance between host and endophytic organisms seems to be controlled in part by chemical factors, for example, herbicidal natural products produced by the fungus versus antifungal metabolites biosynthesized by the host plant.

The term 'endophyte' is derived from the Greek *endon* (within) and *phyte* (plant), and was introduced by Heinrich Anton de Bary in 1866 and applied to any organism found within a plant. Traditionally, the term had been broadly applied to fungi in plant tissue, including the mycorrhizal fungi in plant roots. However, some authors do not include mycorrhizal fungi in this group. The term endophyte has also been adapted to other microorganisms, such as endophytic bacteria. The associations of endophytic organisms with their host plants are varied and complex and we are only starting to understand these interactions. Endophytic microbial organisms often contribute to the normal health and development of their hosts in exchange for a relatively privileged niche. Microbial endophytes have been isolated from tissues of algae, mosses and hepatics, ferns and fern allies, grasses, other herbaceous plants, and trees growing in tropical, temperate, boreal, and arctic environments.

There are vast regions of the world where fungus–plant associations are unknown, and endophytes represent a large reservoir of undiscovered genetic diversity. Species composition of endophyte assemblages and infection frequencies vary according to (1) host species, (2) growth stage of the host, (3) tissue type, (4) the age of host organs and/or tissues, (5) position in the canopy and associated vegetation, and (6) site characteristics, such as elevation, exposure, and latitude, and (7) anthropogenic factors. Usually, one to a few species dominates the endophyte community, while the majority of the species are rare. Distribution of rare and incidental species is influenced more by site than by host, and the number of rare and incidental species isolated is proportional to the intensity of sampling.

Many plant fungal endophytes are transmitted horizontally by production of conidia or other spores on plants that may spread to adjacent uninfected plants, and are not directly inherited through germplasm from the previous generation of the host. Therefore, with each host generation, environmental fungi must compete to colonize the empty niches within new host individuals. Alternatively, some endophytes, including *Epichloë* species and all of their asexual relatives (*Neotyphodium* spp.), grow intercellularly throughout aboveground parts of plants and are transmitted vertically via seeds. These endophytic fungi are spread by hyphae growing into the developing seeds or vegetative parts of infected maternal host plants.

Epibiosis is another type of association of two organisms, the epibiont and the basibiont. The majority of organisms that exist on the surfaces of plants are referred to as epibionts or epiphytes. In general usage the terms epiphyte and epibiont are used interchangeably. The epibiotic organism may be sustained entirely by nutrients and water received nonparasitically from within the host (basibiont) on which it resides. However, the interaction of epibiont with basibiont is usually unknown. In some epibionts water and nutrients are taken up entirely from suspended soils and other aerial sources such as dead host tissues, airborne dust, mist, and rain. Any negative effect on the host, if it occurs, is indirect.

The assemblage of epiphyllic flora on the surface of healthy plant leaves is usually composed of bacteria, yeasts, and filamentous fungi that belong to different systematic categories. The density and the number of epiphyllic microorganisms can rapidly change and it strongly depends on environmental conditions, host species, habitat of the host, and the age and the surface structure of the organ on which the epiphyllic organism resides.

While some fungi are adapted to the plant surface, the community also includes propagules of airborne species, fungi that would not otherwise be considered epibionts. The surface of the plant, especially the leaf is an extreme environment. Though plants may leak nutrients through the cuticle, the surface is dry, waxy, and affected by UV radiation. Epibionts have many obvious characteristics that enable continuation through the stressful conditions of the leaf surface. For example, some epibionts can digest lipidic substrates, and thus may utilize the waxy layer covering the leaf. Epibionts are likely to be melanized, and thus able to resist UV radiation. Epibiotic yeasts are able to multiply during the short periods of appropriate environmental conditions. Although epibiotic organisms differ markedly from endophytes, it has been shown that some endophytic fungi are also able to produce epiphytic stages with fungal reproductive structures on the surface of the host plant.

FUNGAL ENDOPHYTES

Many groups of fungi exist as plant endophytes. The fungal endophytes have commonly been divided into two major groups. The first group comprises a smaller number of specialized fungal species (clavicipitaceous fungi) that colonize some monocotyledonous hosts, and the second group that includes a large number of fungal species (nonclavicipitaceous fungi) with a broad range of host plants. Every plant species examined to date maintains endophytic associations with fungi and this association is a ubiquitous and cryptic phenomenon in nature. Although most of the endophytic fungi belong to the phylum Ascomycota and its anamorphs, Glomeromycota and some Basidiomycota and Zygomycota are also known. The host range, ecology, and geographical distribution of most groups of endophytic fungi are still poorly known. A recent study by Betsy Arnold (University of Arizona) and François Lutzoni (Duke University) indicated that endophyte infection of host plants increases along latitudinal gradient from the Arctic to the Tropics, with less than 1% to more than 99% of tissue segments (2 mm^2) with endophytes, respectively. This study also showed that foliar endophyte assemblages, at both an individual host species and the community level, increases in diversity with decreasing latitude.

Diversity and composition of endophytic fungal communities vary considerably and their detection depends on biotic, abiotic and experimental factors. Endophyte community of almost all plant species examined to date have been assessed mostly by culture-based approaches and have been identified using morphological characteristics with support of molecular analysis for identification of some endophytes that remain sterile in culture. Abundance and diversity of unculturable endophytes is still mostly unknown, limiting our understanding of endophyte infection frequencies, taxonomic composition, and diversity. Many researchers suggest that culture-based methods alone underestimate diversity and misrepresent the taxonomic composition of endophyte communities. Therefore, recently cultivation-independent approaches, such as DNA cloning, denaturing gradient gel electrophoresis (DGGE), or terminal restriction fragment length polymorphism (TRFLP or sometimes T-RFLP), have gained popularity as a means to assess the diversity and composition of uncovering endophytes with obligate host association, that is, fungal species that do not grow on standard media or fungi that grow slowly and are lost during the culturing process in competitive interactions. Regardless some limitations and needs for further improvement, these approaches are gaining popularity along with the culture-based methods as means to assess the entire diversity and composition of endophytic fungal communities present within host plant.

Fungal endophytes of grasses (also known as clavicipitaceous endophytes or e-endophytes) are widely studied; with most work concentrating on fungi of the family Clavicipitaceae (Hypocreales; Ascomycota). The following genera of this family have been identified as grass endophytes or epiphytes: *Atkinsonella*, *Balansia*, *Balansiopsis*, *Echinodothis*, *Epichloë*, *Myriogenospora*, and *Parepichloë*. Clavicipitaceous endophytes, also known as e-endophytes, are widespread in grasses; approximately over 10% of all grass species are estimated to harbor clavicipitaceous fungal endophytes. The genus *Epichloë*, with anamorphs in *Neotyphodium*, is the most studied group of grass endophytes. These endophytes are found growing systemically in the aboveground tissues of some temperate grass species of the cool-season grasses. Based on the relative costs and benefits to the hosts, these grass–fungal endophyte symbioses have been showed to range from pathogenic to mutualistic. Endophytic species from the genus *Epichloë* are often considered to be pathogenic and may cause partial or complete sterilization of hosts due to the production of a fungal stroma on the flowering culms (choke disease) of the host. However, the degree of effect on host reproduction varies with species. In mutualistic associations, that is, in some *Epichloë* spp. and *Neotyphodium* spp., the endophyte grows systematically within its host (Figure 1), including the developing seeds (Figure 2), and is entirely dependent on the survival and growth of the grass host plant for its own growth.

It is believed that fungal endophytes of grasses have two modes of reproduction. The sexual species of genera *Epichloë* and *Balansia* can be horizontally transmitted through development of ascospores. Because *Epichloë* species are obligately outcrossing ascomycetes, development of the sexual spores is dependent upon transfer of spermatia of one mating type to an unfertilized stroma of the opposite mating type occurring on different individuals of the host plants. Transfer of spermatia of *Epichloë typhina* (Pers.) Tul. & Tul. is accomplished by flies of genus *Botanophila* (Anthomyiidae; Diptera), which visit stromata for feeding and oviposition. Immediately after cross-fertilization of the fungus, perithecia begin to develop on the stroma. During flowering of the host plant, the ascospores produced within the perithecia of infected individuals in the population are forcibly ejected. The ascospores, possibly dispersed by air currents, may land on another healthy grass plant and may initiate infection. In contrast, many *Epichloë* species and all of their asexual mutualistic *Neotyphodium* relatives are typically transmitted vertically from maternal plants to their offspring. For most of their life cycle the endophytes inhabit, asymptomatically and systemically, the apoplasts of the aboveground organs of infected host plants, including the embryos of viable seeds, and can be disseminated vertically to successive generations of the host plant. Infected seeds and vegetative tillers of infected host plants are the only known modes of propagation of these endophytes.

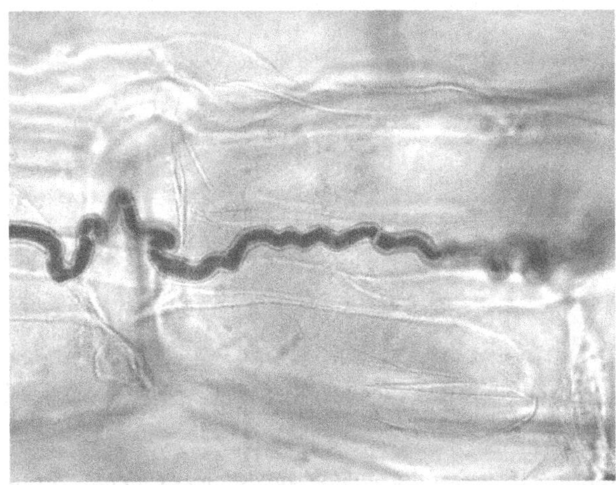

FIGURE 1 Endophytic convoluted hypha of *Neotyphodium coenophialum* in a culm-scraping preparation from *Festuca arundinacea* (×1600).

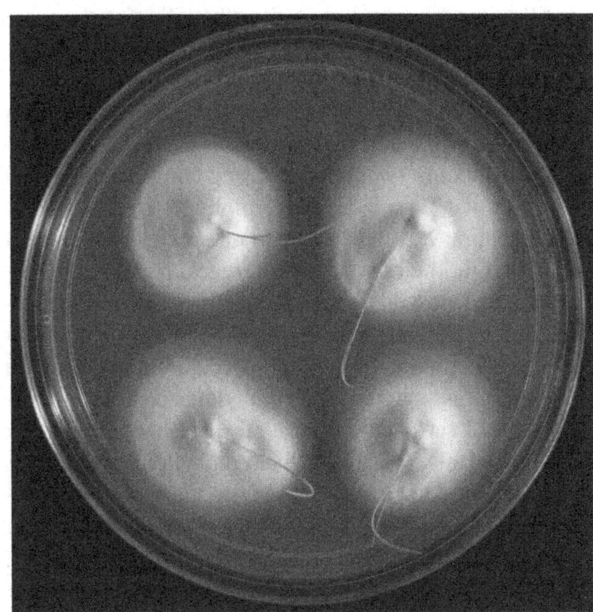

FIGURE 2 Four-week-old culture of *Neotyphodium* sp. from seeds of *Poa ampla* grown on potato dextrose agar at room temperature (×0.7).

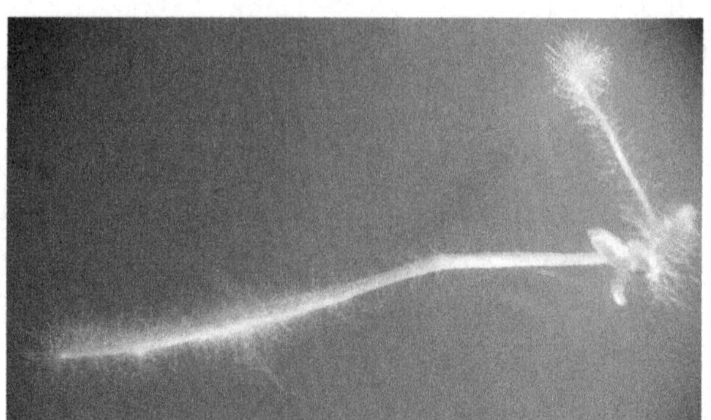

FIGURE 3 Epiphyllous growth of *Neotyphodium* sp. on a 2-week-old *Poa ampla* seedling (×3).

Additionally, several *Epichloë* spp., such as *Epichloë festucae* Leuchtm., Schardl & Siegel, are represented by species that have a remarkable mixed-transmission strategy, that is, horizontal and vertical transmission modes. In these cases, some tillers produce stromata while other tillers on the same plant are asymptomatic and produce normal, vigorous, endophyte-infected seeds.

During the last decade some *Neotyphodium* endophytes have been shown to emerge from plants and produce a network of mycelium, conidiophores, and conidia on the surfaces of plants (Figure 3). This epiphyllous mycelium is most evident on leaf blades of certain species of the Pooideae, including the bent grasses, fescues, forest hedgehog grass, ryegrasses, wild barley, and some blue grasses. Currently, nothing is known regarding the ecology or implications of production of the epiphyllous conidial stages in the life cycles of endophytes. The epiphyllous conidia are viable and spread via water currents, rain splash, or drip splash to adjacent plants. It is likely that epiphyllous conidia are responsible for some of the horizontal transport and parasexual recombination 'hybridization' that may occur in the *Epichloë/Neotyphodium* endophytes.

In addition to the clavicipitaceous endophytes a number of other seed-transmitted grass–fungal endophytes have been described. Certain cool-season grasses harbor nonclavicipitaceous fungal endophytes that are less frequent than clavicipitaceous. These fungi belong to two different groups, that is, p-endophytes and a-endophytes. The p-endophytes include *Gliocladium*- and *Phialophora*-like endophytes ('p' for penicillate disposition of conidiophores, common to *Gliocladium*- and *Phialophora*-like fungal species) that can be isolated from culms and seeds of numerous festucoid grasses, for example, from perennial ryegrass (*Lolium perenne* L.), tall fescue (*Schedonorus arundinaceus* (Schreb.) Dumort.), meadow fescue (*Schedonorus pratensis* (Huds.) P. Beauv.), Arizona fescue (*Festuca arizonica* Vasey), and giant fescue (*Schedonorus giganteus* (L.) Holub). These endophytes are represented by species of the order Eurotiales (Ascomycota). The a-endophytes are a group of grass endophytes found in Italian ryegrass (*Lolium multiforum* Lam.), other annual species of the *Lolium* genus, and *Festuca paniculata* (L.) Schinz & Thell. The a-endophytes belong to parasitic species of *Acremonium* similar to *Acremonium chilense* Morgan-Jones, White & Piontelli (Hypocreales; Ascomycota), an endophyte of orchard grass (*Dactylis glomerata* L.). At present, the ecological and physiological importance of p- and a-endophytes of grasses are not determined or understood.

The warm-season grass, *Trichachne insularis* (L.) Nees, was found to harbor a nonclavicipitaceous, seed-transmitted endophyte identified as *Pseudocercosporella trichachnicola* White & Morgan-Jones, an anamorphic species of genus *Mycosphaerella* (Mycosphaerellaceae; Ascomycota). Another seed-transmitted fungus *Fusarium verticillioides* (Sacc.) Nirenberg (Nectriaceae; Ascomycota) has been isolated from maize (*Zea mays* L.), sorghum (*Sorghum* spp.), and other plants in the Poaceae family as a symptomless endophyte. The horizontally transmitted entomopathogenic fungus *Beauveria bassiana* (Bals.-Criv.) Vuill. (Clavicipitaceae; Ascomycota) is also known to form an endophytic association with maize.

Nonsystemic grass endophytes present another group of nonclavicipitaceous fungal endophytes. Because most studies of grass endophytes have focused on systemic endophytes, at present little is known about nonsystemic endophytic fungi that inhabit grass species. However, some studies indicate that in addition to systemic infections, endophytes that form localized infections in grasses may be also diverse and widespread. Generally, dominant nonsystemic grass endophytes are represented by the genera *Alternaria*, *Cladosporium*, *Epicoccum*, *Fusarium*, *Phoma*, and pathogens typical of grass hosts. Because of their mode of dissemination, diversity and dispersion of this group of grass endophytes is probably more variable than that of systemic grass endophytes and depends on availability and viability of spores of the fungus.

Nonclavicipitaceous endophytes associated with healthy organs of nongrass plants are still poorly known. However, fungal surveys conducted during the last 30 years have reported endophytic fungi from over 100 plant families and demonstrated that tissues of the vast majority of plants are colonized by endophytic microfungi. In general, fungal endophytes within the host may inhabit many different tissues of roots, stems, branches, twigs, bark, leaves, petioles, flowers, fruits, and seeds, including xylem of all available plant organs. Endophytic nonclavicipitaceous microfungi in tissues of plant hosts are usually highly diverse and occur as numerous localized infections that increase in number, density, and species diversity with organ age. It may be because the fungal endophytes associated with nongrass plants generally appear to be transmitted horizontally (i.e., from plant to plant in populations). Most fungal surveys of angiosperms show that nonclavicipitaceous fungal endophytes represent polymorphic assemblages mostly from the Ascomycota or Basisiomycota. Most belong to the Ascomycota with Pezizomycotina (Dothideomycetes, Eurotiomycetes, Leotiomycetes, Pezizomycetes, and Sordariomycetes) that are especially well represented, although some Saccharomycotina are also known. Within the Basidiomycota members of the Agaricomycotina, Pucciniomycotina, and Ustilagionmycotina are known endophytes, although they are reported less frequently than ascomycetous endophytes. Particularly, fungi from the genera *Acremonium*, *Alternaria*, *Chaetomium*, *Cladosporium*, *Cryptocline*, *Cryptosporiopsis*, *Curvularia*, *Fusarium*, *Glomerella*, *Leptostroma*, *Phoma*, *Phomopsis*, *Phyllosticta*, *Physalospora*, and *Trichoderma* are well represented in endophyte assemblages. The profile of fungal endophyte assemblages in specific organs can be completely different from those in other plant organs or tissues. Generally, the main tissues that have been analyzed for the presence of endophytes in woody perennials are the leaves, twigs, branches, and roots; a few studies have looked for endophytes in meristems, flowers, or fruits of woody perennials. However, the profile of fungal endophyte assemblages in flowers and fruits could be radically different than that in other types of plant structures because these organs are young and rapid in development.

Other mostly unknown group of fungal endophytes is represented by marine fungi. Marine fungi are not taxonomically well defined. In general, the marine environment includes obligate marine fungal species, which are considered to be fungi that grow and sporulate exclusively in a marine habitat. Marine fungi that do not germinate in the natural marine habitat are not included in these groups. It also includes facultative marine fungi also called marine-derived fungi, that is, those that can be isolated from both terrestrial and marine environments, and are adapted to and isolated from various marine habitats, like near-shore or estuarine environments. However, not well-documented fungal endophytes harbor by many marine organisms may also protect their host from herbivory, pathogenic and fouling organisms. The marine environment may offer a variety of unexplored epi- and endosymbiotic microorganisms. Due to the complicated nature of marine symbiotic associations and experimental limitations, the ecology and chemistry of these interactions has not been well studied. Only in a few cases a specific microbe–host relationship, for example, symbiosis or permanent association, has been proven. Most of the literature reports describe the sporadic occurrence of variable microorganisms for certain hosts. Mostly bacteria but also fungi are to be found in various marine organisms, including sponges, algae, mangrove plants, and marine animals as epi- and endobionts. Even though metabolite-producing marine-derived fungi are being isolated from sponges, there is still no evidence for the presence of fungal mycelia growing in sponges. One of the better-documented relationships between marine-derived fungi and other marine organisms is the fungal–algal symbiotic association 'mycophycobiosis.' It was estimated that one-third of all known marine fungi are associated with marine algae and these fungi reside inside the algal tissues. Studies of marine-derived fungi indicate the enormous diversity of the fungal community in the world's oceans and their biochemical uniqueness.

SOME COMMON ENDOPHYTE–GRASS ASSOCIATIONS

Tall Fescue–*Neotyphodium coenophialum* Association

Tall fescue (*Schedonorus arundinaceus* (Schreb.) Dumort. = *Festuca arundinacea* Schreb. = *Lolium arundinaceum* (Schreb.) Darbysh.) is one of the best-known examples of a grass with an endophyte that causes toxicity. The grass was brought to the United States from Europe in the late 1800s. It was officially discovered in Kentucky in 1931, tested at the University of Kentucky, and released in 1943 as 'Kentucky 31.' From the mid-1940s it became popular with farmers, spreading quickly throughout the midwestern and southern United States. Today it accounts for well over 16 million hectares of pasture and forage land in the United States. The problem of livestock neurotoxicosis ('fescue toxicosis' also known as 'fescue foot' or 'fescue lameness') became a major concern in the United States. Several studies show that consumption of endophyte-infected tall fescue decreases the feed intake of cattle and therefore lowers animal weight gains. Affected cattle also produce less milk, have higher internal body temperatures and respiration rates, develop a rough hair coat and demonstrate an unthrifty appearance, salivate excessively, have poor reproductive performance, and maintain reduced serum prolactin levels. By the end of the 1970s, the association of this toxicity with the endophyte had been discovered. The endophyte was originally identified as a strain of *E. typhina*, but was later described as *N. coenophialum* (Morgan-Jones & Gams) Glenn, Bacon & Hanlin. *N. coenophialum* produces several alkaloids, particularly, the alkaloid ergovaline that is structurally related to ergotamine, a major factor in ergot poisoning due to ingestion of *Claviceps purpurea* 'ergots' contaminating rye flour.

Perennial Ryegrass–*Neotyphodium lolii* Association

Perennial ryegrass (*L. perenne* L.) is a valuable forage and soil stabilization plant. In New Zealand the neurotoxic disease 'ryegrass staggers' (also known as 'perennial ryegrass staggers') of sheep and cattle has long been reported and attributed to the consumption of perennial ryegrass. In 1898 for the first time the presence of a fungal endophyte in the seeds of *L. perenne* was observed, and 40 years later the endophytic fungus from *L. perenne* was isolated and grown in agar culture. The association between the *L. perenne* endophyte and 'ryegrass staggers' was finally established in 1981. *N. lolii* (Latch, Christensen & Samuels) Glenn, Bacon & Hanlin that infects perennial and hybrid ryegrasses was shown to synthesize several alkaloids. Three of these are known to be particularly important to pasture management, specifically (1) lolitrem B, a tremorgenic molecule responsible for livestock 'staggers,' (2) ergovaline, an ergopeptine that has vasoconstrictive effects and causes heat stress in grazing animals, also responsible for 'fescue toxicosis,' and (3) peramine, a tripeptide that deters some insects, particularly the Argentine stem weevil, from feeding on ryegrass but is not toxic to mammals. Perennial ryegrass staggers is a very serious problem for grazing livestock (sheep, cattle, horses, deer) in New Zealand; however, mortality rates are generally low. Ryegrass staggers occur sporadically in North and South America, Europe, and Australia.

Darnel Ryegrass–*Neotyphodium occultans* Association

Darnel ryegrass = darnel (*Lolium temulentum* L.) is another common endophyte-infected grass species. This species has a long recorded history as a plant poisonous to humans and animals. Darnel was known as a weed and as a poisonous plant in earlier times. The earliest written references to darnel indicating it to be a noxious and toxic weed that caused problems for humans and animals may be found in the Gospel of Matthew of the New Testament and in authors such as Plautus, Virgil, Ovid, Dioscorides, and Shakespeare. Darnel seed from 4000-year-old archeological materials in ancient Egypt contained endophyte mycelium. It was generally understood that human beings would be poisoned by ingestion of flour or baked products containing seeds of darnel as a contaminant. Contamination was also undesirable because of the strong taste of darnel seeds, which, for example, resulted in inferior bread. Darnel seeds were sometimes added to beer as a flavoring. By the end of the 1800s it was discovered that a fungus infects seeds of darnel. Currently we know that darnel seeds contain an endophytic fungus, and this symbiotic fungus is known as *N. occultans* Moon, Scott & Christensen.

The Sleepygrass–*Neotyphodium chisosum* Association

Sleepygrass (*Achnatherum robustum* (Vasey) Barkworth = *Stipa robusta* (Vasey) Scribn.) is a perennial grass forming stout, erected clumps in dry plains, hills, and open woods. The grass is native to North America, and is abundant in southwestern United States. It was found to be toxic and narcotic to grazing animals, that is, to horses, cattle, and sheep. Animals, after consumption of relatively small quantities of the grass, go to sleep for 2–3 days, and then gradually recover. Ergot alkaloids, that is, ergonovine, ergonovinine, lysergic, and isolysergic acid amides, have been identified as the sleep-inducing agents in sleepygrass. These alkaloids are produced by *N. chisosum* (White & Morgan-Jones) Glenn, Bacon & Hanlin, the endophyte isolated from this plant and related species of the genus. However, recent studies suggest that the level of ergot alkaloids in native sleepygrass populations may be highly variable within and among populations despite the level of endophyte infection. Animal toxicity is localized in particular areas where particular strains of the endophyte may predominate.

The 'Drunken Horse Grass'–*Neotyphodium gansuense* Association

Drunken horse grass (*Achnatherum inebrians* (Hance) Keng = *Stipa inebrians* Hance) is a perennial bunchgrass, distributed on alpine and subalpine grasslands of northwestern China and Mongolia. Horses that have grazed *A. inebrians* develop a stagger; the animals walk as if drunk, with some being unable to stand after falling. The symptoms persist for 6–24 h, and usually they appear to be completely recovered after 3 days. Severely affected animals die within 24 h of consumption of the grass. Toxicity of *A. inebrians* was reported by Marco Polo and the Russian explorer Nikolai Mikhaylovich Przhevalsky. In the last decade the presence of *N. gansuense* Li & Nan endophyte in this grass species was confirmed. In the study of alkaloid content it was found that infected *A. inebrians* contains two major toxic alkaloids: ergonovine and lysergic acid amide.

The 'Dronkgras'–*Neotyphodium melicicola* Association

Staggers grass = dronkgras (*Melica decumbens* (L.) Weber) is a tufted perennial, very coarse grass with usually rolled and very rough leaves. *M. decumbens* is endemic to Africa and has a limited distribution in South Africa, being found only in the arid areas of central South Africa. It grows amongst rocks, under trees, and shrubs on hill- and mountainsides, occasionally in areas along roadsides. The name 'staggers grass' comes from the fact that this grass has narcotic effects on cattle, horses, donkeys, and sheep. The tremorgenic neurotoxins produced by the endophyte *N. melicicola* Moon & Schardl were found to be responsible for the narcotic effect of *M. decumbens* on grazing livestock. Usually, consumption of grass is not lethal and the animal recovers. This is probably because the leaves of the grass are very coarse, and animals do not graze it frequently, except when the grass is very young.

Forest Hedgehog Grass–*Neotyphodium* Association

Forest hedgehog grass (*Echinopogon ovatus* (G. Forst.) P. Beauv.) is a tufted perennial, mesophytic grass often found in moist forested areas (in wet sclerophyllic woodlands and by creeks). The grass is endemic to Australia, New Guinea, New Zealand, and Tasmania. Young plants of *E. ovatus* cause 'staggers,' sometimes known as 'wobbles' in stock. The grass was found to harbor endophytes *Neotyphodium aotearoae* Moon & Schardl and *Neotyphodium australiense* Moon & Schardl. The presence of indole-diterpenoid and loline alkaloids in endophyte-infected *E. ovatus* was confirmed.

'Hueću' Grasses–*Neotyphodium tembladerae* Association

Poa huecu Parodi and several other grasses of South America (e.g., *Bromus auleticus* Trin. ex Nees, *Festuca argentina* (Speg.) Parodi, *Festuca hieronymi* Hack., *Poa magellanica* Phil. ex Speg., *Melica stuckertii* Hack., and some other *Poa* spp.) are colonized by the endophytic fungus *N. tembladerae* Cabral & White. The association of the endophyte with *F. hieronymi* and *P. huecu* is probably responsible for toxicosis syndromes, called 'tembladera' or 'hueću' in grazing animals. 'Tembladera' is from the Spanish word that means 'tremble,' and word 'hueću' is from the indigenous Araucanian language of the tribes that lived in the region and means 'intoxicator.' Hueću toxicosis results from consumption of *P. huecu* and is frequently lethal to animals. Studies on *N. tembladerae* suggest that toxicity of the infected grasses to mammals is associated with the ergot alkaloids, lolitrems, and some glycoproteins, produced by the fungus.

ECOLOGICAL IMPACTS OF ENDOPHYTES

Abiotic and biotic stresses due to noninfectious and pathogenic plant diseases, pests, and unfavorable growing conditions are major causes for plant productivity losses. Productivity of cultivated plants relies heavily on high chemical inputs. Recently, natural and biological control of diseases and pests affecting cultivated plants has gained much attention as a way of reducing the use of chemical products in agriculture. Biological control offers an alternative or supplement to chemical pesticides in plant protection. This approach incorporates one or more organisms to maintain another pathogenic organism below a level at which it is no longer an economical problem. Use of endophytic organisms could enhance plant growth and productivity on a worldwide basis.

Some groups of endophytic microorganisms have been described as 'defensive mutualists' that protect host plants against biotic and abiotic stresses. In return, the endophytic symbionts acquire nutrients from their host plants. Endophytic arbuscular mycorrhizal fungi, which live in mutualistic symbiosis with at least 80% of plants, effectively protect host plant from many root diseases, reduce infection by nematodes, increase stress tolerance like drought resistance, tolerance to heavy metals and salinity, and increase phosphorous uptake in phosphorous-deficient soils. Endophytes of upper parts of grasses and other plants also benefit their hosts. The benefits frequently reported include systemic resistance against pathogens, reduced herbivory, increased drought resistance, improved tolerance to heavy metals, and generally enhanced growth and an increase in the plant's fitness to environmental extremes. The fungal endophytes may also influence the plant pathogen assemblages, and reduce their diversity and abundance. For example, in *Schedonorus arundinaceus*, endophyte infection reduced seedling blight caused by *Waitea circinata* Warcup & Talbot and crown rust caused by *Puccinia coronata* Corda relative to endophyte-free plants, but colonization by *N. coenophialum* did not affect rust caused by *Puccinia graminis* Pers. Also, Alternaria leaf spot caused by *Alternaria triticina* Prasada & Prabhu was significantly more common on endophyte-free *Panicum agrostoides* Spreng. than on plants with endophyte. Endophyte-infected grasses may also experience a lower incidence of disease when insects, vectors of plant pathogens, are deterred by endophytes. Although the ecological roles played by endophytic fungi are more diverse and varied, these benefits arise in part from the production of fungal metabolites (usually alkaloids) by the endophyte or endophyte–plant complex. However, in some cases the mechanisms of defense are not fully understood and require further study.

In contrast, in horizontally transmitted endophytes associated with phytosynthetic tissues such as leaves of non-grass plants, the benefits are less clear. Generally, these endophytes are believed to also function as defensive mutualists of host plants; however, in most cases their ecological roles have not been assessed experimentally. Recently, several studies have provided evidence for important roles of class 3 endophytes in enhancing plant defenses against biotic stresses, such as protection from herbivores, pathogenic fungi, nematodes, and neighboring plants' competition, and abiotic stresses. Some studies have shown that the endophyte *Trichoderma ovalisporum* Samuels & Schroers that colonized stems and fruits of *Theobroma gileri* Cuatrec. has the ability to parasitize and antagonize the necrotrophic mycelium of pathogenic *Moniliophthora perniciosa* (Stahel) Aime & Phillips-Mora, and *Moniliophthora roreri* (Cif.) Evans, Stalpers, Samson & Benny. Another study showed that seedlings of *Theobroma cacao* L. inoculated with seven endophytic fungal species and exposed to *Phytophthora* spp., species that are pathogens of the plant, were more resistant than endophyte-free plants.

Leaves of the control endophyte-free plants died in greater numbers and suffered much more pathogen damage than endophyte-infected plants. Also, the relative benefit of endophyte inoculation was higher in older than in younger leaves.

Habitat-adopted symbioses of plants with endophytic fungi from class 2 contribute to and may also be responsible for the adaptation of host plants to environmental stresses. For example, the heat-tolerant perennial panicgrass (*Dichanthelium lanuginosum* (Elliott) Gould) is symbiotic with the fungus *Curvularia protuberata* R.R. Nelson & Hodges. The endophytic fungus was isolated from the grass collected from geothermally heated soils of Yellowstone National Park (Wyoming, Montana, and Idaho, USA) and Lassen Volcanic National Park (California, USA). Geothermal soils of Yellowstone and Lassen Volcanic National Parks may reach temperatures as high as 57 °C, and on an annual basis the grass is exposed to high temperatures as well as periods of drought conditions. Laboratory and field studies conducted by Rusty Rodriguez (US Geological Survey/ University of Washington), Regina Redman (US Geological Survey/University of Washington), and Joan Henson (Montana State University) have shown that the endophytic fungus confers thermotolerance to the host plant and that this fungal–plant association is responsible for survival of both species in geothermal soils. When these organisms were grown asymbiotically under controlled conditions, the maximum growth temperatures of *D. lanuginosum* and *C. protuberata* were 40 and 38 °C, respectively; while with the endophyte plants grew at considerably higher temperatures. A similar effect was observed with *Fusarium culmorum* (W.G. Sm.) Sacc., which colonizes a costal dunegrass (*Leymus mollis* (Trin.) Pilg.). The plant did not survive and the development of the fungus was retarded when both partners were grown separately and they were exposed to levels of salinity that they experienced in their native habitat. Another example is the fungus *Piriformospora indica* Verma, Varma, Rexer, Kost & Franken originally isolated from a spore of mycorrhizal fungus *Glomus mosseae* (T.H. Nicolson & Gerd.) Gerd. & Trappe. Some studies showed that *P. indica* is able to endophytically colonize roots of various crop plants. It has been shown that isolates of *P. indica* in roots of barley plants enhanced development of the host plant and increased grain yield. It was speculated that this endophytic fungus elevated antioxidative status of the infested roots and therefore protected roots from root pathogens like *F. culmorum* and *Cochliobolus sativus* (Ito & Kurib.) Drechsler ex Dastur. Also, the systemic plant response, as a result of induced resistance by this endophyte, causes a reduction of powdery mildew (*Blumeria graminis* (DC.) Speer) infection of barley leaves. The fungus also protects the barley plants from salt stress.

Production of Nontoxic Endophytes

It is well documented that endophytic fungi have been implicated in toxicity of some poisonous plant species. Class 1 endophytes of grasses are a particularly important group of fungi that have been found to naturally produce a range of mycotoxins, alkaloids, and physiologically active chemical compounds. Some of these substances cause problems for livestock and are recognized as, for example, the causative agents of economically important livestock toxicoses, such as 'fescue toxicosis' and 'ryegrass staggers.' It has been found that some endophytes of ryegrasses and some other related grass species have reduced toxicity to grazing livestock while at the same time they enhance tolerance to pests and/or abiotic stresses. It has also been discovered that some toxic and beneficial alkaloids have separate biosynthetic pathways, allowing the selection or development of endophyte strains that show low animal toxicity but still possess anti-insect qualities. Axenic cultures of selected grass–fungal endophytes may be produced and inoculated into seedlings of grasses. The discovery and commercialization of low-toxicity *Neotyphodium* endophytes along with the technology for their reinoculation resulted in the development of elite grass cultivars that are persistent, productive, and enhance animal performance, for example, Greenstone tetraploid hybrid ryegrass with Endosafe endophyte, and tall fescue cultivars with MaxQ endophyte.

ENDOPHYTES AS SOURCES OF BIOACTIVE METABOLITES

Microorganisms demonstrate many unique characteristics. Among these characteristics is the production of a vast range of biologically active substances and secondary metabolites. Microorganisms are one of the richest resources of biologically active substances and secondary metabolites with novel structures and potential activities. Secondary metabolites, also known as idiolites, have been defined as low molecular weight and naturally produced substances that often possess chemical structures quite different from primary metabolites, such as amino acids, organic acids, and sugars from which they are produced. Their functions in the producing organisms are not obvious. They are usually hypothesized to function as chemical defenses for the hosts, but may be potential viable weapons against other organisms. In addition, secondary metabolites may be agents of symbiosis and agents of metal transport. Finally, these metabolites may act as sex hormones, plant growth stimulants, and as effectors of differentiation. For many reasons these compounds may have tremendous economic importance to humankind and an extraordinary impact on the quality of human life. Some fungal secondary metabolites are beneficial (antibiotics such as cephalosporin and penicillin) while others are harmful (carcinogens such as aflatoxin and ochratoxin). The secondary metabolites produced by microbes are generally large and complex chemicals that are not readily synthesized using recombinatorial synthetic approaches. However, a number of drugs derived from fungal metabolites have been developed as their modified analogs. Antibiotics, antifungal, anticancer, immunosuppressive agents, and hypocholesterolemic agents that are derived from fungal compounds have been used for over 50 years.

During the last two decades the fungi living internally in living tissues of plants have been targeted as valuable sources of new bioactive compounds and they have become a mainstay of natural product screening programs. Growth of endophytic fungi within hosts without causing apparent disease symptoms and metabolic interaction of endophytes with hosts may favor the synthesis of biologically active secondary metabolites. According to a study performed by Barbara Schulz (Technical University of Braunschweig) and her colleagues, the proportion of novel structures produced by endophytic isolates (51%) was considerably higher than that produced by soil isolates (38%). Among the fungal genera, endophytic fungi from *Acremonium, Chaetomium, Colletotrichum, Fusarium, Pestalotiopsis, Phoma*, and *Phomopsis* are known to produce bioactive compounds of medical importance. In addition, endophytic fungi growing in axenic culture can also produce biologically active compounds, including several alkaloids, antibiotics, and plant growth-promoting substances. However, the amount and kind of compounds that are produced by a fungus will be affected by factors like temperature, degree of aeration, and the composition of the medium used for culturing. Some endophytic fungi have been identified as sources of anticancer, antidiabetic, and immunosuppressive compounds. For example, anticancer molecule paclitaxel (Taxol, Figure 4), originally discovered in the bark of the Pacific yew tree (*Taxus brevifolia* Nutt.), has also been found to be produced by the endophytic fungus *Taxomyces andreanae* Strobel, Stierle, Stierle & Hess. Another endophytic fungus *Muscodor albus* Worapong, Strobel & Hess has an ability to produce a mixture of volatile organic compounds (VOCs) that were fatal to a wide variety of human- and plant-pathogenic fungi and bacteria.

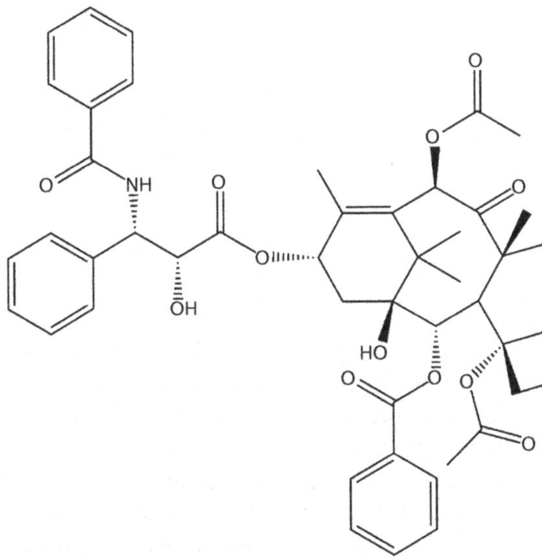

FIGURE 4 Paclitaxel (Taxol).

Endophytic fungi are also more often recognized as a group of organisms capable of providing a source of novel bioactive compounds and secondary metabolites for biological control in agriculture as well as for biotechnology and industrial applications. It was shown that production of herbicidally active substances by endophytic fungi is two and three times higher than phytopathogenic fungi and soil fungi, respectively.

Endophytic fungi play an increasingly important role in the integrated pest management programs in agriculture. Many of the secondary metabolites produced by *Epichloë/Neotyphodium* grass endophytes significantly increase deterrence of vertebrate herbivores, insects, and nematode pests. It has been demonstrated that the secondary metabolites produced by the fine fescue endophyte *E. festucae* are inhibitory to other fungi. It has also been shown that endophyte-infected grasses contain a range of biologically active compounds that either are derived from endophyte or are produced as a result of the association. Four main classes of defensive compounds have been associated with endophyte presence in grasses and the following effects have been described.

Ergot Alkaloids

Ergot alkaloids cause toxicoses in grazing mammals with a range of symptoms from stupor and appetite suppression through reproductive problems and dry gangrene to death. The ergot alkaloids (Figure 5) are produced by a few representatives of filamentous fungi, members of the Trichocomaceae family including the *Aspergillus* and *Penicillium* genera, and Clavicipitaceae family, with the above-mentioned grass endophytes of the genera *Epichloë* with its *Neotyphodium* anamorphs and the genera *Claviceps* and *Balansia*.

Indole-Diterpene Metabolites

Indole-diterpene metabolites produce a number of biological effects, including anti-insect activity (feeding deterrence, modulation of insect receptors functions, toxicity) and mammalian tremorgenic activity (staggers). Indole-diterpenoides (Figure 6) have been reported from *Epichloë* and *Neotyphodium* as well as from some fungi of the genera *Claviceps, Aspergillus*, and *Penicillium*.

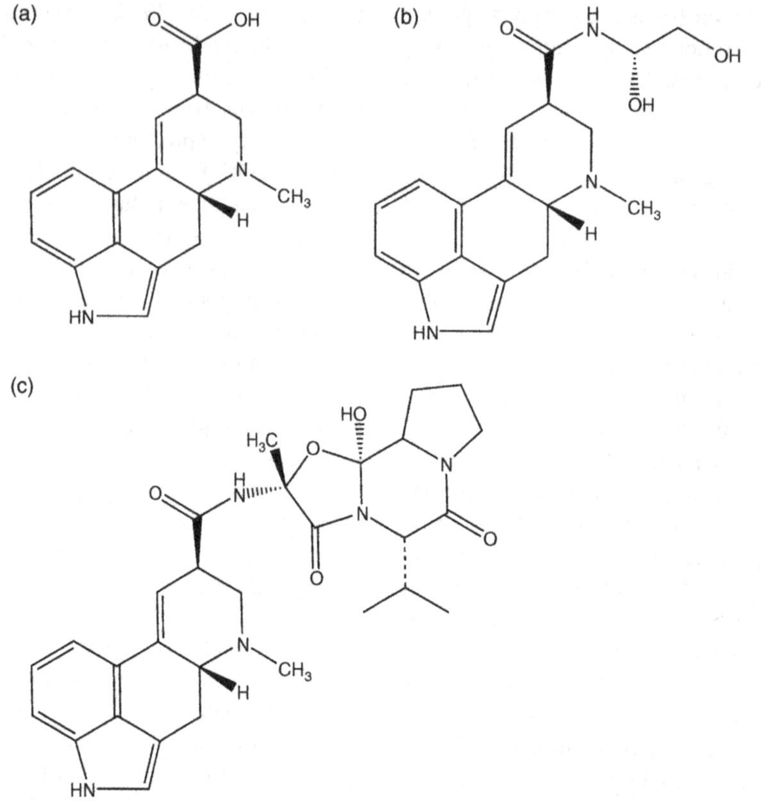

FIGURE 5 Ergot alkaloids: (a) lysergic acid, (b) ergonovine, and (c) ergovaline.

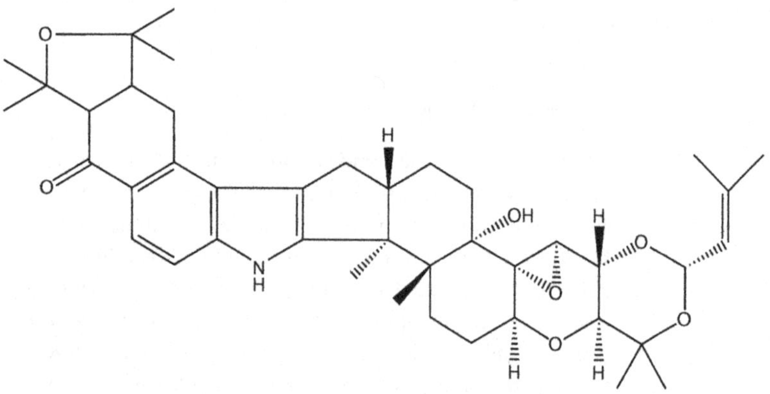

FIGURE 6 Indole-diterpene: lolitrem B.

Peramine

Peramine is an unusual pyrrolopyrazine, and is produced by most *Epichloë/Neotyphodium* endophyte species symbiotic with grasses (Figure 7). It is a metabolite that has specific biological activity as an insect feeding deterrent. Peramine is deterrent to invertebrate herbivores, especially the Argentine stem weevil, and is also active in preventing feeding activity of aphids, but is not toxic to mammalian herbivores.

Lolines

Lolines are classified as pyrrolizidines, a class that also includes plant alkaloids known for their insecticidal activity (Figure 8). These substances are insecticidal alkaloids with insect-deterrent activities, possessing little or no activity against large mammals. Lolines are neurotoxic to a broad range of insects, and when produced by endophytes in plants they have been shown to defend the plants from aphids. Lolines are only known from endophyte-infected grasses,

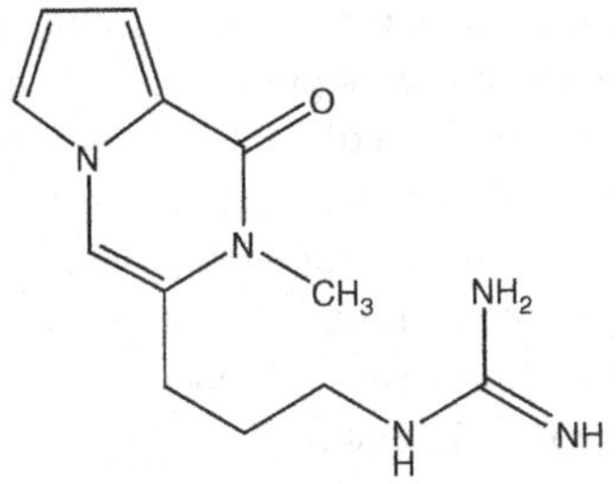

FIGURE 7 Peramine.

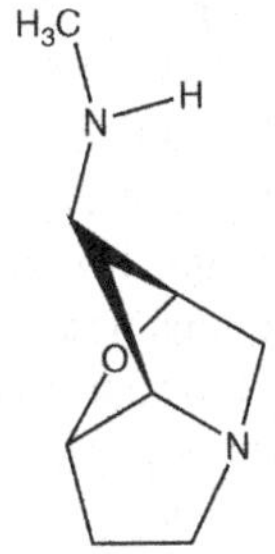

FIGURE 8 Loline.

and plants of the genus *Adenocarpus* (Fabaceae) and *Argyreia mollis* (Burm.f.) Choisy (Convolvulaceae). It is possible that undiscovered fungal symbionts might be responsible for loline production in *Adenocarpus* and *Argyreia* species. In fact, clavicipitaceous epibiotic fungus has recently been discovered on leaf surfaces of *Ipomoea asarifolia* (Desr.) Roem. & Schult. and related plants (Convolvulaceae).

Many, but not all, *Epichloë/Neotyphodium* species produce up to three classes of these alkaloids. In addition to the above, several simple indole alkaloids have been isolated from cultures of endophytic fungi; these include the simple auxins, for example, 3-indoleacetic acid (IAA, Figure 9) and the indole glycerols, for example, 3-indolybutanetriol.

Reports of the secondary metabolites from true marine endophytes are relatively rare. This may be due to the relative difficulty in collecting and culturing of marine endosymbionts and their usually slow growth in laboratory conditions rather than lack of ability to produce secondary metabolites. In recent years, an increasing number of natural products from marine-derived fungal endophytes have been reported. Described below are some examples of secondary metabolites from marine-derived endophytic fungi. Isolated from the green alga *Enteromorpha* sp., the endophytic fungus *Wardomyces anomalus* Brooks & Hansford (Microascaceae; Ascomycota), collected around Fehmarn island in the Baltic Sea, showed (1) antimicrobial effects of the crude extract toward *Microbotryum violaceum* (Pers.) Deml & Oberw. and *Aspergillus repens* (Corda) Sacc. and (2) inhibition of HIV-1 reverse transcriptase (HIV-1-RT). Investigation of the extract yielded several xanthone derivatives. Xanthones are a unique class of biologically active compounds possessing numerous bioactive capabilities, such as antimicrobial, antitubercular, antitumor, antiviral, and antioxidant properties. Xanthone derivatives occur in a number of higher plant families and fungi. Some fungal species are well known as sources of xanthone derivatives, for example, *Penicillium raistrickii* G. Sm., *Phomopsis* sp., *Actinoplanes* sp., *Ascodesmis sphaerospora* Obrist, and *Humicola* sp. The cultivation of the marine fungus *Apiospora montagnei* Sacc. isolated from inner tissue of the North Sea alga *Polysiphonia violacea* Grev. led to the isolation of several new secondary metabolites, where some of them exhibited significant cytotoxicity against human cancer cell lines. From the green alga *Ulva* sp., the endophytic and obligate marine fungus *Stagonosporopsis salicorniae* (Magnus) Died. was isolated. This fungus was found to produce, among others, the unusual tetramic acid-containing metabolites ascosalipyrrolidinones A and B. Ascosalipyrrolidinone A has antiplasmodial activity toward (1) strains K1 and NF 54 of *Plasmodium falciparum* Welch (causing malaria in humans) and (2) general antimicrobial activity. Penostatins, new cytotoxic agents toward leukemic cell lines, were reported from a *Penicillium* sp., inhabiting the marine environment and originally isolated from the marine alga *Enteromorpha intestinalis* (L.) Nees. Antimicroalgal substances, halymecins, were isolated from *Fusarium* and *Acremonium* spp. isolated from a marine alga *Halymenia dilatata* Zanardini.

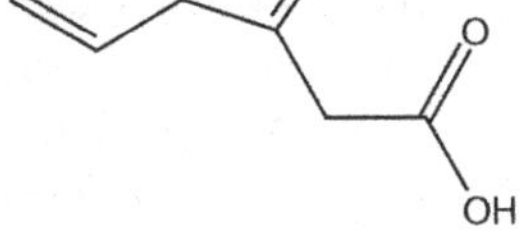

FIGURE 9 Auxin – 3-indoleacetic acid (IAA).

RECENT DEVELOPMENTS

Although in the last few years we noticed significant activity in the study of fungal endophytes associated with host plants, our understanding of their complex and diverse interactions with hosts as well as their ecological functions is still limited. As some researchers have pointed out, if we want to better understand host community dynamics and ecosystem function, we need to better understand the importance of endophytes in host biology. In general, our approaches to study microbe–host interactions are focused on one individual association but individual hosts usually comprise communities of many microorganisms, including viruses, bacteria, microalgae, and fungi that vary in their symbiotic associations with hosts. These fungal–plant associations may be even more complex. In a last decade it has been documented that phylogenetically diverse endohyphal bacteria are associated with living hyphae of several different groups of fungi, including hyphae of some soilborne pathogenic or mycorrhizal fungi in Mucoromycotina,

Glomeromycota, Basidiomycota, and Ascomycota, and have recently been also observed in hyphae of foliar fungal endophytes, members of four classes (Dothideomycetes, Eurotiomycetes, Pezizomycetes, and Sordariomycetes) of filamentous Ascomycota. These bacteria can alter morphology and physiology of the fungal host, and thus fungal interactions with host plants may depend on that association or may be modified in diverse ways. For example, the vertically transmitted bacterium *Candidatus* Glomeribacter gigasporarum colonizes spores and hyphae of the arbuscular mycorrhizal fungi *Gigaspora margarita* Becker & Hall, *Scutellospora castanea* Walker *and Scutellospora persica* (Koske & Walker) Walker & Sanders (Diversisporales, Glomeromycota). Although, *Candidatus* Glomeribacter gigasporarum is not essential to the survival or reproduction of the fungal host, removal of the bacterial partner from the fungal spores suppresses fungal growth and development, altering the morphology of the fungal cell wall, vacuoles, and lipid bodies. In another example, it was reported that a soilborne plant pathogen, *Rhizopus microsporus* Tiegh. (Mucoromycotina) harbors endosymbiotic bacteria *Burkholderia endofungorum* and *B. rhizoxinica*, producers of antimitotic polyketide metabolites, rhizonin and rhizoxin. These toxins are responsible for the pathogenicity of the fungus, are important in securing nutrients for the fungus and protecting the nutrient sources from competitors. In addition to toxin synthesis, the endosymbionts control the ability of the fungal host to reproduce asexually. Fungal mycelia cured of endosymbionts are unable to form asexual spores, which are critical for dispersal of the fungal host. Experimental assessments of such effects will provide key evidence to understanding the degree to which bacterial associates influence the nature of endophytic symbioses. Therefore, complex and multidisciplinary approaches need to be applied to study endophytic fungal–host interactions. Currently, researchers are trying to develop and apply approaches that will take into consideration the host organism as an individual system of various interactions, which is part of the habitat in a specific landscape, and multiple hosts within a single habitat or across different landscapes.

Further Classification of Fungal Endophytes

As previously described, fungal endophytes have frequently been divided into two major groups based on differences in taxonomy, host range, colonization and transmission patterns, tissue specificity, and ecological functions. The first group comprises a smaller number of specialized fungal species (clavicipitaceous endophytes also known as e-endophytes or C-endophytes) that typically colonize some monocotyledonous hosts, and the second group that includes a large number of fungal species (nonclavicipitaceous endophytes or NC-endophytes) with a broad range of host plants. However, complexity and varied phylogenetic origins of the nonclavicipitaceous endopytes, their ecological significance in nature, and continuously growing knowledge about the endophytic associations enabled Rusty Rodriguez (US Geological Survey/University of Washington) and colleagues recently to differentiate endophytic fungi into four separate functional classes. Widespread clavicipitaceous fungal endophytes of grasses of the family Clavicipitaceae (Hypocreales; Ascomycota) are referred as class 1 endophytes. As noted previously, these endophytes are growing systemically in the aboveground tissues and are horizontally and/or vertically transmitted. The second group, the nonclavicipitaceous endophytes, has been divided into three separate functional classes (Class 2, Class 3, and Class 4 endophytes). Class 2 endophytes are represented mostly by the Ascomycota with a minority of Basidiomycota. The class 2 endophytes usually form extensive infections within plants and are able to colonize above- and belowground tissues of a host, but they have low abundance in rhizosphere. They are able to transfer horizontally as well as vertically via seeds, seeds coats, and/or rhizomes. Class 2 endophytes typically have high infection frequencies in plant growing in high-stress habitats. Class 3 endophytes as functional group primarily or exclusively occur in aboveground tissues; they form highly localized limited infections, are horizontally transmitted, and are extremely diverse in host plants. The majority of class 3 endophytes are found in the Ascomycota, but some also belong to the Basidiomycota. Class 4 endophytes also known as dark septate endophytes (DSE) are distinguished from the other two classes of nonclavicipitaceous endophytes based on the presence of darkly melanized septa. Class 4 endophytes are primarily conidial or sterile ascomycetous fungi, and are transmitted horizontally. DSE are restricted to plant roots where they form melanized structures such as inter- and intracellular hyphae and microsclerotia. Class 4 endophytes are associated with mycorrhizal plants, but are also associated with nonmycorrhizal plants. It is assumed that these endophytes have little host or habitat specificity. DSE appear to be ubiquitous and abundant across diverse ecosystems worldwide, and are found to be associated with over 600 plant species. It is almost a hundred years after their first observation and still little is known about the role of DSE in their hosts.

New Endophytic Associations

Clearly, many plants contain, as yet, uncharacterized species and strains of endophytes, and there are vast regions of the world where fungal-plant associations are unknown and where many new fungi await discovery. For example, some endophytic fungi, known as 'endolichenic' fungi were reported to occur within asymptomatic lichens. During the last 3 years another seven new *Neotyphodium* taxa of grass endophytes had been found and described. That includes new species for some *Achnatherum–Neotyphodium* associations previously characterized in this chapter, that is,

Sleepygrass and the 'Drunken Horse Grass' associations. It was found that the North American *Achnatherum robustum* species is associated with a new *Neotyphodium* species, *N. funkii* Craven & Schardl, instead of *N. chisosum* as originally described. For the Asian grass *Achnatherum inebrians*, another *Neotyphiodium* endophyte was described as *N. gansuense* var. *inebrians* Moon & Schardl, which is phylogeneticaly closely related to *N. gansuense* previously reported from this host species.

Novel Mechanism of Hyphal Growth

It is generally believed that fungal vegetative hyphae growth is exclusively apical by extension at the hyphal tip. However, Michael Christensen (AgResearch–Grasslands Research Center) and colleagues presented evidence suggesting that vegetative hyphae of *Epichloë* and *Neotyphodium* species infect elongating grass leaves via a unique and novel mechanism of growth, intercalary hyphal division and extension. According to their observations hyphae within the grass shoot apical meristem invade leaf primordia as they form on the shoot apical meristem. At the beginning, the fungus grows amongst dividing plant cells to form a heavily branched mycelium in leaf primordia. Next, the hyphae attach to adjacent plant cells of an emerging leaf, and when the plant cells divide and then enlarge to form a leaf, hyphal compartments are stretched, which causes intercalary extension of the filament accompanied by cellular division along the length of the filament. These findings suggest that intercalary extension of the hypha in the leaf expansion zone along with increased number of compartments formed during their growth, enable a fungus to grow and keep pace with extended leaf tissues. As a result, the growth of endophytic hyphae is correlated with the host's life cycle; it means that hyphal compartments in grass leaves are of similar age to neighboring cells of the host. Beyond the leaf expansion zone hyphal expansion ceases, suggesting that intercalary growth of the hypha of the endophyte may be activated by stretching. Although endophytic fungi in old, matured leaf tissues no longer are growing, nevertheless they remain metabolically active.

Consequently, these observations might suggest that both modes of growth of vegetative hyphae, that is, apical growth and intercalary hyphal extension, exist in *Epichloë* and *Neotyphodium* species. It may be possible that apical growth accounts for hyphal growth at the beginning of colonization when hyphae are intensively branching among dividing plant cells of leaf primordia in the meristematic zone. In contrast, the intercalary hyphal extension and division is responsible for hyphal expansion later during further development of leaf.

Defensive Mutualism and Mechanisms of Stress Tolerance in Hosts

As previously described in this chapter some grasses from geothermal, hypersaline, and other extreme habitats depend on endophytic fungi to tolerate the high-stress conditions. It has been shown that geothermal endophytes confer heat tolerance but not salt tolerance; coastal endophytes confer salt tolerance but not heat tolerance. Fungal endophytes from agricultural crops conferred disease resistance but not heat or salt tolerance. It was also experimentally shown that the agricultural, coastal and geothermal plant endophytes when inoculated into other plant species conferred the tolerance that was exhibited in the original host. This habitat-specific phenomenon was defined as habitat-adapted symbiosis and it was hypothesized that it is responsible for the establishment of plant communities in high-stress habitats. It is clear that these endophytes provide a mechanism for plant adaptation to habitat stresses but the mechanism responsible for stress tolerance remains unclear. It was recently proposed that the beneficial effects of endophytes on host plants stress tolerance may be the result of the production of reactive oxygen species (ROS) by endophytes. Other investigations have demonstrated that some endophytic fungi, clavicipitaceous endophytes of grasses, produce and secrete ROS to limit host colonization and maintain mutualisms, while other endophytic fungi (class 2 endophytes) reduce ROS production to possibly mitigate the impact of abiotic stresses. In response, the tissues of symbiotic plants produce antioxidants, which increase resistance of plants to oxidative stresses produced by plant pathogens, droughts, heavy metals, and other oxidative stressors. However, this hypothesis that antioxidants are responsible for enhanced stress tolerance in endophyte-infected plants requires further experimental evaluations.

Bioactive Compounds

Recent reviews further demonstrate the importance of endophytic and epiphytic fungi as potential sources of bioactive secondary metabolites as new lead structures for medicine as well as for plant protection. Especially, marine-derived fungi regardless of their origin have shown promising potential and developed into an important source of new and structurally unprecedented metabolites, and more than 330 new metabolites were described between 2002 and 2006. This indicates a growing interest in marine and marine-derived fungi as sources of new bioactive compounds and shows that further progress in sampling, isolation, identification as well as fermentation and extraction was obtained.

New Characterization Methods in the Study of Endophytes

Recently, study of the ecology and evolution of the epiphytic and endophytic interactions between fungi and plants or marine organisms has undergone rapid expansion from using traditional methods to molecular systematic and ecological genomics and further to develop new DNA- and RNA-based tools and associated bioinformatics approaches.

In addition, some biochemical, for example, stable isotope profiling (SIP) and metabolic incorporation of nucleotide analogs such as bromodeoxyuridine (BrdU) technologies are now being applied. Regardless some limitations and needs for further improvement, these techniques would not only help to assess the entire diversity and composition of endophytic fungal communities but also would allow focusing at a co-evolutionary context of epiphyte and endophyte associations at different scales. These new tools may allow us to better understand processes that shape epiphytic and endophytic fungal-plant interactions at local as well as global scales.

CONCLUSION

Endophytic microbes are relatively common in all families of plants from polar to tropical regions. The endosymbionts have in several cases proven to play adaptive and/or defensive roles in the ecology of host plants. Relatively little is known about the relationships between fungi and their hosts especially with respect to chemical ecology. Nonetheless, some observations suggest the importance of the host as well as the ecosystem in influencing the general metabolism of endophytic microbes. Although endophytes are still a poorly investigated group of organisms, with the exception of the clavicipitaceous endophytes of grasses, many studies have proven that they are relatively rich and promising sources of bioactive and chemically novel compounds with a wide variety of potential uses in medicine, agriculture, and industry.

ACKNOWLEDGMENT

This work was partly supported by Fogarty International Center, NIH under U01 TW006674 for the International Cooperative Biodiversity Groups and Specialty Crop Research Initiative, USDA.

FURTHER READING

An, Z. (Ed.), (2005). *Handbook of industrial mycology*. Marcel Dekker: New York.

Arnold, A. E., & Lutzoni, F. (2007). Diversity and host range of foliar fungal endophytes: Are tropical leaves biodiversity hotspots? *Ecology, 88*, 541–549.

Bacon, C. W., & White, J. F., Jr. (Eds.), (2000). *Microbial endophytes*. Marcel Dekker: New York.

Cheplick, G. P., & Faeth, S. H. (2009). *Ecology and evolution of the grass–endophyte symbiosis*. Oxford University Press: New York.

Clay, K., & Schardl, C. (2002). Evolutionary origins and ecological consequences of endophyte symbiosis with grasses. *The American Naturalist, 160*, S99–S127.

Dighton, J., White, J. F., Jr., & Oudemans, P. (Eds.), (2005). *The fungal community: Its organization and role in the ecosystem*. Taylor & Francis: New York.

Gloer, J. B. (1997). Applications of fungal ecology in the search for new bioactive natural products. In D. T. Wicklow & B. E. Soderstrom (Eds.), *Environmental and microbial relationships* (pp. 249–268). *The mycota: A comprehensive treatise on fungi as experimental systems for basic and applied research. Vol. IV* (pp. 249–268). Springer-Verlag: New York.

König, G. M., Kehraus, S., Seibert, S. F., Abdel-Lateff, A., & Müller, D. (2006). Natural products from marine organisms and their associated microbes. *Chembiochemistry, 7*, 229–238.

Kuldau, G., & Bacon, C. (2008). Clavicipitaceous endophytes: Their ability to enhance resistance of grasses to multiple stresses. *Biological Control, 46*, 57–71.

Mueller, G. M., Bills, G. F., & Foster, M. S. (Eds.), (2004). *Biodiversity of fungi: Inventory and monitoring methods*. Elsevier Academic Press: New York.

Rodriguez, R. J., White, J. F., Jr., Arnold, A. E., & Redman, R. S. (2009). Fungal endophytes: Diversity and functional roles. *New Phytologist, 182*, 314–330.

Schardl, C. L., Leuchtmann, A., & Spiering, M. J. (2004). Symbioses of grasses with seedborne fungal endophytes. *Annual Review of Plant Biology, 55*, 315–340.

Schardl, C. L., Scott, B., Florea, S., & Zhang, D. (2009). *Epichloë endophytes*: Clavicipitaceous symbionts of grasses. In: H. B. Deising (Ed.), *Plant relationships* (pp. 275–306). (2nd ed.). K. Esser (Ed.), *The mycota: A comprehensive treatise on fungi as experimental systems for basic and applied research. Vol. V* (pp. 275–306). Springer: Berlin Chap. 15.

Schulz, B., Boyle, C., Draeger, S., Römmert, A.-K., & Krohn, K. (2002). Endophytic fungi: A source of novel biologically active secondary metabolites. *Mycological Research, 106*, 996–1004.

Smith, S. E., & Read, D. J. (2008). *Mycorrhizal symbiosis* (3rd ed.). Academic Press: Boston, MA.

Strobel, G. A. (2003). Endophytes as sources of bioactive products. *Microbes and Infection, 5*, 535–544.

Tadych, M., Bergen, M., Dugan, F. M., & White, J. F., Jr. (2007). The potential role of water in spread of conidia of the *Neotyphodium* endophyte of *Poa ampla*. *Mycological Research, 111*, 466–472.

White, J. F., Jr., Bacon, C. W., Hywel-Jones, N. L., & Spatafora, J. W. (Eds.), (2003). *Clavicipitalean fungi: Evolutionary biology, chemistry, biocontrol, and cultural impacts*. Marcel Dekker: New York.

White, J. F., Jr., & Torres, M. S. (Eds.), (2009). *Defensive mutualism in microbial symbiosis*. CRC Press/Taylor and Francis: Boca Raton, FL.

Wilson, D. (1995). Endophyte – The evolution of a term, and clarification of its use and definition. *Oikos, 73*, 274–276.

Zhang, H. W., Song, Y. C., & Tan, R. X. (2006). Biology and chemistry of endophytes. *Natural Product Reports, 23*, 753–771.

Chapter 5

Microsporidia: A Model for Minimal Parasite–Host Interactions

Catherine Texier[1,2], Cyril Vidau[1,2], Bernard Viguès[1,2], Hicham El Alaoui[1,2] and Frédéric Delbac[1,2]

[1]*Clermont Université, Université Blaise Pascal, Laboratoire Microorganismes: Génome et Environnement, BP 10448, F-63000 Clermont-Ferrand, France*

[2]*CNRS, UMR 6023, LMGE, F-63177 Aubiere, France*

Corresponding authors: Delbac, Frédéric (frederic.delbac@univbpclermont.fr)

Chapter Outline

Introduction	65	Host immune response: from protective to overwhelmed defences	68
The polar tube and the spore wall of microsporidian spores are involved in host cell invasion	66	Conclusion and future directions	70
The reduced and compact microsporidian genomes reflect host dependency	66	References and recommended reading	70

Current Opinion in Microbiology 2010, **13**:443–449

This review comes from a themed issue on Host-microbe interactions: fungi
Edited by Michael Lorenz

Available online 9th June 2010

1369-5274/$ – see front matter

DOI 10.1016/j.mib.2010.05.005

INTRODUCTION

Microsporidia are fungi-related unicellular eukaryotes, all of which are obligate intracellular parasites with more than 1200 species parasitizing a wide range of hosts from insects to most mammal groups [1]. Since their discovery in the 1850s as the causative agent of the silkworm disease pebrine (works of Balbiani and Pasteur) which devastated the silk industry in Europe, these pathogens demonstrated their major economic implication in animal farming with nosemosis in beekeeping (*Nosema apis* and *Nosema ceranae*) and sericulture (*Nosema bombycis*), and microsporidiosis in aquaculture (*Loma salmonae* for salmonids, *Thelohania* spp. for shrimps). Nowadays, *N. ceranae* is considered as an emerging threat to honeybee health and may contribute to massive colony losses of the European honeybee observed worldwide during the last decade [2]. In human, these opportunistic emerging pathogens have become a public health problem since the AIDS pandemic, and were added to the National Institute of Allergy and Infectious Diseases (NIAID) priority pathogen list (category B, Biological Diseases, Food and Waterborne Pathogens). Infections range from asymptomatic to serious diseases and mortality in some cases, in relation to parasite virulence and host immune response efficiency. The fact that human-infecting microsporidia are also pathogens of several animal groups and are found in the environment as resistant spores suggest both zoonotic and environmental potential transmissions to humans. Moreover, microsporidia seem to play an important role in ecosystem functioning, as illustrated by species which can modulate reproduction and behaviour of *Daphnia* populations, the best-understood organisms in the field of ecology [3••]. Microsporidia have also attracted attention for their original host cell invasion system involving a highly specialized organelle called the polar tube, for their extreme reduction at the molecular (smallest eukaryotic genomes), cellular and biochemical level and for their host dependency. These parasites were long time considered as primitive amitochondriate protozoa but recent molecular phylogenetic studies supported a placement of these eukaryotes amongst fungi with a probable zygomycete ancestor [4]. Moreover, they were shown to contain a mitochondria remnant organelle called mitosome for which the only known essential functions are the biosynthesis of Fe–S clusters [5], and maybe the redox balance [6]. This review will focus on microsporidia cell invasion, host dependency in relation to microsporidian genome reduction, and host immune responses.

THE POLAR TUBE AND THE SPORE WALL OF MICROSPORIDIAN SPORES ARE INVOLVED IN HOST CELL INVASION

Microsporidia survive outside of their hosts as environmentally resistant spores [7] protected by a thick two-layered wall (Figure 1a). They contain a highly specialized organelle, the polar tube, which is a unique structure in the eukaryotic world and constitutes one of the most sophisticated infection mechanisms. Under appropriate conditions, the polar tube is expelled from the anterior part of the spore (Figure 1b,c) allowing the transfer and the release of the infectious material (sporoplasm) into the cytoplasm of a target cell where the intracellular development takes place (Figure 1b,d). The great diversity of intracellular developments amongst microsporidia has been already described [8]. Distinct routes can be used by microsporidia to invade a host cell: the polar tube acting as a needle to pierce the host cell membrane, actin-dependant phagocytosis followed by spore digestion or parasite evasion and host plasma membrane invaginations as a consequence of mechanical force of the discharge tube itself [9,10,11•,12]. The signalling pathways that lead to polar tube extrusion (spore germination) *in vivo* remain largely unknown. The current view is that this process is the result of an increase in osmotic pressure inside the spore followed by a rapid influx of water through the cell membrane [13•]. Intra-membranous aquaporin-like channels were postulated to be the mechanism for the influx of water into the activated spore [14] which results in the swelling of both the lamellar polaroplast and the posterior vacuole and then in polar tube extrusion.

In the last decade, several proteins and glycoproteins from both the polar tube (review in [11•]) and the spore wall [15–20] were identified for several microsporidian species. In particular, three families of polar tube proteins (PTP1, PTP2 and PTP3) with no homology within databases were characterized [11•]. A recent yeast two-hybrid study demonstrates that these proteins interact with each other, which is likely essential for assembly and function of the polar tube [21••]. Some components of the polar tube and the spore wall have been suggested to be involved in tissue recognition and thereby in the initiation of host cell invasion. Microsporidian spores are believed to utilize glycosaminoglycans (GAGs) as host cell receptors for cell adhesion and infection [22], through the possible interaction with a major spore wall protein named EnP1 [23]. Moreover the *O*-mannosylated moiety of PTP1 can interact with a host cell mannose receptor [10]. These interactions could be the initial steps leading to polar tube extrusion and cell invasion. Nevertheless more detailed functional characterization of PTPs and spore wall proteins are still needed.

THE REDUCED AND COMPACT MICROSPORIDIAN GENOMES REFLECT HOST DEPENDENCY

Microsporidian genomes may allow comparative analyses that could explain adaptations of these unusual fungi to obligate intracellular parasitism. Compaction and small size are typical of these parasitic genomes. Indeed, genomes range from 2.3 Mbp (*Encephalitozoon intestinalis*) to 24 Mbp (estimated for *Octosporea bayeri*) (Table 1). The complete genome of *Encephalitozoon cuniculi* (2.9 Mbp) is the smallest eukaryotic sequenced genome [24]. Genomics surveys are completed for *Edhazardia aedis* and *Anncaliia algerae* [25]. Draft genomes have also been obtained for *Enterocytozoon bieneusi* [26], *Nosema ceranae* [27•], *O. bayeri* [28] and *Antonospora locustae* (Marine Biological Laboratory, Woods Hole, USA). Intergenic region shortness owing to the extreme genome compaction must have modified transcription, as demonstrated by overlapping mRNAs between contiguous genes for *E. cuniculi* and *A. locustae* [29]. In contrast to *E. cuniculi* and *E. bieneusi* genomes, transposable elements (Table 1) were discovered in *Vittaforma cornea*, in *Nosema bombycis, E. aedis, A. algerae, N. ceranae* and *O. bayeri*. Transposons are usually considered as parasitic DNAs, tracing the origin of these elements will help defining specific lineages within microsporidia.

Compared to other eukaryotes, and especially fungi, microsporidian protein-encoded genes are shorter (~20% less than their yeast orthologues) and the number of proteins is reduced (1997–3632, Table 1). This severe reduction in the gene number is likely due to losses in metabolic capacities as consequences of parasitism [30,13•]. Whilst *E. cuniculi* is considered as the model microsporidia for minimal proteome, *E. bieneusi* seems to be more host-dependant since fewer genes, involved in energy generation, lipid and isoprenoid metabolism, were found [26]. On the contrary, *O. bayeri* is less metabolically dependant since metabolism, cell growth, transcription, protein destination, DNA synthesis and energy production genes are overrepresented in its genome, and substrate transport underrepresented [28]. Finally, microsporidian genomes suggest that core genes have been retained in this group evolution and that differences could be due to adaptation

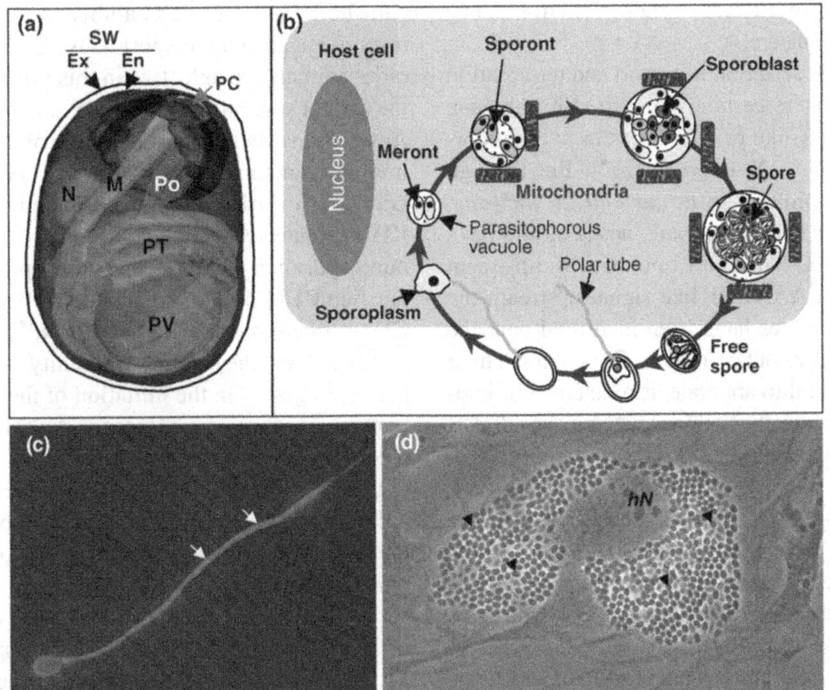

FIGURE 1 The microsporidian spore structure and the intracellular life cycle for *Encephalitozoon cuniculi* **(a)** and **(b)** and *Anncaliia algerae* **(c)** and **(d)**. (a) Three-dimensional reconstruction of the monokaryotic *E. cuniculi* spore with its organelles. The parasite organelle outlines were identified on serial ultrathin sections of *E. cuniculi*-infected human foreskin fibroblast (HFF) cells, and aligned using the Photoshop programme. The 3D model was then built using the IMOD software (http://bio3d.colorado.edu). SW, spore wall; Ex, exospore (outer layer of the cell wall); En, endospore (inner layer of the cell wall); N, nucleus; PC, polar cap; Po, polaroplast (membranous tubules and/or lamellae associated with the tube in the anterior region of the spore); PV, posterior vacuole; PT, polar tube; M, manubrium (anterior and straight part of the polar tube and anchored in the polar cap). (b) Representation of the three distinct phases of the *E. cuniculi* life cycle: first, the infective phase characterized by polar tube extrusion and injection of the sporoplasm within the host cytoplasm; second, the proliferative phase or merogony and third, the spore-forming phase or sporogony during which meronts transform into sporonts, sporoblasts then spores. In *E. cuniculi* the intracellular development occurs within a parasitophorous vacuole of host origin [12,64]. Host mitochondria relocalize to be in close contact with the membrane of the parasitophorous vacuole. (c) The dikaryotic *A. algerae* spore after polar tube extrusion. The tube (in green) is stained with antibodies raised against the polar tube protein PTP1. The spore (in red) is labelled with antibodies directed against a spore wall component. The two nuclei in the lumen of the extruded polar tube are stained with DAPI (in blue) and are indicated by arrows. (d) *A. algerae* intracellular developmental stages (arrow heads) in the cytoplasm of HFF cells. In contrast to *E. cuniculi*, all the intracellular stages are in direct contact with the host cell cytoplasm. *hN*: host nucleus.

TABLE 1 Main features of available microsporidian genomes

Microsporidia	Host	Sequenced amount/ genome size	Coverage	Reference	Gene number	Transposons and retrotransposons	% G + C	Mean intergenic region size (bp)	Gene density
Encephalitozoon cuniculi	Mammals	2.5 Mb/ 2.9 Mb	86%	[24]	1997	No	47	129	1 gene/ 1.025 kb
Enterocytozoon bieneusi	Mammals	3.86 Mb/ 6 Mb	64%	[26]	3632 ?[b]	No	25	127	1 gene/ 1.148 kb
Nosema ceranae	Insects	7.86 Mb/?[a]	–[a]	[27•]	2614	Gypsy, Merlin, Helitron, piggyBac and MULE	26	–[c]	1 gene/ 1.666 kb
Octosporea bayeri	Crustaceans	13.3 Mb/ 24 Mb	55%	[28]	2174	Gypsy, Mariner, Copia	26	429	1 gene/ 4.593 kb

Data concerning genome surveys of *A. algerae* and *E. aedis* [25] are not presented in this table owing to their very low genome coverage.
[a]*Genome size not characterized.*
[b]*Overestimation as 50% of the CDSs are short ORFs.*
[c]*Undetermined.*

to different hosts, and to different genome environments, which set the pace of change.

Consensuses for transcription initiation and termination within microsporidia were certainly modified by genome compaction [31]. TATA-like promoters seem to be important in gene regulation in *N. ceranae* [27•]. But through analysis of ribosomal proteins in *E. cuniculi, A. locustae, E. bieneusi, A. algerae* and *N. ceranae*, novel motifs have been discovered, a CCC-like motif immediately upstream the start codon and an AAATTT-like signal upstream the CCC motif. Such sequences have been identified near the start codon for 1591 genes out of 1997 in *E. cuniculi*. These motifs will be very useful to annotate or re-annotate microsporidian genomes, especially within the MicrosporidiaDB genomics resource (MicrosporidiaDB: The microsporidia genome resource; URL: http://microsporidiadb.org/micro/), which was released on April 2010 as an extension of the Eukaryotic Pathogens Database (EuPathDB). Recently, the *E. intestinalis* genome sequence has been deposited in this database. The *A. algerae* genome sequence is about to be published and the one of *Tubulinosema ratisbonensis*, a *Drosophila* parasite, is currently running at the Génoscope (France) (personal data). Moreover, an ambitious sequencing project encompassing 13 microsporidian genomes is in progress at the Broad Institute Genomic Sequencing Center for Infectious Diseases which was established by the NIAID. This will undoubtedly provide essential data to better understand the host-microsporidia interactions and the evolution of these very specific genomes.

HOST IMMUNE RESPONSE: FROM PROTECTIVE TO OVERWHELMED DEFENCES

Microsporidian infections in immunocompetent mammals are often chronic and asymptomatic whilst immunocompromised hosts develop lethal disease. The deciphering of the protective mammal immune response (Figure 2, left panel) is conducted using mice experimentally infected with *Encephalitozoon* species (for reviews see [32,33••]). CD8+ cytotoxic T lymphocytes (CTL) play a central role in the protection against *E. cuniculi* challenges [34]. This CTL immunity is not activated by CD4+ T cells [35] but strongly depends on the Thl cytokines IFN-γ and IL-12 [36,37]. γδ T cells play a role in its induction probably via an early production of IFN-γ [38]. Dendritic cells (DCs), well-known IL-12-producer and IFN-γ-producer in response to microbial invasion [39], are also implicated in microsporidia-specific cellular immunity. In mice oral infection with *E. cuniculi*, IFN-γ-producing mucosal DCs are critical in priming antigen-specific intraepithelial lymphocytes (IELs) with inflammatory and immunoregulatory properties which lead to a protective immune response to the intestinal parasite [40,41]. The DC ability to trigger a robust T cell response against microsporidia is age-dependant: DCs from older animals are defective in this priming, thus explaining the greater susceptibility to infection of ageing animals [42]. Antigens involved in the elicitation of the protective cellular immunity against *Encephalitozoon* species are largely unknown. Nevertheless, mice immunization with purified PTP1 protein can elicit a strong T cell response against microsporidian spores, and murine splenic DC pulsed with purified PTP1 protein are able to activate IFN-γ-producing CD8+ T cells but not CD4+ cells [43•]. A key role has been attributed to the innate immunity — in particular to macrophages — in the initiation of the protective mammal immune response to microsporidia. Human macrophages recognize *Encephalitozoon* species by a Toll-like receptor (TLR2) resulting in the activation of NF-κB and in the upregulation and secretion of several chemokines which lead to the recruitment of naïve monocytes [44,45] known to be DC precursors [46]. Recently, Lawlor *et al.* demonstrated a strong TLR4-dependant DC activation which was involved in CD8+ T cell immunity against *E. cuniculi* [47]. TLR2 and TLR4 recognize fungal pathogens through phospholipomannans and *O*-linked mannans respectively as PAMP (pattern associated molecular proteins) [48], suggesting that *Encephalitozoon O*-mannosylated proteins [10] and *O*-mannans [49] may act as PAMP. Microsporidia-infected macrophages can either kill the intracellular invader, or fail to do so depending on the pre-activation state of the cell and on the existence of parasite survival mechanisms [33••]. Finally, the mammal immune response also involves not only microsporidia-specific antibodies insufficient to protect the host [32], but also mucosal antimicrobial molecules such as defensins that may play a role in preventing microsporidian infection by reducing spore germination and enterocyte infection [50].

Several wild or laboratory insect models have been used to study the host defence response to microsporidian infections, for example, *Drosophila melanogaster, Aedes aegypti, Apis mellifera* and members of the Lepidoptera and the Orthoptera families. Such a diversity of host-parasite systems gives rise to a diversity of response that may appear as contradictory. Nevertheless, the induction of a cellular immunity was observed in Lepidoptera caterpillars (wax moth and gypsy mox [51,52]), Orthopera (cricket and locust [53,54]) and *Drosophila* [55]. Depending on the model, this induction results in an increase in the total number of haemocytes, phagocytosis, encapsulation of infected tissues, nodulation of melanized parasites and/or melanization which was associated with microsporidian abnormal sporogony. Insect humoral immunity can be also triggered against microsporidia as illustrated by the early increase of honeybee antimicrobial peptides (AMP) after *Nosema apis* infection [56]. Little is known about the molecular regulation of the anti-microsporidian insect immunity. However, some elements which are summarized Figure 2 (right panel), are given by two studies describing the

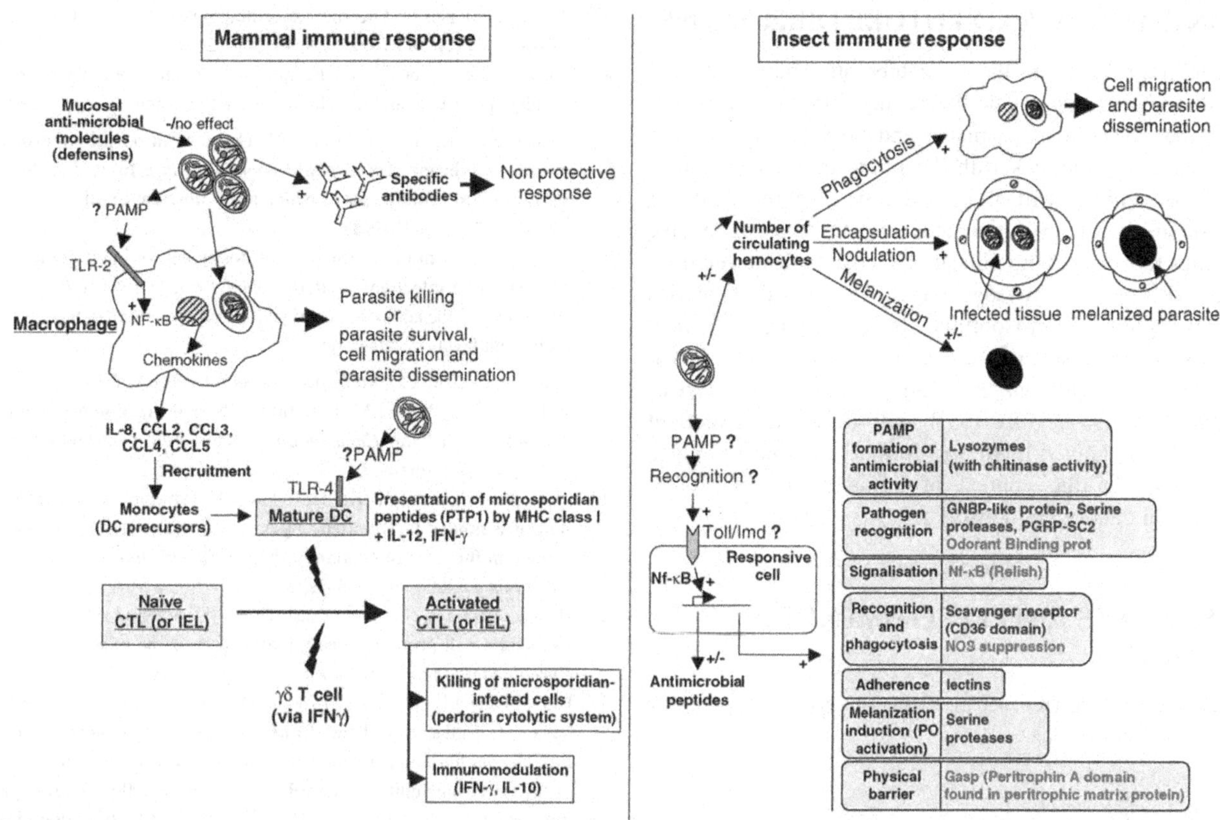

FIGURE 2 Overall picture of mammal (left panel) and insect (right panel) immune responses. Left: the mammal protective immune response against *Encephalitozoon* spp. infection results from the cooperation between adaptive immunity and innate immunity. Whilst the former (cytotoxic T cells primed by dendritic cells) is clearly essential for the clearance of these parasites, the latter (dendritic cell recruitment and activation, macrophages) may ultimately define whether or not the parasite can survive. See text for details and references. Right: the insect cellular immunity against microsporidia has been observed in several wild and laboratory models. Molecular data are either from the *D. melanogaster* adult transcriptomic response against *O. muscaedomesticae* [57] (in blue), or from the *A. aegypti* larvae proteomic response against and *V. culicis* [58] (in red). The Toll and Imd pathways are the may signalization immune pathways described in *Drosophila* [65]. ?: unknown mechanisms or molecules; +: activation; −: repression; CTL: cytotoxic T lymphocytes; DC: dendritic cell; IEL: intraepithelial lymphocytes; MHC: major histocompatibility complex; NOS: nitric oxide synthase; PAMP: pattern associated molecular proteins; PO: phenoloxidase; TLR2 and TLR4: Toll-like receptors 2 and 4.

D. melanogaster adult transcriptomic [57] and the *A. aegypti* larvae proteomic [58] responses against *Octosporea muscaedomesticae* and *Vavraia culicis* respectively. It is noteworthy that *D. melanogaster* infected with microsporidia develops a fundamentally different response by known immunity-related genes as compared to the other microorganisms used for immune challenge (virus, bacteria and fungus). Nevertheless, these defence responses do not prevent the progression of infections which can either be chronic or cause host mortality. Moreover, several data suggest that some microsporidia may possess survival mechanisms and are able to module/suppress the host immunity [52,53,56].

For many fish-infecting microsporidia, the most characteristic feature in host interactions is the formation of well-organized structures corresponding to hypertrophied microsporidia-infected cells and called xenoma [59,60]. The host cell undergoes a complete restructuring which is highly variable according to the host-microsporidia couple. The xenoma offers optimal growth conditions for the parasite including protection against the host immune system [61•], whilst confining it to one cell and preventing its free spread in the host organism. Nevertheless this protection occurs whilst the xenoma is young or growing, and as soon as its wall loses its integrity, macrophages pervade it to digest the parasites, resulting in inflammatory diseases. Whether the parasite or the host induces the formation of the xenoma, remains to be elucidated. Very few molecular data are available about the nature and induction of fish immune response against microsporidia (reviewed in [62]).

Whatever the host is, the resolution of the infection strongly depends on the efficiency of the immunity induced by the microsporidia, which may depend not only on the inner characteristics of the host but also on the parasite manipulation and evasion capacities. Experimental coevolution systems using *Drosophila* and *Daphnia* demonstrated that host resistance towards microsporidia has a fitness cost at the population scale [3••,55], illustrating the host necessary trade-offs between resistance and other biological functions.

CONCLUSION AND FUTURE DIRECTIONS

Accumulating data on the polar tube and spore wall will permit to better understand their composition and architecture, the invasion mechanisms and host cell specificity. Comparative genomics with the microsporidian genomes under sequencing will allow a deeper comprehension of parasitism-driven evolution and help to develop reverse genetics tools to better delineate the host–parasite relationships. This better knowledge may provide new therapeutics targets against microsporidia. In particular, two new microsporidia–host models, *Tubulinosema ratis-bonensis*/*Drosophila melanogaster* and *Nematocida parisii*/*Caenorhabditis elegans* [63•] will constitute excellent and well-recognized models to further study host-parasite interactions in the context of minimal protein–protein interaction eukaryotic networks.

REFERENCES AND RECOMMENDED READING

Papers of particular interest, published within the period of review, have been highlighted as:

- • of special interest
- •• of outstanding interest

1. Keeling PJ, Fast NM: **Microsporidia: biology and evolution of highly reduced intracellular parasites.** *Annu Rev Microbiol* 2002, **56**:93–116.
2. Higes M, Martin-Hernández R, Botías C, Bailón EG, González-Porto AV, Barrios L, Del Nozal MJ, Bernal JL, Jiménez JJ, Palencia PG, Meana A: **How natural infection by *Nosema ceranae* causes honeybee colony collapse.** *Environ Microbiol* 2008, **10**:2659–2669.
3. Ebert D: **Host-parasite coevolution: insights from the *Daphnia*–parasite model system.** *Curr Opin Microbiol* 2008, **11**:290–301.

•• This review describes the role of microsporidia in the functioning of ecosystems through the manipulation of host populations.

4. Lee SC, Corradi N, Byrnes EJ 3rd, Torres-Martinez S, Dietrich FS, Keeling PJ, Heitman J: **Microsporidia evolved from ancestral sexual fungi.** *Curr Biol* 2008, **18**:1675–1679.
5. Goldberg AV, Molik S, Tsaousis AD, Neumann K, Kuhnke G, Delbac F, Vivares CP, Hirt RP, Lill R, Embley TM: **Localization and functionality of microsporidian iron-sulphur cluster assembly proteins.** *Nature* 2008, **452**:624–628.
6. Williams BA, Elliot C, Burri L, Kido Y, Kita K, Moore AL, Keeling PJ: **A broad distribution of the alternative oxidase in microsporidian parasites.** *PLoS Pathog* 2010, **6**:e1000761.
7. Vavra J, Larsson JIR: **Structure of the microsporidia.** In *The Microsporidia and Microsporidiosis*. Edited by Wittner M. ASM Press; 1999:7–84.
8. Cali A, Takvorian PM: **Developmental morphology and life cycles of the microsporidia.** In *The Microsporidia and Microsporidiosis*. Edited by Wittner M. ASM Press; 1999:85–128.
9. Franzen C: **Microsporidia: how can they invade other cells?** *Trends Parasitol* 2004, **20**:275–279.
10. Xu Y, Weiss LM: **The microsporidian polar tube: a highly specialised invasion organelle.** *Int J Parasitol* 2005, **35**:941–953.
11. Delbac F, Polonais V: **The microsporidian polar tube and its role in invasion.** *Subcell Biochem* 2008, **47**:208–220.

• This recent review deals with the molecular architecture of the microsporidian polar tube and its role in host cell invasion.

12. Rönnebäumer K, Gross U, Bohne W: **The nascent parasitophorous vacuole membrane of *Encephalitozoon cuniculi* is formed by host cell lipids and contains pores which allow nutrient uptake.** *Eukaryot Cell* 2008, **7**:1001–1008.
13. Williams BA: **Unique physiology of host-parasite interactions in microsporidia infections.** *Cell Microbiol* 2009, **11**:1551–1560.

• In this review, the author summarizes host–microsporidia interactions at different levels of integration.

14. Ghosh K, Capiello CD, McBride SM, Occi JL, Cali A, Takvorian PM, McDonald TV, Weiss LM: **Functional characterization of a putative aquaporin from *Encephalitozoon cuniculi*, a microsporidia pathogenic to humans.** *Int J Parasitol* 2006, **36**:57–62.
15. Bohne W, Ferguson DJP, Kohler K, Gross U: **Developmental expression of a tandemly repeated, glycine and serine-rich spore wall protein in the microsporidian pathogen *Encephalitozoon cuniculi*.** *Infect Immun* 2000, **68**:2268–2275.
16. Hayman JR, Hayes S, Amon J, Nash TE: **Developmental expression of two spore wall proteins during maturation of the microsporidian *Encephalitozoon intestinalis*.** *Infect Immun* 2001, **69**:7057–7066.
17. Peuvel-Fanget I, Polonais V, Brosson D, Texier C, Kuhn L, Peyret P, Vivares C, Delbac F: **EnP1 and EnP2, two proteins associated with the *Encephalitozoon cuniculi* endospore, the chitin-rich inner layer of the microsporidian spore wall.** *Int J Parasitol* 2006, **36**:309–318.
18. Southern TR, Jolly CE, Lester ME, Hayman JR: **Identification of a microsporidian protein potentially involved in spore adherence to host cells.** *J Eukaryot Microbiol* 2006, **53**:568–569.
19. Li Y, Wu Z, Pan G, He W, Zhang R, Hu J, Zhou Z: **Identification of a novel spore wall protein (SWP26) from microsporidia *Nosema bombycis*.** *Int J Parasitol* 2009, **39**:391–398.
20. Polonais V, Mazet M, Wawrzyniak Y, Texier C, Blot N, El Alaoui H, Delbac F: **The human microsporidian *Encephalitozoon hellem* synthesizes two spore wall polymorphic proteins useful for epidemiological studies.** *Infect Immun* 2010, **78**:2221–2230.
21. Bouzahzah B, Nagajyothi FNU, Ghosh K, Takvorian PM, Cali A, Tanowitz HB, Weiss LM: **Interactions of *Encephalitozoon cuniculi* polar tube proteins.** *Infect Immun* 2010, **78**:2745–2753.

•• This work presents essential data on interactions between the main components of the microsporidian polar tube.

22. Hayman JR, Southern TR, Nash TE: **Role of sulphated glycans in adherence of the microsporidian *Encephalitozoon intestinalis* to host cells in vitro.** *Infect Immun* 2005, **73**:841–848.
23. Southern TR, Jolly CE, Lester ME, Hayman JR: **EnP1, a microsporidian spore wall protein that enables spores to adhere and to infect host cells in vitro.** *Eukaryot Cell* 2007, **6**:1354–1362.
24. Katinka MD, Duprat S, Cornillot E, Méténier G, Thomarat F, Prensier G, Barbe V, Peyretaillade E, Brottier P, Wincker P *et al.*: **Genome sequence and gene compaction of the eukaryote parasite *Encephalitozoon cuniculi*.** *Nature* 2001, **414**:450–453.
25. Williams BA, Lee RC, Becnel JJ, Weiss LM, Fast NM, Keeling PJ: **Genome sequence surveys of *Brachiola algerae* and *Edhazardia aedis* reveal microsporidia with low gene densities.** *BMC Genomics* 2008, **9**:200–209.
26. Akiyoshi DE, Morrison HG, Lei S, Feng X, Zhang Q, Corradi N, Mayanja H, Tumwine JK, Keeling PJ, Weiss LM, Tzipori S: **Genomic**

survey of the non-cultivatable opportunistic human pathogen, *Enterocytozoon bieneusi*. *PLoS Pathog* 2009, **5**:e1000261.

27. Cornman RS, Chen YP, Schatz MC, Street C, Zhao Y, Desany B, Egholm M, Hutchison S, Pettis JS, Lipkin WI, Evans JD: **Genomic analyses of the microsporidian *Nosema ceranae*, an emergent pathogen of honey bees.** *PLoS Pathog* 2009, **5**:e1000466.

• *Nosema ceranae* seems to be involved in a severe bee disease. This work provides data for new insights into its interactions with its host.

28. Corradi N, Haag KL, Pombert JF, Ebert D, Keeling PJ: **Draft genome sequence of the *Daphnia* pathogen *Octosporea bayeri*: insights into the gene content of a large microsporidian genome and a model for host–parasite interactions.** *Genome Biol* 2009, **10**:R106.

29. Corradi N, Gangaeva A, Keeling PJ: **Comparative profiling of overlapping transcription in the compacted genomes of microsporidia *Antonospora locustae* and *Encephalitozoon cuniculi*.** *Genomics* 2008, **91**:388–393.

30. Vivarès CP, Gouy M, Thomarat F, Méténier G: **Functional and evolutionary analysis of a eukaryotic parasitic genome.** *Curr Opin Microbiol* 2002, **5**:499–505.

31. Peyretaillade E, Gonçalves O, Terrat S, Dugat-Bony E, Wincker P, Cornman RS, Evans JD, Delbac F, Peyret P: **Identification of transcriptional signals in *Encephalitozoon cuniculi* widespread among Microsporidia phylum: support for accurate structural genome annotation.** *BMC Genomics* 2009, **10**:607–619.

32. Khan IA, Moretto M, Weiss LM: **Immune response to *Encephalitozoon cuniculi* infection.** *Microbes Infect* 2001, **3**:401–405.

33. Mathews A, Hotard A, Hale-Donze H: **Innate immune responses to *Encephalitozoon* species infections.** *Microbes Infect* 2009, **11**:905–911.

•• This review focuses on the mammal protective immune response against *Encephalitozoon* spp.

34. Khan IA, Schwartzman JD, Kasper LH, Moretto M: **CD8+ CTLs are essential for protective immunity against *Encephalitozoon cuniculi* infection.** *J Immunol* 1999, **162**:6086–6091.

35. Moretto M, Casciotti L, Durell B, Khan IA: **Lack of CD4+ T cells does not affect induction of CD8+ T-cell immunity against *Encephalitozoon cuniculi* infection.** *Infect Immun* 2000, **68**:6223–6232.

36. Khan IA, Moretto M: **Role of gamma interferon in cellular immune response against murine *Encephalitozoon cuniculi* infection.** *Infect Immun* 1999, **67**:1887–1893.

37. Moretto MM, Lawlor EM, Khan IA: **Lack of IL-12 in p40 deficient mice leads to poor CD8+ T cell immunity against *E. cuniculi* infection.** *Infect Immun* 2010, **78**:2505–2511.

38. Moretto M, Durell B, Schwartzman JD, Khan IA: **γδ T cell-deficient mice have a down-regulated CD8+ T cell immune response against *Encephalitozoon cuniculi* infection.** *J Immunol* 2001, **166**:7389–7397.

39. Mellman I, Steinman RM: **Dendritic cells: specialized and regulated antigen processing machines.** *Cell* 2001, **106**:255–258.

40. Moretto MM, Weiss LM, Khan IA: **Induction of a rapid and strong antigen-specific intraepithelial lymphocyte response during oral *Encephalitozoon cuniculi* infection.** *J Immunol* 2004, **172**:4402–4409.

41. Moretto MM, Weiss LM, Combe CL, Khan IA: **IFN-γ-producing dendritic cells are important for priming of gut intraepithelial lymphocyte response against intracellular parasitic infection.** *J Immunol* 2007, **179**:2485–2492.

42. Moretto MM, Lawlor EM, Khan IA: **Aging mice exhibit a functional defect in mucosal dendritic cell response against an intracellular pathogen.** *J Immunol* 2008, **181**:7977–7984.

43. Moretto MM, Lawlor EM, Xu Y, Khan IA, Weiss LM: **Purified PTP1 induce antigen-specific protective immunity against *E. cuniculi*.** *Microbes Infect* 2010. [Epub ahead of print].

• This work provides the characterization of the first microsporidian antigen which is able to trigger a protective immune response, and has implication in vaccine development.

44. Fisher J, West J, Agochukwu N, Suire C, Hale-Donze H: **Induction of host chemotactic response by *Encephalitozoon* spp.** *Infect Immun* 2007, **75**:1619–1625.

45. Fischer J, Suire C, Hale-Donze H: **Toll-like receptor 2 recognition of the microsporidia *Encephalitozoon* spp. induces nuclear translocation of NF-κB and subsequent inflammatory responses.** *Infect Immun* 2008, **76**:4737–4744.

46. Lanzavecchia A, Sallusto F: **Regulation of T cell immunity by dendritic cells.** *Cell* 2001, **106**:263–266.

47. Lawlor EM, Moretto MM, Khan IA: **Optimal CD8 T cell response against *Encephalitozoon cuniculi* is mediated bt TLR4 upregulation by dendritic cells.** *Infect Immun* 2010. [Epub ahead of print].

48. Van de Veerdonk FL, Kullberg BJ, van der Meer JWM, Gow NAR, Netea MG: **Host-microbe interactions: innate pattern recognition of fungal pathogens.** *Curr Opin Microbiol* 2008, **11**:305–312.

49. Taupin V, Garenaux E, Mazet M, Maes E, Denise H, Prensier G, Vivarès CP, Guerardel Y, Méténier G: **Major *O*-glycans in the spores of two microsporidian parasites are represented by unbranched manno-oligosaccharides containing alpha-1,2 linkages.** *Glycobiology* 2007, **17**:56–67.

50. Leitch GJ, Ceballos C: **A role for antimicrobial peptides in intestinal microsporidiosis.** *Parasitology* 2009, **136**:175–181.

51. Vorontsova I, Tokarev I, Sokolova I, Glupov W: **Microsporidiosis in the wax moth Galleria mellonella (Lepidoptera: Pyralidae) caused by Vairimorpha ephestiae (Microsporidia: Burenellidae).** *Parazitologiia* 2004, **38**:239–250.

52. Hoch G, Solter LF, Schopf A: **Hemolymph melanization and alterations in hemocyte numbers in *Lymantria dispar* larvae following infections with different entomopathogenic microsporidia.** *Entomol Exp Appl* 2004, **113**:77–86.

53. Sokolova I, Tokarev I, Lozinskaia I, Glupov W: **A morphofunctional analysis of the hemocytes in the cricket *Gryllus bimaculatus* (Orthoptera: Gryllidae) normally and in acute microsporidiosis due to *Nosema grylli*.** *Parazitologiia* 2000, **34**:408–419.

54. Tokarev YS, Sokolova YY, Entzeroth R: **Microsporidia–insect host interactions: teratoid sporogony at the sites of host tissue melanization.** *J Invertebr Pathol* 2007, **94**:70–73.

55. Vijendravarma R, Kraaijeveld AR, Godfray HCJ: **Experimental evolution shows *Drosophila melanogaster* resistance to a microsporidian pathogen has fitness costs.** *Evolution* 2008, **63**:104–114.

56. Antunez K, Martin-Hernandez R, Prieto L, Meana A, Zunino P, Higes M: **Immune suppression in the honey bee (*Apis mellifera*) following infection by *Nosema ceranae* (Microsporidia).** *Environ Microbiol* 2009, **11**:2284–2290.

57. Roxstrom-Lindquist K, Terenius O, Faye I: **Parasite-specific immune response in adult *Drosophila melanogaster*. a genomic study.** *EMBO Rep* 2004, **5**:207–212.

58. Biron DG, Agnew P, Marché L, Renault L, Sidobre C, Michalakis Y: **Proteome of *Aedes aegypti* larvae in response to infection by intracellular parasite *Vavraia culicis*.** *Int J Parasitol* 2005, **35**:1385–1397.

59. Lom J, Nilsen F: **Fish microsporidia: fine structural diversity and phylogeny.** *Int J Parasitol* 2003, **33**:107–127.

60. Lom J, Dyková: **Microsporidian xenomas in fish seen in wider perspective**. *Folia Parasitol* 2005, **52**:69–81.
61. Sitja-Bobadilla A: **Living off a fish: a trade-off between parasites and the immune system**. *Fish Shellfish Immunol* 2008, **25**:358–372.
- This is an interesting review about the balance between microsporidia and fish immunity.

62. Alvarez-Pellitero P: **Fish immunity and parasite infections: from innate immunity to immunoprophylactic prospects**. *Vet Immunol Immunopathol* 2008, **126**:171–198.
63. Troemel ER, Félix MA, Whiteman NK, Barrière A, Ausubel FM: **Microsporidia are natural intracellular parasites of the nematode *Caenorhabditis elegans***. *PLoS Biol* 2008, **6**:2736–2752.
- This work describes a microsporidian species lethal for the model organism *C. elegans*, and opens the way to a new host–microsporidia model.

64. Fasshauer V, Gross U, Bohne W: **The parasitophorous vacuole membrane of *Encephalitozoon cuniculi* lacks host cell membrane proteins immediately after invasion**. *Eukaryot Cell* 2005, **4**:221–224.
65. Lemaitre B, Hoffmann J: **The host defense of *Drosophila melanogaster***. *Annu Rev Immunol* 2007, **25**:697–743.

Chapter 6

Mycorrhizae

J. Dighton
Rutgers University Pinelands Field Station, New Lisbon, NJ, USA

Chapter Outline

Abbreviations 73
Defining Statement 73
Introduction 73
Types of Mycorrhizae 74
Endomycorrhizae 74
Ectomycorrhizae 75
Orchid Mycorrhiza 75
Other Mycorrhizae 76
Mycorrhizal Function 76
Nutrient Acquisition 76
Water Acquisition 77
Plant Defense 77
Ecology of Mycorrhizae 78
Global Distribution of Mycorrhizal Types and Soil Nutrients 78
Influence of Mycorrhizae on Plant Communities 78
Mycorrhizae as Food 79
Mycorrhizae and Plant Production 79
Agriculture 79
Commercial Forestry 79
Restoration 80
Mycorrhizae and Pollution 80
Acidifying Pollutants 80
Heavy Metal Pollutants 80
Radionuclide Pollutants 81
Organic Pollutants 81
Climate Change 81
Fungal Conservation 81
Recent Developments 82
Further Reading 82
Relevant Website 83

ABBREVIATIONS

MHB Mycorrhizal helper bacteria

DEFINING STATEMENT

Mycorrhizal symbiosis come in a variety of types that influence plant growth and community composition by changing nutrient and water uptake and providing defense from root grazers and pathogens. Mycorrhizal fungi are important food sources for animals. Their interactions with pollutants make them potentially useful for restoration and remediation.

INTRODUCTION

Terrestrial fungi appear to have emerged at about the same time as land plants. In addition to their role as saprotrophs, some fungi became intimately associated with roots of plants, enhancing their abilities to sequester nutrient elements. This became a symbiotic association known as mycorrhizae, which has evolved in a number of directions, forming contrasting morphological changes to root structure and providing different ecological services to different groups of plants. Fossil records show these associations as possibly primitive endomycorrhizae in the Rhynie cherts (410–360 Ma) and as ectomycorrhizae of pines in the Princeton cherts (50 Ma). The term mycorrhiza is derived from the combination of two Greek words – 'mykos' meaning fungus and 'rhizos' meaning roots. Thus, the 'fungus roots' have been distinguished as a specialized adaptation of plant roots, occurring in some 85% of all plants species on this planet. Recent estimates suggest that some 3617 plant species of 263 families have a mycorrhizal association. Thus, it is regarded that the mycorrhizal condition is the most prevalent symbiotic condition on earth.

The efficiency of the mycorrhizal state is related to the enhanced inorganic nutrient uptake by plants. The fungal hyphae emanating from the root surface is able to explore a larger volume of soil than roots and root hairs alone could do. To make this a true symbiosis, the benefit for the fungus

Eukaryotic Microbes.

is a supply of carbohydrates from the photoassimilates of the host plant to support fungal growth. The actual efficiency is defined as the amount of carbon gained by the plant minus the carbon expended to support the mycorrhizal fungal partner. This efficiency is dependent on a variety of abiotic and biotic factors. There is also evidence to suggest that the function of mycorrhizae goes beyond that of pure nutrient acquisition, to include access of less readily labile nutrients (organic nutrient sources), enhanced root access to water, and protection of roots from pathogenic bacteria and fungi and from grazing of soil invertebrates. An additional attribute of mycorrhizae is their ability to provide protection to plants from heavy metals by limiting translocation of metals into aboveground plant parts.

Due to the nutrient uptake-enhancing properties of mycorrhizae, the ability of mycorrhizae to stimulate plant growth has been used in agriculture and forestry. Particularly in nursery conditions, the addition of mycorrhizal fungi at an early stage of plant growth has been shown to yield larger plants in comparison with uninoculated plants. The findings that mycorrhizal inoculation of plants in polluted soils often enhances plant survival and growth have led to the commercial production of mycorrhizal inocula for use in horticulture, agriculture, forestry, and restoration.

TYPES OF MYCORRHIZAE

The mycorrhizal association is made between fungal hyphae and the plant root. Fungal hyphae and fungal spores in soil can act as inocula for roots, and there is evidence to show that the juxtaposition of a root to a spore stimulates spore germination in response to root exudates. Current research is revealing the signaling systems that are involved in host/fungal recognition, where by the plant and fungi recognize their compatibility and, given appropriate environmental factors, will result in the fungal hyphae associating with the root to form the morphological structures associated with that specific symbiosis. In arbuscular mycorrhizae, plant-exuded chemical signals, such as flavonoids, strigolactones, and surface or thigmotropic signals, are recognized by receptor proteins on the fungal plasma membrane. One of these receptors may be the *Gin1* gene, which interacts with ATPase to initiate an internal fungal signaling process allowing the fungus to enter a symbiotic mode. In ectomycorrhizae, it appears that a combination of endogenous rhythms of growth flushes in both shoot and root (carbohydrate supply) and the expression of fungal-stimulated genes that are not present in the root alone stimulates the initiation of the symbiosis. The ability of fungi and plants to form mycorrhizae may be assisted by the presence of, particularly, fluorescent pseudomonad bacteria. These bacteria help to elicit mycorrhizal formation by production of cell wall softening enzymes, possible chelating agents, and chemicals to elicit the plant–fungus recognition system. These bacteria have been termed mycorrhizal helper bacteria (MHB).

There are varying degrees in specificity of plant–fungal association in mycorrhizae and dependency of the plant on mycorrhizal associations. A number of plant species and families will only associate with a limited number of fungal species, leading to specific mycorrhizal types, such as ericoid, arbutoid, orchid, and monotropid mycorrhizal associations being highly specific to limited plant families. A large number of grasses, herbs, and trees form associations with a relatively restricted fungal flora to form arbuscular mycorrhizal associations, whereas a more limited set of plant species (mainly trees) associate with a vast diversity of fungal species in the ectomycorrhizal state. Even within these broad categories of specificity, there is specificity within plant and fungal families. Some plants are heavily or entirely dependent upon mycorrhizal associations for their survival (obligate mycosymbionts), whereas others may only associate with mycorrhizal fungi under times of need (facultative mycosymbionts). Certain fungal species may only associate with one plant species (e.g., the European larch will only associate with *Suillus grevillei*), whereas others may have broad host specificity. The factors determining these degrees of specificity are not clearly understood, but it is likely that a combination of genetics, evolution, and environmental factors are involved.

Endomycorrhizae

As the name suggests, endomycorrhiza is a general term for all mycorrhizal associations where the fungal component is predominantly internal to the root structure, with fungal penetration into host cortical cell walls. Two major groups are the arbuscular and ericoid mycorrhizae, but others, such as the dark septate endophytes, may also be included in this category. In all of these mycorrhizal forms, there is hyphal penetration into the host cells.

Arbuscular mycorrhizae

This group of mycorrhizae are formed between a limited number of fungal species (~ 150) of the phylum Glomeromycota with a very large number of vascular plant species, including grasses, herbs, and tress – particularly tropical tree species.

Fungal hyphae penetrate the epidermal cells by a combination of enzyme activity and hydrostatic force, leaving an appressorium as a hyphal swelling on the root surface where pressure builds up. The hyphae then penetrate through cortical cell walls, pushing aside the plasma membrane and branch into characteristic arbuscules, tree-like structures, to maximize the area of contact between the fungus and host cell contents. This large surface area facilitates nutrient and carbon exchange between the fungal and plant component of the symbiosis. There are classically two types of arbuscule development, the Arum and Paris type. The Arum type develops multibranched hyphal structures that form tree-like arbuscules, whereas the Paris type consists of extensive hyphal coils in the host cells to form their arbuscules.

As a result of colonization of root tissue by the mycorrhizal fungus, root hair development is suppressed as extraradical hyphae (hyphae running from the root surface into soil) effectively take over the role of root hairs to increase the absorptive surface area.

In some fungal genera, excluding *Gigaspora* and *Scutellospora*, vesicles may be formed within the root tissue. These are fungal structures that completely fill the host cell and are stained by lipid stains. The presence of vesicles gave rise to the older name of vesicular-arbuscular mycorrhizae. These vesicles are terminal hyphal swellings that contain many nuclei and lipid bodies. These are thought to be involved with material storage.

Asexual spores may be produced either externally or within the root. Spores of different fungal species are of contrasting size and have characteristic ornamentation and layering of chitin filaments in the spore wall that allow fungal species identification. These spores are dispersed in the air, by water, or by grazing soil animals.

Ericoid mycorrhizae

Ericoid mycorrhizae are a restricted group of fungi associated with a restricted diversity of plant species in the Ericaceae, Epacridaceae, and Empetraceae. *Hymenoscyphus* (*Pezizella*) *ericae* was the first fungal species identified as an ericaceous endosymbiont. However, more recently a number of other fungal genera (*Oidiodendron*, *Myxotrichium*, and *Gymnascella*) have been identified as forming mycorrhizal associations with ericoid plants. In a similar manner to arbuscular mycorrhizae, the fungal hyphae invade cortical cells, usually of very fine roots, in which they form hyphal coils, rather than arbuscules. These fine roots of ericoid plants consist of a vascular bundle and one outer layer of cortical/epidermal cells.

Ectomycorrhizae

As their name suggests, ectomycorrhizae have a significant proportion of their fungal partners biomass external to the root. This comprises two parts – the sheath or mantle of fungal hyphae that wrap around the outside of the root and the extraradical hyphae and hyphal structures that extend into the surrounding soil. The sheath can be of variable complexity, from a loose weft of hyphae to a thick and structured multilayer of cells, which have the appearance of plant parenchyma tissue; termed pseudoparenchymatous. Internal to the root structure, but external to the cortical cells is a layer of hyphal penetration between cortical cells inward to the endodermis. This is termed the Hartig net and is the region of juxtaposition of fungal and plant symbionts where nutrient and carbon exchange occurs. The colonization of root tips by ectomycorrhizal fungi characteristically suppresses root hair development and changes root branching by the induction of altered levels of cytokinins, resulting in increased branching. Branching patterns are varied and can consist of simple bifurcations, through pinnate and pyramidal, with ultimate branching to such an extent that multiple (50 or so) root tips may be enveloped by a continuous hyphal sheath, a condition called coralloid or tuberculate. The fungal hyphae also impart their characteristic color to the mycorrhizal root surface. The degree of branching, surface color, degree of complexity of the sheath, and character of emanating structures (hyphae, rhizomorphs, cystidia, etc.) serve as morphological characteristics by which mycorrhiza may be identified. More recently, molecular analysis of mycorrhizae, together with morphological characteristics, are being used to create a more comprehensive database of mycorrhizal identification.

In the same way as extraradical hyphae increase the surface area of arbuscular mycorrhizal roots, this is taken to a greater degree by many ectomycorrhizae, where extraradical hyphae can extend considerable distances from the root surface. This is particularly the case where congregations of hyphae, rhizomorphs, which are structured entities for rapid, long-distance translocation of nutrients, and water may extend meters from the root surface into soil. In some species and habitats, the hyphal extension is extensive, forming fungal mats, which can alter the soil physical conditions, particularly making them hydrophobic.

The ectomycorrhizal association may be formed by a range of Basidiomycotina, Ascomycotina, and some Zygomycotina, in which about 5500 species have been identified as mycosymbionts. Many of the fungal associates are common forest and woodland mushrooms (e.g., *Russula*, *Hebeloma*, *Cortinarius*, *Lactarius*, *Laccaria*, *Amanita*, and *Lycoperdon*) and truffles (e.g., *Tuber*). These fungi associate with a limited number of tree species in all biomes. Tree genera include most coniferous trees, larch, birch, beech, oak, and eucalypts.

In addition to pure ectomycorrhizae, pines and larches can produce a mycorrhizal type having characteristics of both ectomycorrhizae and arbuscular mycorrhizae, a condition known as ectendomycorrhiza. This association has been attributed to what is called E-strain fungi and is likely to be due to members of the Pezizales (*Wilcoxina* spp. and *Sphaerosporella brunnea*).

Orchid Mycorrhiza

Orchid mycorrhizae may be considered the epitome of plant fungal interdependency in the mycorrhizal world. Some 17 000 species of orchids exist and depend upon their basidiomycete fungal partners for acquisition of nutrients.

In some cases, the dependency of the plant on its fungal partner has become so extreme that fungal propagules are carrier within the seed of the orchid in order to be present at seed germination and enhancing nutrient uptake at an early stage of plant development. This is important where seed size is so small as to limit the nutrient reserve that can be carried in the absence of an endosperm. Initial protocorm development after seed germination is dependent on the symbiotic fungus also for carbohydrate supply, until photosynthetic capacity can be developed. This may take up to a year in some species, and in achlorophyllous orchid species, it never develops. The fungi develop highly coiled arbuscules (peletons) within the cortical cells of the host plant. These fungal coils have a finite lifespan and upon death, deposit cellulose and pectin within the host cell. These cells may be subsequently 'invaded' by new hyphae that can access these carbohydrate and nutrient supplies.

Other Mycorrhizae

Arbutoid

This specific mycorrhizal association occurs between two specific members of the Ericaceae, *Arbutus* and *Arctostaphylos*, and several genera in the Pyrolaceae. A number of ectomycorrhizal fungal species have been established to form these arbutoid relationships, which consist of a very thin fungal sheath surrounding the outside of the root, a paraepidermal Hartig net consisting of fungal penetration between the epidermis, and outer layer of cortical cells into which the hyphae invade to form hyphal coils (intracellular hyphal complexes). It is suggested that there is exchange of nutrients and photosynthates between arbutoid plants and adjacent tree species.

Monotropoid

These mycorrhizal types are restricted to the Monotropoideae in the family Ericaceae, which have largely lost their photosynthetic capacity and live as achlorophyllous plants on the forest floor. At first thought to be parasitic on forest tree species, it is now known that they share mycorrhizal symbionts that are common with their neighboring trees. In this way, these plants obtain carbohydrates from adjacent trees through mycorrhizal bridges between their root systems. Fungal species identified forming these associations include *Tricholoma*, *Russula*, and *Rhizopogon*. On their monotropoid hosts, these fungi form fungal sheaths and Hartig nets in a similar manner to the ectomycorrhizal condition on trees. However, they also produce hyphal pegs from the inner sheath layer of hyphae into the tangential wall of host cortical cells, with one peg per cortical cell. The abundance of peg formations appears to be related to growth and development of the host plant, increasing up to flowering and subsequently declining. The classic work of Björkman in the 1960s, using radiotracers, established that there was movement of photosynthates from trees into *Monotropa*. Subsequent work has shown that this occurs via mycorrhizal bridges between the shared fungal hyphae forming mycorrhizae with both the *Monotropa* plant and the adjacent trees. The nutrient acquisition by monotropoid mycorrhizae has been less well studied, but it may be assumed that it is similar to that of the ectomycorrhizal condition.

Dark septate endophytes

These root endophytes have been recognized in a number of plant species from almost all plant families, particularly those growing in cool, nutritionally poor environments. Many of the fungal species have not been identified, but members of the genera *Chloridium*, *Leptodontidium*, *Phialocephala*, and *Phialophora* have been identified from roots. These fungi enter the root via root hairs or cortical cells and establish runner hyphae between cortical cells from which dense, multibranched structures, called microsclerotia, develop within the host cell. Initially hyaline hyphae frequently deposit melanin in their cell walls resulting in the dark color that gives these fungi their name. Although these mycorrhizal types are frequently found, there is relatively little information on their function. However, given their greater abundance in oligotrophic and climatically limited environments, it is likely that they have the enzymatic competence to access nutrients for inorganic sources (see 'Influence of Mycorrhizae on Plant Communities').

MYCORRHIZAL FUNCTION

The traditional concept of mycorrhizal function is that they enhance inorganic nutrient uptake from soil into the host plant. Indeed this is true, but may be an oversimplification of their function as there is evidence of multiple host benefits of mycorrhizal association. However, most of these functions have been shown under contrived experimental conditions and the magnitude of these effects in natural ecosystems has recently been questioned.

Nutrient Acquisition

Numerous studies have shown that plants grown in the presence of mycorrhizae grow larger than those grown in the absence of mycorrhizae. In an elegant study on arbuscular mycorrhizae, Nye and Tinker used radioactively labeled phosphate in soil and autoradiography to demonstrate that the phosphate uptake capability of mycorrhizal roots was larger than nonmycorrhizal roots. They demonstrated that the nutrient depletion zone developing around a mycorrhizal root system was greater than that for a root with root hairs due to the greater exploratory capacity of multiple, small diameter hyphae taking up phosphorus from the soil pore water. It has been suggested that the energetics of producing extensive and far-reaching hyphae in place of root hairs provides a larger absorptive surface for nutrient acquisition at less expense to the plant. However, the energetic cost–benefit hypothesis has recently been challenged.

Carbon is made available to the fungus by way of photosynthates. In some ectomycorrhizal associations, this may account for some one-third of the plant's photoassimilates. Both the plant and the fungus increase their hexose importer gene function to accomplish this carbohydrate exchange. Sugar-dependent gene expression has been shown to provide a number of physiological functions of ectomycorrhizae, such as defense against faunal grazing and to pathogenic bacterial and fungal attack. In arbuscular mycorrhizae, phosphate acquisition is achieved through membrane integral proteins, including PHT phosphate transporter and the P-type H^+-ATPase. Following uptake, transport within fungal tissues of both arbuscular and ectomycorrhizae is mainly in the form of polyphosphates, which may also be deposited in cortical cells as nutrient reserves in polyphosphate granules. Within arbuscular mycorrhizae, specific *MtPT4* protein production is closely associated with arbuscule formation, and thus with phosphorus uptake. There is similar genetic regulation of nitrogen uptake. In ectomycorrhizal association with *Hebeloma crustuliniforme*, three ammonium transporters and one nitrate transporter have been identified for inorganic nitrogen uptake. Both amino acid and polypeptide transporter genes along with protease and subtilase genes have been identified for acquisition of organic nitrogen.

Enhanced nutrient uptake by mycorrhizae has led to the assumption that the growth benefit of mycorrhization was entirely nutrient driven. Indeed, there is plenty of evidence from experimentation, agricultural plant production, and forest nurseries to show the nutritional benefits of mycorrhizae for improving plant growth (height and stem diameter), foliar nutrient content and mass, and nutrient content of plant products (peanuts, grain, etc.). In addition, it has been shown in a number of cases that the presence of mycorrhizae improves plant survival, especially at the establishment phase. As a result of these growth benefits, mycorrhizae are being used commercially for enhancing growth of agricultural crops, forest trees in nurseries, and in horticulture. This has resulted in a recent increase in the number of companies producing and selling mycorrhizal inoculum for arbuscular mycorrhizae, ectomycorrhizae, and, more recently, ericoid mycorrhizae. However, there are a number of nonnutritional benefits of mycorrhizal associations.

Water Acquisition

The ability of mycorrhizal extraradical hyphae and, in particular, rhizomorphs of ectomycorrhizae to translocate nutrients from soil to plant also enables them to translocate water. This has also been found to be of great importance for plant growth in dry environments. Water uptake and the interaction between water and nutrient acquisition is plant species-dependent. For example, the presence of arbuscular mycorrhizae on *Acacia* did not improve plant growth under drought stress in both the absence and the presence of addition of P fertilizer. However, in another tropical tree *Leucaena*, the presence of mycorrhizae significantly improved plant growth under drought conditions at both levels of P supply. The finding of mycorrhizal associations of *Acacia* roots at depths of 30 m in Senegal is likely to be more associated with water acquisition than that of nutrients and may be linked to enhancing hydraulic lift. Recently it has been shown that the colonization of roots by arbuscular mycorrhizae in xeric systems increases the expression of plasma membrane aquaporin (PIP) genes to increase water flow into host cortical cells.

Plant Defense

The presence of mycorrhizae within the root provides some degree of protection from both grazing soil fauna and plant pathogens. This defense appears to take two forms. One form of defense is the production of a physical barrier to root tissue created by an ectomycorrhizal sheath. The other is a biochemical defense mechanism in which secondary metabolites of fungi defend the mycorrhizal root against pathogenic fungi and bacteria. For example, the presence of *Glomus mosseae* arbuscular mycorrhizae on peanuts significantly improved plant growth (28%), pod production (22%), and seed weight (12%). In the presence of two fungal pathogens, *Fusarium solani* and *Rhizoctonia solani*, the presence of mycorrhizae negated the growth suppression of the pathogens for all parameters and increased production over the pathogen alone by 26% (plant weight), 35% (pods per plant), and 39% (seed weight). In seedling studies, the ectomycorrhizal fungi *Laccaria laccata*, *Paxillus involutus*, *H. crustuliniforme*, and *Hebeloma sinapizans* significantly reduced the effect of the root pathogen *Phytophthora cinnimomi* on the growth of chestnut tree seedlings. These disease prevention attributes of mycorrhizal fungi are important in both the agricultural context and tree nurseries, where both the density of plants and the intensive management practices are favorable for plant pathogen development. The role of mycorrhizae in disease prevention in more natural systems has not been adequately quantified. In a meta-analysis of published data on the interactions between mycorrhizae and fungal pathogens and nematodes, Borowicz, in 2001, showed that mycorrhizal fungi inhibited the pathogen and nematode effects on plants. However, the defense against nematode attack was heavily biased by the large numbers of references to a sedentary nematode, *Heterodera*, suggesting that mycorrhizal defense against other nematode species may be less than reported.

ECOLOGY OF MYCORRHIZAE

Global Distribution of Mycorrhizal Types and Soil Nutrients

Naturally, the distribution of mycorrhizal types follows closely the global distribution of appropriate plant host species. However, Read suggested an underlying factor that linked both plant and mycorrhizal symbionts with soil factors. Using the analogy of an altitudinal cline up a mountain to the variability in soil characteristics from poles to the equator, Read suggested that the dominant mycorrhizal types (ericoid mycorrhizae, ectomycorrhizae, and arbuscular mycorrhizae) were closely related to the nutrient sources and availability in soil and the enzymatic competence of the mycorrhizal community. Organic matter in soil, derived from dead plant and animal remains, provides soluble mineral nutrients in soil as a result of decomposition and mineralization processes carried out by saprotrophic bacteria and fungi, with the help of soil animals. In colder latitudes and higher altitudes, organic matter accumulates. Here, in colder and wetter environments, litter quality is reduced by the presence of complex chemistry and toxic secondary plant metabolites and where the climatic window of opportunity for decomposition is restricted. In these conditions, soil nutrient capital can be high, but availability as soluble nutrients is low and plants are frequently ericaceous with ericoid mycorrhizae. These mycorrhizae benefit their host plant not only by enhancing soluble nutrient uptake, but also by possessing protease enzymes, which allow them to directly access organic forms of nitrogen. In more mesic environments supporting coniferous and mixed coniferous and deciduous forest, the ability of the predominantly ectomycorrhizal fungal community allows access to organic forms of both nitrogen and phosphorus by the production of protease and acid phosphatase enzymes. In most of the arbuscular mycorrhizal-dominated plant communities of deciduous trees and grasslands and at lower altitudes and in the tropics, nutrient mineralization is more rapid, the pool of inorganic nutrients is larger, and the mycorrhizal community has limited enzymatic capabilities. Thus, the link between host plant and mycorrhizal type appears to be a mixture of fungal–host compatibility and environmental limitations.

The interaction between soil conditions and mycorrhizae can be seen during forest growth where changes in dominant ectomycorrhizal fungal species in the community change with tree age. As the forest floor accumulates more tree-derived and recalcitrant woody material, there is a shift from fungal species that efficiently scavenge for inorganic nutrients in soil to those with enzymatic capabilities to decompose organic complexes to access the mineral nutrients within. This short-circuiting of the saprotrophic community mineralizing nutrients has been named the 'direct nutrient cycling hypothesis.' There is still debate in the literature regarding the interactions between mycorrhizal and saprotrophic fungi in this role, where some literature support the idea that mycorrhizal fungi compete against saprotrophs in low inorganic, high organic nutrient conditions and others suggest they work in harmony.

Influence of Mycorrhizae on Plant Communities

Plant establishment and growth is enhanced by the ability of the plant to form mycorrhizal associations. Within a plant community, not all plants have the same capacity to form mycorrhizae, nor the same dependency on mycorrhizae for their survival. Hence, the outcome of competition between plant species in a community is a complex series of interactions with the environment and availability and effectiveness of mycorrhizal fungal symbionts. These interactions have been explored by Bever and Schultz, in 2005, who show that a range of scenarios of plant–mycorrhizal strategy can influence the growth and competitive ability of different species in a community, which results in niche partitioning and the maintenance of diversity within the plant community. The relative dependence of a plant species on mycorrhizal fungi for nutrient uptake may also influence plant community composition depending on availability of nutrient resources, plant competition, and other environmental factors influencing mycorrhizal and plant community assembly rules. Indeed, diversity of mycorrhizal community often increases correspondingly to that of plant species diversity.

Following the classic research of Björkman who used radioactive carbon to trace carbohydrate flow from trees to the forest floor achlorophyllous plant, *Monotropa*, via mycorrhizal linkages between the two plants, interest has arisen about the potential for exchange of information between plants via mycorrhizae. The ability of like species of plants to share arbuscular mycorrhizal bridges between their root systems has been shown to benefit the status of that species in the community by the ability to move photosynthates and nutrients from dying plants to live members of the population. The discovery that a similar sharing of information may occur between plants of different species, in both the arbuscular and ectomycorrhizal plant community, has connotations that change our view on the rules for establishing plant community composition. Given that a 'nurse effect' between plant species can be established by sharing resources via mycorrhizal bridges and 'cheater' effects where plants effectively steal resources from others via mycorrhizal bridges invokes the concept of both competition and synergism being involved in regulating plant communities. This is in contrast to the classical plant community theory of competition alone as being the main driving force influencing plant community assembly rules.

Mycorrhizae as Food

Fungi are rich in important nutrients, particularly nitrogen, phosphorus, minerals, and vitamins, and have been reported to be as good as beef as a food resource. Thus, the fruitbodies of ectomycorrhizal fungi basidiomycete mushrooms are grazed upon by a variety of invertebrates including mollusks and fly larvae. Many fly larval species are even tolerant of the toxic component of *Amanatia* spp., α-amanitin. Additionally, many mushrooms and, particularly, hypogeous fruitbodies (e.g, truffles) are an important food source for small mammals. In Australia, 37 species of native and 4 species of feral mammals exhibit extensive mycophagy, where fungi may comprise more than 25% (by volume) of the diet of brush-tailed potoroo (*Potorus longipes*) at all times of the year. In old growth Douglas fir forests in western Oregon, 11 052–16 753 (some 2.3–5.4 kg ha^{-1} dry mass) fruiting bodies of hypogeous mycorrhizal fungi can be produced per hectare per year. As these fungi have a higher content of nitrogen, phosphorus, potassium, and micronutrients than epigeous fungi (fungi fruiting aboveground), it makes them a high quality food resource for mammals. Indeed, it has been shown that small mammals foraging in adjacent intact forest for ectomycorrhizal fungi carried spores and propagules into refugia in the ash fields of Mount St. Helens to significantly enhance revegetation by trees following the volcanic eruption.

Belowground, the mycelia of both ectomycorrhizal and arbuscular mycorrhizal fungi are important food sources for a variety of soil animals. Populations of nematodes and Collembola (springtails) are maintained by fungal consumption. Not all fungi are, however, equally palatable or provide adequate nutrition, resulting in hierarchies of feeding preference. There continues to be some controversies regarding the influence of soil animal grazing on mycorrhizal effectiveness with suggestions that severing hyphal connections between the root and surrounding soil reduces nutrient influx. It is likely that natural densities of grazing animals have minimal effects.

An indirect effect of arbuscular mycorrhizae on animals can be seen by mycorrhizal enhancement of host plant nutritional status and grazing herbivores. The presence of *Glomus* spp. mycorrhizae on *Lotus corniculatus* significantly reduces larval mortality and increases growth rates of the leaf-grazing larvae of the butterfly *Polyommatus icarus*.

MYCORRHIZAE AND PLANT PRODUCTION

Agriculture

Mycorrhizal growth-enhancing properties have been explored in agricultural settings. However, due to the application of fertilizer in most farming practices, the impact of the fungal inoculum is variable due also to the fact that there is frequently an abundance of native fungal propagules in the soil and that high fertilizer applications suppress mycorrhizal development in roots. The beneficial effects of enhanced plant nutrition often go hand in hand with protective benefits of mycorrhization to protect crops from root-feeding nematodes and bacterial and fungal pathogens. One area in which mycorrhizal associations are of benefit is in high saline soils, especially where irrigation and evaporation lead to salt accumulation. In maize, halotolerance appears to be induced by the mycorrhizae altering root architecture by reducing specific root length at small root diameters (>0.2 mm diameter) and increasing that of larger diameter roots (0.2–0.6 mm). A variety of methods of mycorrhizal delivery systems have been developed for commercial mycorrhizal inoculum production. These include incorporating arbuscular mycorrhizal spores into pellets or producing large numbers of spores in a loosely packed matrix (rockwool, sand, clay-brick granules). Due to demand for mycorrhizal inoculum for both arbuscular and ectomycorrhizal plants, a number of companies have developed, selling their mycorrhizal inoculum to gardeners, horticulturalists, and forest nurseries (e.g., Horticultural Alliance Inc.; Soil Moist; Mycorrhizal Products).

Commercial Forestry

The fact that mycorrhizal inoculation of tree seedlings enhances growth under experimental conditions has stimulated an interest in using ectomycorrhizal and arbuscular mycorrhizal inoculum in commercial forestry. The fungi (especially ectomycorrhizae) have to be the species that are readily culturable and easy to grow in bulk. These may be made commercially available in perlite–peat mixture or in alginate-entrapped mycelial cultures. The growth benefit provided by commercial inocula in the nursery has been demonstrated on many occasions, and the survival and enhanced growth of seedlings planted into harsh environments has been shown. For example, the presence of the ectomycorrhizal fungus *Pisolithus tinctorius* on tree seedlings allows improved survival and growth on heavy metal-contaminated mine spoils. However, generally outplanting tree seedlings that have been inoculated with mycorrhizal fungi often do not provide a significant growth enhancement – a fact that varies by tree species and locality. Part of this problem may lie with the fact that in a nursery setting, nutrients and water are rarely limiting. Hence, plants growing in rich soil conditions with poorly competitive, ruderal, mycorrhizal fungal associates are outplanted into poor soils. It would appear that in many instances, there is a replacement of the nursery-applied mycorrhizae with a native flora of the site of planting. During the time that this mycorrhizal competition occurs, any benefit of mycorrhizal colonization is lost and tree performance is reduced until a new mycorrhizal community has been established, which has the physiological adaptations to the site. A comprehensive review of the literature suggests that the benefits of nursery application of mycorrhizal inoculum to tree seedlings may not be as advantageous as may be thought.

Restoration

The ability of mycorrhizae to promote enhanced growth and establishment of plants in disturbed habitats suggests that mycorrhiza can be important for restoration projects. As we see below, many mycorrhizae have the ability to protect the host plant from pollutants, but in the context of restoration, the ability of mycorrhizal association to improve soil structure, such as aggregate stability, may be of equal importance. First, mycorrhizal fungi assist the establishment of vegetation on, usually, poor soils in a restoration site. This allows the return of plant parts to soil, establishing the organic component of a more healthy soil. Second, the ability of fungi to bind soil mineral particles together either by physical means or by the production of the sticky glycoprotein, glomalin, by some arbuscular mycorrhizal fungal species increases soil stability and prevents soil erosion. The combination of both the factors increases soil aggregation and the storage of carbon as protected organic matter within soil aggregates. Mycorrhizal fungal assisted long-term carbon sequestration in soil aggregates is a potentially valuable resource in a world of increasing atmospheric CO_2.

In some agricultural contexts, particularly in tropical areas, frequent crop irrigation leads to an accumulation of salt and the development of highly saline soils. In these conditions, arbuscular mycorrhizae have been shown to have significant benefit for crop production enabling plants to grow more effectively under these stressed conditions. This allows crop production to continue in areas that would otherwise be abandoned.

MYCORRHIZAE AND POLLUTION

Acidifying Pollutants

It is in the late 1970s when soil ecologists became involved in the research on acid rain (sulfuric acid from dissolved SO_2 in rain) in relation to the 'Waldsturben' effect of the forest die-back in Bavarian forests, where the observations of Ulrich, Huttermann, and Blaschke alerted researchers to the fact that acid rain was affecting root growth and the ectomycorrhizal status of trees. Based on the idea that mycorrhizal formation was affected by both carbohydrate supply and nutrient levels in soil, which in turn influence hormone levels in roots, a two-directional impact model of acidifying pollutants on the development of mycorrhizae was developed. In this model, reduction of mycorrhizal associations of the plant roots occurred (1) via a reduction in photosynthesis in the tree canopy, reducing the energy supply to roots and (2) via acid-induced increase in the availability of toxic metal ions (aluminum, manganese, and magnesium) in soil, resulting in root damage and loss or changes in community composition of ectomycorrhizal fungal symbionts. This model has confirmation from a number of studies showing that ectomycorrhizal community composition changes with increasing acid rain loading and evidence that the photosynthesis is significantly reduced by acid rain.

Using historical mushroom foray records, Arnolds detected changes in the ratio of ectomycorrhizal to saprotrophic basidiomycete fruitbody abundance in The Netherlands over recent years. He attributed this to enhanced nitrogen deposition that acted both as a fertilizer (reducing the effectiveness of mycorrhizae for nitrogen uptake into host plants) and as an acidifying pollutant. Both experimental and observational studies have shown that community shifts in ectomycorrhizae occur with increasing nitrogen loading to forest ecosystems, where the shift from nitrophobic to nitrophilic species could be used as an indicator of nitrogen pollution.

Heavy Metal Pollutants

Heavy metals are known to be toxic to many living organisms. Fungi are no exception to this; however, fungi have a degree of tolerance to heavy metals, which is especially apparent in ectomycorrhizae. In the late 1970s, Don Marx observed enhanced survival and growth of pine trees in mine spoil soils with trees inoculated with ectomycorrhizal fungi. Of particular interest was the fungus *P. tinctorius*, which appeared to be more frequent in these polluted sites than in other habitats. The effect of inoculation with *P. tinctorius* resulted in tree volumes 250% greater than those trees assuming natural inoculum from the site or inoculation with *Thelephora terrestris*. These trees also had higher foliar phosphate levels, but reduced levels of Ca, S, Fe, Mn, Zn, Cu, and Al, suggesting that the effect of this mycorrhizal fungus may reduce the uptake of heavy metals into the host tree. Similar prevention of heavy metal toxicity to host plants afforded by mycorrhization has also been seen in ericaceous plants of the genera *Calluna*, *Vaccinium*, and *Rhododendron*.

The mechanism of plant protection in the ectomycorrhizal system was elucidated by Denny and Wilkins. Using electron microscopy, coupled with X-ray diffraction (EDAX), they identified adsorption of heavy metals onto fungal hyphae in the extraradical hyphal network, the fungal sheath, and Hartig net, preventing translocation of the metal into the host cortex and, particularly, preventing movement into the vascular tissue. Some of this binding capacity is related to extracellular slime formed by the hyphae of the ectomycorrhizal fungus *P. tinctorius* and some to Cu and Zn complexing with polyphosphate granules, which metabolically inactivate the heavy metals within the fungal hyphae. Similar protection has been seen in arbuscular mycorrhizae, where the rate of uptake of heavy metal was increased in the presence of mycorrhizae, but the transfer of metals to the host plant was reduced, effectively locking metals up in the fungal component of the mycorrhizal association. Protection in arbuscular mycorrhizae has been attributed to oxidative stress alleviation, where the production of zinc transporter genes, metallothionien, a 90 kDa heat shock protein, and glutathione have been shown to be induced in extraradical hyphae of *Glomus intraraices* in the presence of heavy metals.

Radionuclide Pollutants

In the same way that fungi can accumulate heavy metals, it seems that they have the capacity to accumulate radionuclides. Accumulation into fruitbodies of basidiomycetes is often higher than surrounding soils. For example, facultative mycorrhizal fungal fruitbodies had ten times the concentration of radiocesium than leaf litter or organic soil horizons in Sweden following the Chernobyl accident, whereas saprotrophic fungal fruitbodies had half that level of accumulation. Many of these hyperaccumulators are ectomycorrhizal species, for example, members of the Cortinariaceae.

The analysis of the isotope ratio of radiocesium from Chernobyl (^{137}Cs:^{134}Cs) in fruitbodies of ectomycorrhizal basidomycete fungi indicates that a large proportion (25–92%) of ^{137}Cs was accumulated that originated from sources occurring prior to the accident at Chernobyl. This suggests long-term accumulation of radionuclides by these fungi. Research suggests that this is achieved by internal translocation within the mycelium and directional transport to fruiting bodies, along with other nutrients.

In both ericoid and arbuscular mycorrhizal host plants, there appears to be a mycorrhizal effect of enhancing radionuclide uptake by the plant, but a significant change in the internal allocation of those radionuclides. Radionuclides appear to be prevented from being translocated into the shoots of plants by these mycorrhizae – a possible defense mechanism.

Given the fact that mycorrhizal fungi, in association with their host plants, have the capacity to absorb and retain high levels of heavy metals and radionuclides, there are suggestions that they may be beneficial in the restoration and remediation of polluted environments. In particular, the high accumulation into basidiomycete fruiting structures in ectomycorrhizal fungi could be used as a harvestable product that could be exported from the site, resulting in a net export of the pollutant chemical.

Organic Pollutants

In a screening of 21 mycorrhizal fungi for PCB decomposition, 14 species were found to be able to decompose some of the PCBs by at least 20%. The ectomycorrhizal fungi *Radiigera atrogleba* and *Hysterangium gardneri* were able to degrade 80% of 2,2′-dichlorobiphenyl and two ericoid mycorrhizae, *Hymenoscyphus ericae* and *Oidiodendron griseum* had decomposer abilities, but they were less effective than the ectomycorrhizal species. The interaction between ectomycorrhizal fungi, rhizospheric bacteria, and the decomposition of organic pollutants is a new area of study.

Climate Change

Climate change can encompass a number of changes in our environment, including increase in atmospheric CO_2 levels, increased temperature, and changes in rainfall amount and distribution pattern (both spatial and temporal). The impacts of these factors on mycorrhizae are in their infancy, but it is clear that changes in CO_2 concentrations in the atmosphere influence the C:nutrient ratio of plant material that enters the decomposition pathways. As a result, the change in nutrient availability causes changes in both the composition of mycorrhizal communities as nutrient content is reduced. The response of mycorrhizae to rainfall, ozone, UV light, as well as CO_2 appears to be unclear from the few studies carried out to date.

Fungal Conservation

Long-term fungal foray records are a potential source of information about long-term trends in abundance of fungal fruitbodies. Using such records, Arnolds discovered that the fungal flora of The Netherlands had changed from being dominated by mushrooms of ectomycorrhizal fungal species to that of saprotrophs. By overlaying information about levels of acidifying pollutants (particularly nitrogen deposition), combined with the knowledge of the effects of acidifying pollutants on mycorrhizae, he produced evidence to suggest that pollution was the cause of this change. As a result, he was the first to construct a red data list of fungal species that he thought were in danger of extinction. Subsequently, fungal conservation has developed both in the United Kingdom and North America, where the discussion relates to the maintenance of habitat for declining fungal species, many of which are ectomycorrhizal. In addition, there is concern in a number of localities about the effects of mushroom harvesting, where research is starting to evaluate if there is a relationship between harvesting rates and fungal species decline. Additionally, the way in which we manage our ecosystems, particularly forest systems, may have a profound effect on the ectomycorrhizal fungal flora.

RECENT DEVELOPMENTS

Over the last 20 years or so, the field of mycorrhizal research has expanded at an amazing rate. The number of published articles in theme-specific journals (e.g., *Mycorrhiza*) and others is staggering, and almost every year a book or two on the subject of mycorrhizae appears. Thus, the review only skims the very surface of the subject, merely providing headings of major areas of interest within the enormous literature.

Molecular techniques have allowed us to fine tune identification of fungal partners with specific plant species and changes in mycorrhizal communities with changes in environmental factors. Advances in methodologies to understand biochemical pathways, physiology of mycorrhizae, and gene expression have allowed us to investigate signaling between the plant root and mycorrhizal fungus to initiate the formation of the symbiosis and to track the metabolic pathways of nutrient acquisition. Unfortunately, these two pathways have emerged somewhat separately so that we can detail community shifts in mycorrhizae, but with the detailed knowledge of function being necessarily restricted to a few specific interactions, it is still not possible to adequately translate changes in community composition to function.

There continues to be interest in the interactions between mycorrhizae metallic, organic, and radionuclide pollutants that demonstrate both plant protection and fungal accumulation of these pollutants. There is speculation that these fungi could be used in restoration and detoxification of polluted sites, but the number of studies that demonstrate successful application is few. This is an area that is probably ripe for commercial intervention given the number of Superfund sites that could potentially be restored to useful agricultural, forest, or recreational land.

A number of commercial organizations produce and market mycorrhizal inoculum for arbuscular, ecto-, and ericoid-mycorrhizal plant species. These inocula are often combined with fertilizers and rhizospehric bacteria, so the efficacy of their mycorrhizal component is not easily testable. Their true utility in enhancing plant growth, plant survival, and nutrient use efficiency still has not been adequately explored. Hence, their potential use in restoration projects is still uncertain.

In natural ecosystems, we are continuing to reveal interesting facets of the ecology and function of mycorrhizae. Recent studies have identified the belowground linkages of plants via mycorrhizal bridges, which has emerged from both controlled experiments and field observations. Although often not completely water-tight evidence, it appears that nutrient and carbon sharing between plants of the same and differing species may occur. This has significantly changed our thinking of plant community assembly rules which were, hitherto, based primarily on the tenet of competition, not synergism. This may be an important subject to bear in mind for land use management and conservation.

Fungal conservation is a relatively new discipline which has evolved as result of increasing land use change, pollution, and climate change. Much of the interest is in the loss of macrofungi (due to their visibility) of which many are ectomycorrhizal and harvested for food. Conservation of these fungi has led us to acknowledge that we probably know less about their ecology and function than we thought we did. A good case in point is the perceived overharvesting of morels in the eastern United States. It transpires that in many cases we do not even know what species, subspecies, or races we have, let alone their ecological niche and mycorrhizal status.

As with other disciplines in the biological sciences, we need our 'model organisms' in which we can obtain detailed physiological and biochemical information; the mycorrhizal community has investigated detailed function of relatively few fungal species. It is my hope that this will expand, so that we have a better understanding of the huge diversity of fungi that are associated with plants and a greater understanding of whether they can be used for restoration and for maintaining sustainable practices in agriculture and forestry.

FURTHER READING

Agerer, R. (1987–2006). *Colour atlas of Ectomycorrhizae*. Einhorn-Verlag: Munich.

Bever, J. D., & Schultz, P. A. (2005). Mechanisms of arbuscular mycorrhizal mediation of plant–plant interactions. In J. Dighton, J. F. White & P. Oudemans (Eds.), *The fungal community: Its organization and role in the ecosystem* (pp. 443–459). (3rd ed.). Taylor & Francis: Baton Rouge, LA.

Castellano, M. A. (1996). Outplanting performance of mycorrhizal inoculated seedlings. In K. G. Mukerji (Ed.), *Concepts in mycorrhizal research* (pp. 223–301). Kluwer: The Netherlands.

Dighton, J. (2003). *Fungi in ecosystem processes*. Marcel Dekker: New York.

Durall, D. M., Jones, M. D., & Lewis, K. J. (2005). Effects of forest management on fungal communities. In J. Dighton, J. F. White & P. Oudemans (Eds.), *The fungal community: Its organization and role in the ecosystem* (pp. 833–855). (3rd ed.). Taylor & Francis: Baton Rouge, LA.

Feddermann, N., Finlay, R., Boller, T., & Elfstrand, M. (2010). Functional diversity in arbuscular mycorrhiza – The role of gene expression, phosphorus nutrition and symbiotic efficiency. *Fungal Ecology*, *3*, 1–8.

Koide, R. T., & Mosse, B. (2004). A history of research on arbuscular mycorrhiza. *Mycorrhiza*, *14*, 145–163.

Peterson, R. L., Massicotte, H. B., & Melville, L. H. (2004). *Mycorrhizas: Anatomy and cell biology*. CABI: Wallingford.

Read, D. J., Lewis, D. H., Fitter, A. H., & Alexander, I. A. (Eds.), (1992). *Mycorrhizas in ecosystems*. CABI: Wallingford.

Smith, S. E., Facelli, E., Pope, S., & Smith, P. A. (2010). Plant performance in stressful environments: Interpreting new and established knowledge of the roles of arbuscular mycorrhizas. *Plant and Soil*, *326*, 3–10.

Smith, S. E., & Read, D. J. (1997). *Mycorrhizal symbiosis* (2nd ed.). Academic Press: San Diego, CA.

van der Heijden, M. C. A., & Sanders, I. R. (Eds.), (2003). *Mycorrhizal ecology*. Springer: Berlin, Germany.

Wang, B., & Qiu, Y. L. (2006). Phylogenetic distribution and evolution of mycorrhizas in land plants. *Mycorrhiza, 16*, 299–363.

Watling, R. (2005). Fungal conservation: Some impressions – A personal view. In J. Dighton, J. F. White & P. Oudemans (Eds.), *The fungal community: Its organization and role in the ecosystem* (pp. 881–896). (3rd ed.). Taylor & Francis: Baton Rouge, LA.

RELEVANT WEBSITE

http://www.hortsorb.com – Horticultural Alliance Inc.
http://www.mycorrhizalproducts.com/ – Mycorrhizal Products.
http://www.soilmoist.com – Soil Moist.

Chapter 7

Lichens

S. Hammer
Boston University, Boston, MA, USA

Chapter Outline

Defining Statement	85	**The Symbiosis**	92
Introduction	85	**Discussion**	94
Historical Note	89	**Further Reading**	94
Lichen Evolution	90		

DEFINING STATEMENT

Lichens represent symbioses between heterotrophic fungi and photosynthetic algae or cyanobacteria. The broadly mutualistic symbiosis involves nutrient exchange, chemical signaling, and recognition mechanisms that are only partially understood. The lichen body is a discrete entity that resembles neither partner, with unique morphology and morphogenetic patterns. Lichens play an important role in community ecology, and are broadly comparable to other fungal organisms that participate in pathogenic and mutualistic activities.

INTRODUCTION

Lichens represent an interspecific symbiosis between a fungus (mycobiont) and one or more species of unicellular algae and/or cyanobacteria (photobionts). Lichen species are based on the taxonomy of the mycobiont, which in most lichens, is the sexually reproducing partner. Heterotrophic lichen fungi obtain carbon-based nutrients from the photosynthetic activity of their photoautotrophic partners. When a cyanobacterium is involved in the symbiosis, as is the case in approximately 10% of lichen species, the fungal partner may also obtain nitrates or ammonium ions that are processed by the bacterium.

Lichens are unique among symbiotic relationships in that they form a distinct, potentially long-lived thallus (Figures 1 and 2). The lichen thallus is a multicellular and usually macroscopic body that in most cases resembles neither the filamentous fungal partner nor the unicellular photobiont. The bulk of the thallus is comprised of heterotrophic fungal cells, but because photobionts are part of the thallus, lichens are by definition photosynthetic. Lichens are exclusively epiphytic, whether on plants, leaf litter, sand, soil, rock, or anthropogenic surfaces such as glass, roofing material, and asphalt. There are no roots or other organs, and lichens lack conducting tissues. Lichen thalli obtain all of their moisture from the ambient environment, either through rain, fog, fog drip, or humidity. The vast majority of lichens are terrestrial, but a few species live in freshwater or tidal marine environments and obtain their moisture from these sources. Lichens are blotter-like (poikilohydric) and when hydrated may increase their weight and size very dramatically. As epiphytes, lichens draw no carbon nutrients from their substrata, but many species prefer particular surface environments. For example, preferences for acidic versus basic (and vice versa) substrata are well documented. In addition, many lichens are found growing on particular plant species, for example, oak trees, which suggests that some trace nutrients are derived from exudates of the host's surface.

Lichens are often considered as 'pioneer' species, some of the first species to invade disturbed habitats. However, this may be a misconception, as many lichens require undisturbed natural microhabitats. Lichens are found in a great diversity of habitats, with species growing in ecosystems as varied old growth forests, endolithic quartzite layers in Antarctica, and desert localities whose only moisture is marine fog banks. As epiphytes, lichens are subject to particular selective pressures, and any consideration of their biology, distribution, dispersal, physiology, or interaction with their environment must take into account the fact that lichens are epiphytes (Figures 3 and 4).

The nature of the mycobiont–photobiont relationship also reflects the epiphytic biology of the thallus. The fitness

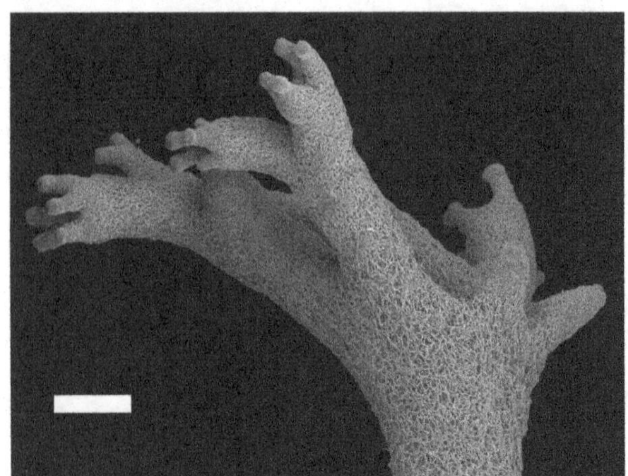

FIGURE 1 The branching lichen thallus of *Cladina rangiferina* ('reindeer lichen'). Scale = ~1 mm.

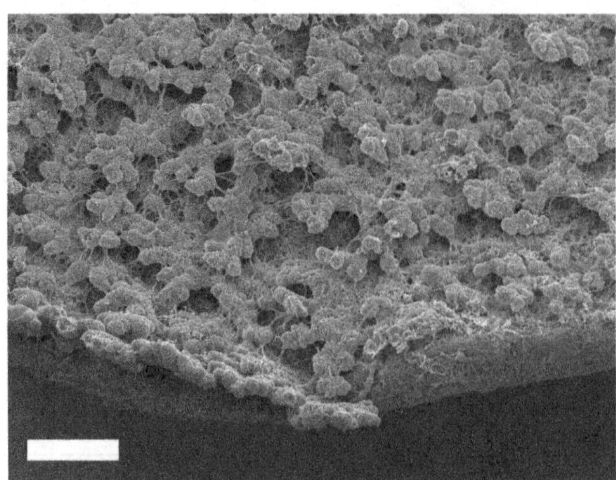

FIGURE 3 Epiphytic lichen on a leaf. Aerial fungal cells (hyphae) connect thallus lobes. Hyphae are also visible along the surface of the leaf. Scale = ~1 cm.

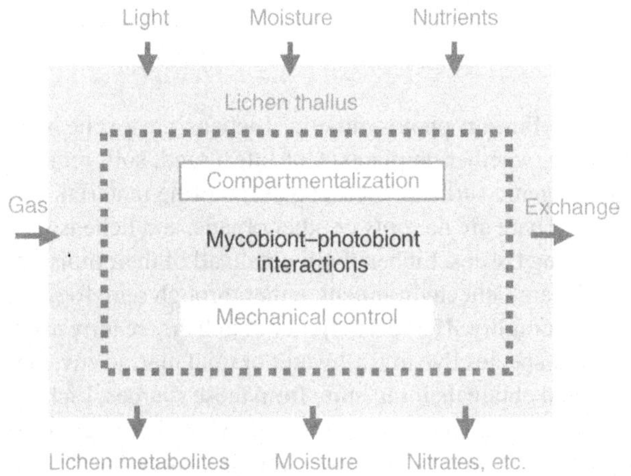

FIGURE 2 Diagram of a hypothetical lichen thallus, highlighting internal and external interactions of the thallus.

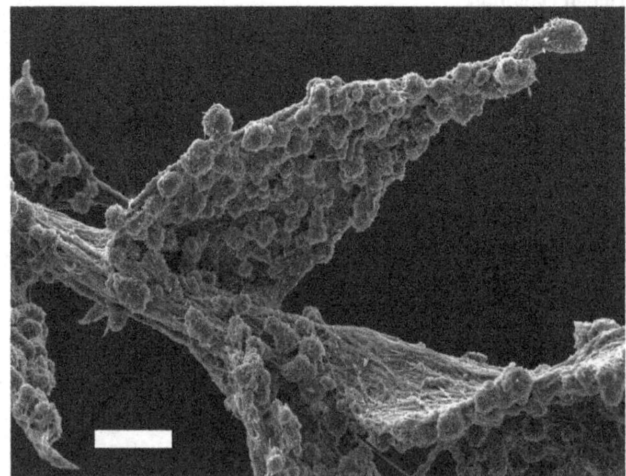

FIGURE 4 Epiphytic lichen on a moss. Loosely organized thallus on leaf surface is connected by fungal hyphae. Scale = ~100 μm.

of both bionts is presumed to be enhanced by their relationship within the thallus, which has led most authors to characterize the lichen as a broadly mutualistic relationship. However, the flow of nutrients is from the photobiont to the mycobiont, suggesting a controlled parasitism in which the photobiont is 'farmed' by the mycobiont. Recent findings indicate that at least some mutual control is exerted between the bionts.

In the broadest sense, lichens bear a superficial resemblance to plants, in which a long-term relationship between plant cells and endosymbiotic organelles such as chloroplasts and mitochondria provide a coordinated, highly organized, and mutually beneficial exchange of nutrients in a discrete multicellular environment such as a leaf. However, lichens are fundamentally different from plants in a number of ways. For example, the lichen photobiont is housed among the fungal hyphae, but unlike an organelle such as the chloroplast, it maintains an independent spatial and genomic presence outside of and apart from the hyphae. In addition to their own genomic structure, both the mycobiont and the photobiont algal partners maintain their own organelle systems, such as chloroplasts in algal cells and mitochondria in both the fungus and the alga. Finally, the anatomical organization of plants (e.g., organs, tissues, conducting systems, and sophisticated gas exchange structures) is much more complex than that of lichen thalli.

In spite of its simplicity, the lichen thallus is a self-sufficient entity characterized by the collaborative physiological activities of the mycobiont and the photobiont. The lichen thallus is mostly fungal tissue comprised of loosely organized to strongly agglutinated hyphae. Photobionts, which are variously distributed among the fungal tissue (Figure 5), are provided with support, display, and protection from extreme light by the fungal partner. In the many

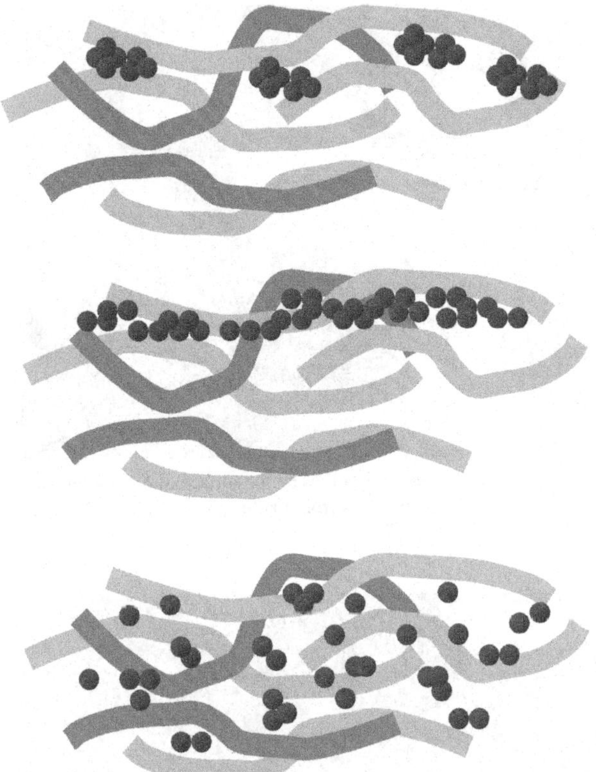

FIGURE 5 The spatial orientation of photobionts varies in different lichen species. From top: clumped, layered, and randomly arranged.

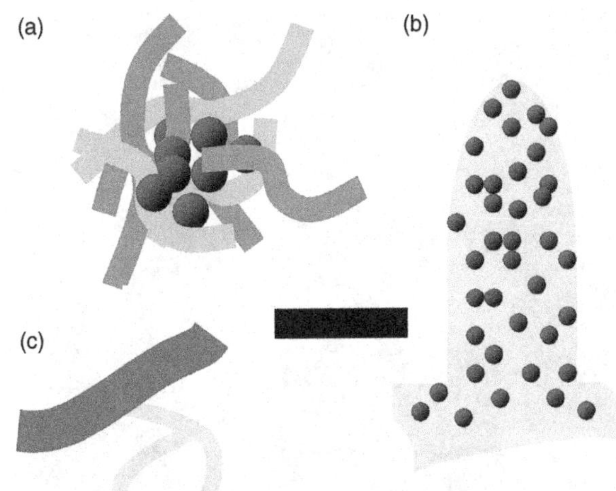

FIGURE 6 Asexual (vegetative) reproduction occurs through variants of thallus fragmentation. Examples clockwise from upper left: (a) soredium, a clump of photobiont cells surrounded by fungal hyphae (scale = ~50 μm); (b) isidium, a minute thallus, usually with discrete cell layers (scale = ~1 mm); (c) macroscopic thallus fragmentation (scale = 1 mm or greater).

lichen species that reproduce asexually, the photobiont is carried along with the mycobiont in vegetative propagules (Figure 6).

Lichen thalli are diffuse to relatively highly organized (Figures 7 and 8). Although they lack well-delineated tissues, they possess various levels of thallus specialization and compartmentalization; for example, cell layering, localized photobionts, and discrete air spaces (Figures 9 and 10). Many lichens possess a cellular or noncellular thickened layer in thallus portions that receive the most sunlight or that have the most interface with the atmosphere. Some lichens maintain enlarged pores that may involve gas exchange. Lichens may employ rhizine-like outgrowths of the thallus that attach them to substrata and potentially serve as rudimentary conducting systems. Thallus organization as well as the production and secretion of complex molecules such as hydrophobins result in considerable internal control over moisture and CO_2 regimes in spite of the fact that poikilohydric lichens experience wide environmental fluctuations. Lichens are not passive entities. They protect themselves from environmental challenges such as ultraviolet light by the production of secondary metabolites. These substances have also been shown to protect lichen thalli from pathogens such as herbivores. Lichens also exert some control over their environment (e.g., participating in soil-building through breakdown of rocky substrata) through

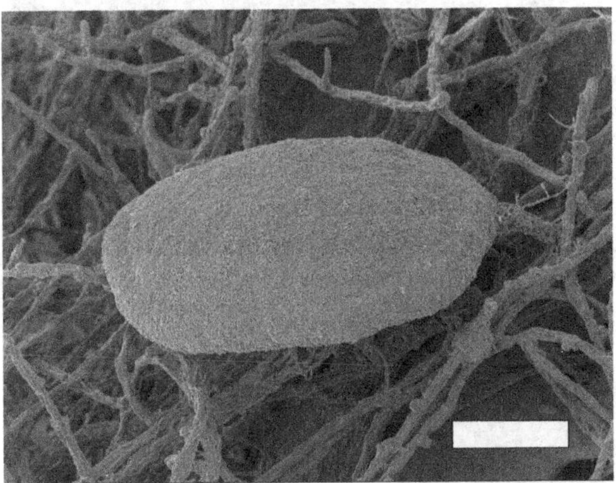

FIGURE 7 The diffuse thallus of *Coenogonim* sp. Large body in the center is the spore-bearing apothecium. Filamentous structures are the vegetative thallus, containing fungal and algal cells. Scale = ~100 μm.

the secretion of unique secondary metabolites. The unique phenolics and other secondary metabolites that characterize many lichen species (Figures 11–13), are primarily secreted by the mycobiont. The analysis of lichen substances has been used widely in the taxonomy of lichens since the late nineteenth century, and in recent decades, sophisticated chromatographic technologies have improved the resolution and subsequent analysis of lichen phenolics.

Of some 15 000–20 000 lichen fungi, most are Ascomycetes. Less than 2% are Basidiomycetes or very rarely Deuteromycetes. Most estimates concur that approximately

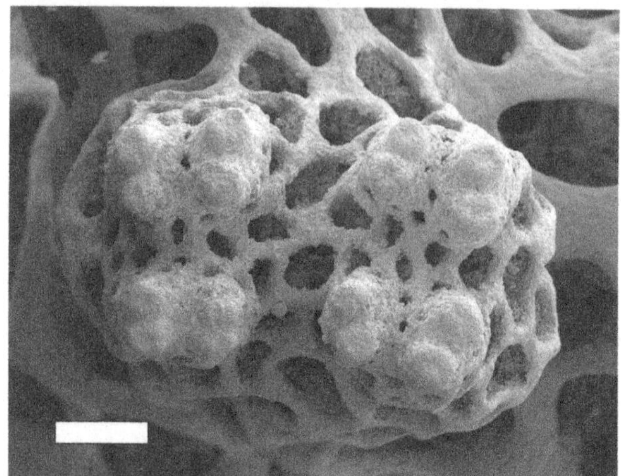

FIGURE 8 The highly organized thallus of *Cladia retipora*. Detail of symmetric branch tips displaying synchronous growth in a roughly helical pattern. Scale = ~ 250 μm.

FIGURE 11 Secondary metabolites on the surface of *Evernia prunastri*. The crystals have been extruded from thallus pores. Scale = ~0.5 μm.

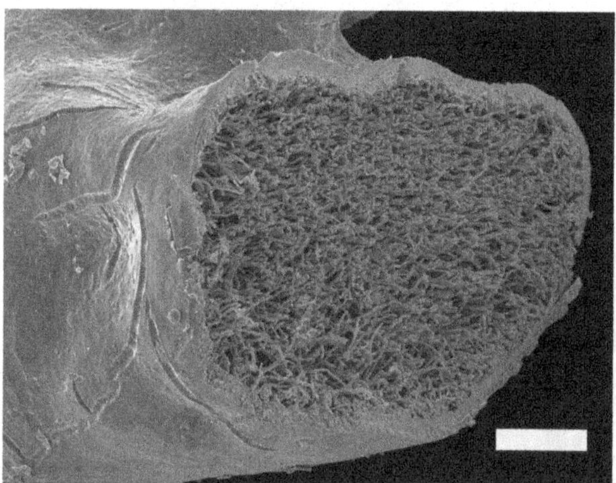

FIGURE 9 Transverse section of a branch of *Siphula* sp. showing a well-developed outer layer (includes photobionts) with parallel-oriented fungal hyphae inside. Scale = ~1 mm.

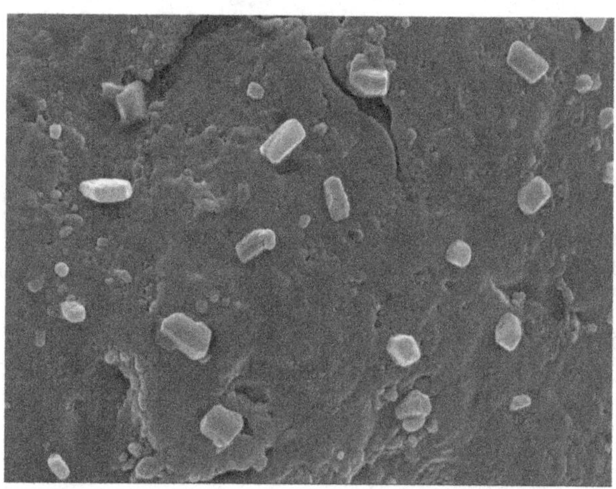

FIGURE 12 Vulpinic acid on the surface of *Letharia vulpina*. Scale = ~0.25 μm.

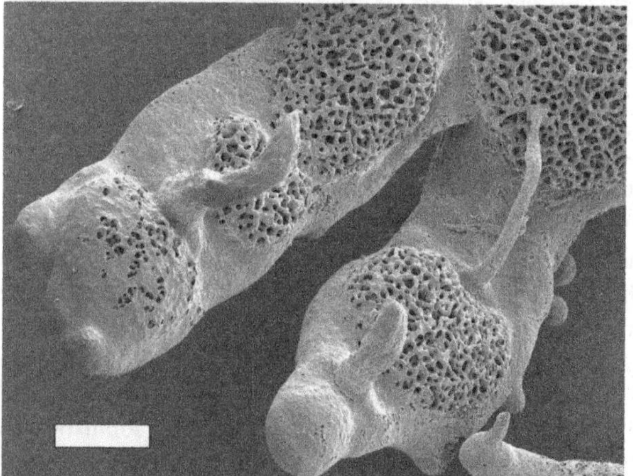

FIGURE 10 Underside of *Pannoparmelia* sp. showing the development of discrete openings into thallus interior, presumably aiding in gas exchange. Note the perpendicular rhizines (elongated structures) extending from thallus bottom. These function to anchor the thallus to its substratum. Scale = ~2 mm.

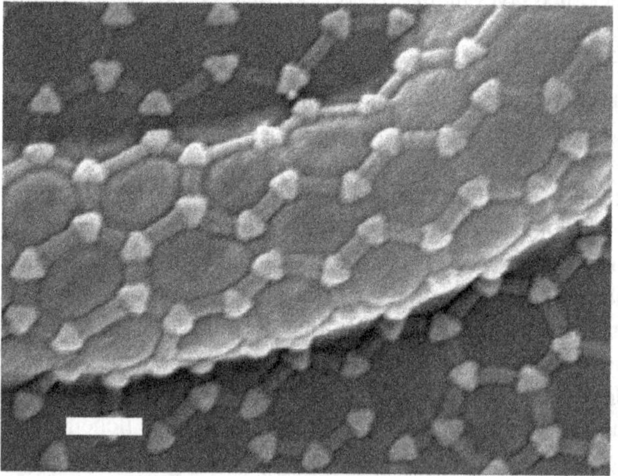

FIGURE 13 Unknown compound crystallized on surface of lichen fungal hypha. Scale = ~0.25 μm.

half of all ascomycetous fungi are represented by lichens. The fact that lichen fungi interact with their photobiont hosts in a long-lived biotrophic symbiosis is remarkable given that Ascomycetes represent some of the most virulent pathogens in the fungal kingdom. These include a number of parasitic species which, while they exert a negative impact on their host, may still be broadly considered as symbiotic. The relationships termed as lichens are generally described as mutualistic, but the natural lichen symbiosis has historically been subject to conjecture, as discussed below. Some authors have characterized the lichen symbiosis as a controlled parasitism, in which the bionts maintain a nonpathogenic relationship and the flow of nutrients is from the photobiont to the mycobiont.

HISTORICAL NOTE

Lichens were known to the ancients and appear in the writings of, among others, Theophrastus, Dioscorides, and Pliny. The English word lichen derives from the Greek *leikhein* – 'to lick.' This may be based on the appearance of certain lobate thalli, which appear to spread across their substrata tongue-like (Figure 14). Some lichens were familiar to the herbalists and found uses through the 'Doctrine of Signatures.' For example, species of *Lobaria* were used to treat lung ailments because of their lung-like resemblance. Lichens have also been used for their chemical components. For example, *Letharia vulpina*, one of the few poisonous lichens, was used as wolf poison for its vulpinic acid constituent. Lichens provide natural dyestuffs in many traditional cultures around the world, including the famous Harris tweed woolens of Scotland. Lichens are also used in the production of perfumes. For example, 'oak moss' (*Evernia* spp.) was used as a component of several high-quality French perfumes in the late nineteenth and early twentieth centuries.

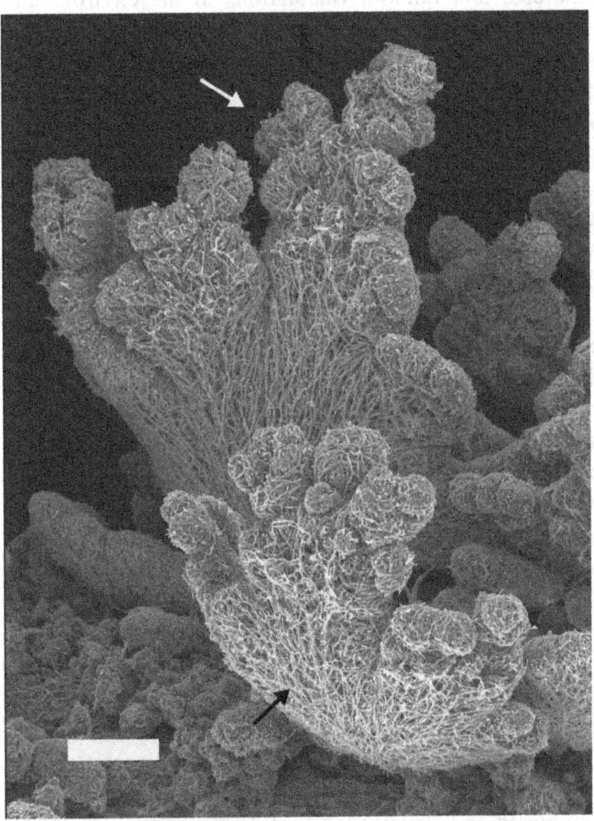

FIGURE 14 A tongue-like lobate portion of the thallus of *Cladonia acervata*. Black arrow indicates filamentous fungal hyphae. White arrow indicates developing lobe tip comprised of a bunch of algal cells surrounded by hyphae. Scale = ~100 μm.

Perhaps because of the unusual and varied forms of lichen thalli, natural historians traditionally struggled to describe the species. The familiar descriptive terms crustose, foliose, and fruticose were used as early as the eighteenth century by the botanist Micheli, who may have garnered them from his reading of the Greek physician and botanist Dioscorides. In general, these commonly used descriptors are useful for field identification of lichen thalli. However, these terms have practically no systematic value, in part because closely related lichens may have different growth forms. Further, in some lichens, the same species may possess different thallus forms at different life stages. Finally, many lichen species cannot be included under any of the categories (Figure 15). The use of the terms crustose, foliose, and fruticose, which have very limited biological relevance, belies the complexity of growth patterns and subsequent mature forms in lichens and should be limited to broad field studies, much like the recent artificial introduction of 'common names' for lichens.

Most lichen terminology, as well as early lichen taxonomy, was adopted before lichen biology and evolution were

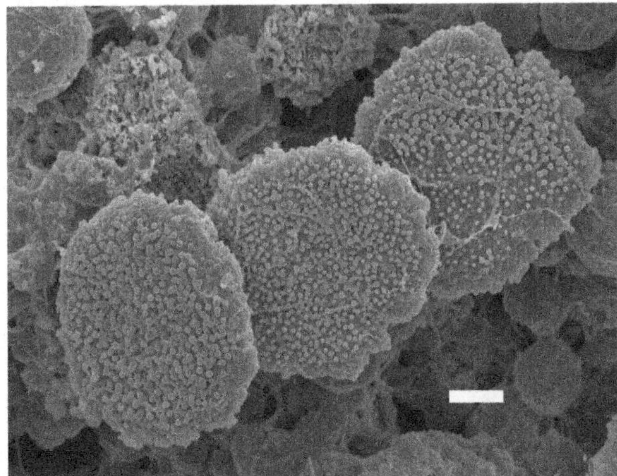

FIGURE 15 Unidentified lichen species from a forest in far Northern Queensland, Australia. This tiny lichen might roughly be termed as 'foliose,' but too little is known of its developmental characteristics and other biological features to pigeonhole it within this simplistic term. Scale = ~100 μm.

understood, even at a rudimentary level. It can be said that early workers who took an interest in lichens approached them from a nonbiological standpoint, considering them very broadly as 'lower plants,' with no understanding of the symbiotic partnership. It is important to note however, that the contemporary concept of fungi was unknown to the early classifiers as well, and that fungi and many other non-plant organisms were also considered as plants. The great eighteenth century classifier Linneaus, whose binomial nomenclatural system still provides the basis for most classification schemes, placed the lichens among algae on the basis of their roughly seaweed-like appearance. It was not until the early nineteenth century that the anatomical characteristics, physiological functions, and geographic distribution of lichens were studied in somewhat greater depth, beginning with the 'father of lichenology,' Erik Acharius. Lichen secondary chemistry was also unknown to the early classifiers and had its rudimentary beginning in the mid-nineteenth century. 'Spot tests' in which a particular reagent applied to the thallus changed its color were devised by the Finnish lichenologist William Nylander, but these techniques and the importance of lichen substances were not widely accepted until much later. Lichen evolution itself was ignored long after Darwin. Late in the twentieth century, questions of lichen evolution began to be considered seriously, and it was only at the dawn of the twenty-first century that lichen systematists began to analyze lichen relationships from a phylogenetic approach.

The signal highlight in the history of lichenology was the 'gonidial hypothesis' (also called the 'dual hypothesis') of Simon Schwendener, in which he posited that lichens are a composite of fungal and algal components. Like many other advances in lichenology, Schwendener's late nineteenth century contribution to lichen biology was not readily accepted and was in fact assailed by the scientific community. Schwendener's hypothesis is universally accepted today, and lichens are recognized as a model for understanding symbiotic phenomena. However, much about the lichen symbiosis and its evolution remains unknown.

LICHEN EVOLUTION

Lichens are considered to be an ancient group, probably extant during the Permo-Triassic (~252 Ma). Presumed lichen fossils from the early Devonian (400 Ma) provide evidence for a much longer time span, although it is important to note that in general, lichens are very poorly and scantily preserved in the fossil record. Remains of so-called subfossilized lichens from regions with recently retreated glaciers, or lichen fragments that have remained frozen until recently in bogs and other habitats provide the most reliable evidence of lichens from the past. However, these remains are usually in the vicinity of tens of thousands, not millions of years old. Lichen species distributions present a fascinating problem. The distribution on distant continental land masses of related or conspecific lichens that lack asexual propagules that would allow long-distance dispersal suggests that ancient populations were separated during vicariance events such as continental drift over the past 250 My. However, these assertions, for instance how closely geographically divergent populations are related, have yet to be tested through comparative genomic analyses. According to popular lore, lichen species are considered to be very widespread, with indistinct distributional patterns. Many lichens are widely distributed, but as species concepts are refined, evidence for discrete and often quite narrow distributional ranges becomes stronger for many of the species.

The lichen thallus is subject to certain unique selective pressures that can be traced to its epiphytic, poikilohydric character. For example, lichens have evolved to cope with unique challenges such as water stress, photooxidative stress, nutrient shortage, dependence on airborne nutrients, and susceptibility to pollutants and airborne toxic metals. At a more general level, lichen evolution is driven by the same forces that affect all life, viz., selective pressures in concert with genetically heritable variations. As in other organisms, these forces are mitigated through phenomena such as sexual selection and reproduction, rare beneficial mutations, and random fluctuations in the environment. Because the lichen is an obligate symbiosis, it is a point of some conjecture which organism (the mycobiont or the photobiont) is affected by selective pressures. Further, the overwhelming evidence supports sexual reproduction in the fungus exclusively, which eliminates the potential for recombinative changes in the photobiont. Many if not most lichens reproduce readily through vegetative reproduction and in many species sexual reproduction through the production and germination of haploid fungal spores is either rare or unknown. How variation is introduced into these species is a matter of conjecture, though some authors have adopted a 'species pair' hypothesis, in which sexually- and asexually-reproducing conspecifics exchange gametes or strains of the mycobiont. It is assumed that many asexual species may become endangered as global climate conditions change. Some evidence of loss of fitness in lichens has been documented in regions where human-induced changes have been dramatic. These qualifications add to the problematic nature of lichen evolution.

Lichen thalli can be short-lived (less than several years) or very long-lived (more than several hundred years). Lichens that are known to be very long-lived have been used to estimate the age of relatively recent geological perturbations, for example, rock slides or glacial retreat. A low, fairly predictable growth rate and relative longevity have encouraged the development of lichenometric dating in geology as well as other disciplines.

Older thalli in particular may include more than one genetic strain of either photobiont or mycobiont as a result of repeated recolonization events by vegetative or sexual propagules. This implies that a photobiont colony in a given thallus is subject to varying levels of control by the mycobiont, allowing new strains of conspecific fungi to establish on a photobiont colony formerly utilized by the initial thallus. Genetic heterogeneity of algal hosts may also relate to the production of asexual propagules such as soredia. The introduction of novel photobiont genomes may increase fitness in lichen species with predominantly asexual reproduction. Some authors have considered the persistence of sexuality in certain lichens as a response to particularly harsh constraints, for example, extreme climate conditions or parasites such as pathogenic licheniicolous fungi.

Little is known about the evolution of collaboration between the bionts. Some fungi and some photobionts are specific in their choice of collaborator, and lichen culture experiments have shown that a fungus may parasitize and kill algal cells that are closely related but not conspecific with its particular photobiont host. There appears to be some specialization among the bionts, although there is ample evidence of collaboration of a given fungus with many species of photobiont and vice versa. There is little evidence of shared genes between myco- and photobionts, although some authors have detected mutual 'gene suppression' during some phases of the lichen symbiosis. Generally, both mycobionts and algal photobionts exhibit anatomical as well as physiological changes in the lichen thallus. For example, algal partners with enlarged chloroplasts, altered membrane permeability, and other ultrastructural changes have been documented.

Both the heterotrophic fungus and its photoautotrophic partner appear to experience increased fitness within the functioning lichen symbiosis (Figure 16). In general, the mycobiont is considered to experience somewhat greater fitness than the photobiont, thanks to the supply of carbon-based nutrients by the photobiont and the fact that sexual reproduction is restricted to the mycobiont. Theoretically, the fitness of both the partners is increased as they attain a greater ecological amplitude than either could experience on its own. This is borne out by estimates that some 8% of the Earth's surface is inhabited by lichens. Almost none of the fungal bionts has been found in nature without its requisite photobiont. Most of the photobionts require a mycobiont although some of the algae (e.g., *Trentepohlia* spp.) are potentially free-living, and many of the bacterial photobionts, for example, *Nostoc* spp., readily inhabit a great variety of environmental niches without the mycobiont. Some authors suggest that lichenization may have been a contributing factor in the invasion of terrestrial environments by green algae.

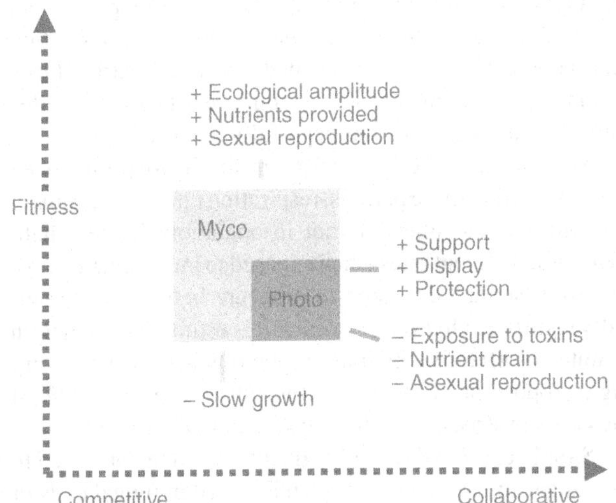

FIGURE 16 A summary of fitness versus collaboration in lichens. Broadly, the mycobiont, which derives nutrients from the photobiont, enjoys higher fitness than the photosynthetic partner.

Lichen photobionts and hence lichen species in general are exquisitely sensitive to airborne pollutants in the ambient environment. This is especially the case when the thallus is moistened, perhaps because of increased membrane permeability of photobionts in the hydrated state. With the rise of fossil fuel combustion and associated pollutants, many industrialized cities became lichen 'deserts' during the twentieth century and extinction seemed imminent for some species. Rising air quality in many postindustrial societies has led to a resurgence of native lichen populations in urban areas, but lichens are severely threatened in developing industrial societies. Because of their sensitivity to air pollution, lichens have been used widely as air quality monitors. Combustion-related airborne pollutants impact lichen survival, but anthropogenic disturbance may threaten lichen communities in a number of other ways. Heavy metals that are released from mines, smelters, and industrial sources are readily absorbed into lichen thalli and may reach toxic levels. Lichens growing downwind of Chernobyl were found to contain very high levels of radioactivity, threatening the lichen species as well as ungulates such as reindeer that use them as food. Airborne pesticides and nutrients from agriculture, lawns, and golf courses also threaten lichen communities.

Habitat fragmentation ranging from logging to compaction along hiking trails change ambient conditions of light, soil, and substratum availability. All of these affect lichen survival on both a local and a global scale. Some lichen species are weedy or invasive, and anthropogenic disturbance may encourage their establishment. Postlogging recovery has been shown to lead to the establishment of common, nonendemic lichen species. Certain opportunistic species may establish along road embankments and other anthropogenic disturbances that provide altered light and moisture regimes. Natural disturbance such as fire appears to be nondetrimental to lichen communities, especially those that inhabit ecosystems that are visited by frequent fire events.

THE SYMBIOSIS

The great majority of lichens (>99%) involve ascomycetes. It is estimated that nearly 50% of all known ascomycetes are lichenized. All of these are so-called inoperculate ascomycetes, a group that includes virulent plant and insect pathogens. Approximately 90% of lichens involve an algal photobiont. Some 25–30 genera of algae, which may represent hundreds of species, are involved in lichen associations. In addition, more than a dozen genera of cyanobacteria participate in lichen symbioses. However, information on photobiont taxonomy is incomplete because in most lichens neither the algal nor cyanobacterial photobionts have been identified to the level of species.

Most lichens involve two partners, a fungus and either an alga or cyanobacterium, but many species include both an algal and a cyanobacterial partner. Generally, the cyanobacterial partner in a tripartite relationship is considered as a secondary partner and is sometimes housed separately on the surface of the thallus. The cyanobacterium in a tripartite symbiosis is most significant for its input of biologically useable nitrogen into the thallus system. Lichens with cyanobacterial partners are significant nitrogen sources for many ecosystems, and net nitrification of soils under lichens without cyanobacterial partners has been documented as well. Biological soil crusts with significant lichen populations contribute to potential nitrogen fixation, reduce nitrogen leaching, and moderate soil microclimate. Moisture and temperature fluctuations beneath lichen thalli are slowed, and certain potentially toxic metals, for example, aluminum, are absorbed by lichen thalli. Lichens ameliorate and improve soil conditions, encourage the establishment of microbial organisms, and produce an environment that supports a diversity of invertebrates. They can be considered as important contributors to pedogenesis (soil-building) in many ecosystems.

Lichen mycobionts are similar to other fungi that associate in a long-term, stable, cell-to-cell collaborative interface with other species. But in other interspecies symbioses that include fungi, for example, mycorrhizal relationships, which affect approximately 90% of plant species, nutrient exchange occurs in both directions between the mycobiont and the photobiont. With all of the nutrients in lichens flowing from the photobiont to the mycobiont, the relationship has been considered as a controlled parasitism, in which the fungal partner maintains a population of carbohydrate-producing photobionts. Anatomically, lichen fungi communicate with photobionts in simple cell-to-cell interfaces or through the extension of haustoria or appressoria, specialized outgrowths of the fungal cells that produce invaginations in the photobiont (Figure 17). In most cases, the relationship is nonpathogenic. Photobiont cell membranes are not penetrated and photobiont cells are not regularly destroyed in a stable lichen thallus. Mutual sustained

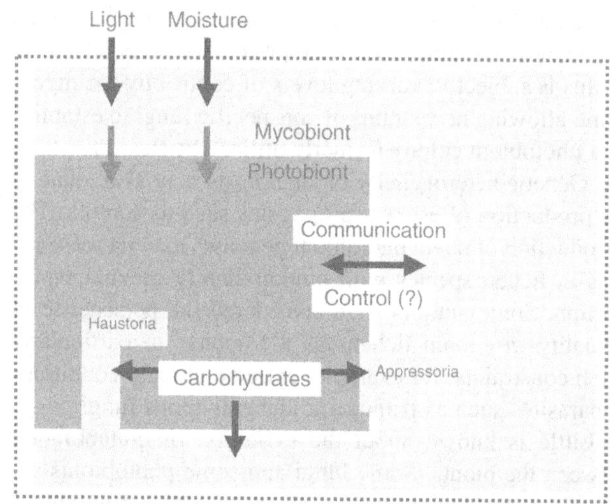

FIGURE 17 Summary of nutrient exchange and mutual control between mycobiont and photobiont in the lichen thallus.

recognition between bionts is essential for thallus success, and some control is apparently exerted by the photobiont. Some authors note that an extended photobiont cell cycle, such as that observed in many lichens, would reduce the number of times the symbiosis must be reestablished on a cell-to-cell basis, minimizing opportunities for pathogenic encounters. The rate at which photobiont cells undergo the cell cycle is apparently under regulation, but that whether and to what extent that regulation is influenced by the fungal partner is unknown. Recognition between the mycobiont and photobiont may be mediated by polysaccharides that are found in both, especially on membrane surfaces. Lectins, which are produced by the mycobiont and are present on the mycobiont cell wall, have been demonstrated to bind to specific ligands on the cell wall of algal photobionts, suggesting a role in communication between the partners. Studies of reestablishment of lichens from axenic culture suggest that certain nonself recognition and defense signals are weakened during the early stages of thallus development. The genomic implications of this mutual 'quieting' have not been fully explored.

Most lichens experience profound drying for pronounced periods. During these periods, respiration rates are extremely low and photosynthesis is not in operation. Lichen thalli hydrate at different rates when exposed to moisture, but most require sustained exposure to moisture before physiologically significant hydration occurs. It is estimated that within minutes of effective hydration, photobionts release a very high proportion of stored photosynthate (up to ~90% of the cell's reserves), which is absorbed by the fungus.

Simple carbohydrates like glucose are then converted to the fungal sugar mannitol, which is stored in fungal cells or converted for use in cellular respiration in the mycobiont. Cell wall building and other cellular functions of both partners occur in the presence of available nutrients and when adequate moisture is available for cellular activity. Hence, thallus growth is limited by the frequency and duration of drying events. These events may be diurnal, for example, in desert lichens that obtain their moisture from morning dew, or in foliicolous (leaf-inhabiting) lichens in rainforests that experience daily rainfall. Drying events may be less frequent but are longer and lasting for several weeks or even months. Frozen water is generally unavailable for photosynthesis and other cellular activities, and lichens may experience many months of inactivity under snow or ice. Many lichens are able to take advantage of microscopic amounts of liquid water that may occur under snow or ice in very cold temperatures. These species can apparently maintain marginal physiological activities in extremely cold conditions. Lichens are notoriously slow growers and generally this has been attributed to the wetting-drying patterns that they experience. However, fungal growth may also be limited by controls exerted by the photobiont host that have yet to be elucidated.

The lichen thallus has evolved in response to potentially very high levels of thallus moisture fluctuation. When dry, the lichen surface may be any shade of opaque white, yellow, brown, orange, greenish, or even purple-black, but when moistened, most lichens appear translucent. Lichens recover relatively rapidly from desiccation, but studies have shown that water saturation can limit photosynthesis and effective gas exchange. The role of hydrophobins, a class of proteins produced in the thallus, have been intensively studied in relation to this problem. Hydrophobins have been implicated in several adaptive strategies, such as allowing water translocation, maintaining water-free interhyphal spaces (for optimal oxygen diffusion), and defining strata within the thallus that slow water loss. Light response curves of net photosynthesis in lichens are highest at relatively low light levels, perhaps as an adaptive response to desiccation that occurs at higher light levels. This lends support to the hypothesis that some photobionts, which reach maximal photosynthetic levels in attenuated light, achieve greater fitness in lichen thalli than in free-living conditions.

Photobiont algae, and to an extent cyanobacterial partners, undergo changes in morphology in the lichen thallus. These changes have been intensively studied. Less is known about physiolgical and morphological changes that the fungus experiences. Lichen fungi grown axenically often appear mold-like, with branching hyphae growing more or less concentrically outward from the point of establishment. This is a profound difference from the appearance of the fungi in lichen thalli. Within thalli, fungal cells that are in contact with photobionts may be thin-walled, presumably as a mechanism for obtaining nutrients from photobionts. This is especially the case with haustoria and appressoria, which are cellular outgrowths of the fungus. In thallus regions where there is no photobiont contact, fungal cells may become quite thick-walled, agglutinating into sclerotium-like bodies that appear hard and sometimes carbonized. Fungal cells organize into pseudotissues within the thallus, stratifying, partitioning, and otherwise organizing its internal space. Agglutinated fungal cells may provide adherence to rocks, tree trunks, and other substrata. They may also participate in water relations of the thallus. The fungal partner has been implicated in directing the overall growth of the thallus. Mersitem-like regions of very tightly packed fungal cells with no direct photobiont content have been demonstrated to undergo coordinated growth that leads to macrophenomena such as thallus branching (Figures 18 and 19).

Although it is still a matter of some controversy, it appears that secondary metabolite production is accomplished mostly if not exclusively by the fungal cells. It is generally assumed that production of these metabolites is

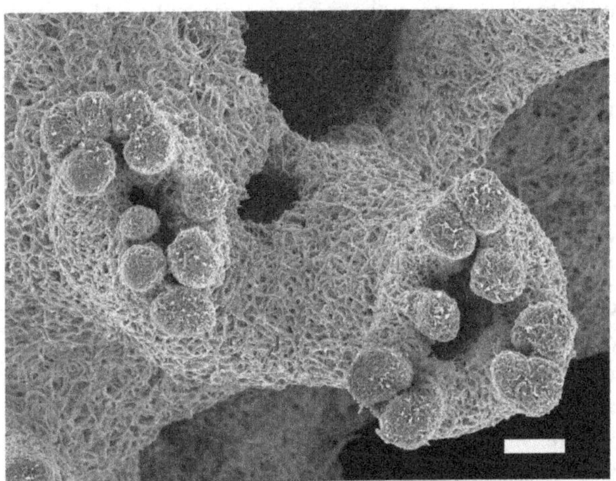

FIGURE 18 Meristem-like fungal regions at growing tip of *Cladina mitis* thallus. Fungal bundles guide branching events as well as subtle torsion of the thallus. Scale = ~100 μm.

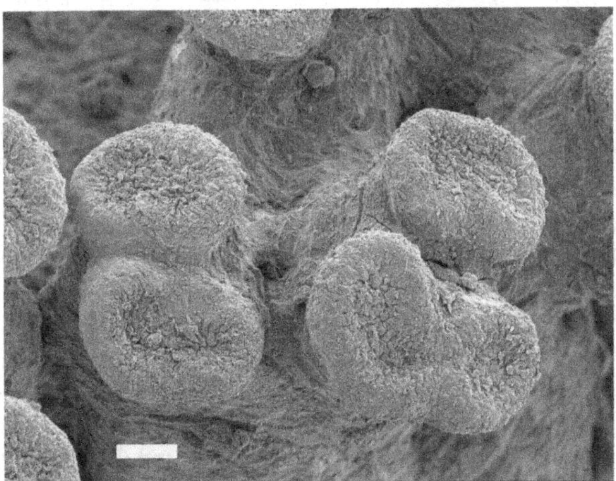

FIGURE 19 Meristem-like fungal regions at very young growing tip of *Notocladonia* sp. Development of these regions, including enlargement, splitting, and torsion is synchronized and is reflected in later thallus development. Scale = ~50 μm.

undertaken in association with (or at least through contact with) photobionts. However, the production of secondary metabolites has been observed in many lichen fungi when grown axenically. The presence of photobionts has also been implicated in inhibiting the production of certain metabolites, perhaps through the secretion of urease inhibitors by algal partners. Generally, secondary metabolites are more varied and abundant in species with algal photobionts than in species with cyanobacterial partners.

DISCUSSION

Lichens are problematic. The thallus, a highly integrated functioning body, is morphologically simple and in many species, indistinct. Its developmental biology and functional anatomy are only obliquely understood. How morphogenetic roles within the thallus are partitioned is not understood. Likewise, the modes of production of secondary metabolites, the partitioning and control over nutrients, and the role of secondary partners such as cyanobacteria are incompletely known. The mycobiont–photobiont relationship provides a model for studying other symbioses. Yet, the state of our knowledge about mutual control between symbionts in lichens is in its very early stages. What is the extent of control? What is the agency of control? Is there a suite of interrelated chemical signals that maintain control? Are associated proteins activated from signals apprehended at the membrane level? What are the feedback mechanisms that modulate immune recognition between partners? How are signals between (or among) symbionts mediated? What causes the profound morphological and physiological changes that both symbionts undergo in the lichen thallus?

The nature of the symbiosis requires further consideration. From one perspective, lichens may be considered as a sort of commensalism. In this framework, the photobiont and mycobiont have coevolved such that the photosynthetic host suffers no damage and the mycobiont is resistant to the host's immune system. However, the loss of sexuality in the photobiont is a significant loss in terms of fitness, and the fact that most myco- and photobionts require the symbiotic thallus to live suggests that in some respects the fitness of both the partners is compromised. Mutualism, a model in which both partners benefit, may better describe lichens. However, the unidirectional flow of nutrients into the fungal partner complicates this model and suggests that parasitism describes the lichen symbiosis more accurately. Some authors have accepted the classification of mutualism on the basis of the increased ecological amplitude of both symbionts in the lichen and the fact that lichen relationships are relatively long-lived while parasitic symbioses are generally shorter in duration. However, the parasitic nature of the fungus on the photobiont cannot be overlooked. It seems that Ahmadjian's (1993) term 'controlled parasitism' is the most likely descriptor for lichens at present. It is ironic that while lichens may be the best known symbiotic relationship, they defy easy classification. Terms such as commensalism, mutualism, parasitism, and others that have been mustered to describe other symbiotic phenomena do not adequately describe the lichen symbiosis.

Lichens make a significant contribution to the health of the biosphere. Most studies that describe lichen activities within ecosystems point to the beneficial effect of lichens as ground cover, forest epiphytes, soil stabilizers, and soil builders. Lichens are an important food for some ungulates and probably they provide nutrition for other organisms as well. Yet lichens do not appear in most macroecological analyses of the biosphere. A great deal more ecological work is needed in order to model the role of lichens in natural ecosystems. Lichens are widely distributed, but many species are highly sensitive to their environment and reflect a narrow and specialized geographic distribution. While all lichen species tolerate a great range of environmental conditions, especially as regards water relations, they are threatened by numerous anthropogenic changes in their environment. There is much to learn about the physiological mechanisms by which lichens function as photosynthetic units in challenging environments.

Lichens have always been difficult to identify on the basis of their variability and human limitations in describing lichen form. Generally, the group has suffered because too few scientists undertake the study of lichens. Much more floristic work is required in order to better document lichen populations worldwide. As lichen habitats are destroyed by human interference, the species will inevitably be lost. Further resources are required, not only to identify and record lichen species but also to explore the numerous physiological phenomena that characterize the symbiosis.

Lichens are subject to evolution like every organism on the planet. Their evolutionary history suggests an ancient lineage that has weathered innumerable changes in the environment. That evolutionary history is complicated by the presence of two or more symbionts in every lichen species. Further research will elucidate not only the evolutionary relationships of lichens but also the nature of their symbiosis.

FURTHER READING

Ahmadjian, V. (1993). *The lichen symbiosis.* Wiley: New York.

Ahmadjian, V., & Paracer, S. (1986). *Symbiosis: An introduction to biological associations.* University Press of New England: Worcester, MA.

Cornelissen, J. H., Lang, S. I., Soudzilovskaia, N. A., & During, H. J. (2007). Comparative cryptogam ecology: A review of bryophyte and lichen traits that drive biogeochemistry. *Annals of Botany, 99*, 987–1001.

Elix, J. A. (1996). Biochemistry and secondary metabolism. In T. H. Nash, III (Ed.), *Lichen biology* (pp. 154–180). Cambridge University Press: New York.

Galun, M. (Ed.), (2000). *CRC handbook of lichenology*. CRC Press: Boca Raton, FL 2 Vols.

Green, T. G. A., Lange, O. L., & Cowan, I. R. (1994). Ecophysiology of lichen photosynthesis: The role of water status and thallus diffusion resistances. *Cryptogamic Botany*, *4*, 166–178.

Grube, M., & Hawksworth, D. L. (2007). Trouble with lichen: The re-evaluation and re-interpretation of thallus form and fruit body types in the molecular era. *Mycological Research*, *111*(Pt 9), 1116–1132.

Hammer, S. (2001). Lateral growth patterns in the *Cladoniaceae*. *American Journal of Botany*, *88*, 788–796.

Hawksworth, D. L., & Pirozynski, K. A. (Eds.), (1988). *Coevolution of fungi with plants and animals*. Academic Press: London.

Hill, D. J. (2004). The cell cycle of the photobiont of the lichen *Parmelia sulcata* (*Lecanorales*, Ascomycotina) during the development of thallus lobes. *Cryptogamic Botany*, *4*, 270–273.

Honneger, R. (1993). Developmental biology of lichens. *The New Phytologist*, *125*, 659–677.

Kershaw, K. A. (1985). *Physiological ecology of lichens*. Cambridge University Press: Cambridge.

Lawrey, J. D., Torzilli, A. P., & Chandhoke, V. (1999). Destruction of lichen chemical defenses by a fungal pathogen. *American Journal of Botany*, *86*, 184–189.

Molnár, K., & Farkas, E. (2010). Current results on biological activities of lichen secondary metabolites: A review. *Zeitschrift für Naturforschung C*, *65*, 157–173.

Piercey-Normore, M. D. (2006). The lichen-forming ascomycete *Evernia mesomorpha* associates with multipe genotypes of *Trebouxia jamesii*. *The New Phytologist*, *169*, 331–344.

Schwenender, S. (1869). Die algentypen der flechtengonidien. *Programm für die Rectorsfeier der Universität Basel*, *4*, 1–42.

Taylor, T., Hass, H., Remy, W., & Kerp, H. (1995). The oldest fossil lichen. *Nature*, *37*(8), 244.

Chapter 8

Plant Pathogens and Disease

Newly Emerging Diseases

J.J. Burdon[1], P.H. Thrall[1] and L. Ericson[2]
[1]CSIRO, Canberra, Australia
[2]Umeå University, Umeå, Sweden

Chapter Outline

Defining Statement 97
Introduction 97
Where Disease Emergence Stems from Simple Ecological Changes in Distribution 98
Where Disease Emergence Stems from Genetic Change in Pathogens 99
Where Disease Emergence Stems from Environmental Change 102
Can We Predict the Identity of Future Emerging Diseases? 102
Countering Invasive Plant Diseases 103
Conclusions 103
Further Reading 104

DEFINING STATEMENT

Ecological and/or genetic changes in the interaction occurring between plant pathogens and their abiotic and biotic environment may lead to severe disease outbreaks that regularly recur where previously there were none. These epidemics may impose significant economic, social, and environmental damage that is hard, if not impossible, to redress.

INTRODUCTION

The evolutionary history of many fungal groups is littered with evidence of jumps in the host range of individual pathogen species. Nowhere is this more apparent than in the rust fungi where many heteroecious rusts alternate between gymnosperm and angiosperm (e.g., *Chrysomyxa* spp. alternating between *Picea* spp. and ericaceous hosts) or angiosperm and angiosperm (e.g., *Puccinia graminis* alternating between *Berberis* spp. and a large number of grass hosts). Within one economically important rust family, the Pucciniaceae, recent molecular evidence points to numerous instances where pathogens have jumped from one host to another taxonomically unrelated but ecologically associated host. At this stage, the causal basis (genetic, environmental, or otherwise) determining such changes in the evolutionary history of the rusts is unknown. However, such examples do provide strong evidence that the issue of new emerging diseases is not entirely one of, or created by, the modern world.

In natural host–pathogen associations, pathogens affect the evolutionary fitness of their hosts by inducing reductions in fecundity or increases in mortality. However, while significant levels of disease may be observed in individual populations, in general the overall outward appearance of the reciprocal battle between host and pathogen is somewhat muted. This is not surprising given that these associations have, for the most part, evolved together over millennia. Examples of contrasting situations occur when pathogens encounter hosts (1) with which they have no evolutionary contact; (2) from which they were separated in the relatively recent past by limited long-distance host dispersal that has resulted in genetically depauperate populations; or (3) that are grown artificially in very large, high-density isogenic monocultures, and recurring severe epidemics lead to major fecundity and or mortality impacts and consequent highly visible changes to the nature of the plant community. In this respect the general consequences of novel pathogen–host contact events, successful biological control programs, or the vulnerability of cereal monocultures in agriculture are essentially the same. Large-scale epidemics result from the genetic uniformity of the target host population.

What is an emerging disease? Emerging diseases have been defined as "any disease that is currently spreading

Eukaryotic Microbes.

within host populations." While it certainly is true that evolutionary change in the pathogen is not a necessary prerequisite for the appearance of an emerging disease, such a broad definition needs to be qualified to avoid inclusion of typical epidemic cycles that may be repeated on a near-annual basis in many agricultural crops and wild plant populations. Notwithstanding this caveat, many diseases of plants, animals, and humans that currently would be generally regarded as emerging have existed for considerable periods of time persisting till now at very low levels in populations of their host or having recently invaded new hosts from reservoir populations of another host.

Emerging diseases may arise in a number of ways, through

1. existing well-recognized pathogens 'catching up' with anthropogenically generated changes in the distribution of existing host species;
2. ecologically based jumps to new, previously unexposed hosts;
3. environmental changes leading to increased aggressiveness in the pathogen or susceptibility in the host; or an expansion of the pathogen's range into parts of its host range from which it was previously excluded; or
4. genetic change leading to an alteration in the host range and/or virulence of an existing pathogen species.

The nature of the first three of these mechanisms – potentially simple changes in the ecology of individual pathogens – reflects the way in which anthropogenic changes in many parts of the world have greatly increased the opportunity for long-distance dispersal with its attendant range of new opportunities. Prior to the advent of increasingly rapid and intense global transport in the last few centuries, the flora of whole continents was essentially quarantined from any possible encounters with exotic pathogens. Elements of the collapse of 'protection through isolation' can be seen in the spread of human diseases like plague in Europe in the thirteenth century, and subsequently in the transmission of diseases from the Americas to Europe and from Europe to the Pacific in subsequent centuries. This increased opportunity for pathogens and naive hosts to come into close contact is a common feature underlying not only the appearance of many animal and human diseases but the increase of many new plant diseases.

WHERE DISEASE EMERGENCE STEMS FROM SIMPLE ECOLOGICAL CHANGES IN DISTRIBUTION

Pathogens 'Catching Up' with Anthropogenically Generated Changes in the Distribution of Their Hosts

Almost all the classic examples of these emerging diseases come from agricultural situations where, over the centuries, crop species have been introduced and grown (often over large areas) in regions far from the original center of diversity of the crop as well as its coevolved pathogens. Over time, selection for traits other than resistance to diseases not present in the new environment, combined with agronomic practices that favor pathogen increase (high nutrition, large, dense, and uniform stands), may make the crop particularly vulnerable to attack. Examples of these situations are found in agricultural (e.g., the Irish potato famine of the 1840s driven by the appearance of *Phytophthora infestans*; the first appearance of stripe rust of wheat caused by *Puccinia striiformis* in Australia in 1979), horticultural (e.g., the collapse of the coffee industry in Ceylon in the nineteenth century due to the impact of *Hemileia vastatrix*), and forestry (Dothistroma blight of *Pinus radiata*) plantings. However, once host and pathogen have been reunited for some time and breeding approaches have begun to deploy genetic approaches to control, differentiating between subsequent periodic failings of resistance deployment strategies and whether it is appropriate to confer the status of a 'new emerging disease' is questionable. Perhaps where the pathogen has been present for some time and causes periodic epidemics, the status of emerging disease should be restricted to special circumstances. An appropriate example might be the threat facing wheat production throughout the Eastern Hemisphere as a novel pathotype of *P. graminis* [Ug99] spreads from its site of origin (Uganda) through the Rift Valley to the Yemen and on to Central and South Asia. Throughout these areas few existing wheat varieties carry appropriate resistance, and major production losses, with accompanying famine, are a significant possibility.

Another example relates to invasive plant species and poses the question about whether pathogens native to an invasive host and that later catch up with the invader in its new area be regarded as new diseases, and if so after how long a time of separation? For example, after its escape from botanical gardens in Europe in 1837, *Impatiens parviflora* from Central Asia and the Himalayas gradually became a dominant species of nutrient-rich forest sites in Europe. About 80 years later its native rust pathogen, *Puccinia komarowii*, showed a dramatic expansion. In this case, although the disease has caught up with the invasive host, and affects host demography, host range expansion still continues.

Jumps to New, Previously Unexposed Hosts

Some of the best-known examples of emerging diseases represent simple situations where existing pathogens have been transported to an environment from which they were previously excluded (nearly always by distance from source populations) and in that new environment came into contact with plant species that lack resistance. In some instances, the new hosts may be close taxonomic relatives of existing host(s) (e.g., *Cryphonectria parasitica* moving from Asian members of the genus *Castanea* to American chestnut *Castanea dentata*), while in others relationships may be somewhat more distant (e.g., *Puccinia lagenophorae* in Australia native on various asteraceous genera such as *Arctotheca* and *Lagenophora* sp. attacking *Senecio vulgaris* in Europe). Similarly, invasive pathogens may be restricted to single new hosts or may be able to attack a very wide range of species in multiple families (e.g., *Phytophthora cinnamomi* attacking members of the Proteaceae, Myrtaceae, and Epacridaceae among others in Western Australia; Table 1). In some cases, spread may be vector-independent, while in others spread is dependent on the prearrival of existing vectors (e.g., Dutch elm disease of amenity trees in New Zealand) or the presence of ecological analogs in the new environment.

Although examples of host jumps are frequently associated with movement of pathogens into new environments, this is not always the case. Indeed, there are many examples in the literature of exotic stands of important forestry species being attacked by pathogens that naturally occur on related native species. Thus in Brazil, plantations of *Eucalyptus* species have been severely and repeatedly attacked by *Puccinia psidii*, a rust native to South America, where it has evolved with a number of Myrtaceous species including guava. In South Africa the native species *Syzygium cordatum* is host to a number of canker and dieback pathogens in the Botryosphaeriaceae that are pathogenic on *Eucalyptus* species; while a range of *Mycosphaerella* species unknown on *Eucalyptus* in Australia cause damage to plantations in Vietnam. Similarly, in Europe, extensive plantations of North American pines have seen host jumps from native pathogens. In the nineteenth century *Pinus strobus* stands were severely damaged by *Cronartium ribicola* native to the Eurasian *Pinus cembra*, and more recently large-scale plantations of *Pinus contorta* in northern Sweden have been damaged by *Gremmeniella abietina*, a pathogen of native conifers.

The importance placed on newly emerging diseases is clearly influenced by their impact. Hence, although *Puccinia malvacearum*, native to Chile, has successfully spread throughout the globe, attacking a broad array of species in the Malvaceae, it is regarded as little more than an oddity as its hosts are neither economically nor ecologically important nor apparently much affected by the disease. A similar case is found in the Australian species *P. lagenophorae* that is now extremely widespread and occurs at high frequency through much of the range of *S. vulgaris* in Europe. Unfortunately, for other introduced newly emerging pathogens this is not the case. Thus, epidemics of *Fusarium circinatum* and *Phytophthora ramorum* that are currently gathering pace in California are already imposing large-scale losses of a range of host species and show no sign of abating. These epidemics are at a very early phase and must be expected to continue to high amplitudes until constrained by the pathogen's environmental envelope and by significant reduction in the frequency and density of highly susceptible host individuals.

In the case of *F. circinatum* and *P. ramorum*, reaching a point on the evolutionary trajectory of these newly established host–pathogen associations at which a new ecological balance is established will be a very long-term process. However, more generally, the speed at which such a balance occurs will be host–pathogen combination, and possibly environment, specific. For shorter-lived species an ecological, if not genetic, balance may be attained within decades. Successful biological control events that reduce host density and population sizes provide significant examples (e.g., the control of *Chondrilla juncea* with *Puccinia chondrillina* in Australia). For longer-lived host species, however, this process could take millennia (e.g., note the continuing major epidemic of white pine blister rust (*Pinus* spp. – *C. ribicola*) in the western United States even though at least one major gene for resistance occurs in some host populations).

A major concern with regard to emerging diseases that attack multiple species or 'keystone' dominants is the impact they have on the community as a whole. Thus, prior to the arrival of *C. parasitica*, *C. dentata* dominated the forests of northeastern United States, consisting of more than 30% of the basal area in many stands. In those situations as the pathogen swept through the community, the loss of chestnut not only resulted in linked extinctions of several specialized natural enemies (or specialized herbivorous insects), but was also often accompanied by an overall increase in diversity of trees and shrubs. Such overarching changes in community structure could have further impacts on the coevolution of native plant–mutualist–pathogen interactions. Furthermore, interplay with changed microclimatic conditions has resulted in an acceleration of successional processes in moist sites and the development of more xeric communities in drier ridge situations. Changes resulting from the introduction of *P. cinnamomi* into Western Australia have been even more dramatic with complex *Banksia*-dominated shrublands being converted into communities with a totally different species mix. It is unlikely that these communities will ever return to a state similar to that prior to the arrival of this pathogen.

WHERE DISEASE EMERGENCE STEMS FROM GENETIC CHANGE IN PATHOGENS

Notwithstanding the discussion above, genetic changes in the nature of pathogens have a very important part to play in the origin of emerging plant diseases. The simplest of genetic changes that can lead to major epidemics are those typically seen in small grain cereal crops where monocultures of the one resistance genotype are often grown over very large areas, thereby imposing very strong selection pressures on the existing pathogen population in

TABLE 1 Some examples of past, present, and potential future emerging plant diseases

Species	Hosts attacked	Origin	New environment	Impact
Past emerging diseases				
Puccinia malvacearum	Wide host range of Malvaceae	Chile	Global	None
Erysiphe alphitoides	Wide host range including Fagaceae, Hippocastanaceae, and Anacardiaceae	First observed in Portugal	Nearly circumglobal	Affects regeneration of *Quercus* spp. in Europe
Phytophthora cinnamomi	Wide host range including Proteaceae, Epacridaceae, and Myrtaceae	?	Western Australia	Major community changes due to loss of multiple species
Ophiostoma novo-ulmi	*Ulmus* spp.		Europe	Major threat to elm forests and ornamental trees
Uncinula necator	*Vitis vinifera*	North America	Global	Earlier devastating epidemics (in Europe)
Sphaerotheca mors-uvae	*Ribes* spp. above all *R. grossularia* and *R. nigrum*	North America or/ and Russia	Europe	Devastating damage on *R. grossularia* (past), now increasingly on *R. nigrum*
Cryphonectria parasitica	*Castanea dentata*	East Asia	Eastern USA	Major community changes associated with loss of community dominant species
Newly emerging diseases				
Phytophthora ramorum	*Quercus* spp.	?	California, Western Europe	Devastating on dominant trees, strong effects on juvenile survival
Discula destructiva	*Cornus florida*	?		Devastation of dominant understory shrub of many northeast American temperate forests
Cronartium ribicola	Five-needled pines	Asia and Eastern Europe	North America	Major effect on dominant white pines, associated with impact on forestry
Puccinia lagenophorae	*Senecio vulgaris*	Australia	Western Europe	Widespread incidence on *S. vulgaris* in northwestern Europe; little apparent population impact
Puccinia komarovii	*Impatiens parviflora*	Central Asia	Europe	Widespread incidence on *I. parviflora* throughout Europe. No visible effect on host expansion
Gremmeniella abietina (European race)	*Pinus resinosa*	Europe	Eastern North America	Devastating epidemics on *P. resinosa*
Phytophthora alni	*Alnus* spp.	Europe	Europe	Major effect resulting in tree die back in particular along slow-flowing watercourses
Future threatening diseases				
Puccinia psidii	Wide host range in Myrtaceae including many *Eucalyptus* species	South America	Australia	Potentially devastating to a wide range of forest communities in Australia
Erysiphe flexuosa	*Aesculus* spp.	North America	Europe	Potential threat to native *A. hippocastanum* on the Balkan peninsula
Dasyscypha willkommii	*Larix decidua*	Europe	Europe	Potential threat to high-altitude forests in the Alps

Note that inclusion in any one category does not exclude the possibility that the disease is still a major problem, or may yet migrate to currently disease-free areas.

favor of novel genotypes with matching virulence. Frequently these changes occur through simple stepwise mutations involving one or two virulence genes as was the case when Australian triticales protected from wheat stem rust attack by the resistance gene *Sr27* were rendered vulnerable following the appearance of a newly virulent pathotype of *P. graminis*.

The process of host-imposed selection is a natural part of all wild host–pathogen systems and is particularly apparent in associations in which gene-for-gene type interactions occur. However, while such selection leads to changes in the structure of wild pathogen populations, in those circumstances the often complex resistance structure of wild host populations leads to epidemic outbreaks that are strictly limited in both time and space. There are no known cases where a simple single mutation increase in virulence in a fungal pathogen has been responsible for major disease epidemics in nonagricultural systems. Despite this, significant genetic changes that give rise to the emergence of new and devastating diseases do occur in wild pathogens. The primary mechanism for this is through hybridization.

Hybridization between related organisms giving rise to a new taxon is increasingly being recognized as a major evolutionary force in changing the adaptability of invasive plants, and has been long identified as a major factor driving recurrent global human influenza epidemics. Fungi are no different in their capability to recombine, and fungal pathotypes with hybrid origin have been known for some time in cereal rusts. Indeed, many fungi have greater genetic flexibility than higher plants and animals with hybridization occurring through two distinct routes – sexual recombination through closely related species or through horizontal gene transfer (somatic hybridization). In somatic hybridization events, genetic exchange occurring via anastomosis of vegetative hyphae is widely regarded as the most likely mechanism, although precisely how exchange occurs is still in question given the existence of fungal nonself recognition systems and that direct observation of this process has rarely occurred. Exchange may result in transfer of whole nuclei, individual chromosomes, and/or of components of the cell's cytoplasm including plasmids and viruses. Where whole nuclei are exchanged, diploidizations, recombination, and subsequent haploidization may or may not subsequently occur (parasexual recombination).

Regardless of the actual mechanism, as the full power of molecular technologies including markers and gene sequencing is applied, it is becoming increasingly clear that hybridization between different pathogen species is a lot commoner than previously thought and plays a significant role in the origin of several of the more important emerging diseases. Thus in Australia, a native wild wheat relative (*Agropyrum scabrum*) that is host to both *P. graminis* f.sp. *tritici* and *P. graminis* f.sp. *secalis* (stem rust of wheat and rye, respectively) provided a common host on which a somatic hybrid naturally developed. The hybrid rust, generated through the exchange of haploid nuclei between the two dikaryotic parent formae speciales, has a different host range to either parent, having acquired through this recombination process virulence on a range of *Hordeum vulgare* lines that are immune to either parent.

Somatic hybrids may also occur between different species, and a number of novel pathogens of trees of agricultural landscapes have arisen in this way. Thus, interaction between two distinct species of poplar rust (*Melampsora larici-populina* and *Melampsora medusae*) in New Zealand, where the sexual stage of these rusts has not been reported, has generated a hybrid rust with a novel combination of virulence specificities. The parents of this hybrid are naturally geographically isolated from each other (Europe and North America, respectively) and this provides a clear example where ecological changes in the distribution of pathogens have brought previously reproductively isolated species together and created the conditions for genetic changes leading to the emergence of a new pathogen. Similar hybrids and even backcrosses have also been found in *Melampsora* species attacking *Populus* species in the northwest United States.

Hybridization among highly specialized taxon groups like the rusts may be of economic concern, but of broader environmental concern and threat to biodiversity are instances of hybridization involving genera housing individual taxa with wide host ranges or pathogens of important dominant species. Probably the best example of the consequences of such events is found in the complex changes that have occurred as two pandemics of Dutch elm disease, the first caused by *Ophiostoma ulmi* and a subsequent epidemic caused by *Ophiostoma novo-ulmi*, swept across much of Asia, Europe, and North America during the twentieth century. These epidemics, and the continuing evolutionary changes occurring in *O. novo-ulmi sensu lato* as a consequence of horizontal gene flow, provide examples of an emerging disease that had its origin in the ecological consequences of geographic spread into naive populations but that has subsequently continued to rage as a result of evolutionary changes made possible by the resultant intimate juxtapositioning of species that were previously allopatric in distribution.

Sexual recombination events also play a role in the origin of a number of emerging and threatening diseases. Recently, this has been documented in the appearance of a number of *Phytophthora* hybrids. For example, in Europe, hybridization between *Phytophthora cambivora* (an introduced pathogen of hardwood trees) and *P. fragariae* (a pathogen of strawberries and raspberries) has given rise to a group of heteroploid hybrid taxa, causing significant destruction to *Alnus* spp. – a host that neither fungal parent is capable of attacking.

WHERE DISEASE EMERGENCE STEMS FROM ENVIRONMENTAL CHANGE

What of the future? With patterns of precipitation and temperature widely predicted to change in current models of future climate, are there likely to be further increases in the incidence of novel disease problems? This question has been the focus of a growing number of studies in recent years. A major problem is to discriminate between epidemics due to more extreme although regularly occurring weather fluctuations (cf. the potato blight famine in Ireland) and long-term climatic trends. Also, relatively small changes may have drastic effects on disease dynamics as recently found for a devastating epidemic of *Dothistroma septosporum* on *P. contorta* in western Canada. Thus, although changing climate will interfere with the complex processes that affect host–pathogen interactions, any forecasts have to be treated with circumspection. Hence, at this stage speculation outweighs fact, but there are a range of reasons to believe that range expansion of existing pathogenic fungi into areas from which they were previously excluded by an inability to cope with harsh environmental conditions will lead to significant and persistent disease epidemics at the margins of distribution of many host species. Thus, there are a range of fungal pathogens whose current distribution appears to be limited by altitude. Where these limitations are imposed by cold, water stress, and season length limitations rather than UV insolation, we might expect a shift in distribution boundaries into previously unchallenged and potentially susceptible host populations. A case in point may become apparent in the European Alps where high-altitude accessions of *Larix decidua* are unable to survive forestry cultivation at lower levels due to epidemics of *Dasyscypha willkommii* – a canker that at high elevations only persists as a saprophyte on fallen twigs. Similarly, sequences of warm winters have been associated with the appearance of overwinter survival of oak mildew *Erysiphe alphitoides* in northern Sweden up to more than 800 km north of its former distribution limit. Another example of dramatic distributional change is the rust *Melampsoridium hiratsukanum* from eastern Asia, which since the first epidemics were observed in Estonia in 1996–1998 and Finland in 1997–1998 has rapidly spread on *Alnus* spp. over most of Europe.

However, there are also likely to be trends in the opposite direction with the impact of pathogens adapted to harsh, cold conditions, particularly, for example, snow-blight of pines (*Phacidium infestans* on *Pinus sylvestris*) and dwarf shrubs (*Eupropolella vaccinii* on *Vaccinium vitis-idaea*), declining as snow cover becomes thinner and less persistent across much of the boreal forest environment.

Another potential driver for new diseases is the radical change in the global nitrogen cycle, which has resulted in accelerating eutrophication of large areas of Europe, eastern North America, and East and South Asia, with consequent dramatic impacts on biota. Whether these changes can be attributed to fungal pathogens has rarely been studied for natural communities. However, for a number of host–pathogen systems, increased concentrations of foliar nitrogen will favor disease, and rapid responses of increased disease severity among biotrophic fungal pathogens have been demonstrated both in agricultural and in natural settings.

Over the past half century particularly dramatic changes have been observed in Europe among the powdery mildews, including range expansions, new host jumps, and increased disease severity of earlier accidentally infected hosts. For example, three mildews have shown rapid range expansions from the east, one in the 1940s (*Microsphaera hypophylla* on *Quercus robur* and *Quercus petraea*), and two in the 1980s (*Microsphaera palcewskii* on *Caragana arborescens*, and *Microsphaera vanbruntiana* var. *sambuci-racemosae* on *Sambucus racemosa*). Another example of an ongoing epidemic is found in *Sphaerotheca erigeronis-canadensis* on *Chamomilla suaveolens*. In Sweden the anamorph of the pathogen was first observed in 1942 and its ascomata in 1978. It is now distributed throughout the country with marked effects on host population dynamics. However, it is not known whether this system represents a host jump due to environmental conditions favorable to the pathogen is the result of evolutionary change in the pathogen population, or a combination of both.

Whether these and similar cases will have long-lasting effects on their respective hosts or whether the epidemics are transient remains to be seen. As demonstrated in the Park Grass Experiment at Rothamsted for the *Blumeria graminis–Anthoxanthum odoratum* system, environments favorable to the pathogen may favor selection for resistance within the host population, thus resulting in low disease levels. However, in situations involving invasive species, rapid evolutionary change like this is dependent on the preexistence of variation for resistance in the host population.

More broadly, the environmental consequences of changing temperature and precipitation regimes and nitrogen loads, as well as their interaction will have different, even opposing, impacts on different fungal pathogens. For some, little if any change in incidence or severity will be apparent; in other cases, however, such changes will alter the environmental envelope of individual species sufficiently to lead to an increase in activity (more frequent and severe epidemics) while in other cases the reverse is likely to be the case.

CAN WE PREDICT THE IDENTITY OF FUTURE EMERGING DISEASES?

As we have seen, there are very many examples of organisms that through new ecological opportunities have jumped hosts or through genetic changes such as recombination have evolved to give rise to major disease epidemics. Understanding the epidemiology and evolutionary biology underlying these differences is a crucial step in understanding the phenomenon of emerging infectious diseases. A major driving force in the epidemiology of any disease association is the interaction of host and pathogen life history features and the way these play out in terms of effective transmission, survival, and reproduction. While each disease has its unique characters, there are also suites of attributes that permit grouping of different host and pathogen types. Thus highly host-specific pathogens and those with a broad host range sit at opposite ends of one particular continuum that determines the amount of host material vulnerable to infection in any situation. Highly specific pathogens with clear boom-and-bust epidemiological patterns may be less of a threat than pathogens with broader host ranges. But how stable are these patterns under different environmental conditions? As host range broadens and pathogens become less specific their impact on any individual host may rise as the total pathogen population size will be determined by the sum of its fecundity on all its hosts – not just on the host of particular focus. In particular, polycyclic pathogens with sexual recombination that are able to respond rapidly upon benign environmental conditions (climatic or nutritional) are likely to play an important role as increased pathogen populations will increase the likelihood for evolutionary change, resulting in new host jumps. However, whether this will be the case is likely to vary broadly even among closely related taxa.

COUNTERING INVASIVE PLANT DISEASES

One of the most frustrating aspects of any study of invasion ecology and its practical application to disease incursion and control strategies is the lack of information as to the frequency with which nascent invasion events occur, the proportion of those that fail to gain a foothold, and even the number that do but remain unrecognized due to their low frequency and a general lack of observers on the ground. If this situation applies in visually obvious organisms like potentially weedy vascular plants, it is substantially worse in considerations of plant disease. Even for diseases of agricultural crops, effective exclusion processes are poor. To paraphrase Thomas Jefferson, "the price of freedom (from disease) is eternal vigilance," but even then for rapidly dispersed diseases of agricultural crops, effective quarantine barriers are potentially the only option, as once entry has been achieved, spread in large extensively planted crop monocultures is often swift and essentially irreversible.

What approaches can then be taken to minimize the arrival of diseases into new environments? The standard preinvasion response is one of maintaining effective quarantine barriers (usually at national boundaries). Such an approach provides a generalized response and has some benefits. However, given the mobility of the world's population and the effective dispersal mechanisms of very many fungi, this approach, while valuable, is essentially only a temporary one that postpones rather than prevents incursion. Where concern is focused on the possible incursion of specific threatening pathogens, matching of the climatic envelope of pathogen distribution in its native range with that potentially available in the target area has been advocated. For example, this approach has been used to estimate geographic areas of highest risk should *P. psidii* arrive in Australia. However, the overall predictive value of climatic matching approaches is generally unproven and where they have been used to predict distribution ranges of invasive weeds the results have often been uninformative.

With postinvasion strategies there is a gulf between the options available for the control of diseases of agriculture and those of native ecosystems. Thus, the appearance of a threatening crop disease would typically precipitate a three-phase response. The first stage of 'contain and destroy' is applied when the pathogen is detected at a very early stage, where the area of ingress is limited and (preferably) isolated (e.g., *Phakopsora vinifera* in northern Australia), and/or the pathogen has limited dispersal capability. During this stage the aim is to eradicate or, failing that, restrict distribution of the pathogen through effective area control. The second stage 'manage and reduce' relies on immediate options provided by agronomic approaches (tillage, rotation cropping, rouging, pesticides) and longer-term ones developed through resistance breeding to ensure the economic impact of the pathogen is limited in most, if not, all years. Finally, the third stage of 'abandonment' occurs where epidemics cannot be controlled and production is subeconomic, leading to crop replacement. Unfortunately, few of these postinvasion strategies are applicable to diseases of plants of native forest and communities. In these situations, if a pathogen successfully circumvents border quarantine barriers and initiates infection, the probability of disease detection early enough to affect eradication is extremely small. The management responses available to agricultural enterprises are not relevant and with the exception of the protection of amenity plantings and sites of special significance, there is little that can be done.

CONCLUSIONS

For a long period the main focus on new emerging diseases has been related to invasive pathogens that have passed oceanic borders. Nowadays, the focus on emerging diseases is above all related to global change issues. Although ongoing and future environmental change is likely to have an impact and may accelerate the emergence of new diseases, it is important to remember that hybridization events and host jumps have regularly occurred during evolutionary time, resulting in the fascinating diversification we now can see and begin to understand. Thus, we foresee a major ongoing challenge as how to prevent the unwanted impacts of pathogens on agriculture, horticulture, and forestry, but in general little, if anything, can be done to hinder them in natural situations.

FURTHER READING

Brasier, C. M. (2001). Rapid evolution of introduced plant pathogens via interspecific hybridization. *BioScience, 51*, 123–133.

Burdon, J. J. (1987). *Diseases and plant population biology*. Cambridge University Press: Cambridge.

Burdon, J. J., Thrall, P. H., & Ericson, L. (2006). The current and future dynamics of disease in plant communities. *Annual Review of Phytopathology, 44*, 19–39.

Collinge, S. K., & Ray, C. (Eds.), (2006). *Disease ecology: Community structure and pathogen dynamics*. Oxford University Press: Oxford.

Gäumann, E. (1951). *Pflanzliche Infektionslehre* (2nd ed.). Birkhäuser: Basel.

Glen, M., Alfenas, A. C., Zauza, E. A., Wingfield, M. J., & Mohammed, C. (2007). *Puccinia psidii*: A threat to the Australian environment and economy – A review. *Australasian Plant Pathology, 36*, 1–6.

Large, E. C. (2003). *The advance of the fungi*. APS Press: St. Paul, USA.

Schrag, S. J., & Wiener, P. (1995). Emerging infectious disease: What are the relative roles of ecology and evolution? *Trends in Ecology & Evolution, 10*, 319–324.

Woolhouse, M. E. J., Haydon, D. T., & Antia, R. (2005). Emerging pathogens: The epidemiology and evolution of host jumps. *Trends in Ecology & Evolution, 20*, 238–244.

Chapter 9

Fungal and Protist Plant Pathogens

A.B. Gould
Rutgers University, New Brunswick, NJ, USA

Chapter Outline

Abbreviations **105**
Defining Statement **105**
Introduction **105**
Fungal Characteristics **106**
Hyphae and Fungal Cells 106
Nutrition 106
Reproduction 108
Dispersal and Survival 109
Fungi and the Environment 110
Symptoms of Plant Disease Caused by Fungi **111**
Abnormal Growth 111
Abscission 111
Host Tissue Replacement 111
Necrosis 111
Permanent Wilt 112
Pathogen Groups **113**
Kingdom Protozoa 113
Kingdom Stramenopila 113
Kingdom Eumycota (Fungi) 116
Diagnosis **123**
Control 124
Further Reading **125**

ABBREVIATIONS

ELISA Enzyme-linked immunosorbent assay
LSD Lysergic acid diethylamide
PARP Pimaricin + ampicillin + rifampicin + pentachloronitrobenzene
RAPD Random amplified polymorphic DNA
RFLP Restriction fragment length polymorphism

DEFINING STATEMENT

An overview of the phyla of plant pathogenic fungi is presented, summarizing hyphal characteristics, reproduction, dispersal, and survival mechanisms, symptoms, fungal diagnostics and control, and definition of commonly used terms. For each pathogen group, examples of common plant pathogenic fungi and the diseases they cause are provided.

INTRODUCTION

The 'fungi' encompass microorganisms that fill a similar ecological niche and yet are distributed among several taxonomic groups, broadly classified as true fungi and pseudofungi (fungus-like). As such, the fungi are examples of convergent evolution, where organisms that are not related evolve similar traits to exploit a similar environment. In their primary role, fungi are extremely important as agents of decay. Many fungi are symbionts; some are parasites and use other organisms (including plants and animals) as a source of food; others are partners with algae to form lichens; some partner with plant roots to form mycorrhizae; and others are endophytes, living within plant tissue without discernible changes in host development. Some fungi are capable of existing in more than one mode.

Although the formal study of fungi, called mycology (Greek for *Mycos* (fungus)+ −*logy* (study)), is about 250 years old, these organisms have had an impact on human social and economic development throughout history. Fungi are a source of food, are used to produce food (cheese, leavened bread, and wine), and can also destroy food at any stage of production, processing, or storage. Many fungi are poisonous; some produce hallucinogens or toxins and others produce enzymes that degrade fabrics, leather goods, and wood products. Societies thatdepended on a single crop for major sustenance have been devastated by fungal disease. For example, the potato crop, a major source of food for Irish peasants in the 1800s, was destroyed by a fungal disease, called late blight of potato, in 1845–1846. The impact of this disease was enormous, resulting in massive starvation and emigration of the Irish people for years to come. The discipline of plant pathology had its origins in the scientific and political controversy

Eukaryotic Microbes.

caused by this disease. Other fungal diseases that have resulted in severe crop loss throughout recorded history include the cereal rusts and smuts, ergot of rye and wheat, brown spot of rice, coffee rust, Sigatoka disease of banana, chestnut blight, and the downy and powdery mildews of grape. Those fungi found consistently in association with a particular plant disease are called pathogens.

FUNGAL CHARACTERISTICS

Fungi are eukaryotic, heterotrophic (lacking photosynthesis) organisms that, in most fungal groups, develop a microscopic, tubular thread called a hypha (pl. hyphae). A group of hyphae is known collectively as a mycelium (pl. mycelia), which makes up the vegetative (nonreproductive) body or thallus of the fungus. In some fungi, the thallus is single-celled (as in the yeasts) or may be plasmodial (without a cell wall, as in slime molds). Although fungi lack a complex vascular system, they can form specialized structures for survival, dispersal, and spore production. Most fungi, except for a few groups, are not motile.

Hyphae and Fungal Cells

Fungal hyphae differ in diameter among species (3–4 to 30 μm or more micrometer wide) and may be septate (with crosswalls) or aseptate (coenocytic, without crosswalls). Septa usually contain small pores to ensure continuity with other cells. Fungal hyphae elongate by apical growth (from the tip). Hyphae grow over a surface stratum, may penetrate it, or may produce an aerial mycelium. In culture, fungi form colonies, which appear as collections of hyphae with or without spores that arise from a central cell or grouping of cells. Fungal cells from different hyphal strands may often anastomose, or fuse, to form a three-dimensional network. This process permits the development of specialized survival or dispersal structures such as rhizomorphs, sclerotia, and fruiting structures, also known as sporocarps.

Hyphal cells are bound by a cell envelope called the plasmalemma (or plasma membrane) that contains the sterol ergosterol. The plasmalemma of fungi differs from that of plants, which contain a phytosterol, and animals, which contain cholesterol. Outside the plasmalemma is the glycocalyx, which appears as a firm cell wall (for most fungi) or as a slimy sheath (as in the slime molds). The glycocalyx is composed chiefly of polysaccharides that differ among fungal groups (Table 1). Primary cell wall polysaccharides of the Ascomycota, Basidiomycota, and Deuteromycota (mitosporic fungi) include chitin (β-(1→4) linkages of *N*-acetylglucosamine) and glucans (long chains of glucosyl residues); in the Zygomycota, chitosan, chitin, and polyglucuronic acid are present; and the Oomycota contain cellulose (β-(1→4) linkages of glucose) and glucans. Cell walls also contain proteins and, in some fungi, dark pigments called melanins. Strong fibers called microfibrils, which lend support to the fungal cell wall, are composed of chitin, glucans, and cellulose.

Fungal cells may be uni- or multinucleate. Nuclei are haploid (most often) or diploid. Cells with genetically identical haploid nuclei are monokaryotic; cells with two genetically different but compatible haploid nuclei are dikaryotic (a characteristic of fungi in the Basidiomycota). Compared to animals and plants, the nuclei of fungi are relatively small with fewer chromosomes or less number of DNA base pairs. Plasmids (extrachromosomal pieces of DNA that are capable of independent replication) are found in some fungi, including the common yeast fungus, *Saccharomyces cerevisiae*. Other organelles found in fungal cells include mitochondria (energy-producing organelles), which vary in size, form, and number; vacuoles, which serve to store water, nutrients, wastes, or enzymes such as nucleases, phosphatases, or proteases; and plastids, which contain pigments and enzymes and may store food. Fungal cells accumulate reserve carbon materials as glycogen, lipids, or low molecular weight carbohydrates such as trehalose.

Nutrition

Fungi lack chlorophyll and, unlike plants, cannot manufacture their own food. Fungal growth requires an external food source, water, and appropriate environmental conditions. Growth ceases when any of these become limiting. Nutrients required by fungi include a source of carbon (sugars, polysaccharides, lipids, amino acids, and proteins); nitrogen (nitrate, ammonia, amino acids, polypeptides, and proteins); magnesium, phosphorus, potassium, and sulfur; and trace elements such as calcium, copper, iron, manganese, molybdenum, and zinc.

Fungi obtain nutrients by using an absorptive mechanism: hyphae secrete digestive exoenzymes into an external food source, and nutrients are carried back through the fungal cell wall and stored in the cell as glycogen. This process requires the presence of free water. The source of carbon used by a given fungus is limited only by the exoenzymes the fungus produces (Table 2).

Food relationships

Most fungi are scavengers or decay organisms that use nonliving sources of organic material as a source of food. These organisms are called saprophytes, and along with bacteria, recycle carbon, nitrogen, and mineral nutrients. At the other end of the spectrum are biotrophs (sometimes called obligate parasites) that derive all nutrients necessary for growth or reproduction only from a living host. Biotrophs, which are highly host specific, do not readily kill their food source, thus ensuring a steady supply of nutrients. Some biotrophs produce haustoria, which are structures designed for

TABLE 1 Classification and characteristics of plant pathogenic fungi

	Kingdom					
	Protozoa	Stramenopila	Eumycota			
	Plasmodiophoromycota	Oomycota	Chytridiomycota	Zygomycota	Ascomycota	Basidiomycota
Habitat	Aquatic	Aquatic	Aquatic/ terrestrial	Terrestrial	Terrestrial	Terrestrial
Form of thallus	Plasmodium	Coenocytic hyphae	Globose or ovoid thallus (no true mycelium)	Coenocytic hyphae; rhizoids, stolons	Septate hyphae; some are single-celled (yeasts)	Septate hyphae
Ploidy	Haploid	Diploid	Diploid	Haploid	Haploid	Dikaryotic, haploid
Chief component (s) of the cell wall	None	Cellulose/ glucans	Chitin	Chitin/chitosan	Chitin/glucans	Chitin/glucans
Motile stage	Zoospores with two anterior, unequal, whiplash flagella	Zoospores with two flagella: one anterior tinsel and one posterior whiplash flagella	Zoospores with one posterior whiplash flagellum	None	None	None
Food relationship	Biotroph	Facultative, biotroph	Biotroph	Facultative	Facultative, biotroph	Facultative, biotroph
Asexual reproduction	Zoospores in zoosporangia	Zoospores in sporangia; chlamydospores	Holocarpic	Sporangiospores in sporangia	Conidia on conidiophores (singly or in fruiting bodies)	Conidia, chlamydospores, oidia
Sexual reproduction	Resting spores, the result of fusion of zoospores to form a zygote	Oospores, the result of fusion between male antheridia and female oogonia	Not confirmed	Fusion of isogametes to produce a zygospore	Formation of ascospores within an ascus	Formation of four basidiospores on a basidium; fusion of pycniospores and receptive hyphae (rusts); fusion of compatible mycelia (smuts)

Reproduced with permission from Gould, A. B. (2008). Plant pathogenic fungi and fungal-like organisms. In R. J. Trigiano, M. T. Windham, & A. S. Windham (Eds.), *Plant pathology concepts and laboratory exercises*, 2nd ed. Boca Raton: CRC Press.

penetration and absorption of nutrients. Although a haustorium may penetrate the host cell wall, it does not penetrate the host plasmalemma, thus essentially remaining outside the host cell while nutrients are transferred into the fungus. Compared to strict saprophytes, biotrophs are relatively rare. Classic examples of pathogenic biotrophs include the powdery mildew fungi and rusts.

Fungi may also be classified as facultative saprophytes or facultative parasites; these organisms are more versatile and utilize both living (as parasites) and nonliving (as saprophytes) sources of carbon. Facultative saprophytes live primarily as parasites; they attack living hosts but then subsist between growing seasons as saprophytes within the normal soil microflora. Alternatively, facultative parasites under normal conditions utilize nonliving sources of carbon (as saprophytes); given fortuitous circumstances, however, these organisms will attack highly susceptible living plant tissue, as parasites.

Other pathogens, called necrotrophs, survive as saprophytes in the absence of a living host, but given the opportunity, will kill a host plant and subsequently feed on the dead plant tissues. Necrotrophs produce secondary

TABLE 2 Function of digestive exoenzymes produced by pathogenic fungi

Enzyme	Substrate	Function
Amylase	Starch (storage polysaccharide in plant cells)	Hydrolyzes starch to glucose, which is used as food by pathogen
Cellulase (cellulase C_1, C_2, C_x, and β-glucosidase)	Cellulose (insoluble, linear polymer of β-(1→4) linkages of glucose); skeletal component of primary and secondary plant cell walls	Softens cell walls, facilitating spread
Cutinase	Cutin (long-chain polymer of C_{16} and C_{18} hydroxy fatty acids); with waxes and cellulose, main component of cuticle	Facilitates direct penetration into host cuticle; enzyme necessary for pathogenicity
Hemicellulase (e.g., xylanase, arabinase)	Hemicellulose (mixture of amorphous polysaccharides); component of primary and secondary cell walls	Role of these enzymes in pathogenesis is unclear
Ligninase	Lignin (complex, high molecular weight polymer made of phenylpropanoid subunits); major component of secondary cell wall and middle lamella of xylem tissue	A few basidiomycetes degrade lignin in nature; brown rot fungi degrade lignin but cannot use it as food; white rot fungi can do both
Lipase	Fatty acid molecules; major components of plant cell membranes; also stored in cells and seeds and found as wax lipids on epidermal cells	Fatty acid molecules used as a source of food by pathogens
Pectinase (pectin methyl esterase, polygalacturonase)	Pectin (chains of galacturonan molecules (α-(1→4)-D-galacturonic acid) and other sugars); main component of the middle lamella and primary cell wall	Degrades middle lamella and primary cell wall, macerating host tissue; facilitates penetration and colonization of host
Proteinase	Protein (major components of enzymes, cell walls, and cell membrane); enzymes hydrolyze protein to smaller peptide fractions and amino acids	Disruption of cell membrane and enzymatic activity affects host cell function; precise role of these enzymes in pathogenesis is unclear

Reproduced with permission from Gould, A. B. (2008). Plant pathogenic fungi and fungal-like organisms. In R. J. Trigiano, M. T. Windham, & A. S. Windham (Eds.), *Plant pathology concepts and laboratory exercises,* 2nd ed. Boca Raton: CRC Press.

metabolites that are toxic to susceptible host cells; cells killed by these toxins are degraded by fungal enzymes, and the cell constituents are used as food. Tissues killed by necrotrophs often appear blackened or sunken. An example of a necrotroph is *Monilinia fructicola*, which causes brown rot of peaches.

Finally, the hemibiotrophs function as both a biotroph and a necrotroph during the life cycle. At first, these fungi subsist as biotrophs, growing between the plasma membrane and the cell wall of living cells, and switch later to a necrotrophic phase, killing all the colonized cells and utilizing the dead tissues for nutrients. Those parasites found consistently with a given plant disease are called pathogens. The soybean anthracnose pathogen, *Colletotrichum lindemuthianum*, is a hemibiotroph.

Reproduction

Most fungi produce spores, which are small, microscopic units consisting of one or more fungal cells. Spores are produced asexually or sexually and provide a dispersal and/or survival function for the fungus. The size, shape, color, and genesis of spores are also used as taxonomic criteria.

Asexual reproduction

Asexual reproduction, the result of mitosis, results in progeny that are genetically identical to the parent. Spores that result from asexual reproduction include conidia, chlamydospores, zoospores, and sporangiospores.

Most true fungi in the Ascomycota, Basidiomycota, and those classified as deuteromycetes (mitosporic fungi) produce conidia (nonmotile, asexual spores) at the tip or side of a supporting structure known as a conidiophore. Conidiophores are arranged singly or in fruiting structures such as synnema (stalks of fused conidiophores), acervuli (flat, saucer-shaped beds of short conidiophores that grow between host tissues and the epidermis or cuticle), and pycnidia (flask-shaped structures lined with conidiophores).

Chlamydospores, found in many groups of fungi, are thick-walled resting spores formed as hyphae cells develop a thick wall and separate. Sporangiospores, produced by fungi in the Zygomycota, are also borne in a sporangium. Zoospores, produced in the Oomycota, Plasmodiophoromycota, and Chytridiomycota, are motile spores with one or more flagella. In the Oomycota, zoospores are borne within a sac-like structure called a sporangium, supported by a stalk called a sporangiophore. Zoospores are specialized for short-distance dispersal, responding to external stimuli, such as root exudates (chemotaxis), for identifying suitable penetration sites, where they encyst and initiate new infections.

Sexual reproduction

Sexual reproduction results in genetic recombination, producing offspring that are genetically different from either parent. In the fertilization process, gametangia produce special sex cells (gametes or gamete nuclei) that fuse to form a zygote. Fertilization is a two-step process: (1) plasmogamy, where the two nuclei join together in one cell; and (2) karyogamy, where these nuclei fuse to form a zygote. In different hyphal groups, karyogamy does not necessarily occur immediately following plasmogamy. Examples of sexual spores include ascospores, basidiospores, oospores, and zygospores.

The thallus of most true fungi is haploid (contains one set of chromosomes); thus the diploid (two sets of chromosomes) zygote that results from karyogamy must undergo meiosis before a haploid thallus can then develop. In contrast, fungus-like species in the Oomycota possess a diploid thallus, and meiosis occurs in the gametangia as the gametes are formed. Homothallic fungi (producing both male and female gametangia on a single mycelium) are self-fertile. These gametangia may be differentiated into male (antheridium) and female (oogonium) structures or may be undifferentiated. In contrast, heterothallic fungi (producing male and female gametangia on separate individuals, or mating types) are self-sterile. Mating types are usually designated by using letters (e.g., A and A′) or as plus (+)/minus (−).

Anamorph–teleomorph relationships

Most fungi exhibit both sexual and asexual reproduction. These fungi are pleomorphic, having more than one form or state. The sexual state of a fungus is called the teleomorph (in the literature, often referred to as the perfect state), and the asexual state is called the anamorph (the imperfect state). The organism in its totality (both anamorph and teleomorph) is the holomorph. The anamorph or teleomorph in some fungi may be lacking or not described. Those fungi that lack a known teleomorph are artificially described as mitosporic fungi, imperfect fungi, or deuteromycetes.

Many true fungi, particularly in the Ascomycota and Basidiomycota, produce sexual and asexual states at different points in the life cycle. As a result, these fungi are associated with two names: one that refers to the anamorph at the time it was described and the other that refers to the teleomorph at the time that it was described. This can be confusing from a taxonomic standpoint. When an anamorph is finally associated with its corresponding teleomorph, the holomorph is referred to by its teleomorph name. Although the fungus may reside in different reproductive states during its life cycle, the genetics of the two states are the same; thus molecular biology has been helpful to link anamorphs to their corresponding teleomorphs. Examples of holomorphs include *Claviceps purpurea* (teleomorph) ana. *Sphacelia segetum* (anamorph) (causal agent of ergot of rye and wheat); *Cochliobolus heterostrophus* ana. *Cochliobolus heterostrophus* (southern corn leaf blight), and *Venturia inaequalis* ana. *Spilocea pomi* (apple scab).

In general, the sexual and asexual reproductive strategies exhibited by many pathogenic fungi may confer different ecological advantages. When food is abundant and dispersal is important, fungi reproduce asexually. Conversely, when food is limited or dispersal is not as critical, fungi reproduce sexually. For example, the fungi that cause powdery mildew are biotrophs and produce easily dispersed conidia (asexual spores) during the growing season when host material is abundant. At the end of the growing season, sexual reproduction is triggered by host and environmental factors. The resulting ascocarp (called a cleistothecium) overwinters in association with plant parts or debris, releasing meiotic ascospores the following growing season.

Dispersal and Survival

Most fungi that cause disease in plants spend a portion of their life cycle as parasites and the remainder as saprophytes, using nutrients found in soil or in plant debris. Hyphal growth ceases during adverse environmental conditions or when nutrients are limiting. Pathogenic fungi must develop mechanisms to disperse to new hosts or to otherwise survive until conditions improve. Spores used for dispersal include conidia and zoospores; spores and other structures that serve a survival purpose include chlamydospores, oospores, sclerotia, teliospores (the rust and smut fungi), and zygospores. Other spores, such as ascospores and conidia, may survive adverse environmental conditions within fruiting structures such as pycnidia, perithecia, pseudothecia, and cleistothecia. Pathogens may also survive and disperse as spores or hyphae in association with infected or infested seed and other plant parts.

Spore liberation and pathogen dispersal

Fungal survival depends on the exploitation of new food sources; thus, liberation from the parent mycelium and subsequent dispersal are critical processes. Spore liberation from the parent mycelium is passive or active. Spores such as conidia may be passively liberated through outside forces, mechanically (splashing water, wind, animal disturbance, or cultivation equipment) or by electrostatic repulsion between the spore and its subtending stalk (sporophore). An example of a pathogen commonly liberated by air currents in greenhouses is *Botrytis cinerea*. Active spore liberation occurs due to mechanisms within the fungus itself, such as changes in water pressure within cells. Ascospores are often forcibly discharged from the ascus (supporting sac) in many ascomycetes, and changes in cell shape in some rust fungi, such as species of *Puccinia*, actively propel aeciospores into the air.

Dispersal is also achieved by passive (most common) or active means. Passively dispersed (vectored) pathogens may, depending on the mechanism, travel short or long distances. Spores are mostly commonly dispersed by water (rain splash and flowing water), air currents, insects, or with seed or plant parts. For example, the sporangia and zoospores of pythiaceous fungi (oomycetes), which cause root diseases of many important crops, are easily dispersed over long distances as water moves through the soil profile. Urediniospores of the stem rust of wheat pathogen, *Puccinia graminis*, are blown hundreds of miles from winter wheat crops in the south to the spring wheat-producing portions of the United States along a corridor known as the 'Puccinia pathway.' Insects are intimately associated with the life cycle and spread of many fungi. For example, *Ophiostoma ulmi*, the pathogen that causes Dutch elm disease, is vectored by bark beetles that spread fungal spores during feeding. This fungus is also dispersed from tree to tree through root grafts. Fungi dispersed with seed, such as *Tilletia caries* (stinking smut of wheat) or *C. purpurea* (ergot of wheat and rye), are conveniently situated near their food source at the beginning of the infection process. Plant pathogens are most effectively dispersed across land masses or vast bodies of water by human activities. For example, both *Cryphonectria parasitica* (chestnut blight) and *Phytophthora parasitica* (late blight of potato) destroyed vast acreage of host plants when moved with infected host plant material from the pathogen site of origin to an exotic location.

Active dispersal in pathogenic fungi is limited. As mentioned previously, zoospores produced by several fungal groups actively seek new root penetration sites, using stimuli such as chemicals, oxygen, or light for direction (taxis). Zoospores do not swim linearly, however, and instead are pushed or pulled by flagellar motion, frequently changing direction when hitting an obstacle in the soil. Zoospores are best for dispersal over short distances.

Rhizomorphs, common in some basidiomycete fungi, are thick strands of somatic hyphae that resemble roots. Pathogens such as *Armillaria mellea* disperse from an existing carbon source, which may be a plant host or woody plant debris to new substrates as the active meristem of the rhizomorph grows through the soil. Rhizomorphs have a dark outer rind that protects an inner cortex of active hyphae. These root-like structures normally penetrate new host roots, usually trees, near the soil line. A new mycelium then proceeds to grow up the trunk just under the bark, killing vascular tissues and causing tree decline or death.

Survival

As mentioned previously, fungi survive adverse environmental conditions as spores (e.g., oospores, zygospores, chlamydospores) or other specialized structures that act as propagules or sources of inoculum. Common to most of these structures is a thick, protective outer cell wall (as in spores) or rind (as in rhizomorphs and sclerotia). Survival structures may be the result of both sexual and asexual reproduction, or in certain groups, may germinate to release sexual (as in *C. purpurea*) or asexual (as in *B. cinerea*) spores. With the exception of rhizomorphs, which also serve a dispersal purpose, most survival structures are passively dispersed, if at all.

Oospores and zygospores are the result of sexual reproduction in the Oomycota and Zygomycota, respectively. An oospore forms when an oogonium (female gamete) is fertilized by an antheridial (male gamete) nucleus; a characteristically thick wall and food reserves help to ensure survival. Zygospores, which form following fertilization of gametes that are morphologically indistinguishable, are protected by a thick wall containing melanin and ornamented by warts. As in the Oomycota, the thick cell wall structure as well as abundant lipid reserves facilitates long-term survival of the fungus. Meiosis does not occur until conditions are sufficient for zygospore germination, which results in the production of a sporangium with zygospores.

Asexually produced chlamydospores are found in many fungal groups. Hyphal cells, either within the strand (intercalary) or at the terminus, accumulate a thick wall and nutrient reserves, round up, and separate. Although chlamydospores are common to soilborne fungi, they are associated with aerial pathogens as well, such as *Phytophthora ramorum*, the causal agent of sudden oak death and ramorum blight.

A common survival, or resting, structure is the sclerotium. Sclerotia are compact masses of mycelium, often spherical or pellet-shaped, that range from 1 mm to 1 cm in diameter and consist of a central core of hyphae with lipid and glycogen reserves protected by a thick-walled rind. Plant pathogens that produce sclerotia or a similar structure, the microsclerotium, tend to persist many years in the absence of a suitable host. Sclerotia are common among species in the Ascomycota and Basidiomycota, especially among those that infect herbaceous plants as a means of surviving between crops.

Fungi and the Environment

Many factors (moisture, a carbon source and other nutrients, and the proper environment) are necessary for fungal growth. Of these, adequate moisture (free water as well as high relative humidity) is highly critical. Moisture is necessary to prevent desiccation of hyphae, to facilitate nutrient uptake, to facilitate germination and penetration of host tissues during the infection process, and for dispersal. Fungi that can adapt to low moisture availability do so by regulating the concentrations of cellular solutes or by producing resting structures with a wall structure that withstands drying.

Most plant pathogenic fungi are aerobic, requiring oxygen for respiration (generation of energy). Some fungi such as the common yeast (*S. cerevisiae*), however, do not need oxygen for respiration and are anaerobic or fermentive. There are a few plant pathogens that derive energy by either oxidation or fermentation; these are called facultative fermentives. Although light is not a requirement for fungal growth, some species produce melanins in the hyphal wall to protect against damage from sunlight.

Fungi must tolerate temperatures that may fluctuate considerably during the day or throughout the year. Most fungi grow well between 10 and 40 °C (with an optimum between 25 and 30 °C) and are considered mesophilic. Species that grow best at 40 °C or higher are thermophilic; those that grow at temperatures of less than 10 °C are psychrophilic. Fungi function best at a pH of 4–7.

SYMPTOMS OF PLANT DISEASE CAUSED BY FUNGI

Symptoms are the visual manifestation of the infection process. Symptoms vary depending on the host, the infected plant part, and the environment. Symptoms caused by fungal pathogens can be similar to those caused by other biotic disease agents, such as bacteria and viruses, as well as abiotic (environmental) disease agents, including extremes in moisture and temperature. Symptoms caused by fungal pathogens are broadly described as abnormal growth, abscission, host tissue replacement, necrosis, and wilt.

Abnormal Growth

Abnormal growth is caused by hypertrophy (excessive cell enlargement), hyperplasia (excessive cell division), and etiolation (excessive elongation). Tissues affected by these processes may be gall- or club-like, misshapen, or curled. Symptoms associated with abnormal growth include the following:

Clubroot – roots are swollen, spindle- or club-shaped. For example, clubroot of crucifers (caused by *Plasmodiophora brassicae*); the clubs form as cells abnormally divide or enlarge.

Etiolation – excessive shoot elongation and chlorosis, induced in poor light or by growth hormones. For example, foolish seedling disease of rice (also called bakanae, caused by *Gibberella fujikuroi*). Study of this disease led to the discovery that plants produce similar compounds called gibberellins.

Gall – many fungi cause galls (enlarged growths, round or spindle-shaped) to form on leaves, stems, roots, or flowers. For example, cedar affected by cedar-apple rust (caused by *Gymnosporangium juniperi-virginianae*); galls consisting of both host and fungal tissue develop in stem tissues. These galls later serve as a source of inoculum.

Leaf curl – leaves are discolored and curled, often at the edge of the leaf. For example, species of *Prunus* affected by *Taphrina deformans* (causal agent of peach leaf curl).

Wart – outgrowth on stems and tubers. For example, potato wart disease caused by the chytrid *Synchytrium endobioticum*.

Witches' broom – profuse branching, resembling a spindly broom, is a common symptom associated with several fungal diseases. For example, witches' broom of cacao (*Crinipellis perniciosa*). *C. perniciosa* is a basidiomycete found wherever cocoa is grown in the western hemisphere. Yields in severely affected plantings may be reduced by as much as 90%.

Abscission

Abscission includes premature defoliation and fruit drop, and shot hole (where affected portions of the leaf blade drop out). Many foliar diseases cause premature leaf or fruit drop. These include apple scab (*V. inaequalis*), shade tree anthracnose (*Apiognomonia* spp.), and ash rust (*Puccinia sparganioides*). Shot-hole forms on leaves of stone fruit when portions of the leaf blade that surround an infection site break down, causing the affected portion to shrivel and drop from the blade.

Host Tissue Replacement

Host tissues, particularly reproductive structures, may be replaced by fungal hyphae and spores. In the disease ergot of rye and wheat, *C. purpurea* invades the grain within the seed, replacing the embryo with a fungal sclerotium (the ergot) that is poisonous to animals and humans. Other examples in this category include the bunts and smuts, characterized as galls or seed heads that are entirely filled with masses of teliospores. Ears of corn affected by corn smut (caused by *Ustilago maydis*) become infected through the silk. The galls that form in place of the kernels enlarge, eventually filling with hyphae that convert to smutted masses of black teliospores.

Necrosis

The most common reaction to fungal infection is necrosis. Cells are killed by fungal enzymes, toxins, or host defense responses; affected tissue is brown or blackened, dry or slimy, sunken, and (in leaves) is often preceded by chlorosis (yellowing, caused by a breakdown of chlorophyll). Symptoms associated with necrosis include the following:

Anthracnose – sunken or blackened lesions appear on leaves (often following leaf veins and/or leaf margins), stems, and fruits. For example, anthracnose (or cane spot) of brambles (*Elsinoë veneta*), a serious disease of purple and black raspberries. In severe cases, fungal lesions girdle the stem to weaken or kill the cane.

Canker – elliptical lesions on branches and stems that destroy vascular tissue; lesions appear cracked, raised, sunken, or associated with resin (conifers); affected branches may wilt, die back, and die. For example, Cytospora canker of spruce (*Cytospora kunzei*, syn. *Leucocytospora kunzei* (anamorph), *Leucostoma kunzei*, syn. *Valsa kunzei* (teleomorph)) is one of the most important diseases of Colorado and Norway spruce in the landscape. Cankers develop on the lower limbs, killing branches as the disease progresses up the tree. As with many canker diseases, fruiting structures are evident in the killed tissue, which is also filled with resin.

Cutting rot – cuttings in propagation beds are blackened from the cut end and up through the stem; ensuing rot rapidly kills the cuttings. For example, Blackleg of geranium (*Pythium* spp.). This cutting rot is common in propagation beds that are overly moist or poorly drained and contaminated with the fungus. Affected plants must be destroyed.

Damping-off – seeds and seedlings are killed before (preemergent) or after (postemergent) they break the soil surface. For example, damping-off of ornamental or vegetable seedlings caused by species of *Fusarium*, *Pythium*, and *Rhizoctonia*. The first indication of damping-off is a bare patch (preemergent damping-off), often circular, within a flat of seedlings. As the disease progresses through the flat, seedlings that have already emerged collapse and die (postemergent damping-off). The fungi attack these seedlings at the root tips or at the soil line.

Dieback – twig or branch necrosis that begins at the tip and progresses toward the twig base. For example, sudden oak death and ramorum blight (*P. ramorum*) is a recently identified disease that, in its canker form (sudden oak death), has killed over 1 million trees in the Fagaceae (oak family) in coastal California. The host list for this pathogen includes many native understory species as well as horticultural crops, such as azalea and rhododendron, where shoot dieback is a predominant symptom (ramorum blight).

Dry rot – a crumbly decay of fleshy plant organs. For example, Fusarium dry rot of potato (*Fusarium sambucinum*) affects tubers and seed pieces, reducing crop establishment.

Leaf spot or blotch – discrete lesions of dead cells on leaf tissue between or on leaf veins; appearance varies as lesion borders and centers differ in color or are target-like; sometimes accompanied by a yellow halo; fungal fruiting structures are often evident in dead tissue. Example of leaf spot is strawberry leaf spot (*Mycosphaerella fragariae*); example of leaf blotch (larger, more diffuse regions of dead tissue) is horse-chestnut leaf blotch (*Guignardia aesculi*).

Needle cast – needles of conifers, very often the previous season's growth, are prematurely cast from the tree. For example, Rhabdocline needle cast (*Rhabdocline pseudotsugae*) affects only Douglas-fir. In the spring, red-brown lesions with fruiting structures form on the previous season's needles. Ascospores are released during damp weather to infect the new growth. Needles are then cast from the tree after spore release, resulting in trees that may be unfit for sale.

Root and crown rot – root necrosis extends from death of feeder roots to the entire root system, often extending into the crown to girdle the base of the stem; symptoms aboveground include dieback, wilt, and death of the canopy. For example, Phytophthora root and crown rot (also known as Rhododendron wilt) (*Phytophthora cinnamomi*, *P. parasitica*, and other species) affects the roots of rhododendron, azalea, and other ericaceous hosts grown in soils with excessive moisture. The fungal pathogens produce zoospores that disperse within the root systems of susceptible species. The resulting root necrosis causes aboveground symptoms of wilt (leaves of rhododendron will roll downward along the midvein), dieback, and die.

Scab – lesions on fruits, leaves, tubers, and other organs become crusty, raised, or sunken. For example, apple scab (*V. inaequalis*) is one of the most serious diseases of apple and crabapple grown commercially or for landscape use. Seriously affected trees defoliate, fruits develop scabbed lesions and may be unfit for sale, and yield is poor.

Soft rot – fleshy plant organs (bulbs, corms, fruits, rhizomes, and tubers) are macerated, becoming water soaked and soft; affected tissues lose moisture, becoming hard or shriveled into mummies. For example, Rhizopus soft rot of papaya (*Rhizopus stolonifer*) is a common disease of papaya that occurs postharvest. Affected papaya fruits in storage rapidly decay, leaving the cuticle intact. The fungus emerges from cracks in the cuticle, spreading to other fruits. Fruits are further decayed by other fungi and bacteria, emanating a sour odor.

Permanent Wilt

Permanent wilting occurs when vascular tissues are blocked or destroyed by fungal growth, toxins, or host defense responses. Tissue connected to affected vessels may wilt and die, and associated leaves may scorch and prematurely abscise. A common disease of both herbaceous and woody plant species is Verticillium wilt (caused by *Verticillium dahliae*). The fungus survives as microsclerotia in the soil, which germinate in the presence of root exudates. The fungus subsequently invades the xylem producing conidia on verticilliate whorls of sporophores. The ensuing vascular dysfunction is accompanied by symptoms of permanent wilt, which, in many cases, results in the death of the plant.

PATHOGEN GROUPS

Fungal taxonomy has been classically based on the morphology and genesis of spores, hyphal and colony characteristics, nutrition, and growth on selective media. Recent advances in molecular techniques that examine fungal genetic sequences or their protein products have become increasingly useful to taxonomists, especially with regard to matching an anamorph with its corresponding teleomorph. Methods such as isozyme analysis, restriction fragment length polymorphism (RFLP), and random amplified polymorphic DNA (RAPD) are used to compare closely related isolates. PCR techniques are also useful for detecting pathogens in culture, host plants, or debris.

Although once classified in the kingdom Planta, the fungi are now placed in three different kingdoms: Protozoa, Stramenopila, and Eumycota (also called kingdoms Mycota or Fungi). Pathogenic fungi are best identified by the phylum to which they belong and are placed in these phyla based on the sexual state of their life cycle. Organisms considered true fungi (Eumycota) are classified in the phyla Chytridiomycota, Zygomycota, Ascomycota, and Basidiomycota. These organisms are referred to as chytrids, zygomycetes, ascomycetes, and basidiomycetes, respectively. Fungus-like organisms, which differ in their evolutionary history, are placed in the Protozoa (the Plasmodiophoromycota or endoparasitic slime molds) and Stramenopila (the Oomycota, or oomycetes; also referred to as water mold fungi). The Plasmodiophoromycota also contains the slime molds (Myxomycota), which are commonly seen saprophytes in the landscape but are not plant parasites. The Chytridiomycota, once considered protozoans, may be regarded as ancestors of the other true fungi. As stated previously, the 'fungi' as a group fill a similar ecological niche, even as their evolutionary backgrounds differ. Basic differences among these groups are described in Table 1. Below are descriptions of the different phyla within their respective kingdoms and examples of plant pathogens for each.

Kingdom Protozoa

Plant pathogenic protozoans are classified within the Plasmodiophoromycota. These endoparasitic slime molds are biotrophs, and represent one of several fungi that produce zoospores, which serve a dispersal (primary zoospores) as well as a reproductive (secondary zoospores) purpose. The thallus of these organisms is a multinucleate, amoeboid plasmodium that lacks a cell wall and is restricted to the host plant cell. Sexual reproduction in this group includes fusion of zoospores (as isogametes) to form a zygote. Species of two genera within this group, *Polymyxa* and *Spongospora*, can also vector plant pathogenic viruses.

Clubroot of crucifers, caused by *P. brassicae*, is one of the most significant diseases of crucifers (cauliflower and cabbage) worldwide. This soilborne fungus invades root cells, inducing hypertrophy and hyperplasia, to form 'clubs' that interfere with normal root function. Affected plants wilt and stunt. *Plasmodiophora* survives in soil for extremely long periods as resting spores (the product of meiosis); thus, fields with a history of this disease cannot be replanted with susceptible host species.

The disease cycle of clubroot of crucifers begins as haploid resting spores in the soil germinate to form primary zoospores. These spores, as agents of dispersal, identify, encyst, and penetrate the root hairs of susceptible hosts. A primary, multinucleate plasmodium forms as the nuclei divide through mitosis. The plasmodium eventually cleaves into multinucleate portions, each developing into a sporangium (zoosporangium) with four to eight haploid secondary zoospores. At this stage of the disease, the impact of the parasite on plant function is slight. Secondary zoospores are discharged to soil through pores dissolved in the root cell wall. These zoospores may act as isogametes to form diploid cells that penetrate roots directly or through wounds. The diploid plasmodia that result from fertilization are much more damaging to the host; the fungus spreads to cells throughout the cortex and vascular system, clubs form as a result of hyperplasia and hypertrophy, and the plants wilt and stunt. Meiosis in plasmodial cells results in the production of resting spores, which are released into the soil as the roots disintegrate; a group of resting spores is called a sorus.

Management of clubroot of crucifers includes crop rotation (crucifers no more than once every 3–5 years) and increasing soil pH to 7.2 to inhibit spore germination. Some cultivars are resistant to some races of the pathogen, but not to all.

Kingdom Stramenopila

Organisms in the Stramenopila are considered protists that include the brown algae, diatoms, and water molds. The flagella in this group exhibit hair-like projections, a characteristic that gives the Stramenopila (Latin: *stramen* (flagella) + *pilos* (hairs)) its name. Water molds are placed in the phylum Oomycota.

Plant pathogenic oomycetes are considered 'water molds' because most species produce motile, biflagellate zoospores that require free water for dispersal. The thallus of the Oomycota consists of diploid, coenocytic hyphae that contain cellulose (a defining characteristic) and glucans in the cell wall. The result of asexual reproduction in this group is the zoospore produced in a zoosporangium; some species produce additional chlamydospores, others reproduce asexually by means of nonmotile sporangiospores. Sex in the Oomycota results in a diploid oospore produced when a female gamete (oogonium) is fertilized by a nucleus from the male gamete (antheridium). Since the thallus is diploid, meiosis must occur as the gametes, which are morphologically differentiated, are formed.

Most plant pathogenic oomycetes are placed in two orders (Table 3). The only significant plant parasites within the Saprolegniales are species of the genus *Aphanomyces*, which causes a root rot of annual plants such as pea and sugar beet. Within the Peronosporales, most pathogenic oomycetes, such as species of *Pythium* and many species of *Phytophthora*, are soilborne and attack roots. Others, such as the downy mildews, the white rusts, and other species of *Phytophthora*, are associated with aerial plant parts.

Pythium diseases are common and can be very damaging to crops during production, transit, post harvest, and at market. *Pythium* species are ubiquitous, causing a damping-off, seed and root rot, or cutting rot of all plant types. A soft rot can also occur when fleshy organs (such as fruits and vegetables) come in contact with the soil. These oomycetes are especially troublesome in greenhouse plants and in turfgrasses, and in general, younger plants are more susceptible than older plants.

The *Pythium* infection process begins when germ tubes from encysting zoospores or from germinating sporangia directly penetrate susceptible plant tissues. The fungus grows between and within host cells, producing pectinase to dissolve pectin in the middle lamella and cellulase to break down cell walls. The result is a rotted mass of macerated tissue. As colonization progresses, sporangia produce a balloon-like vesicle from which zoospores are released. Chlamydospores, as well as oogonia, also form. At the appropriate time, oospores, which serve a survival purpose for the fungus, may germinate by producing a germ tube or by releasing zoospores from a vesicle (Figure 1).

Management of diseases caused by *Pythium* species is very important in propagation operations where highly vulnerable plants may be introduced to fungal propagules. In these cases, care is taken to clean propagation and growing areas to remove any traces of contaminated soil. Pasteurized growing medium (often with inhibitory amendments

TABLE 3 Plant pathogenic oomycetes

Order	Characteristics	Examples
Saprolegniales	Well-developed mycelium, zoosporangium is long and cylindrical	*Aphanomyces euteiches* (pea)
Peronosporales (family)	Well-developed mycelium, zoosporangia oval or lemon-shaped	
Pythiaceae	Sporangia borne on hyphae of indeterminate length	*Phytophthora infestans* (late blight of potato and tomato)
		Phytophthora palmivora (black pod of cacao)
		Phytophthora nicotianae (black shank of tobacco)
		Phytophthora ramorum (sudden oak death, ramorum blight)
		Pythium aphanidermatum (damping-off and root rot of many hosts; root rot of cucumber)
		Pythium debaryanum (damping-off) *Pythium ultimum* (damping-off; black leg of geranium)
Peronosporaceae (downy mildew)	Sporangia borne on hyphae of determinate length	*Bremia lactucae* (lettuce)
		Peronospora destructor (onion)
		Peronospora lamii (coleus)
		Peronospora tabacina (blue mold of tobacco)
		Plasmopara viticola (grape)
		Pseudoperonospora cubensis (cucurbits)
Albuginaceae (white rust)	Sporangia formed in chains on club-shaped sporangiophores of indeterminate length	*Albugo candida* (crucifers)*Albugo ipomoeae-panduranae* (sweet potato)

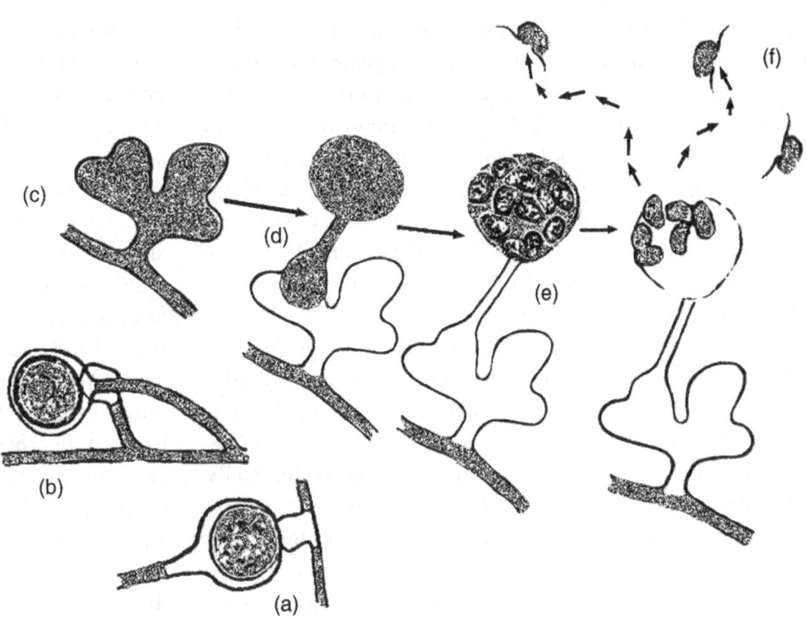

FIGURE 1 Reproduction in *Pythium aphanadermatum* (Oomycota), a root, stem, seed, and fruit rot pathogen with a very wide host range. (a) and (b) Sexual stage. Oogonium (female gamete) with an (a) intercalary or (b) a terminal antheridium (male gamete). (c–f) Asexual stage. (c) Inflated, lobate sporangium. (d) Sporangium with vesicle. (e) Sporangium with vesicle containing zoospores. (f) Biflagellate zoospores. *Reproduced from Shishkoff, N., with permission from CRC Press.*

such as composted tree bark) is used, and overwatering is avoided. Chemical seed treatments and preventive fungicides are available for use, and resistant cultivars of some commercially grown plants may be selected by growers in lieu of more susceptible varieties.

Phytophthora species cause diseases on a variety of hosts, including seedlings, annual plants, herbaceous perennials, shrubs, and fruits and forest trees. These diseases are characterized as damping-off, root rot, rot of the crown, stem, tuber, corm, bud, or fruit, foliar blight, stem canker, and dieback. Species of *Phytophthora* may be host specific or may have a broad host range. For example, a newly described disease caused by *P. ramorum* (sudden oak death or ramorum blight) has, to date, a broad host range of more than 100 species. Symptoms include bole cankers on certain oak species as well as a variety of leaf spots, leaf blights, and shoot blights on many ornamental hosts, some of the natural understory plants associated with susceptible forest trees.

Late blight of potato, a disease caused by *Phytophthora infestans* (meaning 'infectious plant destroyer'), has special historical significance. Controversy over the etiology of late blight (which caused the Irish potato famine in the mid-1840s) eventually led to the first accepted experimental proof that microorganisms, in this case fungi, are the cause of disease and not, as surmised at the time, the result of wet weather or the wrath of God. The story begins in the late 1500s when the potato was brought from its center of origin in Central America to Europe by Spanish conquistadors. By the 1800s, potatoes were widely planted throughout Europe and North America, although genetic variation within the genus was limited. Eventually, a pathogen of potato was introduced with potatoes from South America to potato-growing regions elsewhere. The ensuing disease in the 1840s, the result of a combination of susceptible potatoes grown in monoculture, weather favorable for disease development, and a virulent pathogen, proved catastrophic to cultures, and especially Irish peasants, who depended heavily on the potato for sustenance.

P. infestans affects aerial plant parts (leaves and stems) as well as tubers. This oomycete overwinters within infected tubers. When tubers are used as seed pieces in the spring, the fungus infects the developing seedling, emerging from infected tissues as a white mildew of hyphae. Sporangia produced on hyphal tips are dispersed by air currents to aboveground plant parts or may be washed into the soil. These propagules form a germ tube that penetrates host tissue directly or through stomates. In slightly cooler weather, each sporangium instead releases eight biflagellate zoospores that serve as propagules. Within a few days, necrosis occurs on the affected plant part, shortly accompanied by sporangial development. In this manner, many generations of asexual spores are produced within a single growing season. Foliage affected by late blight becomes brown, water soaked, and covered with white fungal mycelium. Plants may die in the field, severely reducing yield. Tubers infected with the pathogen develop brown and purple splotches that appear water soaked, later drying to become shrunken and firm. Affected tubers may rot in storage and cannot be used as seed pieces the following season.

P. infestans is heterothallic; thus, sexual reproduction requires two mating types (A1 and A2) for production of oospores. Until the 1980s, sexual reproduction was unreported outside of Mexico. Since then, both the mating types have been identified from some potato-growing regions, including the United States, and several more virulent strains of the fungus have emerged. Management of the disease requires the use of seed pieces that are free of the disease as well as preventive fungicides. Field resistance to *P. infestans* among commercial potato cultivars varies; many of the most popular cultivars are susceptible. Environmental conditions have a great impact on disease development; in cool, wet weather, the disease persists, killing plants in a matter of days. In drier conditions, the progress of the disease is diminished.

Downy mildews are caused by a handful of obligate parasites that occur on many cultivated crops. These oomycetes are aerial pathogens that cause necrosis on foliage, stems, and fruits. Infected plants may stunt, defoliate, decline, and die. Younger plants may be systemically infected, resulting in up to 90% crop loss in severe cases. The most significant downy mildews are those that infect cucurbits (*Pseudoperonospora cubensis*), grape (*Plasmopara viticola*), onion (*Peronospora destructor*), and tobacco (blue mold, *Peronospora tabacina*).

The 'downy' growth associated with infected tissues, from which the disease gets its name, consists of characteristic, dichotomously branched stalks (sporangiophores) that emerge through the stomates on the lower leaf surface. Sporangia form at the tips of the sporangiophores and are dispersed by wind or water to new hosts. Sporangia may germinate directly to infect new plants or, in cooler weather, produce zoospores, which rapidly facilitates disease spread. The pathogen overwinters as oospores in infected plant material or in soil, depending on species. Free moisture, high relative humidity, and cool or warm, but not hot, weather enhance disease development.

Downy mildew of grape was introduced from the United States to European vineyards of *Vitis vinifera* in about 1875. Although the fungus does little damage to the grapes native to North America, *V. vinifera* is highly susceptible, and the grape and wine industry throughout France and much of Europe was nearly destroyed. Studies of this disease in France led to the discovery of one of the first fungicides, bordeaux mixture (copper sulfate and hydrated lime). Most varieties of *V. vinifera* are still highly susceptible to the disease, and fungicides remain an important tool for management of this disease.

Kingdom Eumycota (Fungi)

Chytridiomycota

Fungi within the Chytridiomycota, called chytrids, inhabit water or soil and are the oldest known true fungi. Chytrids lack a true mycelium. The thallus is irregularly shaped and the cell wall, as in other true fungi, contains chitin and glucans. Unlike other true fungi, however, chytrids produce motile zoospores that possess a single, posterior, whiplash flagellum.

Members of the Chytridiomycota are mostly saprophytes; the few known pathogens of vascular plants in this group include *Olpidium brassicae* (a root pathogen of cabbage and other hosts), *Physoderma alfalfa* syn. *Urophlyctis alfalfae* (crown wart of alfalfa), *Physoderma maydis* (brown spot of corn), and *S. endobioticum* (black wart of potato). The wart-like (gall) symptoms induced by *Physoderma* and *Synchytrium* occur as cells in affected tissues are stimulated to divide repeatedly. The brown of corn pathogen affects aerial plant parts; severe infection results in stalk rot and lodging in the field.

O. brassicae is a symptomless parasite and is probably most significant because it vectors plant viruses such as tobacco necrosis virus and lettuce big vein virus. The pathogen is endobiotic: the entire thallus resides within a single host cell. The disease cycle begins when the thick-walled resting spores germinate to form zoospores, which during the penetration process dissolve a small pore in the cell wall of root epidermal cells or root hairs. The zoospore protoplast enters the cell and grows to form a zoosporangium. As the zoosporangium forms, nuclei from repeated mitotic divisions are packaged into zoospores; as such, the entire thallus is converted into an asexual reproductive structure (i.e., holocarpic). When the roots are wet, zoospores escape through an exit pore to infect nearby cells. The thallus may also convert into a thick-walled resting structure that can remain viable in the soil for many years. The presence of a sexual cycle in this organism has not been confirmed.

Zygomycota

The Zygomycota are terrestrial fungi with a well-developed, coenocytic, haploid mycelium. The thallus is haploid, and chitin and chitosan are significant constituents of the hyphal cell wall. Asexual reproduction in the zygomycetes results in nonmotile spores called sporangiospores. Sexual spores, or zygospores, are produced when two morphologically similar gametangia of opposite mating types fuse. These fungi are saprophytes or weak pathogens, causing postharvest molds and soft rots. For example, some species of *Mucor* are soil inhabitants that penetrate fruits (through wounds or at the calyx) that have fallen to the orchard floor. Within 2 months of cold storage, the fruits are completely decayed and fungal mycelium emerges in tufts through the cuticle.

Although *R. stolonifer* is best recognized as the common bread mold, under the right circumstances, this ubiquitous saprophyte also causes a soft rot of fleshy fruits and vegetables, bulbs, corms, flowers, and seeds. The thallus consists of hyphal structures known as rhizoids (short branches of hyphae that resemble roots), which penetrate the food substrate, and stolons (longer branches of hyphae) that skip over the substratum surface. The fungus produces asexual sporangiospores (also called mitospores) that form on the swollen tip (columella) of a long aerial sporangiophore. The fungus is heterothallic; as its food supply is depleted, isogametes of opposite mating type fuse to form a zygosporangium containing a single heterokaryotic zygospore. After a 1–3 month period of dormancy, the zygospore undergoes meiosis to form four haploid nuclei (two of each mating type) and germinates to form a sporangium. Sporangiospores are formed by mitosis; these spores are homokaryotic; one-half of the spores are of one mating type, and the other half are of the opposite mating type.

The soft rot disease process begins as sporangiospores, ubiquitous in the air, penetrate through wounds in various plant parts. Cellulase and pectinase enzymes, produced by the fungus, degrade the middle lamella and cell walls of plant tissues, causing a soft, water-soaked rot. The nutrients released are used by the fungus to produce fluffy tufts of a gray/brown aerial mycelium consisting of fungal hyphae and fruiting structures. Eventually, moisture is lost from the degraded tissue, which becomes firm and mummy-like.

Ascomycota

The Ascomycota are some of the best-known true fungi, and at least 30000 different species of ascomycetes are described. The group is very diverse and occupies a variety of niches. A minority of these fungi form partnerships with algae to form lichens; the remainder are saprophytes or symbionts. Parasitic ascomycetes may derive nutrition as biotrophs, necrotrophs, or hemibiotrophs. Although some ascomycetes, such as yeasts, have a single-celled thallus, the thallus of most of these terrestrial fungi consists of a well-developed, septate, haploid mycelium that contains chitin in the cell wall. The ascomycete thallus grows under the substratum surface; only reproductive structures are exposed to the air.

Ascomycetes are named after the ascus, a sac-shaped structure that contains ascospores, the products of meiosis during the sexual reproductive process. Asci are formed when the female sex cell (ascogonium) is fertilized by the male gamete (antheridium). The diploid zygote nucleus undergoes meiosis followed by one mitotic division to form eight ascospores, which remain in the sac until they are discharged and disseminated.

Asci are unitunicate (with a single wall) or bitunicate (with a double wall). For the majority of ascomycetes, asci are produced in fruiting structures called ascomata (or ascocarps). The different types of ascomata are the apothecium (open, cup-shaped with exposed unitunicate asci); cleistothecium (completely closed, lined with one or more unitunicate ascus); perithecium (flask-shaped with an opening, or ostiole, at the tip, lined with unitunicate asci); and pseudothecium (or ascostroma; bitunicate asci are produced in a cavity or locule buried within a stroma of fungal mycelium). (Historically, the ascoma associated with powdery mildews has been called a cleistothecium because, at least initially, the sporocarp is fully closed. Later, however, these structures develop a line of weakness to break open during ascospore release. In other texts, these structures are defined as chasmothecia. Although, developmentally, ascoma produced by powdery mildews may be more similar to perithecia, the previously accepted term 'cleistothecium' is used here.) Those asci that are formed freely without a supporting fruiting structure (e.g., for leaf curl fungi and yeasts) are called naked asci.

Asexual reproduction in ascomycetes is most common as conidia produced on conidiophores; other forms include chlamydospores and reproduction by budding or fission (yeasts).

One of the most notable diseases caused by an ascomycete is ergot of rye and wheat, caused by the perithecial fungus *C. purpurea* (Table 4). Grain in the seed head is replaced by a survival structure of the fungus, called an ergot (sclerotium). Ergots are poisonous to humans and animals and, when ingested with flour made from contaminated rye or wheat, affect the nervous system and restrict blood vessels. Side effects of the disorder in humans, known as ergotism, holy fire, or St. Anthony's fire, include convulsions, gangrene, hallucinations, and miscarriage. Convulsive hallucinations caused by ergot were implicated in the Salem witch trials, but the relationship remains unproven. Interestingly, LSD (lysergic acid diethylamide) was first isolated from the sclerotia of this fungus.

C. purpurea (ana. *S. segetum*) occurs worldwide and is a more serious pathogen of rye than of wheat or other cereals. In the spring, sclerotia associated with fallen seed heads germinate to form many perithecia on stalks along the sclerotium periphery. Each perithecium contains many asci, which in turn contain eight multicellular ascospores. The ascospores are disseminated by wind to infect the ovaries of developing flowers. Within a week, droplets of conidia exude from the infected florets in a matrix of sticky liquid called honeydew. Insects, attracted to the honeydew, carry conidia from flower to flower. Conidia are also dispersed by splashing rain. As the disease develops, ergots form in place of kernels. The ergots mature at the same time as the grain, and are harvested or fall to the ground. Since these sclerotia lose viability after a year, management of this disease includes deep plowing and crop rotation, as well as use of pathogen-free seeds. It is not permissible to mill flour that contains more than 0.3% by weight of sclerotia.

Apple scab, one of the most important diseases of apple, crabapple, and other rosaceous species worldwide, is caused by *V. inaequalis*. Symptoms of this disease include lesions on the foliage (olive-green spots with feathered margins), premature leaf and fruit drop, scabbing and cracking of fruit, poor bud set, and reduced yield. The disease was

TABLE 4 Plant pathogenic ascomycetes

Class	Ascoma	Examples
Archiascomycetes	Naked	*Taphrina caerulescens* (oak leaf blister)
		Taphrina deformans (peach leaf curl)
Leotiomycetes (Erysiphales)	Cleistothecium	Tribes (genus):
		Erysiphae: *Erysiphe* (herbacious plants), *Microsphaera* (lilac), *Uncinula* (grape)
		Phyllactinieae: *Phyllactinia* (shade trees), *Leveillula* (tomato)
		Golovinomyceteae: *Arthrocladiella* (boxthorn), *Golovinomyces* (Asteraceae, coreopsis)
		Cystotheceae: *Podosphaera* (apple, cucurbits, rose)
		Blumerieae: *Blumeria* (cereals)
Pyrenomycetes	Perithecium	*Ceratocystis fagacearum* (oak wilt)
		Claviceps purpurea (ergot of rye and wheat)
		Gaeumannomyces graminis (take-all of wheat)
		Gibberella zeae ana. *Fusarium graminearum* (head blight of wheat)
		Glomerella cingulata (bitter rot of apple and pear)
		Monosporascus cannonballus (root rot and vine decline of melon)
		Nectria spp. (canker of hardwoods)
		Ophiostoma ulmi; *Ophistoma novo-ulmi* (Dutch elm disease)
Loculoascomycetes	Pseudothecium	*Apiosporina morbosa* (black knot of plum)
		Elsinoë ampelina (grape anthracnose)
		Guignardia bidwellii (black rot of grape)
		Mycosphaerella musicola; *M. fujiensis* (Sigatoka disease of banana)
		Venturia inaequalis (apple scab)
Discomycetes	Apothecium	*Diplocarpon rosae* (black spot of rose)
		Lophodermium spp. (needle cast of pine)
		Monilinia fructicola (brown rot of stone fruit)
		Rhabdocline pseudotsugae (Rhabdocline needlecast of Douglas-fir)
		Rhytisma acerinum; *R. punctatum* (tar spot of maple)

likely introduced to North America in the 1600s as European colonies planted infected apple trees and scions. The first botanical description of symptoms was made by Elias Fries in Sweden in 1819, and the pathogen itself was described by Cooke in 1866. Once considered an accepted fact of life, apple scab can be debilitating to susceptible commercial apple and ornamental hosts.

Apple scab first develops soon after a budbreak as ascospores are forcibly ejected from pseudothecia embedded in infected plant material (leaf litter) on the ground. Each pseudothecium contains up to 100 asci. The ascospores are carried by air to the developing leaves and fruits on the tree. Ascospore release (as primary inoculum) continues for a period of up to 9 weeks, through the period the host is most vulnerable to infection. Penetration of new leaves requires a period of leaf wetness that varies from 9 to 28 h, depending on temperature. A fungal germ tube from the germinating ascospore penetrates the cuticle, and a mycelium develops between the epidermis and the cuticle. Conidia are produced on sporophores that push through the cuticle in mats. These asexual spores (as secondary inoculum) are released through the remainder of the growing season. At leaf fall, the fungus grows deeper into the mesophyll of infected leaves, forming pseudothecia. The fungus overwinters in this condition until the following spring, when the disease cycle starts anew.

Management of apple scab is extremely important for commercial growers, and usually requires a multifaceted approach. Cultural management techniques include proper orchard placement and tree orientation, management of humidity and leaf wetness, the use of resistant cultivars where available, removal of fallen leaves if practical, and shredding leaf litter with a mower, adding urea to hasten decomposition. Growers usually rely, however, on fungicide sprays for effective disease control. The target of chemical control is to effectively control the release of primary inoculum (ascospores) at the beginning of the season. Calendar sprays, used in the past, are based on the phrenology of the host tree. Currently, more sophisticated forecasting systems, based on temperature and rainfall, are in use. As a result, fewer better-timed sprays are made.

Powdery mildews are biotrophs, named after the mats of mycelium and spores evident on the plant surface. They are the most common, and probably the best recognized, of all plant diseases. Although all kinds of plants are affected by powdery mildew, the greatest economic impact probably occurs on cucurbits and cereals. Powdery mildew fungi seldom kill their hosts, but diminished photosynthetic capacity resulting in impaired growth and reduced yield may occur in some susceptible species. For example, severe disease caused by powdery mildew of wheat (caused by *Blumeria graminis* f. sp. *tritici*) results in significant yield loss and lodging in the field.

As biotrophs, powdery mildew fungi draw nutrients from epidermal cells via haustoria that penetrate through the cell wall. All mycelial and spore development occurs on the plant surface. Throughout the growing season, powdery mildew fungi produce asexual spores (conidia, some egg-shaped) in chains on short conidiophores on the plant surface. Conidia are disseminated by air currents to new hosts (secondary cycle). When humidity is high, conidia germinate to initiate new infections. At the end of the growing season, host and environmental factors trigger the sexual cycle. Cleistothecia serve as survival structures through the winter; some fungi, such as the powdery mildew of rose fungus, *Podosphaera* (sect. *Sphaerotheca*) *pannosa* f. sp. *rosae*, also overwinter in buds. In this species, primary inoculum is produced in spring, when mycelium in buds infects developing tissue or when ascospores developed within cleistothecia are released.

Powdery mildews are classified in one of the five tribes (Table 4). Characteristics of the anamorph (conidial chains and surface ornamentation) as well as cleistothecial morphology (number of asci per ascocarp, morphology of hyphal appendages on the cleistothecium) are useful taxonomic criteria.

Management of powdery mildews includes the use of resistant cultivars when available. Unfortunately for rose growers, most popular cultivars are susceptible to this disease. The powdery mildew of wheat pathogen is easily adaptable and has many races, so breeders strive to incorporate as many genes for resistance as possible into wheat cultivars to increase their durability. Best practices for this disease include rotating crops, removing volunteer plants, and mixing wheat cultivars with different genes for resistance within a single planting. For most powdery mildews, humidity control through proper spacing and weed control is helpful, as is the use of contact or systemic fungicides. Unlike many other diseases, nontraditional control products (such as oils, potassium bicarbonate, dilute hydrogen peroxide, and biological controls) are useful for powdery mildew management, presumably because the thallus of the fungus is on the plant surface and is easy to eradicate. These products offer attractive alternatives to fungicides for the landscape or residential clientele.

Brown rot of stone fruits (e.g., peaches, cherries, plums, and almond) is caused by several different species of the discomycete *Monilinia* (*M. fructicola*, *M. laxa*, and *M. fructigena*). Losses from this disease occur both in the field and post harvest. Brown rot manifests as a blossom and fruit blight. Cankers may also develop on twigs and branches. Fruits affected by this disease develop brown spots that quickly consume the fruits. Tufts of gray/brown mycelium break through the cuticle, and the fruit eventually loses moisture and mummifies. The sexual stage develops on mummified fruits that drop to the orchard floor and become partially buried. As many as 20 apothecia, each lined with thousands of asci, develop per fruit. Management is best achieved by controlling the blossom blight phase of the disease.

Basidiomycota

The Basidiomycota are an interesting, diverse group of terrestrial fungi that include decay organisms such as white and brown (wood) rot fungi, and symbionts such as parasites and mycorrhizal fungi. Many basidiomycetes often produce large, spectacular fruiting structures; others are known for their hallucinogenic properties. Like ascomycetes, basidiomycetes occupy a variety of niches. Mushrooms are classified in this phylum, as are highly important plant pathogens, the rust and smut fungi (Table 5).

Basidiomycetes have a well-developed septate mycelium that has chitin in the cell wall. These organisms may spend a majority of their life cycle as dikaryotes, where each cell contains two different haploid nuclei. Basidiomycetes are named after the basidium, a club-shaped structure upon which (usually) four haploid basidiospores (the result of karyogamy and meiosis in the sexual reproductive process) are perched. Basidia are often arranged in a single layer, or hymenium. The basidiospores are supported by short pointed stalks called sterigmata. In most taxa, basidiospores are forcibly ejected from the basidium. The form of the basidium varies by taxonomic group; some basidia

TABLE 5 Plant pathogenic basidiomycetes

Order	Characteristics	Examples
Holobasidiomycetes	Basidia club-shaped and single-celled, arranged on an exposed hymenium; includes the gilled mushrooms, boletes, polypores, puff-balls, bird's nest fungi	*Armillaria mellea* (shoestring root rot)
		Crinipellis perniciosa (witches' broom of cocoa)
		Heterobasidion annosum (heart and butt rot of conifers)
Heterobasidiomycetes	Germinating basidiospores can form secondary spores or yeast-like cells	*Rhizoctonia solani*, tel.
		Thanatephorus cucumeris (damping-off, root disease; aerial blight)
		Rhizoctonia cerealis, tel. *Ceratobasidium cornigerum* (sharp eyespot of cereals)
Urediniomycetes	Rust fungi; complex life cycle consisting of up to five different spore stages on two unrelated plant hosts	*Cronartium ribicola* (white pine blister rust)
		Gymnosporangium spp. (cedar-apple rust, quince rust) *Hemileia vastatrix* (coffee leaf rust)
		Puccinia graminis (stem rust of wheat)
Ustilaginomycetes	Smut fungi; basidia produce numerous basidiospores in smutted groups	*Exobasidium vaccinii* (azalea leaf and flower gall)
		Tilletia caries (common bunt)
		Tilletia indica (Karnal bunt)
		Ustilago avenae (loose smut of cereals)
		Ustilago maydis (corn smut)

are produced on fruiting structures called basidiocarps. Asexual reproduction in basidiomycetes occurs as conidia (monokaryotic or dikaryotic) produced on conidiophores, chlamydospores, and oidia (monokaryotic thin-walled spores formed by hyphal fragmentation). Some fungi in the Basidiomycota produce rhizomorphs as structures for survival and dispersal.

The life cycle of basidiomycetes begins as haploid basidiospores develop into a mycelium with one haploid nucleus per cell. These monokaryotic hyphae are the primary mycelium. The dikaryotic hyphal condition is established in basidiomycetes when two primary mycelia fuse in the first step of fertilization (plasmogamy) (karyogamy is delayed until the basidium is formed). In most basidiomycetes, primary mycelia must be of opposite mating type (heterothallic). It is the secondary mycelium that supports the development of the basidium. A frequent characteristic of the secondary mycelium is the presence of clamp connections, a bridge that forms during hyphal development that serves to maintain the dikaryotic condition of each hyphal cell. In addition, the septa of both primary and secondary mycelia contain a special pore (dolipore) so that cytoplasmic continuity is maintained between cells. The dolipore septum is a characteristic unique to basidiomycetes.

The holobasidiomycetes are probably best represented by the classic mushroom. The mushroom basidiocarp consists of a cap (pileus) and a stalk (stipe). The pileus is lined with gills, which in turn are lined with basidia. The stipe of some basidiocarps has a volva (a remnant of the universal veil that once enveloped the entire developing mushroom) at the base. An annulus (a ringed remnant of the partial or inner veil that once enveloped the gills on the underside of the basidiocarp) may also be present on the stipe, just under the cap.

A. mellea is both an effective decay organism in the forest ecosystem and a pathogen, causing shoestring root rot or honey mushroom root rot. The fungus, which is actually a collection of several species, affects hundreds of fruit trees, shrubs, shade trees, and vegetables worldwide. *Armillaria* forms characteristic rhizomorphs and sheets of bioluminescent mycelium (mycelial fans) under the bark near the crown of infected trees. This white mycelium destroys the phloem and cambium of the tree. The disease primarily spreads as rhizomorphs extend to adjacent hosts. Symptoms exhibited by plants with this disease are similar to those caused by other root diseases, including reduced growth, small yellow leaves, dieback, and gradual decline and death. Clumps of basidiocarps (honey-colored mushrooms) are produced at the base of dead or dying trees in early fall. The fungus continues to grow as a saprophyte after the plant is dead.

From a pathogenic standpoint, the heterobasidiomycetes are important because an anamorph of several teleomorphs in this group, *Rhizoctonia* sp., is such a significant pathogen.

Species in this genus (as well as in the genus *Sclerotium*) were classified for many years in an artificial group called 'mycelia sterilia' because sexual or asexual spores were lacking. They are now known to be the anamorphs of several basidiomycetes, as well as a few ascomycetes, teleomorphs.

Species of *Rhizoctonia* occur worldwide. These pathogens affect most crops, causing damping-off, root and stem rot, stem canker, storage rot, and an aerial (foliage) blight that occurs when foliage touches the soil surface. The hyphae of *Rhizoctonia* branch at right angles, a septum forming just beyond the branch. This characteristic facilitates their identification in culture. Some of these fungi are multinucleate (teleomorph *Thanatephorus*) and some are binucleate (teleomorph *Ceratobasidium*). *Rhizoctonia solani* (multinucleate) is actually a collective species consisting of several unrelated strains.

Strains of *Rhizoctonia* are distinguished by anastomosis groupings. Anastomosis occurs when the hyphae of two isolates fuse and undergo plasmogamy. This process permits the development of hyphal networks. Anastomosis is successful only between hyphae of the same anastomosis group. A killing reaction occurs when these strains belong to different anastomosis groups. The anastomosis groups are not entirely host specific, but some tendencies do exist. For example, strains of AG1 cause seed rot and web blight; of AG2 canker of root crops and brown patch of turfgrass; of AG3 potato diseases; and of AG4 hypocotyl rots on angiosperms. Many species of *Rhizoctonia* produce sclerotia, which remain in soil for up to 3 years. This feature makes control of these organisms in the field very difficult.

Brown patch is a common disease of turfgrass caused by *R. solani*. The fungus survives as sclerotia in plant debris, which germinate and grow saprophytically until hyphae can penetrate a suitable host. Circular lesions appear on infected leaves, and purplish patches of diseased plants develop in the turfgrass stand. Management of brown patch includes avoiding excessive nitrogen fertilizer (to reduce development of susceptible, succulent growth) and thatch, increasing drainage, and applying fungicides.

Plant pathogenic rust fungi are Urediniomycetes, placed within the order Uredinales. This order contains some of the most destructive pathogens of vascular plants. Rust diseases have caused famines and ruined the economies of entire civilizations. Crops particularly affected by rust diseases include bean and soybean, grains (barley, oat, wheat), asparagus, cotton, pine, apple, coffee, and a variety of ornamental plants. Rusts are biotrophs, and each species has a relatively narrow host range.

The generic term 'rust' pertains to the rusty spores produced in pustules by some of these fungal species. Rusts most often appear on leaves and stems as spots or lesions that rupture the epidermis, producing rust-colored, orange, yellow, or white spores. Other rusts cause galls or swellings to form. The rust fungi are fabulously complex; their life cycle can consist of up to five different spores (pycniospores, aeciospores, urediniospores, teliospores, and basidiospores) that develop on one of two unrelated plant hosts (alternate hosts). All rust fungi produce at least two spore types: teliospores and basidiospores. Disease cycles that contain all five spore stages are macrocyclic (long cycle); those that contain less than five (usually the urediniospore stage is lacking) are demicyclic; those that produce only teliospores and basidiospores are microcyclic. A summary of these spore stages, as well as the Roman numerals assigned to each stage (a convention used to describe each stage in the rust life cycle), is presented in Table 6.

Spore stages are produced in sequence. For those fungi with two hosts (heteroecious), the principal host is that which supports the development of the dikaryotic (secondary) mycelium (also called the telial host). Spores produced by the secondary mycelium are urediniospores and teliospores. The monokaryotic (or primary) mycelium develops on the alternate host; spores produced on this host (called the aecial host) are pycniospores and aeciospores. Diseases that lack an alternate host are autoecious. Significant diseases caused by heteroecious macrocyclic rusts include stem rust of wheat (*P. graminis*), ash rust (*P. sparganioides*), and white pine blister rust (*Cronartium ribicola*). Heteroecious, demicyclic rusts include those caused by the genus *Gymnosporangium*, such as cedar-apple and quince rusts. Coffee rust (*Hemileia vastatrix*) is an autoecious, demicyclic rust, and pine-pine

TABLE 6 Spore stages associated with rust fungi

Spore stage	Fruiting structure	Spore name	Ploidy	Parent mycelium	Host (if heteroecious)
0	Pycnium (spermagonium)	Pycniospore	(n)	Primary	Alternate, or aecial
I	Aecium	Aeciospore	$(n+n)$	Primary	Alternate, or aecial
II	Uredinium	Urediniospore	$(n+n)$	Secondary	Primary, or telial
III	Telium	Teliospore	$(n+n)$: karyogamy, $(2n)$	Secondary	Primary, or telial
IV	Basidium	Basidiospore	(n)		

gall rust (*Endocronartium harknessii*) is an autoecious, microcyclic rust.

Stem rust of wheat is a major disease of wheat and barley, causing losses of $5 billion per year worldwide. The fungus overwinters as diploid teliospores in telia (III) on infected wheat debris. In spring, teliospores form filamentous basidia (IV). Meiosis occurs, and four haploid basidiospores (produced on short sterigmata) are released into the air. These spores are dispersed several hundred meters by air currents and are deposited on young barberry leaves (alternate, or aecial host), where they germinate, penetrate the epidermis, and grow intercellularly. Within 3–4 days, flask-shaped pycnia (0) develop, the tips of which rupture the leaf surface. Haploid-receptive hyphae (these can be considered female gametes) extend through the opening of the pycnium, accompanied by pycniospores (male gametes), which are extruded through the opening in honeydew. Insects, always attracted to honeydew, visit the pycnia, become smeared with pycniospores, and carry them to the receptive hyphae of other pycnia of opposite mating type (heterothallic). Pycniospores are also spread by rainwater or dew. The receptive hyphae of the opposite mating type are fertilized (plasmogamy), resulting in a dikaryotic mycelium, where each nucleus is of the opposite mating type.

Elsewhere on the leaf during the fertilization process, aecium primordia (aecial mother cells, also haploid) form near the lower leaf surface. Should fertilization occur, the aecium primordia are converted to the dikaryotic state by a process called dikarytization, which forms the next fruiting structure, the aecium (I). Aecia protrude substantially from the lower leaf surface and release dikaryotic aeciospores in chains, which are blown by the wind in late spring to nearby wheat plants. Aeciospores penetrate the stomates of wheat stems, leaves, or leaf sheaths. The mycelium grows intercellularly and forms a mat just below the epidermis. Dikaryotic urediniospores (repeating spores) push through the epidermis on short sporophores to form uredial pustules or uredinia (II). These spores are blown by air currents up to several hundred kilometers from the point of origin to reinfect wheat through stomates in the presence of a film of water or at a high relative humidity. New urediniospores are produced every 8–10 days in a secondary cycle. At plant maturity, uredinia convert into telia (III) and produce dikaryotic teliospores instead of urediniospores. In addition, new telia may develop from recent urediniospore infections. Karyogamy occurs in the teliospores, which then overwinter in the diploid state.

Management of stem rust includes the use of resistant and earlier maturing varieties. Although fungicides such as sterol or demethylation inhibitors can be effective, they are often cost prohibitive to apply and are not used. Another strategy for some rusts is eradication of the alternate (or perhaps, less economically useful) host. For most rust diseases, this strategy is not useful since spores can travel many miles on air currents. In response to a wheat rust epidemic in 1916, however, an eradication program was enacted in the United States in 1918 to remove barberry plants from wheat-growing regions. By 1941, 295 million barberry plants were destroyed. The program should have worked, but later it was found that the fungus overwinters as urediniospores in the southern United States and Mexico. These spores blow up into the wheat-growing regions of the United States yearly in what is known as the Puccinia pathway. Today, barberry species planted are resistant to the disease.

The order Ustilaginales within the Ustilaginomycetes includes the smut and bunt fungi, which are very similar to rusts in that teliospores and basidiospores are produced, but are dissimilar in that basidia produce numerous basidiospores in smutted groups, which are often distasteful to look at and smell. There are about 1200 species of smuts and bunts distributed worldwide. These destructive pathogens, second only to rusts, can severely stunt hosts, reducing yield. Most smuts attack the ovaries of grains and replace the kernel with fungal mycelium and spores. Others attack leaves, stem, or floral parts. The Ustilaginales are biotrophs.

Smut and bunt fungi overwinter as teliospores on contaminated seed, in plant debris, or as mycelium within infected kernels or plants. These fungi are microcyclic, and basidiospores are the only infective propagule. Smuts are monocyclic, producing one generation of inoculum per year.

Corn smut, caused by *Ustilago zeae*, occurs wherever corn is grown. Galls form on aboveground plant parts, including ears, tassels, stalks, and leaves. Losses of susceptible variety in the field can approach 100%. The fungus overwinters as teliospores in crop debris and in soil where it remains viable for several years. In spring and summer, karyogamy occurs and diploid teliospores germinate to produce haploid basidiospores, which are carried by air currents or splashed to young, developing tissue of corn plants. Basidiospores germinate and penetrate epidermal cells directly, forming a fine primary mycelium. Primary mycelia of opposite mating type (heterothallic) must fuse to form a dikaryotic secondary mycelium before parasitism can continue. The mycelium grows into surrounding plant tissues, stimulating nearby cells to enlarge and divide into galls. Just before sporulation, the galls are invaded by the secondary mycelium, which converts completely into teliospores. Galls turn black and rupture, exuding a smutted mass of teliospores.

Management of corn smut often includes removal of infected ears before teliospores are released. Crop rotation, deep plowing of stubble in the fall, and use of resistant varieties is recommended. Although there are field corn varieties with some resistance to corn smut, most commonly used sweet corns are susceptible. Highly susceptible varieties are grown, however, and the galls, which are edible before the teliospores mature, are sold in Mexican markets. This delicacy is known as huitlacoche.

Deuteromycota (mitosporic fungi)

Some fungal anamorphs have no known sexual state in their life cycle and thus remain unclassified. These organisms, also known as mitosporic fungi, are placed in an artificial group called the Deuteromycota or the Fungi Imperfecti. They reproduce mitotically via conidia and are considered anamorphic partners of teleomorphs in the Ascomycota or Basidiomycota. Molecular techniques have helped to place mitosporic fungi with their associated teleomorphs.

Mitosporic fungi share many of the same features as their sexual relatives. Most of the 17 000 described species are terrestrial and survive as saprophytes. Some trap nematodes, others are symbionts (lichens, grass endophytes, mycorrhizae, and weak or primary pathogens). Mitosporic fungi are divided into three informal classes: Agonomycetes (mycelia sterilia, mentioned previously), coelomycetes (spores are produced in fruiting structures called conidiomata), and hyphomycetes (spores are produced on separate conidiophores).

Classification within these groups is based on conidiophore, conidium, and hyphal characteristics. Conidia can be one-celled, two-celled, or multicellular with transverse or oblique septa. They have different shapes, appearing filiform (thread-like), ovoid (egg-shaped), clavate (club-shaped), cylindrical, stellate (star-like), or branched. They may be pigmented or have appendages. Conidia within the hyphomycetes have distinct conidiophores and are produced singly, as synnema (a group of conidiophores are united together to form a stalk), or as sporodochia (clusters of conidiophores that form a cushion of hyphae on the host surface). Coelomycetes produce conidia in fruiting structures such as pycnidia (a flask-shaped structure lined on the inside with conidiophores) and acervuli (conidiophores lay side by side in a flat, saucer-shaped configuration) embedded in host tissue (Figure 2).

Alternaria solani is an example of a pathogenic hyphomycete. The fungus causes tomato early blight, which is one of the most important tomato diseases in the United States. The pathogen overwinters as chlamydospores in plant debris, which serves as a source of inoculum in spring. Lesions develop on leaves, stems, and fruits; conidia, formed on distinct conidiophores, are disseminated by wind, rain, insects, farm machinery, and infected seeds. Entire plants defoliate and die.

Brown spot of soybean is caused by the coelomycete *Septoria glycines*. The fungus overwinters as pycnidia embedded in the debris of stem and leaf tissues. During warm, moist weather, the fungus grows as a saprophyte, eventually invading leaf tissues through the stomates. A flecked pattern of necrosis appears on the leaves, which defoliate, reducing yield. Crop rotation, resistant varieties, and fungicides are used to manage this disease.

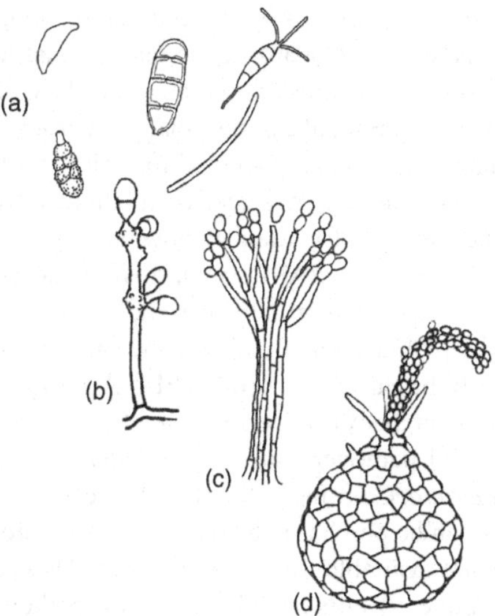

FIGURE 2 Asexual reproduction in mitosporic fungi. (a) Single conidia varying in shape, color (hyaline or brown), and number of cells. (b–d) Asexual fruiting structures. (b) Conidia produced on a distinct conidiophore. (c) Synnema. (d) Pycnidium. *Reproduced from Shishkoff, N., with permission of CRC Press.*

DIAGNOSIS

Classically, disease diagnosis, whether the disease is due to biotic or abiotic causes, is a process where symptoms (host response to disease agent) and signs (pathogen thallus or reproductive structures) are interpreted. Many diseases caused by fungi can be easily identified due to highly characteristic symptoms or signs. For example, powdery mildew fungi are identified by the presence of mycelia and spores (signs) on the surface of susceptible plant tissue. Corn smut is diagnosed by the presence of teliospores (sign) in smutted galls (symptom). For many other diseases, however, the cause is not so obvious, especially in cases where damage due to the primary cause is so advanced that secondary fungi (decomposers or opportunistic organisms) invade the dead or dying tissues.

Some of the standard processes used in diagnosis include microscopic examination of diseased tissues to look for signs such as mycelium or fruiting structures. Samples of diseased tissues are often placed in a moist chamber to encourage the growth of aerial mycelium and spores. A pathogen that is easily coaxed to produce mycelium in this matter is *B. cinerea*, a pathogen of aerial plant parts that produces light, easily dispersed conidia on long conidiophores. Another standard technique is plating small pieces of surface-disinfested symptomatic host tissue on laboratory media. Fungal colonies that grow from the tissue into the agar medium are examined for color, hyphal characteristics, and spores. A selective medium designed to support the growth of certain pathogenic organisms over saprophytes, such as PARP (pimaricin + ampicillin + rifampicin + pentachloronitrobenzene agar, a selective medium for pythiaceous fungi), is especially useful.

Antibody-based tests, such as ELISA (enzyme-linked immunosorbent assay), are used to target certain pathogens, especially soil pathogens such as *Pythium*, *Phytophthora*, and *Rhizoctonia*. Commercially available test kits for many pathogens can be purchased by diagnostic laboratories, and some kits are available for use in the field. These tests can detect the targeted antigen and whether the pathogen is living or not, and may take about 3 h to complete.

Newer molecular diagnostic techniques include standard PCR and real-time PCR. In standard PCR, small bits of DNA from the target organism are extracted and amplified many fold, and the DNA product is visualized as bands on a gel. This technique, although very sensitive, takes 12–24 h to complete and requires a steady hand to appropriately load the gels. In real-time PCR, DNA from the target organism is also extracted, but the amplification process is viewed as it happens. This technique, also very sensitive, is faster than standard PCR and obviates the necessity for gels.

Regardless of the detection technique, a good diagnostician must take other factors into consideration. Most diseases are already described in the literature; identification of the host plant and knowledge of diseases that are common to that host in a particular geographical area are very helpful. In addition, environmental conditions, such as rainfall, temperature, soil conditions, and recent disturbances to the locale, help to pinpoint the causal agent. A correct disease diagnosis is essential if the appropriate control measures are to be applied.

Control

Modern agriculture tips the ecosystem balance to favor the host plant. As a result, growing conditions are artificial: often, the same crops are grown yearly in the same field, providing opportunities for pathogen populations to increase; crops are grown in monoculture, where all plants are genetically similar, thus all of them are susceptible to a virulent pathogen; environmental conditions, especially water management, may be modified to enhance conditions favorable for pathogen development; and high fertilizer inputs may encourage the development of succulent host tissues, which, in many cases, are more susceptible to pathogen attack.

Unlike the earlier concept of disease control, where the goal was to eradicate all pathogen propagules within a given crop system, the concept of disease management seeks to reduce pathogen populations to levels below a certain tolerable, often economic, threshold. Indeed, eradication of pests is seldom even possible. Management strategies are twofold: (1) to reduce initial pathogen inoculum through pathogen avoidance (geographical and temporal changes in planting), exclusion (use of pathogen-free plants and plant parts), and quarantine; and (2) to reduce existing inoculum through pathogen eradication and plant protection.

For example, species of rubber (*Hevea* spp.) are native to the Amazon basin; latex was harvested from these trees in the wild. When American automakers attempted to grow these trees in plantations to supply latex for tires in the 1930s, South American leaf blight, caused by the ascomycete *Microcyclus ulei*, also native to the same region, destroyed the trees. Rubber plantations were moved to the tropical areas of Southeast Asia to avoid the disease. Quarantines are needed to protect these trees from subsequent introduction to the pathogen. The first quarantine enacted in the United States, the Federal Plant Quarantine Act of 1912, was emended in 1915 to include chestnut blight, caused by the exotic pathogen *C. parasitica*. In spite of the quarantine, chestnut blight went on to destroy the American chestnut throughout its range. The quarantine failed because the pathogen was introduced to the Bronx Zoo in 1904, 10 years before the quarantine was enacted. Seed certification programs are useful for diseases caused by fungi that infect the embryo or infest the seed coat, such as *Ustilago nuda*, the cause of loose smut of barley.

Once pathogens are introduced into a cropping system, eradication of fungal propagules may be necessary to ensure sufficient yield. Chemicals, heat, cultural practices (fallow, flooding, rouging), and biological control are used to eradicate pathogens from growing areas, machinery, soil, tools, and plant parts used in propagation (cuttings, tubers, corms, and seeds). As mentioned previously, an eradication program of the alternate host was used to help control stem rust of wheat. This program was only marginally successful because fungal propagules for this disease (urediniospores) are blown to wheat-growing areas from another location.

Biological control, where other microorganisms are used to reduce pathogen inoculum, is useful in some situations where environmental conditions are suitable for development of the biological control agent. An example of a successful biological control program was the use of hypovirulence to manage chestnut blight in Europe. Hypovirulent strains of *C. parasitica* contain a virus that reduces the virulence of the fungal pathogen. Chestnut trees affected with the hypovirulent strain do not develop lethal cases of chestnut blight. The hyphae of mycelia within the same anastomosis group will fuse, passing the virus from one mycelium to another. Hypovirulence has worked well in Europe, where the few genetically different strains there readily anastomose. This has not worked in the United States, where there are many more genetic strains of the pathogen.

Practices that protect plants from pathogens include the use of genetic resistance, cultural practices (barriers, mulches, plant nutrition, management of water and soil edaphic factors), chemicals (fungicides), and biological control. Often, these strategies are more cost-effective than trying to eliminate pathogen propagules from a given production system. The apple scab pathogen, *V. inaequalis*, requires a period of leaf wetness for the spore penetration process to occur. Leaf material that is permitted to dry readily may be protected from the fungal pathogen. The use of seed treatment and protectant fungicides is essential for the management of many fungal diseases, especially for those where the crop value is high and disease management is essential for adequate yield. For example, susceptible apple tissue is sprayed on a protectant basis for apple scab control. In certain locations of the United States, susceptible Douglas-fir in Christmas tree plantations are sprayed with fungicide in the spring to control Rhabodocline needle cast; this disease makes trees unfit for sale where it occurs. Perhaps the wisest management strategy is the use of resistant plant material. For example, since fungicide use is cost-prohibitive for use in many cereals, genetic resistance is necessary for management of stem rust of wheat. Classic plant breeding programs as well as biotechnology and genetic engineering have been used to develop germplasm that is resistant to fungal pathogens of major economic and social importance.

FURTHER READING

Agrios, G. N. (2005). *Plant pathology* (5th ed.). Elsevier Academic Press: Burlington, MA.

Barr, D. J. S. (2001). Chytridiomycota. In D. J. McLaughlin, E. G. McLaughlin & P. A. Lemke (Eds.), *The mycota VII. Part A: Systematics and evolution* (pp. 93–112). Springer: Berlin.

Braselton, J. (2001). Plasmodiophoromycota. In D. J. McLaughlin, E. G. McLaughlin & P. A. Lemke (Eds.), *The mycota VII. Part A: Systematics and evolution* (pp. 81–91). Springer: Berlin.

Carlile, M. J., Watkinson, S. C., & Gooday, G. W. (2001). *The fungi* (2nd ed.). Academic Press: San Diego.

Deacon, J. (2006). *Fungal biology* (4th ed.). Blackwell: Malden, MA.

Hawksworth, D. L., Sutton, B. C., & Ainsworth, G. C. (Eds.), (1983). *Ainsworth & Bisby's dictionary of the fungi*. Commonwealth Mycological Institute: Kew, Surrey.

Schumann, G. L. (1991). *Plant diseases: Their biology and social impact*. APS Press: St. Paul, MN.

Schumann, G. L., & D'Arcy, C. J. (2006). *Essential plant pathology*. APS Press: St. Paul, MN.

Sinclair, W. A., & Lyon, H. H. (2005). *Diseases of trees and shrubs* (2nd ed.). Cornell University Press: Ithaca, NY.

Taylor, J. W., Spatafora, J., & Berbee, M. (2006). *Ascomycota*. http://tolweb.org. Tree of Life Web Project.

Trigiano, R. J., Windham, M. T., & Windham, A. S. (Eds.), (2008). *Plant pathology concepts and laboratory exercises*. (2nd ed.). CRC Press: Boca Raton, FL.

Ulloa, M., & Hanlin, R. T. (2000). *Illustrated dictionary of mycology*. APS Press: St. Paul, MN.

Volk, T. J. (2001a). Fungi. In *Encyclopedia of Biodiversity* (pp. 141–163). Academic Press: New York, Vol. 3.

Volk, T. J. (2001b). *Tom Volk's fungi*. http://botit.botany.wisc.edu; Department of Biology, University of Wisconsin-LaCrosse.

Webster, J., & Weber, R. W. S. (2007). *Introduction to fungi* (3rd ed.). The University Press: Cambridge.

Chapter 10

Entomogenous Fungi

R.A. Humber
USDA-ARS Biological Integrated Pest Management Research Unit, Ithaca, NY, USA

Chapter Outline

Abbreviations	127
Defining Statement	127
Introduction	127
Infection Processes and Pathobiology	128
Practical Uses of Entomogenous Fungi	134
A Look Forward for Entomogenous Fungi	137
Recent Developments	139
Further Reading	140

ABBREVIATIONS

CER Cellular reaction
NCR Noncellular reaction

DEFINING STATEMENT

The biology of the phylogenetically diverse fungi affecting insects and other invertebrates, mainly as pathogens, is summarized with emphasis on the interactions of these diverse fungi and their hosts, and on their uses for biological control and as sources of biologically active compounds.

INTRODUCTION

The term 'entomogenous fungus' has been used historically to refer to almost any type of association between a fungus and an insect. Despite the prefix's reference to insects, the term is also generally understood to include fungi associated with mites, spiders, and other arthropods; even more broadly, 'entomogenous fungi' can also be extended to all fungi associated with (usually as pathogens or parasites of) virtually any free-living microinvertebrates including arthropods, nematodes, tardigrades, rotifers, and protozoans of all sorts. Most studies of these fungi do, however, focus on those associated directly with insects. The emphasis in this treatment is the traditional one, on the fungi affecting arthropods, without further treatment of the many taxonomically diverse fungi from nematodes, tardigrades, rotifers, protozoans, or other nonarthropodous invertebrates.

The types of associations between fungi and arthropods range from wholly benign (e.g., the fungus is only dispersed by the arthropod) to parasitic (either on the cuticle or in the gut but not causing mortality of the host) to pathogenic (resulting in the host's death). It should not be surprising that the diversity of fungi associated with insects is as great as that of the hosts and types of associations that exist between fungus and host.

Spores of nearly any fungus may be present on the surfaces of living arthropod hosts regardless of whether that fungus is saprobic, zoophilic (loosely associated with an animal but not necessarily causing disease), or zoopathogenic, or even phytopathogenic. It is little appreciated, for instance, that fungi that produce their spores in some sort of slime are usually associated with arthropods either as pathogens or parasites, or for the dispersal of those spores. Whether a fungus growing on the cadaver of an arthropod is there as a saprobe or as a pathogen or parasite raises questions that are discussed later.

Fungi are treated here in their historically broad sense. No distinction is made here between true fungi and those entomogenous biflagellate organisms with a tinsel-type flagellum formerly treated as fungi in the oomycetes but now properly placed in the kingdom Straminipila (alternatively referred to as the Chromista). Such a distinction would overlook the fact that the entomogenous organisms involved have common biological problems and act on their hosts in very similar manners despite their highly divergent phylogenetic affinities. It is also worth noting that this discussion does not include the Microsporidia, a very large and diverse group of intracellular parasites of invertebrates (but which also includes a substantial number of pathogens affecting humans and other vertebrates); these fascinating organisms have been variously classified in the past as protozoans, protists, and in other categories, and were for a very long time regarded as extremely ancient eukaryotes devoid of such typical organelles as mitochondria and Golgi bodies and with very small genomes. Recent kaleidoscopic studies on the phylogenetic affinities of these organisms and of phylogenetic studies of an unprecedentedly broad range of fungi made it clear that the microsporidia are highly derived rather than ancestral, and that they do, in a number of proven instances, contain morphologically simplified and highly diminutive structures with the biochemical properties of mitochondria, Golgi bodies, and rough endoplasmic reticulum. The gene sequences of a reasonable spectrum of microsporidians place them among the true fungi in a position basal to the major groups of nonflagellate fungi; the exact point at which the microsporidia should be most accurately grafted onto the Fungal Tree of Life remains uncertain for now although some evidence places them close to but not in the Entomophthorales.

INFECTION PROCESSES AND PATHOBIOLOGY

All infections must begin with an infective unit that contacts its potential host, and for fungal pathogens of invertebrates that infective unit is almost always a spore rather than a hypha. The fungal infective unit may be considered to be a uniflagellate or biflagellate zoospore for chytridiomycetes or oomycetes, respectively, even though those zoospores encyst on the host cuticle before germination and penetration can begin. For the zygomycetes in the Entomophthorales, the infective unit is a primary conidium dispersed directly from an infected host or a secondary conidium that was actively discharged from a primary (or other secondary) conidium or passively dispersed from an elongated capillary-like secondary conidiophore formed on another primary (or secondary) conidium. The conidia of Hyphomycetes or anamorphic ascomycetes are the infective units; for teleomorphic ascomycetes, the infective units are either ascospores or secondary conidia (formed directly from the ascospore, much as in secondary sporulation in the Entomophthorales). The only entomogenous basidiomycetes are the members of the Septobasidiales (primarily *Septobasidium* spp.), and new colonies of these fungi are established by infective basidiospores; established colonies of the fungus, however, constitute an exception to the 'rule' about infective spores since it is hyphae from the fungal colony that penetrate healthy scales underneath the protective roof of the colony and establish their coiled haustorial systems in the hemocoels of affected scales to allow indirect absorption of the nutrients being pumped up from the host plant's phloem through the host's mouthparts. Thick-walled resistant spore stages – resistant sporangia for chytridiomycetes, oospores for oomycetes, and zygo- or azygospores for zygomycetes – are not directly infective but germinate and release infective flagellate zoospores for chytrids or oomycetes or one or more germ conidia produced directly from the thick-walled spore or indirectly on a small germ mycelium.

Among the fungi associated with invertebrates, it is a very general rule that infection of invertebrates proceeds by fungal penetration through the cuticle of the host (discussed in more detail later). However, there are a few notable exceptions where infection either may or must occur through the gut after ingestion of an infective unit: Microsporidia have genomes that are too progressively reduced in their capacities as intracellular parasites to support direct penetrations through the host's cuticle; microsporidia must be ingested (which infect gut epithelial cells first). Among the more traditionally recognized fungi, only three genera are routinely infective through the gut after ingestion of an infective spore: the ascomycetous species of *Ascosphaera* cause chalkbrood disease in bees in which food contaminated with infective ascospores is ingested by larvae after being fed directly by workers in colonies of social bees (Hymenoptera: Apidae) or, in solitary bees (Hymenoptera: Megachilidae and other families), by feeding on contaminated provisions left in their cells during oviposition. *Culicinomyces* spp. (Hypocreales: Clavicipitaceae) are pathogens of the aquatic larvae of such nematoceran flies as mosquitoes (Culicidae) and blackflies (Simuliidae), and infect through the gut after ingestion of the conidia; such a route of entry for the spores is practical for larval blackflies that anchor themselves to rocks or other substrates in rapidly moving water and feed on particulates (including fungal spores) caught on sticky slime nets that are periodically generated and then periodically eaten by the larvae. Germ tubes from the encysted zoospores of the oomycete pseudofungus *Leptolegnia chapmanii* (Saprolegniales) can infect mosquitoes equally well by penetrating either the exoskeleton (from cysts on the exoskeleton) or the gut endothelium (upon the germination of ingested zoospore cysts).

Fungus–Host Interactions During Cuticular Penetration

Unlike with viral, bacterial, microsporidian, and many other microbial pathogens of invertebrates, the infection of healthy hosts by fungi almost always involves penetration of a fungal germ tube through the exoskeleton rather than through the gut after ingestion of the infective propagule. The exoskeleton of arthropods is a complex multilayered structure (Figure 1) providing a sequence of challenges to the entry of a potential pathogen. Fungal units (spores, hyphae, etc.) landing on an arthropod first contact the thin layer of epicuticular waxes that is underlaid by one or more layers of cuticulin and lipoidal material that represent

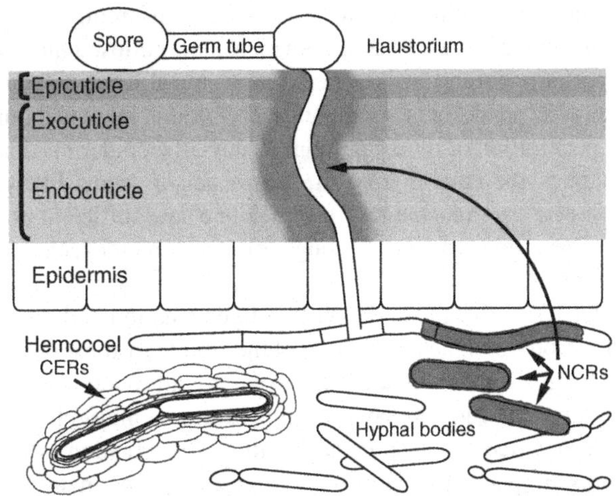

FIGURE 1 Fungal infection processes and host reactions. This diagram summarizes the major steps that may occur in the infection process for a typical fungal pathogen and the major types of host reactions that might occur; in no individual fungus–host pair would all of these interactions be evident. The fungal spore lands on the host's cuticle may germinate to form a germ tube and/or haustorium, and penetrates through the wax and lipoidal layers of the epicuticle and chitinous/proteinaceous layers of the procuticle (tanned exocuticle and untanned endocuticle) before penetrating through the epidermal cells into the hemocoel. Once in the hemocoel, vegetative growth may be as a hypha or, more typically, as short budding hyphal bodies that proliferate rapidly and circulate throughout the blood cavity. Noncellular reactions (NCRs) typically involve strong local melanizations within the cuticle or the deposition of thick melanin coats onto the surface of the invading fungus. NCRs are less common than cellular reactions (CERs) such as phagocytosis by hemocytes or encapsulation and melanization (illustrated here) by multiple layers of hemocytes. Infections are not established until the invading pathogen's development is no longer affected by any host defense reactions.

another waterproofing layer of the exoskeleton. The greatest depth of the exoskeleton is the laminar procuticle in which arrays of chitin microfibrils with changing orientations are embedded in a protein matrix; the exocuticle (outer portion of the procuticle) is usually tanned, while the inner endocuticle is not tanned. Below the cuticle is the cellular epidermis and then, finally, the hemocoel (blood cavity) where most of the development of an infecting fungus occurs.

Beginning at the outer surface of the cuticle, a potentially infective fungal propagule must react to the chemical environment either by not germinating or by germinating to produce either a germ tube or some type of secondary spore for dispersal to another (presumably more suitable) substrate. Oleic acid in the epicuticle is one compound whose presence or absence has been shown either to stimulate or to inhibit the development of entomogenous fungi and may, therefore, act as a key determinant of pathogen specificity. Germinating spores produce a germ tube on the host's cuticle that may immediately begin to grow downward into the cuticle of the potential host on which it forms, or the germ tube may grow across the surface for some distance until it reaches a more chemically suitable location (e.g., the comparatively thinner intersegmental membranes through which more stimulatory or useful compounds may diffuse to support further penetration into the host) or a natural opening into the body such as the mouth, anus, or spiracle (a natural opening into the tracheal system that supports breathing and gas exchange by arthropods). Many (but not all) entomogenous fungi produce a differentiated appressorium at the tip of the germ tube; the cytoplasmic content of the germ tube enters the appressorium, is walled off from the spore and germ tube by a complete septum, and then initiates a penetration peg into the cuticle. A fungus may penetrate more or less straight through the procuticular layers or, especially in comparatively thick cuticles, the penetrating fungus may grow laterally through the procuticle (even possibly branching there) and be subject to noncellular cuticular melanization reactions (Figures 1 and 2). Such melanization reactions may stop or retard the progress of a fungus through the cuticle either by their toxicity to the invading fungus or by making the cuticle a physically stronger barrier to fungal penetration (since few fungi produce enzymes capable of degrading melanin) or that stop or reduce the loss from the host to the fungus of nutrients or water through the chemically intractable melanin barrier.

Invertebrate hosts have immune systems able to mount diverse types of defenses against potential infective agents. While invertebrate immunology is a much younger science

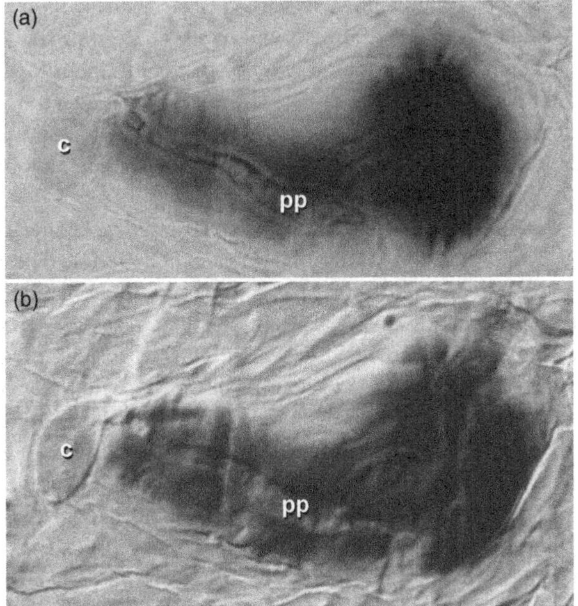

FIGURE 2 Cuticular melanization reaction in alfalfa weevil larva, *Hypera postica* (Coleoptera: Curculionidae) to penetration by germ tube of *Zoophthora phytonomi* (Entomophthorales). (a) Bright field microscopy. (b) Differential interference microscopy. c, infective conidium; pp, penetration peg traveling laterally in cuticle.

than vertebrate immunology, invertebrates' immune responses have successfully overcome most potential challenges over many more millions of years than have those of vertebrates, all of which face similar challenges but with a major dependence on the immunoglobulin-based defenses that invertebrates lack. Compounds constitutively present in the cuticle may be toxic or inhibitory for development by some fungi and may, therefore, help to determine which fungi can or cannot penetrate or infect a specific host. Melanotic or other noncellular immune reactions within the cuticle may bring new compounds to act against an invading nonself organism. Most insects are able to confront an invader that has reached the hemocoel with an array of cellular immune responses mounted by the various classes of hemocytes circulating in the host's blood such as by phagocytosis of small particles (e.g., bacteria) or, with larger invaders, by encapsulation reactions where large numbers of cells converge on the invader, and encase it in a layer many hemocytes deep whose innermost cells in close contact with the nonself material flatten, lyse, and melanize it. In a much smaller diversity of invertebrates, noncellular (humoral) immune reactions may result, among other strategies, in the deposition of thick coats of melanin onto the surface of the invader (Figure 1). Regardless of which defensive strategy or combination of strategies is employed, a microbial invader is able to develop without further restraint only when the host's ability to react becomes ineffective, and the infected host will almost inevitably die. At least one more and later occurring defensive 'trick' may available to cure some invertebrate hosts of actively developing pathogens: Some insects such as infected individual flies, grasshoppers, and locusts may show modified behaviors ('behavioral fevers') that cause them to remain on hot places (e.g., sun-heated rocks) for long enough that their body temperatures rise to 40–45 °C, temperatures that may not affect the host but that may kill infections by fungi or other microbial pathogens.

Fungal pathogens generally secrete a series of lipases among the first steps of germination and penetration. These enzymes clearly aid fungal passage through the epicuticle whether or not any appressorium is formed to help anchor the fungus to the host and in which high hydrostatic pressures may be generated to help push a penetration peg or hypha through the cuticle by mechanical means alone. Nonetheless, fungal penetration of an arthropod exoskeleton seems to depend on a combination of both enzymatic and mechanical mechanisms. The procuticle is composed of innumerable layers of varyingly oriented chitin fibrils embedded in a protein matrix. While it is commonly believed that strong chitinases must be excreted to allow an invading fungus to penetrate this part of the cuticle, enzymatic studies suggest that the production of a battery of proteases to weaken or to remove the protein matrix from around the chitin fibrils may be more effective than production of chitinase to break the reinforcing chitin fibrils; the removal of the protein matrix may allow a fungus to push the chitin fibrils aside by mechanical force without having to break them. Comparatively little is known about the penetration of invading fungi through the epidermis that separates the cuticle from the hemocoel; it seems likely, however, that mechanical forces alone are sufficient for the germ tube to push through or between the cells of this layer.

During all of the early stages of germination and probably throughout the penetration of the cuticle by most fungi, the fungus remains in a germinative growth mode. During this time, the fungus may operate entirely on nutrients that were included in the spore at the time of its creation. Most spores are packaged with stores of glycogen (the main carbohydrate storage polymer for fungi) and oils as sources of essential materials for producing the cell wall on the germ tubes and to provide the required energy for the germination process. When in the germinative mode, the external environment provides oxygen and water but no nutrients to support fungal growth and development. Germinative growth is necessarily determinate because of the dependence on internal resources: Germ tubes may be frequently septate but do not branch while the cytoplasmic volume tends to diminish as it uses those reserves but moves away from the spore by remaining in the tip of the growing germ tube while a series of empty cells accumulate between the cytoplasm and its spore. When the transition to vegetative growth occurs, the fungus must then depend primarily on external nutrient sources for energy and structural components, and indeterminate growth begins. This germinative/vegetative metabolic shift has profound implications for the biology of a fungus, but is usually never seen except with fastidious pathogens that need very special (nutrient and/or physical) conditions to trigger this transition to allow the fungus to continue growing indeterminately by using external sources of nutrients. While many entomogenous fungi easily grow from conidial inoculum on almost any available medium, the most fastidious pathogens may be seen to produce multiseptate germ tubes that stop growing when their reserves are exhausted. If fungal spores fail to make the germinative–vegetative transition despite attempts on many media or in many cultural conditions, then the likelihood of obtaining a growing culture is increased greatly by using vegetative inoculum obtained directly from an infected host that is already supporting vegetative growth of the desired fungus. Vegetative inoculum is already doing everything initially expected of a fungus without its having to go through the series of dramatic metabolic and physiological changes required to initiate germination, to form a germ tube, and then to shift over all cellular processes to accept external sources of nutrients; such inoculum may grow and develop readily on media where spores consistently fail to yield a culture. The germinative mode may not last beyond the earliest stages of germination in species of *Beauveria*, *Metarhizium*, and many other easily culturable genera. Many entomogenous water molds (especially *Coelomomyces* pathogens of mosquitoes), some entomophthoraleans, and even some hypocrealean fungi frustrate every attempt to isolate cultures, even when using specially formulated media and vegetative inoculum.

It may be difficult to know when during the penetration process any given fungus will undergo the germinative–vegetative transition, but it is certain that a successful infection will require the fungus to be vegetative by the time it penetrates the cuticle and epidermis to enter the hemocoel. In the hemocoel, the now vegetative fungus will follow one of two development courses (apart from the need to its interactions with the host's immune responses). The fungus will either continue to grow in a thread-like hyphal form in the blood or begin to fragment into short segments ('hyphal bodies') that may proliferate by further fragmentation or by budding. Hyphal bodies tend to circulate throughout the insect body, even through the wing veins as in Figure 3. These cells may increase the surface area of the fungus in the host's blood (and, thereby, the ability to absorb nutrients) faster than would the growth of hyphae, and also hasten the general colonization of the body.

Hyphal development may result in a more localized growth and less consequent damage to the host tissues. The spatial restriction of the hypha-like protoplastic growth of *Massospora* species (Entomophthorales: Entomophthoraceae) to the terminal three to four abdominal segments (containing the genitalia) of periodical cicadas (Hemiptera: Cicadidae) leaves all other major organ systems largely unaffected and may be a key reason why these fungi can sporulate in and disperse their spores from living hosts; the exoskeleton of the affected segments dissociates and exposes the mass of fungal spores. In a somewhat similar manner, the vegetative growth of *Strongwellsea* species

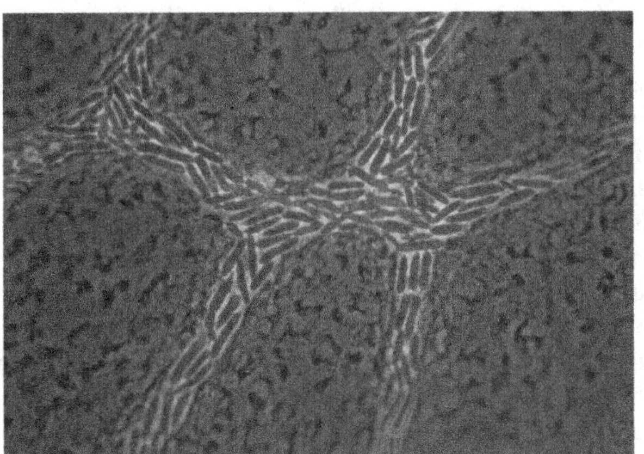

FIGURE 3 Hyphal bodies of a *Hirsutella* species circulating in the wing veins of *Bemisia tabaci* (Hemiptera: Aleyrodidae). Such a high density of fungal cells suggests that the host is moribund (or already dead) rather than in an early or middle stage of postinfection fungal development.

FIGURE 4 *Strongwellsea magna* (Entomophthorales: Entomophthoraceae) sporulating from a living *Fannia canicularis* (Diptera: Fanniidae) showing an abdominal hole underlaid by a cup-like fungal hymenium from which conidia are forcibly discharged through the hole. An accumulation of discharged conidia (arrow) is seen on the lip of the hole.

(Entomophthorales: Entomophthoraceae; Figure 4) is also as protoplastic hyphae confined largely to the abdominal hemocoel where the fungus eventually forms a large, hollow ball that attaches to the thin cuticle of the living fly's abdominal pleuron and, as the ball grows, tears open a large hole through which conidia are then forcibly discharged from a layer of conidiophores covering the inner surface of the ball.

The surface of the typical fungal cell wall incorporates abundant chitin and β-1,3 glucan molecules as integral structural components of the wall. These compounds usually elicit strong polyphenol oxidase (melanization) reactions in most insects, thereby causing some of the strongest host reactions to the penetration of potential fungal pathogens to an insect hemocoel. If, however, a fungus does form a cell wall, it does not secrete these immunological epitopes, and the presence of an invading fungus is more difficult to detect and may not be (or cannot be) rejected by the host's immune system. Growth in the host's blood as wall-less protoplasts is routine for many (but not all) entomopathogens of the Entomophthorales. The protoplasts may be either hypha-like in form or amoeboid and variable in shape but tending to become separated in uninucleate fusoid cells strongly resembling many of the host's hemocytes. Similarly, host immune defenses may be evaded if the surface of an invading fungus either is promptly covered with the host's own antigens or in any way blocks or fails to elicit strong reactions. These latter possibilities are best exemplified by some bacteria or protozoans in other invertebrates or, indeed, in vertebrate hosts; they are not well known for fungal pathogens affecting invertebrates.

The interaction between a fungus and its potential host is a process that involves a series of complex actions and reactions by both organisms. The infection process for any potential pathogen may be stopped or deterred at any of these stages, or it may proceed successfully. No fungal infection is properly established until the host's last cellular or noncellular immune barriers are exhausted (e.g., all available hemocytes having been committed unsuccessfully), overcome (e.g., a fungus may grow through any cellular or noncellular encapsulation reaction), or avoided (by failing to elicit any effective host response). Once a fungus has produced substantial biomass in the host, the generation of cell walls becomes a moot issue since any remaining elements of the host's immune system would be quickly saturated and unable to stop further fungal development.

Fungal Development in the Hemocoel

Established fungal infections tend to develop in the hemocoel and do not immediately attack the host's solid tissues. Some very small hosts may be killed in a day or less, but most fungal infections require two to several days to cause the host's death. The confinement of fungal development to the blood allows the fungus to grow while continuing to depend on the living host for essential life support functions while the host eats, breathes, and excretes. As a fungus grows and begins to absorb an increasingly large proportion of blood-borne nutrients, there is increasing competition for available dissolved oxygen as the fungus mass increases and impedes the circulation of the blood through the affected host. The physiological state of both fungus and host changes rapidly as competition for resources proceeds and host organ systems begin to fail, thus hastening both the host's death and the need for the fungus to begin its reproductive phase. It is usually at this time that the host is suffering significant negative effects of the mycosis and that the fungus begins to secrete large quantities of degradative enzymes that attack the host's muscles, nervous system, Malpighian tubules (excretory system), and other organs. And it is at this time that the destruction of tissues and the loss of excretory and respiratory functions and that a complete shift in the internal environment of the infected host occurs, which contributes strongly to the death of the host and to the initiation of the sporulation by the fungus.

The insect fungi from the Hypocreales are well known to produce a chemically diverse range of toxins that include cyclic peptides (destruxins, beauvericin, bassianolide, oosporein, cyclosporins, etc.), efrapeptins, polyketides, and a wide range of other compounds with diverse biological activities. Deleterious effects of these compounds have been proven against various insect species and also against nontarget invertebrate and vertebrate test species. It has been widely believed that such activities suggest that these compounds are involved in the infection processes and in killing their hosts. Some experimental evidence does, indeed, confirm that these sorts of compounds may play an active role in the infection or pathogenesis of a host.

The toxins and other secondary metabolites produced by entomogenous fungi may play major roles in killing the host, but the main function of these compounds may be to provide a degree of antibiotic protection that has long been overlooked: Pathogenic fungi in hosts with chewing mouthparts meet intense microbial competition when the gut of a moribund mycotized host breaks down and releases its bacteria and fungi into the mass of the pathogenic fungus; without antibiotic protection, the pathogen could be overwhelmed since most bacteria and saprobic fungi grow faster than entomogenous fungi. Insects such as hemipterans and dipterans with piercing/sucking mouthparts have no such complement of potential contaminants in their guts as do chewing insects, and it is notable that scales, aphids, and cicadas (all sucking insects in the Hemiptera) are affected by the highest incidence of mycoses and are also hosts to the greatest taxonomic diversity of fungal pathogens. While production of conidia or other asexual propagules usually occurs very quickly after the host's death, in comparatively few hypocrealean mycoses, the fungus in the cadaver may not produce external conidiophores but forms a metabolically quiescent sclerotium filling the cadaver, and later gives rise to the external erect sexual stroma of a *Cordyceps* species. The total time needed by any *Cordyceps* species to form the sclerotium, then sexual stroma with its perithecia, asci, and ascospores, and then to discharge ascospores from the fruiting structures remains unknown but requires at least several weeks to many months to occur; indeed, the sexual stroma may not even necessarily be formed in the same season in which the host was killed. Until these ascomycetes have exhausted their ability to discharge ascospores, the stromata are vulnerable to microbial degradation and the array of invertebrates and vertebrates that might eat the fungus. The erect stromata of *Cordyceps* species emerge through the soil surface (rotten log, or some other substrate) from buried, sclerotized host cadavers, and these stromata remain surprisingly intact and free of obvious degradation or mycophagy regardless of their age (e.g., even when most perithecia have discharged their ascospores) or, most significantly, even in those environments where the activities of ants, other small animals, or microbial degraders may rapidly clear the soil surface of plant litter or any other readily processed organic matter. Many entomogenous fungi may take long enough to complete their spore dispersal only because of such inherent chemical defenses such as these compounds. As is discussed briefly later, the toxins and metabolites are a growing subject for pharmacological and biochemical research that confirms the historical uses and ascribed properties of fungi such as *Cordyceps sinensis* and *Cordyceps militaris* in East Asian medical traditions.

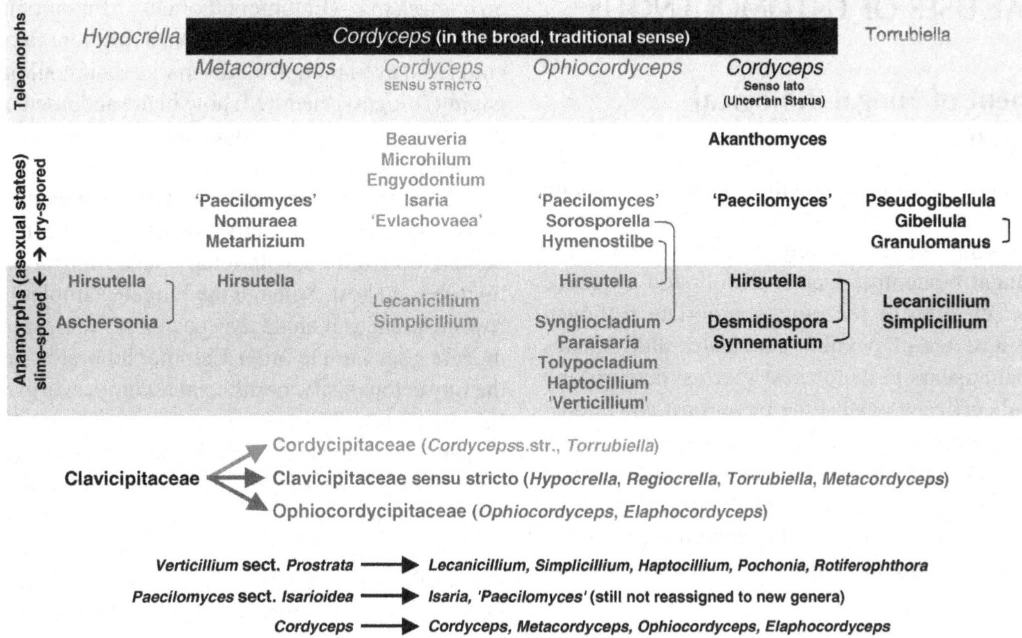

FIGURE 5 The shifting taxonomy for fungal pathogens in the Hypocreales (Sordariomycetes) that are among the most important, most studied entomogenous fungi. At the bottom, entomogenous species from *Verticillium* and *Paecilomyces* have been revised phylogenetically, and all of their invertebrate pathogens reclassified. Above that is an indication of the recent splitting of the family Clavicipitaceae into three families that are listed with their major teleomorphic genera either segregated from *Cordyceps* or otherwise pathogenic for arthropods. The top section shows the major entomogenous teleomorphic genera of this group, how *Cordyceps* has been split in the current classification, anamorphic genera are listed below the teleomorphic genera with which they are associated. Linked synanamorphic genera are joined by brackets; whether the conidia of these anamorphic genera form in dry chains or in slime is noted; the segregate families into which these anamorphic genera are currently placed are indicated by the color of the type (corresponding with the segregate families in the lower section). Note that *Torrubiella* species are being assorted among two different families, and that some anamorphic genera or some species from them (in black type) have not yet been placed into their appropriate segregate families. Status of genera in quotes remains uncertain.

Just as nearly all entomogenous fungi penetrate through the host's cuticle, it is also a virtual rule that their sporulation occurs after the death of the host and outside its mycotized body. The growth of the fungus out through the dead, unreactive exoskeleton of a cadaver before sporulating is less of a problem than was its initial penetration. This emergence, again accomplished by both mechanical and enzymatic means, often occurs first through the weakest parts of the host cuticle (intersegmental membranes, joints of the legs, about the eyes, through the mouthparts, etc.) but is not spatially restricted in most entomogenous fungi. Almost all spores able to transmit a fungus are formed outside the host body; however, the morphologies of those structures vary enormously across the more than 100 genera of entomogenous fungi. The conidia of entomophthoralean fungi that are formed directly on the hosts are actively discharged from the cells on which they are formed and may either be shot to a distance of several millimeters (or even centimeters!) or may become airborne or fall from the elevated positions where hosts infected by these fungi often place themselves during their dying hours downward onto the healthy populations of these fungi. Conidial fungi (hypocrealean anamorphs) show the greatest morphological variations in their reproductive systems but completely apart from the distinctive morphologies of conidia and conidiogenesis in any of the ~25 anamorphic genera for entomogenous hypocrealeans (Figure 5).

Thick-walled spores serve as the major spores for the vertical transmission from generation to generation or season to season and are the major spore types forming inside the cadavers of infected hosts; these environmentally durable spores include the resistant (meio)sporangia of chytridiomycetes, oospores of oomycetes, resting spores (zygo- or azygospores) of zygomycetes, and the chlamydospores of some anamorphic ascomycetes (e.g., *Sorosporella* spp.). Comparatively few entomogenous fungi produce their asexual (mitotically derived) spores inside the body of a host; the exceptions include the species of *Strongwellsea* and *Massospora* for the Entomophthorales, both of which were noted above to form the conidia inside the host body and to disperse those spores actively or passively, respectively, from living hosts rather than from cadavers.

PRACTICAL USES OF ENTOMOGENOUS FUNGI

Development of Fungal Biological Control Agents

Whether any particular fungus can kill a given invertebrate host is a primary concern when trying to find a suitable fungal biocontrol agent. In general, any program to find and to develop a fungal biocontrol agent will follow a sequence that includes selection of the most appropriate pathogen from among a series of possible candidates and isolates of the one that appears to be the best species; determining that pathogen's efficacy against the target host and choosing the most appropriate isolate; optimizing procedures to grow, to formulate, and to apply the pathogen; determining the pathogen's safety for nontarget invertebrates and then for vertebrates; and ultimately seeking governmental registration of a biocontrol product whose active ingredient is the chosen pathogen.

Host and geographical ranges

The host range of any particular fungus may be extremely broad (able to attack hosts from many different orders of hosts), extremely narrow (limited to a single genus or even species of host), or anywhere in between. What is perceived as the host specificity of a fungus is the end result of innumerable complex interactions of a microbe with the environment and its potential host. The most commonly encountered and, arguably, the most important and widely used fungal pathogens of arthropod pests have the broadest host ranges. Nonetheless, recent taxonomic studies based on gene sequence data indicate that *Beauveria bassiana*, *Metarhizium anisopliae*, and the fungus previously known as *Verticillium lecanii* are now shown to be species complexes.

Most entomopathogens, regardless of the order or class to which they belong, show a moderate degree of specificity for hosts usually limited to a single family or to a small group of related families within a single order. An excellent example of this sort of specificity is *Nomuraea rileyi*, a globally distributed anamorphic (conidial) species of the Clavicipitaceae (Hypocreales), whose hosts are exclusively lepidopterans (including many major pest species) from genera in the family Noctuidae; the rare incidences of *N. rileyi* from hosts outside the Noctuidae (e.g., silkworms, *Bombyx mori*) are still from families classified in the superfamily Noctuoidea. *Pandora neoaphidis* is a globally distributed entomophthoralean fungal pathogen affecting aphids (Hemiptera: Aphididae) but shows no apparent preferences for genera or species within this insect family, and is the most common and effective of all fungi affecting these plant-sucking insect pests. The species of *Strongwellsea* (Entomophthorales: Entomophthoraceae) each appear to affect species from a different family of muscoid flies, even though all discharge their conidia through a gaping (fungus-generated) hole in the abdomen of the living host fly from a ball-like hymenium (fertile layer) inside the abdomen (see Figure 4).

The narrowest host ranges, those in which a single microbial species affects only a single host species, appear to be derived from coevolutionary accommodations between microbe and host. Some of the fungal examples of such narrow host ranges include the species of *Massospora*, a zygomycete genus in the order Entomophthorales, and many of the tiny ectoparasitic perithecial ascomycetes in the Laboulbeniomycetes. *Massospora* species have one-to-one relationships with periodical cicada species (Hemiptera: Cicadidae) whose life histories may extend up to 17 years between emergences. In the years between emergences, the fungi remain in their quiescent resting spore stages until the cicada populations begin to construct their emergence tunnels; many emergent cicadas are infected and immediately begin dispersing highly infective conidia to other healthy cicadas, each of which then produces resting spores, the form in which the fungus persists between emergences. The many hundreds of tiny species of the Laboulbeniomycetes are classified among many more than 100 genera; these fungi often show not just extreme host specificities but also site specificities for particular body parts that differ between males and females of suitable hosts; these extreme spatial specificities derive from transmission during copulation of the fungus from infected individuals to the opposite body parts of their healthy mates.

Despite our usually poor understanding of them, highly specific sequences of interactive biological processes control the ability of any given fungus to succeed or to fail to cause disease in a given host. The perceived specificities of a pathogen – the accumulated observations of infections in particular hosts and from particular geographical sites – are treated as the 'host range' and 'geographical range,' respectively. These observed and reported host and geographical ranges may in reality have little to do with the actual biological capacities or distributions of an organism. The broadest host ranges may be those for such conidial fungi as *B. bassiana* and *M. anisopliae* since these are indisputably the most commonly occurring and broadly distributed of all fungi affecting insects, and are each reported to affect literally many hundreds of host species from nearly all major insect orders and families throughout the world. The advance of the systematics of these fungi from its traditional morphological basis to the phylogenetically based approach now revolutionizing our knowledge of organismal relationships has inevitably led to the finding that such 'mega-species' as *B. bassiana* and *M. anisopliae* are species complexes comprising genetically distinct but morphologically similar taxa with narrower host or geographical specificities than those of the more traditionally circumscribed species from which these cryptic taxa are segregated. Despite such taxonomic refinements, most phylogenetically driven revisions of these fungi have tended to break small but homogeneous groups from the margins of the main taxa while still leaving the narrowed circumscriptions of the main species applicable to fungi affecting a huge diversity of hosts from throughout the world.

Despite the inevitable influence of individual collectors, and the extents of their travels and collection activities on our knowledge of host and geographical ranges, some biogeographically significant results bear mention. In the broad traditional sense of the genus, the hundreds of species of *Cordyceps* (Sordariomycetes: Hypocreales) and of the anamorphic (conidial) states connected with these sexual species present a special biological puzzle. In the first place, there may be more morphologically distinct genera of conidial fungi linked to *Cordyceps* than to any other teleomorphic (sexual) fungal genus (see Figure 3). The conidial entomopathogens from the Hypocreales are much more common than the vast majority of the hundreds of teleomorphic species described originally in *Cordyceps* (in the broad traditional sense). While *B. bassiana* and *M. anisopliae* are known from hundreds of host species from throughout the world, no sexual states were confirmed for these species until only very recently. The sexual states of species in these two ubiquitous anamorphic genera are very rarely collected species of *Cordyceps* (in the broad taxonomic sense). Even without considering that *B. bassiana* is a partially resolved species complex, only five individual insects have been collected with a teleomorph appearing to be linked to a conidial state attributable to *B. bassiana*, and only two individual insects infected by a *Cordyceps* have been linked to a fungus identifiable in the *M. anisopliae* species complex. All of these teleomorphs of the two most universally distributed and common conidial entomopathogens came from China, Korea, or Japan. The biogeography of *Cordyceps* species suggests that the greatest diversity of these species occurs in East Asia although there are differing opinions about whether the greatest diversity may be in northeastern Asia (China, Korea, Japan, and eastern Russia) or in southeastern Asia (where the greatest diversity may be in Thailand). The belief that one of these regions of exceptional biodiversity is the center of distribution and the probable center of origin for *Cordyceps* and its many anamorphic forms is gaining wide acceptance.

The fungi most frequently and most widely applied as biocontrol agents against arthropod pests are *B. bassiana*, *M. anisopliae*, fungi previously identified as *V. lecanii* (now identified as *Lecanicillium lecanii*, *Lecanicillium muscarium*, or *Lecanicillium longisporum*), *N. rileyi*, and species such as *Isaria farinosa* and *Isaria fumosorosea* (both of which were previously classified in *Paecilomyces*). These species have become major candidates for use as biocontrol agents mainly because their broad host ranges and easy manipulation in culture made them attractive enough to complete the lengthy development and testing programs, to optimize their production processes and formulations, and to complete the expensive safety testing and governmental regulatory hurdles required by almost all countries before microbial biological control agents are allowed to be produced, sold, or used commercially. Note, however, that the broad host ranges of these fungi (with their comparatively broad possible markets for products based on them) and all that their nonspecificity implies about their biologies also mean that these fungi raise the greatest concerns about the potential nontarget effects against beneficial arthropods and even against humans or other vertebrates.

Uses as biological control agents

The use of fungi as biological control agents against insect pests has been a desired goal in many countries for more than 150 years. There is an inconsistent record of success in attempts to use fungal biocontrol, but there appear to be more successes than prominent failures. Many of the failures have resulted from insufficient appreciation of the overall ecology and interactions of organisms in the ecosystem where an introduction is attempted. For example, the introduction of *Entomophthora sphaerosperma* (=*Zoophthora radicans*; Entomophthorales) against *Plutella maculipennis* (=*P. xylostella*; Lepidoptera: Plutellidae) in South Africa in the early twentieth century strongly depressed populations of both the target pest and its other insect natural enemies so that in the next year the *Plutella* populations were worse than ever in the absence of the usual extra controls from its other natural enemies that had been eliminated. Some notable successes with fungal biocontrol in the twentieth century were the near elimination of brown-tail moth, *Euproctis chrysorrhoea* (Lepidoptera: Lymantriidae), from the northeastern United States by what is now identified as *Entomophaga aulicae* (Entomophthorales) and, in part of the same study, the introduction from Japan of what is now known as *Entomophaga maimaiga* to control gypsy moth, *Lymantria dispar* (Lepidoptera: Lymantriidae). *E. maimaiga* is now widespread throughout the range of gypsy moth in the eastern United States, and is dispersing both by natural means and by artificial introductions. *Z. radicans* was also introduced to New South Wales, Australia, in the 1970s to control spotted alfalfa aphid, *Therioaphis maculata*; the fungus spreads quickly and continues to exert substantial levels of control over this significant pest aphid throughout northeastern Australia.

Each of the examples cited above for successes and failures of fungal biocontrol involved inoculative introductions (the microbial equivalent of classical biocontrol where the control agent is introduced to a target population and allowed to establish and to disperse by itself rather than through constant human intervention). There are many successful programs using either augmentative or inundative application strategies for fungal biocontrol agents either to supplement or to shift the time of activity of natural pathogen populations (the augmentative approach) or to make large-scale applications of a pathogen to induce rapid, significant reductions of pest numbers (the inundative approach). Most augmentative or inundative control programs tend to rely on using a small range of commercialized fungi that can be easily mass produced in liquid fermentations or on solid substrates such as sterilized rice or other grains, and that have been registered by the appropriate regulatory agencies. The fungi used in inoculative biocontrol releases are much more taxonomically diverse and have higher specificities for their hosts than do the commercialized fungi; the presence of so many species of the Entomophthorales among the successful inoculative releases of fungal biocontrol agents reflects the much greater difficulty in controlling these zygomycete fungi both in culture and after their release than is true for so many conidial pathogens such as species of *Beauveria*, *Metarhizium*, *Nomuraea*, *Isaria*/*Paecilomyces*, and *Lecanicillium*/*Verticillium*. Some success has also been had with such conidial fungi as *Aschersonia aleyrodis* (Hypocreales: anamorphic Clavicipitaceae) either through inoculative or augmentative releases or, at the very least, by attempting not to interfere with the recognized presence and activity of the fungus against its whitefly (Hemiptera: Aleyrodidae) hosts in citrus orchards in the United States and China, but this fungus is also commercialized in Europe for inundative sprays in greenhouses against whiteflies on various herbaceous crops.

Is a Fungus a Pathogen or Saprobe?

The cuticle of an insect or other arthropod may carry ungerminated spores of nearly any fungus that are wholly noninfective for invertebrates. Similarly, the growth of a fungus on a dead insect is no indication that the fungus caused the insect's death. Floristic surveys of entomogenous fungi often report the mere presence of fungi merely because they might be found on an invertebrate substrate but rarely distinguish the key functional difference between saprobes (for which cadavers are available, suitable food sources) and facultative or obligatory pathogens or secondary pathogens that gained entry to a living host's body only after the host's immune defenses were destroyed earlier by the actions of a primary fungal, bacterial, viral, or any other type of pathogen. When trying to select a candidate fungus as a biocontrol agent, it is vital to know that the candidate fungus can actively infect and kill significant proportions of a population of healthy target hosts. There is no way to predict the host range, virulence or pathogenicity, or other vital biological properties of any given isolate of a fungus to know whether that isolate might be a good biocontrol candidate in a specific environment against a specific host. This is why all programs to develop microbial biocontrol agents depend on extensive bioassay programs to select among a wide range of possible candidate organisms, and then to focus on optimizing conditions for the production and application of the pathogen against the target host.

One of the greatest microbiological advances in the nineteenth century was the proposition of Koch's postulates and their broad acceptance as an essential tool to determine whether a particular bacterium causes disease. Robert Koch stipulated four steps to ensure that a specific microbe is responsible for a specific disease. The organism in question must always be present in diseased individuals (but is generally absent from healthy individuals). It must then be isolated from a diseased individual and grown in pure culture; this requirement is problematic for fastidious microbes (including many fungal entomopathogens) that resist any culture media and techniques used for their isolation. The cultured microbe (or a concentrated suspension of the microbe if no culture was obtained) must cause the same disease again when introduced into healthy individuals. Finally, the microbe must be reisolated (or reextracted) from experimentally infected individuals and must be proved to be identical to the microbe administered earlier to cause the disease.

When considered for entomogenous fungi, Koch's postulates are accurate but superfluous for all of those fungi that are already well proven to be primary pathogens of their hosts. Paradoxically, for weakly or facultatively pathogenic organisms whose ability to cause disease in small invertebrates may be in doubt, attempts to follow Koch's postulates will often give false-positive conclusions merely because the physical conditions required to conduct bioassays of potential pathogens against arthropods or other invertebrates usually stress the hosts so that their natural ability to resist a microbial challenge may be mildly to severely compromised.

Safety of Fungi as Biological Control Agents

This paradox of false or higher virulence so often shown by bioassay programs requires the recognition that the ecological (natural) and physiological (experimental) host ranges for a microbe are distinctly different. The evaluation of such differences may critically affect regulatory decisions governing the applied use of a pathogen for the biological pest control. For example, cockroaches (Blattodea) and honeybees (Hymenoptera: Apidae) are observed to be subject to very few fungal diseases in natural conditions, whereas bioassay studies indicate that both may be infected by many different fungi. Such a finding for physiological (experimental) host ranges is typical in comparison with observed ecological host ranges. Despite such differences between the natural and experimental host ranges, it is notable that the United States Environmental Protection Agency (whose high standards for the safety of registered products are globally appreciated) has registered *B. bassiana* and *M. anisopliae* – the two most widely used and, arguably, the least highly specific fungi affecting insects – multiple times for uses against a diverse range of target insects pests in fields, greenhouses, and even homes.

Entomogenous fungi have generally proven to have a very strong record of safety for nontarget vertebrates and even invertebrates. While there can be unintended and unexpected effects of introducing fungi for the intentional control of insect pests, there is also much evidence for their successful use as biocontrol agents. Some of the main concerns about the safe use of these organisms include the displacements of nontarget microbes by the introduced biocontrol agent; the allergenicity of the biocontrol agent for humans, livestock, pets, or other nontarget animals; toxicity for nontargets; and pathogenicity for nontarget organisms. Among these, the major concerns with entomogenous fungi have generally been the potential allergenicity of conidial entomopathogens for those who produce or apply them, and the pathogenicity of these fungi for nontarget invertebrates. Adverse allergenic effects can usually be minimized or eliminated through suitable means to produce, to handle, and to apply these fungi. The nontarget pathogenicities are best addressed by suitable bioassay programs to select the isolates best suited for use against specific combinations of target hosts, environments, and application conditions. If the objective is, for example, to use *B. bassiana* against the complex of aphids on cereal grains, the fungal isolate used must be carefully chosen not to endanger other natural enemies of aphids such as ladybird beetles (Coleoptera: Coccinellidae), hoverflies (Diptera: Syrphidae), or lacewings (Neuroptera: Chrysopidae).

A LOOK FORWARD FOR ENTOMOGENOUS FUNGI

Modernization of Systematics and Taxonomy

The recent upheavals in mycology in the wake of a massively multiauthored overview of fungal phylogeny and formal reclassification based on that study have placed the fungi associated with insects and other arthropods in a rather new light since these fungi are understood to have had multiple evolutionary origins and to have demonstrated a remarkable fluidity in their host associations over time. The confirmation that Microsporidia are highly derived fungi rather than extremely primitive eukaryotes represents a huge status change for these organisms that is old news to microsporidiologists (and also to many mycologists) but is still a seismic surprise to those who are unfamiliar with the diverse data supporting this reclassification. This recent phylogenetic reshuffle of the fungi has separated the posteriorly uniflagellate water molds into two phyla (Chytridiomycota and Blastocladiomycota), and fragmented the long-familiar Zygomycota into the Glomeromycota (for arbuscular mycorrhizal fungi) and four other groups recognized at the subphylum level (Entomophthoromycotina, Mucormycotina, Kickxellomycotina, and Zoopagomycotina), which may eventually each be recognized as separate new phyla or grouped in some manner yet undetermined, but these changes involve relatively few fungal entomopathogen. By far, the most radical changes for insect fungi involve the vast majority of conidial (anamorphic) entomopathogens and their sexual (teleomorphic) states that have been recognized to belong to the perithecial ascomycetes (Sordariomycetes) in the order Hypocreales and, mainly, in the family Clavicipitaceae. These fungi notably include the very large and taxonomically complicated genus *Cordyceps*, and the smaller but no less important genera *Torrubiella* and *Hypocrella*. A recent phylogenetically based and long-needed reclassification of the Clavicipitaceae (see Figure 5) has dramatically reworked the taxonomies of the teleomorphs – especially of *Cordyceps* – and split this large family into three smaller ones: Clavicipitaceae *sensu stricto* (primarily for plant-associated fungi, but also incorporating many significant entomopathogens in the teleomorphic genera *Hypocrella*, newly segregated *Metacordyceps*, and *Torrubiella* as well as the conidial genera *Aschersonia*, *Metarhizium*, *Nomuraea*, and some segregate genera formerly incorporated in *Verticillium*), Cordycipitaceae (including primarily entomopathogenic fungi including the newly restricted *Cordyceps* and part of the genus *Torrubiella* along with anamorphic fungi classified in *Beauveria*, *Isaria*, *Lecanicillium*, which was, in turn, the largest and most important genus recently segregated from *Verticillium*), and the Ophiocordycipitaceae (including two more segregates from *Cordyceps* – *Ophiocordyceps* and *Elaphocordyceps* – and a large number of entomogenous conidial genera in comparatively smaller and less well-known genera such as some most species of *Hirsutella*, *Hymenostilbe*, *Tolypocladium*, of the other fungi now segregated from *Verticillium* and *Paecilomyces* after their phylogenetic reclassifications).

Nontraditional and Nonorganismal Uses of Entomogenous Fungi

Evidence is accumulating that the hidden or unsuspected lives of insect-associated fungi – especially those of the pathogenic fungi – may be far more subtle and complex than might have been even imagined only a few years ago by the greatest specialists in these organisms. The discovery of genes in some 'insect' fungi for enzymes that degrade such highly plant-specific compounds as xyloglucans or cellulose and the knowledge that such insect fungi as *B. bassiana* may colonize and exert anti-insect activities in the tissues of diverse herbaceous or woody plants (including maize, opium poppy, banana, and coffee) raise many questions and invite research in previously unforeseen areas to learn more about the 'real' activities and capacities of these fungi in more and wider ecological settings for their actions than was previously thought to be possible.

The extent to which such 'unexpected' capacities of fungi historically believed to affect only arthropods or other invertebrates will remain speculative until any full genomic sequences become available for some of the more important entomopathogens; *B. bassiana* and *M. anisopliae* are expected to be the probable first entomopathogenic fungi to be fully sequenced. Despite the number of organisms whose genomes have now been completely sequenced, no fungal pathogen affecting any insect has yet been sequenced although real-time PCR and differential expression studies have been started to dissect some of the activities of some key entomopathogenic fungi at critically important times such as during their spore germination, penetration of the host cuticle, and early establishment in the host.

Claviceps purpurea, the type of its genus and of the family Clavicipitaceae, is widely known as the causative agent of ergot of rye and also of ergotism, the medically serious syndrome affecting humans or livestock who consume too much ergotized grain. The stromata of *C. purpurea* are the source of a wide range of alkaloids that contribute to the effects of ergotism but that are used medically, and known collectively as 'ergot alkaloids,' to treat a wide range of medical issues from the relief of migraine headaches to the stimulation of smooth muscle contractions (during childbirth and at other times). It may be anticipated that complete sequences of *Beauveria*, *Metarhizium*, or other major fungal entomopathogens will allow unprecedented insights into the extent to which these fungi show 'hidden' metabolic capacities that might allow them to grow and to sporulate in nature on substrates (e.g., as endophytes in plants, or using plant compounds in litter or soil) rather than being limited to living invertebrate hosts. The very many hypocrealean entomopathogens are closely related to *Claviceps* and other plant- and fungus-pathogenic fungi of the Clavicipitaceae and clearly share the abilities to produce biologically active compounds of great pharmacological interest. The tonic and medicinal values of such entomopathogens as *C. sinensis* and *C. militaris* are prized in traditional Asian medicine, and have been used to treat such ailments as tuberculosis, jaundice, drug addictions, blood disorders, and many other ailments up even including various cancers.

Genomic sequencing should also provide clues about how the production of interesting or unusual compounds may be controlled, thereby allowing means by which in vitro production of such compounds may be stimulated and sustained, and also allowing a definitive answer about the long-standing questions about whether the mycotoxins produced by entomopathogenic fungi function primarily during development of pathogenesis in the original arthropod host or work more significantly to protect the slow-growing teleomorphic states of these fungi that might have to withstand weeks to many months (or more) of exposure to decomposer microbes or to mycophages. And there is always the possibility that having complete genomic sequences for some of these organisms may suggest novel and potentially more effective means by which they can be used as agents for the biological control of agricultural pests.

Integration of Multidisciplinary Inputs to Insect Mycology

The continued study of entomogenous fungi (or, indeed, of all organisms) should not become the exclusive province of the genome-based molecular biologists. There is abundant hope that the new biology based on an appreciation and exploitation of the genetic capacities built into the entomogenous fungi needs to be explored, expanded, elaborated, and characterized in a collaborative research environment. The new understandings and data that will continue to be needed for the synthesis of a more comprehensive understanding of the overall biology of these fungi will have to draw upon the best skills and insights that can be assembled from among specialists in many disciplines focused on populations, organisms, cells, and molecules that comprise them all. The teams of specialists advancing our knowledge of these organisms will also include chemists, biochemists, physical scientists, and bioinformaticists.

Insect mycology as a specialized discipline that joins mycology, entomology, invertebrate pathology, and other disciplines in a challenging but fascinating amalgum. Insect mycology is based on foundations initiated more than a century ago with the work of many extraordinarily capable and insightful scientists including Agostino Bassi (whose work confirming that the fungus now known as *B. bassiana* caused a serious disease of silkworms preceded and heavily influenced many of Pasteur's greatest work in disproving spontaneous generation and establishing theories about cells and diseases), Roland Thaxter (a founder of American mycology who was drawn to fungi because of the diseases they caused on the insects on which he had intended to build his career, and who eventually described more than 100 genera and some 1200 species of Laboulbeniales and whose illustrations of these fungi are legendary for their accuracy, beauty, and sheer numbers), Elie Metschnikoff (whose studies of phagocytosis by insect hemocytes made him the founder of immunology, and who first described *M. anisopliae*), Tom Petch (an inexhaustible British botanist/mycologist and civil servant in Ceylon who, when the coffee rust destroyed the main export crop, founded the Tea Research Institute and led the adoption of tea-growing industry there and the rise of tea to become Ceylon's new major export industry, and who over more than 30 years of work on his 'hobby' group of entomogenous fungi accumulated a globally unique and unsurpassable base of knowledge about the taxonomy and incidences of these fungi that, significantly, predated the advent of chemical pesticides), and John Couch (a mycologist at the University of North Carolina who studied of mosquito pathogens, especially *Coelomomyces* spp., for decades and whose 1935 monograph on the scale-parasitic basidiomycetes in *Septobasidium* is universally held to be one of the greatest of all mycological monographs), and many other equally distinguished scientists in many countries and over many decades.

More than ever before, the comprehensive nature of the term 'biology' and all of the subdisciplines included under it can be appreciated as collectively advancing our overall knowledge of the biology of entomogenous fungi. With those advances, the uses of an expanding range of entomogenous fungi as biological control agents against an ever wider range of pests and for other nonorganismal uses in agricultural, pharmaceutical, and biochemical settings will increase. Where most of the major emphasis on entomogenous fungi as biocontrol agents has continued to focus on a very small number of readily manipulated conidial fungi, there should be little doubt that the increasing sophistication of our overall understanding of these fungi and how they operate under field conditions will lead to an increasing dependence on inoculative releases of fungi. Such a smaller scale, less technologically demanding, less bureaucratically complex, and, consequently, less expensive approach to using fungi for pest biocontrol also offers other distinct advantages: Such approaches are also likely to ensure a higher specificity of the biocontrol agents for their target hosts and greater safety for nontarget invertebrates and vertebrates. Despite all the demonstrable difficulties in using fungus-based biocontrol against arthropod pests, the potential benefits of such practices will continue to inspire both hope and continuing research throughout the world.

RECENT DEVELOPMENTS

The phylogenetically based reclassification of fungi is being adopted with a gratifying rapidity and is helping to reshape or to stimulate new mycological research and further taxonomic revisions. Despite the scope and early impacts of this reclassification which strongly affected all major groups of entomopathogenic fungi, there appear to be many surprises dealing with entomopathogenic fungi still to be uncovered, including a much needed reclassification of the Entomophthorales, the most basal group of nonflagellate fungi (zygomycetous fungi, ascomycetes and their conidial states, and basidiomycetes). This general reclassification confirms that the entomopathogenic habit is extraordinarily old and that bidirectional interkingdom host-jumping between animal and plant hosts has occurred often during evolution of many fungal pathogens. The best documented examples of this host-jumping are found in the extraordinarily numerous and diverse entomogenous, phytophagous, mycoparasitic and even some saprobic fungi aggregated in the family Clavicipitaceae (order Hypocreales) in its broadest sense. This group includes the largest, most diverse focus of entomopathogens (hundreds of species in dozens of teleomorphic and anamorphic genera) among all fungi. As part of this phylogenetically driven reclassification of fungi, the Clavicipitaceae was segregated into three families, with the plant pathogens, for example, the ergot-causing *Claviceps* species, and numerous endophytes of grasses remaining in the newly restricted Clavicipitaceae while the entomopathogens and mycoparasites were mostly reclassified into the families Cordycipitaceae and Ophiocordycipitaceae. Despite the long-held assumption that plant-associated fungi (especially phytopathogens and parasites) are evolutionarily older than the pathogens and parasites of insects or other animals, the evidence within the Clavicipitaceae suggests that the ergot and grass-endophytic fungi are derived from ancestors associated with animal (primarily arthropod) hosts.

The presence of *Beauveria bassiana* growing as an endophyte has been noted in an increasing and diverse spectrum of plants. How such an 'insect' fungus can enter and establish itself in the tissues of herbaceous and even woody plants remains unexplained. We do not yet know under what conditions an endophytic *Beauveria* (or any toxins it produces) may affect insect pests feeding on the mycotized host plant tissues even though such endophytic entomopathogens do seem to provide some protection to the host plant. An intriguing but overlooked issue involving fungal entomopathogens and their 'secret' alternative lives inside plant tissues is that *B. bassiana* seems to be virtually the only such fungus routinely found as an endophyte. Despite the large number and taxonomic diversity of entomogenous fungi there is almost no record of *Metarhizium* species, for example, as endophytes even though *M. anisopliae* (which is now recognized as a species complex) is nearly as ubiquitous and important a mortality agent for insects as is *B. bassiana*. The major report of *Metarhizium anisopliae* as an endophyte to date is from the tissues of *Taxus chinensis* (Chinese yew) in China; even more notably, however, this *Metarhizium* is reported to produce an appreciable quantity of the drug taxol (paclitaxol) that is used to treat several types of cancers or, in basic biological research, to inhibit mitoses and to stabilize microtubules.

While the tissues of yew trees (*Taxus* spp.) are the best-known source of taxol, numerous reports since 1993 suggest that many diverse ascomycetes can produce taxol or related compounds.

The phylogenetic segregation of the Clavicipitaceae makes the many incidences of *Beauveria* as an endophyte even more puzzling in view of the conspicuous absence of *Metarhizium* or any other entomopathogenic fungi. *Beauveria* is now classified (together with the entomopathogens in *Isaria* and *Lecanicillium*) in the Cordycipitaceae, whose members are almost exclusively entomopathogens. The entomopathogenic anamorphic species of *Metarhizium*, *Nomuraea*, and *Aschersonia*, however, are among the few entomopathogens remaining in the newly restricted family Clavicipitaceae along all of the phytopathogens and endophytes. Nothing about the biology of *Beauveria*, *Metarhizium*, or any other entomopathogenic fungi explains or even hints at the reasons for *Beauveria*'s preeminence among endophytic entomopathogens. Most fungal entomopathogens such as *Beauveria* and *Metarhizium* are generalist pathogens that show little specificity for particular host taxa and have no clear nutritional specializations that might link them to particular hosts. Another noteworthy paradox raised by entomogenous fungi is that the deeply held concept of exclusion of multiple highly similar species from a single ecological niche is also challenged by recent findings that several genetically distinct *Beauveria* species co-exist within even the field and windrow subhabitats of a small Danish farm. A similar 'violation' of niche exclusion is known for the fungal pathogens of mosquitoes developing in miniscule volumes of water in the sheathing leaf axils of one particular *Colocasia* (Araceae) plant in an Australian rain forest. The mosquitoes in these 3–10 ml volumes of water have been found over time to be infected and killed by two peronosporomycete straminipiles – *Lagenidium giganteum* and *Crypticola clavulifera* – as well as by two clavicipitoid anamorphs – *Culicinomyces clavisporus* and *Culicinomyces bisporalis*; even further, the single plant in question is the type – and only known – locality for both *C. clavulifera* and *C. bisporalis*.

Some common entomopathogens such as *B. bassiana*, *M. anisopliae*, and *Isaria farinosa* can produce enzymes degrading plant-specific substrates such as cellulose, starch, or xyloglucans. Whether more genes for enzymes that degrade only plant rather than animal substrates may occur in entomopathogenic fungal genomes or how extensively such genes may be found in other entomopathogens remains unresolved. However, better data about plant-active degradative enzymes from entomopathogens should be available soon since complete genomic sequences for isolates of *B. bassiana*, two species in the *M. anisopliae* complex (*M. acridum* and *M. robertsii*), and *Cordyceps militaris* are now being generated and annotated. The sequencing of *C. militaris* is especially important for comparative genomics since this teleomorphic species is the only *Cordyceps* that fruits readily in pure culture, and its stromata are widely used in traditional Asian medical practices for the compounds they contain. In sharp contrast, the sexual states for species of *Beauveria* and *Metarhizium* are exceedingly rare in nature and unavailable from cultures. Very little is understood about what stimulates or controls the production of the sexual stromata of *Cordyceps* in nature, so the ability to use genomic tools to study these processes offer a unique opportunity to understand this aspect of the biology of this (and probably many more) clavicipitoid entomopathogens. It also promises opportunities to discover what governs the production of the many biologically active, pharmaceutically useful compounds present in *Cordyceps* fruiting bodies but that are absent or present at only very low titers in the vegetative stages of these fungi.

The capacity of entomopathogens to become endophytic in a wide range of plants is more than a fascinating biological problem. It represents a new strategy for the environmentally benign and economically feasible use of fungi for the biological control of insect pests. It is unfortunate that the applied uses of fungi against insect pests remains nearly negligible in comparison to pesticides and even to other nonpesticidal control strategies. The reasons for the underutilization of fungal biocontrol agents around the world – and, sadly, for their decreasing use in countries such as Brazil where the use of fungal entomopathogens has been widely accepted and desired – is too complex to assess here. Access to a new strategy for using fungi to control arthropod pests provides some encouragement for new alternatives to pesticide-based approaches. This touch of optimism for the practical use of entomogenous fungi comes at a time when increasing numbers of agricultural crop plants are being genetically modified to incorporate *Bacillus thuringiensis* (*Bt*) toxin genes despite widespread concerns and skepticism about the advisability and long-term implications of such practices. Even the more traditional spray applications of *Bt* also seem to be declining and are increasingly restricted to pests affecting forests, orchards, and organic agriculture. Microbe-based biological control of arthropod pests – whether by using viruses, bacteria, microsporidia, or fungi – remains a technical possibility whose tantalizing promise must still conquer many scientific, regulatory, and even societal impediments. Nonetheless, microbial disease outbreaks that reduce arthropod pest populations occur regularly in nature all around us, and they do so often with great efficacy, with no human intervention, and usually without our even being aware of them.

FURTHER READING

Benjamin, R. K., Blackwell, M., Chapela, I. H., *et al.* (2004). Insect- and other arthropod-associated fungi. In G. M. Mueller, G. M. Bills & M. S. Foster (Eds.), *Biodiversity of fungi: Inventory and monitoring methods* (pp. 395–433). Elsevier Academic Press: Amsterdam.

Bidochka, M. J., & Small, C. L. (2005). Phylogeography of *Metarhizium*, an insect-pathogenic fungus. In F. E. Vega & M. Blackwell (Eds.), *Insect-fungal associations: Ecology and evolution* (pp. 28–50). Oxford University Press: Oxford.

Bischoff, J. F., Rehner, S. A., & Humber, R. A. (2009). A multilocus phylogeny of the *Metarhizium anisopliae* lineage. *Mycologia, 101*, 512–530.

Blackwell, M. (1994). Minute mycological mysteries: The influence of arthropods on the lives of fungi. *Mycologia, 86*, 1–17.

Cook, R. J., Bruckart, W. L., Coulson, J. R., *et al.* (1996). Safety of microorganisms intended for pest and plant disease control: A framework for scientific evaluation. *Biological Control, 7*, 333–351.

Couch, J. N. (1938). The genus *Septobasidium*. The University of North Carolina Press: Chapel Hill, NC.

Frances, S. P., Sweeney, A. W., & Humber, R. A. (1989). Crypticola clavulifera gen. et sp. nov. and Lagenidium giganteum: Oomycetes pathogenic for dipterans infesting leaf axils in an Australian rain forest. *Journal of Invertebrate Pathology, 54*, 102–111.

Hajek, A. E., & Leger, R. J., St. (1994). Interactions between fungal pathogens and insect hosts. *Annual Review of Entomology, 39*, 293–322.

Hibbett, D. S., Binder, M., Bischoff, J. F., Blackwell, M., Cannon, P. F., Eriksson, O., *et al.* (2007). A higher-level phylogenetic classification of the Fungi. *Mycological Research, 111*, 509–547.

Hibbett, D. S., Binder, M., Bischoff, J. F., *et al.* (2007). A higher-level phylogenetic classification of the fungi. *Mycological Research, 111*, 509–547.

Humber, R. A. (2000). Fungal pathogens and parasites of insects. In F. Priest & M. Goodfellow (Eds.), *Applied microbial systematics* (pp. 199–227). Kluwer Academic Publishers: Dordrecht.

Humber, R. A. (2008). Evolution of entomopathogenicity in fungi. *Journal of Invertebrate Pathology, 98*, 262–266.

James, T. Y., Kauff, F., Schoch, C., Matheny, P. B., Hofstetter, V., Cox, C. J., *et al.* (2006). Reconstructing the early evolution of fungi using a six-gene phylogeny. *Nature (London), 443*, 818–822.

James, T. Y., Kauff, F., Schoch, C., *et al.* (2006). Reconstructing the early evolution of fungi using a six-gene phylogeny. *Nature (London), 443*, 818–822.

Lacey, L. A. (Ed.), (1997). *Manual of techniques in insect pathology*. Academic Press: San Diego, CA.

Leopold, J., & Samsinakova, A. (1970). Quantitative estimation of chitinase and several other enzymes in the fungus Beauveria bassiana. *Journal of Invertebrate Pathology, 15*, 34–42.

Lewis, E. A., Sullivan, R., & White, J. F., Jr. (2001). Characterization of an extracellular, cellulosome enclosed endoglucanase purified from *Chaunopycnis* spp.. *Phytopathology, 91*(Suppl.), S117.

Lichtwardt, R. W., Cafaro, M. J., & White, M. M. (2001). *The trichomycetes: Fungal associates of arthropods. Rev. ed.* http://www.nhm.ku.edu/~fungi/Monograph/Text/Mono.htm.

Liu, L., Ding, Z., Deng, B., & Chen, W. (2009). Isolation and characterization of endophytic taxol-producing fungi from Taxus chinensis. *Journal of Industrial Microbiology and Biotechnology, 36*, 1171–1177.

Meyling, N. V., Lubeck, M., Buckley, E. P., Eilenberg, J., & Rehner, S. A. (2009). Community composition, host range and genetic structure of the fungal entomopathogen *Beauveria* in adjoining agricultural and seminatural habitats. *Molecular Ecology, 18*, 1282–1293.

Rehner, S. A. (2005). Phylogenetics of the insect pathogenic genus Beauveria. In F. E. Vega & M. Blackwell (Eds.), *Insect-fungal associations: Ecology and evolution* (pp. 3–27). Oxford University Press: Oxford.

Samson, R. A., Evans, H. C., & Latgé, J.-P. (Eds.), (1988). *Atlas of entomopathogenic fungi*. Springer-Verlag, and Utrecht:Wetenschappelijke uitgeverij Bunge: Berlin.

Spatafora, J. W., Sung, G.-H., Sung, J.-M., Hywel-Jones, N. L., & White, J. F., Jr. (2007). Phylogenetic evidence for an animal pathogen origin of ergot and the grass endophytes. *Molecular Ecology, 16*, 1701–1711.

Stierle, A., Trobel, G., & Stierle, D. (1993). Taxol and taxane production by *Taxomyces andreanae*, an endophytic fungus of Pacific yew. *Science (Washington, DC), 260*(5105), 214–216.

Sung, G.-H., Hywel-Jones, J. L., Sung, J.-M., Luangsa-ard, J. J., Shrestha, B., & Spatafora, J. W. (2007). Phylogenetic classification of *Cordyceps* and the clavicipitaceous fungi. *Studies in Mycology, 57*, 5–59.

Tavares, I. I. (1985). Laboulbeniales (fungi, ascomycetes). *Mycologia Memoirs, 9*, 1–627.

Vega, F. E., & Blackwell, M. (Eds.), (2005). *Insect–fungal associations: Ecology and evolution*. Oxford University Press: New York.

White, J. F., Jr., Bacon, C. W., Hywel-Jones, N. L., & Spatafora, J. W. (Eds.), (2003). *Clavicipitalean fungi: Evolutionary biology, chemistry, biocontrol, and cultural impacts*. Marcel Dekker, Inc: New York.

Chapter 11

Fungal Infections, Systemic

J.F. Staab and B. Wong
Oregon Health Science University, Portland, OR, USA

Chapter Outline

Abbreviations	143	Common Fungal Diseases	150
Defining Statement	143	Diseases Caused by True Pathogens	150
Introduction	143	Diseases Caused by Opportunistic Fungi	157
Classification of Pathogenic Fungi	143	Recent Developments	165
Host Defenses against Fungal Infections	145	Summary and Conclusion	166
Molecular Approaches for Studying Fungal Pathogenesis	147	Further Reading	166

ABBREVIATIONS

ABPA Allergic bronchopulmonary aspergillosis
AIDS Acquired immunodeficiency syndrome
CF Complement fixation
CNS Central nervous system
CSF Cerebrospinal fluid
EIA Enzyme immunoassay
HIV Human immunodeficiency virus
ID Immunodiffusion
MTL *MAT*-type-like loci
PAMPs Pathogen-associated molecular patterns
PEG Polyethylene glycol
TLRs Toll-like receptors

DEFINING STATEMENT

Systemic fungal infections are infections of organs or tissues other than the skin or mucosal surfaces. These infections have increased in frequency and importance in recent years. This chapter reviews the fungi that cause most systemic infections in people and the key features of the diseases these fungi cause.

INTRODUCTION

Before the 1980s, serious fungal infections were relatively uncommon, occurring mostly in patients with cancer or other underlying conditions that cause severe immunodeficiency. Since then, several developments have greatly expanded the population at risk for developing these fungal infections. These include advances in the treatment of many forms of cancer, advances in solid organ and stem cell transplantation, the ongoing human immunodeficiency virus (HIV) epidemic, and the widespread availability and use of antibiotics that are effective against a broad range of bacteria. For these reasons, the incidence and importance of serious fungal infections have increased markedly in the past 20 years. For example, data collected from community and teaching hospitals throughout the United States showed that the frequency of fungal bloodstream infections increased by more than threefold during the 1990s, when the frequency of Gram-positive and Gram-negative bacterial bloodstream infections changed little. Similarly, deaths from fungal infections in the United States increased more than threefold between the early 1980s and the mid-1990s. Thus, it is now clear that fungi are very important human pathogens that warrant attention and study.

CLASSIFICATION OF PATHOGENIC FUNGI

Fungi are among the most abundant and diverse of all organisms, but only a few species are responsible for a large majority of serious fungal infections in humans. The most important of these fungal pathogens of humans are listed according to traditional phylogenetic categories in Table 1. These classifications are based on the traditional growth and morphology of vegetative and reproductive fungal forms and also on genetic relatedness as determined by genome sequencing, especially of highly conserved elements like ribosomal DNA sequences. Most of the fungi that cause serious infections in humans are members of the phylum Ascomycota, although a few are members of Basidiomycota or Zygomycota.

Eukaryotic Microbes.

TABLE 1 Phylogeny of the common invasive fungi

Phylum	Subphylum	Class	Order	Genus
Ascomycota				
	Pezizomycotina			
		Eurotiomycetes		
			Onygenales	*Blastomyces*
				Coccidioides
				Histoplasma
				Paracoccidioides
			Eurotiales	*Aspergillus*
				Penicillium
		Chaetothyriomycetes		*Exophiliala* (*Wangiella*)
		Sordariomycetes		
			Ophisostomatales	*Sporothrix*
			Microascales	*Pseudallescheria*
				Scedosporium
			Hypocreales	*Fusarium*
	Saccharomycotina			*Candida*
	Taphrinomycotina			*Pneumocystis*
Zygomycetes			Mucorales	*Rhizopus*
				Absidia
				Mucor
			Endomopthorales	*Basidiobolus*
				Conibiobolus
Basidiomycota		Hymenomycetes		*Cryptococcus*
				Trichosporon
		Ustilaginomycetes		*Malassezia*

Adapted from Taylor, J. W. (2006) Evolution of human-pathogenic fungi: Phylogenies and species. In: J. Heitman & S. G. Filler, J. E. Edwards Jr., & A. P. Mitchell (Eds.) *Molecular principles of fungal pathogenesis* (pp. 113–133, Chap. 8). Washington, DC: ASM Press.

Traditional phylogenetic classification is very useful for analyzing evolutionary relationships and for guiding studies of properties such as sexuality and genetic recombination within populations, but alternative approaches to classification are often more useful when one considers these fungi as human pathogens. As serious fungal infections are often first recognized when fungi are observed in infected tissues or isolated in culture, it is generally useful to divide the pathogenic fungi into groups that grow in tissues or in culture as yeast-like fungi (sometimes with associated elongated forms known as pseudohyphae), as filamentous fungi (or molds), or as dimorphic fungi (which grow as one morphologic form when they are observed in infected tissues or are cultured at 37 °C and as a completely different morphologic form in the environment or when they are cultured at ambient temperatures). Table 2 divides the major human pathogenic fungi into yeast-like fungi (which include both ascomycetes and basidiomycetes), filamentous fungi or molds (which include ascomycetes and zygomycetes), and dimorphic fungi (all of which are ascomycetes).

Some pathogenic fungi cause disease in people with normal host defenses, whereas others cause disease almost exclusively in people whose defenses against infection are seriously compromised. As the adequacy of an individual patient's host defenses is often known well before the etiology of any specific infection is known, it is also useful

TABLE 2 Morphology of the common invasive fungi

	Genus	Usual morphology in tissues
Yeasts	*Candida*	Budding yeast, pseudohyphae, hyphae
	Cryptococcus	Encapsulated budding yeast
	Exophiala (*Wangiella*)	Black yeast, pseudohyphae
	Trichosporon	Pleomorphic yeast, pseudohyphae or hyphae
Molds	*Aspergillus, Fusarium, Pseudallescheria, Scedosporium*	Narrow septate hyphae, branching at 45°
	Rhizopus and other zygomycetes	Wide aseptate hyphae, branching at 90°
Dimorphic fungi	*Blastomyces*	Yeast with broad-based buds
	Coccidioides	Large endosporulating spherules
	Histoplasma	Small budding yeast
	Paracoccidioides	Yeast with multiple buds per cell
	Penicillium	Intracellular yeast
	Sporothrix	Elongated budding yeast
Other	*Pneumocystis*	Trophic forms, and cysts that contain up to 8 internal spores

TABLE 3 Pathogenic potential of the common invasive fungi

True pathogens	*Blastomyces*
	Coccidioides
	Histoplasma
	Paracoccidioides
	Sporothrix
Opportunists	*Aspergillus*
	Candida
	Cryptococcus
	Exophiala (*Wangiella*)
	Fusarium
	Penicillium
	Pneumocystis
	Pseudallescheria and *Scedosporium*
	Rhizopus and other zygomycetes

to divide the human pathogenic fungi into true pathogens that can infect normal hosts and opportunistic pathogens that tend to infect only seriously compromised hosts (Table 3). It should be noted that most of the fungi listed as true pathogens can also act as opportunists because they tend to cause more extensive and severe disease in people with compromised host defenses.

HOST DEFENSES AGAINST FUNGAL INFECTIONS

Humans are continuously exposed to fungi. The majority of fungi are recognized and inactivated within hours by cellular and noncellular innate defense mechanisms. In addition, an important protective mechanism is provided by physical barriers such as intact skin and healthy mucosa. Many fungal infections occur in individuals with predisposing factors that affect or impair the function of skin, mucosa, or airways. A major predisposing factor for *Candida albicans* bloodstream infections, for example, is the placement of an indwelling catheter that provides the fungus with direct access into the bloodstream and internal organs, bypassing the protection provided by intact skin. Another example of a fungus that can take advantage of interruptions in the skin to invade deeper structures is *Malassezia furfur*, a basidiomycete that can contaminate and then proliferate within lipid-containing parenteral nutrition fluids. Filamentous fungi such as the *Rhizopus* can colonize skin that is interrupted by full-thickness burn injuries or by surgical incision, after which the fungi can invade deeper structures and disseminate widely.

Other factors that disturb the integrity of the intestinal mucosa, such as chemotherapy treatments for hematological or neoplastic malignancies, can also allow *C. albicans* to translocate from the gastrointestinal tract into the blood. Besides the mechanical barrier provided by intact skin, secretions that bathe the mucosal surfaces can also influence susceptibility to infection. A group of soluble peptides, defensins, and histatins, found in mucosal secretions and other tissue fluids, have antimicrobial activities, and play an important role in innate immunity. Most of these peptides exert their antifungal activities by interacting directly with the microbial membrane, destabilizing it and leading to cell death.

Protection against fungal infections is afforded by both innate and adaptive immunity. Cellular innate immunity relies upon professional phagocytic cells (tissue macrophages/monocytes and dendritic cells) that are able to ingest and kill certain fungal spores or yeasts, and have an important role in containing fungal growth and dissemination. Another aspect of this first line of defense is the recognition of conserved molecular patterns found on the surface or as part of microorganisms that signal infection to the host. These conserved molecular structures are known as pathogen-associated molecular patterns (PAMPs) that are usually units of cell wall or cell membrane structural components found in large groups of pathogenic microorganisms. PAMPs are 'seen' by host cells through a set of surface proteins called Toll-like receptors (TLRs) and by other membrane-bound proteins (such as Dectin-1, the receptor for 1,3-β-glucan) found on professional and nonprofessional phagocytic cells. Binding of PAMPs by TLRs and other non-TLR proteins signals the recruitment and activation of neutrophils to the site of infection in a manner that is tailored for the killing or containment of a particular pathogen. Recruitment and activation of neutrophils occur by signaling cascades generated by the secretion of chemokines and cytokines by macrophages to produce a beneficial inflammatory response. Inappropriate inflammatory responses have been implicated in allergic types of reactions that do not offer protection against the offending organism.

Adaptive immune responses are also important for protection against fungal infections. This can occur through a cell-mediated mechanism involving T lymphocytes that become activated by binding fungal antigens processed by macrophages, dendritic and other antigen-presenting cells. Dendritic cells are capable of bridging innate and adaptive immunity pathways by virtue of having some TLRs on their surfaces and being able to process fungal antigens to present to T cells in lymphoid tissues. The T cell implicated in cellular immunity to fungal infections is the CD4+ helper T cell. CD4+ T cells dictate the immune response by managing the activity of other lymphocytes. These T cells are biased (Th1 or Th2) to secrete specific cytokines that produce a protective type of inflammatory response or a maladaptive response that may lead to chronic infections or uncontrolled allergic responses. How these cells are programmed to produce an effective immune response is not entirely clear and is a topic of significant research interest.

Antibodies are also involved in immunity to fungi although their protective role is not as well defined as in the case of cell-mediated adaptive immunity. Most adults and children have naturally occurring antibodies to *C. albicans*, *Cryptococcus neoformans*, *Histoplasma capsulatum*, and *Pneumocystis jirovecii*. Interestingly, populations at risk for *Cryptococcus* infections often have impaired B cell or immunoglobulin repertoires. Animal models have established a role for antibodies in the protection against certain fungal infections, implying that antibodies may also have important roles in fighting these infections in man. Several fungal antigens have been used experimentally to establish their usefulness as vaccines.

The host can also defend itself against microbial invasion by sequestering essential nutrients such as iron. Although this metabolic defense mechanism has been studied in more detail in bacteria, there is good evidence that patients with iron overload syndromes caused either by underlying metabolic diseases or by frequent blood transfusion are unusually susceptible to infection by zygomycetes such as *Rhizopus*. In fact, there have been outbreaks of life-threatening or fatal zygomycosis infections among iron-overloaded patients who have received the microbial iron-chelating siderophore desferrioxamine in an effort to reduce heavy metal levels in tissues. Whereas iron sequestered in mammalian tissues by proteins such as transferrin and lactoferrin is not usable by *Rhizopus*, iron chelated to desferrioxamine is readily usable by these fungi. Thus, these outbreaks highlight the importance of nutrient sequestration as a host defense against at least some fungi.

It should be clear from the discussion above that host defenses against fungal pathogen are complex and incompletely understood. Nevertheless, it is generally true that particular opportunistic fungi tend to take advantage of particular host defense abnormalities. Thus, Table 4

TABLE 4 Association of selected pathogenic fungi with host defense abnormalities

Disrupted skin or mucosa	*Candida*
	Mallasezia
	Rhizopus and other zygomycetes
Neutrophil deficiency or dysfunction	*Aspergillus*
	Candida
	Rhizopus and other zygomycetes
Cellular immunodeficiency	*Candida*
	Cryptococcus
	Histoplasma
	Coccidioides
	Penicillium
	Pneumocystis
Diabetes/acidosis/iron overload	*Rhizopus* and other zygomycetes

summarizes the fungi that tend to take advantage of interruptions in the skin or mucosal barriers, abnormalities in or deficient numbers of phagocytic cells, abnormalities in cell-mediated immunity, and inadequate iron sequestration.

The recognition of fungi by the innate immune system utilizes fungal molecules that are similar or are shared across species. Many fungi share characteristics such as cell wall composition or their ability to change their morphology in response to temperature, but not all fungi are able to cause disease. This implies that individual characteristics that give some fungi but not others enhanced infectivity exist. As discussed further, even closely related species are not equally pathogenic. How closely related fungi differ at the genomic level has been used to determine what characteristics are important for causing disease. Molecular tools to analyze genomes, inactivate genes, and introduce heterologous genes have tremendously advanced our knowledge of fungal cellular processes and how these can influence infection of the human host. The section below discusses current molecular approaches to better understand fungal biology and how this information has helped us study how fungi cause disease.

MOLECULAR APPROACHES FOR STUDYING FUNGAL PATHOGENESIS

The sequencing of multiple fungal genomes has significantly advanced the knowledge of comparative genomics. Cellular processes that are common or unique among species can reveal information about shared structural and metabolic components, and individual characteristics that help a particular fungus evade the immune system or actively participate in the establishment of infection. Not only do genome sequences allow for the comparison of paralogous and orthologous proteins, but cross-species genetic information allows for the comparison of the linear order of genes (or synteny), information that may yield clues as to how genetic information has been maintained or lost in different fungi. The genomes of opportunistic fungi appear to differ widely in size and in complexity as seen by the number of chromosomes, genome lengths, and abundance of introns (Table 5). Although genome sequencing has revealed invaluable information, the data have yielded few obvious pan-fungal clues as to why relatively few fungi out of thousands tend to cause most of the systemic infections. This is also true of fungi of the same genus: one or two species tend to cause disease whereas the remaining dozen or more are rarely isolated from patients (*Aspergillus fumigatus* and *C. neoformans* being examples). One shared characteristic is the ability to grow well at the host temperature of 37 °C. How this and other growth attributes coded in the genome affect infectivity and virulence is the subject of much interest, and the ready availability of genome sequencing and molecular tools are important resources for furthering such studies.

A very useful application of genome sequence data of medically important fungi has been the construction of microarray chips composed of oligonucleotides, cDNA, or genomic DNA gene fragments that represent the entire organism's genome. In *H. capsulatum*, genomic DNA microarrays were built in the absence of genomics by a shotgun approach, demonstrating the ability to create arrays even in the absence of complete genome sequence data. Microarrays permit the simultaneous analysis of global gene expression within an organism in response to a perturbation, be it a gene mutation, a difference in growth morphology, or the addition of an experimental drug or other stressor. Previous to microarray technology, gene expression patterns were studied a few genes at a time, a method that is not amenable to understanding how seemingly unrelated genes are linked by global expression patterns. Gene expression interrelationship patterns may provide clues as to the functional relatedness of seemingly disparate genes. Microarrays have been used to understand how *C. albicans* can alter its carbon metabolism to survive inside macrophages, how genes are differentially expressed in the different morphotypes of dimorphic fungi, and how antifungal drugs affect global gene expression, to name a few examples. Microarrys have become an important molecular tool that continues to evolve as better algorithms are written to extract the gene expression data patterns.

Comparative genomics has been helpful in many areas such as for the elucidation of signaling pathways, in the discovery of candidate genes involved in host survival or in virulence traits, for the identification of cell surface proteins perhaps associated with host adhesion, for uncovering mating gene loci, and, in general, for establishing genetic relatedness among fungi. An area of heightened interest has been the demonstration of mating in fungi that appear to exist as asexual organisms. Genetic diversity, as provided by mating, is thought as necessary for maintaining species robustness and fitness. Population genetic studies sometimes suggest that sexual reproduction does occur in fungi in the absence of laboratory demonstration of mating. Genomics has been instrumental in uncovering mating loci by homology searches of fungal genomes. Such has been the case for *A. fumigatus* whose genome harbors loci with good homology to *MAT* genes but has not been documented to undergo mating in the laboratory, and appears to propagate in a clonal fashion. In *C. albicans*, the discovery of *MAT*-type-like loci (*MTL*) has led to the description of a parasexual modality of mating that does not produce haploid progeny. Instead, the tetraploid (4n) progeny undergo a concerted chromosome loss to return to the diploid state. The very low recovery of *MTL* homozygous clinical strains (**a**/**a** or α/α) appears to diminish the importance of mating in infectivity of the human host, and the description of five major distinct genetically similar groups (clades) suggests that *C. albicans* tends to propagate clonally and mating may be a rare event. In *C. neoformans*, mating type has been

TABLE 5 Genome organization of selected fungal pathogens

Classification by morphology	Organism	Genome size (Mb)[a]	Introns?	Number of chromosomes	Ploidy	Mating
Yeasts	*Candida albicans*	16	Few (6% of genes)	8	Diploid	Has *MTL**a**/α*[b] loci; mating documented in vitro
	C. glabrata	12.3	Very few	13	Haploid	Has *MTL1,2,3* loci; mating unknown
	Cryptococcus neoformans (serotype A)	20	Intron-rich	12	Haploid/ diploid	Has *MAT**a**/α* loci; mating documented in vitro
	C. gattii	~18.7	Intron-rich	14	Haploid/ diploid	Has *MAT**a**/α* loci; mating documented by population genetic studies
Molds	*A. fumigatus*	30	Most of the genes have short 1–3 introns	8	Haploid	Has *MAT1-1/1–2* loci; mating unknown
Dimorphics	*Blastomyces dermatitidis*	28	Yes[c]	ND	Haploid/ diploid	Two mating types (+/–)
	C. immitis	~29	Most of the genes have short 1–3 introns	4	Haploid	Has *MAT1-1/1–2* loci; mating documented by population genetic studies
	Histoplasma capsulatum	23–25	Most of the genes have 1–3 introns	7	Haploid/ diploid	Two mating types (+/–)[d]
	P. brasiliensis	29–33	Yes[c]	5	Haploid	Unknown
	P. marneffei	~29.5	Yes[c]	3–6[e]	Haploid	Unknown
Other	*Pneumocystis jirovecii* (*carinii*)	~8	Majority of genes have multiple short introns	15	Haploid/ diploid	Unknown

ND, not determined.
[a]*Mega base pairs.*
[b]*MTL: mating type-like.*
[c]*Genes with introns have been cloned but the overall intron/exon organization of the genome awaits completion of genome sequencing.*
[d]*+ strains are MAT1-1; –, strains are MAT1-2.*
[e]*The number of chromosomes is in question.*

associated with virulence as α strains are more virulent that **a** strains in certain genetic backgrounds. It has long been observed that the α-mating type predominates in clinical and environmental samples, perhaps due to monokaryotic fruiting (a form of sexual reproduction by same mating-type strains) that has been mostly observed in α strains. Whether and how often 'successful' mating between two mating-type strains affects pathogenesis or tissue tropism in medically important fungi is unknown. As more genome sequences are completed and analyzed, together with empiric genetic data, similarities may begin to emerge that point toward characteristics that make certain fungi more apt to infect humans.

An invaluable molecular tool for the study of gene function in association with pathogenesis has been the ability to introduce DNA (transformation) into fungi and to specifically target the disruption of genes of interest (Table 6). Transformation systems can be divided into two steps: (1) DNA delivery into the fungus, and (2) integration of the gene disruption DNA cassette into the chromosome by homologous recombination, preferably at a rate that exceeds the inherent mutational rate. A gene disruption cassette consists of a DNA fragment with homologous DNA of your chosen gene flanking a selectable marker gene. This gene disruption DNA cassette is introduced into the fungal cell by one of various methods, after which homologous recombination at the chromosomal gene locus results in the replacement of the functional gene with a nonfunctional one carried on the disruption DNA fragment. Once a gene knockout strain is identified by molecular methods, it can be used to determine how the loss of function of the gene of interest affects virulence attributes. Many of the selectable

TABLE 6 Molecular genetic tools used to study selected pathogenic fungi

Classification by morphology	Organism	Transformation methods	Selectable markers
Yeasts	*Candida albicans*	Spheroplasting, Li acetate/PEG, eletroporation, *Agrobacterium*	*URA3. ARG4, HIS1, LEU2, ADE2, SAT1*[a], *MPA*[b], *hph*[c]
	C. glabrata	Li acetate/PEG, electroporation, *Agrobacterium*	*URA3, TRP1, HIS3, neo*[d], *ble*[e], *hph*
	Cryptococcus neoformans	Biolistic, electroporation, *Agrobacterium*	*ADE2, URA5, NAT*[a], Neo, Hph
	C. gattii	Biolistic	*URA5, NAT,* Neo
Molds	*Aspergillus fumigatus*	Electroporation, protoplasting, *Agrobacterium*	*hgh, pyrG, argB, lysB*
Dimorphics	*Blastomyces dermatitidis*	Electroporation,*Agrobacterium*	*hph, ura5, sur*[f]
	C. immitis	*Agrobacterium*, protoplasting	*hph, ble*
	Histoplasma capsulatum	Li acetate/PEG, electroporation, *Agrobacterium*	*URA5, hph*
	P. brasiliensis	*Agrobacterium*	*hph*
	P. marneffei	protoplasting	*pyrG*
Other	*Pneumocystis jirovecii* (*carinii*)	NA	NA

NA, not available.
[a]*Nourseothricin marker.*
[b]*Mycophenolic acid marker.*
[c]*Hygromycin B (hygromycin phosphotransferase) marker.*
[d]*G418/Geneticin marker.*
[e]*Zeocin marker.*
[f]*Chlorimuron ethyl resistance (sulfonyl urea resistance).*

marker genes in use restore a nutritional requirement (e.g., *URA3* for uracil or *ADE2* for adenine), which requires the use of a mutated strain that lacks the functional corresponding marker gene. A disadvantage of this methodology is the inability to create gene knockouts in clinical strains that normally are not mutated in your chosen gene marker. To overcome this problem, several dominant selectable markers such as genes coding for antimicrobials (e.g., hygromycin B resistance) have been engineered for use in transformation systems of several fungi. This obviates the need of an auxotrophic strain for performing genetic studies and also allows for the transformation of clinical strains. A very useful modification has been the development of gene disruption cassettes that are recycled such that one can generate multiple gene disruptions within one strain using the same selectable marker gene.

Multiple transformation methods have been designed to introduce DNA into fungi. Yeasts or short germ tubes are often treated with cell wall-digesting enzymes to produce spheroplasts or protoplasts, respectively, which allows for the uptake of DNA. The generation of spheroplasts/protoplasts can be a bit cumbersome and can lead to undesired fusion of spheroplasts/protoplasts; thus this method is usually not preferable over electroporation or lithium acetate combined with polyethylene glycol (PEG) if these methods can serve as alternatives. Other methods have been developed for certain fungi that are not amenable to spheroplasting/protoplasting or electroporation. A few fungi can take up DNA by biolistic delivery using a 'gene gun' to propel DNA-coated tungsten beads into fungal cells. The difficulty of introducing DNA into some filamentous fungi has also been solved by using the Gram-negative plant pathogen bacterium, *Agrobacterium tumefaciens*, to deliver DNA as part of a circular molecule (plasmid) upon infection of the host. The plasmid can be engineered to carry a gene knockout cassette of interest bearing a dominant or nutritional selectable marker. The system has also been used in yeasts including *Saccharomyces cerevisiae* (baker's yeast), further demonstrating the cross-genus utility of this method. The one fungus in which a transformation system has not been developed is *P. jivorecii*. An infected animal model is currently the only means of cultivating the organism in the laboratory. In this case, the *P. jivorecii* genome sequence has been the key to understanding the biology of the organism in the absence of a genetically tractable transformation system.

The usual desired fate of the transformed DNA is the targeted integration at the genomic location of the gene of interest with the concomitant loss of gene function. However, some fungi will preferentially integrate DNA randomly into their genome in the absence of homologous recombination. To overcome this, gene 'knockdowns' in several fungi with a low rate of homologous recombination have been achieved with a different methodology termed RNA interference. This method was first used to generate gene knockdowns in organisms with intractable genetic systems such as in mammalian cells. This technique has been used in *Blastomyces dermatitidis*, *H. capsulatum*, and *A. fumigatus* to determine the function of genes associated with virulence. RNA interference relies on the expression within the fungus of a double-stranded RNA molecule that is sense and antisense to the gene of interest. The double-stranded RNA leads to the destruction of the cognate messenger RNA and subsequent loss of gene expression. The system of gene knockdowns is not perfect, as some genes appear to be more readily inactivated than others, and the degree of messenger RNA destruction sometimes is not 100%. A second caveat is that the system has not been shown to work in all fungi; some yeast species appear to lack RNA interference capabilities. But this has not been a detriment because most yeasts perform homologous recombination at frequencies that permit gene inactivation by other established methods.

Genomic data, microarrays, and targeted gene disruptions are rapidly advancing the study of medically important fungi. The ability to compare cross-genus biological phenomena is a powerful approach for understanding how fungi behave in association with humans in ways that can manifest as disease.

COMMON FUNGAL DISEASES

In the following sections, the characteristics of several of the most important fungal pathogens of humans are summarized, along with information about the epidemiology, clinical and pathologic manifestations, diagnosis, and treatment of the infections that these fungi cause.

Diseases Caused by True Pathogens

Blastomycosis

Causative organism

The causative agent of blastomycosis is *B. dermatitidis*, the asexual form of the ascomycete *Ajellomyces dermatitidis*. *B. dermatitidis* is dimorphic in that it grows as a mold and produces conidia in the environment and in culture at ≤ 30 °C (Figure 1(a)), whereas it grows as yeast-like cells in infected mammalian tissues and when cultured at 37 °C. The yeast cells have thick walls; they are multinucleate and reproduce by budding from a broad base (Figure 1(b)). This fungus grows well in media rich in organic content and therefore it is assumed that they subsist in soil, perhaps in decaying wood or near bodies of water. The exact ecology of *B. dermatitidis* is not well understood because this fungus has only rarely been isolated from soil.

B. dermatitidis exists in two mating types (+ and −), both of which are pathogenic. Mating type + and − strains are found in approximately equal proportions in infected people.

Epidemiology

Most cases of blastomycosis occur in the Central and Eastern United States. Areas of highest prevalence include the lower Mississippi River Valley, the Ohio River Valley, the Carolinas, and regions including the Great Lakes and the St. Lawrence River Valley. There are also microfoci in Central and South America and parts of Africa. Conidia are thought to be the infectious forms of *B. dermatitidis*.

Analysis of sporadic cases indicated that persons at greatest risk in endemic areas include farmers, hunters, forestry workers, and campers. More males than females have historically been diagnosed with blastomycosis, but analysis of several distinct outbreaks has not established

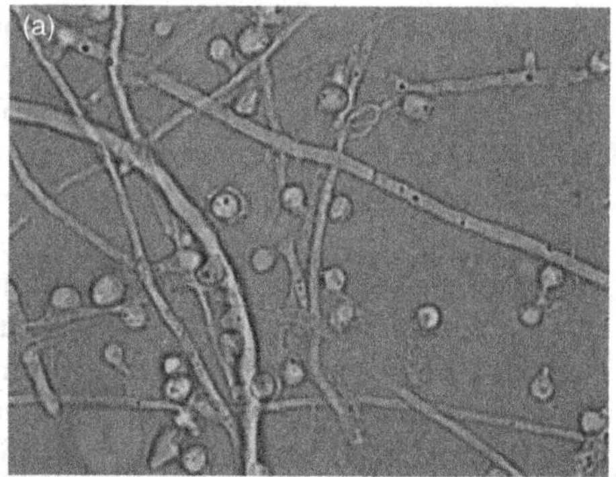

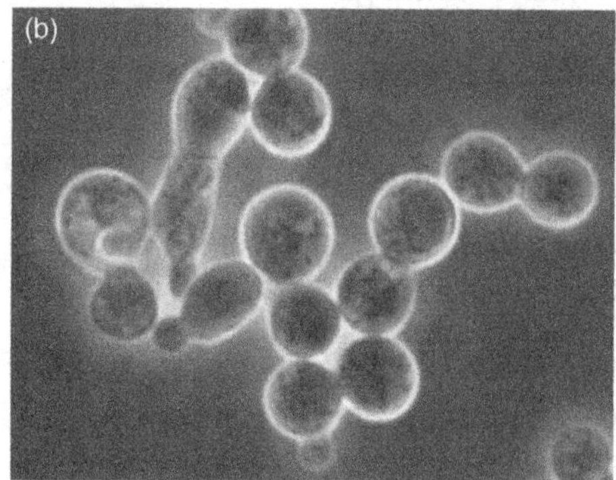

FIGURE 1 *Blastomyces dermatitidis* grows as a mold at 30 °C or below (a) and as thick-walled yeast cells that divide by budding on a broad base (b) in infected mammalian tissues or at 37 °C. *Images courtesy of Bruce Klein, University of Wisconsin.*

any gender or racial differences in susceptibility. Blastomycosis is also common in dogs and other mammals.

Pathogenesis and clinical features

Infectious particles (presumably conidia) are inhaled into the lungs of humans or animals, where they convert into the pathogenic yeast forms. Infection in the lungs elicits an inflammatory reaction with features of both acute inflammation and granuloma formation. Most infections resolve spontaneously and are thus asymptomatic and cause minimal symptoms, but some infections progress to cause progressive pulmonary infection and/or disseminated extrapulmonary disease. Several virulence-associated properties and genes have been studied. One example is a protein on the surface of *B. dermatitidis* (WI-1) that mediates adhesion to host cells.

The ability of an infected host to mount an acute inflammatory reaction and to acquire specific cell-mediated immunity is important for defense against *B. dermatitidis*. Although severe blastomycosis can occur in patients with abnormal immune defenses, *B. dermatitidis* is associated with infection of compromised hosts less often than are most other pathogenic fungi.

Because most pulmonary *Blastomyces* infections are asymptomatic or associated with only mild symptoms, blastomycosis is usually not recognized unless there is persistent or progressive pneumonia or a dissemination to extrapulmonary sites. Symptoms of pulmonary blastomycosis include fever, chills, productive cough, and chest pain. Individuals who do not recover from the early manifestations of infection develop a progressive granulomatous and suppurative pulmonary infection, or the infection may disseminate to other organs. The most common sites of dissemination are the skin and bones, although other tissues can also be involved. The typical skin lesion is a chronic progressive ulcer that is often misdiagnosed as neoplasia.

Diagnosis and treatment

The diagnosis of blastomycosis is based on demonstrating the causative organism in infected tissues or body fluids (e.g., sputum and bronchial washings, other body fluids, skin scrapings, and tissue biopsies), either by microscopy or by culture. The finding of thick-walled, broad-based, unipolar budding yeast-like cells, 8–15 μm in diameter in the collected material or within tissue, is highly suggestive of blastomycosis, and the diagnosis is proven by isolating *B. dermatitidis* in culture. Because the fungi isolated in culture are often molds without distinctive morphologic features, definitive identification of these fungi as *B. dermatitidis* can be accomplished with commercially available DNA probes. Serological tests for antibodies are of limited use because of problems with sensitivity and specificity. An immunologic test for *B. dermatitidis* antigens in serum or urine is commercially available.

Blastomycosis tends to be chronic and progressive, and mortality rates as high as 90% have been reported. Thus, the current consensus is that all people with proven pulmonary or extrapulmonary blastomycosis should receive antifungal therapy. *B. dermatitidis* is susceptible in vitro to the polyene antifungal amphotericin B and to the azole antifungals ketoconazole, itraconazole, and fluconazole. Case series and/or formal treatment trials indicate that all of these treatments are effective. With treatment, most people with blastomycoses recover.

Histoplasmosis

Causative organism

The causative agent of histoplasmosis is *H. capsulatum*, the asexual form of the ascomycete *Ajellomyces capsulatum*. *H. capsulatum* is dimorphic in that it grows as a filamentous mold in the environment and in culture at ≤30 °C (Figure 2(a)), and it grows as small uninucleate budding yeast cells (diameter 3–5 μm) in infected tissues and in culture at 37 °C (Figure 2(b)). The mold form of *H. capsulatum* produces both macroconidia (diameter 8–15 μm) and microconia (diameter 2–5 μm). Microconia are presumed to be the most important infectious particles because their small size allows them to reach the alveoli of the lung. *H. capsulatum* exists in two mating types (+ and −). Approximately equal proportions of the + and − mating types have been isolated from environmental sources such as soil, but most strains isolated from infected people are of the – mating type.

H. capsulatum var. *duboisii* causes histoplasmosis in central Africa. The mold form of this fungus is indistinguishable morphologically from North American *H. capsulatum* strains, but yeast-phase cells of *H. capsulatum* var. *duboisii* are much larger (diameter 10–15 μm).

Epidemiology

Histoplasmosis is endemic in the Eastern and Central United States, specifically in the Ohio River Valley and the lower Mississippi River Valley. Skin test surveys conducted in US military recruits in the 1940s established that as many as 80% of people who live in highly endemic areas have acquired immunity to *H. capsulatum*. The fungus can be found in high concentrations in soil contaminated with bird or bat droppings, caves, areas harboring bats, poultry housing litter, and bird roosts. Outbreaks in endemic areas have been traced to the disturbance of contaminated sites during large-scale excavations or construction projects. Sporadic cases have been diagnosed in people who have neither lived in nor visited endemic areas. *H. capsulatum* var. *duboisii* is endemic in central Africa.

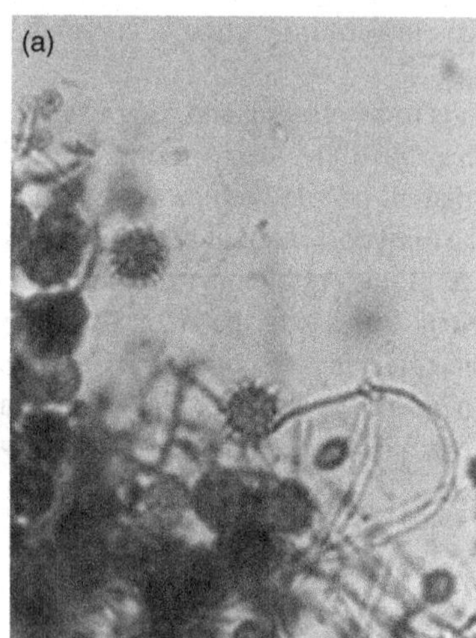

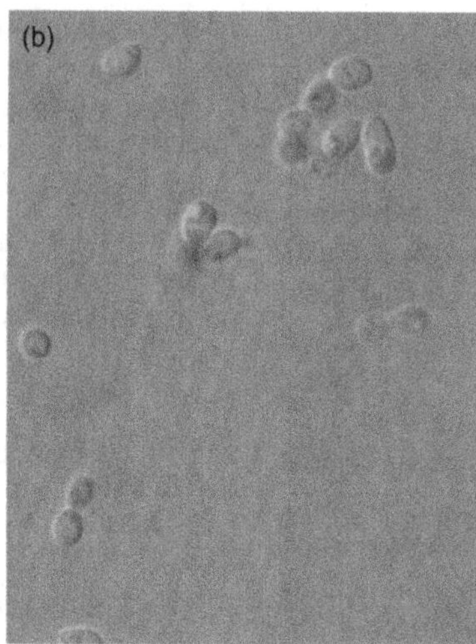

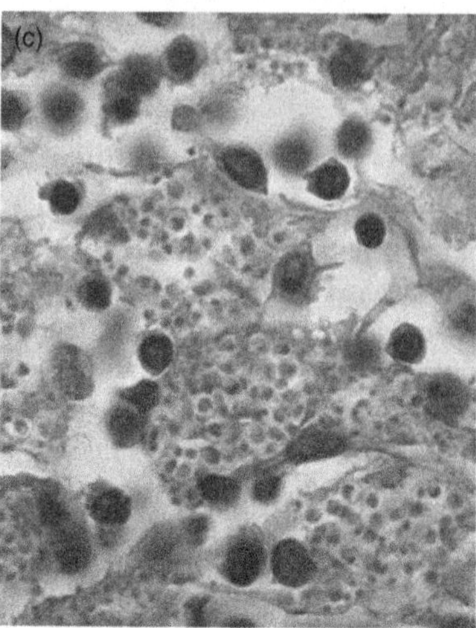

FIGURE 2 *Histoplasma capsulatum* grows as a mold with hyphae, microconidia, and macroconidia at 25 °C (a), and it grows as small budding yeast cells at 37 °C (b). Large numbers of *H. capsulatum* yeast cells can often be demonstrated within mononuclear phagocytic cells of people with disseminated histoplasmosis, as in this specimen of bone marrow (c). *Images courtesy of George Deepe, University of Cincinnati.*

Pathogenesis and clinical features

Inhaled microconidia are deposited in the lungs and are ingested by resident phagocytic cells via receptor-mediated endocytosis. Conversion of microconidia to the yeast-phase form of growth is presumed to occur intracellularly. Once viable fungi are ingested by phagocytes, some are killed, whereas others survive and can replicate within the phagolysosomes of monocytic phagocytes. When the *H. capsulatum* yeast cells proliferate within infected host phagocytes, they eventually kill the cells and go on to infect adjacent phagocytic cells. Also, the yeast cells can disseminate to distant sites within infected host phagocytes. Fungal replication in the lungs or in other organs is eventually controlled by cellular immune mechanisms, which eventually results in the formation of dense granulomata. Acquired T cell immunity to *H. capsulatum* develops and can be demonstrated by skin testing with crude *H. capsulatum* antigen mixtures and by in vitro lymphocyte stimulation studies. Adoptive transfer studies in mice have established that acquired immunity to *H. capsulatum* is protective.

A large majority of people who acquire histoplasmosis by inhalation are asymptomatic or they develop only minimal respiratory symptoms. Acute pulmonary histoplasmosis develops in a small minority of individuals following initial infection, especially when the initial fungal inoculum is large. This form of histoplasmosis is characterized by fever, malaise, chest pain, and nonproductive cough. Chest radiographs demonstrate progressive pneumonia (which is often diffuse and bilateral) and sometimes also regional lymphadenopathy. Most people with pulmonary histoplasmosis recover completely without therapy, so the only evidence of past infection is skin test reactivity and sometimes the presence in radiographs of multiple calcified granulomas in the lungs, spleen, or other organs.

Some people who recover from their initial pulmonary infections develop chronic pulmonary histoplasmosis. As this can occur many years after an individual has left the endemic area, this form of histoplasmosis (which closely resembles pulmonary tuberculosis) presumably results from reactivation of latent infection. Symptoms generally include cough, fever, and weight loss, and these symptoms can progress over the course of months to years. Chest radiographs usually show upper lobe pulmonary disease, often with cavity formation and scarring. Differentiating histoplasmosis from tuberculosis in patients such as these can be very challenging.

H. capsulatum can also cause disease in extrapulmonary sites. In people with profound defects with regard to cell-mediated immunity (e.g., patients with advanced acquired immunodeficiency syndrome (AIDS) or organ transplant recipients) or whose immune systems are immature (e.g., infants and young children), widely disseminated histoplasmosis can develop. This form of histoplasmosis can follow primary infection, or it can develop in previously infected people who develop immune deficiencies. This severe form of disseminated histoplasmosis is characterized by extensive proliferation of *H. capsulatum* yeast cells within phagocytic cells in the blood and in organs such as the spleen, liver, and bone marrow. People with this form of histoplasmosis are usually severely ill with a clinical syndrome that resembles generalized sepsis. Key clinical and laboratory features include fever, chills, enlargement of the liver and/or spleen, hematologic abnormalities including anemia, leukopenia, thrombocytopenia, and disseminated intravascular coagulation. This form of disseminated histoplasmosis is fatal if untreated.

A second form of disseminated histoplasmosis tends to occur in older individuals without known immunodeficiency states. This type of disseminated histoplasmosis varies widely in severity and pace; in some people the only symptoms are gradual weight loss and fatigue and in others there is clear evidence of the involvement of many different extrapulmonary organs (e.g., liver, spleen, meninges, adrenals, and bone marrow). One clue that the problem may be disseminated histoplasmosis is the presence of painful oropharyngeal ulcerations. These lesions usually have raised borders, and they are often mistaken for neoplasms. When biopsied, these lesions show a mix of acute inflammatory reactions and granuloma formation, and special stains generally reveal abundant small budding yeast.

H. capsulatum can also cause disease by eliciting inadequately attenuated host immune responses. In normal individuals, tissues respond to *H. capsulatum* cells or antigens by forming granulomata, which are often dense and fibrotic. In addition to limiting fungal replication and containing infection, this process can lead to the development of pulmonary nodules (histoplasmomas) that sometimes enlarge to several centimeters in diameter. These nodules are often removed because they can be difficult to differentiate from neoplasms. Histopathological examination shows that these lesions are well-organized granulomas with prominent fibrosis; budding yeast cells are generally sparse or absent and cultures are usually negative. Another clinical and pathological entity associated with host responses to *H. capsulatum* antigens is mediastinal granulomatosis and fibrosis. This late complication of histoplasmosis of the mediastinal lymph nodes and adjacent structures is characterized by granuloma formation and dense fibrosis that results in the obliteration and obstruction of key structures, including the airways, esophagus, lymphatics, and blood vessels. Lymphocytic infiltrates and scars in the choroid and retina of the eye occur in people who live in areas in which histoplasmosis is endemic and is presumed to result from cellular immune responses to *H. capsulatum* antigens.

Infection by *H. capsulatum* var. *duboisii* (which causes African histoplasmosis) is presumably also via the lungs, but acute and chronic pulmonary disease is rare. Rather, this organism tends to cause skin, subcutaneous, and bone infections.

Diagnosis and treatment

Histoplasmosis is diagnosed definitively by isolating *H. capsulatum* in culture. Because the filamentous form of *H. capsulatum* cannot readily be distinguished from several other fungi, definitive identification of *H. capsulatum* formerly required its conversion of hyphal-phase organisms to the yeast phase by incubating the cultures at 37 °C or by animal inoculation. However, DNA probes that can definitively identify fungi as *H. capsulatum* are now available commercially. *H. capsulatum* grows slowly; so cultures generally may require several weeks of incubation before growth is detected. Moreover, cultures from some patients are often negative even after prolonged incubation. Thus, it is sometimes necessary to make a presumptive diagnosis of histoplasmosis by microscopy. Histological sections or appropriately stained smears of peripheral blood or other body fluids are examined for the presence of small yeast-like cells that bud from narrow bases. These fungi are generally sparse or absent in pulmonary nodules or in material obtained from patients with mediastinal fibrosis or with ocular histoplasmosis, but they are often easily demonstrated in tissues or body fluids from patients with the acute or chronic forms of disseminated histoplasmosis (Figure 2(c)).

As histoplasmosis can be difficult to diagnose by culture or direct microscopy, serologic testing is an important adjunctive method. Antibodies to yeast and mycelial antigens are generally detected by complement fixation (CF) and immunodiffusion (ID) tests. High or rising antibody titers can provide evidence of active or recent infection, but interpretation of serologic tests is confounded by delayed or absent responses in some people (especially immunocompromised hosts) and by the persistence of detectable antibodies for many years after active infection in some others. Immunologic tests for *H. capsulatum* antigens in serum, urine, bronchial lavage fluid, and cerebrospinal fluid (CSF) have been used to diagnose histoplasmosis in people with acute pulmonary, acute disseminated, and meningeal histoplasmosis. These tests are available in commercial reference laboratories and can provide rapid and specific diagnoses, although some cross-reactions have been described. In AIDS patients or others with disseminated histoplasmosis, serial serum or urine antigen determinations are useful in assessing responses to treatment.

H. capsulatum is susceptible in vitro to amphotericin B and to the azole antifungals ketoconazole, itraconazole, and fluconazole, and all of these drugs have been used with success in people with histoplasmosis. Antifungal chemotherapy is required only in forms of histoplasmosis in which fungal proliferation is a prominent feature, such as severe acute pulmonary histoplasmosis, chronic pulmonary histoplasmosis, and the acute or chronic forms of disseminated histoplasmosis. Antifungal therapy is of no value in patients with stable or enlarging pulmonary nodules, mediastinal granulomatosis and fibrosis, or ocular histoplasmosis.

Coccidioidomycosis

Causative organisms

Coccidioides immitis and *Coccidioides posadasii* are the causative agents of coccidioidomycosis. These are dimorphic ascomycetes that grow as filamentous molds and produce barrel-shaped arthroconidia in the environment and in culture. In infected tissues and in specialized liquid growth media, *Coccidioides* produces thick-walled structures called

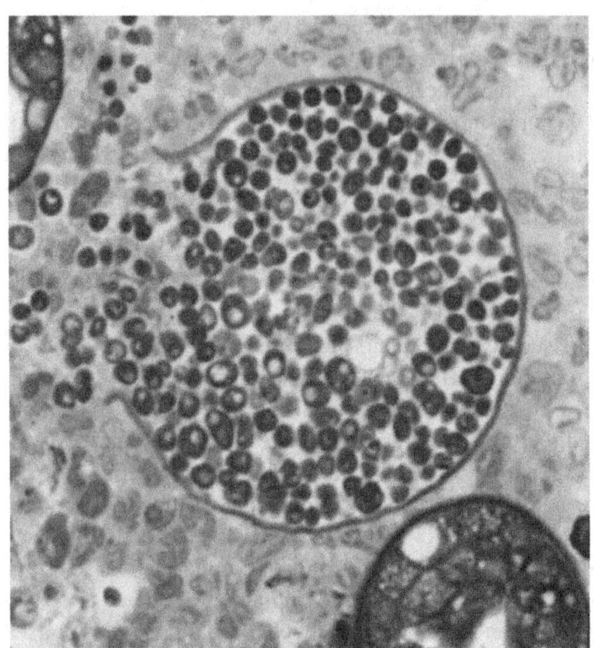

FIGURE 3 *Coccidioides posadasii* spherules filled with endospores are typically seen in infected tissues. The fungus can be induced to form spherules in vitro in special synthetic medium at 37 °C in the presence of CO_2. *Image courtesy of Garry Cole, University of Texas at San Antonio.*

spherules, which reproduce by endosporulation, a process by which the interior of the spherule is divided into as many as several hundred individual cells (endospores) (Figure 3). When the spherule ruptures, the individual endospores are released into the surrounding environment, and each spore is capable of generating a new spherule.

Sexual reproduction in *C. immitis* and *C. posadasii* has not been observed directly, but molecular genetic analyses of isolates collected from infected people and the environment support the view that meiotic recombination probably occurs in the environment.

Epidemiology

Coccidioidomycosis is endemic in arid regions in the southwestern United States, Mexico, and Central and South America. *C. immitis* is limited to the San Joaquin Valley in California and in parts of Arizona, whereas *C. posadasii* is more widespread in geographic distribution. The importance of environmental exposure to airborne spores is illustrated by new infections occurring most frequently in the dry seasons (late summer and early fall) and by large outbreaks of coccidioidomycosis occurring after dust storms, earthquakes, and other events that disturb and aerosolize the soil.

The importance of coccidioidomycosis has increased in recent decades as the population in southwestern United States has grown, especially with the influx of older retirees and immunocompromised individuals.

Pathogenesis and clinical features

The barrel-shaped arthroconidia of *C. immitis* and *C. posadasii* are highly infectious when they are inhaled into the lungs. Once they are deposited in the lung of a susceptible person, the arthroconidia swell and produce spherules that fill with endospores. The endosporulating spherules rupture and release their endospores, each of which can generate a new spherule. This process results in primary pneumonia, which is the most common clinical syndrome associated with coccidioidomycosis.

Approximately 60% of primary *Coccidioides* infections cause no or only minor symptoms and thus are recognizable only later by a positive skin test. Symptoms in the remaining patients with pulmonary infections range from mild to severe. In general, the symptoms are nonspecific and include cough, chest pain, fever, chills, headache, and night sweats. Chest radiographs usually show pneumonia or pleural effusions, with or without hilar or mediastinal lymphadenopathy. Immune-mediated skin lesions such as erythema nodosum or erythema multiforme can also occur in association with pulmonary coccidioidomycosis. Most cases of pulmonary coccidioidomycosis resolve without treatment, although a minority of patients who recover have residual abnormalities such as single or multiple pulmonary nodules or thin-walled cavities on their chest radiographs. That active coccidioidomycosis sometimes develops in people who left the endemic area many years earlier implies that some of these residual lesions contain dormant but still viable fungi that are capable of reactivating when immune surveillance fails.

In a minority of patients, pulmonary coccidioidomycosis progresses to a chronic form of pneumonia that can persist for months or even years. These infections usually resemble pulmonary tuberculosis in that the upper lobes are often involved, and there is often prominent granuloma formation, fibrosis, and cavity formation. Large pulmonary masses and military disease also occur. Histopathological examination of these chronic or persisting lung lesions often shows a combination of suppurative and granulomatous features.

In a small minority of individuals, disseminated coccidioidomycosis develops, either soon after the primary pulmonary infection or much later when latent infection reactivates. Risk factors for the development of disseminated coccidioidomycosis include traditional causes of immunodeficiency such as AIDS, organ transplantation, or therapy with corticosteroids or other immunosuppressive drugs. Disseminated coccidioidomycosis is also more common in people of African American and Filipino ancestry, in young children and the elderly, and in pregnant women. The most common sites of extrapulmonary involvement are the skin, subcutaneous tissues, bones and joints, and meninges, but almost any organ can be involved. Meningeal coccidioidomycosis commonly causes hydrocephalus that requires surgical shunting, and this form of coccidioidomycosis is very difficult to cure.

Diagnosis and treatment

The diagnosis of coccidioidomycosis is established by demonstrating spherules by microscopy in sputum, pleural fluid, or tissues or by isolating *Coccidioides* in cultures of clinical samples. Clinicians should warn laboratory staff when coccidioidomycosis is suspected because cultures of *Coccidioides* are hazardous. A DNA probe that can identify a fungal isolate as *Coccidioides* is available commercially, so it is no longer necessary to convert the hyphal form to spherules in culture or by animal inoculation.

Serologic tests for antibodies to *Coccidioides* antigens are often useful. Enzyme immunoassays (EIA) for IgM antibodies are very sensitive, but also give many false-positive results. EIA tests for IgG and ID tests for IgM and IgG antibodies to *Coccidioides* antigens are less sensitive, but more specific. The CF test is usually positive at titers of 1:16 or higher in patients with disseminated or severe infections and declines with successful therapy. As is the case with most other infectious diseases, patients with profound immunodeficiency (e.g., those with advanced AIDS) are less likely to have measurable humoral immune responses to *Coccidioides* than are patients who are immunologically intact.

Skin testing for delayed-type hypersensitivity to *Coccidioides* is useful mostly as an epidemiologic tool, and the reagents are no longer available commercially.

C. immitis and *C. posadasii* are susceptible in vitro to amphotericin B and to the azole antifungals fluconazole, itraconazole, and voriconazole. Patients with pulmonary nodules, asymptomatic patients with a pulmonary cavity, and patients with acute pneumonia who are not seriously ill can often be observed because many improve without therapy. Patients with severe or progressive pulmonary disease or with extrapulmonary disease are treated with an azole or with amphotericin B. Meningeal coccidioidomycosis is usually treated with high-dose fluconazole or intravenous plus intrathecal amphotericin B; some patients require surgical shunting because of obstructive hydrocephalus.

The prognosis in normal hosts with pulmonary coccidioidomycosis is generally good. The outcome is less favorable in patients with disseminated infection, those who have underlying immunodeficiency states, and those with meningeal coccidioidomycosis.

Paracocciodomycosis

Causative organism

The causative agent of paracoccidioidomycosis is *Paracoccidioides brasiliensis*, a dimorphic ascomycete that grows as a mold in the environment and in culture at ≤25 °C (Figure 4(a)) and as budding yeast in infected tissues and in culture at ≥35 °C. One unique feature of yeast-phase *P. brasiliensis* is that single large mother cells can generate multiple small daughter cells (Figure 4(b)).

Epidemiology

Paracoccidioidomycosis is restricted to Central and South America, between latitudes 23° N and 34° S. Within this region, the disease is common in some areas and almost never observed in others. Favorable environments are characterized by the presence of tropical or subtropical forests, abundant rainfall and water, and moderate temperatures.

Skin test surveys show that approximately equal number of males and females have been exposed to *P. brasiliensis*,

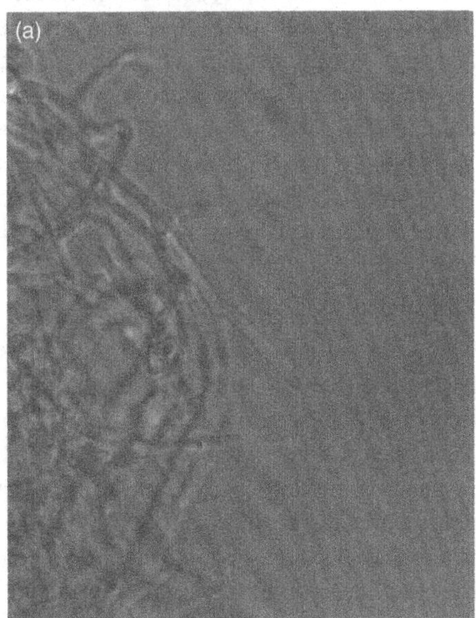

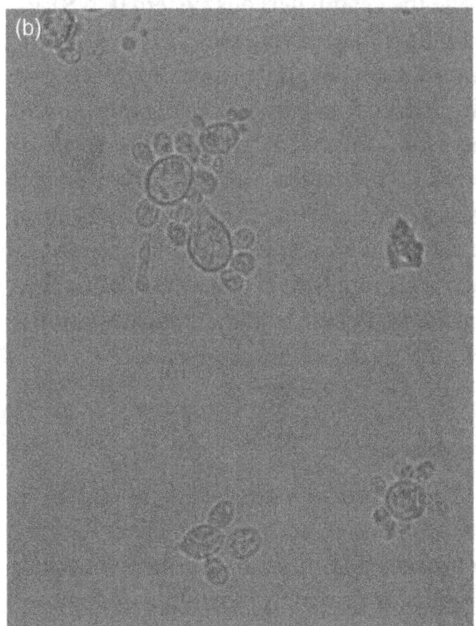

FIGURE 4 *Paracoccidioides brasiliensis* grows as hyphae at 25 °C (a) and as large yeast cells that produce multiple buds per cell at 37 °C (b). *Images courtesy of Paulo Coelho, University of Sao Paulo, Brazil.*

but many more males than females develop clinical disease. Paracoccidioidomycosis is unusual in children; so this is primarily a disease of adult men, most of whom are agricultural workers.

Pathogenesis and clinical features

Infection is presumed to follow the inhalation of conidia produced in the environment by the filamentous form of *P. brasiliensis*. Once the conidia are deposited in the lung, they transform into yeast cells, they begin to proliferate, and they can disseminate to distant sites via the bloodstream and lymphatic system. The fungi initially elicit an intense inflammatory reaction. With time, granulomas form and fungal replication ceases. As some patients develop active paracoccidioidomycosis years after they have left the endemic areas, it is likely that at least some fungi within granulomas or other lesions are dormant but still viable and can thus reactivate when host immunity wanes or becomes insufficient.

Most primary pulmonary infections are asymptomatic or cause only minimal symptoms and are recognized only in retrospect because of positive skin test reactivity. The most common clinical syndrome caused by *P. brasiliensis* is a chronic infection of the lungs and one or more extrapulmonary sites. This form of paracoccidioidomycosis tends to occur in adult males and is characterized by a chronic, progressive pneumonia that usually involves the central and lower portions of both lungs. The most common extrapulmonary sites involved are the skin and oropharyngeal or laryngeal mucosa, both of which develop ulcers characterized by granuloma formation and presence of large numbers of yeast-phase *P. brasiliensis* cells. Lymph nodes, the adrenal glands and the central nervous system (CNS) are also often involved in this form of disease.

A second form of disseminated paracoccidioidomycosis occurs primarily in children and in immunocompromised hosts (e.g., patients with advanced AIDS). This form of paracoccidioidomycosis resembles severe disseminated histoplasmosis in that large numbers of fungi are present in multiple organs such as the liver, spleen, and lymph nodes. Lung disease is much less prominent in patients with this type of paraccocidioidomycosis than in immunologically normal adults with more chronic paraccocidioidomycosis.

Diagnosis and treatment

Yeast-phase cells of *P. brasiliensis* are usually abundant in infected body fluids and tissues; so the diagnosis is usually established by demonstrating typical multiply budding yeast cells in potassium hydroxide preparations of appropriate clinical specimens or in histological sections. *P. brasiliensis* can also be identified in culture, but it generally takes several weeks to obtain hyphal-phase growth and then to convert the mold to yeast-phase cells by shifting to 36–37 °C.

Serologic tests for antibodies are useful in paracoccidioidomycosis. Antibodies can be demonstrated by ID in as many as 95% of patients with active disease. The CF test is somewhat less sensitive, but the ability to quantify the antibody response by generating titers is useful for monitoring responses to treatment. Immunologic tests for *P. brasiliensis* antigens in serum and CSF have been described, but these tests are not widely available.

P. brasiliensis is susceptible to sulfonamides, amphotericin B, and the azoles ketoconazole and itraconazole, and all of these drugs have been effective in clinical studies. Therapy with sulfonamides for 3–5 years is safe, inexpensive, and effective in approximately 70% of cases. Amphotericin B and ketoconazole are also effective in approximately 70% of cases, and itraconazole for 6 months is effective in as many as 95% of cases.

Sporotrichosis

Causative organism

The causative agent of sporotrichosis is *Sporothrix schenckii*, a dimorphic fungus that grows as a filamentous mold in the environment and in culture at 25–27 °C and as small, elongated 'cigar-shaped' yeast cells (3–5 μm) in infected tissues and in culture at 37 °C.

Epidemiology

S. schenckii is found throughout the world, especially in soil and in decaying or live plant matter. Most infections result from inoculation of hyphal-phase fungi through the skin at the sites of minor injury; so infections are most common in people who handle plants or plant products (e.g., gardeners and agricultural workers). *S. schenckii* can also be transferred from soil to people by infected or contaminated animals (e.g., cats). Several outbreaks have been described, including a large outbreak in South African mine workers who handled contaminated timber and outbreaks associated with contaminated plant materials used for gardening.

Pathogenesis and clinical features

The most common form of sporotrichosis is lymphocutaneous disease. This form of sporotrichosis follows inoculation of hyphal material through minor injuries in the skin. The fungus converts to yeast-phase cells, and the host responds by mounting an inflammatory reaction that includes neutrophils and also mononuclear phagocytes. Several days to a few weeks later, a papule or nodule usually develops, and then similar nodules develop over several weeks along the course of the lymphatics that drain the site of initial injury. Some of these nodules ulcerate and drain. The lesions are usually not painful.

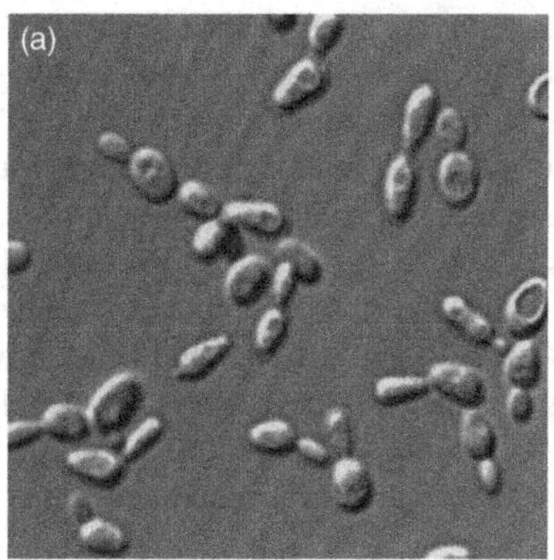

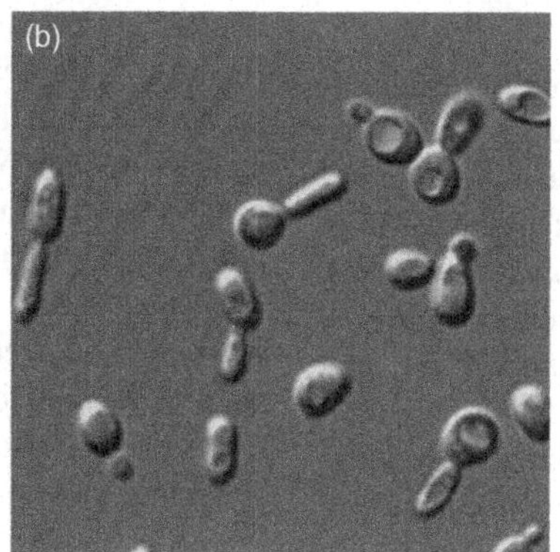

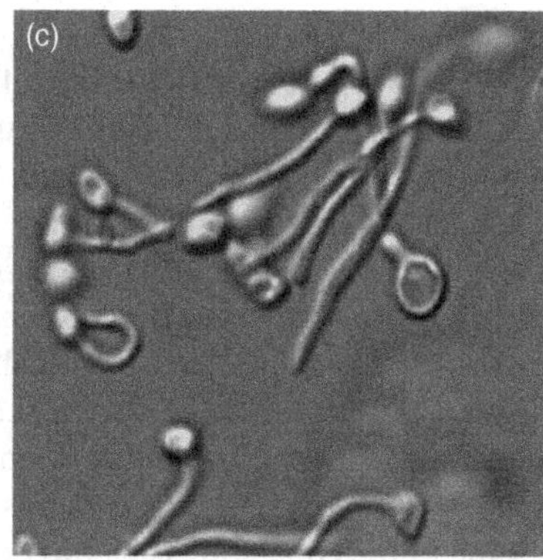

FIGURE 5 *Candida albicans* can grow as budding yeast (a), yeast cells with elongated buds called pseudohyphae (b), or as true hyphae (c). Short hyphal cells shown in (c) are also known as germ tubes.

Less common forms of sporotrichosis include pulmonary infection that closely resembles pulmonary tuberculosis, arthritis or osteomyelitis, and meningitis. Disseminated sporotrichosis has been described in patients with advanced AIDS and in patients who have received tumor necrosis factor antagonists.

Diagnosis and treatment

Definitive diagnosis of sporotrichosis requires that the causative fungus be isolated in culture and then converted from its hyphal phase to the characteristic yeast-phase cells. Because this generally takes several weeks, a presumptive diagnosis can be made if microscopic examination of appropriate fluids or tissues reveals typical elongated budding yeast cells. Unfortunately, these cells can be sparse and difficult to find. Serologic tests for sporotrichosis are generally not available, and there are no tests for *S. schenckii* antigens.

Most patients with lymphocutaneous, pulmonary, or bone and joint sporotrichosis can be treated effectively with itraconazole for at least several months. Potassium iodide by mouth is also effective in lymphocutaneous sporotrichosis; the mechanisms by which this treatment works are unknown. Amphotericin B can be used as initial therapy in people with *S. schenckii* meningitis or in immunocompromised patients with widely disseminated disease.

Diseases Caused by Opportunistic Fungi

Candidiasis

Causative organisms

Candidiasis refers to infections caused by fungi in the genus *Candida*. Together, these organisms are the most common causes of serious fungal infections in people. The most important pathogenic *Candida* species are *C. albicans*, *Candida glabrata*, *Candida tropicalis*, and *Candida parapsilosis*. All of these organisms grow in infected tissues and in culture as budding yeast-like fungi, and all except for *C. glabrata* can also form elongated buds (pseudohyphae). *C. albicans* and some strains of *C. tropicalis* can also form hyphae (Figure 5). The pathogenic *Candida* species are generally differentiated from each other based on (1) the abilities of *C. albicans* (and its close relative *Candida dubliniensis*) to form hyphae (also called germ tubes) when incubated in serum or other protein solutions, (2) the production of distinctive colors on indicator media (e.g., ChromAgar), (3) carbohydrate assimilation and fermentation patterns, and (4) morphologic features when grown in nutrient-limiting media.

C. albicans was once responsible for a substantial majority of serious *Candida* infections, but the importance of other *Candida* species has increased in recent decades. This trend was first noted in the 1970s and 1980s in cancer patients and bone marrow transplant recipients, and it has recently been documented in unselected populations. For example, a recent study of all *Candida* bloodstream infections in Baltimore County, Maryland, and in Connecticut between 1998 and 2000 showed that *C. albicans* was the causal agent in 45% of cases, *C. glabrata* in 24%, *C. parapsilosis* in 13%, *C. tropicalis* in 12%, and other *Candida* species in 6%.

One insight that emerged from sequencing of the genomes of several *Candida* species is that these fungi have mating loci that resemble the mating loci in ascomycetes with known sexual reproductive cycles. This led to the recognition that diploid *C. albicans* strains that were homozygous for the *MTL* **a** or the *MTL* α mating locus can mate with homozygous strains of opposite mating type to form tetraploid cells, which then return to diploidy by chromosome loss. Whether and how often events like this occur in natural populations of *Candida* cells is not known, but these observations raise the possibility that pathogenic eukaryotes that have lost the ability to grow as haploids or to undergo meiosis can acquire alternative mechanisms for genetic recombination.

Epidemiology

The epidemiology of candidiasis differs from the epidemiology of most other fungal diseases because normal people are colonized by *Candida*. Thus, candidiasis is acquired from endogenous sources rather than by exposure to fungi in the environment. Among the major *Candida* species, *C. albicans*, *C. tropicalis*, and *C. glabrata* populate the gastrointestinal tract and vagina of normal individuals, and *C. parapsilosis* is often found on the skin. Consequently, symptomatic infections generally develop when these fungi overgrow at sites of colonization (often as a consequence of suppression of the competing bacterial flora because of antimicrobial therapy) and when the host's defenses against infection are diminished.

Pathogenesis and clinical features

There are two general patterns of candidiasis. In people in whom overgrowth of *Candida* occurs because of suppression of competing bacterial flora by antibiotics or defects in cell-mediated immunity (e.g., corticosteroid recipients and patients with advanced HIV infection), the predominant form of disease is mucosal or cutaneous candidiasis. Typical mucosal lesions consist of white plaques that overlie shallow ulcers on an erythematous base. The ulcers are often filled with inflammatory cells and a mixture of *Candida* yeast cells and pseudohyphae, and the base may bleed if the white plaque is scraped off. Mucosal candidiasis of the mouth and pharynx is referred to as thrush, and similar lesions can occur anywhere in the gastrointestinal tract. Similar lesions also occur in the vagina. Finally, *Candida* can also infect the skin (especially in moist areas where skin maceration has occurred) and the fingernails and toenails (especially in patients with cell-mediated immune deficiency).

The second form of candidiasis is characterized by the fungal invasion of deeper organs and is generally referred to as disseminated or systemic candidiasis. In this form of candidiasis, the fungus usually enters the bloodstream through the gastrointestinal tract or vagina and the major risk factors are loss of mucosal integrity and lack of functional neutrophils. In recent years bloodstream infections occur more commonly via an intravascular device such as an intravenous catheter. *Candida* invades distant organs by (1) adhering to the vascular endothelium, (2) exiting from the vascular space into surrounding tissues, (3) proliferating within those tissues, and (4) eliciting an intense inflammatory reaction. The cardinal feature of this form of candidiasis is the presence of multiple microabscesses within many different tissues. Most patients with disseminated candidiasis are febrile and have signs of generalized infection, but specific clues that the problem is caused by *Candida* are usually lacking.

Hepatosplenic candidiasis is characterized by the onset of fever, abdominal pain, and biochemical evidence of hepatic dysfunction when patients whose bone marrow function is depressed by cytotoxic drugs and/or radiation therapy recover the ability to mount an inflammatory response and thereby respond to *Candida* cells that previously seeded the liver and spleen. In addition to these more generalized syndromes, *Candida* species can also cause deep infections in individual organs, such as bones, meninges, retina and adjacent structures in the eye, and heart valves. Involvement of organs such as these generally develops as a complication of *Candida* fungemia, although the fungemic event may be unapparent or unrecognized.

Diagnosis and treatment

Candida infections can be diagnosed by demonstrating fungal cells with characteristic morphology by microscopy or by culturing the organisms from appropriate sites. Microscopic or microbiologic demonstration of *Candida* species from superficial body surfaces (e.g., skin, mucosal surfaces, and sputum) does not imply infection because *Candida* species are part of the normal commensal flora. Whether *Candida* is causing or contributing to mucosal or cutaneous disease generally must be determined by considering the number of fungi observed, the type of lesions, and relevant host factors. On the other hand, isolation of *Candida* species from body fluids such as blood or CSF or from biopsies of deep tissues constitutes strong evidence of invasive infection. However, at least 50% of patients with disseminated candidiasis have repeatedly negative blood cultures, and there are no reliable nonculture diagnostic methods. Consequently, the diagnosis of disseminated candidiasis is often delayed or missed.

Serologic tests for antibodies to *Candida* species are not useful because normal people have detectable antibodies to this commensal organism. There have been extensive efforts to develop tests for distinctive *Candida* antigens and nucleic acids, but none of these is generally available or used. Several pathogenic *Candida* species produce sufficient amounts of the 5-carbon sugar alcohol D-arabitol in infected tissues and serum and urinary levels of this compound are elevated in a majority of patients with disseminated candidiasis. D-arabitol is used as a diagnostic marker in Japan and Scandinavia, but these tests are generally not available in North America.

Most strains of the medically important *Candida* species are susceptible in vitro to amphotericin B, the azole antifungals (e.g., fluconazole, itraconazole, and voriconazole), and the echinocandins (caspofungin, micafungin, and anidulafungin). Clinical trials conducted mostly in AIDS patients with esophageal candidiasis and in patients with *C.* fungemia have established that antifungals in all of these classes are effective. Resistance to azole antifungals can develop in people who are heavily treated for prolonged periods, and some *Candida* species are inherently more resistant to some classes of antifungals than most others (e.g., *C. glabrata* and azoles, *C. parapsilosis* and echinocandins).

Most patients with mucosal or cutaneous candidiasis can be treated effectively with topical antifungals or systemic azoles. Despite the availability of multiple effective antifungals, disseminated candidiasis remains a serious infection. Attributable mortality rates in the range of 35–50% have been reported, and these have not changed substantially in recent decades.

Cryptococcosis

Causative organisms

Cryptococcosis refers to infections caused by fungi in the genus *Cryptococcus*, which are the asexual forms of basidiomycetes in the genus *Filobasidiella*. A unique feature of *Cryptococcus* is that these are the only medically important fungi that have thick polysaccharide capsules, which consist of chains of α 1,3-linked mannose units with xylosyl and β-glucuronlyl substitutions. Human cryptococcosis is mostly caused by *C. neoformans*, of which there are two varieties (var. *neoformans* (serotype D) and var. *grubii* (serotype A)). *Cryptococcus gattii* (formerly called *C. neoformans* serotypes B and C) is responsible for approximately 5% of all cases of cryptococcosis, especially in several geographically restricted areas (see below).

All pathogenic *Cryptococcus* species grow as encapsulated budding yeast in infected people and in culture. When haploid cells of opposite mating types (**a** or α) are mixed on nutrient-limiting media, mating generates diploid hyphal forms that produce clamp connections and basidia, and haploid basidiospores are produced as a result of meiosis. Filamentation and formation of basidia and basidiospore formation can also be induced by nutrient deprivation in haploid cells of a single mating type. Although mating and basidiospore formation have not been observed in nature, molecular genetic analyses of environmental and pathogenic strains provide strong indirect evidence that sexual recombination occurs.

One surprising observation is that almost all strains isolated from infected people and also a large majority of environmental isolates are of mating type α. Moreover, direct comparison of the virulence in animals showed that a *C. neoformans* mat α strain was much more virulent than a congenic **a** strain. Classical and molecular genetic studies have also established that the ability to grow at 37 °C to form melanin and capsules is required for wild-type virulence.

Epidemiology

C. neoformans is found throughout the world, especially in soil and other material that are heavily contaminated with bird (especially pigeon) feces. Some areas have much higher incidence and prevalence rates than others, which presumably reflects differences in environmental conditions and factors such as bird populations. *C. neoformans* infections were once quite uncommon, tending to occur almost exclusively in people with cellular immunodeficiency states caused by neoplasia, organ transplantation, and administration of immunosuppressive drugs. However, the HIV epidemic has greatly expanded the population at risk for serious *Cryptococcus* infections. Cryptococcosis is now a leading cause of death among people with AIDS in Africa and elsewhere in the developing world.

C. gattii is primarily found in tropical and subtropical areas, often in association with trees (especially eucalyptus) and their surrounding microenvironments. However, a large cluster of cases of human and animal *C. gattii* infections was recently recognized on Vancouver Island in Western Canada, and this epidemic has since spread to the mainland of British Columbia, Washington, and Oregon. One unexpected aspect of this ongoing epidemic is that there is molecular genetic evidence that a hypervirulent epidemic strain may be the product of a sexual cross between a nonvirulent strain endemic and another strain that was recently introduced into this area. *C. gattii* causes serious pulmonary and extrapulmonary disease in immunologically normal hosts; so the epidemiology of *C. gattii* and *C. neoformans* diseases differs substantially.

Pathogenesis and clinical features

Cryptococcus is acquired by inhaling viable fungi in the environment into the lungs. Whether the infectious particles are desiccated yeast cells or basidiospores is not known. Once the fungi are deposited in the lungs, they proliferate as budding yeast cells, and the infection is usually controlled by the combined activities of natural and acquired host defenses. Neutrophils, mononuclear phagocytes, and lymphocytes all participate in cellular defenses against *Cryptococcus*, and phagocytosis and killing are also enhanced by humoral immune molecules such as antibodies to the capsular polysaccharide and its complements.

Most cases of pulmonary cryptococcosis are asymptomatic or result in only minor symptoms, and they resolve spontaneously without sequelae. In a minority of normal people and more often in people with cell-mediated immune deficiency, primary infection with *Cryptococcus* can lead to significant pulmonary disease. Symptoms can include cough, fever, chest pain and dyspnea, and chest radiographs can reveal pneumonitis, pulmonary nodules that may cavitate, pleural effusions, and regional lymphadenopathy.

Dissemination to extrapulmonary sites is common, especially in immunodeficient hosts. Extrapulmonary disease can develop at the same time or shortly after a patient has pulmonary cryptococcosis, but most cases occur without concomitant pulmonary disease. The most important extrapulmonary manifestation of cryptococcosis is meningoencephalitis, which is usually subacute or chronic. Patients with CNS cryptococcosis usually have headache, fever, and signs of meningeal irritation. Mental function abnormalities are common and can range in severity from subtle memory loss or personality changes to profound obtundation and coma. Cranial nerve dysfunction is common and often results from increased intracranial pressure. The pace and severity of CNS cryptococcosis can be highly variable. Some patients progress to coma and death within a few days, whereas others have had untreated chronic cryptococcal meningitis for >10 years.

Other extrapulmonary manifestations of cryptococcosis include mass lesions in the CNS (cryptococcomas), nodular skin lesions, chronic prostatitis, and bone lesions. Diffuse bilateral pulmonary disease that resembles miliary tuberculosis or *Pneumocystis* pneumonia can also occur in profoundly immunodeficient people.

Diagnosis and treatment

Cryptococcosis can be diagnosed by demonstrating encapsulated budding yeast cells in clinical specimens or in histological sections, by isolating the causative fungus in culture, or by demonstrating capsular polysaccharide antigens by immunoassay. As *Cryptococcus* is the only pathogenic fungus with a capsule, microscopic demonstration of encapsulating budding yeast cells in body fluids or tissues is diagnostic of cryptococcosis. The capsule of *Cryptococcus* is easily demonstrated in specimens such as CSF by its ability to exclude india ink (Figure 6). The capsule can appear as a clear halo around yeast cells in histological sections, and it is stained by mucicarmine. *C. neoformans* and *C. gattii* are usually isolated in culture within 2–3 days from CSF, blood, respiratory specimens, or tissue biopsies. Positive cultures can be identified as *Cryptococcus* by demonstrating the polysaccharide capsule, by India ink staining, or by immunoassay. Type-specific sera for differentiating *C. neoformans* from *C. gattii* are generally not available; however, *C. gattii* produces blue pigment on canavanine–glycine–bromothymol blue agar, and *C. neoformans* does not.

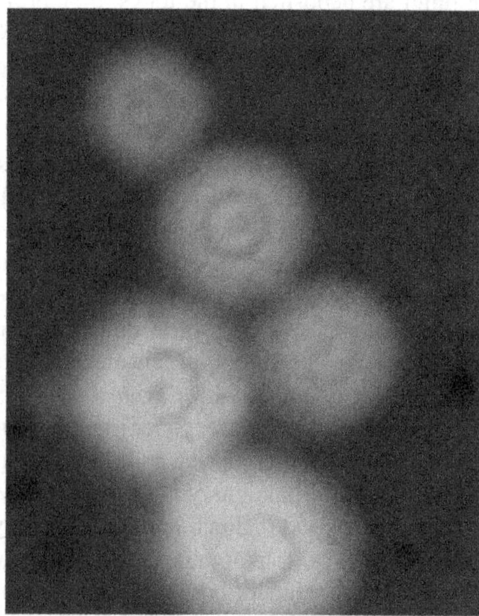

FIGURE 6 *Cryptococcus neoformans* produces a polysaccharide capsule, which is demonstrable as a clear halo surrounding cells in this India ink preparation. *Image courtesy of Kausik Datta, Oregon Health & Science University.*

Several immunoassays are available for demonstrating the capsular polysaccharide of *C. neoformans* and *C. gattii* in body fluids. These tests are now highly sensitive and specific, and they can be used with CSF, blood, and bronchial lavage fluids. Among all the immunoassays for fungal antigens, the test for cryptococcal polysaccharide is the most reliable and widely used.

C. neoformans is susceptible in vitro to amphotericin B, the combination of amphotericin B and flucytosine, and the azole antifungals fluconazole and itraconazole. Clinical trials have established that all of these therapies are effective in patients with cryptococcal meningoencephalitis. In contrast, the echinocandins have no activity against *Cryptococcus* and should not be used. Combination therapy with amphotericin B and flucytosine sterilizes the CSF faster and is associated with fewer late relapses than therapy with amphotericin B or fluconazole alone. Relapses following initial course of therapy are common in AIDS patients and others whose underlying immunodeficiency cannot be corrected; so prolonged maintenance therapy with fluconazole is often required. Marked worsening of intracranial hypertension often occurs shortly after therapy is begun; so careful monitoring for this complication and prompt intervention are mandatory.

No formal treatment trials have compared different therapies in patients with nonmeningeal cryptococcosis. Also, *C. gattii* appears to be less susceptible to azole antifungals than *C. neoformans*, but whether patients with serious *C. gattii* infections should be treated with amphotericin B, amphotericin B plus flucytosine, or an azole antifungal has not been studied.

Penicillosis

Causative organism

Penicillosis refers to diseases caused by fungi in the genus *Penicillium*. The most important *Penicillium* species is *Penicillium marneffei*, which causes serious infections in immunocompromised individuals in Southeast Asia. *P. marneffei* is a dimorphic fungus that grows as hyphae at 25–30 °C and as a yeast at 37 °C. Hyphal-phase *P. marneffei* produces a red pigment that diffuses into agar culture medium. Yeast-phase *P. marneffei* cells divide by fission (i.e., schizogony, as does the model fungus *Schizosaccharomyces pombe*) rather than by budding.

Epidemiology

P. marneffei is endemic in Southeast Asia. Cases of human disease have been reported in Thailand, Laos, Cambodia, Vietnam, Malaysia, Singapore, Burma, Taiwan, and Southern China. The organism has been isolated from healthy bamboo rats and from soil in and around these animals' burrows. Retrospective case–control studies have associated *P. marneffei* infection with exposure to soil and not with exposure to bamboo rats.

P. marneffei causes disease almost exclusively in people with profound abnormalities of cell-mediated immunity. Advanced HIV infection is the most common underlying condition, but cases are also observed in people with hematologic neoplasia, in organ transplant recipients, and in people receiving immunosuppressive drugs. *P. marneffei* infection is the third most frequent serious opportunistic infection among AIDS patients in Northern Thailand, following only tuberculosis and cryptococcosis.

Pathogenesis and clinical features

It is likely that infection is by inhalation of environmental spores, but this has not been proven. Most susceptible people with *P. marneffei* infections develop a subacute to chronic disseminated disease characterized by several weeks of fever, weight loss, and malaise. Many patients also have cough and other symptoms of lung infection, and a majority have skin lesions. These skin lesions are generally located on the face and upper body, and they can include papules (sometimes umbilicated), pustules, ulcers, or abscesses. At autopsy, there is usually widespread involvement of many organs, including the lungs, lymph nodes, bone marrow, liver, spleen, kidney, bowel, adrenals, bone, and meninges. Tissue reactions can include granulomatous and acute inflammatory reactions, often with necrosis or suppuration. Yeast-phase forms of *P. marneffei* are usually abundant in infected tissues.

Diagnosis and treatment

In endemic geographic locales, disseminated *P. marneffei* infection should be considered whenever an immunocompromised patient has a subacute to chronic progressive illness consistent with penicillosis. In nonendemic areas, a history of prior residence or travel to an endemic area should be sought, especially since recrudescence years after departure from an endemic area has been documented.

Once the diagnosis of penicillosis is considered, it is verified by demonstrating the causative organism in body fluids or tissues by microscopy or by culture. The diagnosis can often be established by demonstrating typical fungal forms in smears of skin biopsies, lymph node biopsies, or bone marrow aspirates or biopsies. If there is pulmonary involvement, the fungus can often be demonstrated by microscopic examination of respiratory secretions or bronchoalveolar lavage. *P. marneffei* can be isolated in culture from infected body fluids or tissues, including the blood, bone marrow, skin lesions, lymph nodes, sputum, or bronchoalveolar lavage, or other tissues. If hyphal-phase fungi with morphologic features typical of *Penicillium* species are isolated, identification of the organism as *P. marneffei* is established by conversion to yeast-phase growth by incubating at 37 °C.

Tests for distinctive antibodies to *P. marneffei* or for *P. marneffei* antigens or nucleic acid sequences are under investigation, but they are not generally available. *P. marneffei* has been reported to cross-react with tests for *Aspergillus* galactomannan.

P. marneffei is susceptible in vitro to amphotericin B, itraconazole, ketoconazole, and voriconazole, but not to fluconazole. Seriously ill patients are generally treated with intravenous amphotericin B initially (e.g., for 2 weeks) and then with itraconazole for at least 10 more weeks. Therapy with itraconazole alone is generally reserved for people with less severe disease. Relapses among AIDS patients with penicillosis are common; so long-term suppression with itraconazole or fluconazole is often given. Whether treatment can be stopped if immune reconstitution is achieved with antiretroviral therapy is not known.

Aspergillosis

Causative organisms

The ascomycetous fungi in the genus *Aspergillus* are the causative agents of aspergillosis. These environmental molds are present in abundant amounts in many outdoor and indoor environments. *A. fumigatus* is by far the most important pathogen of humans. *Aspergillus flavus*, *Aspergillus niger*, *Aspergillus terreus*, and *Aspergillus lentulus* also cause serious infections. All of these fungi grow in the environment and in culture as branching hyphae, and they also produce airborne conidia (Figure 7).

Epidemiology

People are exposed to spores of *A. fumigatus* and other *Aspergillus* species on a regular basis; so whether exposure to these fungal spores results in disease is determined primarily by host susceptibility. Cases of invasive aspergillosis have increased markedly in recent decades, which is no doubt due to an increase in the number of people with markedly abnormal host defenses. In one study, cases of aspergillosis per 100 000 people in the United States increased more than fourfold between 1981 and 1997.

Although host susceptibility determines in large measure who acquires aspergillosis, environmental factors are also important. Several studies have documented outbreaks of cases among immunocompromised people

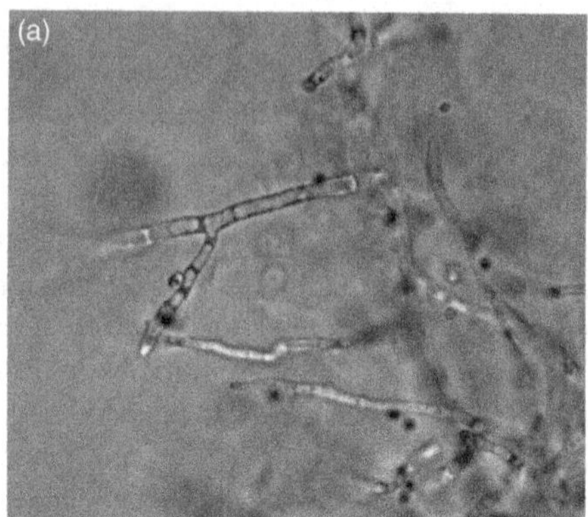

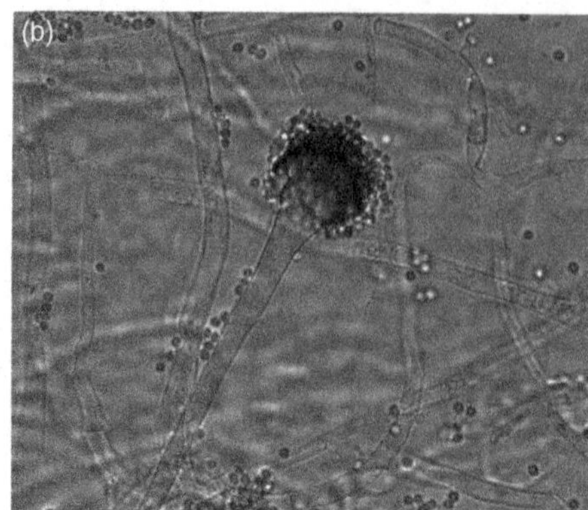

FIGURE 7 *Aspergillus fumigatus* and other *Aspergillus* species grow in infected tissues and in culture as narrow septate hyphae that branch at angles of approximately 45° (a) *A. fumigatus* produces airborne asexual spores (conidia) from columnar fruiting bodies when cultured on agar plates (b) but these structures are seldom seen in infected mammalian tissues.

who were exposed to aerosols generated by construction projects either within hospital buildings or outdoors. It has long been known that removal of airborne particles by high-efficiency air filtration can substantially reduce the incidence of aspergillosis among highly susceptible hosts.

Pathogenesis and clinical features

Aspergillus causes several different types of human disease. Allergic bronchopulmonary aspergillosis (ABPA) is characterized by chronically recurring episodes of bronchospasm associated with sputum production, mucus plugging of airways with resulting atelectasis, sometimes resulting in pulmonary to fibrosis and/or bronchiectasis. This syndrome is caused by hypersensitivity to antigens expressed by fungi colonizing the airways. Most patients with ABPA have high circulating levels of total IgE, demonstrable IgE-mediated responses to *Aspergillus* antigens, and high circulating levels of IgG antibodies to *Aspergillus* antigens. The course is generally chronic and progressive. Patients generally do not improve without therapy.

Aspergillus can also colonize anatomically abnormal airways, thereby initiating a local inflammatory reaction. When this occurs in residual cavities in patients with healed tuberculosis, a mass of fungal hyphae and necrotic debris that eventually fills the cavity is referred to as an aspergilloma or fungus ball. Most aspergillomas cause no symptoms, but they erode the surface of the cavity, which can result in bleeding (hemoptysis) that can be severe. *Aspergillus* can similarly colonize bronchiectatic airways of patients with underlying chronic lung diseases such as cystic fibrosis.

Invasive pulmonary aspergillosis is the form of aspergillosis that has increased markedly in frequency and importance in recent decades. Patients at highest risk include people with prolonged neutropenia due to stem cell transplantation or cytotoxic chemotherapy for neoplasia. Transplant recipients and patients receiving high doses of corticosteroids or other immunosuppressive drugs are also at increased risk. When an immunodeficient host inhales viable *Aspergillus* conidia, the fungi evade ingestion and killing by host phagocytes, they swell and germinate to produce filamentous hyphae, and they invade the surrounding tissues. *Aspergillus* tends to invade and proliferate within blood vessels, thereby causing ischemic injury and widespread necrosis in affected tissues. Patients with invasive pulmonary aspergillosis usually have fever and cough, and pleuritic chest pain is often present. Chest radiographs show pulmonary infiltrates that are often peripheral and wedge-shaped, as would be expected in a lesion in which vascular ischemia is an important contributor. Cavity formation is common and can be a valuable clue to the diagnosis. Disease progression is often very rapid in profoundly immunocompromised individuals, often leading to respiratory failure and death in a few days. In others (especially those with less severe immune deficiencies), the pace of the disease is much slower; chronic progression with formation of multiple cavities can develop over several months.

Aspergillus can also cause necrotizing sinusitis in some profoundly immunocompromised individuals. The initial site of infection is the mucosal surface of a facial sinus or the nasopharynx, which is directly invaded by germinating fungal hyphae. Because *Aspergillus* tends to invade blood vessels and thereby causes extensive tissue necrosis, this process can rapidly extend into the orbit and into the cranium. For unknown reasons, *A. flavus* is especially likely to cause necrotizing sinusitis.

Aspergillus can also disseminate to distant organs via the bloodstream in patients who have primary aspergillosis of the lung or a facial sinuses. Once the fungus reaches these distant organs, it tends again to invade along blood vessels, thereby causing ischemia and tissue necrosis. Virtually any organ can be involved, including the brain, kidneys, liver, spleen, and skin.

Diagnosis and treatment

Aspergillus grows in infected tissues as narrow septate hyphae that branch at 30–45° angles, thus demonstrating fungi with these morphologic features in body fluids of tissues from a person with a compatible clinical syndrome can provide presumptive evidence of aspergillosis. However, the hyphae of several other pathogenic fungi (e.g., *Fusarium*) are similar in morphology; so a definitive diagnosis of aspergillosis requires both demonstration of the fungus in infected body fluids and/or tissues and isolation and identification of the fungus in culture. Interpreting the results of fungal cultures (especially from respiratory sites) can be difficult because (1) *Aspergillus* can colonize the respiratory tract without causing invasive disease and (2) some patients with proven invasive infections have repeatedly negative cultures. It should also be noted that *Aspergillus* is almost never isolated from blood, even in patients with widely disseminated disease.

Patients with ABPA usually have high total IgE levels and high levels of *Aspergillus*-specific IgE and IgG antibodies. Patients with aspergillomas or fungal colonization of bronchiectatic airways often have high levels of IgG antibodies to *Aspergillus* antigens. Serologic testing for antibodies to *Aspergillus* is not useful in invasive pulmonary or extrapulmonary aspergillosis. However, galactomannan derived from the cell wall of *Aspergillus* can be detected by immunoassay in a substantial portion of patients with invasive pulmonary or extrapulmonary aspergillosis, and galactomannan can be detected by immunoassay in bronchoalveolar lavage fluids from a substantial majority of patients with invasive pulmonary aspergillosis.

Growth of *Aspergillus* is inhibited in vitro by amphotericin B, the azoles itraconazole, voriconazole, and posaconazole (but not fluconazole or ketoconazole), and the echinocandins caspofungin, micafungin, and anidulafungin. Prospective clinical trials have established that itraconazole is an effective therapy for ABPA and that voriconazole is superior to amphotericin B for invasive aspergillosis. Uncontrolled studies suggest that caspofungin may be useful in patients who do not respond to or cannot tolerate alternative therapies. Controlled trials showed that posaconazole can prevent invasive aspergillosis in selected high-risk patients. Despite the availability of effective therapies, mortality in severely immunocompromised hosts with invasive aspergillosis remains very high. For this reason, there is considerable interest in combination chemotherapy, adjuvant immunotherapy, and other measures. Also, some patients with extensive necrotizing aspergillosis lesions may benefit from surgical debridement or resection, either to control the infection initially or to reduce the likelihood of relapse.

Zygomycosis

Causative organisms

The term zygomycosis refers to diseases caused by filamentous fungi in the class Zygomycota, the most important of which are in the genera *Rhizopus*, *Mucor*, and *Absidia*. As these fungi are in the order Mucorales, an alternative name for zygomycosis is mucormycosis. These fungi are widespread in nature. They can commonly be found in the soil and on decaying plant and animal matter; for example, zygomycetes include common bread molds. These fungi grow as molds consisting of broad hyphae that usually lack nonseptae. In the environment (but not in infected tissues), fruiting structures (sporangia) develop, which generate large numbers of airborne spores (sporangiospores). Consequently, most people regularly come in contact with airborne zygomycete sporangiospores.

Pathogenesis and clinical features

Most patients with zygomycosis have abnormal host defenses, although these often differ from those that underlie most other opportunistic mycoses. For example, the most important factor associated with zygomycosis is diabetes, especially with ketoacidosis. Other forms of metabolic acidosis (e.g., uremia) also predispose to zygomycosis, as do states associated with availability in tissues of excess amounts of free iron (e.g., hemochromatosis or administration of iron-chelating agents). Corticosteroids, neutropenia, and interrupted skin (e.g., in burns or surgical incisions) are also important predisposing factors.

The most common form of zygomycosis is rhinocerebral zygomycosis, which begins when fungal spores deposit on the mucosal surfaces of the nasopharynx or palate. The spores then germinate into hyphae, which invade and then grow within blood vessels, thereby causing extensive thrombosis, ischemia, and necrosis. Zygomycosis typically breaches anatomical barriers and promptly involves the facial sinuses, the orbit and its contents, and then the cranial contents. Pain, swelling and cranial nerve palsies are often seen early, followed by frank skin necrosis, blindness, stupor, and coma.

In patients such as neutropenic stem cell transplant recipients and leukemia patients, inhalation of zygomyces spores into the lungs can lead to pulmonary zygomycosis, which is very similar in its clinical and pathologic features to pulmonary aspergillosis. As is also observed in rhinocerebral disease, the zygomycetes invade and then grow along blood vessels in the lungs, causing extensive thrombosis, ischemia, and tissue infarction. Most patients have fever, cough, and/or chest pain, and chest radiographs usually show diffuse pneumonia, more focal disease that may cavitate, and/or pleural effusions.

Dissemination via the bloodstream to distant organs occurs in patients with either the rhinocerebral or pulmonary forms of zygomycosis, especially in patients with prolonged neutropenia or in stem cell transplant recipients. The zygomycetes also cause necrotizing skin infections, either as a secondary manifestation of widespread hematogenous dissemination or when viable spores are introduced into burn wounds or surgical incisions via contaminated wound dressing materials. The zygomycetes rarely cause invasive disease of the gastrointestinal tract, especially in malnourished children in the developing world.

Diagnosis and treatment

Most cases of zygomycosis are diagnosed by demonstrating in infected tissues or fluids or in stained histological sections broad hyphae that branch at 90° angles that usually lack cross-septations (Figure 8). The causative fungi can often be demonstrated in potassium hydroxide preparations of material from necrotic eschars in the nasopharynx or palate or from infected sinuses of people with rhinocerebral zygomycosis.

The various zygomycete genera and species are differentiated by morphologic features of fruiting and 'rootlet' structures that develop only in culture. However, it is sometimes very difficult to isolate these fungi in culture, even from specimens that are known to contain abundant hyphae by direct microscopy. Like *Aspergillus*, the fungi that cause zygomycosis cannot be isolated from the blood of infected people. There are no serologic tests for antibodies to the zygomycetes or immunoassays for zygomycete antigens.

Because zygomycosis tends to progress rapidly and is associated with high mortality rates, multiple therapeutic measures are generally required. First, every effort should be made to correct underlying host defense abnormalities such as diabetic ketoacidosis or uremic acidosis as quickly as possible. Similarly, immunosuppressive drugs should be stopped or their dosages reduced, if possible. Second, necrotic tissues should be surgically debrided and removed, even if extensive and disfiguring procedures are required. Next, antifungal therapy should be administered. Amphotericin B inhibits the growth of most of the zygomycetes and has been used with success in many patients with zygomycosis. The newer azole antifungal posaconazole has in vitro activity against many zygomycetes, whereas most other azoles do not, and at least one uncontrolled case series suggested that posaconazole may be effective in patients with zygomycosis who have not responded to or cannot tolerate amphotericin B.

Pneumocystis infections

Causative organism

Organisms in the genus *Pneumocystis* were once thought to be protozoans because of morphology and other characteristics, but genome sequence analysis has established that these are fungi in the phylum Ascomycota. *Pneumocystis* has been found in the lungs of many mammalian species, but it has not been cultivated in vitro. Thus, much of what is known about these fungi has been gleaned from its morphology in infected hosts and from inferences from analyzing their genome sequences. Two morphologic forms are observed in infected lung tissues. Trophic forms are small (diameter 1–5 μm), uninucleate cells that appear to reproduce asexually by fission. Cystic forms are larger (diameter 5–10 μm) spherical structures with thick walls and as many as eight internal spores. As these structures resemble asci, they are presumed to develop as a consequence of mating between trophic cells of opposite mating type.

As different species of *Pneumocystis* appear to infect different mammalian species, the human pathogen that was once called *Pneumocystis carinii* (now the rat pathogen) is now called *P. jirovecii*. Environmental forms of *Pneumocystis* have not been described.

Epidemiology

Serologic surveys have shown that most healthy people develop antibodies to *P. jirovecii* during childhood. Animal studies indicate that *Pneumocystis* is acquired by inhalation, but it is not known if the infectious particles are spores or trophic cells acquired from infected individuals or if they

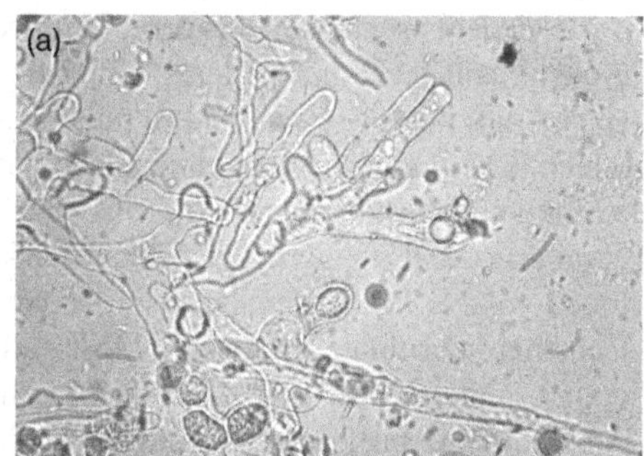

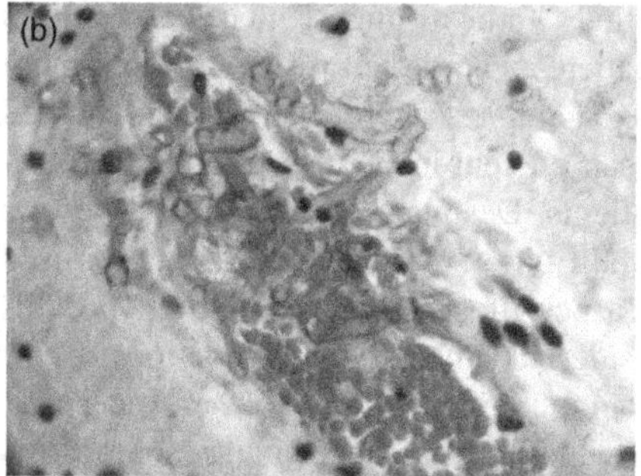

FIGURE 8 *Rhizopus* and other zygomycetes grow as broad nonseptate hyphae that branch at angles of approximately 90°. Typical hyphae can often be demonstrated in potassium hydroxide preparations of infected tissues. (a) Zygomycete hyphae tend to invade within and along blood vessels, as in this section from the brain of a patient with fatal rhinocerebral disease (b).

are produced by an unknown environmental form of this fungus. Clinically apparent disease develops almost exclusively in people with cell-mediated immunodeficiency states associated with conditions such as advanced AIDS, neoplasia, malnutrition, or immunosuppressive drugs.

Pathogenesis and clinical features

The most common clinical manifestation of Pneumocystis infection is bilateral diffuse pneumonia. Most patients have fever, dry cough, and dyspnea. In some patients, the pneumonia can progress to severe disease and death over a few days; in others, the course is indolent over the course of several weeks. Histopathological examination of infected lung tissue generally reveals foamy eosinophilic exudates within affected alveoli. These exudates typically contain abundant *Pneumocystis* trophic cells and cysts. The host inflammatory response is usually mild. Extrapulmonary involvement of many different organs has been documented in patients with advanced AIDS, but is seldom of major importance.

Diagnosis and treatment

Pneumocystis pneumonia is usually diagnosed by demonstrating morphologically characteristic trophic cells and cysts in respiratory specimens or lung biopsies. Immunoassays for *Pneumocystis* antigens in respiratory fluids and polymerase chain reaction tests for *Pneumocystis* DNA sequences have been developed, but these tests are not widely available.

The combination of a sulfonamide and trimethoprim is highly effective for the treatment of *Pneumocystis* pneumonia. Alternative therapeutic include pentamidine (either parenterally or by inhalation), atovaquone, dapsone plus trimethoprim, primaquine plus clindamycin, and trimetrexate. Adjunctive therapy with corticosteroids improves the outcome in AIDS patients with severe *Pneumocystis* pneumonia, presumably by attenuating host inflammatory responses to dead or dying fungi. Chemoprophylaxis against *Pneumocystis* pneumonia is effective in high-risk individuals who have not previously been infected (e.g., AIDS patients with low CD4+ lymphocyte counts, children with acute leukemia) and also in patients who have recovered from an initial episode. Effective chemoprophylactic agents include sulfamethoxazole/trimethorpim, dapsone, atovaquone, and pentamidine administered by inhalation.

RECENT DEVELOPMENTS

Molecular mycology has moved in the direction of genomics in the last few years. It has become feasible to sequence fungal genomes more rapidly and at lesser costs. The generation of genome sequence data has reached a level of ease such that laboratories previously not known for genome work are now capable of sequencing fungi. This has produced sequence data from multiple isolates of a given species, of subspecies, and of obscure and lesser studied fungi. Genome data from lesser studied but related fungi or from multiple isolates may provide important clues as to why certain species or isolates are associated with disease.

Another important use of the DNA data has been for phylogenetic studies. The use of genomic data for the identification of fungi has complicated the field of fungal taxonomy. Multilocus sequence typing has been increasingly used for fungal identification because of the ease of obtaining DNA sequence data that can be compared to ever-expanding DNA databases. One advantage of molecular identification is the quantitative nature of the data. Traditional identification utilizes the morphology of asexual and sexual structures as the basis for classification. This becomes problematic with the identification of atypical isolates that do not display all of the species-defining features. As a consequence, many asexual or imperfect fungi and atypical isolates have been classified together with other fungi that are evolutionarily distant. Genomic data have clarified relationships among asexual fungi but have also challenged the concept of a species. This has been especially true in the genus *Aspergillus*. Various mycologist panels have proposed to utilize a combination of genetic, morphological, physiological, and ecological data to delimit a species of *Aspergillus* to improve upon existing morphological identification.

Another use of multilocus sequence typing is to analyze the evolutionary origins of isolates obtained in the context of new clusters of disease. For example, data generated by these approaches suggest that the *Cryptococcus gattii* strains responsible for the expanding outbreak of infections in British Columbia, Canada, and the Pacific Northwest of the United States may continue to evolve and differentiate from their presumed parents, even over the course of just a few years.

Another advantage of utilizing DNA for molecular taxonomy lies in the ability to extract the nuclei acid from old specimens that are not viable anymore. A second more important use of DNA data is for understanding the biology of an isolate. As more fungal genomes are sequenced and compared, it is likely that organisms will be reclassified into new or existing species based on multilocus sequence typing or whole genome phylogenetic studies.

The accurate classification of medically important fungi is an important undertaking. The use of molecular identification methods has uncovered new and previously described species among clinical isolates that were originally misidentified by morphological features. Some of these species display decreased in vitro antifungal susceptibility profiles and others were unrecognized etiological agents of disease. DNA and genomic data will likely uncover new species and be useful for investigations of the prevalence and significance of newly described and existing fungal pathogens.

SUMMARY AND CONCLUSION

The burden of disease caused by pathogenic fungi has increased markedly in recent decades, primarily because there are now many more people with significant deficiencies in host defenses against infections than there were as few as 20 years ago. Along with the recent increase in the frequency and importance of these diseases, our understanding of the basic biology of the pathogenic fungi and of their interactions with mammalian hosts has also expanded greatly. These advances have resulted in large measure from the sequencing of the genomes of several pathogenic fungi, the application of molecular genetic methods to fungal pathogenesis, and advances in understanding both innate and adoptive immune responses to infection by fungi and other pathogens. There have also been significant advances in recent decades in our ability to diagnose, prevent, and treat several serious fungal infections.

FURTHER READING

Askew, D. S. (2008). *Aspergillus fumigatus*: Virulence genes in a street-smart mold. *Current Opinion in Microbiology*, *11*(4), 331–337.

Butler, G. (2010). Fungal sex and pathogenesis. *Clinical Microbiology Reviews*, *23*(1), 140–159.

Cornely, O. A. (2008). Aspergillus to Zygomycetes: Causes, risk factors, prevention, and treatment of invasive fungal infections. *Infection*, *36*(4), 296–313.

Klein, B. S., & Tebbets, B. (2007). Dimorphism and virulence in fungi. *Current Opinion in Microbiology*, *10*(4), 314–319.

Heitman, J., Filler, S. G., Edwards, J. E., Jr., & Mitchell, A. P. (Eds.), (2006). *Molecular principles of fungal pathogenesis.* ASM Press: Washington, DC.

Hospenthal, D. R., Rinaldi, M. G., & Kauffman, C. A. (Eds.), (2007). *Atlas of fungal infections.* (2nd ed.). Current Medicine: Philadelphia, PA.

Maerterns, J. A., & Marr, K. A. (Eds.), (2007). *Diagnosis of fungal infections.* Informa Healthcare USA: New York.

Wilson, D., Thewes, S., Zakikhany, K., Fradin, C., Albrecht, A., Almeida, R., *et al.* (2009). Identifying infection-associated genes of *Candida albicans* in the postgenomic era. *FEMS Yeast Research*, *9*(5), 688–700.

Chapter 12

Fungal Infections, Cutaneous

C.J. Watts, D.K. Wagner and P.G. Sohnle
Division of Infectious Diseases, Department of Medicine, Medical College of Wisconsin, and the Section of Infectious Diseases, VA Medical Center, Milwaukee, WI, USA

Chapter Outline

Defining Statement	167
Cutaneous Host Defenses	167
Description of the Diseases	169
Recent Developments	173
Further Reading	173

DEFINING STATEMENT

Fungal infections, cutaneous encompass a discussion of the major innate and adaptive host defense mechanisms in the skin and a description of the various superficial and deep fungal infections that occur in this location.

CUTANEOUS HOST DEFENSES

Structure of the Skin

The physical and chemical structure of the skin represents a form of defense against fungal pathogens. The skin surface is relatively inhospitable to fungal growth because of exposure to ultraviolet light, low moisture conditions, and competition from the normal bacterial flora of this site. Therefore, this surface acts as a barrier to the entry of fungi. The stratum corneum is made up of keratin, which most microorganisms cannot use for nutrition. However, *Candida albicans* and the dermatophytic fungi produce keratinases, which hydrolyze this substance and facilitate the growth of these organisms in the stratum corneum itself. This very superficial site of infection may protect the infecting organisms from direct contact with at least some of the effector cells of the immune system. Although neutrophils and small numbers of lymphocytes may enter the epidermis, the major infiltrates of cell-mediated immune responses are generally confined to the dermis.

Keratinization and Epidermal Proliferation

The process by which the stratum corneum is continually renewed through keratinization of the epidermal cells may also present a form of defense against organisms infecting the skin. The basal epidermal cells produce continued growth of the epidermis as they undergo repeated cell divisions that move the resulting daughter cells (keratinocytes) outward, toward the surface. As they mature and differentiate, these cells lose their nuclei and become flattened to form the keratinized cells. This process results in continuous shedding of the stratum corneum, which also may remove infecting fungal microorganisms residing there. Inflammation, including that produced by cell-mediated immune reactions, appears to enhance epidermal proliferation so that rates of transit of epidermal cells toward the stratum corneum are increased. A number of studies have demonstrated that epidermal proliferation is important in the defense against superficial mycosis.

Antifungal Substances

Sebaceous gland secretions and other epidermal lipids, including the sphingosines, fatty acids, polar lipids, and glycosphingolipids, have been shown to be fungistatic against both dermatophytes and *C. albicans*. Unsaturated transferrin is another well-known antimicrobial substance that appears to be active in inflamed skin and inhibits microbial growth by competing for iron. Keratinocytes of the epidermis produce several other antimicrobial proteins, including cathelicidins, human β-defensin 2, and calprotectin. These proteins appear to be generated in greater quantities when the keratinocytes are stimulated by contact with microorganisms or cytokines. A variety of other antimicrobial proteins are also found in the skin, including granzyme B adrenomedullin, antileukoprotease, and melanins. Others are undoubtedly present also, and the

Eukaryotic Microbes.

presence of all these antimicrobial compounds adds to the difficulty that potential pathogens face when trying to invade into the skin.

The Innate Immune System

The innate immune system is composed of cells and mechanisms that defend an organism against nonself entities in a nonspecific manner; it represents a first line of defense against invading pathogens and does not require a preliminary exposure to microbial antigens to generate an effective response. Unlike adaptive immunity, the innate immune system does not confer long-lasting memory or protective immunity against the foreign entity. Rather, innate immunity provides immediate defense against pathogens. This system is especially prominent in those body surfaces that contact the external environment and are colonized by potential pathogens; these areas include the skin, digestive and urinary tracts, and airways. Therefore, cutaneous fungal pathogens must somehow evade these processes when they first try to invade the skin.

All known plants and animals have innate immunity, suggesting that this system appeared early during evolution. The major function of this system seems to be the initial recruitment of leucocytes, such as macrophages and neutrophils, to areas of infection. These cells then phagocytose and kill pathogens while also eliciting continued inflammatory responses. After microbial killing occurs in the phagosomes of macrophages, the organism's components may also be presented to T lymphocytes to activate the adaptive immune system. Thus, there is interplay between these two arms of the immune system.

A critical part of this process is the ability of the host cells to discriminate between themselves and a large number of potential pathogens using a restricted number of germline encoded receptors. The latter recognize particular chemical characteristics of microorganisms, referred to as pathogen-associated molecular patterns (PAMPs). The best described host receptors involved in this process are the toll-like receptors (TLRs), whose genes are present in all eukaryotes. At present there are ten known TLRs in humans, with different PAMPs being recognized by different ones. Once the TLR identifies a PAMP, it begins a cell-signaling cascade that initiates a defensive response directed at eliminating the pathogen. This cascade involves multiple cytokines that promote inflammation, stimulate adaptive immunity, and cause an acute phase response. Factors such as age, nutritional status, and hormone levels may alter innate responses.

The Inflammatory Response

With various kinds of superficial fungal infections, there appears to be an inverse relationship between the degree of inflammation produced by a particular fungal pathogen and the chronicity of that infection. *Malassezia* sp. and the anthropophilic dermatophytes, *Trichophyton rubrum* and *Epidermophyton floccosum*, generally produce little inflammation in their cutaneous lesions and frequently cause infections that persist for long periods. On the other hand, many of the geophilic or zoophilic dermatophytes, for example, *T. verrucosum*, produce highly inflammatory infections that are usually self-limited. Thus, the local inflammatory process may indeed be involved in the defense against this group of pathogens.

Several mechanisms have been described by which inflammatory cells are attracted into sites of cutaneous fungal infections, in addition to the innate immunity processes described above. Fungal organisms are generally capable of activating complement by the alternative pathway to produce chemotactic activity for neutrophils and can also produce low molecular weight chemotactic factors analogous to the ones made by growing bacteria. Keratinocytes themselves can generate chemotactic cytokines that could also be responsible for some of the inflammation in the lesions of cutaneous fungal infections.

Neutrophils and monocytes/macrophages appear to be important in the defense against fungi, including those involved in the cutaneous mycoses. Neutrophils can directly attack pathogens using a variety of microbicidal processes. These processes depend upon either microbicidal oxidants or nonoxidative granule antimicrobial peptides. The latter include defensins, bactericidal/permeability-increasing protein, lactoferrin, lysozyme, and cathelicidin. Most of these compounds have been studied primarily for their ability to kill organisms, although lactoferrin may have both microbistatic and microbicidal effects. Macrophages have an additional antimicrobial mechanism whereby they can use production of nitric oxide to inhibit growth of ingested fungal pathogens. Neutrophils also appear to have significant growth inhibitory activity in their cytoplasm. These cells, as do the keratinocytes themselves, produce large amounts of calprotectin, a calcium- and zinc-binding protein that has potent microbistatic activity against *C. albicans*, the dermatophytes, and other fungi.

The Cutaneous Immune System

Since cutaneous fungal infections are more frequent and more severe in patients with immunologic defects, immune responses to fungal antigens would seem to play an important role in the host defense against these infections. Immunologic host defense mechanisms in normal hosts seem to be effective even when the infections are limited to superficial locations, such as the stratum corneum. A number of studies suggest that the epidermis not only represents a passive barrier against entry of infecting organisms, but also acts as an immunologic organ with some unique elements. A hypothesis regarding the skin-associated lymphoid tissue

(SALT) has been advanced wherein the skin acts as immune surveillance unit. A variety of cell types are believed to have involvement in this cutaneous immune system, including Langerhans cells, dermal dendritic cells, epidermal T lymphocytes, keratinocytes, and microvascular endothelial cells. The mechanisms employed are complex, involving a network of fixed or mobile cells interacting either by the trafficking of the cells themselves from one site to another or by the production of cytokines that influence the function of other cells. Such skin initiated immune responses act against a broad spectrum of foreign antigens including contact allergens, tumors, and transplants, and it is likely that they are also active against the fungal pathogens of interest here. Therefore, this system is probably responsible for initiating immune responses that work to eliminate the infecting organisms in the immune host. In addition, such responses may also produce some of the inflammation that results in much of the symptomatology of these infections.

DESCRIPTION OF THE DISEASES

Superficial Fungal Infections

These infections are limited to the most superficial layers of the epidermis and/or its keratinized appendages, such as the hair and nails. The most common pathogens causing superficial mycoses are listed in Table 1.

Cutaneous candidiasis

Cutaneous candidiasis is an infection of the skin that is generally caused by the yeast-like fungus *C. albicans* and which can be either acute or chronic in nature. *C. albicans* is part of the normal flora of the gastrointestinal tract, rather than that of the skin, although it can be found on the skin on occasion. This organism can grow as either yeast cells or filamentous forms, with mixtures of the two phases generally seen in tissue infections. Cutaneous candidiasis on occasion may be caused by other species of this genus, including *C. parapsilosis*, *C. tropicalis*, or *C. glabrata*, but these are unusual.

Acute cutaneous candidiasis may present as intertrigo, producing intense erythema, edema, creamy exudates, and satellite pustules within folds of the skin. Other infections may be more chronic, as in the feet where there can be a thick white layer of infected stratum corneum overlaying the epidermis of the interdigital spaces. Candida paronychia is marked by infection of the periungual skin and the nail itself, resulting in the typical swelling and redness of this type of candidal infection.

Superficial candidiasis is common in otherwise healthy neonates and young infants, and mainly manifests as either oropharyngeal candidiasis (oral thrush) or candidal diaper dermatitis. In the latter case, moist macerated skin in the diaper area appears to be particularly susceptible to candida invasion. The infection usually starts in the perianal region and may spread to involve the perineum and perhaps the lower abdomen and upper thighs. The lesions are typically

TABLE 1 Superficial cutaneous mycoses and the most common responsible pathogens

Type of infection	Pathogens
A. Cutaneous candidiasis	*Candida albicans*
B. Dermatophytosis	*Trichophyton, Microsporum, Epidermophyton*
1. Tinea pedis	*T. rubrum, T. mentagrophytes, E. floccosum*
2. Tinea cruris	*E. floccosum, T. rubrum*
3. Tinea barbae	*T. rubrum, T. verrucosum, T. mentagrophytes*
4. Tinea unguium (onychomycosis)	*T. rubrum, T. mentagrophytes*
5. Tinea capitus	*T. tonsurans, T. schoenleinii, T. rubrum, M. canis, M.audounii*
6. Tinea corporis	*T. rubrum, T. mentagrophytes, M. canis, T. verrucosum, M. gypseum*
C. Skin and nail infections by nondermatophytes	*Scytalidium dimidiatum, Scopularis brevicaulis, Fusarium oxysporum*
D. Tinea versicolor	*Malassezia* sp.
E. Malassezia folliculitis	*Malassezia* sp.
F. Tinea nigra	*Hortaea werneckii*
G. White piedra	*Trichosporon* sp.
H. Black piedra	*Piedraia hortae*

scaly papules that progress to weeping eroded lesions that often have satellite pustules. Congenital cutaneous candidiasis is a more severe infection of neonates wherein the disease presents within 6 days of birth; there is generally widespread desquamating and/or erosive dermatitis, and a significant risk of systemic and sometimes fatal candida infection. The affected neonates require systemic antifungal therapy.

In some cases, superficial *C. albicans* infections may be particularly severe, persistent, and recalcitrant to treatment, producing the uncommon disorder known as chronic mucocutaneous candidiasis. This condition consists of persistent and recurrent infections of the mucous membranes, skin, and nails, along with a variety of other manifestations. The superficial infections last for years in affected patients unless they are properly treated, although deep candida infections are very rare in this situation. Oral thrush and candida vaginitis are fairly common in patients with chronic mucocutaneous candidiasis. There is often infection of the esophagus, although further extension into the viscera is unusual. Epidermal neutrophilic microabscesses, which are common in acute cutaneous candidiasis, are rare in the lesions of chronic mucocutaneous candidiasis. The oral lesions are generally tender and painful.

A number of other disorders are associated with the syndrome of chronic mucocutaneous candidiasis, including endocrine dysfunction, vitiligo, dysplasia of the dental enamel, congenital thymic dysplasia, thymomas, and certain other infections. The term autoimmune polyendocrinopathy–candidiasis ectodermal dystrophy (APECED) has been used to describe this type of syndrome, and a particular gene, called the autoimmune regulator (AIRE), has been found to be responsible. Otherwise, chronic mucocutaneous candidiasis no doubt represents a group of related syndromes with a variety of predisposing or secondary abnormalities in host defense function, most commonly deficient cell-mediated immune responses against candida antigens.

The diagnosis of superficial candidiasis is usually suspected on clinical grounds and can be confirmed in skin scrapings by demonstrating the organism using potassium hydroxide preparations and/or culture on appropriate antifungal media. Long-term (3–9 months) treatment with azole antifungal drugs can produce good results in chronic mucocutaneous candidiasis, although occasional failures have occurred due to the development of resistant strains of *C. albicans*. Patients who present with chronic mucocutaneous candidiasis should be evaluated for the presence of infection with the human immunodeficiency virus and if presenting as adults, for the possibility of thymoma.

Dermatophytosis

Dermatophytosis is the infection of keratinized structures, such as the nails, hair shafts, and stratum corneum of the skin, by organisms of three genera of fungi termed dermatophytes. Although they are not part of the normal human skin flora, these organisms are particularly well adapted to infecting this location because they can use keratin as a source of nutrients – unlike most other fungal pathogens. The different types of dermatophytes are often classified according to body site, using the word 'tinea,' followed by a term for the particular body site. The major types of dermatophytosis and the most frequent organisms associated with them are listed in Table 1. The degree of inflammation produced in the lesions appears to depend primarily on the particular organism and perhaps also to some extent on the immunological competence of the patient.

Tinea pedis (athlete's foot) is probably the most common form of dermatophytosis. This condition is a chronic toe web infection that can be scaly, vesicular, or ulcerative in form and which can sometimes produce hyperkeratosis of the sole of the foot. Tinea cruris is an expanding dermatophye infection in the flexural areas of the groin and occurs much more frequently in males than females. Dermatophyte infection of the major surface areas of the body is termed tinea corporis. These infections frequently take the classical annular, or 'ringworm,' shape. Involvement of the beard area in men, a condition known as tinea barbae, is often caused by zoophilic organisms such as *T. verrucosum*. Infection of the hair and skin on the scalp is called tinea capitis and is more common in children than adults. Hair infection by dermatophytic fungi can be classified as either endothrix (with organisms present within the hair shaft) or ectothrix (with most of the organisms outside the hair shaft, forming a sheath of spores around it).

Tinea unguium is a form of onychomycosis, or fungal infection of the nails, that is caused by dermatophytes such as *T. rubrum*. This type of infection may be classified into three types by the way the organisms enter the nails: (a) distal subungual onychomycosis, in which the organisms initially enter the distal nail plate and nail bed; (b) proximal subungual onychomycosis, in which the organisms invade proximally in the cuticle area and then migrate distally; and (c) white superficial onychomycosis, in which the fungi invade the superficial layers of the nail plate directly. Nail infections, particularly of the toenails, are among the most difficult types of dermatophytosis to treat.

Diagnosis of dermatophytosis is made in a similar manner to that of cutaneous candidiasis, with examination of skin scrapings by potassium hydroxide preparations and culture on appropriate fungal media. The treatment of this condition has improved markedly in recent years with the development of new antifungal agents for topical application or oral administration. However, certain kinds of dermatophytosis, including widespread infections and those of hair and nails, will often respond poorly to topical therapy and will require prolonged courses of an oral antifungal agent, such as griseofulvin, ketoconazole, itraconazole, fluconazole, and terbinafine.

There is an opportunistic fungal organism, *Scytalidium dimidiatum* (formerly known as *Hendersonula toruloidea*), that can produce conditions clinically mimicking those caused by the usual dermatophyte species, but which does not respond well to conventional antifungal therapy. In addition, some nondermatophytic fungi such as *Scopulariopsis brevicaulis* and *Fusarium oxysporum* can invade the nails to produce the white superficial onychonmycosis pattern of infection. Whereas dermatophytes rarely invade the deep tissues or produce systemic infections, even in severely immunocompromised patients, *Fusarium* spp. can disseminate to produce severe and often fatal infections in patients with inadequate host defenses.

Tinea (pityriasis) versicolor and malassezia folliculitis

Tinea versicolor is a chronic superficial fungal infection of the skin generally affecting the trunk or the proximal parts of the extremities and caused by several species of *Malassezia*, including *M. globosa*, *M. furfur*, and *M. sympodialis*. These organisms are lipid requiring and will not grow on most laboratory media. The lesions resulting from infection with the *Malassezia* sp. are macules that may coalesce into large, irregular patches characterized by fine scaling, along with hypopigmentation or hyperpigmentation. These infections can persist for years unless treated appropriately. *Malassezia* has also been postulated to play a role in certain other diseases, including atopic dermatitis, seborrheic dermatitis, psoriasis, and reticulate papillomatosis. Malassesia folliculitis is a condition that resembles several other cutaneous infections, including acne vulgaris, the macronodular lesions of disseminated candidiasis in immunosuppressed patients, the candidal papular folliculitis of heroin addicts, and graft-versus-host disease in bone marrow transplant patients. The papules of this condition begin as inflammation of the hair follicles, instead of macules typical of tinea versicolor, and may progress to frank pustules.

In tinea versicolor, potassium hydroxide preparations of skin scrapings reveal the typical grape-like clusters of yeast and tangled hyphae of the causative fungus, yielding the diagnosis of this condition. The organism is not usually cultured because of the requirement for specialized media. Tinea versicolor can be treated topically with lotions or creams containing selenium or sodium thiosulfate, specific antifungal agents, or sulfur–salicylic acid shampoo. Oral azole antifungal drugs can also be used for more difficult cases. Malassezia folliculitis can be treated using topical antifungal agents or an oral azole antifungal agent.

Miscellaneous superficial fungal infections

Tinea nigra is a superficial mycosis of the palms that is most often caused by *Hortaea werneckii* (formerly called *Phaeoannellomyces werneckii or Exophiala werneckii*). The lesions are generally dark colored, nonscaling macules that are asymptomatic, but can be confused with melanomas and perhaps result in unnecessary surgery. Tinea nigra is most often seen in tropical or semitropical areas of Central and South America, Africa, and Asia, although some cases do occur in North America. This condition can be treated effectively with either keratinolytic agents or topical azoles. White piedra is an asymptomatic fungal infection of the hair shafts that is caused by various *Trichosporon* species, including *T. ovoides*, *T. inkin*, *T. asahii* and the like (formerly known as *Trichosporon beigelii*). This infection produces light-colored, soft nodules on the hair shafts and may cause the involved hairs to break. Otherwise, this condition appears to be asymptomatic, although the causative fungi can produce serious infections in immunocompromised patients. Black piedra is similar to white piedra in that it is a nodular, generally asymptomatic, fungal infection of the hair shafts. It is caused by *Piedraia hortae* and most commonly affects the scalp hair. Black and white piedra are generally treated by clipping off the affected hairs.

Cutaneous and Subcutaneous Mycoses

The dermis and subcutaneous tissues can be infected by a variety of fungal agents that are directly implanted into the skin by punctures with sharp objects contaminated by the organisms. Organisms causing the deep cutaneous and subcutaneous mycoses are listed in Table 2.

Sporotrichosis

This condition is generally caused by accidental implantation of the causative fungus *Sporothrix schenkii* into the skin. The lesions most often consist of cutaneous and subcutaneous nodules extending up the limb from the site of inoculation. However, spread may occur through the lymphatics or blood vessels, causing infections of the bones, joints, or other organs. It is also possible to develop lesions in the lungs by inhalation of the fungal elements. The causative organism is a dimorphic fungus that exists as either hyphae or elongated yeast cells.

The most common reservoir of the fungus in nature is on vegetation, although it may also be found in the soil. Rose bushes often have the organism, and inoculation by rose thorn punctures is common. The site of implantation may develop into a papule or pustule and cutaneous nodules may then develop proximally in a linear fashion. If the fungus is inhaled, it may cause a granulomatous pneumonitis that can cavitate and produce a clinical picture similar to tuberculosis. Immunosuppressed patients are more likely to develop disseminated disease.

Demonstration of the characteristic small, cigar-shaped yeast cells is diagnostic but often difficult. Multiple sections may have to be examined. Asteroid bodies are stellate,

TABLE 2 Deep cutaneous and subcutaneous mycoses and the most common responsible pathogens

Type of infection	Pathogens
A. Sporotrichosis	*Sporothrix schenckii*
B. Chromoblastomycosis	*Fonsecaea pedrosoi, F. compacta, Phialophora verrucosa, Cladophialophora carrionii*
C. Eumycotic mycetoma	*Scedosporium* sp., *Madurella mycetomatis, M. grisea, Acremonim* sp., *Leptosphaeria senegalensis, Aspergillus nidulans*
D. Endemic fungal infections	*Blastomyces dermatitidis, Coccidioides immitus, Histoplasma capsulatum, Paracoccidioides braziliensis*
E. Infections with immunosuppression	*Cryptococcus neoformans, Trichosporon* sp., *Aspergillus* sp., *Zygomycetes* (*Rhizopus* sp. and *Mucor* sp.), *Fusarium* sp., *Blastoschizomyces capitatus, Penicillium marneffei*

periodic acid-Schiff (PAS) positive eosinophilic material that surround the organisms. The diagnosis of sporotrichosis is best made by culture of material from the lesions on appropriate fungal media. Isolation of this organism is usually indicative of sporotrichosis in that the fungus is not part of the normal flora of humans. Iodides may be given for cutaneous sporotrichosis, with oral intraconazole or fluconazole being used if these measures fail. For disseminated disease, either amphotericin B or itraconazole is generally effective, although relapse is common.

Chromoblastomycosis

Certain species of the dematiaceous (darkly pigmented) fungi can cause chronic cutaneous and subcutaneous infections. A number of genera can be involved, but *Fonsecaea*, *Phialophora*, and *Cladophialophora* are most common. The dark pigment of these organisms is dihydroxyphenylanine melanin also associated with *Cryptococcus neoformans*. Dematiaceous fungi can also cause mycetoma. Chromoblastomycosis is characterized by the presence of sclerotic (muriform) bodies in the tissues. When yeast-like cells, pseudohyphae, or hyphae of the dematiaceous fungi are present in the tissues, the term 'phaeohyphomycosis' is used. Chromoblastomycosis usually results from implantation of the organisms during local trauma, generally to the feet or legs. Usually, the first lesion is an erythematous papule, followed by scaling and crusting, with eventual development into a warty structure. The pathology is characteristic of a suppurative granuloma, often with overlying pseudoepitheliomatous hyperplasia. The distribution of cases is worldwide, although most come from Central and South America.

The finding of the characteristic cross-walled, pigmented sclerotic bodies is pathognomonic of chromoblastomycosis. However, since those formed by all the relevant dematiaceous fungal species are similar, culture of the infecting organism on fungal media containing cycloheximide and antibiotics is necessary to identify it. Treatment may be difficult in that the organisms may not be sensitive to antifungal agents. Surgery or local heat may be other options in the early stages of the disease.

Mycotic mycetoma

Mycetomas are swellings with draining sinuses and grains. They usually affect the feet, legs, or hands and begin with direct implantation of the causative organisms. The latter are either actinomycetes (actinomycotic mycetoma) or the true fungi (eumycotic mycetoma). About half of the cases of mycetoma are caused by true fungi, including the genera *Madurella*, *Leptosphaeria*, *Scedosporium*, *Acremonium*, plus several others. Initially, pain and discomfort develop at the implantation site, followed weeks or months later by induration, abscess development, granulomas, and draining sinuses. The lesions may extend to bone and cause severe bony destruction. Eumycotic mycetomas are rare in the United States, although those caused by *Scedosporium* sp. do occur there.

Specimens of exudates or biopsy material should be examined by the naked eye for the presence of grains. The latter can be gram-stained and examined microscopically to differentiate actinomyctic from eumycotic mycetoma. The grains should be washed with saline containing antibacterial compounds and then cultured on Sabouraud's dextrose agar containing chloramphenicol and cycloheximide, as well as on media for bacterial and actinomycotic organisms. Identification of the organisms is based on gross colonial morphology, pigmentation, and mechanism of conidiogenesis. Treatment of eumycotic mycetoma is often unsatisfactory because the causative organisms generally show poor sensitivity to available antifungal agents. Amputation of an infected limb or surgical debridement of infected tissue may be necessary. Amphotericin B or antifungal drugs such as terbinafine, itraconazole, or posaconazole can be used if the particular fungal strain is sensitive. If not treated effectively, mycetomas may progress for years and produce marked tissue damage, deformity, and even death.

Systemic Mycoses with Cutaneous Manifestations

A number of deep fungal infections may produce cutaneous lesions as part of a disseminated disease process. In these infections, the portal of entry is usually the lung, with development of a pneumonia and spread to other organs. Dissemination is most likely to occur in patients with compromised host defenses. This process is different from that discussed in the last section wherein direct implantation into the skin or subcutaneous tissues is the usual mode of entry. The organisms causing systemic infections with spread to the skin are listed in Table 2; they include both the endemic fungi and opportunistic fungal organisms that infect immunosuppressed patients.

RECENT DEVELOPMENTS

Our ability to understand and manage cutaneous fungal infections has been enhanced by advances in a number of areas, particularly taxonomy of the causative organisms, cutaneous immunology, and development of new antifungal drugs.

Fungal taxonomy, like that of bacteria, has been greatly improved by DNA sequence analysis. This technique is helping to identify the causative agents of infectious diseases accurately, facilitating diagnosis and treatment. For example, tinea (pityriasis) versicolor was previously thought to be caused by a single organism called *Malassezia furfur* (also known as *Pityrosporum orbiculare*). However, recent studies using both DNA sequence analysis and advanced biochemical tests have shown that there are at least 11 species within this genus. *M. globosa* now appears to be the major agent of tinea versicolor, although *M. furfur* and *M. sympodialis* are also involved in some cases. *Malassezia* sp. also appear to be involved in certain other cutaneous conditions including Malassezia folliculitis, seborrheic dermatitis, neonatal cephalic pustulosis, and possibly some cases of confluent and reticulated papillomatosis and atopic dermatitis. It will be interesting to see if future studies can relate particular Malassezia species to these conditions as well.

Similar changes have also been made in the classification of the genus *Trichosporum*. Fungi from this genus cause white piedra and some serious disseminated infections in immunocompromised patients. Previously there were thought to be only a few species in this genus, and that *T. beigelii* (also known as *T. cutaneum*) was the major cause of human infections. However, the genus has been extensively revised using new morphological, biochemical, and DNA sequence criteria. *T. beigelii* has been replaced by a number of other species, and it has been found that white piedra and disseminated infections in immunocompromised patients are caused by *Trichosporon* species such as *T. ovoides*, *T. inkin*, and *T. asahii*. With respect to the disseminated infections, it will be important to determine if the different *Trichosporon* species produce different clinical syndromes and if they vary in their susceptibility to antifungal therapy.

There have also been recent advances in our understanding of cutaneous immunology, and in some cases this work may help us to explain the increased susceptibility of certain persons to cutaneous fungal infections. For example, chronic mucocutaneous candidiasis appears to be a complex syndrome in which a variety of immunological defects lead to chronic superficial infections with *C. albicans*. Clearly, mutations in the autoimmune regulator gene (AIRE) cause many cases of a particular variety of this condition called autoimmune polyendocrinopathy candidiasis ectodermal dystrophy (APECED). Recently, patients with chronic mucocutaneous candidiasis have also been demonstrated to have a deficiency in IL17 cytokine production, which is the function of a recently described subdivision of T cells (Th17 cells) with important proinflammatory and host defense functions. In addition, a family with mucocutaneous candidiasis was found to have a mutation in the CARD9 (caspase recruitment domain-containing protein 9) gene that was associated with low numbers of Th17 cells in the affected family members. The Th17 pathway appears to be important in protecting against chronic superficial candida infections and it may be that in a significant number of patients with such infections the disease is due to genetic abnormalities of this pathway. The same may be true for other chronic fungal infections of the skin or other sites.

Some of the cutaneous fungal infections have been very difficult to treat because causative organisms have innate resistance to many of the available antifungal drugs. These resistant infections generally involve the organisms causing chromoblastomycosis and mycetoma. In some cases, debridement of infected tissue or amputation of an infected limb may be necessary. Although amphotericin B or itraconazole may have activity against the fungal isolates, the newer triazole posaconazole has been found to have increased efficacy in some cases. Posaconazole's potential for treating these difficult infections is quite encouraging, although long treatment courses still seem to be necessary.

The intensity of study on the immunology and treatment of different cutaneous fungal infections varies greatly, partly because the causative fungi are sometimes found only in restricted geographical areas of the world. With less common cutaneous fungal infections, the challenge will be to apply knowledge gained from other fields to understanding and managing these conditions.

FURTHER READING

Body, B. A. (1996). Cutaneous manifestations of systemic mycoses. *Dermatology Clinics, 14*, 125–135.

Braff, M. H., Bardan, A., Nizet, V., et al. (2005). Cutaneous defense mechanisms by antimicrobial peptides. *Journal of Investigative Dermatology, 125*, 9–13.

Elewski, B. E. (1998). Onychomycosis – Pathogenesis, diagnosis, and management. *Clinical Microbiology Reviews*, *11*, 415–429.

Hackett, C. J. (2003). Innate immune activation as a broad-spectrum biodefense strategy. *Journal of Allergy and Clinical Immunology*, *112*, 686–694.

Larone, D. H. (2002). *Medically important fungi – A guide to identification* (4th ed.). Washington, DC: American Society for Microbiology.

Lilic, D., & Gravenor, I. (2001). Immunology of chronic mucocutaneous candidiasis. *Journal of Clinical Pathology*, *54*, 81–83.

McGinnis, M. R., Sigler, L., & Rinaldi, M. G. (1999). Some medically important fungi and their common synonyms and names of uncertain application. *Clinical Infectious Diseases*, *29*, 728–730.

Morishita, N., & Sei, Y. (2006). Microreview of pityriasis versicolor and *Malassezia* species. *Mycopathologia*, *162*, 373–376.

Nestle, F. O., Di Meglio, P., Qin, J. Z., & Nickoloff, B. J. (2009). Skin immune sentinels in health and disease. *Nature Reviews Immunology*, *9*, 679–691.

Pappas, P. G., Kauffman, C. A., Andes, D., Benjamin, D. K., Jr., Calandra, T. F., Edwards, J. E., Jr., et al. (2009). Practice guidelines for the treatment of candidiasis – 2009 Update by the Infectious Diseases Society of America. *Clinical Infectious Diseases*, *48*, 503–535.

Rinaldi, M. G. (2000). Dermatophytosis – Epidemiological and microbiological update. *Journal of the American Academy of Dermatology*, *43* (Suppl. 5), S120–S124.

Vander Straten, M. R., Hossain, M. A., & Ghannoum, M. A. (2003). Cutaneous infections dermatophytosis, onychomycosis, and tinea versicolor. *Infectious Diseases Clinics of North America*, *17*, 87–112.

Wagner, D. K., & Sohnle, P. G. (1995). Cutaneous defense mechanisms against dermatophytes and yeasts. *Clinical Microbiology Reviews*, *8*, 317–335.

Walsh, T. J., Groll, A., Hiemenz, J., et al. (2004). Infections due to emerging and uncommon medically important fungal pathogens. *Clinical Microbiology and Infection*, *10*(Suppl. 1), 48–66.

Warnock, D. W. (1998). Fungal infections in neutropenia – Current problems and chemotherapeutic control. *Journal of Antimicrobial Chemotherapy*, *41*(Suppl. D), 95–105.

Part II

Protists

13. Amitochondriate Protists (Diplomonads, Parabasalids, Oxymonads) 175
14. Amoebas, Lobose 191
15. Ciliates 213
16. Secretive Ciliates and Putative Asexuality in Microbial Eukaryotes 227
17. Coccolithophores 235
18. The Glass Menagerie: Diatoms for Novel Application in Nanotechnology 249
19. Dinoflagellates 263
20. Dyctiostelium 279
21. Foraminifera 291
22. Euglenozoa 311
23. Protozoan, Intestinal 323
24. Leishmania 335
25. Oomycetes (Water Mold, Plant Pathogenic) 347
26. Picoeukaryotes 355
27. Stramenopiles 373
28. Toxoplasmosis 385
29. Trypanosomes 397
30. Sleeping Sickness 411
31. Secondary Endosymbiosis 427
32. Algal Blooms 435
33. Food Webs, Microbial 451

Chapter 13

Amitochondriate Protists (Diplomonads, Parabasalids, and Oxymonads)

A.G.B. Simpson[1] and I. Čepička[2]
[1]*Dalhousie University, Halifax, Nova Scotia, Canada*
[2]*Charles University in Prague, Prague, Czech Republic*

Chapter Outline

Abbrevations	**177**	Diplomonads	182
Defining Statement	**177**	Parabasalids	183
Introduction	**177**	Oxymonads	185
Amitochondriate Protists and Eukaryote Evolution	**178**	**Hydrogenosomes and Mitosomes**	**185**
Systematics	**179**	**Genetics and Genomics**	**187**
Habitats	**182**	**Important Pathogenic Species**	**188**
Cell Organization	**182**	**Further Reading**	**189**

ABBREVATIONS

PFO Pyruvate:ferredoxin oxidoreductase
TCA Tricarboxylic acid
TEM Transmission electron micrograph

DEFINING STATEMENT

This chapter covers the diversity, systematics, life history, and cell organization of diplomonads, parabasalids, oxymonads, and their relatives. The importance of these amitochondriate protists in broad-scale eukaryote evolution is discussed. The mitochondrion-like organelles and genomes of diplomonads and parabasalids are overviewed as are some important diseases that they cause.

INTRODUCTION

Mitochondria are often considered to be a defining feature of eukaryotic cells; however, it is now clear that many eukaryotes lack canonical mitochondria. These 'amitochondriate' organisms do not form a single taxonomic group, and they are referred to collectively as 'amitochondriate eukaryotes' or 'amitochondriate protists.' They have become adapted to environments where little or no free oxygen is present, and the primary function of canonical mitochondria – oxidative phosphorylation using oxygen as a terminal electron acceptor – is not feasible. Well-studied amitochondriate eukaryotes instead of mitochondria have 'mitochondrion-like organelles' that are bounded by two membranes but that lack characteristic features of mitochondria, most notably those associated with oxidative phosphorylation. These often missing features include cristae (infoldings of the inner bounding membrane), most components of the tricarboxylic acid (TCA) cycle and of the electron transport chain, and the mitochondrial F_0F_1 ATPase (ATP synthase). Some mitochondrion-like organelles also display positive properties that are not found in canonical mitochondria, such as the production of molecular hydrogen. Until recently, it was thought that some amitochondriate eukaryotes lacked any sort of mitochondrion-like organelle, but the number of possible cases is dwindling rapidly.

Amitochondriate organisms are scattered across the eukaryote tree of life, and many have well-established evolutionary relationships with mitochondriate taxa. For example, there are amitochondriate species within the ciliates and the heterolobosean amoebae, while one major subgroup of Amoebozoa (Archamoebae: pelobionts and entamoebae) is entirely amitochondriate. There are also amitochondriate fungi, the most prominent of which are certain chytrids that inhabit the rumen of ruminant animals, and the microsporidia, which are intracellular parasites that

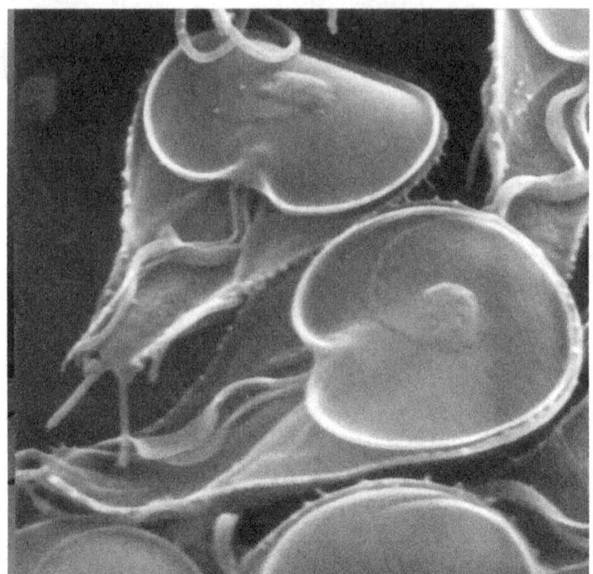

FIGURE 1 Scanning electron micrograph of *Giardia intestinalis*, an amitochondriate eukaryote belonging to the group Diplomonadida and a significant human intestinal parasite. The cells are seen from the ventral face, much of which forms a large, flanged disk that is used for adhesion to the intestinal wall. Several of the flagella (there are eight in total per cell), including two that bear vanes and beat within a ventral channel posterior to the disk, are also visible. *Image courtesy of K. Vickerman.*

were recognized as fungi only relatively recently. The best-known collection of amitochondriate eukaryotes, however, is the assemblage made up of diplomonads, parabasalids, oxymonads, and their relatives. This assemblage, sometimes called 'metamonads,' is the most diverse in terms of cell organization and life history. It is these amitochondriate eukaryotes – diplomonads, parabasalids, and oxymonads – that are the major focus of this chapter.

Diplomonads, parabasalids, and oxymonads collectively contain a broad range of symbiotic and parasitic species. These include the important human parasites *Trichomonas vaginalis* and *Giardia intestinalis* (Figure 1). These organisms are also especially important from an evolutionary point of view. Diplomonads and parabasalids have often been proposed as the earliest diverging branches among living eukaryotes, and thus thought to hold important clues to early events in eukaryote evolution. Although these ideas are now being questioned, these organisms remain among the most unusual types of eukaryotes alive today, and are fascinating subjects for evolutionary cell biologists.

AMITOCHONDRIATE PROTISTS AND EUKARYOTE EVOLUTION

The mitochondrion was derived from a prokaryotic cell, specifically an α-proteobacterium. At some point, perhaps 1.5–2 billion years ago, this bacterium entered into a symbiotic association with a eukaryotic or 'proto-eukaryotic' host cell. The symbiont became a resident of the host cell, and over time, most genes were lost from the symbiont genome. Many of these genes were dispensable in the new environment, and therefore lost entirely, while others were transferred to the host nuclear genome. The protein products of some transferred genes (together with many genes of nonsymbiont origin) are now imported back into the organelle via a complex protein import apparatus that evolved to enable this transport. The evolution of this system cemented a total integration of the erstwhile symbiont into the eukaryotic cell as an organelle.

In the 1980s, it was proposed that some of the living groups of amitochondriate eukaryotes lacked mitochondria because they never had them. It was thought that these taxa might be early branches from the 'main line' of eukaryotic descent that had diverged prior to the acquisition of mitochondria, making them 'primitively amitochondriate.' In groups with a conspicuous mitochondrion-like organelle, this structure was usually interpreted as the result of a different endosymbiosis involving a different prokaryotic symbiont. The general idea of late acquisition of mitochondria seemed to be supported in the late 1980s by molecular phylogenies that placed several groups of amitochondriate protists, including diplomonads and parabasalids (as well as Microsporidia), as the very earliest branches of the eukaryotic tree (Figure 2(a)). This substantially increased interest in the cell biology and molecular biology of amitochondriate protists.

By the end of the last century, however, this 'mitochondria late' hypothesis had been essentially rejected. Several studies of species that were proposed to be primitively amitochondriate showed that these organisms did, in fact, contain genes that seemed to have originated from the genome of the symbiont that had become the mitochondrion. An actual mitochondrion-like organelle has now been found in nearly all of these taxa. In several cases, the products of nucleus-encoded genes of symbiont origin are known to be targeted to mitochondrion-like organelles, further supporting the idea that these organelles do indeed stem from the same endosymbiotic event as canonical mitochondria.

Furthermore, there is now considerable doubt that amitochondriate organisms such as diplomonads and parabasalids are, in fact, early branches on the eukaryotic tree. Under certain circumstances, molecular sequences that have experienced very large amounts of evolutionary change can cluster together artificially in a phylogenetic tree through a phenomenon known as 'long-branch attraction.' Fast-evolving sequences are often drawn toward 'outgroup' sequences, which frequently act as 'long branches.' The fast-evolving sequences may thus appear 'deeper' in the phylogenetic tree than they actually belong. Genes from diplomonads and parabasalids, as well as Microsporidia, appear to be evolving rapidly, in general, and several studies suggest that their deep evolutionary positions could be a long-branch attraction artifact.

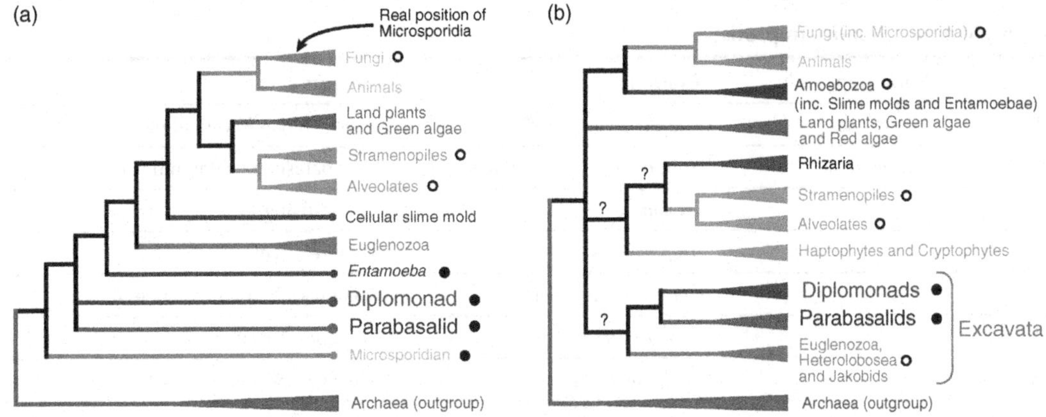

FIGURE 2 Competing views of the evolutionary position of amitochondriate protists, especially diplomonads and parabasalids. (a) Diplomonads and parabasalids as deep-branching eukaryotes. This diagram summarizes a recent phylogeny of 61 ribosomal proteins reported by Morrison, H. G., McArthur, A. G., & Gillin, F. D., *et al.* (2007). Genomic minimalism in the early diverging intestinal parasite *Giardia lamblia*. *Science*, *317*, 1921–1926, (supplementary material), but the recovered tree is remarkably similar to many phylogenies of small subunit ribosomal RNA genes or elongation factor proteins from the late 1980s and early 1990s. Diplomonads and parabasalids form two of the three deepest branches within eukaryotes. Note, however, that the third very deep branch, Microsporidia, is misplaced in this phylogeny, likely due to 'long-branch attraction,' and is placed robustly with Fungi by other data and analyses. The deep placement of diplomonads and parabasalids may be similarly artifactual. (b) Diplomonads and parabasalids as members of Excavata. This diagram summarizes results from several molecular-phylogenetic studies, and other molecular and morphological data. Diplomonads and parabasalids, together with oxymonads, are placed within the supergroup Excavata, along with mitochondriate organisms such as Euglenozoa and Heterolobosea. The monophyly of all Excavata, however, remains somewhat contentious. In both trees, the positions of major 'amitochondriate' groups are shown by black circles. Open circles indicate that only a minority of taxa within a group lack canonical mitochondria (e.g., a few subgroups of ciliates, in the case of alveolates).

An early-branching position for Microsporidia has been clearly disproved, with strong evidence now showing that they are highly derived fungi. The situation is complicated for diplomonads and parabasalids, as there are no strong alternative hypotheses for their broader evolutionary placement (except that they seem to be closely related to each other). Some researchers continue to regard diplomonads and parabasalids as very early-branching eukaryotes. An increasingly popular view, however, is that diplomonads and parabasalids may be members of a larger grouping of eukaryotes called Excavata. Other proposed members of Excavata include Euglenozoa (e.g., *Euglena* and *Trypanosoma*), the heterolobosean amoebae (e.g., *Naegleria*), the oxymonads (which like diplomonads and parabasalids are amitochondriate), and several more obscure groups. The assignment to Excavata would make diplomonads and parabasalids no more 'early branching' than Euglenozoa and Heterolobosea (Figure 2(b)), for example. Nonetheless, they remain some of the strangest organisms in nature, and are still a fascinating group to study for understanding the evolutionary history of life and the range of capabilities of eukaryotic cells.

SYSTEMATICS

Diplomonads (Diplomonadida) are flagellated cells, most of which share the characteristic feature of having two nuclei and two near-identical clusters of flagella per cell (see section 'Cell Organization'). There are two major subgroups of diplomonads: Giardiinae and Hexamitinae. The Giardiinae (*Giardia*, *Octomitus*) are all parasites or commensals. The best known is *G. intestinalis* (synonyms: *G. lamblia*, *G. duodenalis*), which causes a highly prevalent diarrheal disease in humans – giardiasis or 'beaver fever' (see section 'Important Pathogenic Species'). The hexamitines (e.g., *Spironucleus*, *Hexamita*, *Trepomonas*) are more diverse and consist of a mixture of free-living, parasitic, and commensal species (Table 1). Some species (e.g., members of the genera *Enteromonas* and *Trimitus*) have a single nucleus and one flagellar cluster per cell, and look somewhat like half of a typical diplomonad cell. These uninucleate species were traditionally referred to as 'enteromonads' and were considered as ancestral to diplomonads; however, recent molecular phylogenetic studies strongly suggest that 'enteromonads' are actually part of the hexamitine branch within diplomonads.

Parabasalids (Parabasalia or Parabasala) are a diverse assemblage, containing a few hundred described species, almost all of which are parasites, commensals, or beneficial symbionts of animals. There are two basic morphological types of parabasalids. Parabasalids of the first type have a single nucleus associated with a single cluster of a few (usually 4–6) flagella. These are mostly relatively small cells ($>20\ \mu m$), although there are some much larger forms, and they are found in different types of associations with a diversity of animal hosts. There are also a few free-living species. Parabasalids of the second type have dozens to thousands of flagella. These cells are usually large, and

TABLE 1 Example amitochondriate protists

Group	Subgroup	Example	Comments
Diplomonadida	Giardiinae	*Giardia intestinalis*[a]	Prevalent human intestinal parasite
	Hexamitinae	*Spironucleus salmonicida*	Virulent parasite of salmonid fish
		Hexamita	Many free-living
		Trepomonas	Most/all free-living
		Enteromonas hominis	Enteromonad; human commensal (?)
Retortamonadida		*Chilomastix mesnili*	Human commensal
Carpediemonas		*Carpediemonas*	Free-living, marine
Dysnectes		*Dysnectes brevis*	Free-living, marine
Parabasalia	Trichomonadida	*Trichmonas vaginalis*	Prevalent human urogenital parasite
		Trichomonas tenax	Commensal (?) of human oral cavity
		Tritrichomonas suis[b]	Causes spontaneous abortion in cattle
		Histomonas meleagridis	Poultry pathogen ('blackhead')
		Dientamoeba fragilis	Amoeba; human intestinal commensal (?)
	Cristamonadida	*Mixotricha paradoxa*	Termite symbiont; with motility symbionts
		Calonympha	Termite symbiont; with numerous karyomastigonts
	Trichonymphida	*Trichonympha*	Termite/wood-eating roach symbiont; multiflagellated
	Spirotrichonympha	*Spirotrichonympha*	Termite symbiont; multiflagellated
Oxymonadida		*Monocercomonoides*	Small; diverse vertebrate and insect hosts
		Oxymonas	Medium-large; with holdfast and rostellum
		Saccinobacculus	With actively flexing axostyle
		Pyrsonympha	Large, attaches with holdfast
Trimastix		*Trimastix*	Free-living, marine, and freshwater

[a]*Synonyms:* Giardia lamblia, Giardia duodenalis.
[b]*Synonym:* Tritrichomonas fetus.

are found almost exclusively in the hindguts of wood-eating insects, especially termites, where they act (or are suspected to act) as beneficial symbionts. Traditionally, most of the multiflagellated species were referred to as 'hypermastigids' or 'hypermastigotes,' while the simple forms were referred to as 'trichomonads.' Molecular phylogenetic studies confirm, however, that neither hypermastigotes nor trichomonads represent natural monophyletic groups. It is likely that the large multiflagellated forms arose from simpler parabasalid ancestors on several occasions in the evolutionary history of the group.

As of 2009, the high-level classification of parabasalids was based on a combination of morphological characters and molecular trees, and was still not satisfactory. The most widely used contemporary scheme recognizes four major subgroups, namely, Trichonymphida, Spirotrichonymphida, Cristamonadida, and Trichomonadida. The Trichonymphida are all multiflagellated (e.g., *Trichonympha*), as are the Spirotrichonymphida (e.g., *Spirotrichonympha*). Cristamonadida includes often quite large cells with only a few flagella (e.g., *Devescovina*), as well as multiflagellated species, some of which have numerous small clusters of flagella, each usually connected to a nucleus (e.g., *Calonympha*). Finally, Trichomonadida is a diverse collection of simple forms (e.g., *Trichomonas*, *Tritrichomonas*, *Histomonas*). The first three subgroups are, with very few exceptions, symbionts of wood-eating insects. Trichomonadida is technically paraphyletic, as at least two of the other three groups are descended from within it (Figure 3). Trichomonadida contains many parasitic species, including the causative agents of human trichomoniasis (*T. vaginalis*) and of several diseases of domestic animals (see section 'Important Pathogenic Species'). A major reclassification of parabasalids was proposed in 2010 (see Cepicka et al., 2010).

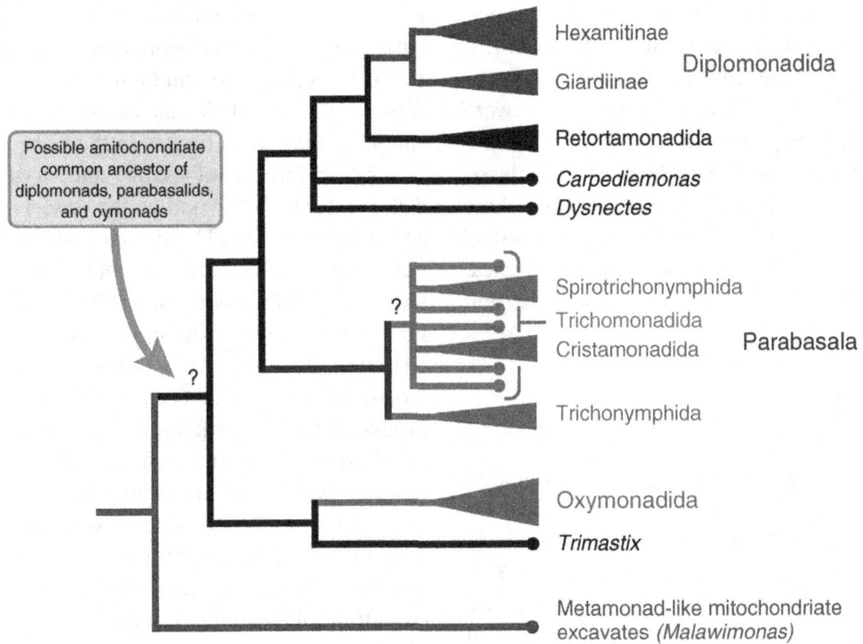

FIGURE 3 Phylogenetic diagram illustrating the evolutionary positions of diplomonads, parabasalids, and oxymonads relative to one another and to related minor taxa of amitochondriate protists. Also shown are the major subtaxa of Diplomonadida and Parabasalia as discussed in the text.

Oxymonads (Oxymonadida) are not immediately related to parabasalids, but share several superficial similarities. They are all commensals or perhaps beneficial symbionts of animals, and they range in morphology from simple species around 10-μm long to elaborate cells up to 200 μm. The moderate-to-large-sized forms, which constitute the bulk of the described species, are found only in wood-eating insects. A few of the largest oxymonads are also multinucleated and multiflagellated, but most species have only one nucleus and four flagella. Examples of oxymonads include the small, simple *Monocercomonoides*, and larger organisms such as *Oxymonas*, *Saccinobaculus*, and *Pyrsonympha*.

The three groups described previously are related to several more obscure lineages of amitochondriate eukaryotes. The immediate relatives of diplomonads are the retortamonads (Retortamonadida), which are flagellates with two or four flagella that are (with one exception) parasites or commensals of animals. Retortamonads are divided into two genera: *Retortamonas*, with two flagella, and *Chilomastix*, with four. Meanwhile, *Carpediemonas* and *Dysnectes* are small free-living cells with two flagella, which have been shown to be related to diplomonads and retortamonads mainly through molecular phylogenies (see Figure 3). A different group of free-living amitochondriate flagellates, *Trimastix*, turns out to be the closest relative of oxymonads. Despite its name, *Trimastix* has four flagella. We will not consider these obscure lineages in detail here.

The evolutionary relationships among these amitochondriate eukaryotes are not completely resolved and are the subject of some debate. There are strong molecular data that demonstrate that parabasalids and diplomonads (plus the obscure relatives of diplomonads) are closely related to each other. Some molecular phylogenies suggest that oxymonads and *Trimastix* represent the immediate sister group of the diplomonad–parabasalid assemblage. If so, it is possible that diplomonads, parabasalids, and oxymonads all descended from a common ancestor that already lacked canonical mitochondria (Figure 3).

Recent phylogenetic analyses based on sequences from >100 genes provide additional support for the hypothesis that diplomonads, parabasalids, and oxymonads branch inside a single large 'amitochondriate' group of organisms – metamonads. This indicates that diplomonads, parabasalids, and oxymonads indeed all descend from a common ancestor that lacked canonical mitochondria. These analyses also provide further support for a monophyletic Excavata supergroup that would include metamonads as a major subgroup. For this and other reasons, it is becoming more accepted that diplomonads (e.g., *Giardia*), parabasalids, and oxymonads are not uniquely early-branching eukaryotes.

Recent studies of suboxic marine samples have uncovered a large number of previously unrecognized major lineages related to diplomonads. Most of these organisms superficially resemble *Carpediemonas* and *Dysnectes*, but are highly distinct at the level of gene sequences. These discoveries confirm that diplomonads, parabasalids, and oxymonads are just three of many major lineages of 'metamonads,' and also suggest that the early ancestors of diplomonads were *Carpediemonas*-like organisms.

The formal taxonomic system of parabasalids that was used until recently was, for the most part, 30 years old. It did not reflect many relationships estimated by molecular-phylogenetic studies, and contained taxa that were clearly paraphyletic or even polyphyletic, the taxon Trichomonadida in particular. In 2010, Parabasalia was reclassified, and the former Trichomonadida was split into three distinct lineages. The new system thus divides parabasalids into six major taxa: Trichonymphea, Spirotrichonymphea, Cristamonadea, Trichomonadea, Tritrichomonadea, and Trichomitea.

HABITATS

Amitochondriate eukaryotes are characteristically found in habitats with little free oxygen. Many do not withstand high oxygen levels, but some at least tolerate or even grow in microoxic conditions and are therefore not strict anaerobes. Many, including *Giardia* and especially *T. vaginalis*, are probably microaerophiles (i.e., achieve optimal growth at low, but nonzero, oxygen levels).

Free-living amitochondriate protists are frequently reported from highly eutrophic (nutrient-enriched) sites where high levels of microbial activity rapidly deplete any free oxygen. Natural environments for these species include freshwater swamps, many marine sediments, and intertidal mudflats. Similar but artificial environments in which amitochondriate eukaryotes are observed include sewage sludges and waste ponds from certain industrial processes such as sugar refining. Amitochondriate eukaryotes also occur in noneutrophic anoxic water masses, such as those that can form at depth in deep fjords due to poor mixing with surface water. Most free-living forms eat prokaryotes.

Many commensal and parasitic amitochondriate eukaryotes live within the intestinal tracts of animals. *Giardia* in humans is one of the many examples. Some parasites, however, have other sites of infection. The parabasalid *T. vaginalis*, for example, infects the human urogenital system, while a related commensal species, *Trichomonas tenax*, lives in the human mouth. Some species of *Spironucleus* that infect fish can cause systemic infections that spread to numerous tissues in the host.

Certain large parabasalids and oxymonads are beneficial symbionts of wood-eating insects, namely, termites and certain roaches. These symbionts live in the insect hindgut, together with a diverse biota of prokaryotes, as well as with other amitochondriate protists such as small parabasalids, in some cases. The larger species produce cellulases, enzymes that degrade cellulose, a major component of woody plant material. These cellulases, together with those produced by the prokaryotic symbionts, are a significant supplement to the endogenous cellulases produced by the termite, which are most important before the hindgut. The larger parabasalids and oxymonads in particular are able to engulf large particles of wood by phagocytosis, and partially digest them. Under the generally anaerobic conditions of the hindgut, some end products of their metabolism (e.g., acetate and other organic acids) can then be absorbed and utilized by the insect for aerobic energy production. Other products of their metabolism are substrates for various prokaryotes, which in turn release products useable by the insect.

CELL ORGANIZATION

Diplomonads

Diplomonads are small cells, usually around 10 μm in length and often less. With the exception of the enteromonad organisms (see section 'Systematics'), diplomonads have a characteristic 'doubled' organization. Each cell has two identical-sized nuclei located alongside each other in the anterior half of the cell (Figure 4). Associated with each nucleus is a flagellar apparatus consisting of a single cluster of flagella and a series of 'microtubular roots' formed by parallel microtubules that originate in

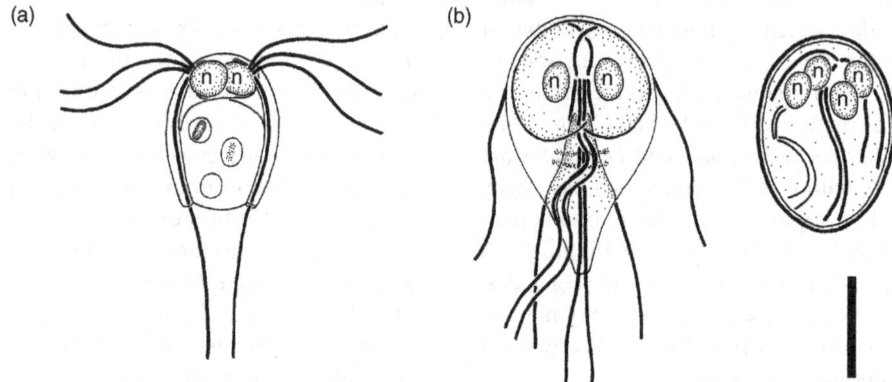

FIGURE 4 Representative diplomonads. (a) *Hexamita* (Hexamitinae). (b) *Giardia* (Giardiinae) in trophozoite form (seen from the ventral side) and in cyst form. *n*, nucleus (note that the *Giardia* cyst contains four nuclei). Scale bar = 5 μm for all images.

association with the basal bodies of the flagella and extend into the cytoplasm in different directions. Each flagellar cluster and associated cytoskeleton is sometimes referred to as a 'mastigont,' and the complex of the mastigont and its associated nucleus is referred to as a 'karyomastigont.' There are usually four flagella per cluster, one of which is directed posteriorly, another of which is directed more or less laterally, and the final two may be directed anywhere from laterally to posteriorly (Figure 4). There are three basic microtubular roots recognized in diplomonads: one that travels posteriorly, usually alongside the posteriorly directed flagellum, and two that pass over the top of and underneath the nucleus, although the last of these roots is absent in Giardiinae (see below). The microtubular roots determine much of the basic shape of the cell. In all species except *Giardia* spp. (see below), the two flagellar apparatuses are more or less identical to each other, and the cell has a rotational symmetry.

The other conspicuous cell component in many diplomonads is the endoplasmic reticulum, which often forms stacks within the cell. Hexamitine diplomonads may also contain food vacuoles, which tend to lie in the posterior half of the cell. A classical stacked Golgi apparatus is not observed by electron microscopy, although individual vesicles appear in encysting *Giardia* cells (see below), and these are involved in Golgi-like posttranslational modifications of cyst wall proteins. To date, mitochondrion-like organelles have been observed in only a couple of species, and these organelles are always tiny compared with canonical mitochondria. In *Giardia*, where the mitochondrion-like organelles are ~ 100 nm in size and are definitively identified as 'mitosomes' (see section 'Hydrogenosomes and Mitosomes'), they are generally located in the center of the cell, posterior to the nuclei, and between the flagellar apparatuses. Often, there are conspicuous glycogen granules in the cytoplasm of diplomonads.

As in most flagellate eukaryotes, each flagellar apparatus duplicates prior to mitosis, and the flagellar apparatuses serve as the microtubular organizing centers for the mitotic spindle. Mitosis in diplomonads is 'semi open,' meaning that the nuclear envelope remains largely intact, but the spindle microtubules penetrate through holes in the envelope to attach directly to the chromosomes. In diplomonads, of course, there are two nuclei, and each must undergo mitosis before actual cytokinesis (cell division). Like many unicellular eukaryotes, some diplomonad species form cysts that are resistant to chemical attack and/or desiccation (Figure 4(b)). In some parasitic forms (e.g., *Giardia*), the cyst is the infectious form that is transmitted between hosts.

In hexamitine diplomonads, the two flagellar apparatuses are located apart from each other, usually toward the 'outside' of the cell. The nuclei, or at least parts of each nucleus, are located between the two flagellar clusters (Figure 4(a)). Each posterior flagellum is either associated with an open groove on the cell surface or is enclosed by a tube that opens near the posterior end of the cell. The grooves or tubes are feeding structures where material (e.g., prokaryotic prey) is phagocytosed.

In Giardiinae, the two flagellar apparatuses are located close to each other, more or less in the center of the cell, and between the two nuclei (Figure 4(b)). The proximal part of each flagellum is intracytoplasmic, that is, it runs some distance through the cytoplasm before emerging as a distinct structure surrounded by a flagellar membrane (this arrangement is also seen in some hexamitines, but to a much more limited extent). In particular, one flagellum from each flagellar apparatus runs down the central axis of the cell and emerges only at the extreme posterior end. The other flagella emerge at various locations around the cell. There are no distinct feeding-related structures, and the cells do not perform phagocytosis.

The best-known member of Giardiinae is *Giardia*. *Giardia* is unusual among diplomonads because the cell has a superficial bilateral symmetry, rather than rotational symmetry. The 'ventral' side of the cell is flattened and the anterior half of the ventral face is shaped into a large flanged disk that the cell uses to attach to the intestine wall of its host (Figures 1 and 4(b)). Two of the flagella (one from each flagellar apparatus) lie within a wedge-shaped channel that extends posterior to the disk. These flagella bear vanes (Figure 1) and beat within the channel during attachment. The exact method by which this parasite attaches to its host is unclear and may involve several mechanisms. It is often assumed that the disk acts like a suction cup. Alternatively, or in complement, contraction within the cytoskeleton that supports the disk may allow the rim of the disk to mechanically grip the intestine wall. The disk flange also appears to be capable of actual adhesion to surfaces.

Parabasalids

As discussed earlier, there is a wide range of cell complexity in parabasalids. The smaller, simpler cells include *Trichomonas* and the other species currently classified in the taxon Trichomonadida (Figure 5(a)). These cells almost always have a single nucleus which is located near the anterior end of the cell. The nucleus is associated with a flagellar apparatus, usually in the form of a single cluster of 4–6 flagella and a cytoskeletal system (i.e., the cell has a single karyomastigont). One flagellum is directed posteriorly, while the others are directed anteriorly or laterally. In many species, the posteriorly directed flagellum is unusual in that the flagellar membrane is physically connected to the cell membrane for some of its length. The cell membrane is extended as a flap or broad projection along this zone of attachment (in some cases, the attached flagellum is also expanded). This creates a fin-like region that moves with the flagellum as it beats, and is referred to as an 'undulating membrane.'

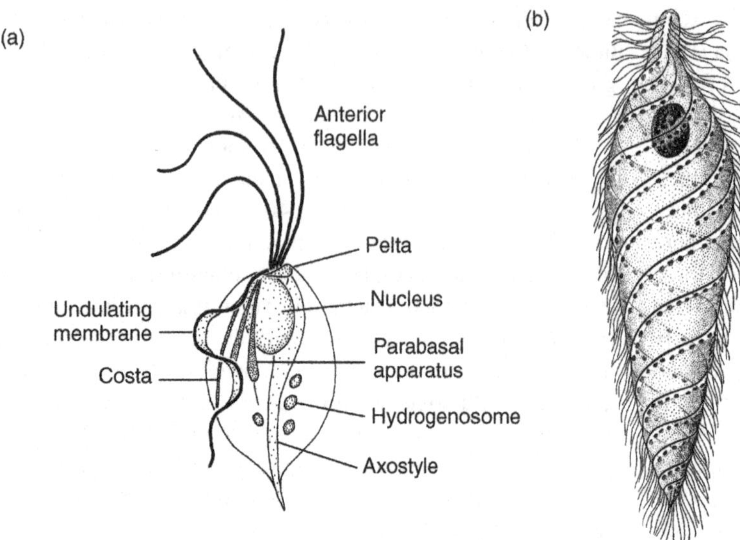

FIGURE 5 Representative parabasalids. (a) *Trichomonas vaginalis* (Trichomonadida). (b) *Spirotrichonympha* (Spirotrichonymphida), a moderately large multiflagellated parabasalid. Scale bar = 5 μm for (a) and 20 μm for (b). (b) After Duboscq, O., & Grassé, P.-P. (1933). L'appareil parabasal des Flagellés, avec des remarques sur le traphosponge, l'appareil Golgi, les mitochondries et le vacuome. *Archives de Zoologie Expérimentale et Générale*, *73*, 381–621.

There are two main microtubular structures within the parabasalid cell – the pelta and the axostyle – which are both associated with the flagellar basal bodies and the nucleus. The pelta is a sheet of spaced microtubules that curves over the anterior side of the nucleus and gives structure to the anterior end of the cell. The axostyle is a column of microtubules that forms the central supportive axis of the cell, and usually extends from the posterior end of the cell into a point. There are also conspicuous nonmicrotubular elements to the cytoskeleton such as the striated 'parabasal fibers' that extend posteriorly from the flagellar apparatus into the cytoplasm. These parabasal fibers are associated with the Golgi apparatus which is conspicuous and shows the classical 'stacked' organization of membrane sacs. The fibers and the Golgi collectively form the 'parabasal apparatus,' which is characteristic for the group. Some species also have a 'costal fiber' or 'costa,' a striated fiber that runs under the cell membrane, in association with the undulating membrane. The costal fibers vary in appearance, and in some cases, they may be contractile. There are a few species (e.g., *Histomonas meleagridis*, *Dientamoeba fragilis*) in which the flagellar apparatus and cytoskeleton are reduced or completely absent; these species are highly variable in shape and usually behave as amoebae.

Simple parabasalids usually phagocytose host cellular debris, prokaryotic cells, and even entire host cells in some cases (*T. vaginalis*). Thus, in addition to endoplasmic reticulum, many cells also contain food vacuoles. The parabasalid mitochondrion-like organelles are referred to as hydrogenosomes (see section 'Hydrogenosomes and Mitosomes'). They are roughly the same size as mitochondria, are quite numerous, and are usually found loosely associated with the axostyle and/or costal fiber. Granules of glycogen also preferentially accumulate in association with the axostyle. Curiously, the formation of cysts seems to be uncommon in parabasalids, though some small forms do produce cysts with a distinct extracellular wall. *T. vaginalis* and some other species form somewhat resistant 'pseudocysts,' which are compact, but lack a cyst wall.

Large parabasalids are quite different in general appearance. In some species, each cell has many individual flagellar apparatus–cytoskeleton complexes, each usually attached to a nucleus (i.e., the cell has dozens to thousands of karyomastigonts). The karyomastigonts may be 'bundled together' by their axostyles. In most species, however, there is a single large nucleus. The numerous flagella are then either organized in a single field near the anterior end of the cell (some cristamonadids) or formed into two or more plates (trichonymphids) or rows (spirotrichonymphids) that originate at the anterior end of the cell, and run down most of its length, often in spirals (Figure 5(b)). The parabasal apparatuses may be aligned to these rows. Cells generally phagocytose objects such as wood particles at their posterior end. Many of the large multiflagellated species are more than 100 μm in length, and some reach 0.5 mm. The complement of cell organelles is similar to that of simple parabasalids, although undulating membranes are absent, and the form of the pelta and axostyle can be modified. The parabasal apparatus may have numerous branches and takes many complex forms.

In parabasalid mitosis, the nuclear envelope does not break down, and the mitotic spindle is formed external to the nucleus, with noncentriolar regions called 'attractophores' serving as the microtubular organizing centers for the spindle. The form of the spindle is distinctive. The attractophores begin mitosis next to each other on one side of the nucleus. Some of the spindle microtubules attach to the chromosomes indirectly through the intact nuclear envelope, while others form a solid rod directly between the attractophores. It is this rod that pushes apart the two attractophores, thereby separating the attached chromosomes.

Some larger parabasalids have substantial numbers of prokaryotic ectosymbionts. One remarkable example is the unusual termite-inhabiting parabasalid *Mixotricha paradoxa* (Cristomonadida), which has only four flagella, yet is very large – up to 0.5 mm. Most of the cell surface is covered by several species of epibiotic bacteria, including spirochetes that attach to distinct docking sites on the surface of the parabasalid cell. The spirochetes are motility symbionts, and *Mixotricha* is propelled through its environment by the locomotory action of these attached spirochetes rather than by its own flagella. Prokaryote–eukaryote motility symbiosis is also seen in another cristamonad, *Caduceia*.

Oxymonads

Oxymonads generally have a single nucleus located in the anterior portion of the cell. This nucleus is associated with four flagella arranged as two separated pairs connected by a broad sheet-like microtubular root, called the 'preaxostyle.' The preaxostyle is also the site of origin of a two-dimensional array of linked microtubules, called the axostyle (Figure 6), that runs down the longitudinal axis of the cell. This structure is only superficially similar to the axostyle of parabasalids and is likely to have evolved independently. In the oxymonad *Saccinobaculus*, the axostyle is highly motile: active sliding of microtubules within the axostyle causes the structure to flex dramatically, and as a result, the entire cell performs a squirming motion.

In many species, the extreme anterior end of the cell forms a microfibrillar structure called a holdfast. This is used to attach the cell to the gut wall of its host. In some groups, this holdfast is located at the end of a long columnar structure supported by microtubules, called the rostellum (Figure 6(b)). The oxymonad cell contains conspicuous endoplasmic reticulum, while a discrete Golgi apparatus is not observed. Mitochondrion-like organelles have not yet been identified with certainty in oxymonads, but possible candidate structures have been observed in at least one species. Most oxymonads can feed by phagocytosis, and cells of smaller species typically ingest prokaryotes, while larger species may ingest wood particles, as per large parabasalids. Many species have prokaryotic symbionts either as endosymbionts within the cytoplasm or as ectosymbionts that attach to the cell surface, as in some large parabasalids. Some oxymonads are known to produce cysts which are presumably used in transmission to new hosts.

The smallest oxymonads are less than 10-μm long, while the largest uninucleate forms can be up to 200 μm in length, and may have eight flagella rather than the usual four (*Pyrsonympha*). A few equally large species have numerous nuclei and flagella. In these cases, each nucleus within the cell is connected to a cluster of four flagella, in a manner analogous to some cristamonadid parabasalids (see section 'Parabasalids').

HYDROGENOSOMES AND MITOSOMES

There are two basic types of mitochondrion-like organelles recognized at present: hydrogenosomes and mitosomes (Figure 7). Hydrogenosomes were first identified in parabasalids, where they seem to be universal. Similar organelles are also found in some amitochondriate eukaryotes not considered in detail in this chapter. These include anaerobic/microaerophilic ciliates, some rumen-dwelling fungi, and some heteroloboseans.

The hydrogenosomes of parabasalids are about the same size as mitochondria and, like mitochondria, are redox organelles that generate energy for the cell. However, they lack

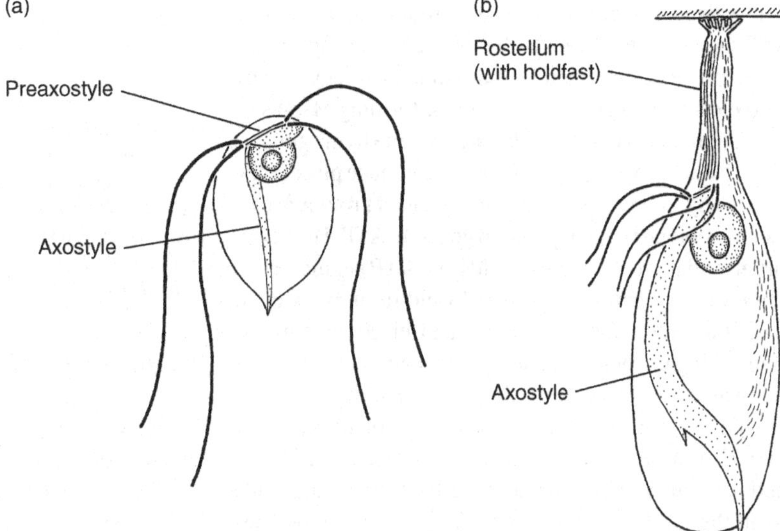

FIGURE 6 Representative oxymonads. (a) *Monocercomonoides*, a small free-swimming form. (b) *Oxymonas*, an attached form. Scale bar = 5 μm for (a) and 10 μm for (b).

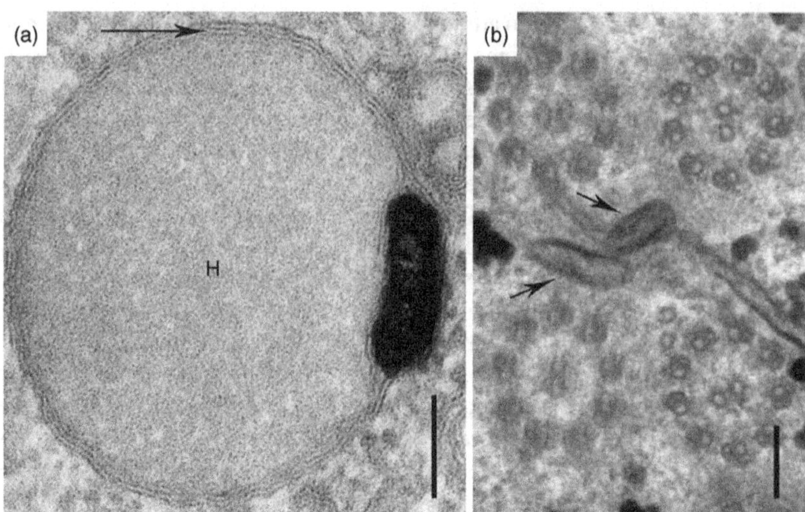

FIGURE 7 Hydrogenosomes and mitosomes. (a) Transmission electron micrograph (TEM) of a hydrogensome of the trichomonad parabasalid *Tritrichomonas suis.* Note the presence of two closely adpressed bounding membranes (see arrow), but the lack of cristae. (b) TEM showing two mitosomes of *Giardia intestinalis* (arrows), located in the center of the cell, between the intracytoplasmic portions of some of the flagella. Scale bars = 100 nm. *(a) Image courtesy of M. Benchimol, from Benchimol, M. (2008). Structure of the hydrogenosome. In J. Tachezy (Ed.)* Hydrogenosomes and mitosomes: Mitochondria of anaerobic eukaryotes. *Berlin: Springer. Copyright 2008 Springer Berlin Heidelberg, reproduced with kind permission of Springer Science and Business Media. (b) Image reprinted with Tovar, J., Leon-Avila, G., Sanchez, L. B.,* et al. *(2003) Mitochondrial remnant organelles of* Giardia *function in iron–sulphur protein maturation.* Nature, 426, *172–175. Copyright 2003, with permission from Macmillan Publishers Ltd: Nature.*

most components of the mitochondrial electron transport chain, they do not have a functioning TCA cycle, and they do not use oxygen as an electron acceptor when generating energy. They do, however, contain several enzymes that are absent in canonical mitochondria, including pyruvate:ferredoxin oxidoreductase (PFO) and [FeFe] hydrogenase.

The core of hydrogenosomal metabolism is depicted in Figure 8. Like mitochondria, hydrogenosomes can utilize pyruvate ($C_3H_4O_3$) as a primary substrate. Within hydrogenosomes, PFO catalyzes the decarboxylation and oxidation of pyruvate, forming acetyl-CoA and carbon dioxide. The electrons lost from pyruvate are accepted by the carrier ferredoxin. CoA is recycled via a succinate-to-succinyl-CoA cycle, releasing acetate as a waste product and yielding one ATP molecule by direct (substrate-level) phosphorylation in the process. Oxidized ferredoxin is then recycled by donating electrons to H^+ ions (protons), forming H_2 gas. It is this reaction that is catalyzed by the enzyme hydrogenase.

All the classical hydrogenosomal reactions take place in the matrix of the hydrogenosome, and movement of protons across the inner membrane is not required to generate ATP. However, this partial oxidation of pyruvate yields one ATP per molecule, many times lower than the theoretical yield from pyruvate in mitochondria with a full TCA cycle and electron transport chain, and using oxygen as the terminal electron acceptor.

The waste products of hydrogenosomal metabolism, acetate and H_2, can be utilized by some prokaryotes. For example, H_2 is a substrate for many methanogenic Archaea, and methanogens and hydrogenosome-bearing parabasalids coexist in the hindguts of termites. Other hydrogenosome-

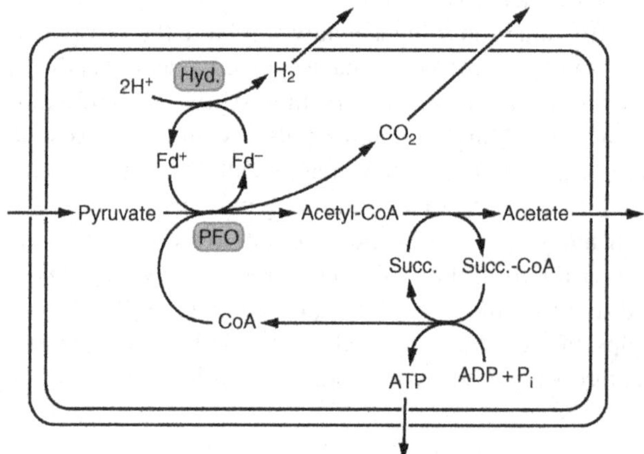

FIGURE 8 Diagram showing the core metabolic pathway of the parabasalid hydrogenosome. Pyruvate is partially oxidized, yielding acetate and carbon dioxide. Electrons are ultimately accepted by protons (rather than by oxygen, as is typical of mitochondria), yielding hydrogen gas. 1 ATP is generated per molecule of pyruvate processed. PFO, pyruvate:ferredoxin oxidoreductase; Hyd., hydrogenase; Fd, ferredoxin; Succ., succinate; Succ.-CoA, succinyl-CoA. *Redrawn from several sources, including Hrdy* et al. *(2008), Metabolism of trichomonad hydrogenosomes. In Tachezy J (ed.)* Hydrogenosomes and Mitosomes: Mitochondria of Anaerobic Eukaryotes. *Berlin: Springer. Copyright 2008 Springer Berlin Heidelberg, reproduced with kind permission of Springer Science and Business Media.*

bearing eukaryotes (e.g., some anaerobic ciliates) harbor endosymbiotic methanogens that reside within the eukaryote's cytoplasm, and these prokaryotes are usually situated immediately alongside the hydrogenosomes.

The second kind of mitochondrion-like organelle is the 'mitosome.' Mitosomes are small and have no known role in energy generation, even under anaerobic conditions. The extremely small (~100 nm) mitochondrion-like organelles of *Giardia* (Figure 7(b)) are classed as mitosomes, and it is generally assumed that mitosomes are present in other diplomonads. Organelles classed as mitosomes are also found in Microsporidia and *Entamoeba*, neither of which is closely related to each other or to diplomonads and parabasalids (see Figure 2).

At present, relatively little is known about the function of mitosomes. Canonical mitochondria perform several important functions in eukaryotic cells that are not directly related to energy generation, and it is possible that amitochondriate protists could rely on their mitochondrion-like organelles for one or more of these other functions. One such process that is performed by canonical mitochondria is the synthesis of iron–sulfur clusters which are components of a number of enzymes in eukaryotic cells. Enzymes from the iron–sulfur cluster synthesis pathway have in fact been localized to the mitosome of *Giardia*, and their activity has been demonstrated in vitro. This indicates that this function is indeed retained by this highly reduced mitochondrial organelle. Iron–sulfur cluster synthesis also seems to be a function of *Trichomonas* hydrogenosomes.

Canonical mitochondria still contain a small genome that is a remnant of the genome of the original α-proteobacterial symbiont. This genome encodes, at a minimum, some ribosomal RNAs and a few proteins, in particular, certain components of the electron transport chain. By contrast, both parabasalid hydrogenosomes and *Giardia* mitosomes have completely lost their genomes, and presumably, their translation machinery. Thus, all of the proteins that function in these hydrogenosomes or mitosomes must be encoded by the nuclear genome and imported to the organelle posttranslationally.

Recent studies have bolstered the evidence that diplomonad mitosomes are indeed highly modified mitochondria. In particular, an outer membrane translocase that is distinctive for mitochondria, Tom40, was found to be encoded by the *Giardia* genome, and can be localized to mitosomes. Research continues into the functions of mitosomes and other types of mitochondrion-related organelles. So far, however, iron–cluster biosynthesis remains the only positively identified function of diplomonad mitosomes.

GENETICS AND GENOMICS

Our knowledge of the genome organization of amitochondriate eukaryotes comes primarily from the draft genome sequences of the parabasalid *Trichomonas vaginalis* and the diplomonad *G. intestinalis* strain WB, both of which were published in 2007. The genome of *Giardia* is small and compact for a eukaryote. At ~12 Mb in size and with ~5000–6500 protein-coding genes, it is similar in size and coding capacity to the genome of the brewer's yeast, *Saccharomyces cerevisiae*. The intergenic regions in *Giardia* are short, and adjacent genes sometimes even overlap. This compactness may be representative of diplomonads in general, since a similar organization was found in genomic surveys of the hexamitine diplomonad *Spironucleus salmonicida*. The *Trichomonas vaginalis* genome is much larger (~160 Mb), with at least 25 000 genes interspersed with numerous repetitive elements, and there is evidence of recent and dramatic genome expansion. Interestingly, spliceosomal introns are very sparse in both diplomonads and parabasalids; there are only ~65 in *Trichomonas* and just four identified in *Giardia*. Both the genomes seem to have been shaped significantly by lateral gene transfer. The bulk of the transferred genes are metabolic genes that have been acquired from prokaryotes, probably more than 100 in the case of *Giardia*. Most of these are probably stable long-term transfers, as a large fraction of the genes of foreign origin in diplomonads are shared by *Giardia* and *Spironucleus*. A smaller number of these gene transfers can be traced back to the common ancestor of diplomonads and parabasalids, and these shared gene transfers represent some of the best evidence that these two groups are specifically related (see Figure 3).

One issue of particular interest with regard to diplomonads is whether or not their two nuclei are genetically equivalent. Prior to cell division, both nuclei undergo mitosis, and one copy of each of the parent nuclei is passed on to each daughter cell. In other words, each nucleus could, in principle, act as a separate lineage and diverge in sequence from the other over evolutionary time. Nonetheless, most of the evidence to date from *Giardia* suggests that the nuclei are, in fact, equivalent, although karyotypic differences have been recorded in one study. Recently, a process of internuclear genetic transfer was documented in *Giardia* cysts, providing a possible mechanism for homogenization of the nuclear genomes. If diplomonads are truly sexual (see below), this may also act as a homogenization mechanism.

The presence of sexuality in many protists is a matter of long-standing debates. Various lines of evidence indicate that at least some diplomonads, parabasalids, and/or oxymonads may be facultatively sexual. Over the 1940s–1960s, L.R. Cleveland examined and described sexual cycles in several large parabasalids and oxymonads from wood-eating insects. These light-microscopy-based accounts record gamete formation, meiosis, fertilization events, and autogamy, that is, fusion of gametic nuclei within a single cell. However, these phenomena await reexamination with modern microscopic techniques. Recent attention has focused on small parasitic forms. The nuclei of *G. intestinalis* appear to be diploid, but strain WB, from which the genome sequence was derived, is almost completely homozygous. This is not what would be expected from a long history without recombination, over which the homologous chromosomes should diverge from one another in sequence. *Giardia* and *Trichomonas* both have genes that encode proteins required for meiosis, and recent evidence suggests that there was genetic recombination in the ancestry of various contemporary strains of *G. intestinalis*.

The first genome sequence of a diplomonad was from a clone of *G. intestinalis* strain WB, which belongs to 'Assemblage A.' Recently, a draft genome sequence has been determined for a second strain, GS. Strain GS belongs to 'Assemblage B,' the other *G. intestinalis* assemblage that is found in humans. The gene complements of WB and GS are similar (only a few predicted proteins are unique to one or other genome), but sequence divergence between their genomes is quite high, with predicted protein sequences showing <80% identity on average.

These genome comparisons did not provide evidence of recent recombination between Assemblages A and B. However, recent studies comparing multiple loci across larger numbers of isolates have bolstered previous findings of recombination within and between *G. intestinalis* assemblages. It is still unclear whether the recombination that has occurred between assemblages was due to a sexual process (albeit an occasional one), or resulted from other phenomena, such as genetic transfer by viruses. For these and other reasons, it is disputed still whether *G. intestinalis* should be considered as one 'species' or several.

IMPORTANT PATHOGENIC SPECIES

Diplomonads, parabasalids, and their relatives are mostly harmless commensals or beneficial symbionts found in the digestive tracts of both vertebrates and invertebrates. However, some species are pathogenic and cause various diseases of the intestine, urogenital tract, or other internal organs. The following is a brief discussion of some of the diseases caused by these organisms, especially those affecting humans and domestic animals.

G. intestinalis (synonyms: *G. duodenalis*, *G. lamblia*) is a parasite of the small intestine of humans and many animals, and several genetic lineages (assemblages) exist. Two of these lineages, assemblages A and B, are anthropozoonotic; humans may be infected from a wide variety of both domestic and wild animals. In the developing world, there is also extensive person-to-person transmission due to sewage contamination of drinking water, and prevalence is high (~20–30%). The disease caused by *G. intestinalis*, giardiasis, is also called beaver fever or backpacker's diarrhea. Humans contract giardiasis by ingestion of water or food that contains *Giardia* cysts. Occasionally, transmission may be via fecal–oral or sexual routes.

After the passage of four-nucleated cysts through the stomach, the binucleated trophozoites are released. These adhere to enterocytes of the small intestine, thereby blocking nutrient absorption. Following an incubation period of one to a couple of weeks, the main symptom of giardiasis manifests – bloodless and nonfestering slimy diarrhea that may be accompanied by stomach ache, nausea, vomiting, and loss of appetite. As lipids are not effectively absorbed by the infected small intestine, the diarrheic feces are often white and oily, and float in water. The disease usually abates spontaneously in a few weeks, but may persist for years without treatment in some cases (although a proportion of infections are asymptomatic). Giardiasis is treated with 5-nitroimidazole derivates (metronidazole, tinidazole). *G. intestinalis* also causes an intestinal disease of some domestic animals that is similar to human giardiasis. Three nonpathogenic or slightly pathogenic relatives of *Giardia* also live in the human large intestine: *Enteromonas hominis* (an enteromonad species belonging to Hexamitinae), *Chilomastix mesnili*, and *Retortamonas intestinalis* (the latter two are retortamonads).

Spironucleus is a genus of hexamitine diplomonads that includes several species that cause enteritis, skin disease, or systemic infections of fish, birds, and rodents. *S. salmonicida* has caused serious mortality in farmed salmonid fish.

Trichomonas vaginalis is a pathogen of the human urogenital tract and causes human urogenital trichomoniasis, perhaps the most prevalent sexually transmitted disease. There are estimated to be more than 170 million infections worldwide. The incubation period lasts from a few days to a few weeks. The infection is asymptomatic or accompanied by only mild symptoms in most men and approximately half of women, and these people serve as carriers of the disease. Rarely, epididymitis and prostatitis, or even sterility, may develop in some men. The infection is typically more severe in women, and can include inflammation of the vagina and womb accompanied by 'strawberry cervix.' In these cases, the normal vaginal microbiota is perturbed and a fetid discharge develops. Although the symptoms usually disappear spontaneously after some time, *Trichomonas* cells persist in the vagina and inflammations may reappear in the future. Acute trichomoniasis during pregnancy may cause premature birth, and in some long-term untreated cases, it has caused sterility. Persons with trichomoniasis are also considerably more susceptible to HIV infection due to inflammation of the genital mucosa. Human urogenital trichomoniasis is treated with 5-nitroimidazole derivates (usually metronidazole or ornidazole), and it is necessary to treat both sexual partners even if the symptoms have developed in only one of them. A closely related species, *T. tenax*, lives in the human oral cavity and is usually considered to be nonpathogenic, although its harmlessness has been disputed recently by some workers. Two trichomonad parabasalids live in the human large intestine. Of these, *Pentatrichomonas hominis* is probably a nonpathogenic commensal, whereas the aflagellated, amoeboid *D. fragilis* may be involved in some intestinal disorders.

Tritrichomonas suis, more commonly known by its synonym *Tritrichomonas fetus*, lives in the nasal cavity of pigs and the urogenital tract of cattle. Although harmless to pigs, *T. suis* is highly pathogenic to cattle and causes bovine urogenital trichomoniasis. The epidemiological role of pigs has not been elucidated yet, and transmission between pigs is probably fecal–oral. In cattle, however, *T. suis* is transmitted sexually. In infected bulls, the disease has only mild symptoms, and they serve as lifelong carriers of the parasite. In cows, *T. suis* causes serious inflammation of the vagina and womb, and the infected cow may abort and/or become sterile. Bovine urogenital trichomoniasis was effectively treated with metronidazole; however, the application of metronidazole to domestic animals has been prohibited in several countries, including the United States and European Union countries. As a result, the disease is now virtually untreatable and the infected animals have to be destroyed. Artificial insemination is an effective prevention.

Histomonas meleagridis and *Trichomonas gallinae* are pathogens of poultry. The uniflagellated *H. meleagridis* lives in the large intestine of chickens and turkeys and is transmitted between birds by the eggs of the intestinal nematode *Heterakis gallinarum*. *T. gallinae* occurs in the beak and crop of pigeons and is transmitted by water or by predation (to birds of prey). Under some circumstances, these parasites are capable of invading the intestinal mucosa (*H. meleagridis*) or oral mucosa (*T. gallinae*) and may then be transported via the blood to the internal organs. The resulting diseases are then often lethal for the hosts.

FURTHER READING

Ankarklev, J., Jerlström-Hultqvist, J., Ringqvist, E., Troell, K., & Svärd, S. G. (2010). Behind the smile: Cell biology and disease mechanisms of *Giardia* species. *Nature Reviews Microbiology*, *8*, 413–422.

Cacciò, S. M., & Sprong, H. (2010). *Giardia duodenalis*: Genetic recombination and its implications for taxonomy and molecular epidemiology. *Experimental Parasitology*, *124*, 107–112.

Carlton, J. M., Hirt, R. P., Silva, J. C., *et al.* (2007). Draft genome sequence of the sexually transmitted pathogen *Trichomonas vaginalis*. *Science*, *315*, 207–212.

Cepicka, I., Hampl, V., & Kulda, J. (2010). Critical taxonomic revision of parabasalids with description of one new genus and three new species. *Protist*, *161*, 400–433.

Embley, T. M., & Martin, W. (2006). Eukaryotic evolution, changes and challenges. *Nature*, *440*, 623–630.

Franzén, O., Jerlström-Hultqvist, J., Castro, E., *et al.* (2009). Draft genome sequencing of *Giardia intestinalis* assemblage B isolate GS: Is human giardiasis caused by two different species? *PLoS pathogens*, *5*, e1000560.

Hampl, V., Hug, L., Leigh, J., Dacks, J. B., Lang, B. F., Simpson, A. G. B., & Roger, A. J. (2009). Phylogenomic analyses support the monophyly of Excavata and robustly resolve relationships among eukaryotic "supergroups" *Proceedings of the National Academy of Sciences of the United States of America*, *106*, 3859–3864.

Kolisko, M., Silberman, J. D., Cepicka, I., Yubuki, N., Takishita, K., Yabuki, A., *et al.* (2010). A wide diversity of previously undetected relatives of diplomonads isolated from marine/saline habitats. *Environmental Microbiology*, *12*, 2700–2710.

Kreier, J. P. (Ed.), (1978). *Parasitic protozoa*. Academic Press: New York; Vol. 2.

Lasek-Nesselquist, E., Welch, D. M., Thompson, R. C., Steuart, R. F., & Sogin, M. L. (2009). Genetic exchange within and between assemblages of *Giardia duodenalis*. *Journal of Eukaryotic Microbiology*, *56*, 504–518.

Lee, J. J., Leedale, G. F., & Bradbury, P. (Eds.), (2002). *An illustrated guide to the protozoa*. (2nd ed.). Allen Press: Lawrence.

Morrison, H. G., McArthur, A. G., Gillin, F. D., *et al.* (2007). Genomic minimalism in the early diverging intestinal parasite *Giardia lamblia*. *Science*, *317*, 1921–1926.

Simpson, A. G. B., Inagaki, Y., & Roger, A. J. (2006). Comprehensive multigene phylogenies of excavate protists reveal the evolutionary positions of 'primitive' eukaryotes. *Molecular Biology and Evolution*, *23*, 615–625.

Tachezy, J. (Ed.), (2008). *Hydrogenosomes and mitosomes: Mitochondria of anaerobic eukaryotes*. Springer: Berlin.

Tovar, J., Leon-Avila, G., Sanchez, L. B., *et al.* (2003). Mitochondrial remnant organelles of *Giardia* function in iron–sulphur protein maturation. *Nature*, *426*, 172–175.

Chapter 14

Amoebas, Lobose

A. Smirnov
St. Petersburg State University, St. Petersburg, Russia

Chapter Outline

Abbreviation	191	Morphotypes of Gymnamoebae	200
Defining Statement	191	Diversity	200
Introduction	191	Amoeboid Movement	207
Some Noteworthy Dates	192	Biology and Ecology	209
Systematics and Phylogeny	192	Importance	210
Morphology	193	Further Reading	211

ABBREVIATION

MTOC Microtubule-organizing center

DEFINING STATEMENT

Naked lobose amoebae (gymnamoebae) are characterized by wide, smooth, nonanastomosing cytoplasmic projections (lobopodia), driven by an actomyosin cytoskeleton. There are 206 described species and there is little evidence so far of sexual reproduction. They are abundant in all types of habitats and play an important role in freshwater, marine, and soil ecosystems.

INTRODUCTION

Long ago, all amoeboid organisms were considered as members of a large taxonomic group (Rhizopoda), which, among other organisms, included lobose amoebae (Lobosea) and filose amoebae (Filosea). Within each group, there were naked forms and testate forms grouped into different taxa (e.g., Gymnamoebia vs. Testacealobosia). Nowadays, rhizopods are no more taxonomic group; its members are dispersed between a number of supergroups of eukaryotes.

The naked lobose amoebae (as well as testate ones) are members of the eukaryotic supergroup Amoebozoa, which also includes the cellular and acellular slime molds (Dictyostelia and Myxogastria, respectively) testate lobose amoebae pelobionts, entamoebids and a number of flagellated protists (*Phalansterium* and *Multicilia*). These amoebae are only very distantly related to the naked and testate filose which are now placed in the Rhizaria. Both groups are even more distantly related to the heterolobosean amoebae, which are now placed in the tentative eukaryote supergroup Excavata. The naked lobose amoebae (also known as gymnamoebae, the Latinized form of the same name) are among the best-known lobosean amoebas, which are one of the least studied groups of protists.

As organisms without any differentiated locomotive organelles and with variable body shape, amoebae have long been considered as the simplest, most ancient eukaryotic cell type. Only recently has it become clear that different types of amoeboid cell organization (lobosean, rhizarian, and heterolobosean) have arisen independently in various phylogenetic groups. The switch to an amoeboid organization requires complex specialization of the entire cell and, especially, of its cytoskeleton. Amoeboid cells have a dynamic shape and few stable morphological characters, which makes them very hard to identify. Thus it is not surprising that we still count only slightly more than 200 recognizable amoebae species, which is 10–100 times less than the number of species of ciliates or flagellates. Nonetheless, amoebae are everywhere. They are among the most widely distributed eukaryotes, and it is hard to imagine any ecotope containing no lobose amoeba. Their abundance may reach millions of individuals per cubical centimeter of soil or sediment, and they are among the primary consumers of bacteria in the majority of soil, freshwater, and marine habitats. Some amoebae species are important pathogens of invertebrate and vertebrate hosts, including humans. It is now clear that these organisms, which were long out of the mainstream of protistological studies, are among the most interesting groups of protists with many unique adaptations and interesting morphological and molecular characteristics.

SOME NOTEWORTHY DATES

1755: The first amoeba species observed and documented by Rösel von Rosenhorf in his paper entitled *Der kleine Proteus*

1766: A species, now known as *Amoeba proteus*, discovered by P.S. Pallas (as *Volvox proteus*); properly described by Leidy in 1878

1767: The first genus of naked amoebae still valid today (*Chaos*) is established by C. Linnaeus

1822: The genus *Amoeba* established by Bory de St. Vincent (as *Amiba*)

1854: Naked amoeba are separated from testate amoebae by M. Schulze

1879: 'Freshwater Rhizopods of North America' published by J. Leidy, the first comprehensive regional survey of amoebae diversity

1902: *Faune rhizopodique du bassin du Leman* – fundamental monograph by E. Penard, with description of 33 species; the total number of properly described species after his monograph is about 59

1905: The first volume of *The British Freshwater Rhizopoda and Heliozoa*. A series of monographs in five volumes by James Cash, John Hopkinson, and George Wailes; last volume in 1921. First comprehensive study of British rhizopods

1914: Probably the first instance of microphotography application to document a lobose amoeboid organism, *Gephyramoeba delicatula*, by A. Goodey

1926: *Taxonomy of the amebas with description of thirty-nine new marine and freshwater species* – the monograph by A.A. Schaeffer. First recognition of 'locomotive form' as a primary criterion for amoebae description. Number of 'known' amoebae species is given as around 200, of which many are probably synonyms

1926: The first comprehensive model of amoeboid movement by S.O. Mast, titled *Plasmasol–plasmagel interconversion theory*

1953: E. Chatton attempts to classify amoebae based on their nuclear division patterns. This was probably the first attempt to use a single nonmorphological character for classification of amoebae

1956: Probably the first application of electron microscopy to study amoebae, by G.D. Pappas

1961: *Frontal zone contraction* model of amoeboid movement by R.D. Allen

1965: System of amoeboid protists, based on the mechanism of locomotion by T. Jahn and E. Bovee. This was revolutionary because it reunited naked and testate amoebae. Despite many shortcomings, it appears to be the closest to the modern phylogenetic system

1974: The first comparative electron-microscopic study of lobose amoebae, by C.J. Flickinger

1979: *Generalized cortical contraction* model of amoeboid movement by A. Grebecki, subsequently developed in his further publications

1985: Morphological system of amoebae by E. Bovee, the last classification system based solely on light microscopy and pseudopodial pattern

1987: Morphological system of amoebae by F.C. Page – still the most comprehensive classification of lobose amoebae based on combined light- and electron-microscopic data. The number of valid species is about 176

2004: First attempt to create a congruent molecular and morphological systematics of amoebae, by T. Cavalier-Smith, E. Chao, and T. Oats

SYSTEMATICS AND PHYLOGENY

Naked lobose amoebae are among the most difficult protists to differentiate. Because they are believed to be agamous (clonal) organisms, the biological species concept, which involves defining species based on their reproductive isolation, is not applicable. The general consensus is that for such taxa the morphospecies concept is the only one practically available. However, analysis of the morphological differences between amoeboid protists is rather difficult, and conclusions are often unreliable, especially for closely related species. This is partly because the shape of an amoeba is dynamic; in stained preparations after fixation and dehydration specimens are often no longer representative. So, there is no way to preserve a type specimen of an amoeba – a holotype, so important in traditional biological systematics. Many amoebae species are culturable, and therefore type strains can be deposited in culture collections. However, this practice became widely used only after the 1960s and there are still many examples where strains deposited with the culture collections were lost. So, until the advent of microphotography, the only tools to document amoebae species were line drawings and text descriptions, both of which tended to be rather author-specific. For example, despite careful descriptions provided by E. Penard in his fundamental monograph published in 1902, and a large number of stained preparations left by him, many of his 'species' are now unrecognizable.

Recognition of the importance of locomotive morphology for adequate characterization of amoebae by A.A. Schaeffer in 1926 and careful attention to the conformations of moving amoeba, combined with the morphology of uroidal structures noted earlier (by G.C. Wallich in 1863) for distinction of gymnamoebae, allowed more reliable species description. Even so, for example, only 13 species of the 39 described in Schaeffer's fundamental monograph from 1926 are recognized today. The use of microphotography for amoebae identification improved the situation only slightly, until it became widely distributed in journals in the early 1960s. However, the problem of adequate species descriptions persisted.

Thus it is not surprising that attempts to construct a morphological system of amoebae have long resulted in variable, often changing classification schemes. However, even with the relatively poor light microscopy of the early twentieth century it was possible to outline groups of similar species or genera. This is the reason why the genus has always been, and still remains, the most solid taxon in amoebae classification. But it was never really clear how to combine genera into higher taxa because it was very hard to weigh morphological characters and to establish shared features. Attempts were made to use single but very fundamental characters, such as nuclear division patterns, for grouping amoebae into higher taxa (E. Chatton and N.B. Singh), but these were never widely accepted. Likewise, attempts to use biochemical characters for classification of naked amoebae, like the taxonomic serology of the genus *Amoeba* by C.T. Friz developed during 1970–1990, also did not result in a practical system except for the identification of *Acanthamoeba* isolates.

An approach to the classification of amoeboid protists, based on the analysis of their mechanisms of locomotion, was developed by T. Jahn and E. Bovee in the 1960s. This proved very successful for the construction of a higher-level classification of amoebae, although these authors included too many dubious species and genera to create a practical system. Nonetheless, some of their basic ideas were later confirmed by molecular phylogeny. Attempts to construct a solely light-microscopic system of amoebae were probably finalized by E. Bovee in 1985, but this system was soon replaced by a combined light- and electron-microscopic one by F.C. Page, in 1987.

The development of electron microscopy led to the discovery of differences in the cell coatings of gymnamoebae, in the nuclear lamina, and other characteristics that helped investigators to delimit amoebae genera. Most of these ultrastructurally delimited genera are now confirmed by molecular studies. However, electron microscopy did little for species resolution because the ultrastructure of amoebae in the same genus is rather uniform. Similarly, electron microscopy did not reveal any shared characters that could help to clarify the composition of amoebae families and higher taxa, except for the shape of mitochondrial cristae. This latter character supported the separation of lobose amoebae with smooth cytoplasmic flow from those with eruptive flow of the cytoplasm; Page and Blanton in 1985 used this criterium and helped to establish the amoeboid class Heterolobosea (now placed as a member of the tentative eukaryotic supergroup Excavata). As a result, the morphological system of amoebae is, in fact, a classification that is convenient, comprehensive, and logical but does not pretend to be a phylogenetic (evolutionary) reconstruction. Nonetheless, this is still the system used in all available keys and guides to gymnamoebae, so it remains popular and widely distributed (Table 1).

As with so many groups of microorganisms, molecular phylogeny revolutionized the evolutionary systematics of amoebae, particularly at higher taxon levels. Molecular studies indicate that all naked lobose amoeba belong to the supergroup Amoebozoa Cavalier-Smith 1998, but do not form a monophyletic clade within it, being distributed between several ribogroups. One remarkable finding was that certain testate lobose amoebae are very closely related to naked lobose amoebae, as previously suggested by T. Jahn and E. Bovee based on the analyses of the mechanisms of amoeboid movement. However, molecular phylogeny of amoebae is still limited, being based mostly on the sequences of a single gene (nuclear-encoded small subunit ribosomal RNA), and includes only a fraction of known species. Therefore, attempts to construct a practical, congruent molecular and morphological system of gymnamoebae are still in progress. This is further complicated by the fact that many amoebae are known only from ancient descriptions and type cultures are no longer available, if they ever were in the first place. Hence, these taxa cannot be properly placed in the molecular tree. Table 2 provides a current version of the modern combined molecular and morphological systems, where many genera are placed on the basis of only the morphological evidence or left *incertae sedis*, is provided here.

MORPHOLOGY

A stationary or slowly moving amoeba may acquire different conformations, but an active, continuously moving cell – the locomotive form – adopts a dynamically stable shape. This is the morphology used as the basis for the description of the amoeba. When an amoeba detaches from the substratum and starts to float (i.e., to move passively with the water currents), it is called the floating form. This floating morphology is often very specific and helpful in genus and species level distinction.

The leading edge of a locomotive amoeba consists of an optically transparent region of cytoplasm referred to as a hyaloplasm, which can take the form of a frontal hyaline area, anterolateral hyaline crescent or anterior hyaline cap (Figure 1(a), 1(c), and 1(e), respectively). The rest of the cytoplasm is filled with various granules, crystals, and other inclusions and is referred to as the granuloplasm (Figure 1). Note, the terms 'hyaloplasm' and 'granuloplasm' are descriptive terms and are not entirely equivalent to the widely used terms 'ectoplasm' and 'endoplasm,' which refer,

TABLE 1 Morphological system of naked lobose amoebae by F.C. Page, updated with later described genera

Class Lobosea Carpenter 1861	
Subclass Gymnamoebia Haeckel 1866	
Order Euamoebida Lepşi 1960	
Family Amoebidae (Ehrenberg 1838) Page 1987	
	Genera: *Amoeba, Chaos, Polychaos, Parachaos, Trichamoeba, Hydramoeba*, and *Deuteramoeba*
Family Thecamoebidae (Schaeffer 1926), (Smirnov and Goodkov 1994)	
	Genera: *Thecamoeba, Sappinia, Stenamoeba, Dermamoeba, Paradermamoeba, Pseudothecamoeba, Parvamoeba*, and *Thecochaos*
Family Hartmannellidae (Volkonsky 1931) Page 1974	
	Genera: *Hartmannella, Saccamoeba, Cashia, Glaeseria*, and *Nolandella*
Family Paramoebidae (Poche 1913) Page 1987	
	Genera: *Mayorella, Korotnevella*, and *Paramoeba*
Family Vexilliferidae (Page 1987)	
	Genera: *Vexillifera, Pseudoparamoeba, Neoparamoeba*
Family Vannellidae (Bovee 1970) Page 1987	
	Genera: *Vannella*[a], *Pessonella, Clydonella, Lingulamoeba, Ripella*
Family Pellitidae Smirnov et Kudryavtsev 2005	
	Genus: *Pellita*
Order Acanthopodida Page 1976	
Family Acanthamoebidae Sawyer and Griffin 1975	
	Genera: *Acanthamoeba, Protacanthamoeba*
Order Leptomyxida (Pussard and Pons 1976) Page 1987	
Suborder Rhizoflabellina Page 1987	
Family Flabellulidae (Bovee 1970)	
	Genera: *Flabellula, Paraflabellula*
Family Leptomyxidae (Pussard and Pons, 1976) Page 1987	
	Genera: *Rhizamoeba, Leptomyxa*
Suborder Leptoramosina Page 1987 Family Gephyramoebidae Pussard and Pons 1976	Genus: *Gephyramoeba*
Family Stereomyxidae (Grell 1966)	
	Genera: *Stereomyxa, Corallomyxa*
Insertae sedis	
Balamuthia, Cochliopodium, Comandonia, Echinamoeba, Filamoeba, Flamella, Gocevia, Ovalopodium, Paragocevia, Stygamoeba	

[a]*The genus* Vannella *now includes all members of the former genus* Platyamoeba *– see Smirnov* et al. *(2007).*

TABLE 2 Combined morphological and molecular system of gymnamoebae

Class Tubulinea (Smirnov *et al.*, 2005) (=Lobosea Cavalier-Smith *et al.*, 2004)			
	Order Tubulinida Smirnov *et al.* (2005)		
	Superfamily Amoeboidea Cavalier-Smith *et al.* (2004)		
		Family Amoebidae (Ehrenberg, 1838) Page (1987)	
			Genera: *Amoeba, Chaos, Polychaos, Parachaos, Trichamoeba, Deuteramoeba, Hydramoeba*
		Family Hartmannellidae (Volkonsky, 1931) Page (1974)	
			Genera: *Glaeseria, Cashia, Hartmannella, Saccamoeba, Nolandella*
	Superfamily Echinamoeboidea (Cavalier-Smith *et al.*, 2004) Smirnov *et al.* (2008)		
		Family Echinamoebidae (Page, 1975) Smirnov *et al.* (2008)	
			Genus: *Echinamoeba*
		Family Vermamoebidae Smirnov *et al.* (2008)	
			Genus: *Vermamoeba*
	Order Arcellinida (Kent, 1880) 18 families, not listed		
	Order Leptomyxida (Pussard and Pons, 1976) Page (1987)		
		Family Leptomyxidae (Pussard and Pons, 1976) Page (1987)	
			Genera: *Leptomyxa, Rhizamoeba*
		Family Flabellulidae (Bovee, 1970) Page (1987)	
			Genera: *Flabellula, Paraflabellula*
		Family Gephyramoebidae Pussard and Pons (1976)	
			Genus: *Gephyramoeba*
		Order Copromyxida[a] Cavalier-Smith (1993)	
		Family Copromyxidae Olive and Stoianovich (1975)	
			Genera: *Copromyxa, Copromyxella*
Class Discosea (Cavalier-Smith *et al.*, 2004) Smirnov *et al.* (2008)			

Continued

TABLE 2 Combined morphological and molecular system of gymnamoebae—cont'd

	Subclass Flabellinia (Smirnov *et al.*, 2005) Smirnov *et al.* (2008)		
	Order Dactylopodida Smirnov *et al.* (2005)		
		Family Paramoebidae (Poche, 1913) Smirnov *et al.* (2008)	
			Genera: *Paramoeba, Korotnevella*
		Family Vexilliferidae Page (1987)	
			Genera: *Vexillifera, Neoparamoeba, Pseudoparamoeba*
	Order Vannellida Smirnov *et al.* (2005)		
		Family Vannellidae Bovee (1979)	
			Genera: *Vannella*[b], *Clydonella, Lingulamoeba*, Pessonella, Ripella
	Order Himatismenida Page (1987)		
		Family Cochliopodiidae De Saedeleer (1934)	
			Genus: *Cochliopodium*
	Order Stygamoebida Smirnov *et al.* (2008)		
		Family Stygamoebidae Smirnov *et al.* (2008)	
			Genus: *Stygamoeba*[c]
	Order Trichosida Moebius (1889)		
		Family Trichosidae Moebius (1889)	
			Genus: *Trichosphaerium*
	Order Pellitida Smirnov and Cavalier-Smith (2008)		
		Family Pellitidae Smirnov and Kudryavtsev (2005)	
			Genus: *Pellita*
Subclass Longamoebia Smirnov *et al.* (2008)			
	Order Dermamoebida (Cavalier-Smith *et al.*, 2004) Smirnov *et al.* (2008)		
		Family Mayorellidae (Schaeffer, 1926) Smirnov *et al.* (2008)	
			Genus: *Mayorella*
		Family Dermamoebidae Smirnov *et al.* (2008)	

TABLE 2 Combined morphological and molecular system of gymnamoebae—cont'd

			Genera: *Dermamoeba, Paradermamoeba*
	Order Thecamoebida Smirnov *et al.* (2008)		
		Family Thecamoebidae (Schaeffer, 1926) Smirnov *et al.* (2008)	
			Genera: *Thecamoeba, Sappinia, Stenamoeba, Parvamoeba*
	Order Centramoebida (Rogerson and Patterson, 2002) Cavalier-Smith *et al.* (2004)		
		Family Acanthamoebidae Sawyer and Griffin (1975)	
			Genera: *Acanthamoeba, Protacanthamoeba*
		Family Balamuthiidae Cavalier-Smith *et al.* (2004)	
			Genus *Balamuthia*
		Discosea incertae sedis: *Pseudothecamoeba, Thecochaos, Gocevia, Paragocevia Janickia*	
Class Variosea (Cavalier-Smith *et al.*, 2004) Smirnov *et al.* (2008)			
	Order Varipodida Cavalier-Smith *et al.* (2004)		
		Family Filamoebidae Cavalier-Smith *et al.* (2004)	
			Genera: *Filamoeba, Flamella*
		Family Acramoebidae Smirnov *et al.* (2008)	
			Genus: *Acramoeba*
	Order Phalansteriida Hibberd (1983)		
		Family Phalansteriidae Kent (1880/1881)	
			Genus: *Phalansterium*
	Order Holomastigida (Lauterborn, 1895) Cavalier-Smith (1997)		
		Family Multiciliidae Poche (1913)	
			Genus: *Multicilia*
		Amoebozoa incertae sedis: *Stereomyxa, Corallomyxa*	

[a]*Assignment of these orders to classes needs corroboration.*
[b]*Includes all members of the former genus* Platyamoeba.
[c]*Genus* Vermistella *Moran, Anderson, Dennett, Caron et Gast 2007 perhaps is a junior synonym of the genus* Stygamoeba.

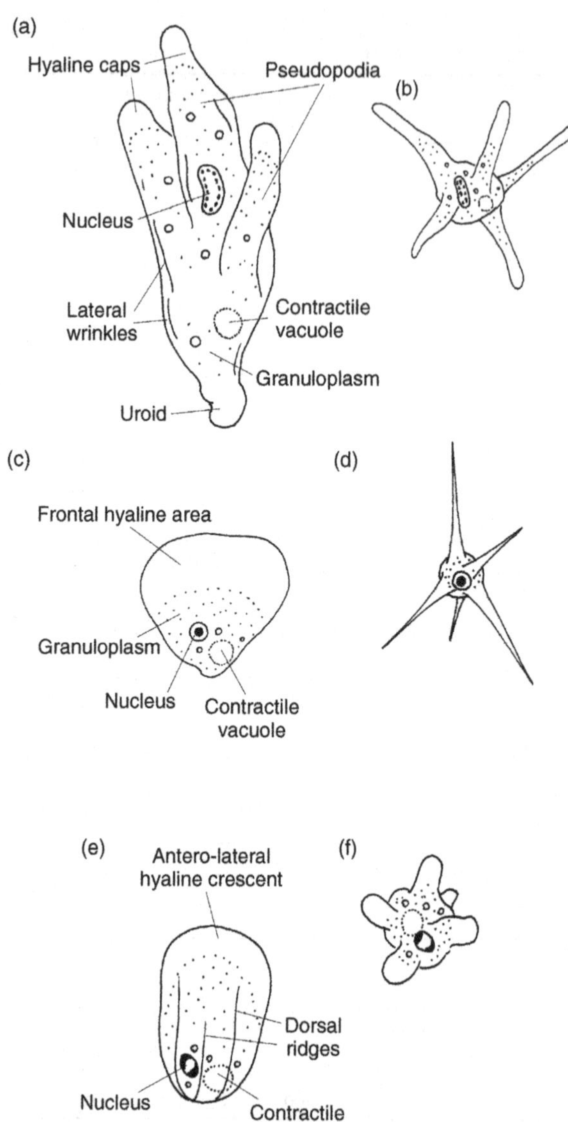

FIGURE 1 Morphology of amoebae. Locomotive and floating forms of *Amoeba proteus* (a, b), *Vannella simplex* (c, d), and *Thecamoeba striata* (e, f), respectively, as shown but not drawn to scale.

respectively, to the 'gel' and 'sol' viscosity states of the cytoplasm used in models of amoeboid movement.

Many lobose amoebae, especially the larger ones, produce lobose pseudopodia (lobopodia). These are defined here as variable cell projections that are smooth in outline with rounded tips, which participate in the relocation of the main cytoplasmic mass of the cell and include both the granuloplasm and the hyaloplasm. The cytoskeleton of lobopodia consists of actomyosin filaments. In fact, however, few amoebae (mostly the species of the family Amoebidae) form discrete pseudopodia and thus are polypodial in locomotion. Small- and medium-sized species usually move 'as a whole,' that is, without forming discrete pseudopodia. If such a cell is oblong, wormlike or avate in locomotion, it is often called 'monopodial' or 'limax-like' (although most of the historical group of 'limax amoebae,' which included the smallest monopodial amoeboid protists, are now assigned to the Heterolobosea).

A more detailed classification of the locomotive forms of amoebae of the family Amoebidae was suggested by A. Grebecki in 1991. In his system, polypodial cells with several actively growing pseudopodia are called polytactic. Cells that are monopodial in locomotion and bear numerous lateral wrinkles are called orthotactic, while monopodial and smooth ones are called monotactic. Floating amoebae in this classification are heterotactic. This classification reflects the physiological state of cells as the same amoeba, for example, *A. proteus*, may be polytactic at the beginning of locomotion, orthotactic during actively directed locomotion, and monotactic in an old, dying culture.

Together with pseudopodia formation or independently of it, most lobose amoebae can produce subpseudopodia. These are hyaline projections of different shape, usually anteriorly directed, which do not take part in the relocation of the main cell mass. The function of these structures remains unclear. Their shape may be very characteristic and these are widely used in systematics (dactylopodia, acanthopodia, and echinopodia) (Figure 2). The posterior formations of an amoeba are called uroidal structures, and these may be of different types (Figure 3). Some amoebae have folds or wrinkles on the dorsal surface of the locomotive form, which is especially characteristic of the genus *Thecamoeba*, where it is a species-specific character.

Most of the amoebae species are mononucleate, but a few are multinucleate, containing from 10 to 15 to several thousands of nuclei. Amoebae nuclei vary in morphology and may be vesicular (single, central, or eccentric nucleolus), granular (many small nucleoli), or they may have a complex nucleolar structure (e.g., *Polychaos annulatum*, *Polychaos fasciculatum*, or *Thecamoeba striata*) (Figure 4). Freshwater species normally have one or several contractile

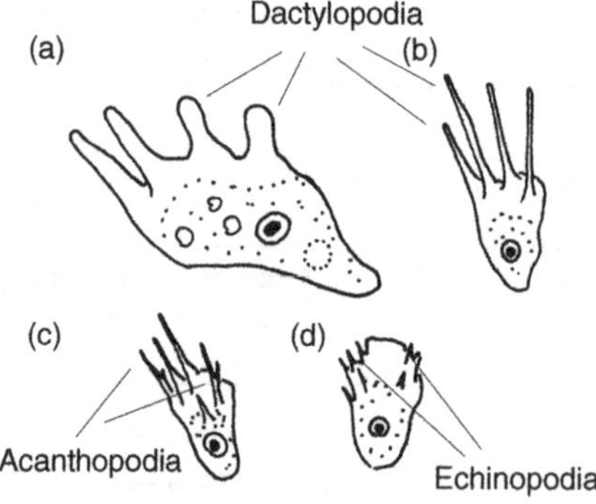

FIGURE 2 Various subpseudopodia types are shown, including the dactylopodia of *Korotnevella* (a); the dactylopodia-like projections of *Vexillifera* (b); the narrow, tapering, and sometimes furcating acanthopodia in *Acanthamoeba* (c); and the short, spineolate echinopodia of *Echinamoeba* (d). Drawings are not to scale.

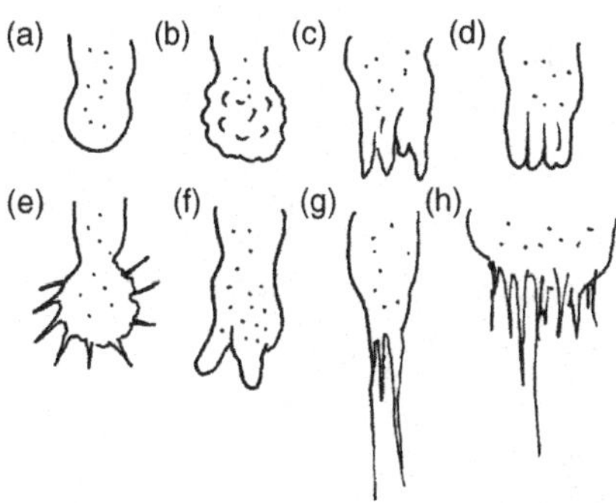

FIGURE 3 The main types of uroidal structures in lobose amoebae. These are (a) bulbous, (b) morulate, (c) fasciculate, (d) plicate, (e) villous-bulbous, (f) posterior hyaline lobes, and (g) and (h) adhesive. Not to scale.

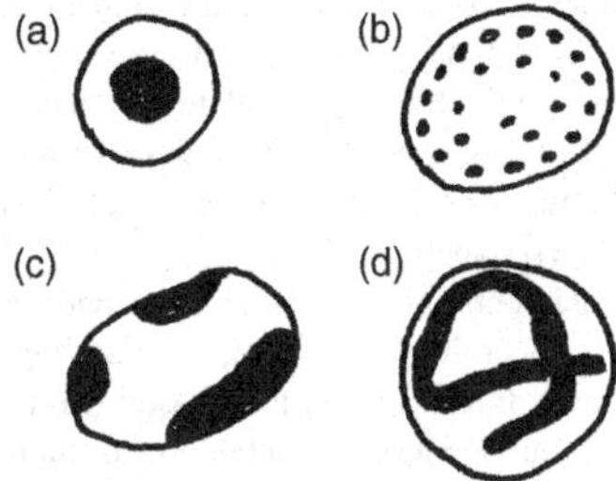

FIGURE 4 Basic types of nuclei in lobose amoebae. (a) Vesicular nucleus with central nucleolus; (b) granular nucleus; (c) nucleus with peripheral arrangement of nucleolar material; and (d) nucleus with complex nucleolar structure. Not to scale.

vacuoles. Marine species lack contractile vacuoles under normal salinity conditions, but may form them when the salinity decreases. The cytoplasm of an amoeba often contains opaque granules, crystals, and other cytoplasmic inclusions, and these may sometimes be characteristic and species-specific.

The cell coat of an amoeba consists of a highly differentiated glycocalyx. This may be amorphous, of different thicknesses (e.g., *Thecamoeba*, *Flabellula*, and *Hartmannella*), or filamentous, in the latter case consisting of a layer of radiating filaments of different thicknesses over a thin amorphous layer (e.g., *Amoeba*, *Chaos*, and *Polychaos*). The glycocalyx may be organized into glycostyles, pentagonal in some species of *Vannella*, hexagonal in *Vexillifera*, and blister-like in *Pseudoparamoeba* (Figure 5). The thick multilayered cell coat of *Mayorella* and *Dermamoeba* is often referred to as a 'cuticle.' Amoebae of the genera *Korotnevella* and *Paramoeba* are covered with scales. Amoebae of the genus *Cochliopodium* are also covered by scales, but unlike the latter two genera, they cover only the dorsal surface of the locomotive cell. This kind of a cell surface structure is called a tectum (Figure 5).

Other characteristic organelle morphologies of amoebae concern the nuclear membrane, mitochondria, and Golgi complex. A characteristic structure observed inside the nuclei of some amoebae species is the nuclear lamina. This can be honeycomb-like, as in *A. proteus*, or filamentous, as in *Saccamoeba limax*. The functional role of this layer is unclear; the honeycomb-like layer probably fulfills some regulatory role in nuclear–cytoplasmic transport because in studied species a single nuclear pore is located at the bottom of every hexagonal cell. The mitochondria of lobose amoebae have tubular cristae, with the exception of the genus *Stygamoeba*. The Golgi complex is organized as a dictosome, that is, with stacks of flattened saccules. Most amoebae species also possess endobionts, which are mostly bacteria. A few species, such as *Mayorella viridis*, have algal symbionts, but their role in amoebae-feeding and biology remains poorly known.

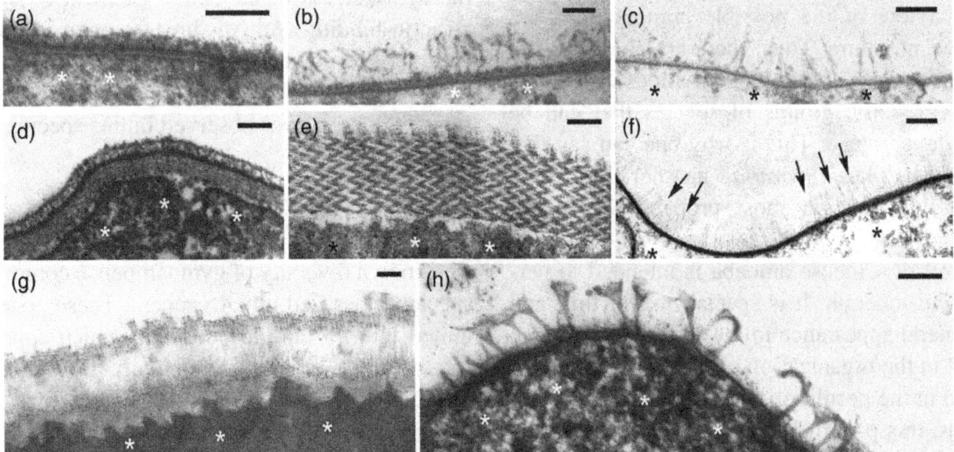

FIGURE 5 Variety of cell surface structures in amoebae. (a) Amorphous glycocalyx of *Thecamoeba*; (b) filamentous cell surface of *Amoeba proteus*; (c) fine filaments in the glycocalyx of *Polychaos annulatum*; (d) cuticle of *Mayorella vespertilioides*; (e) cell coat of *Paradermamoeba valamo*; (f) glycostyles of *Vannella* (arrowed); (g) cell coat of *Pellita catalonica*; and (h) scales of *Korotnevella nivo*. Asterisks mark the cytoplasm of the cell, which is not perfectly fixed in order to achieve the best view of the glycocalyx. Scale bar = 100 nm.

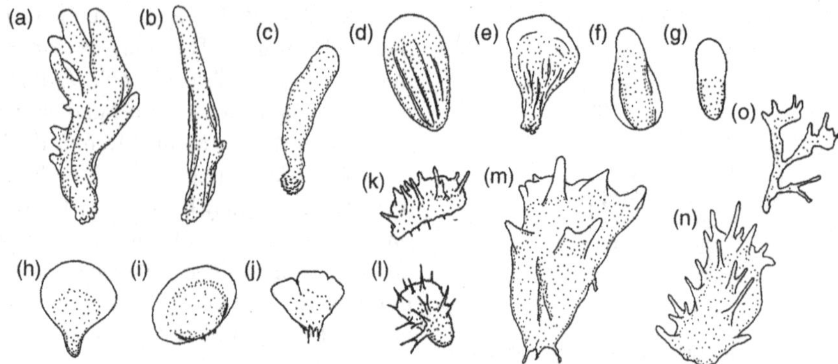

FIGURE 6 Basic morphotypes of gymnamoebae. A single drawing is used to illustrate each morphotype; for more comprehensive plate and text definitions of morphotypes see Smirnov and Brown (2004). (a) Polytactic (polypodial, with several distinctly separated pseudopodia of different size that are formed from the anterior part of the body); (b) orthotactic (body elongated with bell-like cross section, lateral wrinkles always present); (c) monotactic (monopodial, body subcylindrical with circular cross section, no lateral wrinkles); (d) striate (flattened, ovoid, or oblong with regular outline and several nearly parallel dorsal folds); (e) rugose (flattened, ovoid, or oblong, with more or less regular outline, dorsal and/or lateral folds and wrinkles that are usually irregular); (f) lanceolate (flattened, lancet-like, with distinct lateral flatness, no dorsal folds or wrinkles); (g) lingulate (flattened, oblong, or even spineolate, with more or less regular outline, without any dorsal and lateral folds and wrinkles); (h) fan-shaped (flattened, semicircular, fan-shaped, or spatulate, with regular outline; distinct separation into anterior hyaline area and posterior granuloplasmic region; anterior edge entire; no subpseudopodia); (i) lens-like (rounded cells with lens-like profile, its dorsal surface is covered with a rigid envelope); (j) flabellate (flattened, irregularly flabellate, prominent anterior hyaloplasm with uneven frontal edge, no subpseudopodia, trailing adhesive uroidal filaments); (k) flamellian (flattened, irregularly flabellate, prominent anterior hyaloplasm with short narrow subpseudopodia and lobes, trailing adhesive uroidal filaments); (l) acanthopodial (flattened, somewhat irregular in outline, numerous short, slender, flexible, tapering subpseudopodia, sometimes furcating near their base); (m) mayorellian (flattened, irregularly triangular or oblong, anterior hyaline border with a few blunt, conical, hyaline subpseudopodia); (n) dactylopodial (flattened, irregularly triangular or of variable shape, wide anterior hyaline zone with distinct frontal or fronto-lateral hyaline dactylopodia); and (o) branched (flattened and expanded, branching or reticulate with loboreticulopodia). Not to scale.

MORPHOTYPES OF GYMNAMOEBAE

The locomotive form of an amoeba can be characterized by a set of discrete features like the general outlines, profile in cross section, presence of folds and wrinkles, and morphology of the uroidal structures. Taken in combination, these features are sufficient to generate a descriptive statement for all known species. Indeed, the number of theoretically possible combinations of features is immense and a broad range of locomotive forms can be envisaged. However, only a few of the possible combinations are actually observed in nature. Thus, there are characteristic 'patterns of organization,' that is, combinations of features that characterize specific groups of species that can be recognized with experience. This is why one can say that one organism 'looks like *Mayorella*,' another 'looks like *Chaos*,' and a third one is most probably a 'rugose thecamoebian.'

The entire cell of a lobose amoeba is intended to perform a locomotive function. It is specialized for this, and its shape and general appearance follow from this function. This is reflected in the organization of the cytoskeleton, of the cell coat and in the peculiarities of the cell–substratum interaction. Thus, this pattern is a combined characteristic of an amoeba, synthesizing many basic features. These general patterns of the morphodynamic organization of a locomotive form of an amoeba were termed morphotypes by Smirnov and Goodkov in 1999, although a species may belong to more than one morphotype. In particular, many members of the family Amoebidae can adopt two basic morphotypes during locomotion, polytactic and orthotactic, and some species also become monotactic under unfavorable conditions. Figure 6 shows the basic morphotypes of lobose amoebae, which gives an overall impression of the general morphological diversity of these organisms.

There is interesting evidence demonstrating how the morphotype of an amoeba may reflect the organization of the cytoskeleton of the cell. Treatment of *A. proteus* with an actin-binding agent disrupts the organization of the cell's contractile system. The result is a drastic change of amoeba morphotype with the cell adopting a fan-shaped morphotype, which is never observed in this species under normal conditions.

DIVERSITY

The known diversity of gymnamoebae comprises 206 valid species, classified into 46 genera. These genera are further united into 13 families, some of which appear to be very robust, while others are relatively arbitrary and probably artificial. The latter is indicated by the differences in composition of these families between the morphological and the molecular systems (Table 1 vs. Table 2). The following is a brief review of amoebae families using the morphological system by F.C. Page, as this is the most compact and comprehensive system currently available.

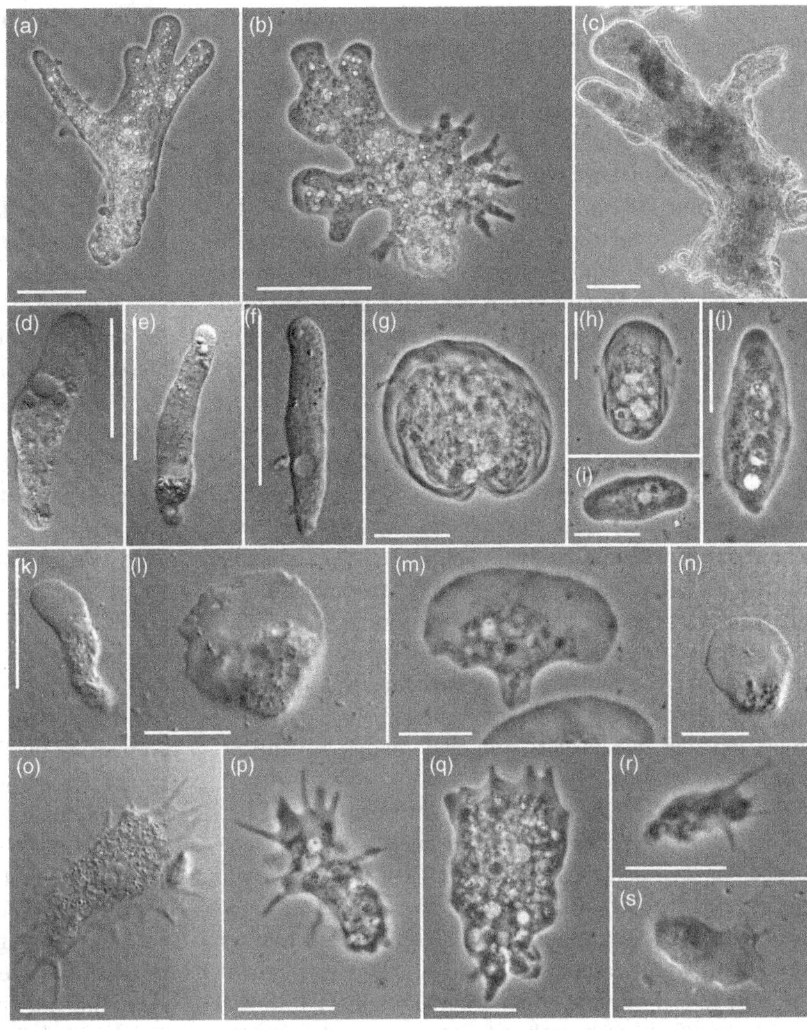

FIGURE 7 Examples of diversity of amoebae. Family Amoebidae includes the following: (a) *Amoeba borokensis* CCAP 1503/7, (b) *Polychaos fasciculatum* CCAP 1564/1, and (c) *Chaos carolinense*, strain form Carolina Biological Supply Company. Family Hartmannellidae includes the following: (d) *Saccamoeba limax* CCAP 1534/6, (e) *Glaeseria mira* CCAP 1531/1, and (f) *Hartmannella canrabrigiensis* CCAP 1534/11. Family Thecamoebidae includes the following: (g) *Thecamoeba similis* CCAP 1583/8, (h) *Thecamoeba striata* CCAP 1583/4, (i) *Dermamoeba algensis* (strain by Dr. A. Goodkov), (j) *Paradermamoeba valamo* (Geneva strain), and (k) *Stenamoeba stenopodia* CCAP 1565/8. Family Vannellidae includes the following: (l) *Vannella calycinucleolus* CCAP 1565/6, (m) *Vannella simplex* (Geneva strain), and (n) *Ripella platypodia* CCAP 1589/2. Family Paramoebidae includes the following: (o) *Paramoeba eilhardi* CCAP 1560/2, (p) *Korotnevella stella* CCAP 1547/6, and (q) *Mayorella cantabrigiensis* CCAP 1547/7. Family Vexilliferidae includes the following: (r) *Vexillifera bacillipedes* CCAP 1590/1 and (s) *Pseudoparamoeba pagei* CCAP 1568/1. Scale bars: 100 μm (a–c) and 25 μm (d–s).

Family Amoebidae: This is a family of amoebae with polytactic, orthotactic, or monotactic morphotypes (7 genera, 20 valid species; Figure 7(a)–7(c)). Nuclei are granular or with different arrangements of the nucleolar material, but never vesicular. All species are mononulceate with the exception of the genera *Chaos* and *Parachaos*, which contain from several up to several thousand nuclei per cell. These amoebae have clear hyaline caps at the tips of their pseudopodia and always contain lots of cytoplastmic crystals, which sometimes are very characteristic (e.g., in *Deuteramoeba mycophaga*). Uroidal structures are fasciculate, morulate, or bulbous. The cell surface is covered with a filamentous or amorphous glycocalyx. A nuclear lamina is present in some species, where it is organized into the honeycomb layer. The genera *Amoeba* and *Chaos* include species with dimorphic mitochondria, where two types of mitochondria exist simultaneously in the same cell – some with electron-dense matrices and others with electron-transparent matrices and a slightly different morphology of their cristae. These are freshwater or soil-inhabiting amoeba, some of which are known to form cysts.

Family Hartmannellidae: These are amoebae with purely a monotactic morphotype and a single vesicular nucleus (5 genera, 15 valid species; Figure 7(d)–7(f)). Cells are wormlike or clavate in locomotion, forming discrete pseudopodia only when changing their direction of movement. A pronounced frontal hyaline cap is seen in most species. These amoebae have bulbous, willous-bulbous, morulate uroidal structures, or posterior hyaline lobes. The glycocalyx is amorphous or consists of tightly packed layers of cup-shaped structures. The nuclear lamina is known in some species, where it represents a layer of filaments lying under the nuclear membrane but not arranged into honeycomb-like structures. The group includes freshwater, soil, and marine amoebae, most of which are known to form cysts.

Family Thecamoebidae: This is a rather diverse family, and molecular data indicate that the family is polyphyletic (8 genera, 22 valid species; Figure 7(g)–7(k)). These are amoebae with a striate, rugose, lingulate, and lanceolate morphotype. Members of the genera *Pseudothecamoeba* and *Thecochaos* are polytactic, but it is not certain that these two genera belong in this family. Differentiated uroidal structures are usually absent but a bulbous, plicate, or morulate uroid may be present in some species. The usually single nucleus is vesicular in most species, but in some the nucleolar material is dispersed into several peripheral fragments or has a more complex structure, while the nuclei of *Pseudothecamoeba* and *Thecochaos* are close to the granular type. Amoebae of the genus *Thecochaos* are multinucleate, but this genus is known only from the stained preparations of E. Penard. Amoebae of the genus *Sappinia* have a pair of closely apposed nuclei, resembling a diplokaryon. For *Sappinia diploidea* binary fusion of cells during encystment is described, but the details of this process as well as the presence of meiosis have never been documented. Nonetheless, it is widely believed to be the only sexual amoeba. The cell coat differs between genera; it can be amorphous (*Thecamoeba*, *Parvamoeba*, and *Stenamoeba*), or have an extra structured layer (*Sappinia*), a thick 'cuticle' (Dermamoeba), or very thick layer of densely packed helical glycostyles (*Paradermamoeba*), or it can be filamentous (*Pseudothecamoeba*). The latter genus is also unique in having a highly vacuolated cytoplasm, surprisingly resembling the structural vacuoles of *Pelomyxa palustris*. The group includes freshwater, soil, and marine organisms, some of which form cysts.

Family Vannellidae: These are amoebae with a fan-shaped morphotype, sometimes semicircular, crescent-shaped, or spatulate, in locomotion (5 genera, 41 valid species; Figure 7(l)–7(n)). There is a single vesicular nucleus in most species, and a few with nucleoli that are peripheral. No differentiated uroidal structures are known. The glycocalyx consists of an amorphous layer followed by a layer of either densely packed hexagonal prismatic structures or pentagonal glycostyles. Until recently, the genus *Vanella* was thought to be distinguished by the presence of pentagonal glycostyles, and the genus *Platyamoeba* by the presence of prismatic structures in the cell coat. However, molecular phylogeny clearly indicates that the cell coat structure cannot be used to distinguish genera in this case and all members of the genus *Platyamoeba* were transferred recently to the genus *Vannella*. These are freshwater, soil, and marine organisms, a few of which are cyst-forming.

Family Paramoebidae: This is a rather diverse family of amoebae with dactylopodial or mayorellian morphotype (3 genera, 17 valid species; Figure 7(o)–7(q)). Molecular data again indicate that the taxon is polyphyletic. Amoebae of this family form subpseudopodia, which are dactylopodia (*Paramoeba* and *Korotnevella*) or short conical pseudopodia that are sometimes accompanied by dactylopodia (*Mayorella*). Most lack differentiated uroidal structures, although the mayorellas may form a plicate or bulbous uroid and temporary longitudinal dorsal folds. There is a single vesicular nucleus, and the cell coat is a multilayered 'cuticle' in *Mayorella* and has scales in *Korotnevella* and *Paramoeba*. Amoebae of the latter genus contain one or several parasomes, a specific DNA-containing self-replicating organelle, which is probably the remnant of a very ancient kinetoplastid symbiont. The group consists of freshwater, soil, and marine organisms, none of which are known to form cysts.

Family Vexilliferidae: This is probably another polyphyletic collection that is split in molecular trees (3 genera, 15 valid species; Figure 7(r) and 7(s)). These amoebae have dactylopodial and flamellian morphotypes, and the locomotive amoebae form (1) long dactylopodia-like projections directed anteriorly (*Vexillifera*), (2) short dactylopodia (*Neoparamoeba*), or (3) short conical pseudopodia (*Pseudoparamoeba*). A bulbous uroid may be found in species of *Neoparamoeba*. There is a single vesicular nucleus, and amoebae of the genus *Neoparamoeba* may contain one or several parasomes. The cell coat is amorphous or has glycostyle-like structures (*Neoparamoeba*), and may consist of short prismatic hexagonal structures (*Vexillifera*) or hexagonal blister-like structures (*Pseudoparamoeba*). This is a group of freshwater, soil, and marine organisms, none of which are known to form cysts.

Family Pellitidae: These are amoebae of flamellian morphotype that lack subpseudopodia (one genus, two valid species; Figure 8(a)). There is a single vesicular nucleus and no differentiated uroidal structures. The cell coat is very thick (up to 800 nm) and consists of densely packed tulip-like glycostyles. For locomotion, these amoebae form short ventral hyaline projections that extend through the cell coat; during phagocytosis amoeba capture prey by liberating part of the cell membrane, 'pushing away' the glycostyles. This may be an adaptation to the presence of a very thick layer of glycostyles that cannot mediate cell adhesion or be embedded in the food vacuoles together with the food object. The group includes freshwater and marine organisms and no cysts are known.

Family Acanthamoebidae: This family includes a number of well-known pathogens of animals and humans, as well as the molecular model organism *Acanthamoeba castellanii* (2 genera, about 24 valid morphospecies). These are amoebae with an acanthopodial morphotype, meaning the presence of short, tapering, sometimes furcating subpseudopodia, referred to as acanthopodia. There are no differentiated uroidal structures, and 'centriole-like bodies' (cytoplasmic microtubule-organizing centers (MTOCs)) are found near the single vesicular nucleus. The cell coat is amorphous, all species are cyst-forming; cysts are double-walled with pores (*Acanthamoeba*) or are simpler with single walls (*Protacanthamoeba*). Species are known from freshwater, soil, and marine habitats.

Family Flabellulidae: These amoebae have a flabellate or flamellian morphotype (two genera, eight valid species; Figure 8(b)). In rapid locomotion they may become temporarily monotactic. Characteristic adhesive uroids are found

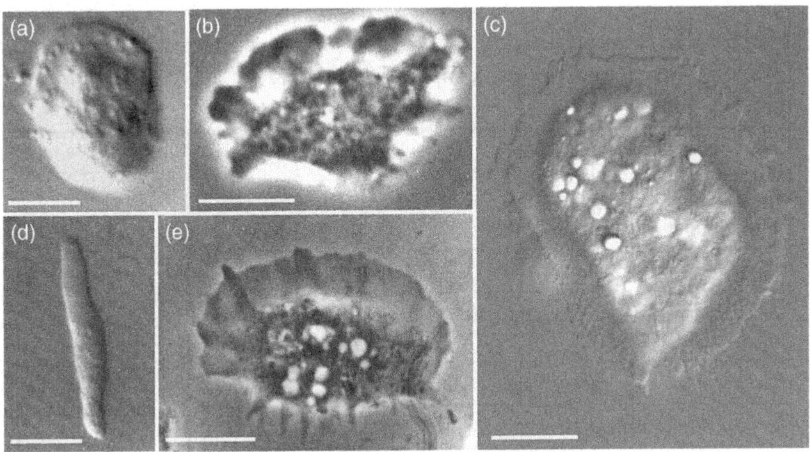

FIGURE 8 Diversity of amoebae. (a) *Pellita catalonica*, (b) *Flabellula baltica*, (c) *Cochliopodium bilimbosum*, (d) *Stygamoeba regulate*, and (e) *Flamella lacustris*. All photos are from the type strains, except for *C. bilimbosum*. Scale bar = 10 μm.

in all species. Nuclei are vesicular and vary in number between species. *Flabellula baltica* is capable of agamous cell fusions. These are freshwater and marine species, and no cysts are known.

Family Leptomyxida: These sometimes striking amoebae have a monotactic (*Rhizamoeba*) or branched (*Leptomyxa*) morphology, the latter becoming monotactic in rapid locomotion or when disturbed (two genera, nine valid species). Both uninucleate and multinucleate species occur, and *Leptomyxa reticulata* is a plasmodial organism with hundreds of nuclei. The glycocalyx is amorphous. These are both freshwater and marine species and several types of cysts (resting, digestive, and multiplication) are found in *Leptomyxa*.

Family Gephyramoebidae: This is a monotypic family consisting of a single described species, *Gephyramoeba delicatula*. This amoeba has a branched morphotype, which, during rapid movement acquires a nearly monotactic form (as deduced from the available drawings). There is a single vesicular nucleus, although A. Goodey, the author of the description of this species notes the presence of a 'few supernumerary nuclei' in some. The ultrastructure is unknown for this soil-inhabiting species, but the cysts are known to be single-walled.

Family Stereomyxidae: These are amoeboid organisms with a branched morphotype (two genera, five valid species). Cells are uninucleate, and no monotactic form is known. Cytoplasmic MTOCs have radiating mictotubules and are found near the dictyosome. The glycocalyx is thin and amorphous. All known examples are marine species, and no cysts are known. The morphology and biology of the genera *Stereomyxa* and *Corallomyxa* are very different; they hardly belong to the same family.

Gymnamoebia *insertae sedis*. In the morphological system of Page, there are a large number of amoebal groups for which the taxonomy is still highly uncertain. Some of them have found a home in the molecular system (Table 3), but here they are listed according to Page's system for convenience.

Family Filamoebidae: Molecular data indicate that this taxon is polyphyletic (three genera, six valid species). These are all amoebae with acanthopodial morphotype, with spineolate subpseudopodia that are sometimes furcating. None have differentiated uroidal structures, although amoebae of the genus *Echinamoeba* can become monotactic in active locomotion or under oxygen deficit. The glycocalyx is thin and amorphous. There are both marine and freshwater species, all cyst-forming.

Family Cochliopodiidae: These are amoebae with a lens-like morphotype, is some species they form a kind of fasciculate uroid or adhesive uroidal filaments in some species (1 genus, 14 valid species; Figure 8(c)). The nucleus is single and usually vesicular. Cells are covered with a tectum, which is a layer of scales enveloping the dorsal surface such that the ventral surface is free for locomotion and phagocytosis. There are freshwater, soil, and marine species, and cysts are known in some species.

Genus Stygamoeba: These are amoebae with a lingulate or branched morphotype that often are spineolate in locomotion (three valid species; Figure 8(d)). A bulbous uroid occurs in some species. The single nucleus is vesicular, and the glycocalyx is amorphous in the only studied species. Cytoplasmic MTOC with radiating mictotubules occur near the dictiosome. Unlike the rest of the lobose amoebae, mitochondria in these species have flattened cristae. Taken together, these are a unique combination of characteristics of lobose amoebae. All known species are marine, with single-walled cysts.

Genus Gocevia: These lens-like amoebae are covered with a thick cuticle covering the dorsal surface only, similar to *Cochliopodium* (see above; three probably valid species). The moving cell produces a hyaline border from under this cuticle, which may be covered with adhering foreign particles. There are freshwater and marine species.

Genus Paragocevia: These are similar to *Gocevia*, except that cells produce an expanded frontal lamellipodium with numerous short projections on their ventral surface

TABLE 3 Checklist of currently known gymnamoebae species. Species listed are (1) all taxonomically valid species and (2) species where descriptions contain enough data for reliable reisolation. Some comments are provided on the most debated genera

Class Lobosea Carpenter 1861 Subclass Gymnamoebia Haeckel 1866 Order Euamoebida Lepşi 1960 Family Amoebidae Ehrenberg 1838
1. *Amoeba proteus* (Pallas, 1766) Leidy 1878
2. *Amoeba borokensis* Kalinina, Afon'kin, Gromov, Khrebtukova et Page 1987
3. *Amoeba leningradensis* Page et Kalinina 1984
4. *Amoeba amazonas* Flickinger 1974
5. *Chaos carolinense* (Willson, 1900) King et Jahn 1948
6. *Chaos illinoisense* (Kudo, 1950) Bovee et Jahn 1973
7. *Chaos nobile* (Penard, 1902) Bovee et Jahn 1973
8. *Chaos glabrum* Smirnov et Goodkov 1997
9. *Polychaos fasciculatum* (Penard, 1902) Schaeffer 1926
10. *Polychaos annulatum* (Penard, 1902) Smirnov et Goodkov 1998
11. *Polychaos dubium* (Schaeffer, 1916) Schaeffer 1926
12. *Polychaos nitidubium* Bovee 1970
13. *Deuteramoeba algonquinensis* (Baldock, Rogerson et Berger, 1983) Page 1987
14. *Deuteramoeba mycophaga* (Pussard, Alabouvette et Pons, 1980) Page 1988
15. *Trichamoeba sinuosa* Siemensma et Page 1986
16. *Trichamoeba myakka* Bovee 1972
17. *Trichamoeba cloaca* Bovee 1972
18. *Trichamoeba osseosaccus* Schaeffer 1926
19. *Parachaos zoochlorellae* (Willumsen, 1982) Willumsen, Siemensma et Suhr-Jessen 1987
20. *Hydramoeba hydroxena* (Entz, 1912) Reynolds et Looper 1928
Family Hartmannellidae (Volkonsky, 1931) Page 1974
1. *Saccamoeba stagnicola* Page 1974
2. *Saccamoeba limax* (Dujardin, 1841) Page 1974
3. *Saccamoeba lucens* Frenzel 1892
4. *Saccamoeba wakulla* Bovee 1972
5. *Saccamoeba limna* Bovee 1972
6. *Saccamoeba wellneri* Siemensma 1987
7. *Saccamoeba marina* Anderson, Rogerson et Hannah 1997
8. *Cashia limacoides* (Page, 1967) Page 1974
9. *Glaeseria mira* (Glaeser, 1912) Volkonsky 1931
10. *Hartmannella cantabrigiensis* Page 1974
11. *Hartmannella vermiformis* Page 1967
12. *Hartmannella lobifera* Smirnov 1997
13. *Hartmannella abertawensis* Page 1980
14. *Hartmannella vacuolata* Anderson, Rogerson et Hannah 1997
15. *Nolandella hibernica* (Page, 1980) Page 1983
Family Thecamoebidae (Schaeffer, 1926) Smirnov et Goodkov 1994
1. *Thecamoeba striata* (Penard, 1890) Schaeffer 1926
2. *Thecamoeba quadrilineata* (Carter, 1856) Lep¸i 1960
3. *Thecamoeba munda* (Schaeffer, 1926) Smirnov 1999
4. *Thecamoeba orbis* Schaeffer 1926
5. *Thecamoeba hilla* Schaeffer 1926
6. *Thecamoeba sparolata* Fishbeck et Bovee 1993
7. *Thecamoeba sphaeronucleolus* (Greef, 1891) Schaeffer 1926
8. *Thecamoeba verrucosa* (Ehrenberg, 1838) Schaeffer 1926
9. *Thecamoeba terricola* (Greef, 1866) Lep¸i 1960
10. *Thecamoeba similis* (Greef, 1891) Lep¸i 1960
11. *Thecamoeba pulcra* (Biernacka, 1963) Page 1977
12. *Thecamoeba hoffmani* Sawyer, Hnath et Conrad 1974
13. *Sappinia diploidea* (Hartmann et Naegler, 1908) Alexieff 1912
14. *Dermamoeba granifera* (Greef, 1866) Page et Blakey 1979
15. *Dermamoeba minor* (Pussard, Alabouvette et Pons, 1979) Page 1988
16. *Paradermamoeba valamo* Smirnov et Goodkov 1993
17. *Paradermamoeba levis* Smirnov et Goodkov 1994
18. *Pseudothecamoeba proteoides* (Page, 1976) Page 1988
19. *Parvamoeba rugata* Rogerson 1993
20. *Thecochaos album* (Greef, 1891) Page 1981
21. *Thecochaos fibrillosum* (Greef, 1891) Page 1981
22. *Stenamoeba* (former Platyamoeba) *stenopodia* (Page, 1969) Smirnov, Nassonova, Chao et Cavalier-Smith 2007
Family Vannellidae (Bovee, 1970) Page 1987
1. *Ripella* (former *Vannnella*) *platypodia* (Glaeser, 1912) Smirnov, Nassonova, Chao et Cavalier-Smith 2007
2. *Vannella simplex* Wohlfarth-Bottermann 1960
3. *Vannella lata* Page 1988

TABLE 3 Checklist of currently known gymnamoebae species. Species listed are (1) all taxonomically valid species and (2) species where descriptions contain enough data for reliable reisolation. Some comments are provided on the most debated genera—cont'd

4. *Vannella cirifera* (Frenzel, 1892) Page 1988
5. *Vannella miroides* Bovee 1965
6. *Vannella persistens* Smirnov et Brown 2000
7. *Vannella peregrinia* Smirnov et Fenchel 1996
8. *Vannella devonica* Page 1979
9. *Vannella aberdonica* Page 1980
10. *Vannella caledonica* Page 1979
11. *Vannella arabica* Page 1980
12. *Vannella anglica* Page 1980
13. *Vannella septentrionalis* Page 1980
14. *Vannella sensilis* Bovee et Sawyer 1979
15. *Vannella mira* (Schaeffer, 1926) Smirnov 2002
16. *Vannella ebro* Smirnov 2001
17. *Vannella crassa* Schaeffer 1926
18. *Vannella danica* Smirnov, Nassonova, Chao et Cavalier-Smith 2007
19. *Vannella* (former Platyamoeba) *placida* (Page, 1968) Smirnov, Nassonova, Chao et Cavalier-Smith 2007
20. *Vannella* (former Platyamoeba) *schaefferi* (Singh et Hanumaiah, 1979) Smirnov, Nassonova, Chao et Cavalier-Smith 2007
21. *Vannella* (former Platyamoeba) *plurinucleolus* (Page, 1974) Smirnov, Nassonova, Chao et Cavalier-Smith 2007
22. *Vannella* (former Platyamoeba) *calycinucleolus* (Page, 1974) Smirnov, Nassonova, Chao et Cavalier-Smith 2007
23. *Vannella* (former Platyamoeba) *bursella* (Page, 1974) Smirnov, Nassonova, Chao et Cavalier-Smith 2007
24. *Vannella* (former Platyamoeba) *australis* (Page, 1983) Smirnov, Nassonova, Chao et Cavalier-Smith 2007
25. *Vannella* (former Platyamoeba) *mainensis* (Page, 1971) Smirnov, Nassonova, Chao et Cavalier-Smith 2007
26. *Vannella* (former Platyamoeba) *flabellata* (Page, 1974) Smirnov, Nassonova, Chao et Cavalier-Smith 2007
27. *Vannella* (former Platyamoeba) *pseudovannellida* (Hauger, Rogerson et Anderson, 2001) Smirnov, Nassonova, Chao etCavalier-Smith 2007
28. *Vannella* (former Platyamoeba) *nucleolilateralis* (Anderson, Nerad et Cole, 2003) Smirnov, Nassonova, Chao et Cavalier-Smith 2007
29. *Vannella* (former Platyamoeba) *murchelanoi* (Sawyer, 1975) Smirnov, Nassonova, Chao et Cavalier-Smith 2007
30. *Vannella* (former Platyamoeba) langae (Sawyer, 1975) Smirnov, Nassonova, Chao et Cavalier-Smith 2007
31. *Vannella* (former Platyamoeba) *weinsteini* (Sawyer, 1975) Smirnov, Nassonova, Chao et Cavalier-Smith 2007
32. *Vannella* (former Platyamoeba) *douvresi* (Sawyer, 1975) Smirnov, Nassonova, Chao et Cavalier-Smith 2007
33. *Vannella* (former Platyamoeba) *oblongata* Moran, Anderson, Dennett, Caron et Gast 2007
34. *Vannella* (former Platyamoeba) *contorta* Moran, Anderson, Dennett, Caron et Gast 2007
35. *Vannella* (former Platyamoeba) *epipetala* Amaral-Zettler, Cole, Laatsch, Nerad, Anderson et Reysenbach 2006
36. *Clydonella rosenfeldi* Sawyer 1975
37. *Clydonella sindermanni* Sawyer 1975
38. 38. *Clydonella wardi* Sawyer 1975
39. *Clydonella vivax* (Schaeffer, 1926) Sawyer 1975
40. *Lingulamoeba leei* Sawyer 1975
41. 41. *Pessonella marginata* Pussard 1973
Family Paramoebidae (Poche, 1913) Page 1987[a]
1. *Mayorella viridis* (Leidy, 1874) Harnisch 1968
2. *Mayorella cantabrigiensis* Page 1983
3. *Mayorella vespertilioides* Page 1983
4. *Mayorella penardi* Page 1972
5. *Mayorella bigemma* (Schaeffer, 1926) Schaeffer 1926
6. *Mayorella augusta* Schaeffer 1926
7. *Mayorella pussardi* Hollande, Nicolas et Escaig 1981
8. *Mayorella kuwaitensis* (Page, 1982) Page 1983
9. *Mayorella gemmifera* Schaeffer 1926
10. 10. *Mayorella dactylifera* Goodkov et Buryakov 1986
11. *Korotnevella stella* (Schaeffer, 1926) Goodkov 1988
12. *Korotnevella bulla* (Schaeffer, 1926) Goodkov 1988
13. *Korotnevella diskophora* Smirnov 1999
14. *Korotnevella nivo* Smirnov 1997
15. *Korotnevella hemistilolepis* O'Kelly, Peglar, Black, Sawyer et Nerad 2001
16. *Korotnevella monacantholepis* O'Kelly, Peglar, Black, Sawyer et Nerad 2001
17. *Paramoeba eilhardi* Schaudinn 1896
Family Vexilliferidae Page 1987[b]
1. *Vexillifera bacillipedes* Page 1969
2. *Vexillifera granatensis* Mascaro, Osuna et Mascaro 1986

Continued

TABLE 3 Checklist of currently known gymnamoebae species. Species listed are (1) all taxonomically valid species and (2) species where descriptions contain enough data for reliable reisolation. Some comments are provided on the most debated genera—cont'd

3. *Vexillifera lemani* Page 1976
4. *Vexillifera minutissima* Bovee et Sawyer 1979
5. *Vexillifera armata* Page 1979
6. *Vexillifera expectata* Dykova, Lom, Machakova et Peckova 1998
7. *Vexillifera telmathalassa* Bovee 1956
8. *Vexillifera aurea* Schaeffer 1926
9. *Vexillifera browni* Sawyer 1975
10. *Vexillifera ottoi* Sawyer 1975
11. *Pseudoparamoeba pagei* (Sawyer, 1975) Page 1983
12. *Neoparamoeba pemaquidensis* Page 1970
13. *Neoparamoeba aestuarina* Page 1970[c]
14. *Neoparamoeba branchiphila* Dykov´, Nowak, Crosbie, Fiala, Peckov´, Adams, Mach´kov´ et Dvo´kov´ 2005
15. *Neoparamoeba perurans* Young, Crosbiea, Adamsa, Nowaka et Morrison 2007

Order Leptomyxida (Pussard et Pons, 1976) Page 1987
Suborder Rhizoflabellina Page 1987
Family Flabellulidae Bovee 1970

1. *Flabellula citata* Schaeffer 1926
2. *Flabellula calkinsi* (Hogue, 1914) Page 1983
3. *Flabellula demetica* Page 1980
4. *Flabellula trinovanica* Page, 1980
5. *Flabellula baltica* Smirnov 1999
6. *Paraflabellula kudoi* (Singh et Hanumaiah, 1979) Page 1983
7. *Paraflabellula reniformis* (Schmoller, 1964) Page 1983
8. *Paraflabellula hoguae* (Sawyer, 1975) Page 1983

Family Leptomyxidae (Pussard et Pons, 1976) Page 1987

1. *Leptomyxa reticulata* Goodey 1914
2. *Leptomyxa fragilis* (Penard, 1904) Siemensma 1987
3. *Rhizamoeba polyura* Page 1972
4. *Rhizamoeba saxonica* Page 1974
5. *Rhizamoeba australiensis* (Chakraborty et Pussard, 1985) Page 1988
6. *Rhizamoeba schnepfii* Kuhn 1997
7. *Rhizamoeba clavarioides* (Penard, 1902) Siemensma 1980
8. *Rhizamoeba flabellata* (Goodey, 1914) Cann 1984
9. *Rhizamoeba caerulea* (Schaeffer, 1926) Siemensma 1980

Suborder Leptoramosina Page 1987
Family Gephyramoebidae Pussard et Pons 1976

1. *Gephyramoeba delicatula* Goodey 1914

Family Stereomyxidae Grell 1966

1. *Stereomyxa ramosa* Grell 1966
2. *Stereomyxa angulosa* Grell 1966

Order Acanthopodida Page 1976
Family Acanthamoebidae Sawyer et Griffin 1975[d]

1. *Acanthamoeba tubiashi* Lewis et Sawyer 1979
2. *Acanthamoeba astronixis* (Ray et Hayes, 1954) Page 1967
3. *Acanthamoeba comandoni* Pussard 1964
4. *Acanthamoeba lugdunensis* Pussard et Pons 1977
5. *Acanthamoeba castellanii* (Douglas, 1930) Volkonski 1931
6. *Acanthamoeba rhysodes* (Singh, 1952) Griffin 1972
7. *Acanthamoeba mauritaniensis* Pussard et Pons 1977
8. *Acanthamoeba polyphaga* (Pushkarew, 1913) Volkonsky 1931
9. *Acanthamoeba griffini* Sawyer 1971
10. *Acanthamoeba quina* Pussard et Pons 1977
11. *Acanthamoeba divionensis* Pussard et Pons 1977
12. *Acanthamoeba triangularis* Pussard et Pons 1977
13. *Acanthamoeba hatchetti* Sawyer, Visvesvara et Harke 1977
14. *Acanthamoeba palestinensis* (Reich, 1933) Page 1977 (syn. Acanthamoeba pustulosa)
15. *Acanthamoeba royreba* Willaert, Stevens et Tyndall 1978
16. *Acanthamoeba culbertsoni* (Singh et Das, 1970) Griffin 1972
17. *Acanthamoeba lenticulata* Molet et Ermolieff-Braun 1976
18. *Acanthamoeba gigantean* Schmoller 1964
19. *Acanthamoeba jacobsi* Sawyer, Nerad et Visvesvara 1992
20. *Acanthamoeba stevensoni* Sawyer, Nerad, Lewis et McLaughlin 1993
21. *Acanthamoeba pearcei* Nerad, Sawyer, Levis et McLaughlin 1995
22. *Acanthamoeba healyi* Moura, Wallace et Visvesvara 1992
23. *Protacanthamoeba caledonica* Page 1981
24. *Protacanthamoeba invadens* (Singh et Hanumaiah, 1979) Page 1981

Gymnamoebia incertae sedis
Family Echinamoebidae Page 1975

1. *Echinamoeba exudans* Page 1975

TABLE 3 Checklist of currently known gymnamoebae species. Species listed are (1) all taxonomically valid species and (2) species where descriptions contain enough data for reliable reisolation. Some comments are provided on the most debated genera—cont'd

2. *Echinamoeba silvestris* Page 1975
3. *Echinamoeba thermarum* Baumgartner, Yapi, Gr¨bner-Ferreira et Stetter 2003
4. *Filamoeba nolandi* Page 1967
5. *Filamoeba siniensis* Dykov´, Peckov´, Fiala et Dvor´kov´ 2005
6. *Comandonia operculata* Pernin et Pussard 1979
Genus *Flamella* Schaeffer 1926
1. *Flamella citrensis* Bovee 1956
2. *Flamella magnifica* Schaeffer 1926
3. *Flamella tiara* Fishbeck et Bovee 1993
4. *Flamella lacustris* Michel et Smirnov 1999
5. *Flamella aegyptia* Michel et Smirnov 1999
Rhizopoda incertae sedis **Genus *Stygamoeba* (Sawyer, 1975) Smirnov 1995**
1. *Stygamoeba polymorpha* Sawyer 1975
2. *Stygamoeba regulata* Smirnov 1995
3. *Stygamoeba antarctica* (Moran, Anderson, Dennett, Caron et Gast, 2007)[e]
Family Entamoebidae Chatonn 1925
The only probably free-living member of this family is *Entamoeba moshkovskii* Chalaja 1941 All other species are obligate parasites
Genus Cochliopodium Hertwig et Lesser 1874
1. *Cochliopodium bilimbosum* (Auerbach, 1856) Leidy 1879
2. *Cochliopodium actinophorum* (Auerbach, 1856) Page 1976
3. *Cochliopodium barki* Kudryavtsev, Brown et Smirnov 2004
4. *Cochliopodium larifeili* Kudriavtsev 1999
5. *Cochliopodium minutoidum* Kudryavtsev 2006
6. *Cochliopodium minus* Page 1976
7. *Cochliopodium gulosum* Schaeffer 1926
8. *Cochliopodium vestitum* (Archer, 1871) Archer 1877
9. *Cochliopodium gallicum* Kudryavtsev et Smirnov 2006
10. *Cochliopodium spiniferum* Kudryavtsev 2004
11. *Cochliopodium kieliense* Kudryavtsev 2006
12. *Cochliopodium maeoticum* Kudryavtsev 2006
13. *Cochliopodium clarum* Schaeffer 1926
14. *Cochliopodium granulatum* Penard, 1890
Genus Gocevia Valkanov 1932
1. *Gocevia binucleata* De Saedeleer 1934
2. *Gocevia fonbrunei* Pussard 1965
3. *Gocevia obscura* Penard 1890
Genus Paragocevia Page 1987
1. *Paragocevia placopus* (H¨lsmann, 1974) Page 1987

[a]*E. C. Bovee named 19 more* Mayorella*-like and* Korotnevella*-like amoebae as belonging to the genus* Mayorella *and four more genera (*Flagellipodium *Bovee 1961;* Oscillosignum *Bovee 1951;* Subulamoeba *Bovee 1951; and* Trienamoeba *Bovee 1970); all these species need reisolation and study before they are recognized as a valid ones and assigned to any genus; the same is true for all the four genera mentioned. The generic name* Korotnevella *is used instead of Page's 'Dactylamoeba,' Korotneff, 1880.*
[b]*There are two little-known genera of* Vexillifera*-like amoebae –* Boveella *Sawyer, 1975 and* Trienamoeba *Bovee, 1970. Representatives of both must be studied with EM to make a decision about the validity and position of these genera.*
[c]*The 'subspecies'* Neoparamoeba aestuarina antarctica*, Moran, Anderson, Dennett, Caron et Gast 2007, is probably an independent species.*
[d]*Species composition of the genus* Acanthamoeba *is questionable, and nonmorphological methods are required for distinguishing some of its species. There are serious disagreements between various authors about the validity of species and synonyms. In contrast with other sections of this checklist, only approximate species composition is given here, that is, species that are recognizable at the morphological level and are generally accepted.*
[e]*Genus* Vermistella *Moran, Anderson, Dennett, Caron et Gast 2007 is perhaps a junior synonym of the genus* Stygamoeba.

(one described species). A single freshwater species, *P. placopus* has been described and cysts are unknown.

Genus Flamella: These are amoebae with a flamellian morphotype that never acquire a monotactic form (five valid species; Figure 8(e)). They may have one or many vesicular nuclei. Uroidal structures are usually adhesive, and the glycocalyx is thin and amorphous. The trophozoites of *Flamella lacustris* are known to undergo agamous fusions. There are both freshwater and marine species, and cysts are known in some.

AMOEBOID MOVEMENT

The naked lobose amoebae do not have any specialized locomotive organelles, so the entire cell performs the locomotive function, combining the activity of cytoskeleton and adhesion to the substratum. The cytoskeleton is relatively well studied in large lobose amoebae and *Acanthamoeba*, but very poorly, if at all, in other species. Basically, it probably includes all three components typical of eukaryotic cells – microtubules, microfilaments, and (perhaps) intermediate filaments. The system of microtubules, in contrast with most other cells, is poorly developed, while the system of microfilaments (actin–myosin filaments) is highly developed and plays the main role in amoeboid locomotion. There are little data available on the presence (or not) of intermediate filaments in the cytoplasm of naked lobose amoebae. In the vast majority of species MTOCs are noticeable only in mitosis; there are no morphologically pronounced cytoplasmic MTOCs, although separate microtubules have been noted in the cytoplasm. In contrast, *Acanthamoeba*, *Stygamoeba*, and members of the order Stereomyxida appear to have a microtubular network, as well as a persistent cytoplasmic MTOC. This suggests that the mechanism of locomotion in these latter species may differ in some detail from that in other naked amoebae.

Attempts to explain the phenomenon of amoeboid movement date back to Dujardin who suggested in 1835 that the cytoplasm of a cell possesses a 'contractivity.' Delinger, in 1906, suggested that all kinds of amoeboid movement are based on the contraction of the semipermanent filamentous network in the cytoplasm of the cell (surprisingly foreseeing the modern models of the amoeboid movement). Later observers of moving amoebae noted the difference between the liquid (sol) cytoplasm flowing inside the cell (endoplasm) and the gel-like peripheral cytoplasm bordering these flows (ectoplasm). On the basis of these observations, Hyman in 1917 attempted to develop a model of amoeboid movement based on the local liquefaction of the cytoplasm at the tip of the growing pseudopodium. In 1922, Fürth proposed a model based on the local pH changes in the moving cell. He suggested that the local change of pH caused by a special substance 'lactacidogen' results in the decrease in the surface tension at the tip of the growing pseudopodium and the increase in the volume of cytoplasm, which he considered as a protein colloid. However, these and other models created to that time considered only the growth of separate pseudopodia and did not explain the entire mechanism of amoeboid locomotion.

The first, relatively comprehensive model of amoeboid movement was proposed by C.F. Pantin in 1923 and further developed by S.O. Mast in 1926, who considered the plasmasol–plasmagel interconversion as a primary mechanism in amoeboid locomotion. According to this model, the gel–sol conversion is accompanied by the contraction of an ectoplasm generating a force that pushes liquid endoplasm forward, toward the front of the cell. Meanwhile, in the frontal area, the endoplasm would fountain to the walls of the cell and convert to the gel-like ectoplasm. Thus, an amoeba could be represented essentially as a tube of ectoplasm permanently growing at the anterior edge and degenerating at the posterior, resulting in the locomotion of the cell. This 'hydrostatic' model was modified to the 'tail conversion' model by R.J. Goldacre and I.J. Lorch in 1950, who suggested that the contraction site was limited to the posterior part of the cell. This caudal contraction was supposed to create the pressure difference necessary to promote the cytoplasmic flow.

Alternatively, R.D. Allen suggested in 1961 that the amoeba's locomotive force is generated at the anterior end of the growing pseudopodium (the frontal zone contraction theory). According to this model, the sol-like endoplasm converts to a gel-like ectoplasm at the frontal part of the cell. Sol–gel conversion is accompanied by the reduction of cytoplasmic volume, which generates the force pulling forward everything behind it. This model rejects active contraction of the cortical ectoplasm and any role for hydrostatic pressure in the transmission of the motive force. However, one advantage of this theory is that the control and steering of locomotion in this model is localized to the anterior region of the cell, where the growth of the pseudopodium occurs. Both these models explained some aspects of amoeboid movement, but not all. As a result, in 1977 L.N. Seravin suggested that more than one mechanism could be involved and that features of amoeboid locomotion might depend on the activity and combination of these mechanisms in the given cell.

The first model explaining most of the known aspects of movement in large lobose amoebae was proposed by A. Grebecki in 1979 in his 'generalized cortical contraction' model. This suggested that the amoeboid cell contains a polarized cortical layer of microfilaments. At the tip of the growing pseudopodium, this contractile layer detaches from the membrane and retracts inward forming a transverse network of actin filaments, while the membrane lifts outward by the pressure of endoplasm filtered through this network. This flow of the cytoplasm brings molecules of actin that polymerize on the membrane, forming a new cortical filament layer. This newly formed layer contracts, detaching from the plasma membrane and forming a new transverse network, while the former network moves back and degenerates. The cycle repeats every 2 s on an average, resulting in the advancement of the pseudopodium. The same mechanism is responsible for the retraction of pseudopodia; in this case the actin layer at the tip of the pseudopodium contracts without detaching from the membrane (Figure 9).

The progress of the moving cell is impossible without proper cell–substrate interactions, which are poorly understood. Large lobose amoebae adhere to the substratum by means of local contacts, called minipodia, which are surface microprotrusions that cover mainly the middle–anterior area of the ventral cell surface. Minipodia are grouped into 'rosette contacts,' which look like cauliflower papillae with crowns of minipodia emerging from them. This pattern of adhesion seems to be common among amoeboid organisms, as the well-studied amoeboflagellate *Naegleria gruberi* (Heterolobosea) adheres to the substratum in a similar manner. This organism forms close cell–substratum contacts, with a 20 nm gap separating the cell and the substratum, which are linked to much larger platforms about 100 nm

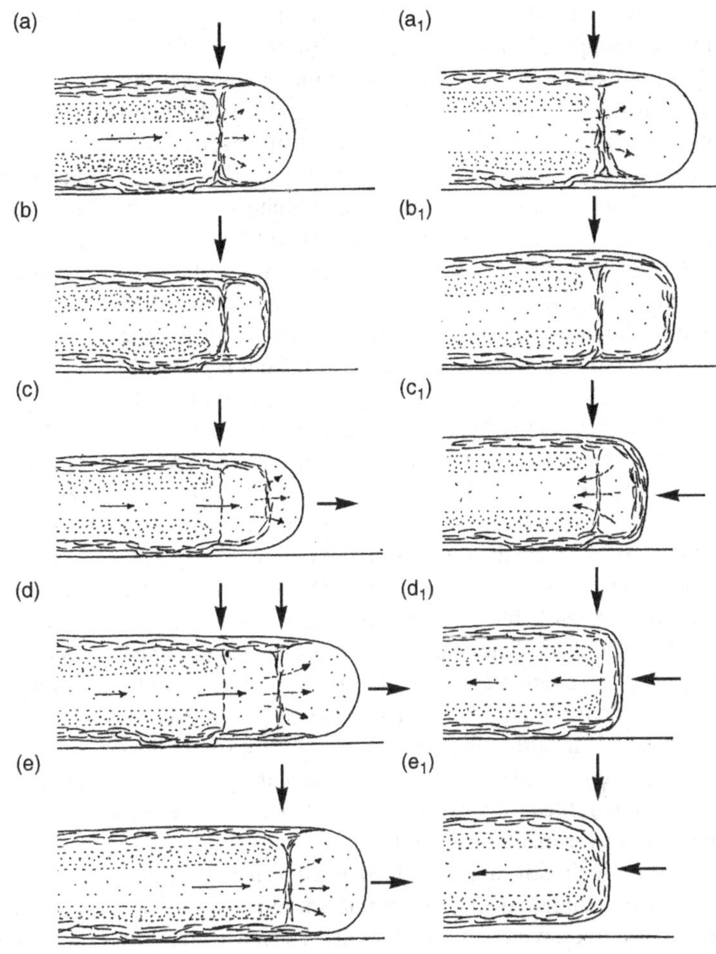

FIGURE 9 Amoeboid movement, scheme illustrating the model of generalized cortical contraction. Stages (a)–(e) illustrate subsequent phases of the elongation of the pseudopodium; stages (a_1)–(e_1) illustrate the retraction of the pseudopodium. The position of the transverse actin network border is marked with vertical arrows. Horizontal arrows illustrate the flow of the cytoplasm and the direction of the pseudopodial growth/contraction. *Modified from the scheme by A. Grebecki from 1990.*

apart. For moving amoebae, the substratum need not necessarily be a hard surface; amoebae can move along the layers of biofilms, on mucose slime floating in the water, and even along the surface water film. Besides the usual means of locomotion, where the cell is flattened on the substratum, the large amoebae demonstrate a 'walking' locomotion (first described by Dellinger in 1906). The cell moving in this way forms a pseudopodium directed up and forward that adheres with its tip to form a bridge-like structure. The posterior of the cell then separates from the surface, causing the amoeba to essentially 'walk' upon the substratum.

These models fairly thoroughly describe the mechanism of movement and pseudopodial formation of amoebae that are cylindrical or subcylindrical in cross section (class Tubulinea). The locomotory mechanisms of flattened, flabellate amoebae (class Discosea) are probably similar, but these have not yet been analyzed in detail.

BIOLOGY AND ECOLOGY

The life cycle of gymnamoebae includes the trophozoite and cyst stages. The former refers to any active amoeba, whether locomotive, resting, or floating, while the latter is a facultative resting stage. Cysts are single- or double-walled, in the latter case often with pores at the sites of contacts of the two cyst walls (endocyst and ectocyst, respectively). The cyst pore is the site of future excystment. Some amoeba species round up but do not produce any external envelope and persist like this for some time (pseudocyst). However, the ability to form cysts may also be lost during laboratory cultivation. In members of the genera *Leptomyxa* and *Gephyramoeba*, there are several functional groups of cysts intended for digestion of phagocytized food, reproduction of the organism, and finally for rest. These different cyst types may be formed consecutively in the life cycle of the same organism.

Reproduction is by binary division or by the fragmentation of multinucleate cells (characteristic of the order Leptomyxida). The trophozoites of some species (e.g., *Flabellula baltica* and *Flamella lacustris*) are capable of agamous fusion, and members of the genus *Leptomyxa* exist only as multinucleate plasmodia with a few or up to several hundreds of nuclei.

All naked amoebae are aquatic organisms; nominal 'soil species' actually inhabit the layer of capillary water surrounding soil particles and fill the pores between them.

On the other hand, there are a number of clearly distinguished groups of marine versus freshwater/soil species. Although experiments show that some marine morphospecies, especially those from boundary, brackish-water habitats, can persist at very low salinity or even under freshwater conditions, molecular studies show that at least some of such seemingly euryhalinous morphospecies may be genetically distinct (i.e., different species).

Difficulties in species identification have so far prevented an extensive study of amoebae biogeography. Most of the valid amoebae species were described from Europe and North America, and a few are known from the Middle East, India, Australia, or Japan. Surprisingly, one relatively well-studied region is Antarctica, while the amoebal biogeography in most regions of Asia, South America, and Africa is virtually unknown. Thus we do not have reliable data with which to deduce species distribution patterns on a global scale. Current data suggest that some amoeba species have a worldwide distribution, although these are mostly small or unremarkable species, where misidentification is possible, and many other amoebae species have been found only once in a single geographic location. Another complication in determining if amoebae morphospecies have a worldwide distribution is evidence that at least some morphospecies consist of several genotypes, and the geographic distribution of these genotypes is probably complex, as has been shown for many other protists.

Theoretically, the ability to form resting stages capable of persisting under adverse conditions should allow amoebae to colonize a wide range of habitats. The small size and low weight of amoebae and the high resistance of their cysts should facilitate easy dispersal with local- and global-scale mass flows, like air and water currents. This suggests that there should be a ubiquitous dispersal of cysts and trophozoites, resulting in global-scale metapopulation for most species. Thus, any suitable habitat should contain a very wide diversity of amoebae. However, this is a matter of considerable debate for protists in general, complicated by the unknown level of cryptic species in nearly all groups and a lack of the molecular data needed to test it. Regardless, most of amoebal diversity is expected to be passive, that is, in the form of resting cysts that do not participate in mass and energy cycling. Thus, the composition of an active community, that is, the fraction of species actually participating in the ecosystem functioning, is expected to depend on the current microhabitat structure of the environment, as well as the relative abundance of the species.

One side effect of ubiquitous dispersal is that if indirect methods of species recovery are used to explore diversity, such as methods that do not distinguish between cysts and trophozoites (e.g., enrichment cultivation or DNA extraction), a species may be recovered from a habitat where it never actually exists in an active state. This fraction of the passive diversity of amoebae could be a significant problem for ecological studies.

The local-scale distribution of amoebae is very heterogeneous, with pronounced patchiness and vertical structure, especially in boundary habitats, like aquatic bottom sediments and top layers of soil. This may be related to the microhabitat structure of the environment, and this distribution may change rapidly, following microhabitat changes. Amoebae can show local flashes of abundance following an increase in the number of appropriate microhabitats and seeding of the habitat with large numbers of cysts after the favorable conditions have passed. Cysts can persist, presumably for long periods of time, only gradually dying off, until conditions improve. This latter strategy may be the mode of population dynamics for most amoebae species.

The primary food source for amoebae is bacteria. Most amoebae species are polyphagous and in culture can feed on bacteria or other protists of suitable size. Some species are algivorous, for example, species of the genus *Polychaos* or *Dermamoeba algensis*, or they may be carnivorous, such as species of *Amoeba*, *Mayorella*, and *Paradermamoeba*. There also appear to be specialized fungi-feeding amoebae species in soil, and dissolved organic substances consumed by pinocytosis may play an important role in nutrition of the smallest amoebae species.

All the known gymnamoebae species are aerobic and contain mitochondria. However, some species, for example, *Vannella peregrinia*, may persist at very low or virtually zero oxygen concentrations. None of the known lobose amoebae species lack functional mitochondria.

IMPORTANCE

Due to their feeding activity, the amoebae play an important role as grazers of bacteria. They are probably among the main controllers of bacterial populations because of their fast response to increases in bacterial numbers. It seems that in some habitats naked amoebae are specifically significant as primary grazers of attached bacteria that are generally unavailable to other micrograzers. The largest amoebae species, especially those of the genera *Vannella* and *Cochliopodium*, can also mechanically damage a biofilm, making its remnants available for other predators. As decomposers of organic matter and chemical substances, amoebas enhance nutrient cycling and, together with other protozoa, they stimulate carbon and nitrogen cycling in the environment. On the other hand, amoebae are a valuable source of food for metazoan organisms, for example, nematodes, and are known to be highly abundant in the plant rhizosphere.

There are a number of medically important amoebae. Species of *Acanthamoeba* and *Balamuthia* can cause fatal infections in animals and humans and may act as vectors of various pathogenic organisms. A binucleate amoeba identified as *Sappinia diploidea* has been added recently to this list. Species of *Acanthamoeba*, *Vannella*, and *Hartmannella* have been isolated from the eye surface of patients with contact lens keratits, although their role in the infection process is unclear. Amoebae have also been found in the tissues of lobsters and fish and are believed to be infectious agents causing diseases of these organisms.

FURTHER READING

Adl, S. M., Simpson, G. B., Farmer, M., *et al.* (2005). The new higher level classification of eukaryotes with emphasis on the taxonomy of protists. *Journal of Eukaryotic Microbiology*, *52*, 399–451.

Bovee, E. C. (1985). Class lobosea carpenter, 1861. In J. J. Lee, S. H. Hunter & E. C. Bovee (Eds.), *An illustrated guide to the protozoa* (pp. 158–211). Allen University Press: Kansas.

Grebecki, A. (1982). Supramolecular aspects of amoeboid movement part I. In *Progress in Protozoology. Proc. VI Int. Congr. Protozool. Acta Protozool.v. 1*, (pp. 117–130). (special issue).

Grebecki, A. (1990). Dynamics of the contractile system in the pseudopodial tips of normally locomoting amoebae, demonstrated by *in vivo* video-enhancement. *Protoplasma*, *154*, 98–111.

Page, F. C. (1976). *An illustrated key to freshwater and soil amoebae*. Institute of Terrestrial Ecology: Cambridge.

Page, F. C. (1983). *Marine gymnamoebae*. Institute of Terrestrial Ecology: Cambridge.

Page, F. C. (1987). The classification of 'naked' amoebae (phylum rhizopoda). *Archive für Protistenkunde*, *133*, 199–217.

Page, F. C. (1988). *A new key to freshwater and soil gymnamoebae*. Freshwater Biological Association: Ambleside.

Page, F. C. (1991). In *Nackte rhizopoda. Nackte Rhizopoda Und Heliozoea (Protozoenfauna, Band 2)* (pp. 3–170). Gustav Fisher Verlag: Stuttgart, New York.

Preston, T. M., & King, C. A. (1978). Cell-substrate associations during the amoeboid locomotion of *Naegleria*. *Journal of General Microbiology*, *104*, 347–351.

Rogerson, A., & Patterson, D. J. (2000). The naked ramicristate amoebae (gymnamoebae). In J. J. Lee, G. F. Leedale & P. Bradbury (Eds.), *An illustrated guide to the protozoa* (pp. 1023–1053). (2nd ed.). Society of Protozoologists: Kansas.

Smirnov, A., & Brown, S. (2004). Guide to the methods of study and identification of soil gymnamoebae. *Protistology*, *3*, 148–190.

Smirnov, A., Nassonova, E., Chao, E., & Cavalier-Smith, T. (2007). Phylogeny, evolution and taxonomy of vannellid amoebae. *Protist*, *158*, 295–324.

Visvesvara, G. S., Moura, H., & Schuster, F. L. (2007). Pathogenic and opportunistic free-living amoebae: *Acanthamoeba* spp., *Balamuthia mandrillaris*, *Naegleria fowleri*, and *Sappinia diploidea*. *FEMS Immunology and Medical Microbiology*, *50*, 1–26.

Chapter 15

Ciliates

D. Lynn
University of Guelph, Guelph, ON, Canada

Chapter Outline

Abbreviations 213
Defining Statement 213
Introduction 213
Morphological Features 215
Origin of Ciliates 215
Evolution of the Ciliate Cortex 216
Evolution of Nuclear Dimorphism 217
Major Clades 218
Ultrastructure and the Structural Conservatism of the Ciliate Cortex 219
Molecular Systematics and the Major Clades 219
Diversity at the Class Level 219
The Two Subphyla 219
The 11 Classes 219
Probing Diversity at the Species Level 224
Morphological Diversity at the Species Level 224
Molecular Techniques at the Species Level 224
Further Reading 226

ABBREVIATIONS

ITS Internally transcribed spacer
mtDNA Mitochondrial DNA
RAPD Randomly amplified polymorphic DNA
SSU rDNA Small subunit ribosomal RNA

DEFINING STATEMENT

The ciliates are defined as a group based on three major features – nuclear dualism, structure of somatic kinetids, and conjugation. Ciliates evolved from a dinoflagellate-like ancestor and diversified into 2 subphyla and 11 classes, which are briefly characterized. Species of ciliates can be recognized using a large variety of morphological features, and more recently, the sequences of genes, such as the mitochondrial cytochrome *c* oxidase subunit 1 gene.

INTRODUCTION

The ciliated protozoa are undoubtedly one of the most easily recognized groups of protists as they are typically highly motile and, in microscopic terms, fairly large. Ciliates are often 100 μm in length. However, some planktonic forms in the genera *Strombidium* and *Strobilidium* can be as small as 10 μm in length, while some benthic karyorelicteans can be ribbon-like and over 4500 μm. This typically large size suggests that the ciliates were once on top of the food chain, probably more than 1 billion years ago when all life was microbial. While they still feed on those organisms that they evolved to depredate, such as bacteria, flagellates, and smaller ciliates, they are now themselves a prey to metazoans (animals), which evolved after the ciliates and achieved larger body sizes through multicellularity, and so themselves have become top of the food chain.

We know for certain that ciliates existed in the Ordovician, some 400–500 Ma, because marine planktonic forms, called tintinnids, which secrete a lorica around themselves, have left fossilized loricae that can be identified as contemporary genera and families. There are even fossils of 'recent' ciliates that are soft bodied and have been preserved in amber as they fed and grew in small drops of water on the bark of ancient trees. Some of these ciliates are well enough preserved that we can place them in contemporary species, even though they are tens of thousands of years old. However, it is very probable that ciliates evolved at least one and perhaps 2 billion years ago, if we can trust calculations that we make using a 'molecular clock' based on the rate of divergence of the small subunit ribosomal RNA gene (SSU rRNA).

The ciliates are easily characterized and despite their tremendous diversity are, without a doubt, a monophyletic group. First, all ciliates are heterokaryotic: they have two kinds of nuclei, a feature called nuclear dualism or nuclear

Eukaryotic Microbes.

dimorphism. The macronucleus is the larger of the two, often contains many copies of the genome, and typically divides by amitosis in which discrete chromosomes are not formed prior to karyokinesis (nuclear division), since the genome has been processed into subchromosome-sized pieces (except in the protocruziids). The macronucleus is the physiologically active nucleus, performing most if not all transcription. The second ciliate nucleus, the micronucleus, is smaller, as its name suggests, and typically presumed to be diploid. The micronucleus divides by an endomitosis in which discrete chromosomes align on a metaphase plate, but the nuclear envelope does not break down at any time during the process. The micronucleus is very probably rarely involved in transcription; while its removal by microsurgery negatively impacts such cell processes as cell division and growth, micronuclear mRNA transcripts have never been isolated. It serves primarily as the germ line reserve of the ciliates; it can be likened to the nuclei of spermatogonia or oogonia in animals. Moreover, some ciliates, like *Tetrahymena* and *Paramecium*, can grow and reproduce without a micronucleus, but they cannot survive without a macronucleus.

The second characteristic feature of ciliates is that they typically propel themselves through the medium by cilia, which are essentially identical to eukaryotic flagella, but present in abundance and arranged in rows. Only one group of ciliates, the suctorians, lack cilia in their 'adult' stage, but their dispersing swarmer stage does have cilia. The basal bodies or kinetosomes of these cilia are distinguished by having three rootlets that together make up a complex infraciliature, which is part of the ciliate cytoskeleton that also includes cortical filamentous systems. These three ciliary rootlets include a striated rootlet and two microtubular rootlets (Figure 1). The character and arrangement of these three fibrillar structures varies from one major group of ciliates to another (Figure 2), and it was these differences in pattern that enabled electron microscopists in the 1980s to realign many ciliates into more natural or monophyletic groups.

The third characteristic that distinguishes ciliates is their sexual process, referred to as conjugation. This is typically a temporary fusion of cells, during which each cell can donate and receive a gametic nucleus. The gametic nuclei are derived by meiosis from the micronucleus and are thus haploid. A complex cytoskeletal structure called the conjugation basket forms in some species to enable the transfer of gametic nuclei from one partner to another. In order to conjugate, ciliates must be sexually mature, a process that may take hundreds of fissions from the time of the last conjugation. After conjugation, the 'new' ciliates are essentially immature again and must go through many cell divisions before they are able to secrete pheromones that stimulate other cells to enter conjugation with them or until they are able to express mating-type proteins on their cell surfaces, which are recognized by cells of a complementary mating type and so enable cell fusion and gametic nuclear exchange.

The ciliates display tremendous morphological diversity, and over 8000 morphological species or morphospecies have been described, both extinct and extant. In fact, the number of 'true biological' species is likely to be two or three times that number. Early assessments of the genetic diversity of morphospecies suggest that there is a great amount of 'hidden' (cryptic) diversity in ciliates. The overall diversity of the ciliates may also have been underestimated because little work has been done on this group outside of Europe and western Asia. As ciliatologists explore more of the world and diverse habitats in other continents, many more morphospecies are being discovered. Based on current trends in the discovery of new species by examination of new locations and habitats, some estimate the existence of 30 000 or more species of ciliates.

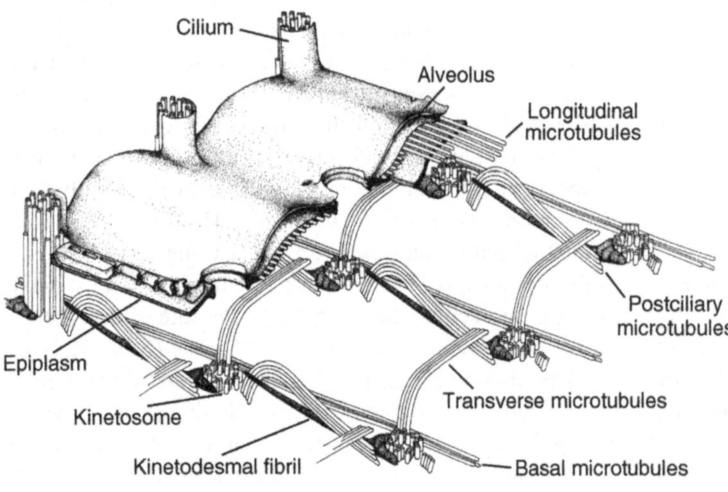

FIGURE 1 A diagrammatic representation of the ciliate cortex centered around the kinetid with its cilium and three rootlets. The striated rootlet is called the kinetodesmal fibril, which originates in the anterior right quadrant of the kinetosome and extends anteriorly or right in the cortex. The two microtubular rootlets are called the postciliary microtubular ribbon and the transverse microtubular ribbon: first, because their microtubules are lined up in a ribbon-like arrangement; and second, because the postciliary ribbon originates from the posterior right quadrant of the kinetosome and extends posteriorly in the cortex, while the transverse ribbon originates from the left anterior quadrant and extends transversely in the cortex.

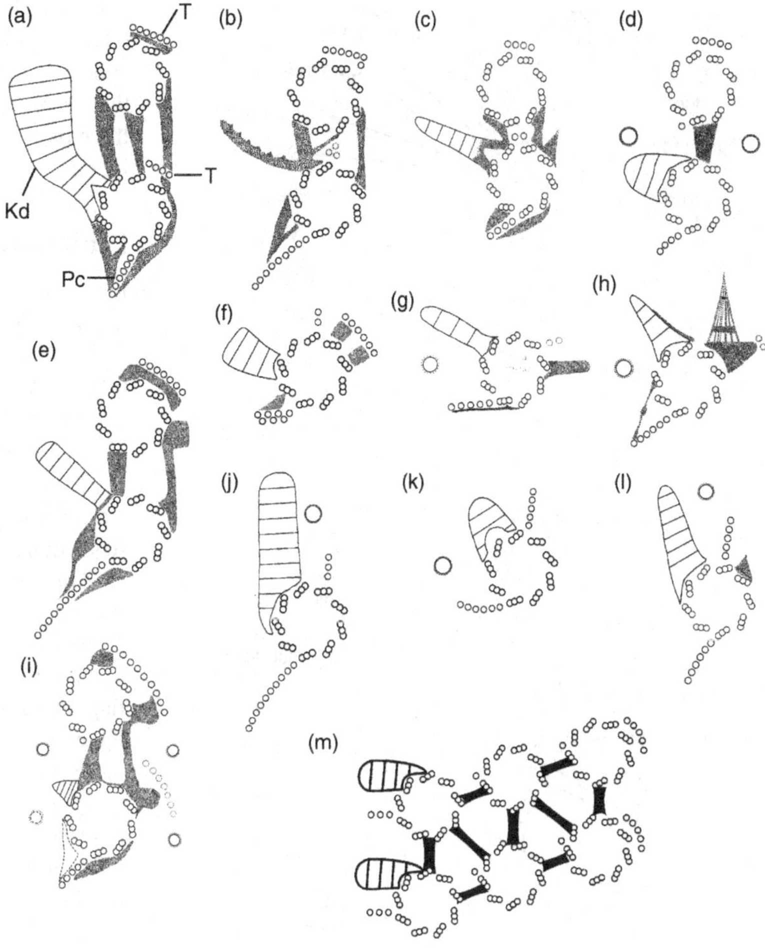

FIGURE 2 The structure of the somatic kinetids of ciliates is representative of the major clades or classes within the phylum. (a) *Loxodes* – Class Karyorelictea; (b) *Blepharisma* – Class Heterotrichea; (c,d) *Protocruzia* (c), *Euplotes* (d) – Class Spirotrichea; (e) *Metopus* – Class Armophorea; (f) *Balantidium* – Class Litostomatea; (g) *Chilodonella* – Class Phyllopharyngea; (h) *Obertrumia* – Class Nassophorea; (i) *Colpoda* – Class Colpodea; (j) *Plagiopyla* – Class Plagiopylea; (k) *Holophrya* – Class Prostomatea; (l) *Tetrahymena* – Class Oligohymenophorea; (m) *Plagiotoma* – Class Spirotrichea. Kd, kinetodesmal fibril; Pc, postciliary microtubular ribbon; T, transverse microtubular ribbon.

Morphological Features

The 'classic' approach to uncovering diversity among ciliates has been the examination of their morphology by cytological methods, both as living organisms and stained by a variety of techniques. A recent description of these techniques was published by William Foissner in 1991. Examining ciliates for any length of time will reveal to the attentive observer a wealth of characters, and these have all been used at one time or another to describe new species or redescribe those discovered by past microscopists (Figure 3).

Useful somatic structures include characteristics of the ciliature (the total number of somatic kineties and lengths of somatic and caudal cilia), of secretory organelles, such as pigmentocysts, trichocysts, and mucocysts (size and shape), and of other organelles, such as the contractile vacuole and its pore, and the collecting canals and spongiome that 'feed' the vacuole (numbers, positions, and other characteristics) (Figure 3). Structures related to feeding can be highly significant such as paroral kinetosomes and paroral cilia (number and arrangement), oral polykinetids (number and arrangement), circumoral dikinetids and circumoral cilia (number and pattern), oral palps and brosse kinetids (number and length), toxicysts (presence and length), oral ribs (number and arrangement), oral nematodesmata (lengths and number), and cytostome and cytopharynx (position); the size of phagosomes or food vacuoles and their contents as an indication of prey preferences; and cytoproct or cell anus (position and length) (Figure 3). Features of the nuclei like the number, positions, and dimensions of the micronucleus and macronucleus can also be assessed. And, this is not an exhaustive list, as the two examples in Figure 3 still miss much of the diversity in the phylum.

ORIGIN OF CILIATES

Ciliates belong to the eukaryote supergroup Alveolata (alveolates). These protists are characterized by the presence of unit membrane-bound sacs or alveoli, underlying their plasma membrane (Figure 1). There are three major groups within alveolates – the dinoflagellates, the apicomplexans, and the ciliates. Since the dinoflagellates, apicomplexans, and ciliates are 'closely related' to one another and sister to a variety of related alveolate flagellates, like the

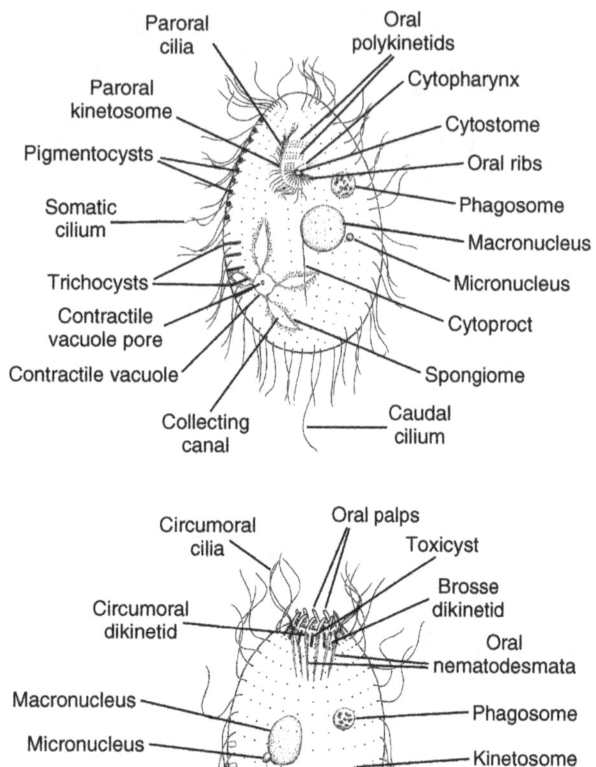

FIGURE 3 Features of ventrostome (top) and prostome (bottom) ciliates that can be used to characterize morphological species or morphospecies. These two examples, however, only provide a small sampling of the variety of features that have been used by morphologists to describe new taxa.

colpodellids, it is most parsimonious to conclude that the ciliates evolved from a flagellate ancestor. How this happened is a matter of conjecture, and there are two theories that seem reasonable to explain the evolution of two of the three major features of ciliates: the complex cortex covered by cilia and nuclear dimorphism.

Evolution of the Ciliate Cortex

Earlier theories on the evolution of the ciliate cortex proposed that the rows of somatic ciliature (files of kinetosomes or kineties) evolved first. After the cell surface was covered by these kineties, simple oral ciliature evolved and this oral ciliature became more complex as ciliates diversified. The problem with these theories is that oral structures of the heterotrophic ciliates appeared to have evolved anew from a presumably heterotrophic ancestral flagellate: these theories required ciliates to have 'reinvented' the oral apparatus.

The theory that seems most plausible today was proposed by Klaus Eisler in 1992. Eisler started with a protociliate-flagellate that had a cytostome (oral apparatus) to the right of which was a file of multiple dikinetid (paired flagella) units (Figure 4), similar to those of modern dinoflagellates such as *Polykrikos*. The two kinetosomes of these dikinetids separated: the right-hand kinetosomes formed somatic kinety 1 (K_1, Figure 4(a)) and dikinetids reformed to the right of the cytostome. This kinety of dikinetids comprised the first paroral for ciliates, and the repetition of this replication process around the cell covered the protociliate with '*n*' somatic kineties (K_n, Figure 4(b)). Ultimately, Eisler argued, oral structures to the left of the cytostome, the so-called adoral organellar complexes like membranelles, developed by the replication of kinetosomes to the left of the cytostome (Figure 4(c)).

Eisler's theory assumes that the cytostome of the protociliate was on one surface of the cell, which is operationally defined as ventral. Thus, ventrostomial ciliates represent the ancestral condition in the phylum. This is contrary to

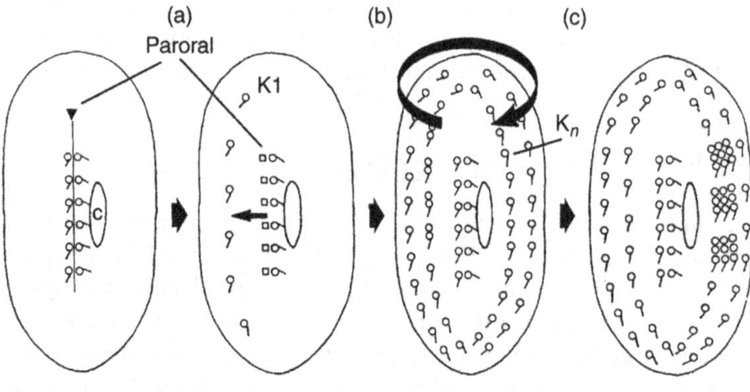

FIGURE 4 A diagrammatic representation of the evolution of the ciliate cortex from a dinoflagellate-like ancestor. The kinetosomes of the paroral dikinetid separate (a) to produce the first ciliary row or kinety (K_1). This process is repeated until the entire cortex of the protist is covered by *n* kineties (b, K_n). In the final step, (c) adoral structures are differentiated within kineties to the left of the oral region. *Modified from Eisler, K. (1992). Somatic kineties or paroral membrane: Which came first in ciliate evolution?* BioSystems, 26, *239–254.*

the 'classical' view of the ciliate cortex, which imagined that prostomial ciliates – those with a simple, anterior oral region – were ancestral. Using Eisler's theory, it must be concluded that the several times that prostomy evolved among ciliates – in the classes Karyorelictea, Litostomatea, and Prostomatea – must have occurred independently.

Evolution of Nuclear Dimorphism

The polyploid macronucleus of ciliates might be more properly called ampliploid because it contains typically many copies of all genes but not many entire copies of the genome. This allows for high rates of transcription for genes essential to support a large cytoplasm and may explain how ciliates achieved large cell size without becoming multicellular. The differentiation of multiple nuclei is common in unicellular eukaryotes, sometimes allowing evolution of extremely large cells such as in mycetozoa and plasmodiophorids.

Orias in 1991 suggested a link between nuclear dimorphism and the sexual cycle of ciliates in which new macronuclei are differentiated in the postzygotic period from the division products of the zygotic nucleus. This often results in more macronuclei than is typical of a particular ciliate species. For example, a species might typically have one micronucleus and one macronucleus in its asexual phase, but the postzygotic cell might have four macronuclei.

Orias also argued that the protociliate had a division-less macronucleus (Div-Mac), like modern ciliates in the class Karyorelictea, and that macronuclear division evolved several times independently within the ciliates. Orias noted that macronuclear division among ciliates occurs in quite diverse ways: in the class Heterotrichea, microtubules assemble outside the macronuclear envelope (extramacronuclear microtubules) and form 'supporting rods' outside the nuclear envelope to elongate the macronucleus, while ciliates of the subphylum Intramacronucleata divide the macronucleus using intramacronuclear microtubules (Table 1).

TABLE 1 Classification of the phylum Ciliophora

Phylum CILIPHORA Doflein, 1901	Phylum CILIPHORA Doflein, 1901
Subphylum *POSTCILIODESMATOPHORA* Gerassimova and Seravin, 1976	Order Tintinnida Kofoid and Campbell, 1929
Class KARYORELICTEA Corliss, 1974	Order Choreotrichida Small and Lynn, 1985
Order Protostomatida Small and Lynn, 1985	Subclass Stichotrichia Small and Lynn, 1985
Order Loxodida Jankowski, 1978	Order Stichotrichida Fauŕ-Fremiet, 1961
Order Protoheterotrichida Nouzar'de, 1977	Order Sporadotrichida Fauŕ-Fremiet, 1961
Class HETEROTRICHEA Stein, 1859	Order Urostylida Jankowski, 1979
Order Heterotrichida Stein, 1859	Subclass Oligotrichia Bütschli, 1887
Subphylum *INTRAMACRONUCLEATA* Lynn, 1996	Order Strombidiida Petz and Foissner, 1992
Class SPIROTRICHEA Bütschli, 1889	Class ARMOPHOREA Jankowski, 1964[a]
Subclass Protocruziidia de Puytorac, Grain and Mignot, 1987	Order Armophorida Jankowksi, 1964
Order Protocruziida Jankowski, 1980	Order Clevelandellida de Puytorac and Grain, 1976
Subclass Phacodiniidia Small and Lynn, 1985	Class LITOSTOMATEA Small and Lynn, 1981
Order Phacodiniida Small and Lynn, 1985	Subclass Haptoria Corliss, 1974
Subclass Licnophoria Corliss, 1957	Order Haptorida Corliss, 1974
Order Licnophorida Corliss, 1957	Order Pleurostomatida Schewiakoff, 1896
Subclass Hypotrichia Stein, 1859	Order Cyclotrichida Jankowski, 1980 *incertae sedis*
Order Kiitrichida Nozawa, 1941	Subclass Trichostomatia Bütschli, 1889
Order Euplotida Small and Lynn, 1985	Order Vestibuliferida de Puytorac et al., 1974
Subclass Choreotrichia Small and Lynn, 1985	Order Entodiniomorphida Reichenow in Doflein and Reichenow, 1929

Continued

TABLE 1 Classification of the phylum Ciliophora—cont'd

Phylum CILIPHORA Doflein, 1901	Phylum CILIPHORA Doflein, 1901
Order Macropodiniida Lynn, 2008[a]	Order Sorogenida Foissner, 1985
Class PHYLLOPHARYNGEA de Puytorac et al., 1974	Class PROSTOMATEA Schewiakoff, 1896
Subclass Cyrtophoria Fauŕ-Fremiet in Corliss, 1956	Order Prostomatida Schewiakoff, 1896
Order Chlamydodontida Deroux, 1976	Order Prorodontida Corliss, 1974
Order Dysteriida Deroux, 1976	Class PLAGIOPYLEA Small and Lynn, 1985[a]
Subclass Chonotrichia Wallengren, 1895	Order Plagiopylida Jankowski, 1978
Order Exogemmida Jankowski, 1972	Order Odontostomatida Sawaya, 1940 *incertae sedis*
Order Cryptogemmida Jankowski, 1975	Class OLIGOHYMENOPHOREA de Puytorac et al., 1974
Subclass Rhynchodia Chatton and Lwoff, 1939	Subclass Peniculia Fauŕ-Fremiet in Corliss, 1956
Order Hypocomatida Deroux, 1976	Order Peniculida Fauŕ-Fremiet in Corliss, 1956
Order Rhynchodida Chatton and Lwoff, 1939	Order Urocentrida Jankowski, 1980
Subclass Suctoria Claparéde and Lachmann, 1858	Subclass Scuticociliatia Small, 1967
Order Exogenida Collin, 1912	Order Philasterida Small, 1967
Order Endogenida Collin, 1912	Order Pleuronematida Fauŕ-Fremiet in Corliss, 1956
Order Evaginogenida Jankowski in Corliss, 1979	Order Thigmotrichida Chatton & Lwoff, 1922
Class NASSOPHOREA Small and Lynn, 1981	Subclass Hymenostomatia Delage & Hrouard, 1896
Order Synhymeniida de Puytorac et al., 1974	Order Tetrahymenida Fauŕ-Fremiet in Corliss, 1956
Order Nassulida Jankowski, 1967	Order Ophryoglenida Canella, 1964
Order Microthoracida Jankowski, 1967	Subclass Apostomatia Chatton & Lwoff, 1928
Order Colpodidiida Foissner, Agatha and Berger, 2002 *incertae sedis*	Order Apostomatida Chatton & Lwoff, 1928
Class COLPODEA Small and Lynn, 1981	Order Astomatophorida Jankowski, 1966
Order Bryometopida Foissner, 1985	Order Pilisuctorida Jankowski, 1966
Order Bryophryida de Puytorac, Perez-Paniagua and Perez-Silva, 1979	Subclass Peritrichia Stein, 1859
Order Bursariomorphida Fernández-Galiano, 1978	Order Sessilida Kahl, 1933
Order Colpodida de Puytorac et al., 1974	Order Mobilida Kahl, 1933
Order Cyrtolophosidida Foissner, 1978	Subclass Astomatia Schewiakoff, 1896
	Order Astomatida Schewiakoff, 1896

[a]*A taxon based on molecular phylogenetics, but still lacking a morphological synapomorphy.*

Ultimately, there are three major kinds of macronuclear categories in the ciliates with the vast majority of species dividing by intramacronuclear microtubules (subphylum Intramacronucleata). There is one unusual ciliate that requires deeper investigation, the macronuclear nodules of the spirotrich *Protocruzia* (subclass Protocruziidia, Table 1). In its macronuclear division, two large, apparently composite chromosomes condense in each nodule and 'slide' past each other by some unknown mechanism to segregate to the two daughter macronuclei.

MAJOR CLADES

Uncovering the history of diversification has occupied ciliate systematics since the nineteenth century. The first ideas about diversity assumed that the species with 'simpler' cell forms were more ancestral, while the more 'complex' cell forms were derived. The former consisted of those ciliates with an anterior cytostome, unadorned by complex ciliary organelles, the 'prostomial' forms (Figure 3). These were considered ancestral to taxa with a ventral cytostome surrounded by many complex ciliary organelles. These ideas, although modified by new evidence, went basically unchallenged until the last half of the twentieth century.

Ultrastructure and the Structural Conservatism of the Ciliate Cortex

In the mid-twentieth century, invention of the electron microscope revolutionized investigations of the ultrastructure of cells. These images provided light microscopists with a whole new perspective on the 'fine' structure of cells, revealed the detailed nature of organelles whose functions had been guessed at since the nineteenth century, and uncovered a diversity of structural details that could be used to test ideas about relationships formulated on the basis of the light microscopic perspective.

For ciliates, the arrangements of the fibrillar structures associated with the kinetosomes turned out to be highly suggestive of major clades. Some clades had two kinetosomes linked by fibers, while others had only a single kinetosome, and the size, direction, and number of fibers could vary from group to group (Figure 2). The problem with this new set of information was that it grouped taxa that were not considered to be closely related based on light microscopy. In 1976, Lynn argued that ultrastructural evidence of affinity was more reliable than the evidence based on light microscopy. That is, patterns of variation of somatic structures, like the arrangement of somatic kineties and their number, could be quite variable in a recognized natural assemblage (monophyletic group) at the level of light microscopy. However, when these structures were examined by electron microscopy, there was virtually no variation in fine structural-level arrangement of the supporting fibrillar structures. Lynn demonstrated this fact for a number of known monophyletic groups, and formulated the principle of structural conservatism: the conservation of biological structure is inversely related to the level of biological organization. This led to the use of the somatic kinetid structure as one of the primary means to recognize clades or monophyletic groups of ciliates.

In 1981, Eugene Small and Denis Lynn proposed a new view of the diversification of ciliate groups, recognizing eight classes instead of three, primarily by aggregating taxa whose somatic kinetids showed similar structural features (Figure 2). This macrosystem was essentially a hypothesis about phylogenetic relationships among taxa. In the 1980s, an approach to systematics using gene sequences – molecular systematics – was becoming increasingly efficient. Since the molecular level represents an even 'lower' level in the biological organization, its results could be expected to provide a robust test of phylogenetic relationships based on ultrastructure – a 'higher' level, based on the structural conservatism hypothesis.

Molecular Systematics and the Major Clades

Molecular systematics provided an objective means for testing the robustness of the major clades of ciliates established based on ultrastructural features. The gene of choice for ciliates has been the SSU rRNA gene, which specifies the RNA skeleton of the small subunit of the ribosome. This molecule includes both highly conservative and more variable regions, and thus can be used to test both deep (ancient) phylogenetic relationships and more recent ones.

Now, there are hundreds of SSU rRNA sequences for ciliates, including multiple representatives for all the major subdivisions in the phylum. Analyses of these data have confirmed the basic integrity of the eight major classes of Small and Lynn, with some changes in assigning subclasses (Table 1). There is now good evidence that there are at least 11 major clades or classes of ciliates (Figure 5).

DIVERSITY AT THE CLASS LEVEL

The ciliates are a morphologically and genetically diverse group. In fact, some classes are as genetically different from one another, based on the SSU rRNA gene, as higher plants are from higher animals. Yet, based on the SSU rRNA gene, the ciliates are undoubtedly a strongly supported monophyletic group, confirming the conclusions of microscopists since the nineteenth century. Our current understanding is that they can be divided into 2 subphyla and 11 classes.

The Two Subphyla

The early phylogenies based on the SSU rRNA gene suggested that the ciliates were divided into two deeply divergent clades, and the subsequent research on this gene and others has confirmed this result. One of the groups also has a very strong unifying cortical feature, that is, their postciliary microtubular ribbons are extremely well developed and extend posteriorly to overlap each other in a complex ribbon (postciliodesma). These microtubules are crucial in reelongating these highly contractile cells after they have shortened their bodies using complex microfilamentous myonemes. This subphylum has therefore been named the Postciliodesmatophora. The other major clade encompasses the vast majority of ciliate diversity and is distinguished by the division of the macronucleus using intramacronuclear microtubules. Therefore, this subphylum is named Intramacronucleata.

The 11 Classes

The 11 classes are divided into the 2 subphyla as follows: subphylum Postciliodesmatophora – classes Karyorelictea and Heterotrichea; and subphylum Intramacronucleata – classes Spirotrichea, Armophorea, Litostomatea, Phyllopharyngea, Prostomatea, Colpodea, Nassophorea, Plagiopylea, and

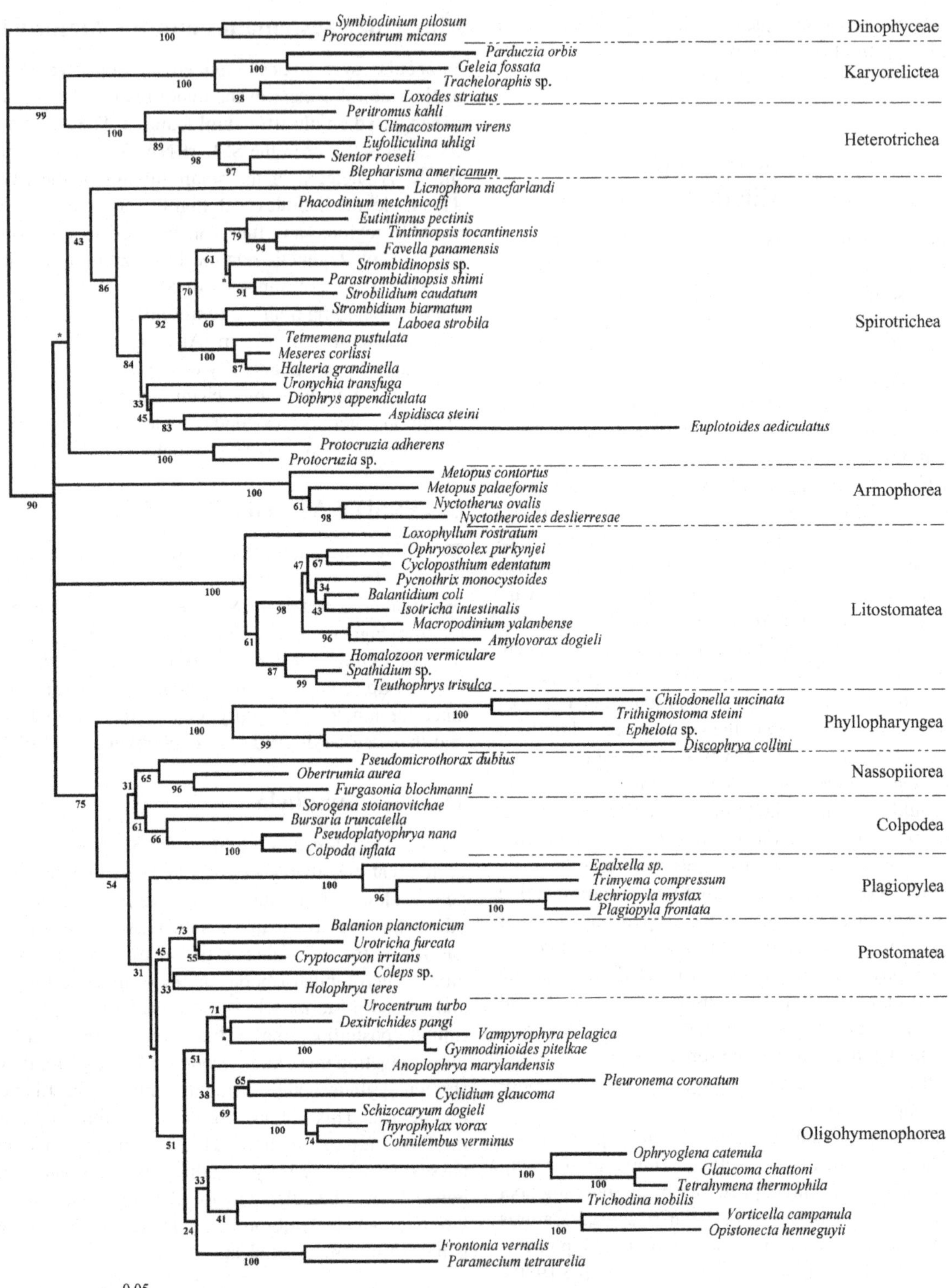

FIGURE 5 A phylogeny of the phylum Ciliophora based on small subunit rRNA (SSU rRNA) gene sequences and using the profile-neighbor-joining method implemented in Profdist ver. 0.9.6.1. There are two deep subdivisions in the phylum, now recognized as the subphyla Postciliodesmatophora and Intramacronucleata (Table 1). These 2 subphyla have diversified into 11 major clades or classes (Table 1).

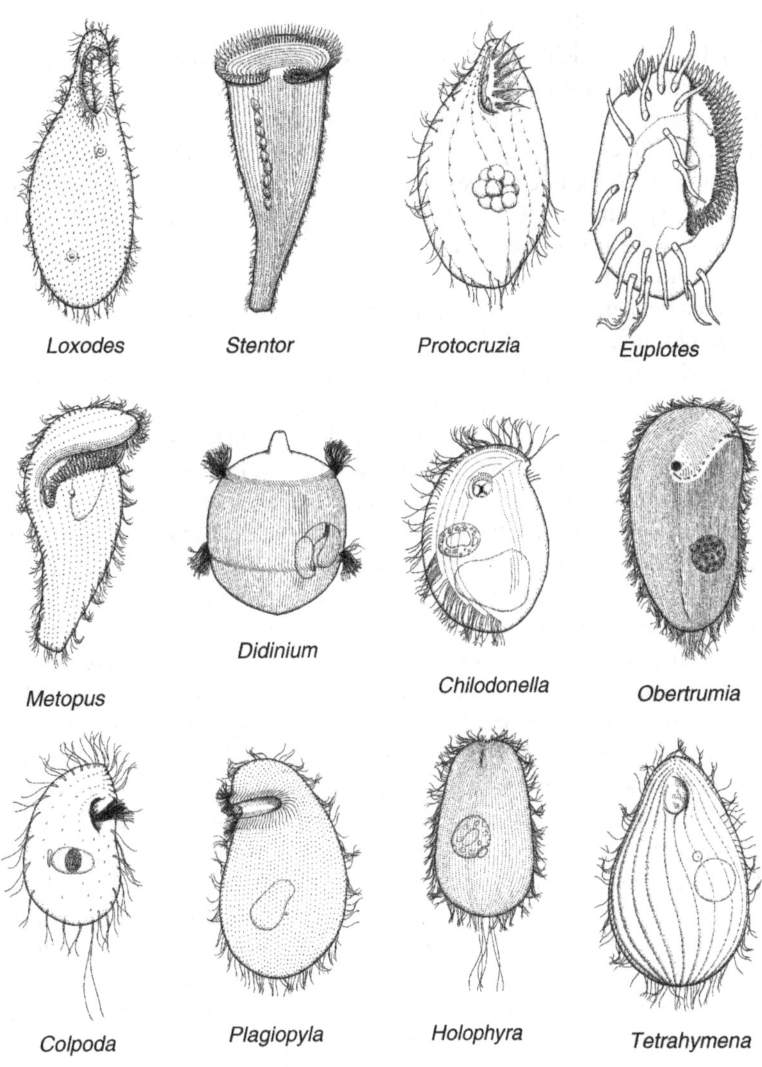

FIGURE 6 Illustrations of genera representative of each of the 11 classes in the phylum Ciliophora. *Loxodes* – Class Karyorelictea; *Stentor* – Class Heterotrichea; *Protocruzia*, *Euplotes* – Class Spirotrichea; *Metopus* – Class Armophorea; *Didinium* – Class Litostomatea; *Chilodonella* – Class Phyllopharyngea; *Obertrumia* – Class Nassophorea; *Colpoda* – Class Colpodea; *Plagiopyla* – Class Plagiopylea; *Holophrya* – Class Prostomatea; and *Tetrahymena* – Class Oligohymenophorea.

Oligohymenophorea (Table 1). Typical genera are illustrated for each of these classes and they are briefly characterized in Figure 6.

The class Karyorelictea (e.g., *Loxodes*; Figure 6) includes 'karyological relicts' (karyon = nucleus; relictos = ancestor). This is because they have macronuclei that do not replicate, and this is presumed to be the ancestral state in the ciliates. As members of the subphylum Postciliodesmatophora, these species have postciliodesmata associated with their somatic dikinetids. However, these ribbons differ from those of the heterotrichs in that the postciliary microtubules are arranged in a 2 + ribbon + 1 pattern. Karyorelicteans are commonly found in marine interstitial sand habitats, often with low oxygen concentrations. *Loxodes* is unique in the class, as it is the only genus found in freshwater. However, it is still typically found in the benthos, unless the water column goes anoxic; in these circumstances, *Loxodes* migrates upward, becoming planktonic. Karyorelicteans vary in size, and this determines the prey they feed on, which vary from bacteria to rotifers.

Members of the class Heterotrichea (e.g., *Stentor*; Figure 6) were once thought to be highly derived species. They are now placed along with the karyorelicteans in the subphylum Postciliodesmatophora because of the presence of postciliodesmata and strong support from SSU rRNA gene phylogeny (Figure 5). The postciliodesma of heterotrichs, however, have microtubules arranged in a ribbon+1 microtubule pattern. Furthermore, the macronuclei of heterotrichs divide, although they do this using extramacronuclear microtubules. Heterotrichs are large and conspicuous ciliates, typically free-swimming with a conspicuous adoral zone of oral polykinetids or membranelles. These conspicuous oral structures give the heterotrichs their name (heteros = different; trichos = hair): the oral ciliary structures are conspicuously different from the somatic ones. These species are found in both benthic and planktonic habitats and in freshwater and marine ecosystems. Heterotrichs are also conspicuously pigmented by cortical vesicles called pigmentocysts. These pigments have two properties: they apparently confer light sensitivity on these ciliates and their extrusion appears to deter some predators from feeding on them.

The Spirotrichea (e.g., *Protocruzia* and *Euplotes*; Figure 6) were also once considered to be highly derived species and were also related to the heterotrichs. It now appears that this group may be quite ancient, as it typically appears as the earliest diverging clade in the subphylum Intramacronucleata (Figure 5). As their name suggests, the spirotrichs (spiros = spiral; trichos = hair) have a spiraling set of ciliary structures – the adoral zone of oral polykinetids or membranelles, which is similar to that of the heterotrichs (Figure 6). Spirotrichs typically have reduced somatic ciliature, and in a number of subclasses, these cilia are differentiated into complex 'leglike' organellar complexes called cirri (e.g., *Euplotes*; Figure 6). The only morphological synapomorphy (shared-derived) character for the spirotrichs is the replication band, a structure that passes through the macronucleus during S phase (DNA synthesis). However, the replication band is absent in *Protocruzia*, the only genus in the subclass Protocruziidia (Figure 6) and the subclass of phacodiniids. It is likely that this feature was lost during evolution of the phacodiniids, as they appear to be derived members of the spirotrich clade (Figure 5). However, protocruziids may be the earliest branch of spirotrichs and are only weakly associated with other members of this class based on molecular characters. Protocruziids, in addition to lacking replication bands, also have unusual macronuclei whose chromosomes appear to remain throughout the vegetative cycle, condensing at karyokinesis. This suggests that protocruziids could actually be the only representatives of a 12th ciliate class. Spirotrichs are found in diverse habitats throughout the world and are especially conspicuous in the plankton of the world's oceans: the loricate tintinnid ciliates and their aloricate relatives often dominate marine ciliate planktonic communities where they consume bacteria, diatoms, and dinoflagellates.

The class Armophorea (e.g., *Metopus*; Figure 6) is one of the newest identified classes, sometimes referred to as a 'riboclass,' because support for the clade comes only from SSU rRNA phylogeny and, now, that of several other genes (Figure 5). Thus, there are no shared morphological features for the two included orders – the Armophorida and Clevelandellida (Table 1). Armophoreans do share the ecological trait of being anaerobes, and they have hydrogenosomes rather than true mitochondria. Often, these organelles are associated with methanogenic endosymbionts that utilize the hydrogen produced by the hydrogenosomes, turning it into methane that is released into the environment. In fact, the methane coming indirectly from these ciliates may contribute a significant amount of this potent greenhouse gas to the atmosphere.

The class Litostomatea (e.g., *Didinium*; Figure 6) includes many species that were long considered to represent early branches of the ciliates, as they typically have simple oral ciliature (litos = simple; stomos = mouth). However, it is now believed that the oral structures of litostomes have been secondarily evolved from somatic kinetids. These species are united by a strong morphological synapomorphy: the somatic kinetid has two transverse microtubular ribbons, a unique feature in the phylum (Figure 2). The Litostomatea include two classes that are ecologically quite different. The Haptoria includes predatory ciliates that use toxic secretory organelles or extrusomes (toxicysts) to immobilize and capture prey. *Didinium* is a classic example of this strategy, which it uses in attacking *Paramecium* species. Species in the subclass Trichostomatia lack toxicysts and are all endosymbionts of animals, ranging from fish to marsupials and even to humans. In fact, the only ciliate parasite of humans is the trichostome *Balantidium coli*, which is typically found in people who live closely with pigs. Balantidiiasis is characterized by diarrhea and in severe cases degradation of the intestinal epithelium by the ciliate. Most trichostomes, however, are commensals that feed on bacteria and other protists in the gut environment of the animal host, and some species that are found in herbivores have adapted to feeding on plant tissues ingested by the host. Trichostomes often have hydrogenosomes rather than mitochondria, and may be indirectly responsible for a significant amount of the methane production of ruminants whose digestive system also harbors methanogens, some of which are dependent on the ciliates to produce hydrogen.

The class Phyllopharyngea (e.g., *Chilodonella*; Figure 6) is the first branch in a clade comprising the last six classes of ciliates. We have no significant morphological synapomorphy for this group of six classes, although it is often recovered in molecular phylogenies of various genes (Figure 5). The phyllopharyngeans are distinguished by an array of microtubular ribbons that form a flower-like arrangement supporting the cytopharynx (phyllos = leaf; pharynx = throat). The somatic kinetids are also characteristic of this group, underlain by ribbons of subkinetal microtubules. The four subclasses are morphologically very diverse. Members of the subclasses Chonotrichia and Suctoria have reduced or no cilia in their 'adult' stages. In fact, for many years, these two groups of ciliates were distinguished from the other 'true' or euciliate groups because of this absence. Suctorians are even more unusual in that they have groups of tentacles emerging from the cell body. These tentacles are used in feeding, and so suctorians have been referred to as polystomatous (many-mouthed). Members of these latter two subclasses are often found as symbionts on other organisms, often crustaceans, where they do not harm their hosts but benefit from being carried around by them. The fourth subclass, the Rhynchodia, comprises 'endosymbionts,' typically found in the mantle cavity of bivalve mollusks where they attach to the mantle or gill tissue and feed on the cell contents. *Chilodonella* can also be found on the gills and epithelia of fish, and can become especially problematic in aquaculture situations. Fish need to be washed with various chemical solutions, such as a mild formaldehyde solution, to remove the ciliates.

The class Nassophorea (e.g., *Obertrumia*; Figure 6) was once considered a 'pivotal' group in the evolution of the 'higher' ciliates, like the heterotrichs and spirotrichs. This was because nassophoreans, like *Obertrumia*, had a few to sometimes many poorly developed oral polykinetids, presaging the complex and conspicuous adoral zones of the 'higher' ciliates. Gene sequences now refute this hypothesis and demonstrate a close relationship with other ciliate groups that typically have few oral membranelles, such as some colpodeans and oligohymenophoreans (Figure 5). Nassophoreans have a complex cytopharyngeal apparatus composed of nematodesmata or large rods of microtubules that form a basket-like structure (nasse = basket; phoros = bear). The morphological trait that is shared by all members of the group is a special ribbon of microtubules called the X-lamella associated with these microtubular bundles. Some species also have special extensions of the alveoli called 'alveolocysts' that extend into the cortical cytoplasm. Nassophoreans typically consume filamentous cyanobacteria or blue-green algae, which they 'suck' down their cytopharynx using arms on the microtubules of the X-lamellae.

The class Colpodea (e.g., *Colpoda*; Figure 6) was the first class to be radically revised based on Lynn's structural conservatism hypothesis. The group now includes species once classified as prostomes, trichostomes, hymenostomes, and heterotrichs. Colpodea is now united by a strong morphological synapomorphy of the somatic kinetid, that is, dikinetids whose posterior kinetosome has a transverse microtubular ribbon that extends posteriorly along the kinety, forming the so-called transversodesma (Figure 2). Colpodeans are often twisted and bulgy cells (colpos = breast) and are typically only freshwater or terrestrial ciliates. Terrestrial species often inhabit the microlayer of water found on terrestrial plants, grasses, and mosses and between soil particles. They feed on the bacteria and small algae found there. Since they must survive the sporadic drying conditions that often occur in terrestrial habitats, colpodeans are able to encyst, holding the 'ciliate world record' for surviving in this cryptobiotic state – almost 40 years!

The class Plagiopylea (e.g., *Plagiopyla*; Figure 6) is another 'riboclass,' as its monophyly and apparent distinctness are based solely on SSU rRNA phylogeny (Figure 5). The class includes two morphologically highly distinct orders – Plagiopylida and Odontostomatida (Table 1). The class name refers to the obliquely oriented oral opening of the plagiopylid ciliates (plagios = oblique; pylon = gate) (Figure 6). All are found in anaerobic to microaerophilic habitats, and both trimyemids and plagiopylids have hydrogenosomes rather than mitochondria that are often accompanied by methanogenic bacteria, as many as 3000 per cell. Plagiopyleans are typically bacterivorous, but since they are rarely highly abundant in their preferred habitats, they have relatively little impact on the bacterial community.

The class Prostomatea, exemplified by the genus *Holophrya* (formerly *Prorodon*; Figure 6), includes ciliates that were 'classically' considered ancestral because their oral apparatus is simple and anteriorly located (pro = before; stomos = mouth). Their main morphological synapomorphy is the simple arrangement of oral ciliature (the brosse), which borders the oral region. Prostomes can be common in the plankton of marine and freshwaters, with genera such as *Urotricha* and *Tiarina* reaching abundances of over 10 000 cells l^{-1}. The parasite of marine fishes, *Cryptocaryon irritans*, has recently been included in this group. Interest in this species is increasing, as it is becoming increasingly problematic in marine aquaculture.

The last class, but by no means the least, is the Oligohymenophorea, exemplified by the genus *Tetrahymena* (Figure 6). This class is comparable to the Spirotrichea in its overall diversity, both in terms of numbers of subclasses and numbers of species. In addition to *Tetrahymena* – the 'white mouse' of the ciliate world, it includes *Paramecium* – the 'lab rat' because it is a larger cell. Species from these two groups are widely studied model organisms because they have been amenable to laboratory culture since the 1920s and genetic manipulation since the mid-twentieth century. Thus, complete or nearly completed genome sequences are available for several of these species, and more are being planned or in progress. Oligohymenophoreans lack a strong morphological synapomorphy, although a paroral and three oral polykinetids or membranelles are shared by virtually all species except those in the subclass Astomatia, which as 'mouthless' ciliates, are presumed to have lost them. The diversity of the group may explain why they are often not recovered as a monophyletic group in gene trees (Figure 5).

There are, in total, six subclasses of Oligohymenophorea: Peniculia, Scuticociliatia, Hymenostomatia, Apostomatia, Peritrichia, and Astomia. Free-living species can be very abundant, especially in the plankton where scuticociliates have been recorded to exceed 40 000 cells l^{-1} and peniculines, such as *Frontonia* and *Stokesia*, have been recorded, but rarely, to exceed 10 000 cells l^{-1}. There are also many symbiotic oligohymenophoreans. Peritrichians are often found as ectosymbionts on animals and plants. Scuticociliatians can be found in the mantle cavities of bivalve mollusks and even as significant parasites of aquacultured fish. Apostomatians and astomatians are all symbionts, the former typically in and on crustaceans and the latter in oligochaete worms. Possibly, the most disreputable parasite among the ciliates belongs to this class – *Ichthyophthirius multifiliis* – the causative agent of white spot disease or 'Ich' in freshwater fish, belongs to the Hymenostomatia. 'Ich' causes problems in freshwater aquaculture, but there is only one report in the literature of a natural mass mortality of fish caused by this ciliate. Oligohymenophoreans were, along with the nassophoreans, thought to be a 'pivotal' group in the evolution of the 'higher' ciliates, that is, species with many membranelles. The molecular phylogenies now suggest that they are probably a derived radiation (Figure 5).

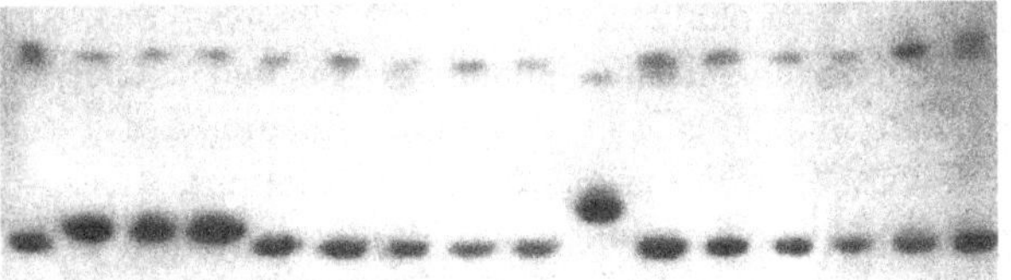

FIGURE 7 An isozymogram of the enzyme isocitrate dehydrogenase from the stichotrich *Stylonychia mytilus* (B, Europe; C, D, North America), *Paraurostyla weissei* (J), and *Stylonychia lemnae* (A, L, O, Europe; EI, K, M, N, P, North America). Note that the isozyme for each species migrates to a slightly different position on the gel. *Reproduced from Ammermann, D., Schlegel, M., & Hellmer, K.-H. (1989). North American and Eurasian strains of* Stylonychia lemnae *(Ciliophora, Hypotrichida) have a high genetic identity, but differ in the nuclear apparatus and in their mating behavior.* European Journal of Protistolology, 25, *67–74.*

PROBING DIVERSITY AT THE SPECIES LEVEL

Understanding the adaptive diversification of ciliates in its broadest sense has relied recently on the technologies of electron microscopy and molecular biology. At the species level, light microscopy is still an extremely important tool for the measurement of morphological variability in ciliates. Nevertheless, molecular tools are becoming increasingly important for rapid large-scale species surveys and for discovering a great deal of hidden diversity.

Morphological Diversity at the Species Level

Morphologists have discovered that taxa can be readily differentiated morphologically from each other, based on a variety of these characters (Figure 3). Nevertheless, morphology only scratches the surface. There are taxa that are clearly 'true' biological species because they do not conjugate, and yet they can only be distinguished, if at all, using multivariate morphometric techniques and highly controlled culture conditions. In *Tetrahymena* and *Paramecium*, we know there are dozens of these so-called cryptic biological species or sibling species. However, the use of mating tests to distinguish species is extremely laborious. Not only does a laboratory have to have a presumed full complement of all the mating types of all the species in a given potential cryptic species complex, but they all have to be sexually mature at the time one discovers a new isolate if a rapid identification of the new isolate is to be made. These problems have, fortunately, been largely solved with molecular genetic approaches.

Molecular Techniques at the Species Level

One of the first molecular techniques to be effectively applied to resolve the identity of cryptic species was isozyme variation. Isozymes are variants of an enzyme that can be distinguished by their differences in electrophoretic mobility. Different isozymes appear as separate distinct bands on electrophoretic gels (electropherograms), and these differences can be precisely quantified (Figure 7). Isozymes effectively distinguish all the cryptic species in the *Paramecium aurelia* complex, as shown by Sonneborn in 1975. However, while informative, isozyme analyses have several drawbacks. They require a significant biomass of cells, which is problematic with unicellular species that can often be difficult to culture; isozyme patterns can sometimes be ambiguous; and there is no easy way to standardize the results except to run samples repeatedly as reciprocal references, which requires even more biomass.

Thus, DNA techniques have mostly replaced isozyme analyses to identify species. One widely useful DNA

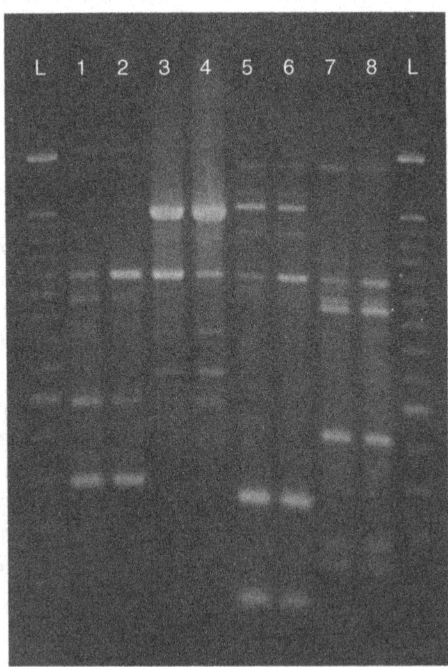

FIGURE 8 A photograph of a gel of DNA fragments derived from a random amplified polymorphic DNA (RAPD) experiment to explore the genetic diversity of species of the ciliate genus *Euplotes*. Three species of *Euplotes* have been examined here: (1, 2) *Euplotes aediculatus* Strain 17, Marseille, France; (3, 4) *E. aediculatus* Strain 18, Ohio, USA; (5, 6) *Euplotes woodruffi*; (7, 8) *Euplotes octocarinatus*. L – molecular weight marker. *Photo courtesy by J¨rgen Kusch, University of Kaiserslautern.*

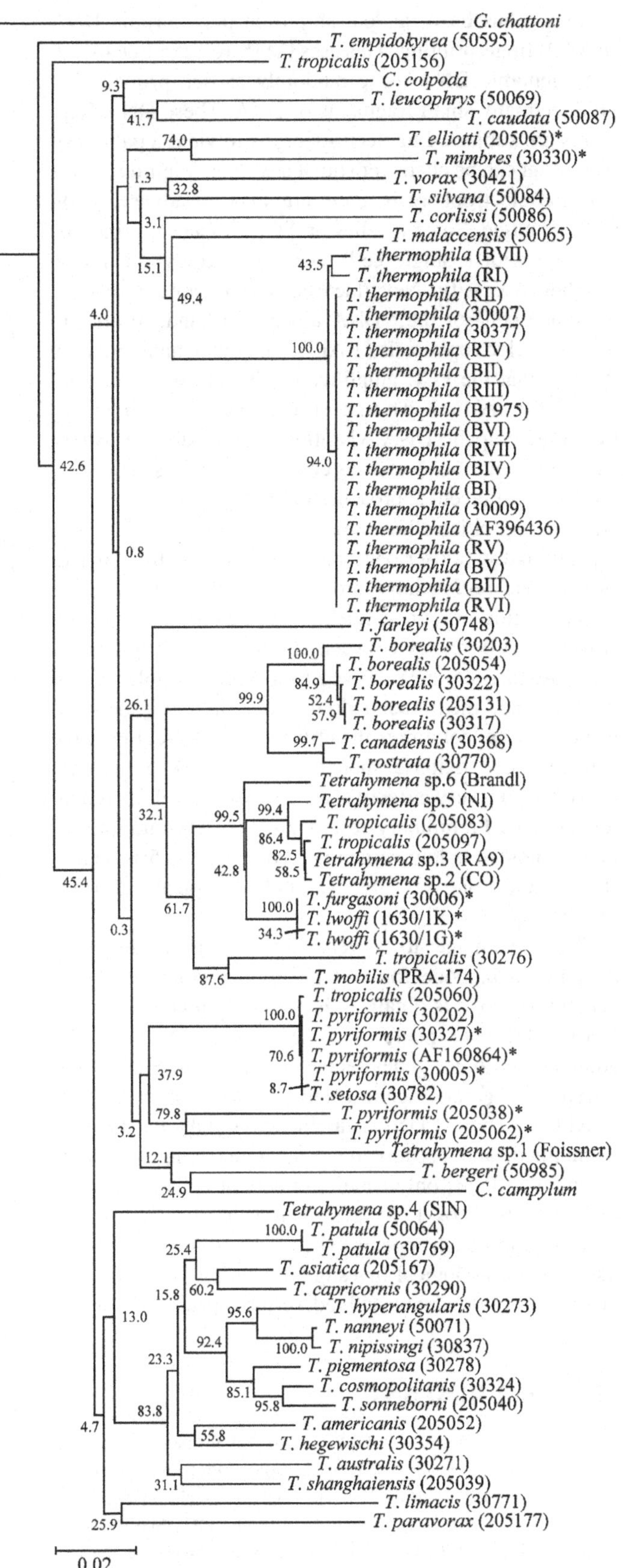

FIGURE 9 A neighbor-joining tree based on 689 bp of the cytochrome *c* oxidase subunit 1 mitochondrial gene sequence, the 'bar code' region, of species of *Tetrahymena* and several related hymenostomes. The 78 isolates cluster out into distinct groups with species represented by tight clusters, except for those assigned to *Tetrahymena pyriformis* and *Tetrahymena tropicalis* (bold taxa), which appear not to be monophyletic groups. Scale bar = 0.02 nt substitutions per site. *Reproduced from Chantangsi, C., Lynn, D. H., Brandl, M. T., Cole, J. C., Netrick, N., & Ikonomi, P. (2007). Barcoding ciliates: A comprehensive study of 75 isolates of the genus* Tetrahymena. International Journal of Systematic and Evolutionary Microbiology, *57, 2412–2425.*

technique has been random amplified polymorphic DNA (RAPD) fingerprinting. This uses the PCR technique to amplify genomic DNA using randomly chosen primers that produce many small fragments of DNA. These DNA fragments are subjected to electrophoresis to yield a pattern of bands that can be characteristic of a species (Figure 8). Like isozyme electrophoresis, there are some disadvantages to this technique. Since it relies on PCR, it can, in principle, be done with a single cell, so mass cultures are not an absolute necessity. Nevertheless, it is important to have reference cells or their DNA always on hand. Similar to isozyme electrophoresis, RAPDs are also hampered by the fact that the PCR amplification is not always equally efficient every time. Thus, even with the same primer and the same template DNA, variations in the banding pattern can occur, and this decreases confidence in these results.

The most recent approaches to resolving species boundaries among ciliates have used gene sequencing. For this, the internally transcribed spacers (ITSs) of the nuclear rRNA gene region have provided useful results such as distinguishing cryptic species in the *P. aurelia* complex. However, differences between species are still not large. Thus, researchers have moved to a more variable part of the 'genome,' the mitochondrial DNA (mtDNA), and work is underway to determine the utility of a small region of the cytochrome *c* oxidase subunit 1 (cox-1) gene that has proved to be very effective as a species-level 'barcode' for animals. Preliminary results are very encouraging, as they demonstrate low variability (typically $>0.5\%$) within species and relatively high variability (typically $>10\%$) between species (Figure 9).

The advantage to a gene sequencing technique is that, like a barcode for a commercial product, once the genetic bar code is obtained, it should be relatively stable (invariable) over many years. Moreover, DNA can be archived and other genes sequenced if it is decided that a different 'bar code' gene should be used in the future. Provided sufficient DNA is archived, cultures need not be maintained, and because the gene sequencing technique begins with PCR, only small numbers of cells are needed to obtain the cox-1 gene sequence in the first place. Thus, this approach appears to be a very promising complement to morphological investigations in providing a complete description of new and redescribed species of ciliates.

FURTHER READING

Asai, D. J., & Forney, J. D. (Eds.), (2000). Tetrahymena thermophila *(methods in cell biology).* (Vol. 62) Academic Press: New York.

Beisson, J. (2008). Preformed cell structure and cell heredity. *Prion, 2*, 1–8.

Chantangsi, C., Lynn, D. H., Brandl, M. T., Cole, J. C., Netrick, N., & Ikonomi, P. (2007). Barcoding ciliates: A comprehensive study of 75 isolates of the genus *Tetrahymena. International Journal of Systematic and Evolutionary Microbiology, 57*, 2412–2425.

Coleman, A. W. (2005). *Paramecium aurelia* revisited. *The Journal of Eukaryotic Microbiology, 52*, 68–77.

Corliss, J. O. (1974). The changing world of ciliate systematics: Historical analysis of past efforts and a newly proposed phylogenetic scheme of classification for the protistan phylum ciliophora. *Systematic Zoology, 23*, 91–138.

Dan, X. M., Li, A. X., Lin, X. T., Teng, N., & Zhu, X. Q. (2006). A standardized method to propagate *Cryptocaryon irritans* on a susceptible host pompano *Trachinotus ovatus. Aquaculture, 258*, 127–133.

Eisler, K. (1992). Somatic kineties or paroral membrane: Which came first in ciliate evolution? *BioSystems, 26*, 239–254.

Eisler, K., & Bardele, C. F. (1983). The alveolocysts of the nassulida: Ultrastructure and some phylogenetic considerations. *Protistologica, 19*, 95–102.

Elliott, A. M., & Nanney, D. L. (1952). Conjugation in *Tetrahymena. Science, 116*, 33–34.

Foissner, W. (1991). Basic light and electron microscopic methods for taxonomic studies of ciliated protozoa. *European Journal of Protistology, 27*, 313–330.

Gentekaki, E., & Lynn, D. H. (2009). High genetic diversity but no population structure of the peritrichous ciliate *Carchesium polypinum* in the Grand River Basin (North America) inferred from nuclear and mitochondrial markers. *Applied and Environmental Microbiology, 75*, 3187–3195. doi:10.1128/AEM.00178-09.

Görtz, H. D. (Ed.), (1988). *Paramecium*. Springer: Berlin.

Juranek, S. A., & Lipps, H. J. (2007). New insights into the macronuclear development in ciliates. *International Review of Cytology, 262*, 219–251.

Lynn, D. H. (1976). Comparative ultrastructure and systematics of the colpodida: Structural conservatism hypothesis and a description of *Colpoda steinii* maupas, 1883. *The Journal of Protozoology, 23*, 302–314.

Lynn, D. H. (1981). The organization and evolution of microtubular organelles in ciliated protozoa. *Biological Reviews, 56*, 243–292.

Lynn, D. H. (2004). Morphology or molecules: How do we identify the major lineages of ciliates (phylum ciliophora)? *European Journal of Protistology, 39*, 356–364.

Lynn, D. H. (2008). *The ciliated protozoa: Characterization, classification, and guide to the literature* (3rd ed.). Springer: Dordrecht.

Lynn, D. H., & Small, E. B. (1997). A revised classification of the phylum Ciliophora doflein. *Revista de la Sociedad Mexicana de Historia Natural, 47*, 65–78.

Orias, E. O. (1991a). Evolution of amitosis of the ciliate macronucleus: Gain of the capacity to divide. *The Journal of Protozoology, 38*, 217–221.

Orias, E. O. (1991b). On the evolution of the karyorelict ciliate life cycle: Heterophasic ciliates and the origin of ciliate binary fission. *BioSystems, 25*, 67–73.

Small, E. B., & Lynn, D. H. (1981). A new macrosystem for the phylum ciliophora doflein, 1901. *BioSystems, 14*, 387–401.

Sonneborn, T. M. (1937). Sex, sex inheritance and sex determination in *Paramecium aurelia. Proceedings of the National Academy of Sciences, 23*, 378–385.

Sonneborn, T. M. (1975). The *Paramecium aurelia* complex of fourteen sibling species. *Transactions of the American Microscopical Society, 94*, 155–178.

Strüder-Kypke, M. C., & Lynn, D. H. (2009). Comparative analysis of the mitochondrial cytochrome c oxidase subunit I gene in ciliates (Alveolata, Ciliophora) and evaluation of its suitability as a marker for barcoding. *Biodiversity and Systematics, 8*, 131–148. doi:10.1080/14772000903507744.

Chapter 16

Secretive Ciliates and Putative Asexuality in Microbial Eukaryotes

Micah Dunthorn[1,*] and Laura A. Katz[1,2]

[1]*Organismic and Evolutionary Biology, University of Massachusetts, Amherst, MA 01003, USA*

[2]*Department of Biological Sciences, Smith College, Northampton, MA 01063, USA*

*Corresponding author: *Dunthorn, M. (dunthorn@rhrk.uni-kl.de) Present address: Department of Ecology, University of Kaiserslautern, 67653 Kaiserslautern, Germany.*

Chapter Outline

Putative Asexual Microbial Eukaryotes	227	**Concluding Remarks and Outstanding Questions**	231
The Powerlessness of Observation	228	**Acknowledgements**	231
An Improved Phylogenetic Framework	229	**References**	232
Reversing the Loss of Sex?	229		
Ancient Asexuality?	231		

PUTATIVE ASEXUAL MICROBIAL EUKARYOTES

Based on the broad distribution of meiotic sex, the ancestor of extant eukaryotes was probably facultatively sexual [1,2]. Although there is a cost to being sexual [3,4], numerous advantages have been proposed for its maintenance in plants and animals (Box 1). While most lineages have remained sexual, but asexual species are found scattered throughout the eukaryotic tree of life, primarily at the tips [4]. When plants and animals are found that are putatively asexual [particularly if they are putatively ancient asexuals (see Glossary)] they are studied intensively, as their existence goes against established theory [3–5]. By contrast, when asexuality is purported in microbial eukaryotes they are ignored or are judged not to be a theoretical problem [6]. But are microbial eukaryotes asexual as often assumed [7–9] and if so, do they pose a problem to our theories and expectations of the distribution and maintenance of sex?

In this paper, we discuss the putative lack of sex in microbial eukaryotes by concentrating on the Colpodea as an exemplary lineage. The Colpodea, one of 11 major ciliate clades, consists of about 200 described species with similar somatic but diverse oral morphologies [10,11] (Figure 1). Like all ciliates, a clade that is sister to the apicomplexans (e.g. *Plasmodium* sp., the causative agent of malaria) and dinoflagellates (e.g. *Symbiodinium* sp., the photosynthetic symbiont of corals), colpodean ciliates posses two types of nuclei within each cell: a 'germline' micronucleus and a 'somatic' macronucleus [12]. Colpodean ciliates are found in numerous habitats, some are fungivores, and at least one species has a multicellular slime mold-like life stage [10]. Much is known about their biology through morphological and molecular analyses [10,11,13–15].

Sex in colpodeans has been observed only in *Bursaria truncatella*, a relatively large species that can be up to 0.5 mm in length, and preys on other ciliates [10,16]. The extent of sexuality in the rest of the colpodeans is debated; Foissner [10] proposed that they are asexual, whereas Dunthorn *et al.* [13] suggested that they are secretively sexual.

In this article we first present the observational evidence for and against sexuality in the Colpodea, then discuss their putative lack of sex in relation to the theoretical problems of reversing the loss of complex traits and ancient asexuality. We conclude that most, if not all, colpodean ciliates are likely to be secretively sexual, particularly if macro-organismic expectations of sexuality apply to microbial eukaryotes. The arguments here are applicable to other putatively asexual microbial eukaryotes, many of which are also likely to be secretively sexual.

Box 1 Sex and its maintenance in macro-organisms

There are many definitions of sex. In this paper, we have restricted 'sex' to the definition used by Normark *et al.* [5]: 'meiosis followed by the fusion of meiotic products from different individuals'. Sex and reproduction are often intimately linked in macro-organisms, but new cells or progeny are not produced during sex in all eukaryotes. Recombination in archaea and bacteria is not covered under this definition, as it is non-meiotic [54,55].

Numerous theories exist for the maintenance of sex in macro-organisms. These include theories that deal with changing environments (e.g. Red Queen, Tangled Bank) and the accumulation and elimination of harmful mutations (e.g. Muller's ratchet, Kondrashov's hatchet), which are extensively reviewed elsewhere [3,4,35,42–46,56,57]. Most macro-organismic species are sexual during their life cycle or after a certain number of generations. The loss of sex is often thought to lead to rapid extinction with little chance of speciation [3,5,42].

THE POWERLESSNESS OF OBSERVATION

Most of the evidence for the purported asexuality in colpodean ciliates derives from the lack of observing sex. But sex is easily missed for three reasons: sex in ciliates is facultative, sex might not occur because of inappropriate laboratory conditions, and there are no obvious morphological features for sex in these ciliates.

GLOSSARY

Ancient asexuals: a group of species that have remained asexual over longer periods of evolutionary or even geological time than is predicted by theory.

Eukaryote: the group of organisms diagnosed by the presence of a nucleus.

Microbial eukaryote: eukaryotes that are not plants, animals or fungi. Also known as protists, protoctistas, algae and protozoa.

Molecular clock: estimate of the time of divergence among taxa using a molecular phylogeny and one or more calibration points such as fossils.

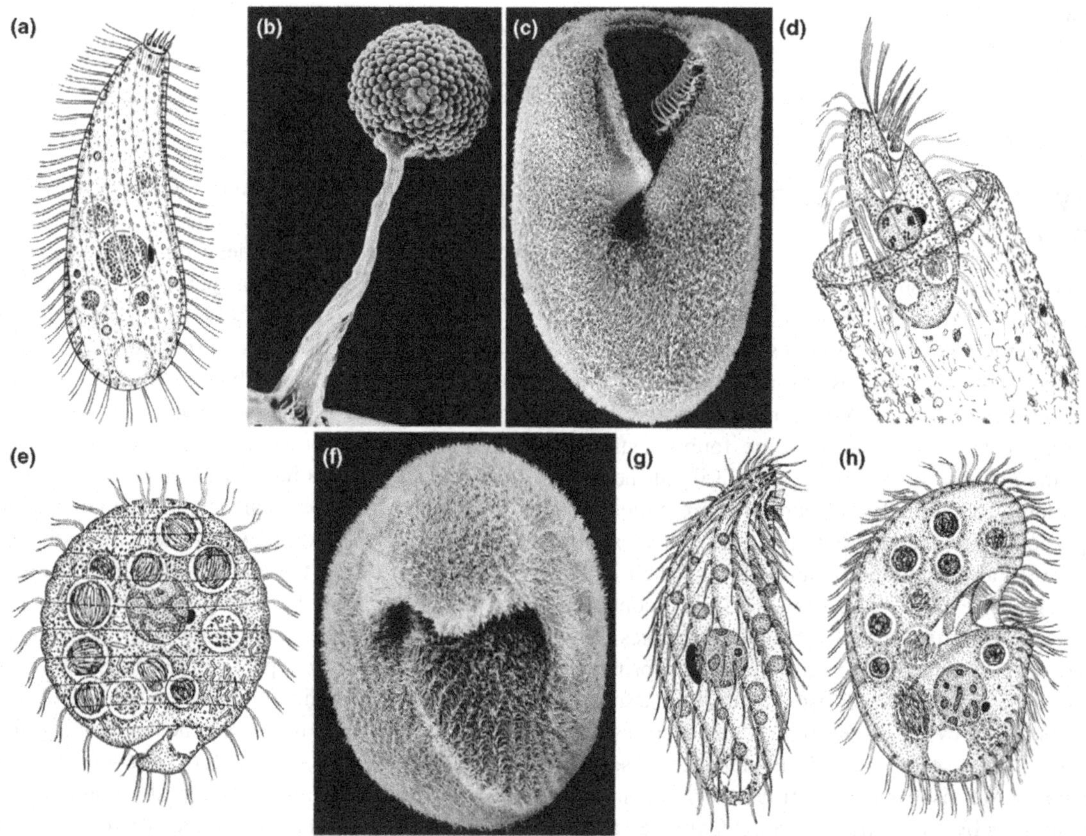

FIGURE 1 Morphological diversity within the putative asexual Colpodea ciliates. **(a)** *Platyophrya vorax* (right lateral view). **(b)** *Sorogena stoianovitchae*, slime mold-like multicellular sorocarp. **(c)** *Bursaria truncatella* (ventral view) the only known sexual species in the Colpodea. **(d)** *Cyrtolophosis mucicola* (left lateral view) in a mucilaginous dwelling-tube. **(e)** *Ilsiella palustris* (dorsal view) with posterior oral structure. **(f)** *Bresslauides discoideus*, (oblique apical view). **(g)** *Grossglockneria acuta* (right lateral view), oral structure modified into a tube used to perforate and feed on fungi. **(h)** *Colpoda cucullus* (right lateral view). *Pictures by Professor Dr Wilhelm Foissner.*

Reproduction: creation of offspring cells or individuals. Can be independent of sex.

Sex: meiosis, followed by the fusion of meiotic products from different individuals. This working definition restricts sex to eukaryotes.

Observing sex in colpodean ciliates in specific and in all ciliates generally, is not easy because sex has remained facultative [17]. The probability, then, that a researcher will observe sex occurring can be extremely low, depending on its rate in nature, which can vary dramatically among species [18,19]. This rate can be further decreased over time in the laboratory [20]. Reproduction in ciliates is via asexual cell division that can continue uninterrupted for thousands of generations, although the rate of division decreases over time without a sexual event [11,21]. In ciliates, sex occurs during conjugation, with haploid micronuclei products being mutually exchanged between complementary cells [11]. Sex is assumed to occur in almost all ciliate clades, although details and direct observations for most species are lacking [11]. There are known derived asexual strains of sexual species in other ciliate clades that have lost their micronuclei and are thus unable to conjugate [11,21]. Although most colpodean ciliates will divide in laboratory cultures, they have never been observed to have sex, despite the many observations that have taken place over many years (with the exception of *B. truncatella*) [10]. This might be because the facultative events of sex are so rare that they are easily missed. Nevertheless, all of these observations could have overlooked that one rare and facultative event between two lonely, but secretively sexual cells.

Observing sex in colpodeans is not necessarily easy because laboratory culture conditions might not be appropriate. For colpodeans, and even for most ciliates, we do not know what the right conditions for inducing sex are, and until we mimic their environment we might not observe their secretive sexuality. Like the Colpodea, the fungal pathogen *Aspergillus fumigatus* was long thought to be asexual because there were no laboratory observations of sex. Only when *A. fumigatus* was finally cultured in conditions that mimicked its natural environment was a sexual life-cycle stage demonstrated [22].

Observing sex in colpodean ciliates is also not necessarily easy because there is a lack of what Schurko *et al.* [23] describe as 'organismal signs of sex', such as sex-specific morphologies and organs, or gender differences, as found in plants and animals. In macro-organisms, if functional organismal signs of sex are observed in an individual it is safe to assume that it belongs to a sexual species, although there are exceptions [23]. Even if organismal signs of sex do exist in nature, they might not be easy to find; for example, darwinulid ostracods were long thought to be asexual until males were found after more than a century of searching [24]. Most microbial eukaryotes such as colpodeans lack morphological signs of sex, and although there might be differences in the mating types of ciliates [25], these differences are not outwardly visible. There is one potential sign of sex in colpodeans: they contain germline micronuclei [10], although these might be incapable of meiotic division. Because there is little reason for an asexual ciliate to maintain micronuclei, as they are only used for sex and for modeling the next generation's somatic nucleus [12], the colpodeans might be secretively sexual.

Given these difficulties in observing sex in colpodean ciliates, the lack of observation does not mean that sex is not occurring. A lack of data is not evidence for the hypothesis of ancient asexuality. There is a history of microbial eukaryotes being long thought to be asexual because of the powerlessness of observation, which were later to be shown either to have sex or the genetic signatures of sex, such as *Giardia duodenalis* [26], *Leishmania major* [27], *Naegleria lovaniensis* [28,29] and *Trichomonas vaginalis* [30]. Putative asexuality in colpodean ciliates (and in other microbial eukaryotes in which there have been a lack of observations of sex) might represent a similar situation: given enough time and requisite conditions their secretive sexuality might be observed.

AN IMPROVED PHYLOGENETIC FRAMEWORK

While acknowledging that observational artifacts might be the root of putative asexuality, an improved phylogenetic framework can allow us to understand the distribution and maintenance of sexuality better in larger groups of microbial eukaryotes. Using earlier estimates of the eukaryotic tree of life, Dacks and Rogers [1] suggested that, based on the distribution of known sexual lineages, facultative sexuality is probably the ancestral condition for all eukaryotes. With the data from recent molecular work that includes an increased number of sampled taxa and increased resolution of relationships, we can make further and more exact inferences about the distribution of sex within and among microbial eukaryotes such as colpodean ciliates, beyond issues relating to observational artifacts.

Reversing the Loss of Sex?

Recent molecular phylogenetic work has shown that the one known sexual colpodean species, *B. truncatella*, nests within the larger clade of putatively asexual colpodeans [13] (Figure 2). If colpodeans are truly asexual, then we would have to hypothesize a reversal of the loss of sex after many millions of years of asexuality.

The general ability of a species to reverse the loss of any complex characteristic is open to debate [31–34]. Complex characteristics can either be lost phenotypically or genotypically; the ability to regain a characteristic depends

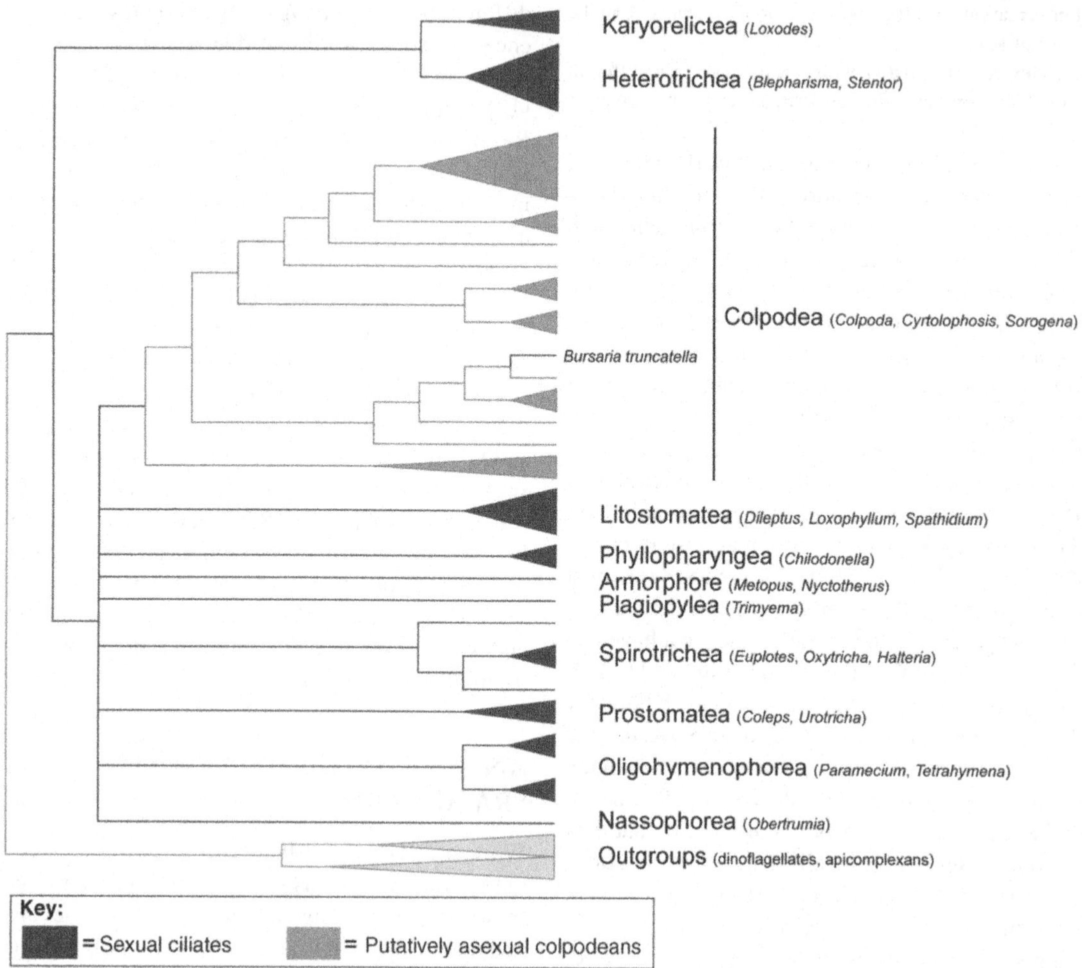

FIGURE 2 Distribution of sex in ciliates. The 11 major ciliate taxa are shown, along with their commonly known subclades. Most of these clades are largely sexual (blue). The Colpodea are putatively asexual (orange), except for one derived sexual species, *B. truncatella*. *Modified from Dunthorn* et al. *[14] and Lynn [11].*

on which of these levels was involved and the amount of time intervening between loss and reversal [31]. Can the loss of sex be reversed? It is a formal possibility that sex might have been lost and regained along lineages throughout the eukaryotic tree of life, although we would not know this given the current distribution of sex in extant species [35]. We do know of two putative cases of regaining sexuality, both at relatively shallow nodes. In multiple populations of the plant *Hieracium pilosella* [36], reversal from asexuality to sexuality entailed returning to homozygosity (and tetraploidy) for a recessive allele [36]. In this case, although the phenotype of sex was lost, sexual genes remained in the population. In oribatid mites, the case for reversing the loss of sex depends on likelihood character state reconstructions [37]. However, these reconstructions could be fundamentally flawed because of incorrect model specifications of both ancestral states and transitions rates, leading to the false acceptance of reversal [32]. The case of regaining sex in oribatid mites is thus ambiguous, and might merely represent multiple independent losses of sex. There is also no current evidence on whether genes involved in sex have been lost in oribatid mites. We also note that had Domes *et al.* [37] applied their method to other macro-organisms (especially in other insect clades) they might have increased the number of putative cases of reversing the loss of sex, although these additional likelihood character state reconstructions would also be fundamentally flawed because of the same model problems.

If genes involved in meiosis or other sexual processes are shown to be retained in the putative asexual relatives of *B. truncatella*, then a case for reversing the loss of sex in colpodean ciliates might be made, although the problem of reactivation of silenced genes increases over time as they could be mutated or lost [5,31,38]. However, retention of sex genes is more consistent with the retention of secretive sexuality. If we assume that both the phenotype and the genotype of sex were lost in the colpodeans, then our expectations for reversing the loss of sex would be much smaller to none. On the other hand, the problem of hypothesizing the reversal of the loss of sex in colpodeans is removed if they are secretively sexual.

Ancient Asexuality?

Phylogenetic analyses coupled with molecular clock estimates suggest that colpodean ciliates would be extremely ancient asexuals (up to 900 MYA) [39]. There are also fossils of derived lineages in 93 MYA-old amber [40]. Colpodeans are thus at least as old, or perhaps even an order of magnitude older, than the putatively ancient asexual bdelloid rotifers that date to about 100 MYA [41].

There are other putative ancient asexuals lineages of plants and animals. Although asexuals are known to exist at the tips of the tree of life of macro-organisms, ancient asexuals are generally thought to be unlikely because asexuality is believed to lead to rapid extinction [3,5,42]. For example, without sex, macro-organisms might become more susceptible to increased mutational load and retrotransposon invasion/expansion, might not be able to adapt to a changing environment, or might not be able to escape predators and parasites [3,4,43–46]. Most claims of ancient asexuals have not been supported [5,47], except possibly those for the bdelloid rotifers [41,48–52]. However, this low expectation of ancient asexuality is based on theory and on observations of macro-organisms [5]. If these low expectations of ancient asexuality in macro-organisms do not apply to colpodean ciliates, then a case for asexuality in colpodeans can be made. For example, it is possible that large effective population sizes combined with non-canonical genetic systems could potentially allow these ciliates to remain asexual over long periods of evolutionary or geological time, beyond that predicted from plants and animals. However, if the expectations from macro-organisms do apply, then colpodean ciliates are likely to be secretively sexual.

CONCLUDING REMARKS AND OUTSTANDING QUESTIONS

Most of our theoretical and empirical understanding of evolution derives from our familiarity and observations of plants and animals; our understanding of the distribution and maintenance of sex is just one example. Although originally thought to be closely related, plants and animals are on disparate branches of the eukaryotic tree of life [53]. Hence, as theoretical work on sex applies to both of these non-sister groups, this suggests that the theory might apply more broadly to other eukaryotic lineages. However, if additional research indicates that our theories do not apply to microbial eukaryotes, then our basis for understanding evolution would be restricted to distantly related macro-organisms in terminal branches in the eukaryotic tree of life – a very limited view indeed.

Despite macro-organismic theory, many microbial eukaryotes are often assumed to be asexual. Undoubtedly some of them truly are, although their age and distribution is unknown. But for the rest, the way in which most species in the laboratory are observed could prevent us from actually observing their secretive sex. For most microbial eukaryotes we are far from understanding aspects such as the facultative nature and timing of their sexual phases, the appropriate laboratory conditions to induce sex, and the organismal signs of sex at the morphological, biochemical and molecular levels. Given these problems, putative asexuality in our exemplary clade of colpodean ciliates is likely to be an observational artifact. This powerlessness of observation might also apply to all other putative asexual microbial eukaryotes, and many of these are secretively sexual as well.

For those microbial eukaryotes that are either anciently asexual and/or have known derived sexual species demonstrated by molecular phylogenetic analyses, a further claim against putative asexuality can be made using macro-organismic theory, such as in colpodean ciliates. Do these expectations also apply to microbial eukaryotes? We do not yet know. A more generalized understanding of ancient asexuality that takes into account microbial eukaryotes, and a better understanding of the possibility to regain the loss of sex in both macro-organisms and microbial eukaryotes, is needed. If our macro-organismic theories of the low expectations of ancient asexuals and the problems of reversing the loss of sex are shown not to apply, then colpodean ciliates and other microbial eukaryotes might be asexual, but if these expectations do apply, then we would expect most to be secretively sexual.

Beyond taking into account observational biases and developing a broader theoretical understanding of the distribution and maintenance of sex in microbial eukaryotes, there are numerous potential tests that can be performed to look for genomic signatures of sex or asexuality (which are reviewed elsewhere [5,23]). For example, the inventory of meiosis-related genes [30] or transposable elements [43] can be assessed. Any of these potential tests can and should be applied to colpodean ciliates and other putative asexual microbial eukaryotes as the next step in understanding the distribution of sex among diverse microbial eukaryotes.

ACKNOWLEDGEMENTS

We thank Norman Johnson, Dan Lahr, David Lahti, George McManus, Ben Normark and Laura Parfrey, Cesar Sanchez, Gail Teitzel and three anonymous reviewers for discussions and suggestions. Professor Dr Wilhelm Foissner kindly provided the pictures for Figure 1. Funding came from postdoctoral fellowships from the Faculty of Biology of the University of Kaiserslautern and from the Alexander von Humboldt Foundation to M.D., and NSF Grant DEB 0816828 to L.A.K.

REFERENCES

1. Dacks, J. and Roger, A.J. (1999) The first sexual lineage and the relevance of facultative sex. *J. Mol. Evol.* 48, 779–783.
2. Ramesh, M.A. *et al.* (2005) A phylogenomic inventory of meiotic genes: evidence for sex in *Giardia* and an early eukaryotic origin of meiosis. *Curr. Biol.* 15, 185–191.
3. Maynard Smith, J. (1978) *The Evolution of Sex*, Cambridge University Press.
4. Bell, G. (1982) *The Masterpiece of Nature: the Evolution and Genetics of Sexuality*, University of California Press.
5. Normark, B.B. *et al.* (2003) Genomic signatures of ancient asexual lineages. *Biol. J. Linn. Soc.* 79, 69–84.
6. Ekelund, F. and Rønn (2008) If you don't need change, maybe you don't need sex. *Nature* 453, 587.
7. Fenchel, T. and Finlay, B.J. (2006) The diversity of microbes: resurgence of the phenotype. *Phil. Trans. R. Soc. B* 361, 1965–1973.
8. Schlegel, M. and Meisterfeld, R. (2003) The species problem in protozoa revisited. *Europ. J. Protistol.* 39, 349–355.
9. Sonneborn, T.M. (1957) Breeding systems, reproductive methods, and species problems in protozoa. In *The Species Problem* (Mayr, E., ed.), pp. 155–324, American Association for the Advancement of Science.
10. Foissner, W. (1993) Colpodea (Ciliophora). *Protozoenfauna* 4/1, i–x, 1–798.
11. Lynn, D.H. (2008) *The Ciliated Protozoa: Characterization, Classification, and Guide to the Literature*, (3rd edn), Springer.
12. McGrath, C.L. *et al.* (2006) Genome evolution in ciliates. In *Genomics and Evolution of Eukaryotic Microbes* (Katz, L.A. and Bhattacharya, D., eds), pp. 64–77, Oxford University Press.
13. Dunthorn, M. *et al.* (2008) Molecular phylogenetic analysis of class Colpodea (phylum Ciliophora) using broad taxon sampling. *Mol. Phylogenet. Evol.* 48, 316–327.
14. Dunthorn, M. *et al.* (2009) Phylogenetic placement of the Cyrtolophosididae Stokes, 1888 (Ciliophora; Colpodea) and neotypification of Aristerostoma marinum Kahl, 1931. *Int. J. Syst. Evol. Microbiol.* 59, 167–180.
15. Lynn, D.H. *et al.* (1999) Phylogenetic relationships of orders within the class Colpodea (phylum Ciliophora) inferred from small subunit rRNA gene sequences. *J. Mol. Evol.* 48, 605–614.
16. Raikov, I.B. (1982) *The Protozoan Nucleus: Morphology and Evolution*, Springer-Verlag.
17. Bell, G. and Koufopanou (1991) The architecture of the life cycle in small organisms. *Phil. Trans. R. Soc. B.* 332, 81–89.
18. Doerder, F.P. *et al.* (1995) High frequency of sex and equal frequencies of mating types in natural populations of the ciliate *Tetrahymena thermophila*. *Proc. Natl. Acad. Sci., U. S. A.* 92, 8715–8718.
19. Lucchesi, P. and Santangelo, G. (2004) How often does conjugation in ciliates occur? clues from a seven-year study in marine sandy shores. *Aquat. Microb. Ecol.* 36, 195–200.
20. Dini, F. and Nyberg, D. (1993) Sex in ciliates. *Adv. Microb. Ecol.* 13, 85–153.
21. Bell, G. (1988) *Sex and Death in Protozoa: the History of an Obsession*, Cambridge University Press.
22. O'Gorman, C.M. *et al.* (2009) Discovery of a sexual cycle in the opportunistic fungal pathogen *Aspergillus fumigatus*. *Nature* 457, 471–474.
23. Schurko, A.M. *et al.* (2009) Signs of sex: what we know and how we know it. *Trends Ecol. Evol.* 24, 208–217.
24. Smith, R.J. *et al.* (2006) Living males of the 'ancient asexual' Darwinulidae (Ostracoda: Crustacea). *Proc. R. Soc. Lond. B* 273, 1569–1578.
25. Phadke, S.S. and Zufall, R.A. (2009) Rapid diversification of mating systems in ciliates. *Biol. J. Linn. Soc.* 98, 187–197.
26. Lasek-Nesselquist, E. *et al.* (2009) Genetic exchange within and between assemblages of *Giardia duodenalis*. *J. Eukaryot. Microbiol.* 56, 504–518.
27. Akopyants, N.S. *et al.* (2009) Demonstration of genetic exchange during cyclical development of *Leishmania* in the sand fly vector. *Science* 324, 265–268.
28. Hurst, L.D. *et al.* (1992) Covert sex. *Trends Ecol. Evol.* 7, 144–145.
29. Pernin, P. *et al.* (1992) Genetic structure of natural populations of the free-living ameba, *Naegleria lovaniensis*. evidence for sexual reproduction. *Heredity* 68, 173–181.
30. Malik, S-B. *et al.* (2008) An expanded inventory of conserved meiotic genes provides evidence for sex in *Trichomonas vaginalis*. *PLoS One* 3, e2879.
31. Collin, R. and Miglietta, M.P. (2008) Reversing opinions on Dollo's Law. *Trends Ecol. Evol.* 23, 602–609.
32. Goldberg, E.E. and Igic, B. (2008) On phylogenetic tests of irreversible evolution. *Evolution* 62, 2727–2741.
33. Gould, S.J. (1970) Dollo on Dollo's Law: irreversibility and the status of evolutionary laws. *J. Hist. Biol.* 3, 189–212.
34. Teotónio, H. and Rose, M.R. (2001) Perspective: reverse evolution. *Evolution* 55, 653–660.
35. Williams, G.C. (1975) *Sex and Evolution*, Princeton University Press.
36. Chapman, H. *et al.* (2003) A case of reversal: the evolution and maintenance of sexuals from parthenogenetic clones in *Hieracium pilosella*. *Int. J. Plant Sci.* 164, 719–728.
37. Domes, K. *et al.* (2007) Reevolution of sexuality breaks Dollo's law. *Proc. Natl. Acad. Sci. U. S. A.* 104, 7139–7144.
38. Marshall, C.R. *et al.* (1994) Dollo's law and the death and resurrection of gene. *Proc. Natl. Acad. Sci. U. S. A.* 91, 12283–12287.
39. Wright, A-DG. and Lynn, D.H. (1997) Maximum ages of ciliate lineages estimated using a small subunit rRNA molecular clock: crown eukaryotes date back to the paleoprotoerozoic. *Arch. Protistenk.* 148, 329–341.
40. Martín-González, A. *et al.* (2008) Morphological stasis of protists in lower Cretaceous amber. *Protist* 159, 251–257.
41. Mark Welch, D. *et al.* (2008) Evidence for degenerate tetraploidy in bdelloid rotifers. *Proc. Natl. Acad. Sci. U. S. A.* 105, 5145–5149.
42. Lynch, M. *et al.* (1993) Mutational meltdowns in asexual populations. *J. Hered.* 84, 339–344.
43. Arkhipova, I. and Meselson, M. (2004) Deleterious transposable elements and the extinction of asexuals. *Bioessays* 27, 76–85.
44. Burt, A. (2000) Sex, recombination, and the efficacy of natural selection — was Weisman right? *Evolution* 54, 337–351.
45. Hamilton, W.D. (2001) In *Narrow Roads of Gene Land, the Evolution of Sex* (Vol. 2), Oxford University Press.
46. Kondrashov, A.S. (1993) Classification of hypotheses on the advantage of amphimixis. *J. Hered.* 84, 372–387.
47. Judson, O.P. and Normark, B.B. (1996) Ancient asexual scandals. *Trends Ecol. Evol.* 11, A41–A46.

48. Arkhipova, I. and Meselson, M. (2000) Transposable elements in sexual and ancient asexual taxa. *Proc. Natl. Acad. Sci. U. S. A.* 97, 14473–14477.
49. Mark Welch, D. and Meselson, M. (2000) Evidence for the evolution of bdelloid rotifers without sexual reproduction or genetic exchange. *Science* 288, 1211–1215.
50. Mark Welch, D. *et al.* (2004) Divergent gene copies in the asexual class Bdelloidea (Rotifera) separated before the bdelloid radiation or within bdelloid families. *Proc. Natl. Acad. Sci. U. S. A.* 101, 1622–1625.
51. Mark Welch, J.L. *et al.* (2004) Cytogenetic evidence for asexual evolution of bdelloid rotifers. *Proc. Natl. Acad. Sci. U. S. A.* 101, 1618–1621.
52. Wilson, C.G. and Sherman, P.W. (2010) Anciently asexual bdelloid rotifers escape lethal fungal parasites by drying up and blowing away. *Science* 327, 574–576.
53. Simpson, A.G.B. and Roger, A.J. (2004) The real 'kingdoms' of eukaryotes. *Curr. Biol.* 14, R693–R696.
54. Gogarten, J.P. and Townsend, J.P. (2005) Horizontal gene transfer, genome innovation and evolution. *Nat. Rev. Microbiol.* 3, 679–687.
55. Vos, M. (2009) Why do bacteria engage in homologous recombination? *Trends Microbiol.* 17, 226–232.
56. Neiman, M. *et al.* (2009) What can asexual linegae age tell us about the maintenance of sex? *Ann. N. Y. Acad. Sci.* 1168, 185–200.
57. West, S.A. *et al.* (1999) A pluralist approach to sex and recombination. *J. Evol. Biol.* 12, 1003–1012.

Chapter 17

Coccolithophores

R.W. Jordan
Yamagata University, Yamagata, Japan

Chapter Outline

Abbreviations	**235**
Defining Statement	**235**
Morphology	**235**
General Features of the Coccolithophore Cell	235
Coccolith Morphology	237
Coccolith Production	238
Functions of Coccoliths	238
Taxonomy	**238**
From the Chrysophyceae to the Haptophyceae	238
Haptophyte Taxonomy	239
Taxonomic Concepts Based on Morphology	239
Taxonomic Concepts Based on Molecular Genetics	240
Collection Methods	**240**
Biogeography and Ecology	**241**
General Distribution	241
Seasonality and Depth Preferences	241
Blooms	**242**
Detection of Coccolithophore Blooms	242
Coccolithophore Viruses	244
Impact on the Regional Climate and Environment	244
Carbon Release and Sinking of Blooms	245
Evolution	**245**
First Appearance of the Coccolithophores	245
Coccolithophore Diversity and Extinctions	246
Past Coccolithophore Blooms	246
Biostratigraphic Use	247
Further Reading	**247**

ABBREVIATIONS

CZCS Coastal Zone Color Scanner
DMS Dimethyl sulfide
DMSP Dimethylsulfoniopropionate
LPZ Lower photic zone
MPZ Middle photic zone
SeaWiFS Sea-viewing Wide Field of view Sensor
SEM Scanning electron microscope
TEM Transmission electron microscope
UPZ Upper photic zone

DEFINING STATEMENT

Coccolithophores are ubiquitous, calcareous scale-bearing marine microalgae, often dominating phytoplankton assemblages, and producing huge blooms that may affect regional and global climate. Since they first appeared in the Late Triassic, coccolithophores have provided an almost continuous fossil record of their evolution and distribution throughout the last 200 My.

MORPHOLOGY

General Features of the Coccolithophore Cell

Coccolithophores are mostly marine, photoautotrophic unicellular algae, ranging from 2.0 to 75.0 μm in cell diameter, including cell coverings. These cells may be characterized by the following ultrastructural features (Figure 1): (1) an emergent or residual haptonema, (2) two smooth flagella, when present, (3) chloroplasts with no girdle lamella, (4) peculiar Golgi bodies, and (5) an extracellular covering of unmineralized scales and overlying calcified scales (coccoliths).

The haptonema is a unique organelle found in only one group of algae, collectively called the haptophytes, to which the coccolithophores belong. Although superficially similar in structure to a flagellum, the main function of the haptonema is probably food capture. This is best studied in the noncoccolithophore species *Chrysochromulina hirta*, where the haptonema captures particles and gathers them into an aggregate at a particular point on the haptonema. This aggregate then moves back to the haptonema tip,

Eukaryotic Microbes.

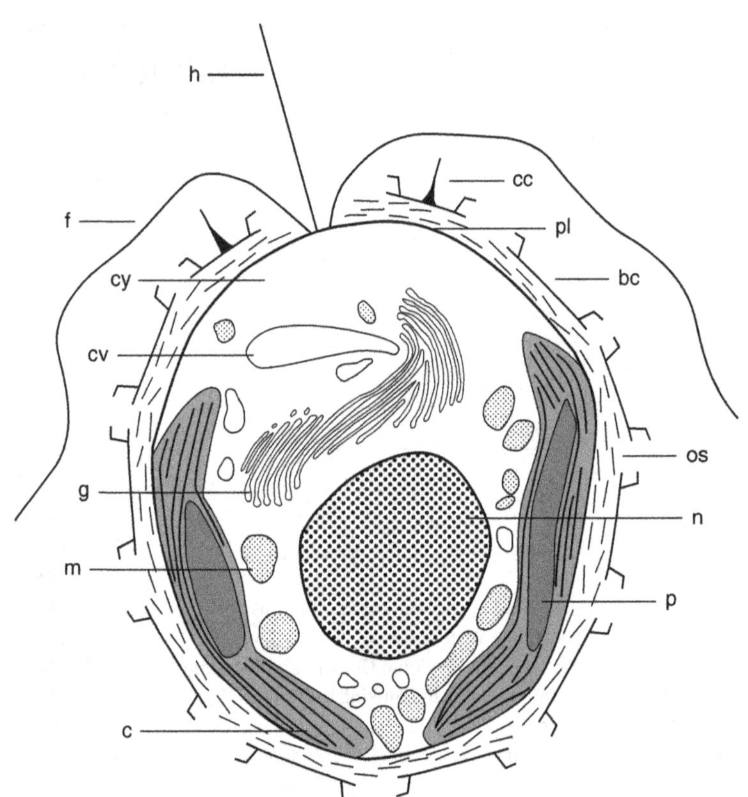

FIGURE 1 Longitudinal section through a generalized coccolithophore cell; b, body coccolith; c, chloroplast; cc, circumflagellar coccolith; cv, coccolith vesicle; cy, cytoplasm; f, flagellum; g, Golgi body; h, haptonema; m, mitochondrion; n, nucleus; os, organic scale; p, pyrenoid; pl, plasmalemma+endoplasmic reticulum. Figure loosely based on that of *Syracosphaera pulchra.* Reproduced from Inouye, I., & Pienaar, R. N. (1988). Light and electron microscope observations of the type species of *Syracosphaera*, *S. pulchra* (Prymnesiophyceae). *British Phycological Journal*, *23*(3), 205–217.

before being deposited into a food vacuole at the posterior end of the cell. It is generally presumed that bacteria caught in this way are used to supplement photosynthesis (i.e., mixotrophy), especially in waters where light levels are low, such as in the lower photic zone (LPZ) or at high latitudes. Haptonemata are variable in length, and in some species they may be nonemergent, with only traces remaining inside the cell. A long haptonema can be coiled, whereas shorter ones tend to be stiff. Normally, flagellated haptophytes employ a forward swimming motion with the flagella and the haptonema in front of the cell. On contact with other objects the cell may change direction by using a rapid backward swimming motion with the flagella and haptonema (coiled, if the haptonema is long) behind the cell. However, the opposite scenario has been reported recently for *Algirosphaera robusta*, in which the haptonema is directed forward when the cell swims rapidly, but is directed backward when the cell swims slowly.

Many coccolithophores are motile in the haploid phase, but often nonmotile in the diploid phase. Coccolithophore flagella tend to be long, equal in length, and smooth (i.e., do not possess tubular hairs), and are usually present in at least one phase of the life cycle. Since only flagellated cells possess an emergent haptonema, diploid phases often lack this specialized organelle. In old cultures, nonmotile diploid cells have sometimes been known to revert into haploid motile phases if the nutrient supply becomes low. This has prompted speculation that the switch from a nonmotile to a motile phase in the natural environment is also related to nutrient supply, since flagellated cells can seek out more favorable waters and also utilize their haptonema to feed.

Most, if not all, coccolithophores are photosynthetic, and are traditionally thought of as typical chlorophyll $a+c$ algae. However, it has recently been shown that their pigments are highly diverse, with the pigment distribution closely mirroring the phylogeny and ecology of the coccolithophores. Thus, coastal species generally contain chlorophyll c_1, while oceanic species possess 19′-hexanoyloxyfucoxanthin. Furthermore, pigment content is maintained throughout the life cycle, indicating that it is a stably inherited trait, rather than a reaction to changing environmental conditions.

In coccolithophores, a layer of organic body scales usually covers the plasma membrane, regardless of whether or not coccoliths are present on top of them. These organic scales as well as the coccoliths are generally (but not always) manufactured by the Golgi apparatus. The stack of flattened cisternae constituting the coccolithophore Golgi body is arranged like an asymmetrical fan, close to the basal bodies of the flagella. This is the main assembly site for scales and coccoliths, which are produced in cisternae with swollen ends. The presence of these dilated cisternae (or vesicles) give the Golgi apparatus an unusual appearance (hence the term, 'peculiar Golgi'), compared to those of other algae.

The morphology of the organic body scales differs between life cycle phases. In general, the body scale of the diploid ($2n$) phase is circular with a prominent rim and with the same pattern on both sides. Meanwhile in the haploid (n) phase the scale is more elliptical, rimless, and with a pattern of radiating ridges arranged in four quadrants on the proximal side and one of concentric fibers on the distal side. In some species, smaller haptonematal scales have also been found.

Coccolith Morphology

While unmineralized scales are used as the primary character for separating most noncalcifying haptophytes (e.g., species of *Chrysochromulina*), it is the diverse morphology of the coccoliths that has proven to be the most useful character for the identification and classification of the coccolithophores. In fact, surprisingly little is known about the organic scales in coccolithophores, due to the paucity of species maintained in culture and the use of the light microscope or the scanning electron microscope (SEM) instead of the transmission electron microscope (TEM) as the preferred tool for detailed observation. Coccoliths may be separated into two main groups: holococcoliths (Figure 2(a) and 2(b)) and heterococcoliths (Figure 2(c)–2(f)). Holococcoliths are composed of individual rhombohedric crystals, while in heterococcoliths the crystals have undergone some degree of growth, forming a variety of structures such as radiating laths, wall elements, plates, shields, spines, and tubular processes. The diverse range of coccoliths and the structures they possess has resulted in the creation of hundreds of descriptive terms. Coccolith terminology has been approached in two different ways, using taxon-specific terms such as helicoliths for coccoliths of the Helicosphaeraceae, or using more general terms like muroliths for any coccolith lacking shields but possessing a raised rim.

Coccoliths are usually arranged around a single cell in one of three ways; they abut, overlap, or interlock. However, some species produce large coccospheres that surround more than one cell (e.g., *Umbilicosphaera sibogae*), while others may possess only one coccolith per cell (e.g., *Ceratolithus cristatus*). The coccospheres of some species consist of 50–200 coccoliths (e.g., *Florisphaera profunda*), whereas species like *Gephyrocapsa oceanica* (see Figure 2(d)) have >20.

In some species (e.g., *Emiliania huxleyi*, see Figure 2(e)) the coccoliths are all identical (monomorphic), while in others more than one coccolith type may be present. Many species have dimorphic coccospheres bearing specialized coccoliths around the flagellar pole (circumflagellar coccoliths) in addition to the body coccoliths, although other combinations occur such as specialized coccoliths at the antapical pole (*Ophiaster* spp.), both poles (*Calciosolenia murrayi*), or around the coccosphere equator (*Scyphosphaera apsteinii*). In other species slight morphological differences in the coccoliths may appear somewhat randomly over the coccosphere (e.g., *Rhabdosphaera clavigera*). Polymorphic coccospheres also exist, for example, in *Michaelsarsia* and *Ophiaster* spp., with up to four different coccolith types. In *Syracosphaera* spp. a second layer of morphologically

FIGURE 2 Scanning electron micrographs of (a) *Calyptrolithina multipora*, (b) *Syracolithus schilleri*, (c) *Coccolithus pelagicus*, (d) *Gephyrocapsa oceanica*, (e) *Emiliania huxleyi*, (f) *Braarudosphaera bigelowii*. All scale bars = 1 μm.

different coccoliths completely or partially cover the body coccoliths, resulting in a dithecate or pseudo-dithecate coccosphere, respectively.

Coccolith Production

Over the last 20 years a lot of research has been carried out on the process of coccolith formation, with some species producing one coccolith at a time (e.g., *E. huxleyi*), while others produce both organic scales and coccoliths in a continuous production line (e.g., *Pleurochrysis carterae*). In general, heterococcoliths are formed in specialized vesicles (coccolith vesicles) originating from the Golgi body. Within these vesicles, calcite crystals are initially laid down on an organic base plate (the base plate scale), and after the coccolith is completed, it is usually excytosed through the flagellar area, and then rhythmically incorporated into the coccosphere. Holococcoliths, on the other hand, are formed extracellularly in the space between the plasmalemma and the outer 'skin.' This apparent lack of cell control during holococcolith formation may be the reason why holococcoliths are composed of undifferentiated rhombohedric crystals. However, despite this, holococcolith morphology is extremely diverse and exceedingly intricate (see Figure 2(a) and 2(b)).

In the case of heterococcoliths, the calcite crystals are initially laid down on the base plate scale in a ring, known as the protococcolith ring, with alternating subvertical (V-unit) and subradial (R-unit) *c*-axis orientations. These units then grow within the coccolith vesicle to form particular parts of the mature coccolith. For instance, in *Coccolithus pelagicus* (see Figure 2(c)) an initial V-unit crystal grows into a distal shield element and lower tube element, while an initial R-unit crystal grows into the upper and lower proximal shield elements and the upper tube element. In contrast, only the R-units are developed in *E. huxleyi*, forming the proximal and distal shield elements as well as the inner and outer tube cycle elements. Research has shown that the V-/R-model works for many coccolithophore taxa; however, the radial lath elements in *Syracosphaera pulchra* represent an additional crystal type with subtangential *c*-axis orientation (T-unit). The V-/R-/T-model is likely to be found throughout the Syracosphaeraceae, Rhabdosphaeraceae, and Calciosoleniaceae, since all the species bear radial laths. Thus, the distribution of these three crystal units and the structures they eventually form are strongly related to coccolithophore phylogeny. Since these *c*-axis orientations are visible in cross-polarized light, this confirms the reliability of coccolith identification by light microscopy.

Functions of Coccoliths

Ever since scientists began observing coccolithophores with the light microscope, they have wondered what purpose the coccoliths serve. The diversity and intricate nature of their structures would suggest that the organisms are not merely disposing of waste products. Instead, the coccoliths probably play multiple roles, for example, as (1) light capture or light reflection devices, (2) ballast, (3) a defense mechanism, (4) aids in food capture, or (5) a buffer zone, protecting the cell membrane against external conditions.

As calcite is a birefringent mineral, the coccoliths of species living in the upper photic waters may act as reflectors protecting the cell from photoinhibiting light levels or UV damage. Alternatively, in dimly lit deeper photic waters, the coccoliths may act as light collectors (or enhancers). The successful deep photic species *F. profunda* has multiple layers of coccoliths covering only half the cell surface, thus giving the coccosphere a bowl shape. Some deeper-living species like *S. apsteinii* have very large coccoliths, supporting the idea that they aid in buoyancy control. Many species, particularly those in the upper photic waters, have spiny coccoliths that may prevent herbivorous zooplankton from eating them. Coccoliths are continuously being shed and replaced, which may also prevent attacks from protozoa, bacteria, or viruses, or prevent them from merely growing on the coccolith surface. Another possible function of spiny coccoliths, although as yet unconfirmed, may be as an aid in food capture, since it is known that bacteria may adhere to the spiny organic scales of the haptophyte *Chrysocampanula spinifera*. The captured bacteria are then transported by the haptonema to a food vacuole. Since the involvement of the haptonema appears essential for this mode of nutrition, one may assume that only motile, spiny coccolith-bearing coccolithophores can utilize the bacteria captured in this way. Of those coccolithophores that do possess spiny coccoliths, most are indeed motile, and many only have spiny coccoliths close to the flagellar area, where the haptonema is located (e.g., *Acanthoica* and *Syracosphaera*). Finally, the coccoliths may be acting as a buffer zone between the delicate cell membrane, which is responsible for gaseous exchange and the transport of various materials in and out of the cell, and the surrounding seawater. This is supported by the fact that almost all coccoliths are perforated. In fact, this role could be widespread among related algal groups, since diatoms, chrysophytes, and Parmales have perforated valves, scales, and plates, respectively.

TAXONOMY

From the Chrysophyceae to the Haptophyceae

The coccolithophores are currently assigned to the algal division Haptophyta, a group of taxa possessing the unique organelle, the haptonema, or remnants of one. However, the taxonomic history of the haptophytes (including the coccolithophores) is rather complicated, and so only some of the key events are given here. At the beginning of the twentieth century, the coccolithophores were placed in their own family, the Coccolithophoridae, as they possessed coccoliths. Meanwhile, marine flagellates bearing two smooth equal flagella were placed in the order Isochrysidales, and included in the class Chrysophyceae largely due to their yellow-brown pigmentation. This was despite the fact that the other chrysophytes possessed two unequal flagella, with the longer one bearing tubular hairs. Since some of the coccolithophores bear two, albeit equal flagella, the Coccolithophoridae was initially included in the Isochrysidales. Later, another group of similar flagellates seemingly bearing three flagella (now known to be two equal flagella and the haptonema) were placed in the family Prymnesiaceae.

Around the middle of the century, two important discoveries were made. First, a new genus (*Pavlova*) of marine flagellate was found, which possessed unequal flagella but also the so-called 'third flagellum' of the Prymnesiaceae. In the following 20 years more species were added to this genus and a new family – the Pavlovaceae – was eventually erected for them. The second significant event in the mid-twentieth century was the elucidation of the structure of the haptonema, and the confirmation of its presence in a wide spectrum of flagellates, including some living in brackish and freshwater habitats. Since all these haptonema-bearing flagellates belonged to the Isochrysidales, Prymnesiaceae, and Pavlovaceae, a new class of algae, the Haptophyceae, was created to unite them. Further research on cell ultrastructure has led to the haptophytes being assigned to their own division, the Haptophyta, and subsequent molecular studies have shown that this division is monophyletic.

Haptophyte Taxonomy

The division Haptophyta may be divided into two classes: the Pavlovophyceae containing species with unequal flagella and the Prymnesiophyceae containing species with equal flagella. In addition, *Pavlova* and related genera may have an eyespot, and although lacking organic body scales or coccoliths, they may possess knob-like bodies (knob-scales) on the plasma membrane. This separation of two classes based on cell ultrastructure has been consistently supported by molecular studies. These studies have also supported the separation of the Prymnesiophyceae into a number of orders, several of which contain extant coccolithophore taxa. Within the Haptophyta, the coccolithophores represent the largest group, comprising 15 extant families as well as about 30 fossil families, while other noncalcifying haptophyte taxa comprise just four families. This disparity has led some researchers to resurrect the old class name Coccolithophyceae, which has priority over the name Prymnesiophyceae; however, for the sake of nomenclatural stability the latter name should be retained, since it has been used for the last 30 years.

Taxonomic Concepts Based on Morphology

At the beginning of the twentieth century, coccolithophores were assigned to families based largely on their cell shape and coccolith arrangement. Later, as coccolith morphology became recognized as a strong taxonomic character, the families were redefined with redistribution of the member taxa. With the advent of commercial TEM in the 1950s and the SEM in the 1970s, cell contents and mineralized components were determined more accurately, allowing rapid improvement of coccolithophore taxonomy. Coccolithophore classification, like many other protist groups, is built around the morphology of the cell covering, with each family possessing a unique coccolith type or a unique combination of several coccolith types. In hindsight this has proved to be a good taxonomic concept, as it has now been corroborated by molecular genetics.

The latest classification schemes of coccolithophores reflect recent advances in our knowledge of their life cycles, through observations of either cultured or wild material, and of molecular phylogeny. For instance, in past schemes coccolithophores bearing either holococcoliths or heterococcoliths were presumed to be separate taxa belonging to different families. However, it is now known that holococcolith-bearing cells represent the haploid phases of diploid, heterococcolith-bearing cells. Illustrations and photographs of so-called combination cells bearing both holococcoliths and heterococcoliths were once assumed to be the result of mere agglutination, but are now recognized as examples of cells changing from one life cycle phase to another. Surprisingly, in the last hundred years, most of these examples have occurred in the Mediterranean Sea.

Other than the example given above, a number of other coccolithophore life cycles exist. Species in the Pleurochrysidaceae have a diploid heterococcolith-bearing phase and a haploid non-coccolith-bearing phase that is benthic and pseudofilamentous, with both phases possessing organic body scales. Species in the Hymenomonadaceae have a life cycle similar to that of the Pleurochrysidaceae, but the haploid phase is not pseudofilamentous and bears dimorphic organic body scales, one of which has species-specific ornamentation. In *E. huxleyi* the nonmotile heterococcolith-bearing diploid phase lacks organic scales (in fact, only *E. huxleyi* and *Umbilicosphaera foliosa* are known to lack organic scales in their diploid phase), while the motile haploid phase bears only organic scales. The motile phase of *G. oceanica* and members of the Isochrysidaceae bear identical organic scales to *E. huxleyi*, which appear to be different from those of other families.

It is generally assumed that most haptophytes have either coccoliths or unmineralized scales. However, siliceous cysts have been found in the noncoccolithophore genus *Prymnesium*. Just recently a siliceous scale-bearing haptophyte, *Hyalolithus neolepis*, found in Japan was identified and cultured (Figure 3). This species has been known for

FIGURE 3 Collapsed coccosphere of the siliceous haptophyte, *Hyalolithus neolepis*, collected off the coast of Western Australia. Scale bar = 5 μm. Photo by Maiko Tanimoto.

decades but was always assumed to belong to another protist group. The presence of a haptonema in *H. neolepis* and its position on the phylogenetic tree close to *Prymnesium* confirmed its true identity. The siliceous scale formation process is very similar to that of coccolithogenesis. In addition, these outer scales are underlaid by an inner layer of unmineralized scales that are of haptophyte design. This discovery demonstrates that past concepts of what constitutes a haptophyte need to be revised, and strongly suggests that other siliceous scaly protists (e.g., *Petasaria heterolepis*) should be reinvestigated in more detail.

Taxonomic Concepts Based on Molecular Genetics

There have been two main approaches to the molecular studies of the haptophytes, with the first and most obvious of these being the utilization of cultured material. However, of the approximately 250 species of haptophytes currently described, perhaps only 25% have been cultured successfully in the last 80 years. Despite this, huge advances in culturing techniques have been made recently, resulting in an increased knowledge of haptophyte ultrastructure, pigments, and life cycles. Molecular studies on these cultures have demonstrated that the largest haptophyte genus, *Chrysochromulina*, is actually paraphyletic, while other cryptic genera (e.g., *Braarudosphaera* and *Chrysoculter*) have been confirmed as *bona fide* haptophytes.

Coccolithophore species that are easily cultured, such as *E. huxleyi* and *P. carterae*, are among the most extensively used eukaryotes in laboratory experiments. *E. huxleyi* in particular has attracted a great deal of attention by a wide spectrum of scientists, including those interested in the general mechanisms of biomineralization and more specifically those interested in the medical implications of calcification (e.g., osteoporosis). Thus *E. huxleyi* is the first haptophyte to have its mitochondrial and plastid genomes sequenced, and a high-quality draft sequence of its nuclear genome is now available. Based on pigmentation, notably chlorophyll *c*, the haptophytes belong to the chromoalveolates, a group comprising the chromists (haptophytes, cryptophytes, and heterokonts/stramenopiles) and the alveolates (dinoflagellates, ciliated ciliates, and the apicomplexa). The plastids of these groups are thought to be derived from the red algal lineage, following a single secondary endosymbiosis event about 1300 Ma.

The second approach to molecular studies has been to investigate picoeukaryote assemblages in seawater samples. One such study revealed numerous novel sequences, some of which were affiliated with the haptophytes. In fact it has been suggested that as much as 35% of these picoeukaryotes are haptophytes. Investigation of these haptophytes, which are less than 3 μm in cell diameter, may revolutionize the traditional method of identifying and classifying haptophytes solely by morphology.

COLLECTION METHODS

Over the last century, coccolithophores have been collected in numerous ways, with the research aims determining the collection method. Perhaps the oldest method has been the use of plankton nets, either neuston nets for surface tows or vertically hauled nets for collecting plankton through a range of water depths (Figure 4). For phytoplankton samples, nets with a mesh opening of 30–60 μm are generally preferred, to prevent clogging, although finer mesh sizes (with openings as small as 10–20 μm) may be used in clearer waters or when the target organisms are small. In the past, the nets (and the continuous plankton recorder) were made of fine silk, but these days the nets are usually made of nylon, since the mesh size can be accurately maintained in water. Plankton nets are generally used when

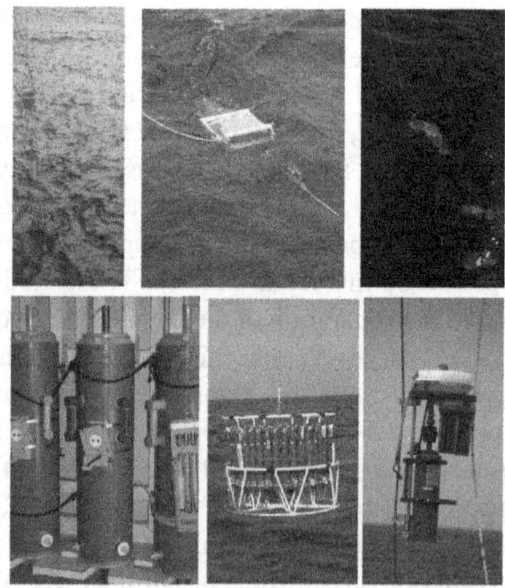

FIGURE 4 Some of the equipment used to collect marine phytoplankton. Upper left to right: buckets, neuston net, vertically hauled nets. Lower left to right: large volume water bottles, rosette of water bottles attached to a CTD rig, large volume water sampler.

seeking to isolate cells for culture, but the relatively large mesh openings preclude its use for accurate cell count determinations.

For those interested in coccolithophore ecology and absolute (or even relative) species abundances the preferred method is collection by water bottles. On oceangoing research vessels the water bottles are often attached to a conductivity, temperature, depth (CTD) rig, which records various physicochemical parameters in real time as the rig is lowered by winch through the water column (Figure 4). The bottles are closed electronically at selected depths and brought back to the surface, where they are sampled. Depending on the subsequent method of sample treatment, the plankton assemblages may be (1) filtered onto filter papers, dried, and stored, (2) concentrated and preserved with a dilute preservative (e.g., formalin or glutaraldehyde), or (3) concentrated and added to a culture medium in preparation for single-cell isolation. In shallow waters such as lagoons where large research vessels and the traditional CTD method cannot be used, some compromises need to be made. Recent methods include using a scuba diver to acquire water bottle samples at known depths (0–40 m), while simultaneously deploying hydrographic equipment (including a miniature CTD) to acquire in situ readings and/or real-time data.

To study coccolithophores on seasonal or annual timescales one needs to use a time-series sediment trap, which is often moored at a water depth a kilometer or more below the euphotic layer. Sinking coccolithophore assemblages fall into the large funnel of the sediment trap, which is equipped with a series of small water bottles. After a designated time period (perhaps 10 days or so), the water bottle directly underneath the funnel is replaced with a new bottle. In this way the sediment trap may collect samples for 6 months or 1 year, before needing to be recovered and subsequently redeployed. While this is still the only way to logistically collect samples from an open ocean location over a long time period, it may not be a truly accurate record of surface production, since a large percentage of the coccolithophore assemblage (especially those bearing holococcoliths or with fragile coccoliths) is recycled within the euphotic zone. Furthermore, intermediate or subsurface currents may transport and deposit different assemblages into the sediment trap. However, assemblages collected by sediment traps can still provide a good comparison with the coccolith assemblages found in the surface sediments.

BIOGEOGRAPHY AND ECOLOGY

General Distribution

Coccolithophores are globally distributed, exhibiting their greatest diversity in subtropical–tropical waters (often >100 species in total within the euphotic zone), with diversity decreasing toward the high latitudes (usually >10 species in subpolar waters). The spatial distribution of coccolithophore communities can be used to distinguish distinct biogeographic zones, which are closely related to those of zooplankton and to the circulation pattern of surface water currents (see Figure 5). However, the positions of these zones should not be thought of as fixed, since the frontal zone of converging surface water currents is a dynamic feature and often exhibits latitudinal and seasonal variation. Furthermore, these zones do not extend into shallow waters, as coastal processes such as upwelling, river outflow, and harbor pollution strongly influence the phytoplankton composition.

While most coccolithophores are planktonic and live in the open ocean, specialized coastal communities also exist, some of which have benthic life cycle stages, and at least one species has been recorded in freshwater. Each of the biogeographic zones shown in Figure 5 is characterized by a specialized assemblage (Table 1).

Seasonality and Depth Preferences

In temperate waters overlying the continental shelf, diatoms are usually numerically dominant in the spring (also occasionally in the autumn). This is when high nutrient concentrations, especially silicates, are present above the seasonal thermocline. However, the silicates are quickly used by diatoms and other siliceous phytoplankton (e.g., the Parmales and silicoflagellates), and once concentrations fall below about 1 μmol^{-1} these algae are unable to make their valves, plates, or skeletons. By late spring/early summer, the diatoms become stressed by these decreasing nutrient concentrations and increasing light intensity, and so sink out of the surface waters. Thus, nonsiliceous microplankton like coccolithophores and dinoflagellates are often more abundant in the summer, since they do not require high concentrations of silicates and can utilize the remaining nutrients such as nitrates (Figure 6). If new nutrients enter the surface waters at this time, rapid growth of opportunistic species (sometimes referred to as r-selected species) may occur resulting in dinoflagellate red tides or mesoscale coccolithophore blooms. In the autumn occasional storm mixing and decreasing light intensities favor the return of the siliceous phytoplankton, and so diatoms may produce autumn blooms, albeit on a smaller scale than those in spring. During the winter months, stratification breaks down and the conditions in the upper water column become homogenous, allowing so-called winter mixing to occur. The low water temperatures, low light intensities, and winter mixing prevent significant phytoplankton growth at this time.

In temperate areas where stratification of the upper water column may persist through the summer or in subtropical/tropical areas where a permanent thermocline exists, coccolithophores may be vertically distributed in depth-related

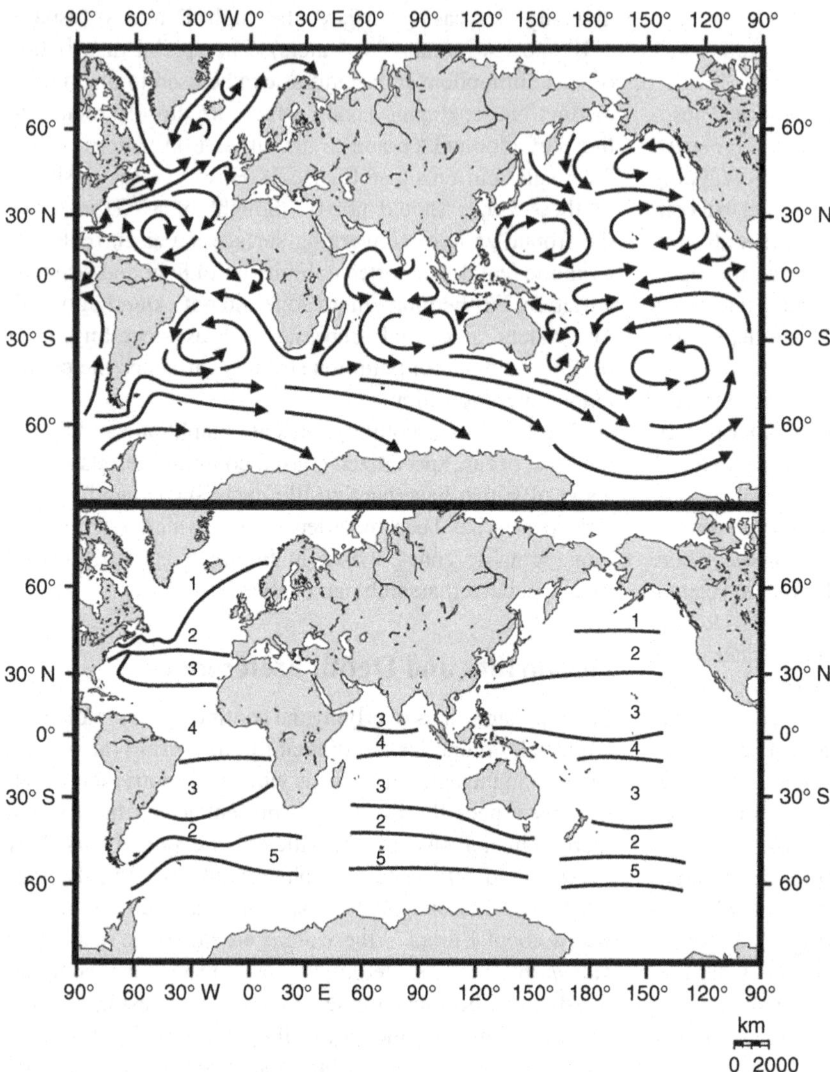

FIGURE 5 (a) Simplified circulation pattern of surface water currents during the Northern Hemisphere winter. Modified from a diagram that appeared in 'Ocean Circulation,' Open University/Pergamon Press, 1989. (b) Biogeographic zones based on phytoplankton distributions (where 1, subarctic; 2, temperate; 3, subtropical; 4, tropical; 5, subantarctic). Modified from Figure 6 in Jordan, R. W., & Chamberlain, A. H. L. (1997). Biodiversity among haptophyte algae. *Biodiversity and Conservation*, *6*, 131–152.

layers within the photic zone. These layers are often referred to as the upper, middle, and lower photic zones (or UPZ, MPZ, and LPZ, respectively), with each one possessing a characteristic assemblage (Figure 7 and Table 2). Some researchers doubt the existence of the MPZ, but it is a relatively thin layer (perhaps >20 m thick) and likely to be missed if the photic zone is sampled at low resolution.

With a lack of in situ observations and a paucity of species in culture, most coccolithophores are assumed to be photosynthetic. However, those species inhabiting the LPZ, where light levels are below 1% of the surface irradiance, could be heterotrophic, possibly supplementing their diet by engulfing bacteria. Alternatively, there is reasonable evidence to suggest that one of the LPZ species, *Reticulofenestra sessilis*, may have a symbiotic relationship with a diatom (*Thalassiosira* sp.), since the two are always found together. It is interesting to note that the coccospheres of *R. sessilis* are usually positioned around the girdle area of the diatom, which is where cell leakage of organic carbon is most likely to occur. This phenomenon has previously been documented for the epiphytic flagellate *Solenicola setigera* and the diatom *Leptocylindrus mediterraneus*.

Most coccolithophore species inhabit subtropical/tropical regions, where nutrient levels are low and species diversity is high. Since the photic zone conditions in these regions are relatively stable all year round, most species are considered to be specialists (or K-selected species), living in the UPZ, MPZ, or LPZ.

BLOOMS

Detection of Coccolithophore Blooms

One of the most spectacular features of coccolithophores is their ability to form massive blooms. Coccolithophore blooms often occur annually in the same geographic

TABLE 1 Characteristic coccolithophore assemblages of each biogeographic zone

Subarctic
Algirosphaera robusta, Balaniger balticus, Calciarcus alaskensis, Calciopappus caudatus, Coccolithus pelagicus, Emiliania huxleyi, Papposphaeraceae, *Quaternariella obscura, Syracosphaera borealis, Wigwamma* spp.
Temperate
Balaniger balticus, Calcidiscus leptoporus, Coccolithus pelagicus, Emiliania huxleyi, Ericiolus spp., *Gephyrocapsa* spp., *Helicosphaera* spp., *Reticulofenestra* spp., *Syracosphaera* spp.
Subtropical
Alisphaera spp., *Alveosphaera bimurata, Canistrolithus valliformis, Emiliania huxleyi, Florisphaera profunda, Gephyrocapsa oceanica, Gladiolithus* spp., *Hayaster perplexus, Helicosphaera* spp., *Navilithus altivelum, Oolithotus fragilis,* Pontosphaeraceae, *Reticulofenestra sessilis,* Rhabdosphaeraceae, *Solisphaera* spp., Syracosphaeraceae, *Tetralithoides quadrilaminata, Turrilithus latericioides, Umbellosphaera* spp., *Umbilicosphaera* spp., *Vexillarius cancellifer*
Tropical
Alisphaera spp., *Calcidiscus leptoporus, Emiliania huxleyi, Gephyrocapsa* spp., *Florisphaera profunda, Gladiolithus* spp., *Reticulofenestra sessilis*
Subantarctic/Antarctic
Calciarcus alaskensis, Ericiolus spp., Papposphaeraceae, *Quaternariella obscura, Wigwamma* spp.
Coastal
Braarudosphaera bigelowii, Cruciplacolithus neohelis, Hymenomonas spp. (except *H. roseola*), *Jomonlithus littoralis, Ochrosphaera neapolitana, Pleurochrysis* spp.
Freshwater
Hymenomonas roseola

Modified from Table 3 in Jordan, R. W., & Chamberlain, A. H. L. (1997). Biodiversity among haptophyte algae. *Biodiversity and Conservation, 6*, 131–152.

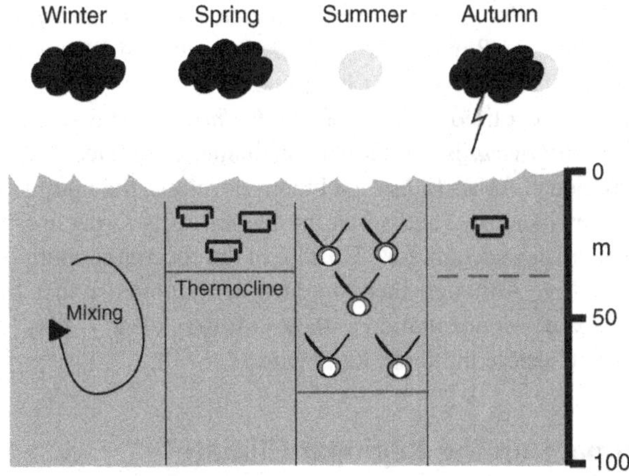

	Spring	Summer	Autumn
Nutrients High	Gradually decreases	Low	Occasionally increases due to storm
Light Low	Gradually increases	High	Gradually decreases
Temperature Low	Gradually increases	High	Gradually decreases

= Siliceous phytoplankton (especially diatoms)

= Flagellates (especially dinoflagellates and coccolithophores)

FIGURE 6 Schematic diagram showing the seasonal variation of some physicochemical parameters and phytoplankton groups in a temperate photic zone.

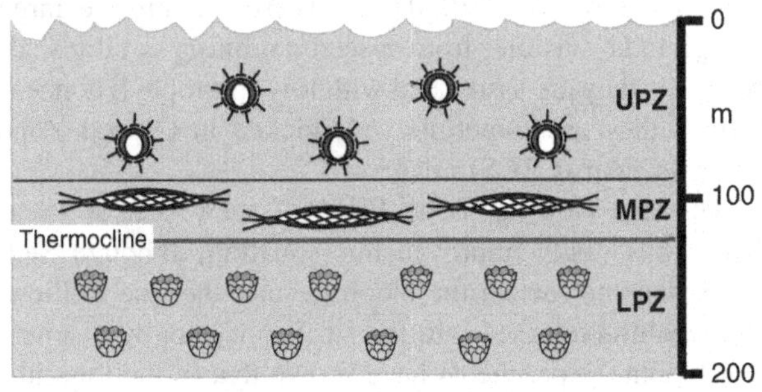

	Nutrients	Light	Temperature
UPZ	Low (<1 μ mol^{-1})	High (100%)	Warm (>20 °C)
MPZ	Higher than UPZ	Lower than UPZ	Cooler than UPZ
LPZ	High (>1 μ mol^{-1})	Low (<1%)	Cool (<20 °C)

= UPZ coccolithophores (e.g., *Rhabdosphaera*)

= MPZ coccolithophores (e.g., *Calciosolenia*)

= LPZ coccolithophores (e.g., *Florisphaera*)

FIGURE 7 Schematic diagram showing the vertical distribution of coccolithophores within a subtropical/tropical photic zone and some physicochemical parameters associated with each layer (where UPZ, MPZ, and LPZ represent the upper, middle, and lower parts of the photic zone, respectively).

TABLE 2 Vertical distribution of coccolithophores within the subtropical/tropical photic zone

Upper photic zone (UPZ)
Alisphaera spp., *Ceratolithus cristatus*, Rhabdosphaeraceae, Syracosphaeraceae, *Umbellosphaera* spp.
Middle photic zone (MPZ)
Calciopappus rigidus, *Calciosolenia* spp., *Michaelsarsia* spp., *Ophiaster* spp., *Picarola margalefii*, *Placorhombus ziveriae*
Lower photic zone (LPZ)
Algirosphaera robusta, *Alveosphaera bimurata*, *Canistrolithus valliformis*, *Florisphaera profunda*, *Gladiolithus* spp., *Hayaster perplexus*, *Navilithus altivelum*, *Oolithotus fragilis*, *Reticulofenestra sessilis*, *Solisphaera* spp., *Syracosphaera anthos*, *Tetralithoides quadrilaminata*, *Turrilithus latericioides*, *Umbilicosphaera anulus*, *Vexillarius cancellifer*
No depth preference
Emiliania huxleyi, *Gephyrocapsa* spp., *Helicosphaera* spp. (although their holococcolith-bearing phase inhabits only the UPZ)

Modified from Table 4 in Jordan, R. W., & Chamberlain, A. H. L. (1997). Biodiversity among haptophyte algae. *Biodiversity and Conservation, 6*, 131–152.

locations, suggesting that their formation is due to a combination of interrelated factors such as permanent bathymetric features (e.g., continental shelves, shallow seas, and fjords), recurrent favorable water conditions (during the summer months), constant 'seeding' from adjoining oceanic areas, and a fast growth rate associated with an opportunistic species. Most of these blooms persist for 3–6 weeks and are produced by *E. huxleyi*, although *G. oceanica* blooms have been recorded from South Africa, Japan, and Australia. Coccolithophore blooms contain at least 1 million cells/liter and often cover >100 000 km^2. These blooms are large enough to be 'visible' from space by orbiting satellites, although as they are associated with low chlorophyll concentrations they are sometimes overlooked in Coastal Zone Color Scanner (CZCS) images.

The Sea-viewing Wide Field of view Sensor (SeaWiFS) has vastly improved this situation, although SeaWiFS does not detect the blooms, but rather the millions of coccoliths released into the surface waters by a senescent bloom. Experiments have shown that in the logarithmic growth phase of a bloom, there are equal numbers of attached versus detached coccoliths, but in aging blooms detached coccoliths exceed attached ones by a factor of 6. Since coccoliths are composed of birefringent calcite, the detached coccoliths act like tiny mirrors reflecting as much as 30% of the surface irradiance back into space. From a ship these bloom waters, laden with millions of discarded coccoliths, often appear milky-turquoise. Since the amount of chlorophyll in the water can subtly alter the color of the water and chlorophyll levels decrease as the bloom declines, the colored satellite images can provide information on the relative age of the bloom. Furthermore, as individual coccoliths sink exceedingly slowly out of the upper water column, this high-reflectance water may persist much longer than the bloom itself, and act as a tracer of past bloom events.

Coccolithophore Viruses

About 40 years ago it became clear that virus-like particles were present inside the cells of all the major groups of marine algae. Later on, it was realized that these marine viruses were species-specific, targeting most of the abundant or bloom-forming algae, with those associated with phytoplankton termed phycodnaviruses. At present, viruses are known to infect and lyse a number of noncoccolithophore haptophytes, including *Prymnesiovirus* on *Chrysochromulina brevifilum* and *PpV* and *PgV* on *Phaeocystis pouchetii* and *Phaeocystis globosa*, respectively. At least one virus has been found to attack coccolithophores, notably *E. huxleyi*. The so-called *Coccolithovirus* is 170–200 nm in diameter, with icosahedral symmetry. As an *E. huxleyi* bloom develops, the number of viruses increases, and when the bloom starts to decline the viruses may account for 25–100% of the coccolithophore cell mortality. Although their production is limited under low phosphate concentrations, these viruses clearly play an important role in bloom termination.

Impact on the Regional Climate and Environment

Some blooms of the noncoccolithophore haptophyte genera *Chrysochromulina* and *Prymnesium* are renowned for producing powerful toxins, particularly those in Scandinavian coastal waters that have resulted in massive fish kills. It has also been demonstrated recently that the coastal coccolithophore genera *Jomonlithus* and *Pleurochrysis* can also produce toxins strong enough to kill nauplii of the brine shrimp *Artemia*. These toxic coccolithophores often form blooms in saline lakes or ponds used for aquaculture of fish or shellfish, so they can have a devastating effect on the local economy. However, the two oceanic bloom producers, *E. huxleyi* and *G. oceanica*, have tested negative for toxins.

Like a number of other phytoplankton species, blooms of coccolithophores and noncoccolithophore haptophytes may release large amounts of dimethylsulfoniopropionate (DMSP), which in the surrounding seawater hydrolyzes into two compounds: dimethyl sulfide (DMS) and acrylic acid. DMS has been implicated in acid rain formation, since in the atmosphere it may be oxidized to form sulfur dioxide, and then sulfuric acid. It is also presumed that sulfate and methane sulfonate aerosols act as cloud condensation nuclei, and since clouds increase the planet's albedo, the aerosols may partly control the amount of solar radiation reaching the earth's surface. The termination of marine phytoplankton blooms is often associated with species-specific viruses, and it has been shown that DMSP release increases dramatically during lysis of infected cells.

The other by-product of DMSP cleavage, acrylic acid, is an antibacterial agent and so may reduce the threat of bacterial infection during the coccolithophore bloom period. There is also some evidence to suggest that acrylic acid affects herbivorous copepods by suppressing grazing.

Carbon Release and Sinking of Blooms

It is well known that aging algal blooms often release huge amounts of organic carbon in the form of mucilage (mucopolysaccharides), particularly when these blooms become stressed due to nutrient limitation. Mucilage released by blooms of the noncoccolithophore haptophyte *Phaeocystis* is often associated with foam production and the clogging of fishermen's nets. Coccolithophore blooms may also release carbon compounds in the form of coccolith polysaccharides. These polysaccharides are involved in the coccolith formation process, and so are excreted along with the coccolith upon completion. However, it has been demonstrated that even naked cells possess an extracellular layer of coccolith polysaccharides, implying that calcification is not essential for their production. Since these polysaccharides are water soluble, a proportion of them must become incorporated into the dissolved organic carbon content of the surrounding seawater. These polysaccharides may also be involved in the aggregation of senescent bloom cells, thus allowing the bloom to sink out of the surface waters when unfavorable conditions prevail.

Since individual coccoliths have exceedingly slow settling rates, and hence would be dissolved long before reaching the seabed, there are only two mechanisms for quickly transporting bloom material from the sea surface to the seabed. One of these mechanisms is via mass aggregates (sometimes referred to as 'marine snow'), which may sink at approximately 100 m day^{-1} through the water column to the seabed, where they are subsequently eaten by the benthos. The other process is via fecal pellets, notably those of herbivorous calanoid copepods. Due to the acid digestion of the copepod gut, only 27–50% of the ingested coccolith material is subsequently egested. However, it is not uncommon to recover live phytoplankton cells from these pellets, implying that the copepods do not crush their food during intake. Like the mass aggregates, the fecal pellets have sinking rates >100 m day^{-1}. The pellet contents are usually protected from dissolution by the organic coating around the pellet, but this often gets degraded in the waters above the seabed where bacterial numbers are high.

Comparisons of the surface water assemblages with those in sediment traps (below 1000 m water depth) or in the surface sediments show that about 70% of the coccolithophore diversity is lost within the upper water column, with only around 50 taxa common to all three assemblages. Those coccolithophores bearing holococcoliths or weakly calcified or easily broken heterococcoliths are generally rare or absent in sediment assemblages. However, holococcoliths do have a geological record with some of the earliest forms in Jurassic sediments (about 185 Ma), in intervals reflecting excellent preservation conditions.

EVOLUTION

First Appearance of the Coccolithophores

There are potentially three ways of dating the first appearance of the coccolithophores in the geological record. The traditional method using microscopy targets the mineralized parts (i.e., the coccoliths), for which the oldest reliable records date back to the Late Triassic (225 Ma), when low-diversity assemblages occurred worldwide. Coccoliths of earlier periods are not available; however, another approach involving group-specific biomolecules may be able to extend this record. Over the last few decades a lot of work has been carried out on long-chain carbon compounds such as alkenones, which are produced exclusively by some members of one coccolithophore order, the Isochrysidales. Their ancestors can be traced back to the Cretaceous–Tertiary boundary (65 Ma), while their alkenones have been found in Cretaceous shales (105 Ma). Unfortunately, the first appearance of these alkenones clearly postdates the coccoliths from the Late Triassic and so additional coccolithophore-related biomolecular compounds need to be discovered in families with much longer stratigraphic records (e.g., the Braarudosphaeraceae and Calciosoleniaceae). The third method involves constructing a molecular clock using the genetic material of living haptophytes (and other algae), with which to estimate the timings of past events. A recent study suggested that the first appearance of the haptophytes probably occurred around 850 Ma (in the Precambrian), while the divergence of the two subclasses was about 281 Ma, just before the Permian–Triassic boundary (250 Ma). If true, this may suggest that the coccolithophores first appeared sometime in the Triassic, thus corroborating the fossil coccolith record.

Coccolithophore Diversity and Extinctions

Since the Late Triassic the diversity of the coccolithophores reached a maximum on at least two occasions, once in the Late Cretaceous (80–68 Ma) and again in the Oligocene (29–26 Ma). Although diversity declined during the Pliocene and Quaternary, about 200 taxa of coccolithophores are currently known to be extant. As with other groups, extinctions and more gradual turnover events have characterized the evolution of the coccolithophores, with extinctions at the Triassic–Jurassic and Cretaceous–Tertiary boundaries. The latter example in particular resulted in 93% of taxa becoming extinct, followed by a gradual recovery during the Paleocene and the subsequent appearance of many new families (see Figure 8 for a summary of these events).

The success and dominance of these algae in past oceans is clearly evidenced by the presence of thick chalk deposits, such as those along the English south coast (Figure 9). These deposits, often hundreds of meters thick, were laid down during the Jurassic and the Cretaceous.

FIGURE 9 The chalk cliffs along the Dorset coast of England.

Past Coccolithophore Blooms

Usually, only detached coccoliths are preserved in the sediment record, but occasionally whole coccospheres are also present (Figure 10). In exceptional cases the coccospheres are so abundant and in monospecific assemblages that they are likely to represent the remains of a past bloom. These deposits are often associated with past anoxic bottom waters, which prevented organic degradation and disturbance by benthic communities. Anoxia usually occurs in isolated eutrophic basins, and results in the deposition of black shale-like sediments. These assemblages are extremely useful in providing information on the arrangement and number of coccoliths on the cell, evidence of dimorphism or polymorphism, and the ontogenetic sequences of coccolith formation. The Kimmeridge Clay (a geological formation in southern England) contains abundant coccospheres of the Late Jurassic species, *Watznaueria fossacincta*, whereas more diverse (>30 species) coccosphere-containing deposits have been found at Cretaceous/Tertiary (K/T) boundary sections in the Netherlands. In the latter deposits it was found that some of the post-K/T boundary species were actually present in low

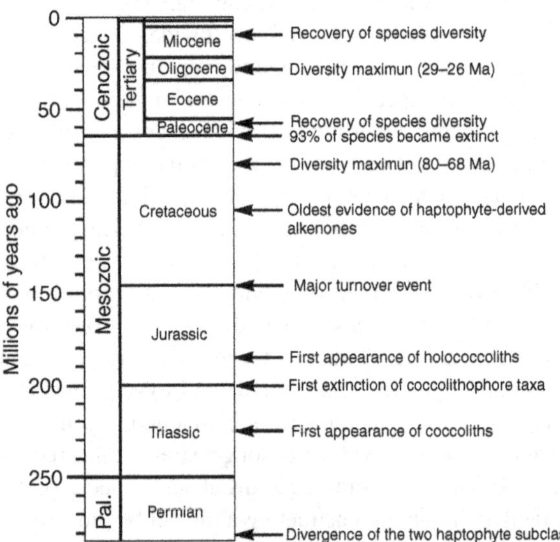

FIGURE 8 A summary of some of the key events in coccolithophore evolution. Modified from Medlin, L. K., Kooistra, W. C. H. F., & Schmid, A. M. M. (2000). A review of the evolution of the diatoms – A total approach using molecules, morphology and geology. In A. Witkowski & J. Sieminska (Eds.), *The origin and early evolution of the diatoms: Fossil, molecular and biogeographical approaches* (pp. 13–35). Cracow: W. Szafer Institute of Botany, Polish Academy of Sciences and Bown et al. in Thierstein, H. R., & Young, J. R. (Eds.) (2004). *Coccolithophores: From molecular processes to global impact*. Berlin Heidelberg New York: Springer.

FIGURE 10 An Early Oligocene coccosphere of *Reticulofenestra umbilica* from the Falkland Plateau. Scale bar = 5 μm.

numbers before the event, and so the number of K/T boundary 'survivors' may be higher than previously thought.

Biostratigraphic Use

Coccoliths have become one of the most powerful tools in biostratigraphy, because (1) they can be easily observed and quickly identified using light microscopy, (2) it takes only a toothpick-sized sample and a few minutes to make a permanent slide for observation, (3) they have a worldwide distribution and are abundant in most marine sediments, and (4) coccolithophores evolved rapidly over a relatively short period of geological time, making them useful for geological dating. Like the other major microfossil groups, the coccolith biostratigraphic schemes employ a coded zonation finely tuned to the chronologies provided by the paleomagnetic and oxygen isotope records. Besides the overall coccolith assemblage, the first and last appearance and occurrence data of selected taxa (usually abbreviated to FAD, LAD, and FO and LO, respectively), as well as acme events, provide extra tie points for dating marine sediments.

In general, land sections have been used to produce the Mesozoic schemes, especially for the Jurassic and the Triassic, although a number of deep-sea drilling sites have yielded excellently preserved Cretaceous sediments. For the Cenozoic schemes, the cores obtained around the world by drilling ships have provided an immense amount of good material, while the Quaternary schemes have benefited from the large number of piston cores taken by nondrilling research vessels. Over the last 50 years, coccolith biostratigraphy has been helpful in understanding a variety of scientific problems. Perhaps the most important role of fossil coccoliths has been as dating tools for the hydrocarbon exploration industry and for scientific ocean drilling research projects. These coccolith biostratigraphic schemes have been constantly improved and tested, not just in terms of reliability and finer resolution, but also for pinpointing regional differences. This has led to the development of biostratigraphic schemes for particular provinces or zones (such as the Boreal Province or North Sea zonation scheme) and a resolution for the Cenozoic of between 1 million and 600 000 years.

Coccolith biostratigraphy has also been used in other ways, for instance, in dating turbidite units and in archeology. Since turbidites are sediments that have been deposited by turbidity currents, and the latter are known to have strong relationships with changing sea level and earthquakes, knowing the timing of these turbidite units can provide useful information about past oceanic events. Using coccolith biostratigraphy, an age range can be assigned to a turbidite unit by comparing its coccolith assemblage with that of known coccolith mixtures prepared in the laboratory. Archeology is another field of study that uses coccolith assemblages to date materials, since ancient potters utilized coccolith-bearing clays, marls, and limestones to make their ceramics. The identification and dating of these coccolith assemblages can help to locate the local strata the potters once used.

FURTHER READING

Baumann, K.-H., Andruleit, H., Böckel, B., *et al.* (2005). The significance of extant coccolithophores as indicators of ocean water masses, surface water temperature, and palaeoproductivity: A review. *Paläontologische Zeitschrift*, *79*(1), 93–112.

Bown, P. R. (Ed.), (1998). *Calcareous nannofossil biostratigraphy*. Chapman & Hall: London.

Bown, P. R., & Young, J. R. (1997). Proposals for a revised classification system for calcareous nannoplankton. *Journal of Nannoplankton Research*, *19*(1), 15–47.

Cros, L., & Fortuño, J.-M. (2002). Atlas of Northwestern Mediterranean coccoliths. *Scientia Marina*, *66*(Suppl. 1), 1–186.

Cros, L., Kleijne, A., Zeltner, A., Billard, C., & Young, J. R. (2000). New examples of holococcolith-heterococcolith combination coccospheres and their implications for coccolithophorid biology. *Marine Micropaleontology*, *39*(1–4), 1–34.

Green, J. C., & Leadbeater, B. S. C. (1994). In: *The haptophyte algae*. *The Systematics Association* Special Vol. 51. Clarendon Press: Oxford.

Jordan, R. W., & Chamberlain, A. H. L. (1997). Biodiversity among haptophyte algae. *Biodiversity and Conservation*, *6*, 131–152.

Jordan, R. W., Cros, L., & Young, J. R. (2004). A revised classification scheme for living haptophytes. *Micropaleontology*, *50*(Suppl. 1), 55–79.

Jordan, R. W., Kleijne, A., Heimdal, B. R., & Green, J. C. (1995). A glossary of the extant haptophyta of the world. *Journal of the Marine Biological Association of the United Kingdom*, *75*, 769–814.

Kleijne, A., & Cros, L. (2009). Ten new extant species of the coccolithophore *Syracosphaera* and a revised classification scheme for the genus. *Micropaleontology*, *55*(5), 425–462.

Medlin, L. K., Kooistra, W.C. H. F., & Schmid, A. M. M. (2000). A review of the evolution of the diatoms – A total approach using molecules, morphology and geology. In: A. Witkowski & J. Sieminska (Eds.), *The origin and early evolution of the diatoms: Fossil, molecular and biogeographical approaches* (pp. 13–35). W. Szafer Institute of Botany, Polish Academy of Sciences: Cracow.

Thierstein, H. R., & Young, J. R. (Eds.), (2004). *Coccolithophores: From molecular processes to global impact*. Springer: Berlin, Heidelberg, New York.

Triantaphyllou, M. (2004). Advances in the biology, ecology and taphonomy of extant calcareous nannoplankton. *Micropaleontology*, *50* (Suppl. 1), 1–170.

Winter, A., & Siesser, W. G. (Eds.), (1994). *Coccolithophores*. Cambridge University Press: Cambridge.

Young, J. R., Bergen, J. A., Bown, P. R., *et al.* (1997). Guidelines for coccolith and calcareous nannofossil terminology. *Palaeontology*, *40*(4), 875–912.

Young, J. R., & Bown, P. R. (1991). An ontogenetic sequence of coccoliths from the Late Jurassic Kimmeridge Clay of England. *Palaeontology*, *34*(4), 843–850.

Young, J. R., Thierstein, H. R., & Winter, A. (2000). Nannoplankton ecology and palaeoecology. *Marine Micropaleontology*, *39*(1–4), 1–318.

Chapter 18

The Glass Menagerie: Diatoms for Novel Applications in Nanotechnology

Richard Gordon[1], Dusan Losic[2], Mary Ann Tiffany[3], Stephen S. Nagy[4] and Frithjof A.S. Sterrenburg[5]

[1]*Department of Radiology, University of Manitoba, Winnipeg MB R3A 1R9, Canada*

[2]*Ian Wark Research Institute, University of South Australia, Mawson Lakes, Adelaide SA 5095, Australia*

[3]*Center for Inland Waters, San Diego State University, 5500 Campanile Drive, San Diego CA 92182, USA*

[4]*Montana Diatoms, PO Box 5714, Helena MT 59604, USA*

[5]*Stationsweg 158, Heiloo 1852LN, The Netherlands*

Chapter Outline

Why Diatoms?	249
Diatom Silica Structure	251
Silica Biomineralization and Diatom Genomics	251
Diatom Biophotonics	255
Microfluidics Within Diatoms	256
Diatoms for Drug Delivery	257
Selective Breeding of Diatoms Using a Compustat	258
Computing With Diatoms	258
Conclusions	258
Acknowledgements	258
References	259

WHY DIATOMS?

Why have grown men and women spent lifetimes, often unpaid and while pursuing other careers (Box 1), examining one division of single-celled algae over the course of more than two centuries? The answer lies in their inordinate beauty: the shells around each cell of Bacillariophyta [1], the diatoms, are made of amorphous, clear silica glass, more ornate [2] than the finest delicate crystal that human artisans have crafted [3]. Indeed, when designing buildings and aircraft, architects and engineers have applied the same structural principles in their work as diatoms use to create their shells [4–7], and now nanotechnologists are turning to diatoms to build a variety of devices [8]. There are ~250 living diatom genera with more than 200 000 estimated species classified by their unique morphologies [9] (Figure 1). Diatoms are also remarkable living creatures with significant biogeochemical [10] and ecological roles on this planet, including '~20–25% of the world net primary production' [11]. Their extraordinary diversity might be due to in part to rapid rates of horizontal gene transfer with many bacteria [12]. Here we will provide an update on our previous reviews published in this journal [13,14] and a compendium [15] that covered the status of the field as of 2005.

The basic advantage of diatoms for nanotechnology over standard photolithography methods (microelectromechanical systems [MEMS]) [16] is that diatoms grow in exponentially increasing numbers on surfaces [17] or in solution [18], whereas MEMS are manufactured in numbers that grow linearly with time. With MEMS, we build to our own design. With diatoms, we either select from available species or attempt to modify their morphogenesis. Doing the latter requires that we understand how diatoms build themselves. Generally, we expect industry to be utilizing basic research, but a counterintuitive consequence of the thrust to use diatoms industrially is an enormous industrially motivated growth in the basic science of diatoms [19]. For example, because diatoms are much like the rest of eukaryotic life in their fundamental biology, diatom nanotechnologists are inadvertently contributing to the solution of one of the major outstanding problems of biology, namely the possibly reciprocal relationship [20] between the one-dimensional, linear, sequenced genotype [12,19,21,22] and the chemistry and physics of the multidimensional phenotype. Diatoms provide a crucial testbed for the reductionist concept of 'specific gene products (proteins) guiding these biomineralization processes' [23]. Given the significant intraspecific variability

Box 1 Gentlemen diatomists of independent means

There was not a single paid professional 'diatomist' on Earth until ~1930. Diatom studies are the classic example of the gentleman amateur scientist. For instance, the leading British diatomist around 1855, William Smith [137], was a reverend. Adolf Schmidt, the man who started the world-famous gigantic Schmidt Atlas in 1874 [138], should be given the honour due to him by referring to him as 'the Archidiaconus [archdeacon] Schmidt' – certainly the most impressive title a diatomist could have.

Towering giants of the Victorian period, when the entire basis of diatom studies was founded, include the coauthors Albert Grunow (an Austrian naturalist and phycologist) and Per Teodor Cleve (a Swedish chemist) [139] and the brothers Hippolyte and Maurice Peragallo [140]. The latter two were 'anciens élèves de l' Ecole Polytechnique' – a famous institute, but not in any way connected to the life sciences.

But the most outstanding example is Henri van Heurck [141,142], a Belgian industrialist of the late Victorian era, who between 1860 and 1908 literally spent a fortune dabbling in diatoms. He was very wealthy indeed, so much so that he could ask Messrs. Zeiss to compute, design and construct a one-off special oil immersion objective for his diatom hobby. Also, he privately published books on microscopy and diatoms. Then he had a handy tool for his diatom studies: his own steam-yacht, completely fitted out as a laboratory. And finally, he spent astronomical (for that time) sums on acquiring materials and every new optical gadget that was being invented. His collections – far from intact, unfortunately – are now in Brussels.

Living examples of diatomists who are not paid to work specifically on diatoms include the NASA astrobiologist Richard B. Hoover [37] and three of the authors of this article: science consultant F.A.S.S. [2], psychiatrist S.S.N. [2,37,143,144] and 'armchair diatomist' R.G., who is a theoretical biologist in a medical school and has sought no grants for his hobby.

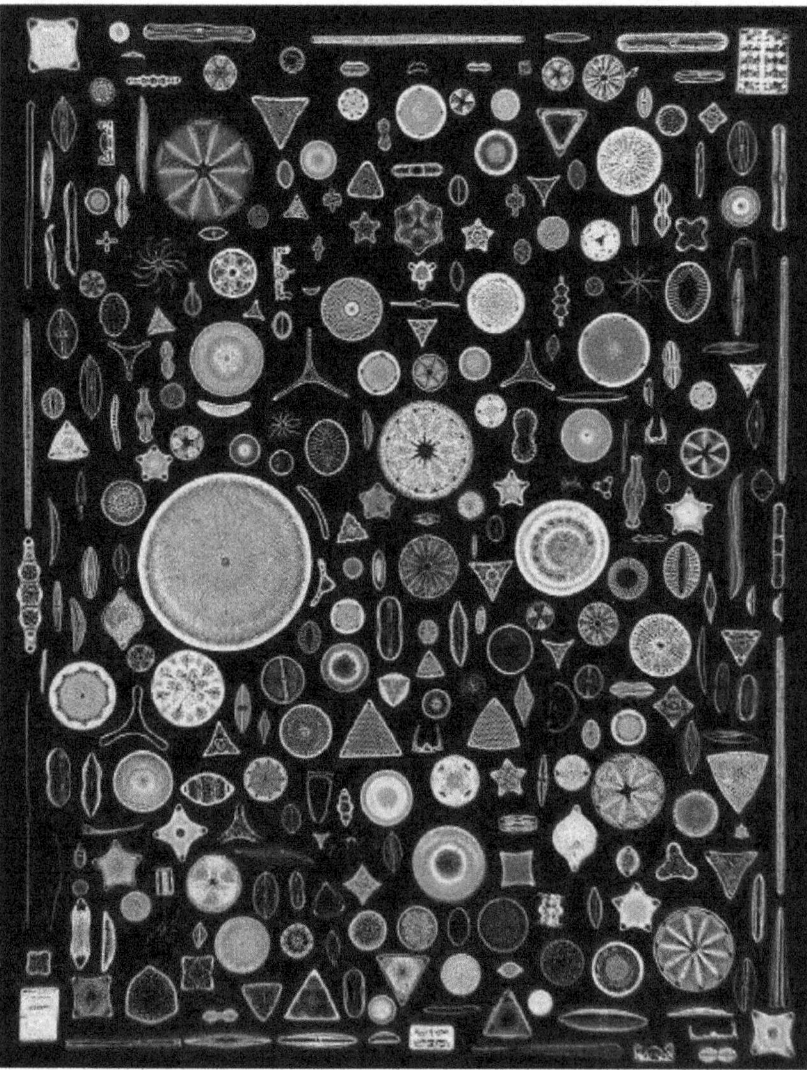

FIGURE 1 Three hundred diatoms mounted individually by hand. The examples shown include recent and fossil and freshwater and marine diatoms. They originate from the UK, Holland, France, New Zealand, Sulawesi, Caribbean, Indian Ocean, Florida, Maryland, Oregon, Montana, Nevada, British Columbia, California, Alaska, Honolulu and Russia. The array is 1.78 × 2.30 mm.

of diatoms [24], the presumption that they are 'under precise genetic control' [25] might be an exaggeration, but a testable one.

DIATOM SILICA STRUCTURE

Diatoms are microscopic (2 μm to 2 mm [26], cf. Figure 1), and species are classified mostly by the shapes and patterns of their hard silica parts, so the foci of diatom taxonomists and nanotechnologists coincide. The silica shell, or 'frustule', consists of two overlapping valves joined with girdle bands [1], much like a Petri dish (Figure 2). There are two major groups that are separated based on valve symmetry [1]. The pennate diatoms are elongate, usually with bilateral symmetry. In the class of centrics, diatoms have radial symmetry (Figure 3a). A proper group theory analysis of diatom symmetries has yet to be done, but the centrics might be said to have n-fold two-dimensional (2D) rotational symmetry, with $n = 3$ on up, approaching full circular symmetry. The pennates are placed into two classes depending on whether or not they have slits in the valves called raphes [1] (Figure 2), which are involved in gliding motility [27,28].

The general structure of a valve can be summarized as follows: lines of silica called costae diverge and occasionally branch (Figure 2, Figure 3b) from a nucleation site, the linear midrib in pennate diatoms or the circular midring in centric diatoms [1,29]. As we shall see, this scenario might include honeycomb structures (Figure 3b). Each valve possesses a three-dimensional (3D) and hierarchical organization of porous plates and solid walls with pore diameters that range from nanometres to micrometres (Figure 3) and with enormous structural diversity of their patterns and shapes. Recent studies using high resolution atomic force microscopy, scanning electron microscopy and time-lapse light microscopy have revealed a diversity of new nano-and meso-scale silica morphologies of diatoms, including the presence of 50 nm spherical silica particles (cf. 'colloidal silica' [29]), which allow better understanding of biosilica formation and valve morphogenesis [30–32]. The general assumption is that the silica itself remains amorphous in all these detailed structures, despite their organic components and 'templating' surrounds [33], but this needs to be directly tested by electron diffraction (cf. [34]).

SILICA BIOMINERALIZATION AND DIATOM GENOMICS

Diatom structures are presumed to be replicated from generation to generation by a genetically controlled biomineralization process that takes place at levels from the molecular to the nano- and micro-scale. However, cytoplasmic inheritance, such as occurs in the also hierarchically patterned surfaces of ciliates [35,36], might have a role, considering the existence of pennate diatoms with significantly differing valves (heterovalvy) despite there being just one cell nucleus, as well as the transmission of shape aberrations or complementary geometry of valves of

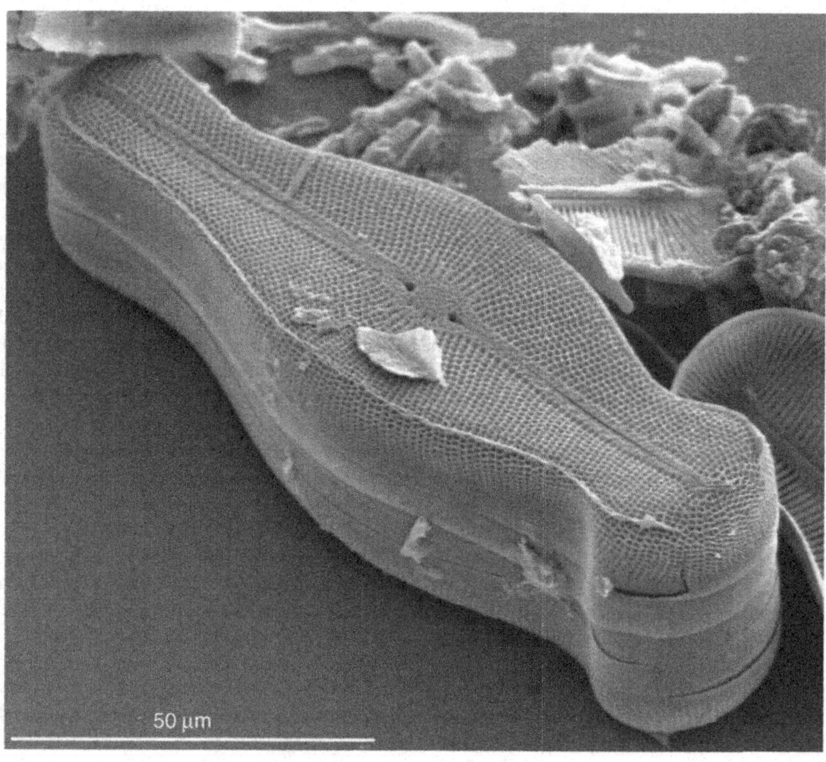

FIGURE 2 Scanning electron micrograph (SEM) of a pennate diatom, *Didymosphenia geminata* (Lyngbye) Schmidt from Cache la Poudre River in Colorado, USA. The two slits along the midline are the raphes, which are involved in motility. The branching silica precipitation of costae proceeded from this midline to the periphery and down the sides; this is more clearly visible on the inner view of the valve on the right. Girdle bands can be seen between the two valves. The scale bar represents 50 μm.

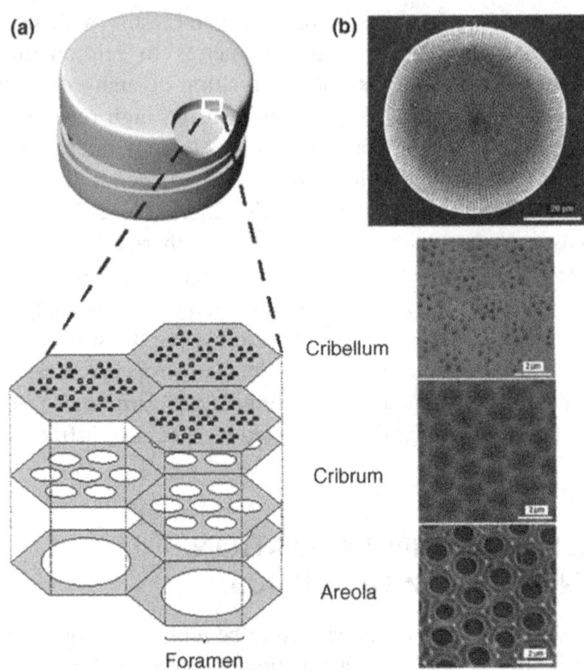

FIGURE 3 Diatom structure. **(a)** Schematic of a centric diatom frustule with cross-sectional three-dimensional (3D) profile of the silica wall based on SEM data. The inner layer contains honeycomb-like vertical chambers called areolae. The large hole in the floor of an areola is known as a foramen. The roof of the areolae is called the cribrum, which contains a regular pattern of pores. The layer over the cribrum is a thin siliceous membrane known as the cribellum, which consists of small pores. **(b)** SEM image of a *Coscinodiscus* sp. with corresponding layers [46,133,134]. *Reproduced from [134] with permission.*

daughter cells [1]. We can anticipate a role of gravity because microtubules (MTs) are undoubtedly involved [37], and manipulations of pattern via environmental changes have begun in earnest [38–40].

There are two basic forms of morphogenesis: (i) patterns that form spontaneously as symmetry-breaking phenomena and (ii) patterns that are guided by prepatterns, so-called 'structure-directing templates' or 'scaffolds' [41]. Prepatterns lead to an infinite explanatory regress, as in the long-defunct idea of Leeuwenhoek of the homunculus in sperm [42], the inflatable 'little man' who contains sperm that each contain another smaller homunculus with sperm, etc., back to Adam and Eve. At this point we have several pieces to the puzzle of diatom shell morphogenesis, but not the whole picture. The morphological evidence suggests that silica precipitates in at least five stages: (i) formation of small silica spheres of 30 to 50 nm diameter [29], perhaps inside membrane-bound silica transport vesicles (STVs) (Figure 4); (ii) transport of these vesicles to the periphery of a flat membrane bag called the silica-lemma (SL), into which the silica spheres are released; (iii) 2D precipitation of the spheres onto the growing valve inside the SL, starting from a nucleating structure and taking but a few minutes [23,43]; (iv) pore formation [44–46]; and (v) thickening of the valve, taking hours and often accompanied by further pattern formation in the third dimension [29,43,44,47]. These steps are generally not distinguished in the current literature on silica biomineralization, giving the impression that the molecular key(s) to morphogenesis have been found, without accounting for the multiple physical and

FIGURE 4 Diatom morphogenesis. Schematic view of a cross section of a pair of daughter diatom cells after cell division while the new valves are forming inside, each in a silica deposition vesicle (SDV) consisting of a bilayer membrane, called the silicalemma (SL), and its contents. The upper cell shows the rapid two-dimensional (2D) phase of valve formation, which only takes minutes. The cell nucleus might be torus-shaped, and bundles of microtubules (MTs) extend through the nuclear hole [85]. MTs emanating from a microtubule-organizing centre (MTOC) are on the inner face of the SL, which is a flat, membrane-bound bag at this stage. The SL contains a nucleating centre where silica precipitation starts. This centre is where colloidal silica spheres of ~50 nm diameter, which are probably the diffusing and precipitating entities, initially adhere. They then stick to the already precipitated silica spheres. The SL also contains the mother liquor, the fluid remaining after silica precipitation and sintering [29]. The mother liquor might have two or more immiscible liquid components [44–46]. The MTs on the inner face of the SL might mechanically counteract contraction of the microfilament (MF) ring around the perimeter of the SL [86]. This tensegrity structure presumably keeps the SL thin and flat, thereby allowing a 2D pattern of precipitated silica to form within it [29]. Three possible routes of entry of monomeric silica, $Si(OH)_4$, into the cell are shown: route **(a)** proceeds via adsorption to spines and migration along the organic casing [1]; route **(b)** is through the silica shell pores; and route **(c)** is via the gap that is formed between the daughter cells and that follows through or past the possibly more permeable girdle bands [31,135] (not shown here, but depicted in Figure 2). Transport of silica within the cell might involve hypothesized silica transport vesicles (STVs), which could be formed at the cell membrane by a clathrin mechanism, then labelled with a trafficking signal and transported to the SL margin by motor molecules on MTs [58]. At the SL, each deposits its membrane and silica sphere. It has been assumed that the silica within the STVs is solid (but cf. [26]). Silica-binding organic and inorganic molecules and proteins [23,53] that are present in the mother liquor, including nickel [38], germanium [39], H^+ [136] and salts [40], might influence the pattern of precipitation and might become incorporated into the silica. The daughter cell depicted in the lower half of the figure is at a later stage of cell division and shows the slow, 3D, thickening phase of valve formation, which typically takes hours. Most cellular details that are the same, such as nucleus and mother liquor, are identical and are not shown for clarity. Instead, only events that differ from the earlier, rapid phase of valve formation are depicted. The MF ring and MTs are no longer in mechanical opposition, so the SL is free to thicken. Fusion of STVs to the face of the SL rather than its perimeter might occur at this valve-thickening stage [29]. Note that because the valves fit within one another, on exocytosis of the new valves, the bottom daughter cell will be smaller than the top one. In addition to this size difference, the epitheca and hypotheca might also have significantly different morphologies (heterovalvy [1]). Considerably less is known about morphogenesis of the girdle bands [1], spines [26] and other silica attachments.

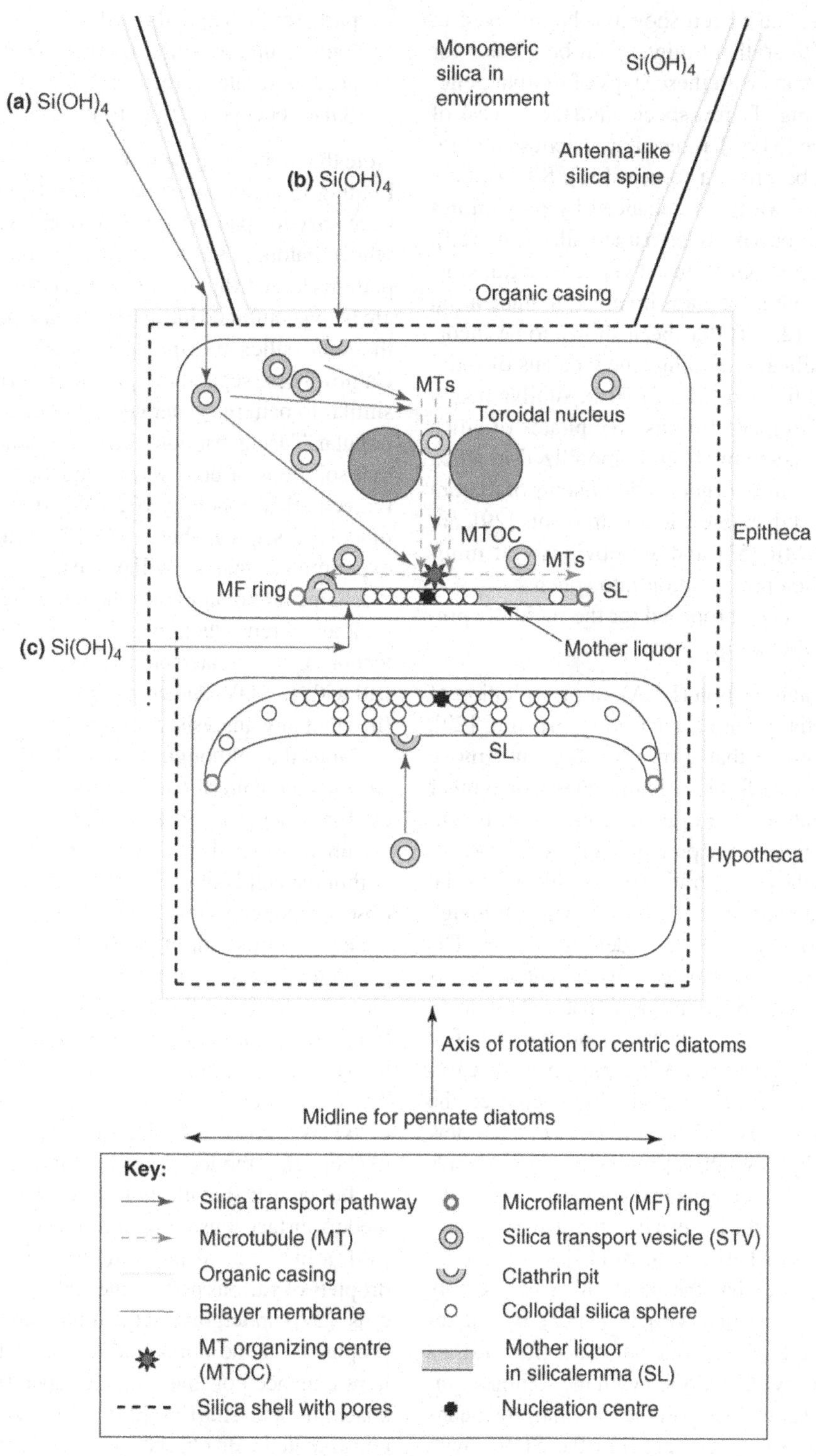

FIGURE 4–cont'd

time scales at which it actually occurs, let alone integrating them into one coherent theory. Of course, we need not follow nature's steps in synthesis of diatom-like structures [48,49], although a deliberate effort to do so might prove rewarding. Ultimately, the proof of a theory of morphogenesis will be its step by step quantitative matching to a computer simulation and/or real-world synthesis, rather than the current approach of shopping for diatoms whose mature valves roughly match a given simulation [29,44].

Seventy-five genes have been shown to be involved in silica metabolism [50], so there might soon be a basis for correlating their functions with these steps of morphogenesis. With full sequencing of a few species and the success of genetic transformation [51,52], molecular genetics and bioinformatics can now be brought to bear [51]. Silicasphere formation has been shown to be enhanced by polyamines associated by specific polypeptides named silacidins [53]. Green fluorescent protein (GFP) fused via genetic transformation to a silaffin protein (perhaps involved in nucleation of silica precipitation [23,54]) has been shown to be incorporated into diatom silica, providing a new means of functional protein immobilization [55]. *In vitro*, similar fusion proteins can be used to make fibrous precipitates of silica [56]. Many enzymes have now been immobilized in silica [41], and old observations of organic components of diatom silica, which we at first dismissed as contaminants [29], are now confirmed by NMR [57] and are obviously at minimum catalytic for silica precipitation into spheres.

Three models have been proposed for the rapid 2D precipitation phase of valve morphogenesis:

- Diffusion limited aggregation (DLA), or precipitation of silica spheres, initially onto a nucleating structure [29], involves a solid phase that grows but does not move except for sintering [29] and a liquid phase, or mother liquor, that concentrates the organic matter in it, potentially changing details of the precipitation, such as keeping pores open, as the concentration of non-silica material increases [29,58] and that is capable of flowing (although flow has been ignored in DLA modelling so far). The DLA model was of limited success when sintering was not invoked, imitating irregular costae patterns of some centric diatoms and aberrant pennate diatoms [29]. The addition of radially organized MTs, presumed to carry silica spheres to the perimeter of the SL, increased the range of diatom patterns simulated [58]. Perhaps a new approach using 'slippery' DLA [59] specific for colloids in water would lead to more realistic patterns.
- In a two-liquid model, the pattern is explained as a phase separation that occurs between them [44].
- It has been suggested that the solid silica forms only within the silica deposition vesicle (SDV) by 'aquaporin-induced syneresis' (extrusion of water from a silica gel) after transport by 'STVs filled with the soluble complex of oligosilicates with polyamines… [that] by means of fusion, discharge their content into the SDV', with long-range order attributed to the cytoskeleton and 'branching due to arrival of new microtubules' [26]. This model has only been simulated for a cross section of an SL, which does not permit detailed comparison with diatom valves. In this proposal, the STVs contain no silica, and are '… simply transport vesicles (TVs) that deliver constituents of cellular origin; most probably membrane parts for the expanding SDV and polypeptides (silaffins, long-chain polyamines) for which it is expected that they accelerate silica precipitation' [40], and pinocytosis of $Si(OH)_4$ occurs only via route C in Figure 4.

Note that none of these models invokes a prepattern of silica binding to supramolecular scaffolds, as has been presumed necessary to span the size gap from the 50 nm sphere to the whole diatom [53,60]. The first two models yield convincing patterns for a different range of diatom valve patterns, so they are being combined into a three-phase model: two liquids plus the solid silica precipitate (in collaboration with Philip J. Camp). These separation–precipitation patterns are uncannily similar to patterns generated by more complicated, but more popular, Turing reaction–diffusion equations [61], suggesting that solutions of both sets of equations share a fundamental mathematical topology [62]. We might find that all of the models discussed here 'work', and that step by step experimental analysis will prove necessary to get at the actual mechanism of diatom morphogenesis. Synthesis, both by experimental reproduction of diatom patterns [63], using nanotechnology to create some semblance of an artificial diatom cell with an SDV, and computer simulation is needed to confirm that any analysis of morphogenesis is sufficient [64].

Note that although purported STVs have not yet been shown to contain silica, contrary to earlier beliefs [1], present evidence suggests that the only detectable form of silica within diatom cells is precipitated silica [65]. Thus, transport within the cell is of solid silica, not $Si(OH)_4$ (Figure 4). This observation seems to contradict the syneresis model [26]. We are equally ignorant of the state of organic matter inside the purported STVs, even whether any of the silica-binding proteins are in them. Such organic macromolecules could be trapped within the silica spheres during STV formation, or between them when they sinter inside the SDV, or both. There is a clear need for a definitive study of the contents of STVs using SIMS (secondary ion mass spectrometry) [37] or other modern analytical tools.

For centric diatoms with hexagonal patterns, once attributed to surface tension, vibrational or electromagnetic forces [66], a hierarchical phase transition model invoking liquid droplets of various polyamines (cf. 'condensed protein spheroids' [29]) undergoing sequential phase separations has been proposed for the pores-within-pores structure [45,46], perhaps enhanced or made species-specific by peptides named silacidins and silaffins [23]. This model has not yet been consolidated with observations that centric diatoms with hexagonal patterns instead have branching patterns emanating from the midring when silica-starved [29], nor with the 3D layered structure shown in Figure 3, and computer simulations are desirable because the varying diameters of the silica spheres obtained [60] do not seem to form any long-range order of closely packed structures, let alone result in 'self-assembly … of structures or patterns at various length scales without external guidance' [60]. If close packing of silica nanospheres or liquid droplets is involved in the sometimes highly regular, long-range order hexagonal patterns of diatoms (Figure 3b), then a mechanism must be found, such as endocytosis via clathrin-coated pits [67,68] (Figure 4) or other means [49,69–71], by which their uptake size might become more monodisperse, and the actual spectrum of diameters of silica colloidal particles within live diatoms needs to be observed and quantified.

Little progress has been reported on the intracellular pathway by which silica makes its way from outside the cell into the SL [23,72], although tools for tracking silica are being developed, including ^{29}Si NMR spectroscopy with confocal laser fluorescence microscopy [65] and dyes [73]. An old standby is germanium (Ge, below Si on the periodic table), which is presently being used to alter photonic properties of diatom valves [39,74,75]. Viral particles of 100-nm diameter have been tracked in live cells [76], so similar work in diatoms should be possible. Perhaps silica-coated nanospheres containing luminescent nanocrystals [77], quantum dots [78,79] or gold [80] would be taken up by diatoms. It seems from earlier work [29] that silica is not stored to any significant extent [72] but rather is taken up from the environment as silicic acid [81] during valve formation, as shown by the time course of silica uptake [82]. Silica transporter (SIT) proteins can 'cycle between the plasma membrane and intracellular vesicles' [82], and an 'organic pentaoxo-azo-silicon complex' might be involved [83]. However, the intracellular silica is likely to be condensed [65], presumably in STVs, shortly after incorporation from the medium as monosilicic acid, $Si(OH)_4$, because the latter is not detected inside cells [23]. Perhaps the external surface of the shell is not only an 'antenna' for silica [29] but also a catalytic surface for condensation. This might be another role of the silica-embedded proteins, whose relationship to the 'organic casing, which coats all the siliceous components' [1] has yet to be investigated, although the casing was found to be chemically removable [60].

Therefore, more studies are required to understand silica trafficking during the process of morphogenesis. It is not known: (i) whether colloidal size particles are ever taken in, which would seem to be more efficient when they are available in the environment; (ii) whether the valves and silica spines have roles in providing adsorption 'antennae' (another DLA problem) and surface migration 'funnels' to bring the external silica to the cell membrane, a form of dimensionality reduction analogous to the cell nucleus and its pores [84]; (iii) whether pinocytosis, clath-rates, etc. are involved in taking up the silica, or where on the cell surface this occurs; (iv) whether silica is packaged into vesicles at the cell membrane or in the Golgi apparatus, or how STVs [23,29] are delivered to the surface of the SL; (v) whether STVs are deposited to spatiotemporally localized positions on the SL [58]; (vi) how their contents are exocytosed into the SDV through its membrane, the SL; and (vii) whether the growing surface area of the SL depends on the STV membranes for its increase as the nascent valve grows inside. There is a need to understand the roles of the cytoskeleton in all this [58], and intriguing hints have been provided by the MT bundles that extend through a hole in a torus-shaped nucleus [85] impinging on the SL to a microtubule-organizing centre (MTOC) on the external surface of the SL, as well as a microfilament ring that might be involved in keeping the SL flat as it grows [86] (Figure 4). The flatness of the SDV might be essential to keep the initial pattern formation confined to an essentially 2D space, as is required for pattern formation by bacterial colonies [87]. Basically, to link the fine molecular genetics that is being done [50] to the morphogenesis of the valve will require that we now turn our attention to the trafficking of silica inside diatoms.

DIATOM BIOPHOTONICS

Because of their similarity to opals, diatoms are often referred to as 'jewels of the sea' or 'living opals'. In fact, diatom shells are opal material made from silica nanoparticles, and diatoms can be described as a living cell inside a glass house [88]. Thus, it is not surprising that within the glass menagerie of diatoms, we can find outstanding examples of multifunctional structures based on interaction with light, so that the beauty of diatoms lies not only in the artistry of their forms and structures but also in the equally valuable optical properties of their transparent silica structures. The strong interaction with light produces stunning structural colours with intense diffraction and interference effects when diatoms are observed under a light microscope (Figure 5). Displaying a play of lustrous

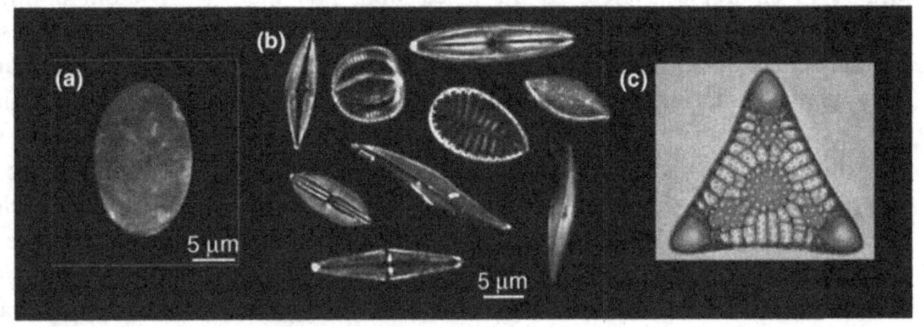

FIGURE 5 Optical and photonic properties of diatoms. **(a)** Opal is shown as an example of a photonic crystal with a characteristic play of colours. Reproduced with permission of S. Ely. **(b)** Light microscopy images of several pennate diatom species with characteristic colours as a result of light interference and diffraction from their silica structure. **(c)** Digitally enhanced Jamin-Lebedeff micrograph of the marine centric diatom fossil *Triceratium morlandii*. Cell width ~120 μm.

colours like those of the rainbow, this phenomenon is caused by multiple reflections from multilayered, semi-transparent surfaces in which phase shift and interference of the reflections modulates the incident light by amplifying or attenuating some frequencies more than others: the precise interference effect depends on the angle at which light strikes the surface, hence the diatom seems to change colour as it or the observer moves position, as with a thin film of oil on water. Such iridescent effects [89] are used widely in colour cosmetics products and personal care packaging, and there is great potential for using diatoms in this industry [90].

Jamin-Lebedeff interferometric microscopy, using polarizing optics to separate specimen and reference beams, gives particularly spectacular colour effects involving transmitted light. Although the raw image is quite washed out, digital post-acquisition image processing allows reduction of haze, the intensification of colour saturation and additional sharpening (Figure 5c). Whereas in transmission mode light diffraction is more important, in reflectance mode the result is similar to iridescent colours. Further study is warranted, especially the possibility of inferring diatom structure via visible light computed tomography methods [91,92], which would then, for example, permit time-lapse of the 3D phase of valve morphogenesis *in vivo*.

Photonic crystals are materials with spatially ordered and periodic nanostructures that can control the propagation of light, only allowing certain wavelengths to pass through the crystal (similar to the propagation of electrons in a semiconductor crystal) [93]. They are able to control photons, producing remarkable effects that are impossible with conventional optics, and have the potential to revolutionize existing electronic and computing technologies [93]. The photonic crystal properties of diatoms' girdle band structures were recently confirmed [38,90,94], suggesting that diatoms are living photonic crystals. Diatom nanotechnology now allows us to grow a huge variety of biophotonic crystals. This extraordinary discovery raises questions about the biological relevance of the photonic properties of diatoms and their practical exploitation. Butterflies, beetles and many other organisms have been using photonic crystals for ages [89,95]. Their function varies from communication, camouflag, and thermal exchange to UV protection. Diatoms' photosynthetic receptors are located in chloroplasts close to the silica wall, and the light-channelling and -focussing [96] properties of their silica structure could help the transmission and collection of more light into the photoreceptors to improve their photosynthetic efficiency [90].

Another optical surprise that comes from diatoms is photoluminescence [38,97,98]. A visual luminescence effect from diatoms is clearly seen after exposing the silica structures to UV light with a broad blue luminescence peak in the visible region (450 nm). This effect was found to be similar to the photoluminescence of artificially fabricated porous silicon. Diatom photoluminescence is strongly species dependent, and it is based on both their frustule structure and the surrounding environment. These characteristics were elegantly exploited [98] to create the first photoluminescence gas-sensing devices based on diatoms [99]. Ultra-sensitive detection (sub ppm) of a series of organic vapours (ethanol, acetone, xylene and pyridine) and gases (nitrogen dioxide, methane, carbon monoxide) has been demonstrated [100]. Based on these findings and the diversity of diatoms available with different photoluminescence characteristics toward different gases, one can predict the development of a universal photoluminescence gas-sensing platform with an array of different diatoms for toxic gas detection or air-pollution monitoring.

MICROFLUIDICS WITHIN DIATOMS

The gliding motility of pennate diatoms is intriguing because the cell does not change shape and there are no moving parts, contrary to our common experience with amoeboid, ciliate and bacterial motility. Two models for the role of microfilaments parallel to the raphes have long been in contention, without definitive resolution: one model identifies the microfilaments as the motor [28] and the other identifies them as controllers of the direction of motility [101]. A fibrous fluid, the 'diatom trail', is left on the surface a diatom traverses.

Biophotonics and motility might be intertwined in the colonial diatom *Bacillaria paradoxa*, in which the chained cells move back and forth against one another [102,103]. Whether the coupling of these autonomously oscillating [104] *B. paradoxa* cells is local, or global via the elasticity or anomalous viscosity [8] of the diatom trail slime, remains to be determined. The resting stage involves alignment of all the cells in a stack with no obvious mechanical stop [103]. Given that there is a photosensitive region at the distal ends of a pennate diatom, which causes the diatom to respond to a 'light wall' by reversing direction [105], we would like to hypothesize that a light pipe forms between these photosensitive regions when the cells are stacked. It might be responsible not only for the aligned resting stage but also for the partial synchrony of movement of the cells.

Diatom motility utilizes active flow of an adhesive fluid through a narrow slit, the raphe (Figure 2), which suggests a new branch of nanotechnology that might be called 'self-propelled microfluidics', compared to so-called 'active' microfluidics, in which the liquid is passively moved by an external force [106]. (Cytoplasm is a more complex self-propelled fluid.) A downside of diatoms is their adhesion to man-made surfaces under water, leading to biofouling [107], although this might lead to new commercially or medically important bioadhesives [108].

DIATOMS FOR DRUG DELIVERY

Nanotechnology is currently opening new therapeutic opportunities for agents that cannot be used effectively as conventional drug formulations owing to poor bioavailability or drug instability. The diatom silica shell possesses a combination of structural, mechanical, chemical and optical features that might both overcome challenges associated with conventional delivery of therapeutic agents and have advantages over existing microparticle delivery systems. The pill-box structures, micro- and nanoscale porosity, enormous surface area (100 m^2/g for unheated, fresh diatom shells [29,109–112]) and biocompatibility and biodegradability of amorphous diatom silica make them a promising biomaterial for drug-delivery applications. They can be easily functionalized, protected and designed for controlled drug release through nano-sized pores or by embedding in the silica [55]. Even though diatoms can be easily cultivated, a large and even less expensive source of diatom silica is diatomite or diatomaceous earth, which is formed by the fossil siliceous frustules of diatoms. The preparation of the ultra-high-purity and fraction-free silica capsules from raw diatomite material (diatomaceous earth) is possible using simple separation procedures (Figure 6). These diatom microcapsules are proposed as excellent natural porous materials for drug-delivery applications. Diatom structure provides flexibility for the design of complex drug-delivery vehicles through functionalization with sensing biomolecules or immunotargeting bioreceptors, optically active dyes

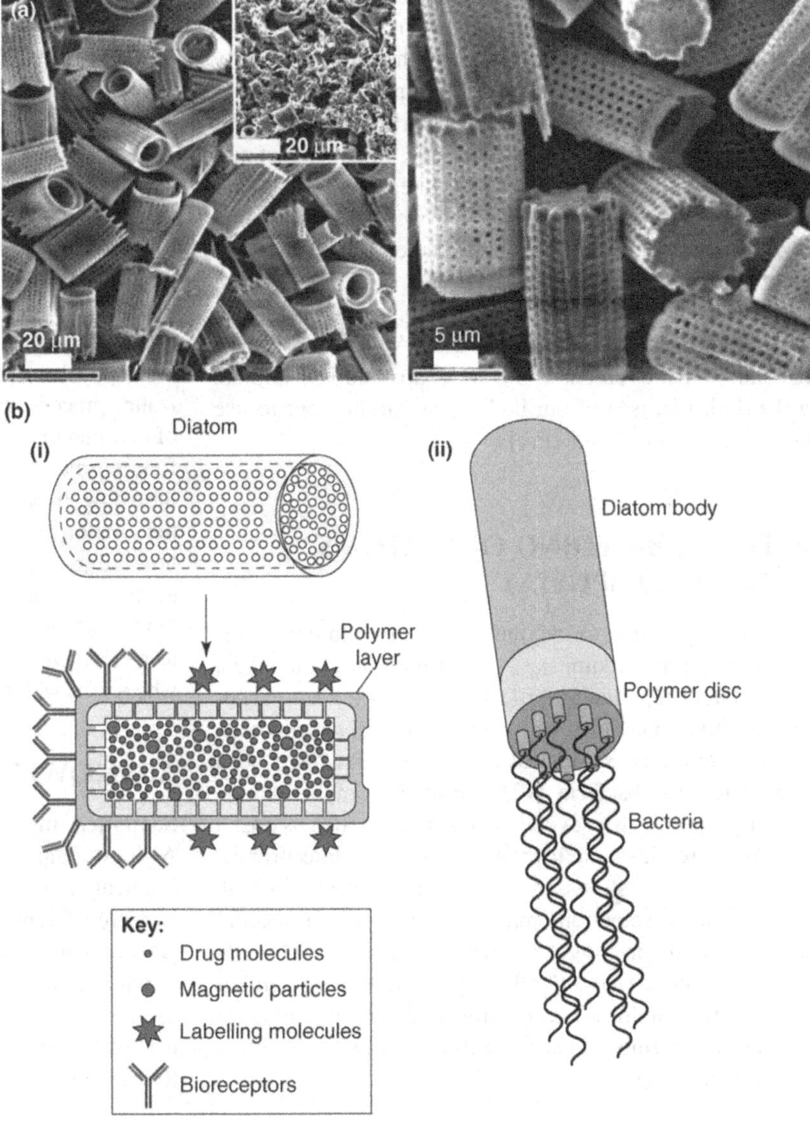

FIGURE 6 Diatoms for drug delivery. Panel (**a**) shows exemplary SEM images of purified diatoms with whole, fraction-free frustules in comparison with raw diatomaceous earth (inset image). Panel (**b**) shows a multifunctional diatom-based drug-delivery system (i) and a model of a self-propelled drug carrier with diatom and attached bacterial biomotors (ii).

(for imaging) and/or magnetic nanoparticles (for controlled movement to target diseased tissue or cancer cells) (Figure 6).

More sophisticated drug-delivery systems, such as self-propelled swimming microrobots, could also benefit from the unique properties of diatoms' frustule structure. Although not quite like science fiction [113], scientists have talked for quite some time about microdevices that can travel inside the human body and carry out a range of complex medical procedures, such as monitoring, drug delivery and cell repair [114]. Recent developments in micro- and nanoscale engineering have led to the realization of various miniature mobile robots, but we have an intriguing opportunity to integrate whole biological organisms or their parts [115,116]. In regards to their physical and structural properties, porous silica capsules of diatoms are ideal microscale bodies for designing these future robotic devices for medical applications. However, the self-propelled function is missing here, and to introduce mobility, we could attach bacteria to the diatom (Figure 6) in addition to, or instead of, the gliding motility of diatoms themselves. Many bacteria propel themselves along in a fluid by rotating their corkscrew-like tails, called flagella, at relatively high speeds, and as robust machines, such flagella can easily be integrated with other microscopic components and do not need to be purified or reconstituted [116–118]. The bacteria motors work using a simple chemical energy source (glucose) and are naturally sensitive to the environment (e.g. metal ions, ethylene diamine tetra-acetic acid [EDTA]), which means that nanobot movement could be controlled. Of course, in the dark recesses of our bodies, we might want to use motile apochlorotic diatoms [119].

SELECTIVE BREEDING OF DIATOMS USING A COMPUSTAT

It might be possible to manipulate the morphology of diatoms by use of a compustat [120], which functions like a chemostat, except that the criteria for survival are morphological rather than nutritional. While this still has not been done for diatoms, somewhat similar devices have been constructed for bacteria [121]. With a compustat, one could try to select for several visual criteria, such as costal spacing, pore sizes, shell shape and biophotonic properties. Curiosity questions, such as whether we can, through such artificial selection, make one species of diatom look like another, might be worth pursuing. Selection could also be based on oil droplet sizes and number (T.V. Ramachandra *et al.*, unpublished) or detection, via absorption or fluorescence spectra, of other important bioactive materials.

COMPUTING WITH DIATOMS

Perhaps the most sophisticated dream for diatoms to date is the hope to grow large numbers of 3D nanocomputers or computer components from them (M.R. Sussman [2008] 'In diatom, scientists find genes that may level engineering hurdle', http://www.eurekalert.org/pub_releases/2008-01/uow-ids011808.php). The idea of growing a computer goes back at least to the experiments of toy maker Robert Stewart on iron dendrites in nitric acid [122,123], which he derived from Lillie's iron wire model for nerve cell signal propagation [20,124]. The achievement so far is to transform the 3D shape of the amorphous silica of a diatom into silicon, preserving the morphology of the original diatom [125]. A combination of 3D diatom nanotechnology with 3D DNA nanotechnology [126,127] and the electronic properties of DNA [128] might be particularly rewarding. DNA binding directly or indirectly to silica [129] and silicon [130,131] has already been demonstrated, and one should keep in mind the silica-dependent nature of DNA replication in diatoms [132].

CONCLUSIONS

Diatom research is rapidly moving from the underappreciated domain of taxonomists and skilled amateur scientists into high nanotechnology and big business. Diatom bionanotechnology, a new interdisciplinary area, has successfully emerged over the past several years into a dynamic and productive research area with hundreds of papers to date. We have witnessed significant progress in understanding diatom structural, mechanical, genomic, optical and photonic properties and the silica biomineralization process, leading to the nanofabrication and engineering of new materials and devices based on diatom silica. Biofuels, food, cosmetics and pharmaceutical products might soon be in the offing. In basic research, diatoms are likely to contribute to the solution of one of the major unsolved biological problems: how the genome is involved in creation of form, and how form evolves. We now have a few competing hypotheses for diatoms, and there is a great need for intracellular observation to resolve what is going on. Their beauty inspires everyone who works with them.

ACKNOWLEDGEMENTS

Supported in part by grants from the Canadian Space Agency, Manitoba Institute of Child Health, Manitoba Medical Service Foundation, University of Manitoba Research Grant Program, MITACS, University of South Australia and Australian Research Council. We would like to thank Philip J. Camp, Ille C. Gebeshuber, Chung-Yuan Mou and the reviewers for critical readings and great suggestions. Dedicated to the memory of diatomists Charles W. Reimer and Ralph A. Lewin.

REFERENCES

1. Round, F.E. *et al.* (1990) *The Diatoms, Biology & Morphology of the Genera.* Cambridge University Press.
2. Sterrenburg, F.A.S. *et al.* (2007) Diatoms: living in a constructal environment. In *Algae and Cyanobacteria in Extreme Environments. Series: Cellular Origin, Life in Extreme Habitats and Astrobiology* (Vol. 11) (Seckback, J., ed.), In pp. 141–172, Springer.
3. Williams, T. (1945) *The Glass Menagerie.* James Laughlin.
4. Bach, K. and Burkhardt, B. (1984) *Diatomeen I, Schalen in Natur und Technik [Diatoms I, Shells in Nature and Technics]*, Cramer Verlag.
5. Schmid, A.M.M. (1984) Schalenmorphogenese in Diatomeen [Valve morphogenesis in diatoms]. In *Diatomeen I, Schalen in Natur und Technik [Diatoms I, Shells in Nature and Technics]* (Bach, K. and Burkhardt, B., eds), pp. 300–317, Cramer Verlag.
6. Aldersey-Williams, H. (2004) Toward biomimetic architecture. *Nat. Mater.* 3, 277–279.
7. Sterrenburg, F.A.S. (2005) Crystal palaces – diatoms for engineers. *J. Nanosci. Nanotechnol.* 5, 100–107.
8. Gebeshuber, I.C. (2007) Biotribology inspires new technologies. *Nanotoday* 2, 30–37.
9. Mann, D.G. and Droop, S.J.M. (1996) Biodiversity, biogeography and conservation of diatoms. *Hydrobiologia* 336, 19–32.
10. Street-Perrott, F.A. and Barker, P.A. (2008) Biogenic silica: a neglected component of the coupled global continental biogeochemical cycles of carbon and silicon. *Earth Surf. Process. Landf.* 33, 1436–1457.
11. Werner, D. (1977) Introduction with a note on taxonomy. In *The Biology of Diatoms* (Werner, D., ed.), pp. 1–17, Blackwell Scientific Publications.
12. Bowler, C. *et al.* (2008) The *Phaeodactylum* genome reveals the evolutionary history of diatom genomes. Nature 456, 239–244.
13. Parkinson, J. and Gordon, R. (1999) Beyond micromachining: the potential of diatoms. *Trends Biotechnol.* 17, 190–196.
14. Drum, R.W. and Gordon, R. (2003) Star Trek replicators and diatom nanotechnology. *Trends Biotechnol.* 21, 325–328.
15. Gordon, R. *et al.* (2005) A special issue on diatom nanotechnology. *J. Nanosci. Nanotechnol.* 5, 1–4.
16. Saliterman, S. (2006) *Fundamentals of BioMEMS and Medical Microdevices.* SPIE Press.
17. Umemura, K. *et al.* (2007) Regulated growth of diatom cells on self-assembled monolayers. *J. Nanobiotechnology* 5, 2.
18. Mirón, A.S. *et al.* (1999) Comparative evaluation of compact photobioreactors for large-scale monoculture of microalgae. *J. Biotechnol.* 70, 249–270.
19. Chepurnov, V.A. *et al.* (2008) In search of new tractable diatoms for experimental biology. *Bioessays* 30, 692–702.
20. Gordon, R. (1999) *The Hierarchical Genome and Differentiation Waves: Novel Unification of Development, Genetics and Evolution.* World Scientific & Imperial College Press.
21. Armbrust, E.V. *et al.* (2004) The genome of the diatom *Thalassiosira pseudonana*: ecology, evolution, and metabolism. *Science* 306, 79–86.
22. Martens, C. *et al.* (2008) Whole-genome analysis reveals molecular innovations and evolutionary transitions in chromalveolate species. *Proc. Natl. Acad. Sci. U. S. A.* 105, 3427–3432.
23. Sumper, M. and Brunner, E. (2008) Silica biomineralization in diatoms: the model organism *Thalassiosira pseudonana. ChemBioChem* 9, 1187–1194.
24. Genkal, S.I. and Popovskaya, G.I. (2008) Morphological variability of *Cyclotella ocellata* from Lake Khubsugul (Mongolia). *Diatom Res.* 23, 75–91.
25. Brzezinski, M.A. (2008) Mining the diatom genome for the mechanism of biosilicification. *Proc. Natl. Acad. Sci. U. S. A.* 105, 1391–1392.
26. Grachev, M.A. *et al.* (2008) Silicon nanotechnologies of pigmented heterokonts. *Bioessays* 30, 328–337.
27. Molino, P.J. *et al.* (2006) Utilizing QCM-D to characterize the adhesive mucilage secreted by two marine diatom species in-situ and in realtime. *Biomacromolecules* 7, 3276–3282.
28. Heintzelman, M.B. (2006) Cellular and molecular mechanics of gliding locomotion in eukaryotes. *Int. Rev. Cytol.* 251, 79–129.
29. Gordon, R. and Drum, R.W. (1994) The chemical basis for diatom morphogenesis. *Int. Rev. Cytol.* 150, 243–372 421–422.
30. Hildebrand, M. *et al.* (2008) Application of AFM in understanding biomineral formation in diatoms. *Pflugers Arch.* 456, 127–137.
31. Losic, D. *et al.* (2007) AFM nanoindentations of diatom biosilica surfaces. *Langmuir* 23, 5014–5021.
32. Kaluzhnaya, O.V. and Likhoshway, Y.V. (2007) Valve morphogenesis in an araphid diatom *Synedra acus subsp radians. Diatom Res.* 22, 81–87.
33. Lobel, K.D. *et al.* (1996) Computational model for protein-mediated biomineralization of the diatom frustule. *J. Marine Biol.* 126, 353–360.
34. Holzhüter, G. *et al.* (2005) Silica structure in the spicules of the sponge *Suberites domuncula. Anal. Bioanal. Chem.* 382, 1121–1126.
35. Grimes, G.W. and Aufderheide, K.J. (1991) Cellular Aspects of Pattern Formation: The Problem of Assembly. *Karger.*
36. Frankel, J. (1992) Genes and structural patterns in ciliates: Vance Tartar and the 'cellular architects'. *Dev. Genet.* 13, 181–186.
37. Gordon, R. *et al.* (2007) Diatoms in space: testing prospects for reliable diatom nanotechnology in microgravity. *Proc. SPIE* 6694, V1–V15.
38. Townley, H.E. *et al.* (2007) Modification of the physical and optical properties of the frustule of the diatom *Coscinodiscus wailesii* by nickel sulfate. *Nanotechnology* 18, 295101 10.1088/0957-4484/18/29/295101.
39. Qin, T. *et al.* (2008) Biological fabrication of photoluminescent nanocomb structures by metabolic incorporation of germanium into the biosilica of the diatom Nitzschia frustulum. *ACS Nano* 2, 1296–1304.
40. Vrieling, E.G. *et al.* (2007) Salinity-dependent diatom biosilicification implies an important role of external ionic strength. *Proc. Natl. Acad. Sci. U. S. A.* 104, 10441–10446.
41. Betancor, L. and Luckarift, H.R. (2008) Bioinspired enzyme encapsulation for biocatalyis. *Trends Biotechnol.* 6, 566–572.
42. Hall, A.R. (1954) *The Scientific Revolution, 1500–1800: The Formation of the Modern Scientific Attitude.* Beacon Press.
43. Hazelaar, S. *et al.* (2005) Monitoring rapid valve formation in the pennate diatom *Navicula salinarum* (Bacillariophyceae). *J. Phycol.* 41, 354–358.
44. Lenoci, L. and Camp, P.J. (2008) Diatom structures templated by phase-separated fluids. *Langmuir* 24, 217–223.
45. Sumper, M. and Lehmann, G. (2006) Silica pattern formation in diatoms: species-specific polyamine biosynthesis. *ChemBioChem* 7, 1419–1427.

46. Sumper, M. and Brunner, E. (2006) Learning from diatoms: Nature's tools for the production of nanostructured silica. *Adv. Funct. Mater.* 16, 17–26.
47. Tiffany, M.A. (2008) Valve development in *Aulacodiscus. Diatom Res.* 23, 185–212.
48. Fujiwara, M. *et al.* (2006) Silica hollow spheres with nano-macroholes like diatomaceous earth. *Nano Lett.* 6, 2925–2928.
49. Tu, H.L. *et al.* (2008) One-step synthesis of ordered mesostructural organic/silica nanocomposites with tunable fluorescence surfactants. *J. Mater. Chem.* 18, 1771–1778.
50. Mock, T. *et al.* (2008) Whole-genome expression profiling of the marine diatom *Thalassiosira* pseudonana identifies genes involved in silicon bioprocesses. *Proc. Natl. Acad. Sci. U. S. A.* 105, 1579–1584.
51. Kroth, P. (2008) Molecular biology and the biotechnological potential of diatoms. *Adv. Exp. Med. Biol.* 616, 23–33.
52. Sakaue, K. *et al.* (2008) Development of gene expression system in a marine diatom using viral promoters of a wide variety of origin. *Physiol. Plant.* 133, 59–67.
53. Wenzl, S. *et al.* (2008) Silacidins: highly acidic phosphopeptides from diatom shells assist in silica precipitation *in vitro. Angew. Chem. Int. Ed. Engl.* 47, 1729–1732.
54. Sumper, M. and Kröger, N. (2004) Silica formation in diatoms: the function of long-chain polyamines and silaffins. *J. Mater. Chem.* 14, 2059–2065.
55. Poulsen, N. *et al.* (2007) Silica immobilization of an enzyme through genetic engineering of the diatom *Thalassiosira pseudonana. Angew. Chem. Int. Ed. Engl.* 46, 1843–1846.
56. Marner, W.D., II *et al.* (2008) Morphology of artificial silica matrices formed via autosilification of a silaffin/protein polymer chimera. *Biomacromolecules* 9, 1–5.
57. Tesson, B. *et al.* (2008) Contribution of multi-nuclear solid state NMR to the characterization of the *Thalassiosira pseudonana* diatom cell wall. *Anal. Bioanal. Chem.* 390, 1889–1898.
58. Parkinson, J. *et al.* (1999) Centric diatom morphogenesis: a model based on a DLA algorithm investigating the potential role of microtubules. *Biochim. Biophys. Acta* 1452, 89–102.
59. Seager, C.R. and Mason, T.G. (2007) Slippery diffusion-limited aggregation. *Phys. Rev. E Stat. Nonlin. Soft Matter Phys.* 75, 011406.
60. Gröger, C. *et al.* (2008) Biomolecular self-assembly and its relevance in silica biomineralization. *Cell Biochem. Biophys.* 50, 23–39.
61. Sanderson, A.R. *et al.* (2006) Advanced reaction-diffusion models for texture synthesis. *J. Graphics Tools* 11, 47–71.
62. Thom, R. (1989) *Structural Stability and Morphogenesis: An Outline of a General Theory of Models.* Addison-Wesley.
63. Schultze, M.J.S. (1863) On the structure of the valve in the Diatomacea, as compared with certain siliceous pellicles produced artificially by the decomposition in moist air of fluo-silicic acid gas (fluoride of silicium). *Q. J. Microsc. Sci.* 3, 120–134.
64. Jacobson, A.G. and Gordon, R. (1976) Changes in the shape of the developing vertebrate nervous system analyzed experimentally, mathematically and by computer simulation. *J. Exp. Zool.* 197, 191–246.
65. Gröger, C. *et al.* (2008) Silicon uptake and metabolism of the marine diatom *Thalassiosira pseudonana*: solid-state ^{29}Si NMR and fluorescence microscopic studies. *J. Struct. Biol.* 161, 55–63.
66. Thompson, D.W. (1942) *On Growth and Form.* Cambridge University Press.
67. Lu, C.W. *et al.* (2007) Bifunctional magnetic silica nanoparticles for highly efficient human stem cell labeling. *Nano Lett.* 7, 149–154.
68. Rejman, J. *et al.* (2004) Size-dependent internalization of particles via the pathways of clathrin- and caveolae-mediated endocytosis. *Biochem. J.* 377, 159–169.
69. Knecht, M.R. *et al.* (2005) Size control of dendrimer-templated silica. *Langmuir* 21, 2058–2061.
70. Laulicht, B. *et al.* (2008) Evaluation of continuous flow nanosphere formation by controlled microfluidic transport. *Langmuir* 24, 9717–9726.
71. Hartlen, K.D. *et al.* (2008) Facile preparation of highly monodisperse small silica spheres (15 to >200 nm) suitable for colloidal templating and formation of ordered arrays. *Langmuir* 24, 1714–1720.
72. Martin-Jézéquel, V. *et al.* (2000) Silicon metabolism in diatoms: implications for growth. *J. Phycol.* 36, 821–840.
73. Desclés, J. *et al.* (2008) New tools for labeling silica in living diatoms. *New Phytol.* 177, 822–829.
74. Jeffryes, C. *et al.* (2008) Electroluminescence and photoluminescence from nanostructured diatom frustules containing metabolically inserted germanium. *Adv. Mater.* 20, 2633–2637.
75. Jeffryes, C. *et al.* (2008) Two-stage photobioreactor process for the metabolic insertion of nanostructured germanium into the silica microstructure of the diatom *Pinnularia sp. Mater. Sci. Engin. C-Bio. Supramolecular Syst.* 28, 107–118.
76. Jouvenet, N. *et al.* (2008) Imaging the biogenesis of individual HIV-1 virions in live cells. *Nature* 454, 236–240.
77. Chan, Y. *et al.* (2004) Incorporation of luminescent nanocrystals into monodisperse core-shell silica microspheres. *Adv. Mater.* 16, 2092–2097.
78. Zhang, T. *et al.* (2006) Cellular effect of high doses of silica-coated quantum dot profiled with high throughput gene expression analysis and high content cellomics measurements. *Nano Lett.* 6, 800–808.
79. Dembski, S. *et al.* (2008) Photoactivation of CdSe/ZnS quantum dots embedded in silica colloids. *Small* 4, 1516–1526.
80. Park, Y.S. *et al.* (2007) Concentrated colloids of silica-encapsulated gold nanoparticles: colloidal stability, cytotoxicity, and X-ray absorption. *J. Nanosci. Nanotechnol.* 7, 2690–2695.
81. Del Amo, Y. and Brzezinski, M.A. (1999) The chemical form of dissolved Si taken up by marine diatoms. *J. Phycol.* 35, 1162–1170.
82. Thamatrakoln, K. and Hildebrand, M. (2007) Analysis of *Thalassiosira pseudonana* silicon transporters indicates distinct regulatory levels and transport activity through the cell cycle. *Eukaryot. Cell* 6, 271–279.
83. Kinrade, S.D. *et al.* (2002) Silicon-29 NMR evidence of a transient hexavalent silicon complex in the diatom *Navicula pelliculosa. J. Chem. Soc., Dalton Trans.* 2002, 307–309.
84. Peters, R. (2005) Translocation through the nuclear pore complex: selectivity and speed by reduction-of-dimensionality. *Traffic* 6, 421–427.
85. Pickett-Heaps, J.D. (1991) Cell division in diatoms. *Int. Rev. Cytol.* 128, 63–108.
86. Gordon, R. and Brodland, G.W. (1987) The cytoskeletal mechanics of brain morphogenesis. Cell state splitters cause primary neural induction. *Cell Biophys.* 11, 177–238.
87. Tyson, R. *et al.* (1999) A minimal mechanism for bacterial pattern formation. *Proc Biol Sci* 266, 299–304.
88. Siver, P.A. (2005) Diatoms: life in glass houses. *J. Phycol.* 41, 720.
89. Berthier, S. (2007) *Iridescences: The Physical Colors of Insects.* Springer.
90. Parker, A.R. and Townley, H.E. (2007) Biomimetics of photonic nanostructures. *Nat. Nanotechnol.* 2, 347–353.

91. Vest, C.M. (1985) Tomography for properties of materials that bend rays: a tutorial. *Appl. Opt.* 24, 4089–4094.
92. Debailleul, M. *et al.* (2008) Holographic microscopy and diffractive microtomography of transparent samples. *Meas. Sci. Technol.* 19, 074009.
93. Hall, N. and Ozin, G. (2003) The photionic opal – the jewel in the crown of optical information processing. *Chem. Commun. (Camb.)* (21), 2639–2643.
94. Fuhrmann, T. *et al.* (2004) Diatoms as living photonic crystals. *Appl. Phys.* B 78, 257–260.
95. Srinivasarao, M. (1999) Nano-optics in the biological world: beetles, butterflies, birds, and moths. *Chem. Rev.* 99, 1935–1961.
96. De Stefano, L. *et al.* (2007) Lensless light focusing with the centric marine diatom *Coscinodiscus walesii. Opt. Express* 15, 18082–18088.
97. Butcher, K.S.A. *et al.* (2005) A luminescence study of porous diatoms. *Mater, Sci. Engin. C-Bio. Supramolecular Syst.* 25, 658–663.
98. De Stefano, L. *et al.* (2005) Marine diatoms as optical chemical sensors. *Appl. Phys. Lett.* 87, 233902.
99. Setaro, A. *et al.* (2007) Highly sensitive optochemical gas detection by luminescent marine diatoms. *Appl. Phys. Lett.* 91, 051921 10.1063/1.2768027.
100. Lettieri, S. *et al.* (2008) The gas-detection properties of light-emitting diatoms. *Adv. Funct. Mater.* 18, 1257–1264.
101. Gordon, R. (1987) A retaliatory role for algal projectiles, with implications for the mechanochemistry of diatom gliding motility. *J. Theor. Biol.* 126, 419–436.
102. Schmid, A-MM. (2007) The "paradox" diatom Bacillaria paxillifer (Bacillariophyta) revisited. *J. Phycol.* 43, 139–155.
103. Ussing, A.P. *et al.* (2005) The colonial diatom *'Bacillaria paradoxa'*: chaotic gliding motility, Lindenmeyer model of colonial morphogenesis, and bibliography, with translation of O.F. Müller (1783), 'About a peculiar being in the beach-water'. *Diatom Monographs* 5, 1–140.
104. Drum, R.W. *et al.* (1971) On weakly coupled diatomic oscillators: Bacillaria's paradox resolved. *J. Phycol.* 7, 13–14.
105. Cohn, S.A. *et al.* (1999) High energy irradiation at the leading tip of moving diatoms causes a rapid change of cell direction. *Diatom Res.* 14, 193–206.
106. Khatavkar, V.V. *et al.* (2007) Active micromixer based on artificial cilia. *Phys. Fluids* 19, 083605.
107. Molino, P.J. and Wetherbee, R. (2008) The biology of biofouling diatoms and their role in the development of microbial slimes. *Biofouling* 24, 365–379.
108. Coulthard, P. *et al.* (2004) Tissue adhesives for closure of surgical incisions. *Cochrane Database Syst. Rev.*, CD004287.
109. Lewin, J.C. (1961) The dissolution of silica from diatom walls. *Geochim. Cosmochim. Acta* 21, 182–198.
110. Lewin, R.A. (1962) *Physiology and Biochemistry of Algae*. Academic Press.
111. Iler, R.K. (1979) *The Chemistry of Silica: Solubility, Polymerization, Colloid and Surface Properties, and Biochemistry*. John Wiley & Sons.
112. Hurd, D.C. (1983) Physical and chemical properties of siliceous skeletons. In *Silicon Geochemistry and Biogeochemistry* (Aston, S.R., ed.), pp. 187–244, Academic Press.
113. Asimov, I. *et al.* (1966) *Fantastic Voyage:* A Novel. Houghton Mifflin.
114. Chrusch, D.D. *et al.* (2002) Cytobots: intracellular robotic micromanipulators. In *Conference Proceedings, 2002 IEEE Canadian Conference on Electrical and Computer Engineering: 2002 May 12–15; Winnipeg* (Kinsner, W. and Sebak, A., eds) (Vol. 3), pp. 1640–1645, IEEE.
115. Zhang, Y. *et al.* (2006) Development of a micro swimming robot using optimised giant magnetostrictive thin films. *Applied Bionics Biomech.* 3, 161–170.
116. Behkam, B. and Sitti, M. (2007) Bacterial flagella-based propulsion and on/off motion control of microscale objects. *Appl. Phys. Lett.* 90, 023902 10.1063/1.2431454.
117. Kim, M.J. and Breuer, K.S. (2008) Microfluidic pump powered by setf-organizing bacteria. *Small* 4, 111–118.
118. Behkam, B. and Sitti, M. (2006) Towards hybrid swimming microrobots: bacteria assisted propulsion of polystyrene beads. In *Proceedings of the 28th IEEE EMBS Annual International Conference: 2006 August 30–September 3,* New York. pp. 2421–2424, IEEE.
119. Lauritis, J.A. *et al.* (1968) Studies on the biochemistry and fine structure of silica shell formation in diatoms IV. Fine structure of the apochlorotic diatom *Nitzschia alba Lewin and Lewin. Arch. Mikrobiol.* 62, 1–16.
120. Gordon, R. (1996) Computer controlled evolution of diatoms: design for a compustat. *Nova Hedwigia* 112, 213–216.
121. Balagadde, F.K. *et al.* (2005) Long-term monitoring of bacteria undergoing programmed population control in a microchemostat. *Science* 309, 137–140.
122. Stewart, R.M. (1965) Progress in experimental research on electrochemical adaptive systems. In *Biophysics and Cybernetic Systems, Proceedings of the Second Cybernetic Sciences Symposium: 1964 October 13; University of Southern California* (Maxfield, M. and et, al., eds), Los Angeles. pp. 25–36, Spartan Books.
123. Stewart, R.M. (1965) Fields and waves in excitable cellular structures. *Prog. Brain Res.* 17, 244–250.
124. Lillie, R.S. (1918) Transmission of activation in passive metals as a model of the protoplasmic or nervous type of transmission. *Science* 48, 51–60.
125. Bao, Z. *et al.* (2007) Chemical reduction of three-dimensional silica micro-assemblies into microporous silicon replicas. *Nature* 446, 172–175.
126. Seeman, N.C. (1998) DNA nanotechnology: novel DNA constructions. *Annu. Rev. Biophys. Biomol. Struct.* 27, 225–248.
127. Goodsell, D.S. (2004) *Bionanotechnology, Lessons from Nature*. Wiley-Liss.
128. Chakraborty, T. (ed.) (2007) *Charge Migration in DNA: Perspectives from Physics, Chemistry and Biology*, Springer.
129. Rosi, N.L. *et al.* (2004) Control of nanoparticle assembly by using DNA-modified diatom templates. *Angew. Chem. Int. Ed. Engl.* 43, 5500–5503.
130. Fritz, J. *et al.* (2000) Translating biomolecular recognition into nanomechanics. *Science* 288, 316–318.
131. Keren, K. (2002) Sequence specific molecular lithography on single DNA molecules. *Science* 297, 72–75.
132. Okita, T.W. and Volcani, B.E. (1978) Role of silicon in diatom metabolism. IX. Differential synthesis of DNA polymerases and DNA-binding proteins during silicate starvation and recovery in *Cylindrotheca fusiformis. Biochim. Biophys. Acta* 519, 76–86.
133. Losic, D. *et al.* (2006) Pore architecture of diatom frustules: potential nanostructured membranes for molecular and particle separations. *J. Nanosci. Nanotechnol.* 6, 982–989.

134. Losic, D. *et al.* (2007) Atomic force microscopy (AFM) characterisation of the porous silica nanostructure of two centric diatoms. *J. Porous Mater.* 14, 61–69.
135. Hogan, H. (2007) AFM maps the diatom exoskeleton. *Biophotonics International* (June) 14–15.
136. Telford, R.J. *et al.* (2006) How many freshwater diatoms are pH specialists? A response to Pither & Aarssen (2005). *Ecol. Lett.* 9, E1–E5.
137. Smith, W. (1856) *Synopsis of British Diatomaceae.* John Van Voorst.
138. Schmidt, A. (1874) *Atlas der DiatomaceenKunde.* Koeltz Scientific Books.
139. Cleve, P.T. and Grunow, A. (1880) Beiträge zur Kenntniss der Arctischen Diatomeen [Contributions to knowledge of Arctic diatoms]. *Kongliga Svenska Vetenskaps-Akademiens Handlingar* 17, 1–122 + 127 plates.
140. Peragallo, H. (1897) *Diatomées marines de France et des districts maritimes voisins [Marine Diatoms of France and the Close Maritime Districts]* (2 volumes), *J. Tempere.*
141. Van Heurck, H. (1896) *A Treatise on the Diatomaceae, Containing Introductory Remarks on the Structure, Life History, Collection, Cultivation and Preparation of Diatoms, and a Description and Figure of Every Species Found in the North Sea an Countries Bordering It, Including Great Britain, Belgium, etc.* William Wesley & Son.
142. Frison, E. (1954) *L'évolution de la partie optique du microscope au cours du dix-neuvième siècle* (*Communication No. 89*), Rijksmuseum voor de geschiedenis der natuurwetenschappen.
143. Patwardhan, S.V. *et al.* (2005) On the role(s) of additives in bioinspired silicification. *Chem. Commun.* (*Camb.*) 9, 1113–1121.
144. Lopez, P.J. *et al.* (2005) Prospects in diatom research. *Curr. Opin. Biotechnol.* 16, 180–186.

Chapter 19

Dinoflagellates

M.-O. Soyer-Gobillard
Université Pierre et Marie Curie Paris 6, Laboratoire Arago, Banyuls-sur-Mer, France

Chapter Outline

Abbreviations	**263**	*Noctiluca scintillans* McCartney	268
Defining Statement	**263**	*Prorocentrum micans*	269
Introduction	**263**	**Mixotrophic Dinoflagellates**	**270**
Dinoflagellate Evolution	**263**	**Evolution of the Mitotic Apparatus**	**271**
Dinoflagellate Diversity	**265**	**Conclusions**	**274**
Crypthecodinium cohnii Biecheler	265	**Further Reading**	**277**

ABBREVIATIONS

FeSODs Iron SODs
SOD Superoxide dismutase

DEFINING STATEMENT

The diversity of dinoflagellate morphology and lifestyles is presented with an emphasis on model organisms. These include heterotrophic (*Crypthecodinium cohnii*, *Noctiluca scintillans*), autotrophic (*Prorocentrum micans*), mixotrophic (*Blastodinium* sp.), and parasitic (*Syndinium* sp.) species. Cellular and molecular analyses, including some puzzling questions regarding dinomitosis, eyespots, and an antioxidant enzyme, are discussed.

INTRODUCTION

With their large distribution throughout all the seas and freshwater bodies of the world, dinoflagellates play a prominent role in the trophic chain and in water ecology. They constitute a major group of unicellular eukaryotic microbes and can be autotrophic (about half are photosynthetic), heterotrophic, parasitic and/or mixotrophic, or symbiotic. Current classification recognizes approximately 161 genera, 48 families, and 17 orders or a total of ~6000 described species. Cells are typically biflagellated, and generally, the two flagella are oriented perpendicular to one another. This results in a characteristic spiral locomotion from which the group derives its name (*dino*, in Greek means 'to turn'). Most autotrophic dinoflagellates possess a characteristic carotenoid pigment, peridinin, in association with chlorophylls *a* and *c* and other pigments such as β-carotene, diadinoxanthin, and dinoxanthin.

One of the most striking features of dinoflagellates is their nucleus (dinokaryon). Although the pattern of dinoflagellate DNA synthesis is typically eukaryotic, they are the only eukaryotes in which the chromatin is totally devoid of histones and nucleosomes. Instead, it contains specific DNA-binding basic proteins. The only known exception to this is the plasmodial genus *Syndinium* in which basic nuclear proteins (histone-like) have been detected using alkaline fast green staining.

Dinoflagellate diversity has inspired not only numerous scientific studies but also artistic representations. These include the wonderful drawings of Ernest Haeckel (1834–1919) and Edouard Chatton. For example, Figure 1 shows a drawing of several free-living dinoflagellate of the genus *Pyrocystis*, which is still actively studied for its bioluminescent properties (www.mcb.harvard.edu/).

DINOFLAGELLATE EVOLUTION

The great protistologist and cell biologist Edouard Chatton (1883–1947) was the first to distinguish between prokaryotic and eukaryotic microbes and to propose a tentative protist classification. However, it was Haeckel who coined

Eukaryotic Microbes.

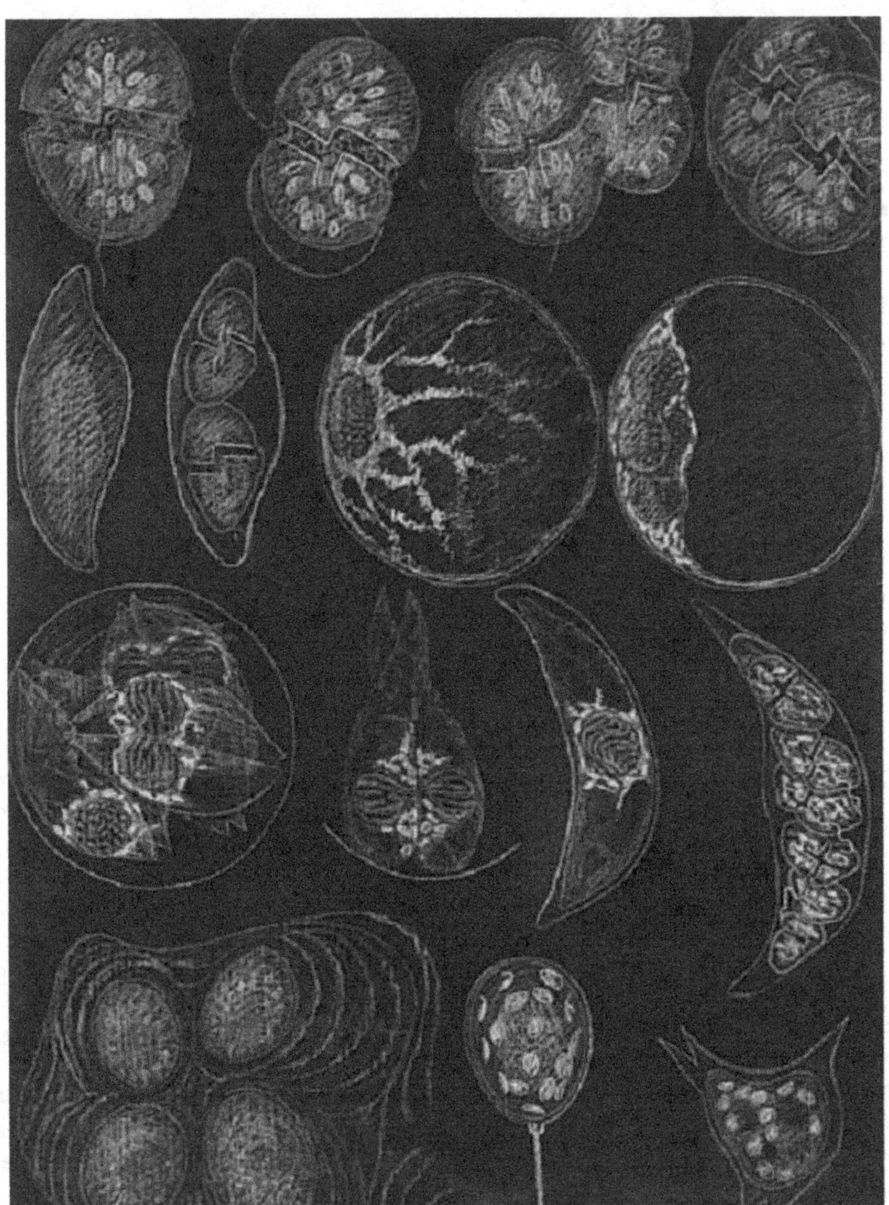

FIGURE 1 Several free-living bioluminescent dinoflagellates of the genus *Pyrocystis* Murray ex Haeckel, 1890. The species *Pyrocystis pseudonoctiluca* is shown in the second row at the right-hand side; *Pyrocystis lunula* in the third row at the right-hand side. The drawing is from a course board for students drawn by Edouard Chatton (1930). Copyright is courtesy of 'Les Archives du Laboratoire Arago,' bequest LWOFF, F-Banyuls-sur-mer, and all rights are reserved for potential further use.

the term 'protist' in the nineteenth century, although he also included bacteria in that group. Dinoflagellates are now recognized as one of the major divisions of the eukaryotic supergroup Alveolata, which is further included in the supergroup Chromalveolata (Figure 2). Like other members of the chromalveolates, phototrophic dinoflagellates have acquired their plastids by secondary endosymbiosis. This appears to have occurred multiple times during dinoflagellate evolution and at least two cases of tertiary endosymbiosis have now been documented in this group.

The first and undisputable fossil dinoflagellate cyst is from the Triassic (~200 Ma). However, biogeochemical evidence suggests the presence of dinoflagellate ancestors as far back as the early Cambrian. This is based on the discovery of triaromatic dinosteroid, sedimentary biomarkers characteristic of dinoflagellates. This suggests that these protists are much older than their fossils suggest. The major diversification of modern dinoflagellates occurred during the Jurassic radiation based on the presence of numerous cysts and thecal plates, which were first formed by photosynthetic dinoflagellates. However, these dates

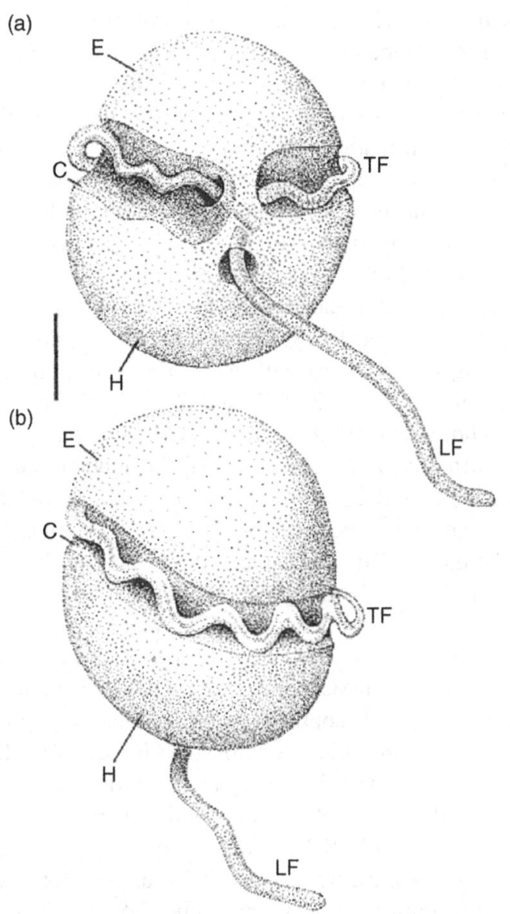

FIGURE 2 Schematic representation of the *Crypthecodinium cohnii* cell. The diagram is drawn from previously published SEM views (Perret, E., Albert, M., Bordes, N., et al. (1991). Microtubular spindle and centrosome structures during the cell cycle in a dinoflagellate *Crypthecodinium cohnii* B.: An immunocytochemical study. *BioSystems, 25*, 53–65). (a) Ventral view and (b) dorsal view. Also indicated are the episome (E), hyposome (H), longitudinal flagellum (LF), and cingulum (C). *Reproduced from Perret, E., Davoust, J., Albert, M., et al. (1993). Microtubule organization during the cell cycle of the primitive eukaryote dinoflagellate* Crypthecodinium cohnii. Journal of Cell Science, 104, *639–651, with permission of the Company of Biologists Limited.*

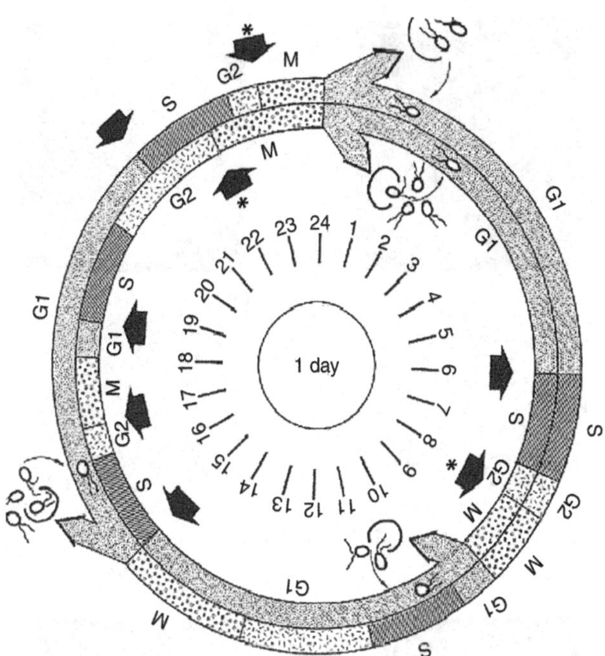

FIGURE 3 Diagram of some successive *Crypthecodinium cohnii* cell cycles over 24 h. In this particular example, one vegetative cell performed two successive complete cell cycles (16 h) and released four daughter cells. One of these new swimming cells released two daughter cells 10 h later (external circle of the diagram). During this time, other swimming cells gave an inverse alternation (in alternance with the other cycle) (internal circle). Different diagrams could be possible with other alternations. Transition points G1S ('start' point) are represented by arrows and G2M by arrows plus star. *Reproduced from Bhaud, Y., Barbier, M., & Soyer-Gobillard, M. O. (1994). A detailed study of the complex cell cycle of the dinoflagellate* Crypthecodinium cohnii *Biecheler and evidence for variation in histone H1 kinase activity.* Journal of Eukaryotic Microbiology, 41, *519–526, with permission of the Company of Biologists Limited.*

based on morphological and cell biology data are not yet fully supported by all the molecular phylogenetic analyses.

DINOFLAGELLATE DIVERSITY

Three species of free-living dinoflagellates and two mixotrophic species belonging to the genera *Blastodinium* and *Syndinium* have been extensively studied as model organisms, and much of what is known about dinoflagellate cell structure is based on this work. These organisms are the heterotrophic species *Crypthecodinium cohnii* and *Noctiluca scintillans*, the photosynthetic species *Prorocentrum micans*, and the mixotrophic species of the genera *Blastodinium* and *Syndinium*.

Crypthecodinium cohnii Biecheler

C. cohnii Biecheler is a marine dinoflagellate whose vegetative cells are devoid of plastids but rich in starch. Two orthogonally disposed flagella are present (Figure 2), and these disappear just before cell division (dinomitosis), which precedes a complex reproductive cycle. Since 1887, two forms of *C. cohnii* swimming cells and cysts that lack flagella have been recognized. Detailed *in vivo* studies published in the 1990s described the complex cell cycle of this dinoflagellate, which lasts 26 h and results in four daughter cells (Figure 3). The organism has been used as a model system to study the cytoskeleton and its role in mitosis, the superoxide dismutase (SOD) evolution.

The cytoskeleton of *C. cohnii* has been intensively studied in our lab, particularly the role of β-tubulin in the maintenance of cell shape (Figure 4) and its importance in dinomitosis. It was also shown in this species that cortical microtubules do not depolymerize during mitosis, and now this has also been shown in the species *Oxyrrhis marina*. Microtubules of cytoskeleton are accompanied by heat

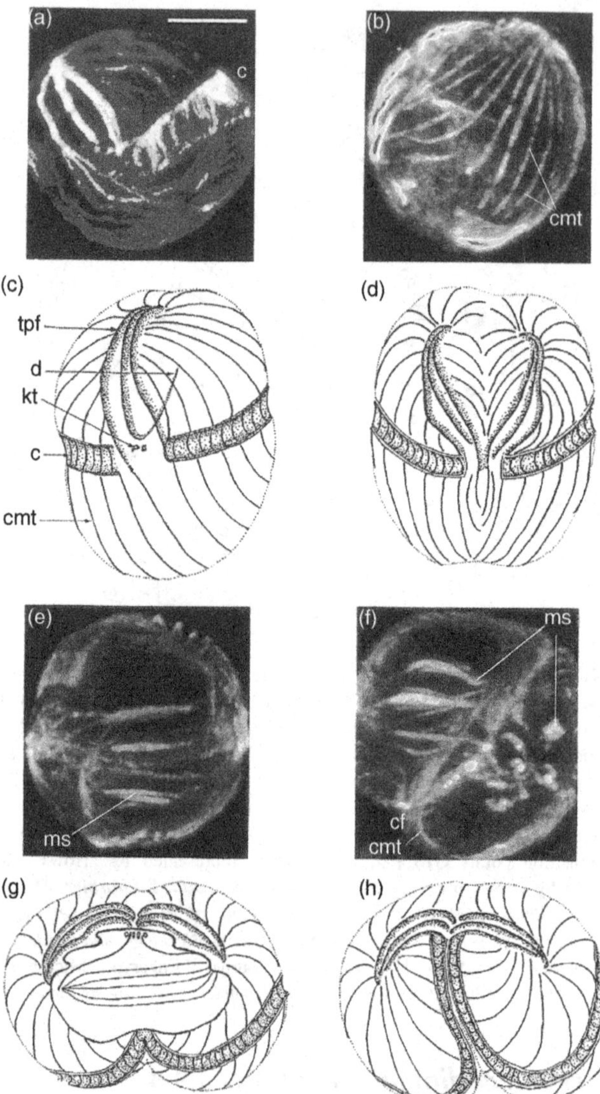

FIGURE 4 Reconstructions (each of 16 confocal laser scanning sections) of the free-living heterotrophic dinoflagellate *Crypthecodinium cohnii* Biecheler. Cells are labeled with anti-β-tubulin antibody. (a) In the cortical region of *C. cohnii*, the cingulum (c) in which the transverse flagellum lies (data not shown) and microtubular bundles are visible (in blue color) (scale bar = 10 μm). (b) Organization of the cortical microtubular rows (cmt) of *C. cohnii*. (c) Cytoskeleton of a nondividing *C. cohnii* cell in G1 phase. Close to the two kinetosomes (kt), three important cortical microtubular bundles are merging at the apical pole. (d) At the beginning of the division, duplicated microtubular bundles migrate. (e) In this dividing (anaphase) cell, microtubules of mitotic spindle (ms) and of the cortex are both visible. (f) Two dividing daughter cells in a cyst, showing two orthogonally oriented mitotic spindle (ms) and cleavage furrow (cf). (g and h) Peeled back views of the same dividing cell are shown. The diameter of a dividing cell is about 20 μm. Drawn by M. J. Bodiou. *Reproduced from Soyer-Gobillard, M. O., Besseau, L., Géraud, M. L., et al. (2002). Cytoskeleton and mitosis in the dinoflagellate* Crypthecodinium cohnii*: Immunolocalization of P72, an HSP70-related protein.* European Journal of Protistology, 38, *155–170, with copyright permission from Elsevier.*

shock protein P72. This may play a protective role for the cortical microtubules, for example, by preventing microtubule depolymerization during mitosis. This well-conserved protein could also play an important role in the regulation of microtubule formation.

SODs, which catalyze the degradation of toxic superoxide radicals, have been studied only recently in dinoflagellates. A multicopy gene family (about 20 genes) encoding iron SODs (FeSODs) was discovered and characterized in a heterotrophic dinoflagellate, while MnSOD and FeSOD activities were detected in *Symbiodinium* sp., a dinoflagellate living in symbiosis with the sea anemone *Anemonia viridis*, and FeSOD, MnSOD, and Cu/ZnSOD were found in the photosynthetic dinoflagellate *Lingulodinium polyedrum* (formerly *Gonyaulax polyedra*). In another photosynthetic dinoflagellate, *Karenia brevis*, both MnSOD and FeSOD have been identified but only by Western blotting. It is thought that these enzymes might play an important role in the survival strategy of this Florida red tide microorganism. Enzymatic assays of SOD activity indicate the presence of two different FeSOD activities in *C. cohnii* lysates, whereas MnSOD and Cu/ZnSOD were not detected in this species. This contrasts with the situation previously characterized in other photosynthetic dinoflagellates. Phylogenetic analyses including 110 other dimeric FeSODs from a wide range of bacteria and eukaryotes showed that the *C. cohnii* SODs form a monophyletic group (Figure 5). These recent results (2008) suggest that all dinoflagellates primitively had a cytosolic FeSOD of bacterial origin, while the MnSOD of mitochondrial ancestry was lost and replaced by a second copy of FeSOD.

Eyespot

The ocellus of dinoflagellates, an unusual photosensitive organelle, is the most complex algal eyespot known. It is present in several heterotrophic dinoflagellates such as *Nematodinium*, *Warnowia*, *Erythropsis*, and several Woloszynskioids. Eyespot is considered as an important phylogenetic marker in dinoflagellates. More often located toward the posterior end of the cell, the most sophisticated structure measures about 25 μm long by 15 μm wide. It is composed of a pigment cup, retinoid, and lens and it functions as a photoreceptor. As suggested by W. J. Gehring in 2004, this photosensitivity probably arose first in cyanobacteria, and was later acquired by the first photosynthetic eukaryotes (archaeplastida) via the cyanobacterial endosymbiosis that gave rise to the plastid. Dinoflagellates would then have acquired this photosensitivity from a red alga via secondary chloroplasts. In some dinoflagellate species, this plastid evolved into an elaborate and sophisticated photoreceptor organelle, which is found in some dinoflagellate species that lack chloroplasts, such as those cited previously. Because some dinoflagellates such as zooxanthellae are

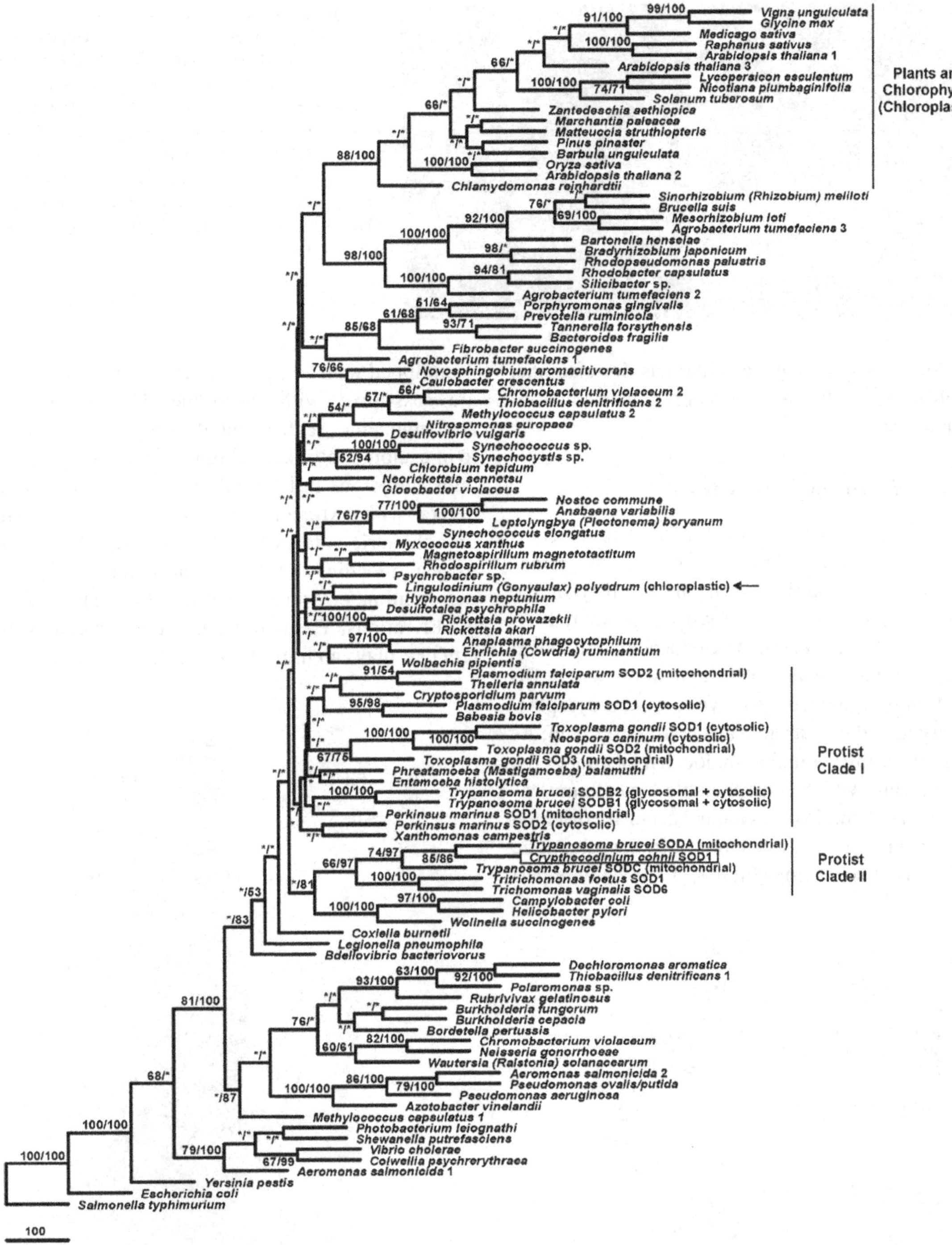

FIGURE 5 Neighbor-joining tree based on protein sequences of dimeric FeSODs and closely related SODs. Accession numbers or references of the sequences retrieved from genome sequencing projects are given in Dufernez, F., Yernaux, C., Gerbod, D., et al. (2006). The presence of four iron-containing superoxide dismutase isozymes in Trypanosomatidae: Characterization, subcellular localization, and phylogenetic origin in *Trypanosoma brucei*. *Free Radical Biological Medicine*, *40*, 210–225. The full-length *Crypthecodinium cohnii* SOD1 sequence obtained in this study is boxed, whereas the chloroplast sequence from *Lingulodinium (Gonyaulax) polyedrum* is indicated by an arrow. If experimentally determined, the subcellular localization of protist SODs is indicated in parentheses. The scale bar corresponds to 0.1 substitution (corrected) per site. *Reproduced from Dufernez, F., Derelle, E., Noël, C., et al. (2008). Molecular characterization of iron-containing superoxide dismutases in the heterotrophic dinoflagellate* Crypthecodinium cohnii. Protist, 159, *223–238, by copyright permission from Elsevier GmbH.*

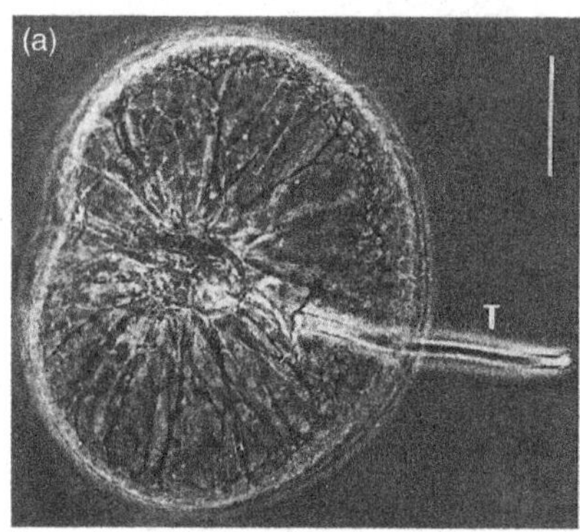

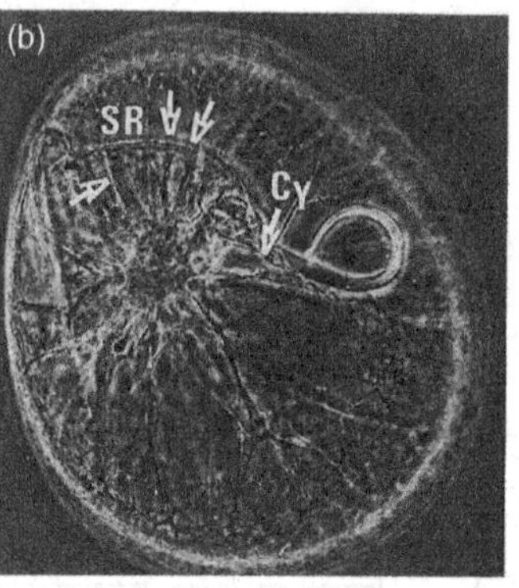

FIGURE 6 *Noctiluca scintillans* McCartney. (a and b) Interference phase-contrast views. (a) Trophozoite with its tentacle (T) seen relaxed and (b) contracted. Tracts of myonems (arrows) are anchored between the cytostome (Cy) and the supporting rod (SR). Scale bar = 200 μm (Photos J. Lecomte). Magnification ×200. *Reproduced from Métivier, C., & Soyer-Gobillard, M. O. (1988). Organization of cytoskeleton during the tentacle contraction and cytostome movement in the dinoflagellate* Noctiluca scintillans *McCartney.* Cell and Tissue Research, 251, *359–370, by copyright permission from Springer-Verlag.*

commonly found as symbionts in cnidarians, for example, these organisms could then have transferred their photoreceptor to cnidarians.

Noctiluca scintillans McCartney

N. scintillans McCartney is a model for studying the complex heterotrophy found in some dinoflagellates. Cytoskeletal elements that participate in the motility of the *N. scintillans* trophozoite are also involved in related nutritional functions. These elements are located at the level of the tentacle (Figure 6) and of the cytostome where filaments are found organized into myonems and cytoplasmic fibrils organized into striated and contractile strip. As shown in Figure 8, microtubules are located just under the peripheral ectosarc, which is responsible for calcium exchange by means of epiplasmic channels, and are oriented parallel to the major axis of the tentacle. In the schematic drawing of transverse section (Figure 7), the convex part is equipped with a large number of microtubule rows that are cross-linked with one another (Figure 8, arrows). Striated myonems are distributed along the tentacle or linked to one another by a knot (Figure 7). The tentacle also contains numerous mitochondria. This complex system together with extracellular calcium ions sequestered in mitochondria is responsible for this dinoflagellate's motility and feeding. Tentacle contraction involves ectosarc deformation, myonem contractility, and microtubule modifications. The major contractile protein of this system that plays an important role in determining cell shape

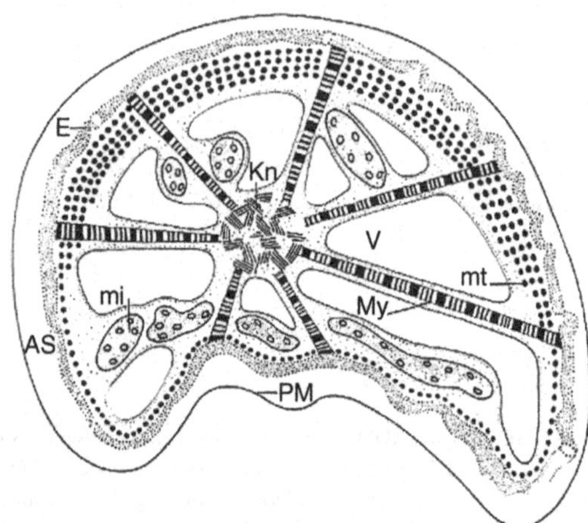

FIGURE 7 Schematic drawing of a transverse section of the *Noctiluca scintillans* tentacle. Note the distribution of the three characteristic elements of the cytoskeleton: the ectosarc (E), microtubules (mt), and myonems (My). Also indicated are the knot (Kn), plasma membrane (PM), vacuoles (V), mitochondria (mi), and alveolar space (AS). *Reproduced from Métivier, C., & Soyer-Gobillard, M. O. (1988). Organization of cytoskeleton during the tentacle contraction and cytostome movement in the dinoflagellate* Noctiluca scintillans *McCartney.* Cell and Tissue Research, 251, *359–370, with copyright permission from Springer-Verlag.*

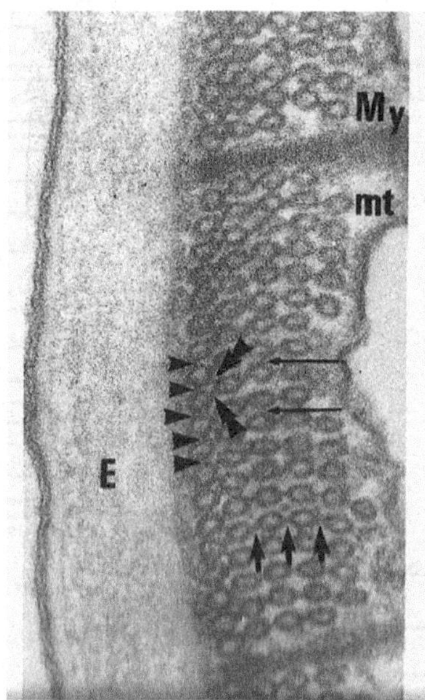

FIGURE 8 Ultrastructural organization of the tentacle of *Noctiluca scintillans*. High magnification of a part of the tentacle on its convex side. Myonems (My) between the microtubules (mt) on the ectosarc (E) of the first row are cross-linked with each other by bridges (single arrowhead), links with the second row result in a Y bond (double arrowheads). In the other rows, links are always observed between each row (thick black arrows); they are sometimes visible between microtubules of the same row (thin black arrows). Scale bar = 0.1 μm. Magnification ×140000. *Reproduced from Métivier, C., & Soyer-Gobillard, M. O. (1988). Organization of cytoskeleton during the tentacle contraction and cytostome movement in the dinoflagellate* Noctiluca scintillans *McCartney.* Cell and Tissue Research, 251, *359–370, by copyright permission from Springer-Verlag.*

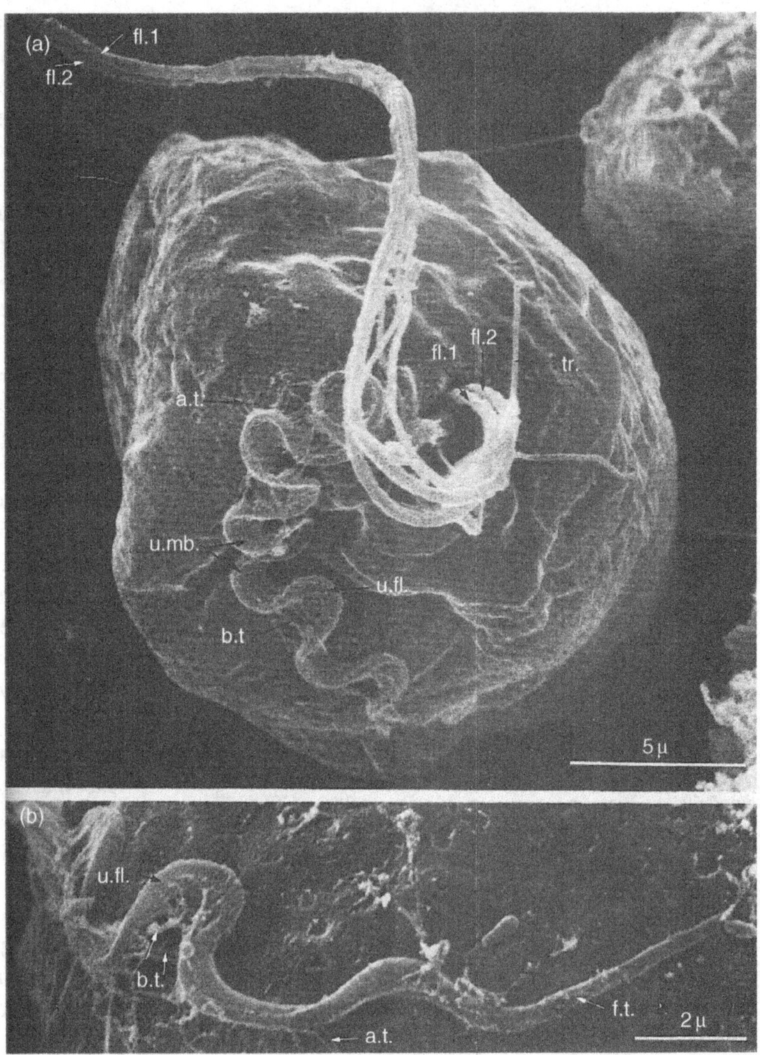

FIGURE 9 (a) *Prorocentrum micans* autotrophic cell apical view observed with scanning electron microscope. Observe the peripheral polysaccaridic 'skin.' Two longitudinal flagella of this predividing cell run more or less parallel to each other. One undulating flagellum arises from the same opening of the epitheca as the longitudinal ones, but adheres to its outer layer; anchoring threads strengthen this adhesion, and bridging threads connect every second wave of the undulating membrane. On the epitheca also lie adsorbed extruded fibrous trichocysts and undefined filaments, probably of bacterial origin. (b) Terminal portion of the undulating flagellum. There are no anchoring or bridging threads on its free tip. Magnification ×11 500. a.t., anchoring thread; b.t., bridging thread; fl1, fl2, longitudinal flagella; f.t., free tip; tr. trichocyst; u.fl., undulating flagellum; u.mb., undulating membrane. *Reproduced from Soyer-Gobillard, M. O., Prévot, P., De Billy, F., et al. (1982).* Prorocentrum micans *E., one of the most primitive dinoflagellates. I. The complex flagellar apparatus as seen in scanning and transmission electron microscopy.* Protistologica, XVIII, *289–298.*

is the epiplasmic actin protein, P45. Cytochemical investigations about the presence of histone-like proteins in the trophocyte nucleus, using in particular alkaline fast green, were negative. At the molecular level, the unusual nucleotide containing the base 5-hydroxymethyluracil has been found in the nuclear DNA of *N. scintillans*. In the dinoflagellate species investigated up to now, this unusual base replaces 12–68% of the thymines, which constitutes a relatively important level of replacement amount.

Prorocentrum micans

P. micans is a model for autotrophic dinoflagellates. In this species, the pellicle covering the whole surface of the theca (the epitheca) is a kind of polysaccharidic 'skin.' This fragile structure surrounding the entire cell was observed only under SEM with careful preservation (Figure 9). The cell cycle of *P. micans* is typically eukaryotic despite the unusual structure of the dinoflagellate chromatin when compared with that of most eukaryotes. The population doubles every 6 days and the S phase of DNA synthesis lasts 4 h. From the end of replication to the end of mitosis (G2+M phases), the time taken on average is 8 h. Phases S+G2+M take less than 12 h and the G1 phase lasts 120 h (Figure 10). This is despite the fact that *P. micans* has more than 100 chromosomes of 1 μm in diameter and a large DNA content of 42 pg cell^{-1}.

In *P. micans*, vegetative cells are haploid and reproduce by binary division. Sexuality can be experimentally induced by changing the culture medium (switching from Erd Schreiber's to Provasoli's medium) or by subjecting cultures of this photosynthetic species to unfavorable conditions such as darkness and low temperature (4 °C).

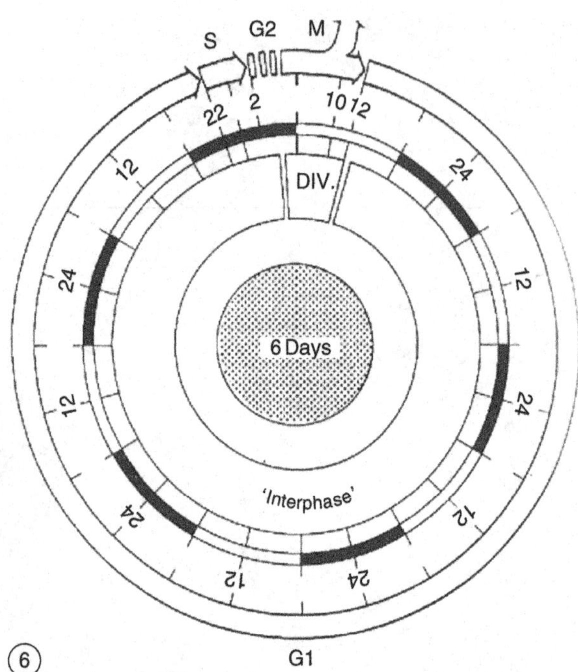

FIGURE 10 Cell cycle and mitotic clock of *Prorocentrum micans* cell. S = 4 h; G2 is short or nonexistent, duration undetermined; G2 + M = 8 h; S + G2 + M = 12 h; G1 = 120 h. *Reproduced from Bhaud, Y., & Soyer-Gobillard, M. O. (1986). DNA synthesis and cell cycle of a primitive dinoflagellate* Prorocentrum micans *Ehr.* Protistologica, XXII, *23–30.*

Mating in dinoflagellates occurs by conjugation (Figure 11). This begins with the formation of a conjugation tube between the two conjugants, followed by passage of one nucleus via the tube. Then, this tube enlarges and the sexual cells fuse, leading to the formation of a dikaryon. There is commonly an absence of fundamental morphological differences between vegetative cells and gametes. The only visible differences are a small variation in pigmentation and cell size. Many different sexual processes occur in dinoflagellates (Table 1), but *P. micans* is unique in having a fertilization tube that permits the transfer of all genetic material (male gametic nucleus) and in the formation of a partition between the empty sexual cells and the dinokaryon. Dinoflagellates appear to be highly specific in terms of environmental factors that induce sexuality. This is probably related to the extreme diversity in their lifecycles and way of life.

Mixotrophic Dinoflagellates

Mixotrophic (parasitic) dinoflagellates were first described by Chatton in his magisterial doctorate thesis, published in 1920. He was particularly interested in *Blastodinids*, which he first discovered in the digestive tract of pelagic copepods. As shown in Figures 12 and 13, blastodinids are binucleated cells. Two kinds of cells are present: the trophocyte, which remains temporarily undivided while it increases in size, and the gonocyte, which produces sporocytes in which the chromatin is progressively more and more condensed after successive divisions. Layers of sporocytes lead to the formation of biflagellate swarming dinospores, which are later released in seawater. Trophocytes can also divide longitudinally and produce other parasites into the same digestive tract. In Figure 13, the two nuclei are seen clotted in anaphase with 'plasmodendrites' (cytoplasm) passing through the nuclei. In 1971, Soyer (Gobillard) showed with transmission electron microscopy that the *Blastodinium* mitotic spindle is located in these channels and linked to the spindle poles, which lack centrioles. During mitosis, the chromosomes of the sporocytes become increasingly compacted and are attached to the nuclear envelope. The latter has differentiated cytoplasmic channels that allow the microtubules of the mitotic spindle to pass through the nuclear membrane (Figure 14).

Chatton also described the genus *Syndinium*, whose various species parasitize marine copepods or various radiolarians. On the basis of the morphology of the spores released by their free-living swarmers, Chatton considered *Syndinium* a 'specialized' dinoflagellate. This was also due to the apparent simplicity of *Syndinium* cell division and the fact that the species had only five chromosomes (Figure 15). On this basis, Chatton proposed it as a model for dinoflagellate mitosis, specifically referring to it as 'syndinian mitosis' (Figure 15).

Later, ultrastructural studies showed this to be very particular peridinian mitosis. The mechanism consists of a permanent closed nuclear membrane outside of which lies an extranuclear mitotic spindle with centrioles and kinetochores that attach to the inner surface of the nuclear membrane (Figures 17 and 18). This is in contrast to other dinoflagellates, which have no centriole. Moreover, typical dinoflagellate chromosomes lack basic proteins such as histones, but instead have specific basic nuclear protein associated with their DNA. Two exceptions to this are *Syndinium* sp. and the symbiotic zooxantellae, which have appreciable amounts of basic proteins of the histone family associated with their DNA. Further biochemical and molecular studies are needed to analyze more thoroughly and to characterize these basic nuclear proteins.

Very recently (2010), a detailed description of a new planktonic free-living mixotrophic dinoflagellate, *Paragymnodinium shiwhaense*, was published: its morphology, pigments, and ribosomal DNA gene sequence analysis suggested that it is a new species into a new genus. The distinctive feature of this species is the presence of both chloroplasts and nematocysts, organelles used for the prey capture as also present in the polynucleated *Polykrikos* dinoflagellate, in addition to fibrous trichocysts, which are detoxication organelles. In addition to the photosynthesis, *P. shiwhaense* sucks the preys through its peduncle.

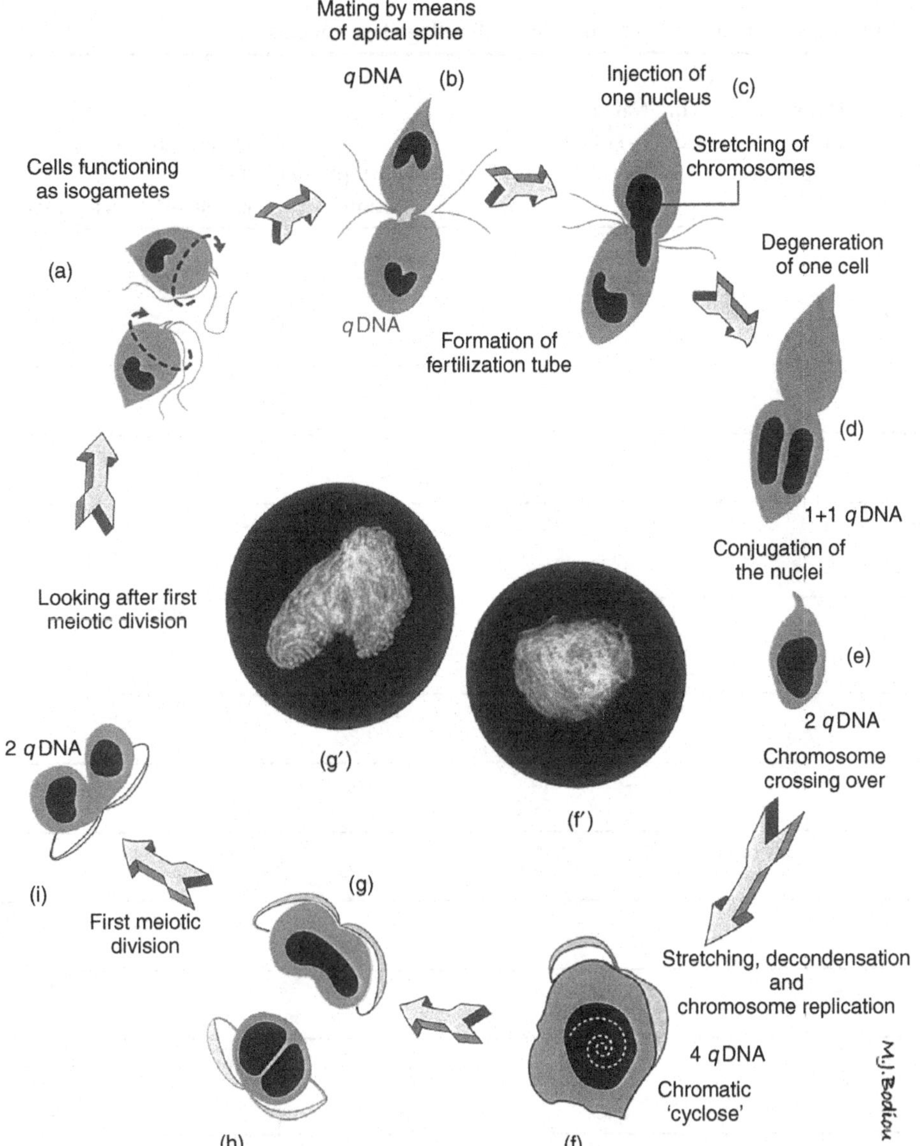

FIGURE 11 (a–i) Diagram of the partial course of sexual reproduction phases of the dinoflagellate *Prorocentrum micans* from *in vivo* and TEM observations of the nuclei after exposure of the cultured cells to low temperature (4 °C) and darkness. Vegetative cells functioning as isogametes (a) and containing *q* DNA (Bhaud, Y., Soyer-Gobillard, M. O., & Salmon, J. M. (1988). Transmission of gametic nuclei through a fertilization tube during mating in a primitive dinoflagellate, *Prorocentrum micans* Ehr. *Journal of Cell Science*, *89*, 197–206) become paired by means of their respective apical spines (b), then the donor cell injects its nucleus with stretched chromosomes into the receiver cell (c). After the two nuclei conjugate (d) in the zygote containing 2*q* DNA (e), crossing over of the chromosomes occurs. Chromosomes become completely unwound, stretched, and replicated. In the nucleus containing therefore 4*q* DNA (f), chromatin begins to spin round (chromatic cyclose) giving a round shape to the nucleus as shown in the photograph (f′). Only one meiotic division occurred (g, g′, h), leading to an incomplete separation of the 2*q* DNA-containing cells (i). *Reproduced from Soyer-Gobillard, M. O., Bhaud, Y., & Saint-Hilaire, D. (2002). New data on mating in an autotrophic dinoflagellate* Prorocentrum micans *Ehrenberg*. Vie Milieu, 52, *167–175, by copyright permission from Vie Milieu.*

Several photosynthetic species of Dinophyceae as *Dinophysis acuminata or D. norvegica* were the first (1994) to be described as mixotrophic with the presence of food vacuoles into the cell. They are predators for other dinoflagellates or other unicellular planktonic organisms and can have an important grazing impact on algal populations.

EVOLUTION OF THE MITOTIC APPARATUS

Most dinoflagellates have a closed mitosis in which the nucleus is surrounded by a persistent nuclear envelope. Within this nucleus, the chromosomes are maintained in a quasi permanently compacted state. Chromosome separation is

TABLE 1 Some properties of sexual reproduction in 25 dinoflagellate species

Species	F or M[a]	Different from vegetative cells[b]	Iso (+) or aniso (−) gamy	Naked (−) or thecated (+)	Fertilization tube[c]	Protoplasmic fusion[d]	Planozygote (nb. flagellae)[e]	Hypnozygote[f]	Authors
Peridinium cinctum	F	+	−	−	?	+	+(2)	+	Pfiester (1975)
idem	F	+	−	−+	+	+	+(4)	+	Spector et al. (1981)
Peridinium willei	F	+	−	−	−	+	+(2)	+	Pfiester (1976)
Peridinium gatunense	F	+	−	+	+	+	+(?)	+	Pfiester (1977)
Peridinium volzii	F	+	−	−+	+	+	+(?)	+	Pfiester and Skvarla (1979)
Peridinium limbatum	F	+	−	+	−	+	+(?)	+	Pfiester and Skvarla (1980)
Peridinium inconspicuum	F	+	−	+	?	+	+(?)	+−	Pfiester et al. (1984)
Peridinium cunningtonii	F	+	−	−+	−	+	+(?)	+	Sako et al. (1984)
Gymnodinium excavatum	F	+	−	?	?	?	+(?)	+	Von Stosch (1972)
Gymnodinium paradoxum	F	+	−	?	?	?	+(?)	+	Von Stosch (1972)
Gy. pseudopalustre	F	+	−	?	+	+	+(?)	+	Von Stosch (1973)
Amphidinium carteri	M	−	−	?	+	+	+(?)	+	Cao Vien (1967, 1968)
Ceratium cornutum	F	+	+	+	+[g]	+[h]	+(?)	+	Von Stosch (1972)
Ceratium horridum	M	+	+	+	+[g]	+[h]	+(?)	+	Von Stosch (1979)
Oxyrrhis marina	M	+	−	?	?	?	+(?)	?	Von Stosch (1972)
Crypthecodinium cohnii	M	+	+−	?	?	+	+(?)	+	Tuttle and Loeblich (1975)
idem	M	?	+−	+	+	+	+(?)	+	Pfiester (1984)

TABLE 1 Some properties of sexual reproduction in 25 dinoflagellate species—cont'd

Species	F or M[a]	Different from vegetative cells[b]	Iso (+) or aniso (−) gamy	Naked (−) or thecated (+)	Fertilization tube[c]	Protoplasmic fusion[d]	Planozygote (nb. flagellae)[e]	Hypnozygote[f]	Authors
Gonyaulax tamarensis	M	+	+	?	−	+	+(4)	+	Turpin et al. (1978)
idem	M	+	?	?	?	?	+(?)	+	Anderson and Lindquist (1985)
Go. excavata	M	?	?	?	?	?	?	+	Anderson and Wall (1978)
Go. monilata	M	+	−	+	−	+	+(4)	+	Walker and Steidinger (1979)
Woloszynskia apiculata	F	+	−	+	+	+	+(4)	+	Von Stosch (1973)
Protogonyaulax catenella	M	+	−	+	−	+	+(?)	+	Yoshimatsu (1981)
Gymnodinium breve	M	+	−	+	−	+	+(4)	−[i]	Walker (1982)
Gyrodiniumun catenatum	M	+	+	+	+	+	+(4)	+	Coats et al. (1984)
Helgolandicum subglobosum	M	+	+	?	?	?	+(?)	−	Von Stosch (1972)
Noctiluca miliaris	M	+	−	−	−	+	+(0)	?	Zingmark (1970)
Prorocentrum micans	M	−	−	+	+	−	+(2)	−	Bhaud and Soyer-Gobillard (1988).

Reproduced from Bhaud, Y., Soyer-Gobillard, M. O., & Salmon, J. M. (1988). Transmission of gametic nuclei through a fertilization tube during mating in a primitive dinoflagellate, *Prorocentrum micans* Ehr. *Journal of Cell Science, 89*, 197–206, with permission of the Company of Biologists Limited.

[a]*F, freshwater; M, marine.*

[b]*Differences observed between vegetative cells and gametes are often morphological variations such as variations in cell volume and color; gametes are generally smaller and less pigmented than vegetative cells; +, different; −, the same.*

[c]*Fertilization tube or similar structure: +, present; −, absent.*

[d]*Protoplasmic fusion of the two gametes: +, fusion; −, no fusion.*

[e]*Planozygote: +, present; −, absent.*

[f]*Hypnozygote: +, present; −, absent.*

[g]*The female engulfs the male.*

[h]*Protoplasm from female and male is misled.*

[i]*No hypnozygote in the culture.*

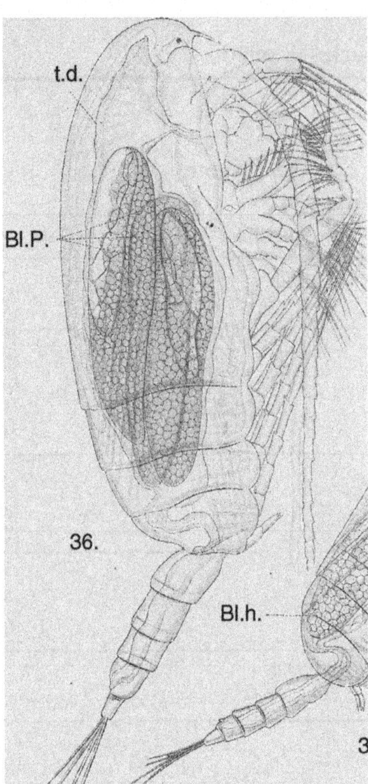

FIGURE 12 A pelagic copepod (*Clausocalanus arcuicornis*) in which stomach is distended by three *Blastodinium pruvoti* Chatton. Total length is 1.3 mm. t.d., digestive tract. *From an original plate (Pl. IV) of the doctoral Chatton's thesis (1920)* Archives de Zoologie Expérimentale et Générale, 59, *1–475*.

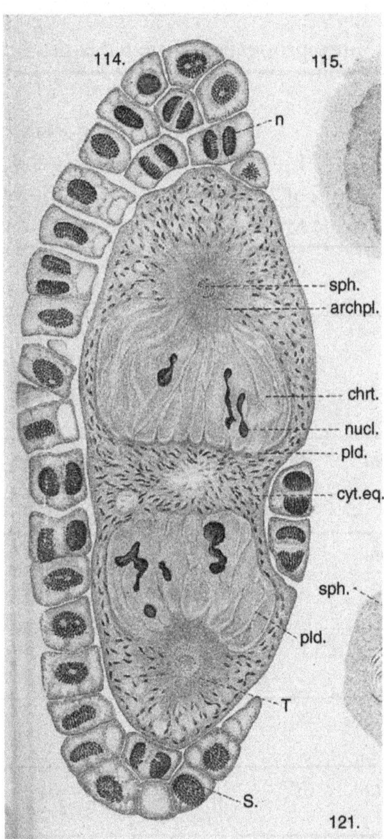

FIGURE 13 Cytology of *Blastodinium crassum* Chatton, a parasitic and mixotrophic dinoflagellate living in the digestive tract of pelagic copepods (*Clausocalanus arcuicornis*). Longitudinal section of the vegetative binucleated cell, the depicted 'trophocyte' (T) containing no functional etioplastids, surrounded by several layers of numerous dividing sporocytes (s) containing functional plastids. The centrosome region-like or 'centro spheres' (sph) is surrounded by archoplasm, specialized cytoplasm containing Golgi bodies. chrt., chromatin; chrs, chromosomes; nucl., nucleolus; pld, plasmodendrites, cytoplasmic channels. Length of the trophocyte is 300 μm. *From an original plate (Pl. X) of the doctoral Chatton's thesis (1920)* Archives de Zoologie Expérimentale et Générale, 59, *1–475*.

accomplished by an extranuclear microtubular mitotic spindle located in cytoplasmic channels that pass through the nuclear membrane during mitosis. Two centrosome regions are linked to the kinetosomes by two microtubular structures called 'desmoses' (Figure 16). There appears to be a great deal of homogeneity in terms of mitotic apparatus and cell division in dinoflagellates. Exceptions include rare species such as the coelomic parasite *Syndinium* sp. (see below) and *N. scintillans*, which show considerable variations when compared with the well-known *C. cohnii* model of mitosis.

Syndinium and related species that parasitize invertebrates were classified as dinoflagellates by Chatton and others on the basis of the morphology of their zoospores. However, the biochemistry of their chromosomes (presence of specific basic proteins) and electron microscopic observations demonstrated fundamental differences in their mitotic apparatus as compared with the rest of the dinoflagellates. In *Syndinium* and relatives, the four V-shaped chromosomes are permanently attached at their apex to a specific area of the nuclear membrane through a kinetochore-like trilaminar disk inserted into an opening in the membrane (Figure 17). Microtubules connect the dense layer of each kinetochore to the base of the two centrioles located in an invagination of the nuclear envelope (Figure 18). In these species, centrioles or basal bodies are intimately involved in chromosome segregation. This is not the case in *Crypthecodinium*, where the basal bodies are indirectly involved with the microtubular spindle, or in *Blastodinium*, where the nuclear envelope functions as the motor for chromosome segregation.

CONCLUSIONS

Despite the great diversity of dinoflagellates in terms of their physiology, lifestyle, and cell cycle, they are remarkably similar in their mitosis except for *Syndinium* spp. and *Oxyrrhis marina*. The system of cytoplasmic channels passing through the intact nucleus indicates that microtubules

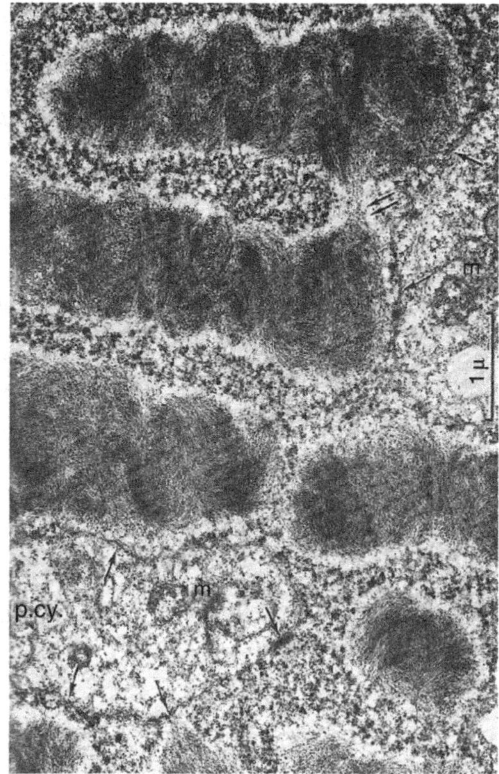

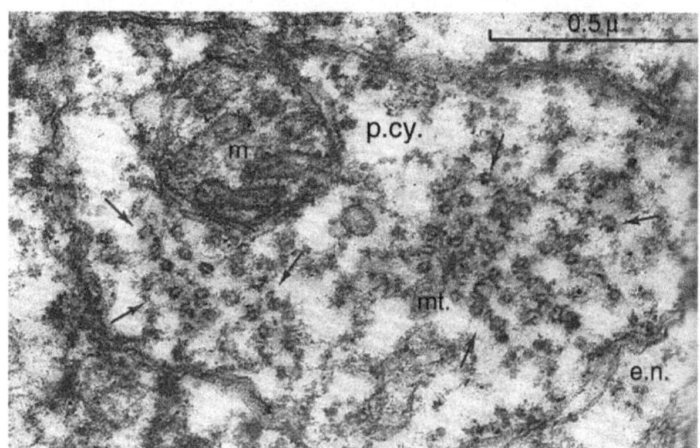

FIGURE 14 Part of a *Blastodinium contortum* Chatton dividing nucleus. Several chromosomes are attached to the nuclear envelope (arrows) and two dividing chromosomes are yet linked by DNA fibers (double arrow; upper panel). Detail of a transversally sectioned cytoplasmic channel (p.cy.) passing through the nucleus. Numerous microtubules (mt.) of the mitotic spindle and mitochondriae (m) are visible (lower panel). *Reproduced from Soyer-Gobillard, M. O. (1971). Structure du noyau des* Blastodinium *(dinoflagellés parasites). Division et condensation chromatique.* Chromosoma, 33, *70–114, by copyright permission from Springer-Verlag.*

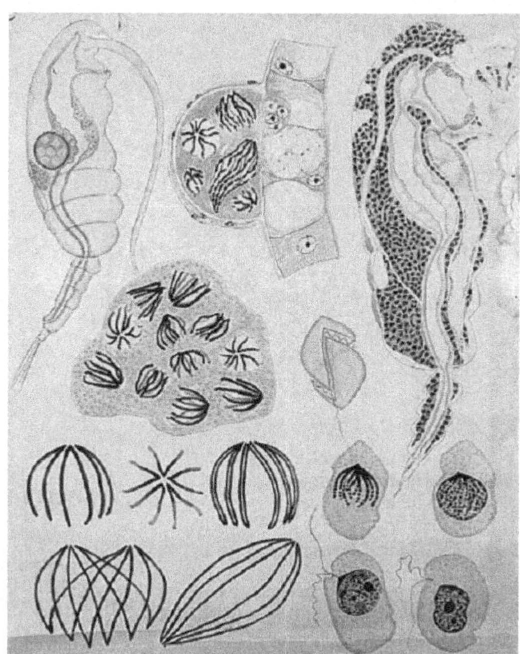

FIGURE 15 Course board (1.60 m by 1 m) drawn by Edouard Chatton (1930). Chatton discovered the genus *Syndinium* Chatton in 1910 in the sea, close to Banyuls sur mer (France). Detail of the syndinian mitosis: *Syndinium turbo* is a plasmodial coelomic and parasitic dinoflagellate living in the general cavity of pelagic copepods (*Paracalanus parvus, Clausocalanus arcuicornis*, or *Corycoeus venustus*). During metaphase, the five chromosomes of *S. turbo* congress to an equatorial plate before anaphase segregation. The motile biundulipodiated zoospores, later released, are of gymnodinian (free-living dinoflagellate) type. Note at the lower right the release from the coelomic mass of the newly formed karyomastigont during zoospore maturation. At the lower right, the characteristic signature of Edouard Chatton (E and C interlaced). *From personal collection of the author, bequest Lwoff, with permission of Dr M.O. Soyer-Gobillard. All rights reserved for potential further use.*

are never in direct contact with the chromosomes but are always separated from them by the persistent nuclear envelope. Despite the absence of histones and nucleosomes in their chromatin, dinoflagellates still have a typical eukaryotic cell cycle.

There is still too little molecular data to sufficiently resolve relationships among the diverse dinoflagellates. Thus, morphological and cell biological analyses will continue to constitute a vital tool for dinoflagellate phylogeny. The first evidence of dinoflagellates in the fossil record is the biogeochemical analyses of early Cambrian sediments (~520 Ma), which detect the presence of dinoflagellate-specific dinosterols.

In addition to their fundamental interest for cell and evolutionary biologists, autotrophic dinoflagellates constitute part of the enormous energy potentially available from the phytoplankton. The heterotrophic *C. cohnii* is also intensively cultivated as a source of maternal milk due to the presence of long chain fatty acids in these organisms. Dinoflagellates are also important as symbionts of corals, where they play important role(s) in the health and survival of their hosts. The impact of toxic dinoflagellates such as species of *Pfisteria, Alexandrium, Gymnodinium, Dinophysis, Karenia*, and others on fish and shellfish industries is substantial and possibly is still significantly underestimated. Indeed, toxic incidents resulting from these different dinoflagellate species are being reported frequently

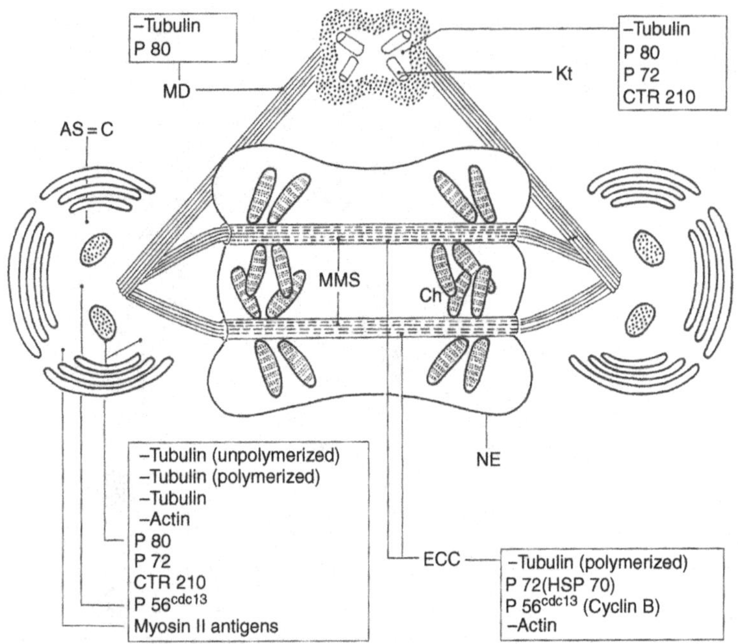

FIGURE 16 Diagrammatic representation of a dinoflagellate (except *Syndinium*) mitotic apparatus schematized in anaphase with listed and framed the up-to-now detected associated proteins. Microtubular mitotic spindles lie throughout the nucleus, pass into the archoplasmic spheres, and are linked to the two pairs of kinetosomes. AS, archoplasmic sphere (containing Golgi bodies); C, centrosome (without any centriole); ECC, extranuclear cytoplasmic channel; MMS, microtubular mitotic spindle; MD, microtubular desmose; Kt, kinetosomes; N, nucleus; NE, nuclear envelope (permanent); Ch, chromosomes. *Reproduced from Ausseil, J., Soyer-Gobillard, M. O., & Géraud, M. L. (2000). Dinoflagellate centrosome: Associated proteins old and new.* European Journal of Protistology, 36, *1–19, by copyright permission from Elsevier.*

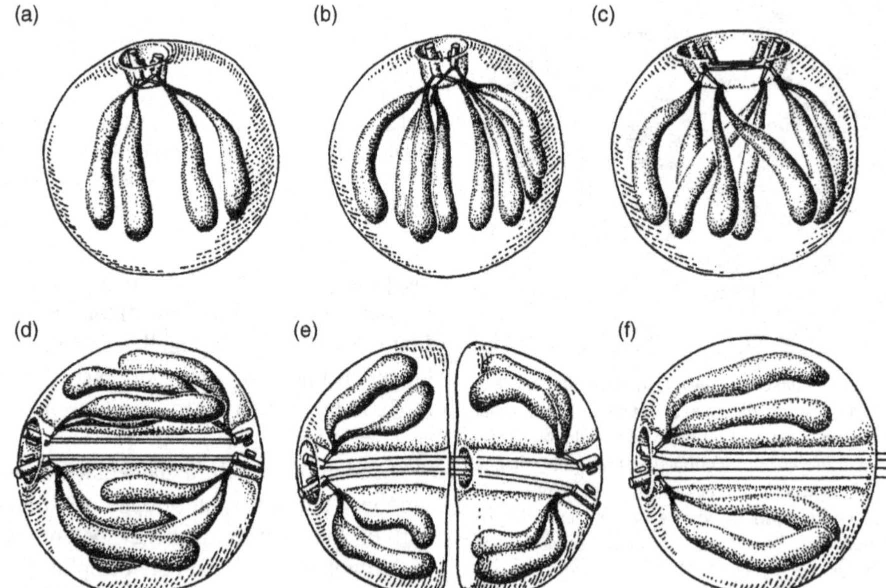

FIGURE 17 Diagrammatic representation of nuclear division in *Syndinium* sp. (a) Interphase. (b) Early division, kinetochores and chromosomes have duplicated. (c) Early stage of chromosome segregation. Central spindle between separating centrioles. (d) Late stage of chromosome segregation. Central spindle in cytoplasmic channel throughout the nucleus. (e) Division of nucleus. (f) Early daughter nucleus with persisting channel and microtubules. *Reproduced from Ris, H., & Kubai, D. F. (1974). An unusual mitotic mechanism in the parasitic protozoan* Syndinium *sp.* Journal of Cell Biology, *60, 702–720, by copyright permission from The Rockefeller University Press.*

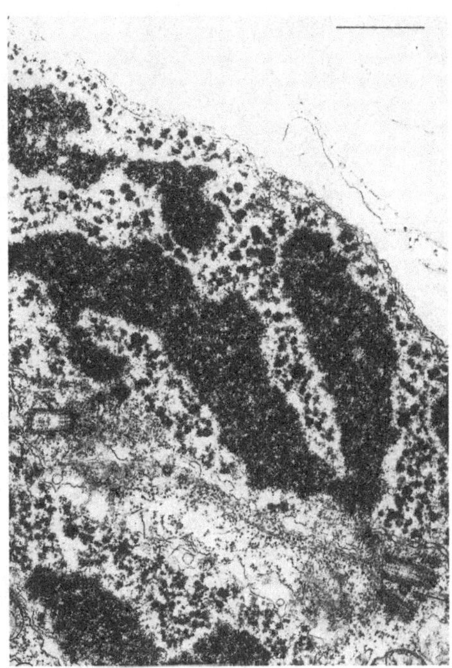

FIGURE 18 Ultrathin section through a nucleus of *Syndinium* sp. in an intermediate stage of division. The central microtubular spindle lies in a central channel and is linked with two pairs of centrioles. The 'V' shape of one of the chromosomes is clearly visible in this micrograph (as previously shown by Chatton, see Figure 15). Scale bar = 1 μm. *Reproduced from Ris, H., & Kubai, D. F. (1974). An unusual mitotic mechanism in the parasitic protozoan* Syndiniumsp. Journal of Cell Biology, 60, *702–720, by copyright permission from The Rockefeller University Press.*

throughout the world, particularly near the coasts. This increase of toxic blooms seems linked to the eutrophization of coasts and the spreading of species due to an increase in commercial exchanges. Thus, the importance of these species, which produce some of the most potent neurotoxins known to man, should not be underestimated.

FURTHER READING

Bhaud, Y., Soyer-Gobillard, M. O., & Salmon, J. M. (1988). Transmission of gametic nuclei through a fertilization tube during mating in a primitive dinoflagellate, *Prorocentrum micans* Ehr. *Journal of Cell Science*, *89*, 197–206.

Biecheler, B. (1952). Recherches sur les Péridiniens. *Bulletin Biologique de la France et de la Belgique*, *36*(Suppl.), 1–149.

Breidbach, O., Eibl-Eibesfeldt, I., & Hartmann, R. P. (1998). *Ernst Haeckel: Kunstformen der Natur.* (pp. 62–63). Prestel-Verlag: München, NY.

Chatton, E. (1920). Les Péridiniens parasites. Morphologie, reproduction, ethologie. *Archives de Zoologie Expérimentale et Générale*, *59*, 1–475.

Delwiche, C. F. (2007). The origin and evolution of dinoflagellates. In: P. G. Falkovski & A. H. Knoll (Eds.), *Evolution of primary producers in the sea* (pp. 191–205). Elsevier: Boston, MA.

Dufernez, F., Derelle, E., Noël, C., *et al.* (2008). Molecular characterization of iron-containing superoxide dismutases in the heterotrophic dinoflagellate *Crypthecodinium cohnii*. *Protist*, *159*, 223–238.

Gehring, W. J. (2005). New perspectives on eye development and the evolution of eyes and photoreceptors. *The Journal of Heredity*, *96* (3), 171–184.

Kang, N. S., Jeong, H. J., Moestrup, O., *et al.* (2010). Description of a new planctonic mixotrophic dinoflagellate paragymnodinium shiwhaense n.gen., n.sp. from the coastal waters of western Korea: Morphology, pigments, and ribosomal DNA gene sequence. *Journal of Eukaryotic Microbiology*, *57*, 121–144.

Margulis, L., Soyer-Gobillard, M. O., & Corliss, J. (1984). *Evolutionary protistology: The organism as cell.* D. Reidel: Dordrecht/Boston, MA (pp. 205–215).

Raikov, I. B. (1982). *The protozoan nucleus: Morphology and evolution. Cell biology monographs* (Vol. 9). Springer: Wien (pp. 1–474).

Ris, H., & Kubai, D. F. (1974). An unusual mitotic mechanism in the parasitic protozoan *Syndinium* sp. *The Journal of Cell Biology*, *60*, 702–720.

Skovgaard, A., Massana, R., Balagué, V., & Saiz, E. (2005). Phylogenetic position of the copepod-infesting parasite *Syndinium turbo* (Dinoflagellata, Syndinea). *Protist*, *156*, 1–13.

Soyer-Gobillard, M. O. (2006). Edouard Chatton (1883–1947) and the dinoflagellate protists: Concepts and models. *International Microbiology*, *9*, 173–177.

Soyer-Gobillard, M. O. (2008). Methods for studying the nuclei and chromosomes of dinoflagellates. In R. Hancock (Ed.), *The nucleus: Vol. 1. Nuclei and subnuclear components. Methods in molecular biology* Vol. 463(pp. 93–108). Humana Press: Totowa, NJ.

Soyer-Gobillard, M. O., Bhaud, Y., & Saint-Hilaire, D. (2002). New data on mating in an autotrophic dinoflagellate, *Prorocentrum micans* Ehrenberg. *Vie et Milieu*, *52*(4), 167–175.

Spector, D. L. (1984). *Dinoflagellates.* Academic Press (Harcourt Brace Jovanovich): Orlando, NY (pp. 1–545).

Taylor, F. J. R. (1987). *The biology of dinoflagellates. Botanical monographs* (Vol. 21). Blackwell Scientific: London (pp. 1–785).

RELEVANT WEBSITES

www.mcb.harvard.edu/ – Department of Molecular and Cellular Biology, Harvard University.

Chapter 20

Dictyostelium

C. Scott and P. Schaap
University of Dundee, Dundee, UK

Chapter Outline

Abbreviations 279
Defining Statement 279
Taxonomy, Evolution, and Ecology 279
Genomics and Tractability 281
The *D. discoideum* Genome 281
Experimental Tractability 282
The Developmental Program of *D. discoideum* 283
Morphogenesis 283
Gene Expression and Cell Differentiation 286
Recent Developments in the *Dictyostelium* Field 288
The Sporulation Cascade 288
Phylogeny Wide Genome Sequencing 288
Evolutionary History of cAMP Signaling 288
Conclusions 289
References 289
Further Reading 289

ABBREVIATIONS

ALC Anterior-like cell
asp Aspartate
cAMP Cyclic AMP
cAR cAMP receptor
CMF Conditioned medium factor
CRAC Cytosolic regulator of adenylate cyclase
his Histidine
PdsA Phosphodiesterase A
PH Pleckstrin homology
PIP2 Phosphatidyl inositolbisphosphate
PIP3 Phosphatidyl inositoltrisphosphate
PKA-C Single catalytic subunit
PKA-R Single regulatory subunit
PSF Prestarvation factor
REMI Restriction enzyme-mediated integration
SDF-2 Spore differentiation factor 2

DEFINING STATEMENT

Dictyostelid social amoebas are popular model systems for studying problems in cell and developmental biology. They also offer unique opportunities for understanding how multicellular life evolved.

TAXONOMY, EVOLUTION, AND ECOLOGY

The Dictyostelids are one of only a few groups of protists that can live either as unicellular predators or as interactive community members. They start life as amoebas that feed on bacteria in decaying vegetation and dung. Starvation triggers social behavior and amoebas move together to form colonies. Here the best-fed amoebas enter a dormant spore stage and the rest construct a pedestal to bear the spore mass aloft.

Of the 100 known social amoebae species, *Dictyostelium discoideum* is the most intensely studied. When *D. discoideum* cells suffer food shortage, a few cells start to secrete pulses of cyclic AMP (cAMP), a molecule that in most organisms regulates events inside the cell. When receiving a cAMP pulse, neighboring cells migrate toward the signal source in a process called chemotaxis. In turn these cells secrete their own cAMP pulse. The pulse thus propagates as a wave through the cell population, allowing up to 1 million cells to come together to form a mound. Subsequent cell–cell communication regulates both cell migration and cell differentiation. Together these processes enable the formation of a fruiting body, which consists of cellulose-encased stalk cells and thickly walled spores. This structure is uniquely well designed to allow relocation of the spores to a more nutrient-rich habitat.

Eukaryotic Microbes.

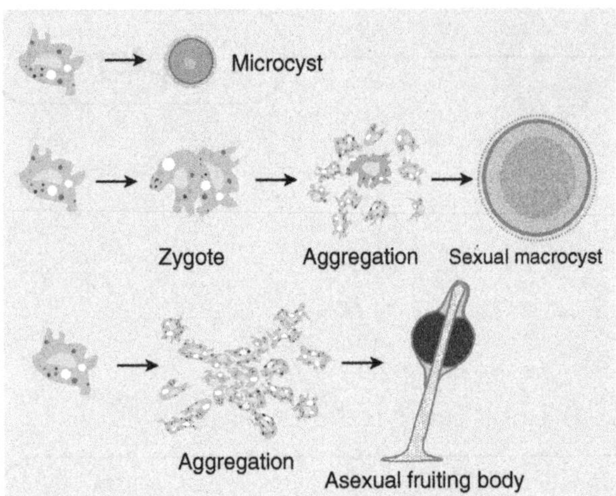

FIGURE 1 Surviving starvation. Amoebas can adopt one of three survival strategies when faced with starvation. Solitary amoeba can encapsulate individually to form microcysts, two amoebas can fuse to form a zygote, which then attracts other cells to form a sexual macrocyst, or amoebas can aggregate to form fruiting bodies, where part of the cells die to form a stalk and the others survive as highly resistant spores.

In addition to fruiting body formation, amoebae have two other choices when faced with starvation. Some species can encyst individually and form dormant microcysts (Figure 1). Microcysts are less dehydrated than spores and have a thinner two-layered cell wall instead of the thick three-layered spore coat. Alternatively, cells can form sexual macrocysts. This often requires the presence of cells with opposite mating types, although homothallic mating is also common. Macrocyst formation begins when two cells fuse to form a zygote. The zygote then chemotactically attracts other starving cells and cannibalizes them. Eventually, the zygote synthesizes a highly resistant thick cell wall and enters a long period of dormancy. Macrocysts are more commonly formed under dark and wet conditions that are not favorable for fruiting body formation.

The Dictyostelids owe their common name of 'cellular slime mold' to the fact that their fruiting structures are similar to those of some of the smaller species of fungi. This also highlights the confusion about the position of Dictyostelids in the tree of life. All amoeba-like protists that formed some type of spore-bearing structure were at one time grouped together within the Myxomycota, placed in the kingdom Fungi. These myxomycota were subdivided into three classes, the Protostelids with fruiting structures of one to four cells, the true slime molds or Myxogastrids that form fruiting bodies from a multinucleate syncytium, and the Acrasiomycetes that form fruiting bodies from cell aggregates. The Dictyostelids were placed as a subclass of the Acrasiomycetes, and subdivided into three genera: *Dictyostelium* with simple or branched cellular stalks, *Acytostelium* with acellular stalks, and *Polysphondylium* with regular whorls of side branches. The other subclass of Acrasiomycetes – the Acrasids – differs from the Dictyostelids in the morphology of their amoebas and aggregates, and the lack of cellulose in spore-bearing structures.

However, it is now clear that this system does not reflect true evolutionary relationships among the different groups. Modern taxonomy based on gene or protein sequence comparisons shows an entirely different view. The Protostelids, Myxogastrids, and Dictyostelids are members of the supergroup Amoebazoa, and this group is separate from but closely related to the opisthokonts, the group containing the animals and fungi. Most importantly, the Acrasids are members of the unrelated supergroup Discicristates. Molecular phylogeny has now clearly demonstrated that the Dictyostelids are subdivided into four major divisions. None of these correspond to the three traditional genera. In fact, one group, group 2, appears to include members of all three (Dictyostelids, Polysphondylids, and Acytostelids). The molecular phylogeny also indicates that the polysphondylium morphology has evolved at least twice. Similarity of fruiting body morphology is evidently not a good marker for genetic similarity (Figure 2).

However, there are some morphological characters that are group specific. Species in groups 1–3 generally form small clustered and branched fruiting structures from a single aggregate (Figure 3). In the most derived group 4, species usually form a single robust unbranched structure. Most species in groups 1–3 can still form microcysts, the dormant stage of their ancestors, the solitary amoebas. However, group 4 species have lost this survival strategy. This suggests that when the process of fruiting body formation became more robust, encystation was no longer necessary. However, the sexual cycle, resulting in macrocyst formation, occurs in all taxon groups. Group 4 species also stand out by using cAMP as chemoattractant, with a variety of other compounds being used by the other groups (Figure 3).

Dictyostelids have been found all over the world on every continent and adapted to many climates. Forest soils contain the largest variety of species, but also cultivated soils and even deserts harbor at least some species. Some species seem happy to grow under almost any conditions, coping with a wide variety of changes in humidity and altitude. However, others are far more particular about their choice of residence. According to their distribution pattern, species can be subdivided into one of four groups: cosmopolitan, disjunct, restricted, and pantropical. Cosmopolitan species grow almost anywhere and include the species *Dictyostelium mucoroides* and the exquisite *Polysphondylium pallidum*. Species with disjunct distributions are found over widely separated areas, but tend to be found in similar habitats and similar climates. The model organism *D. discoideum* falls into this group. Restricted species are only found in specific locations and pantropical species are confined to tropical forest soils around the world. Generally speaking, the tropics contain the highest number of

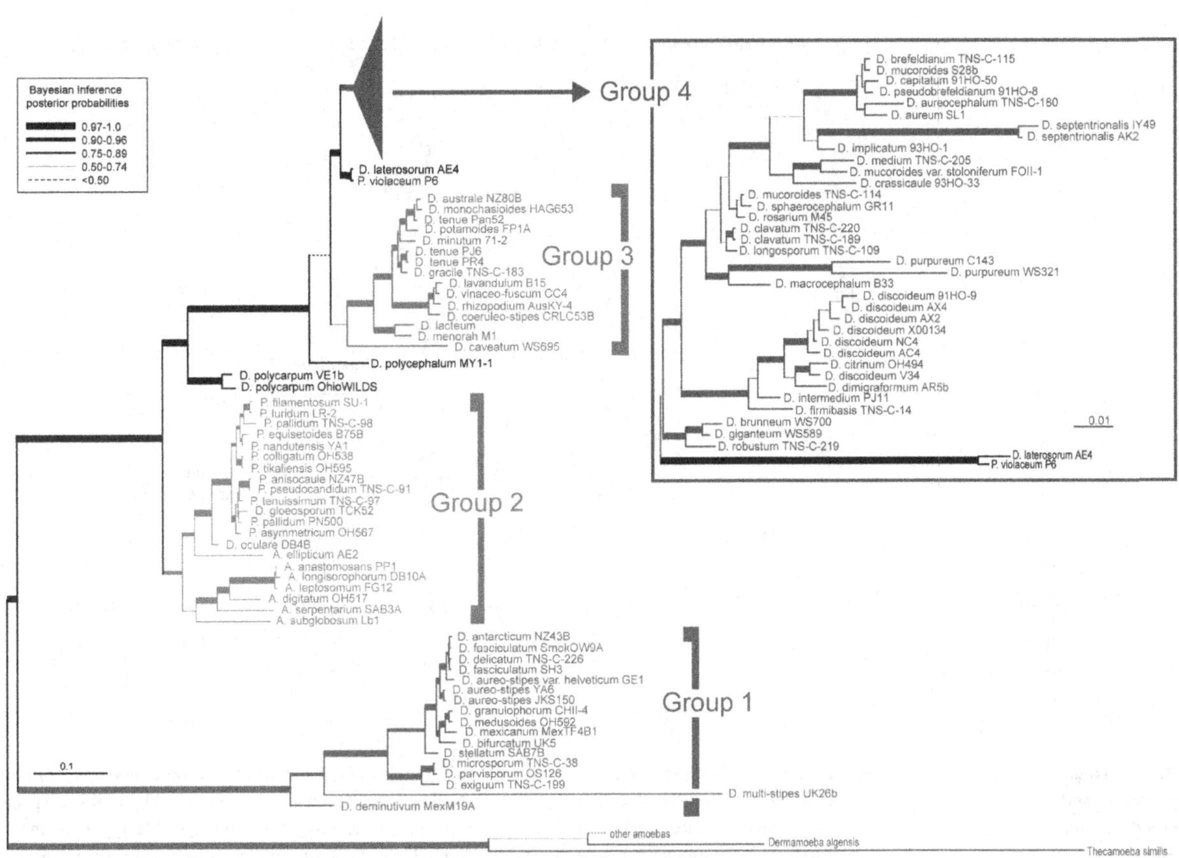

FIGURE 2 The four groups of Dictyostelids. A phylogeny based on molecular data (here SSU rRNA) subdivides the Dictyostelids into four major groups. Note that group 4 contains the model organism *Dictyostelium discoideum* and group 2 contains all three former genera of Dictyostelia – the Acytostelids, the Polysphondylids, and the Dictyostelids. At least one Polysphondylid is also found outside group 2 at the base of group 4. Reproduced from Schaap, P., Winckler, T., Nelson, M., et al. (2006). Molecular phylogeny and evolution of morphology in the social amoebas. *Science*, 314, 661–663, with permission.

species and species variety decreases as latitude and altitude increase. To date there are around 100 social amoeba species characterized, with a further 20 or so still under investigation. This highlights the increased interest in the biodiversity of Dictyostelids, compared to only a decade ago when a mere 65 species had been identified.

GENOMICS AND TRACTABILITY

The *D. discoideum* Genome

D. discoideum is a popular system for studying many processes in cell biology such as chemotaxis and cell migration, phagocytosis and vesicle trafficking, and cell signaling and cytokinesis. Its developmental processes including gene regulation, cell type proportioning, and pattern formation have also attracted a great deal of interest. More recently, Dictyostelids are being used to study genes involved in social behavior and in host–pathogen interactions. All these research areas have received a considerable boost by the completion of the *D. discoideum* genome sequence in 2006.

The *D. discoideum* genome is 34 Mb, arranged over six chromosomes, which is about 80 times smaller than the human genome. However, it contains a 12500 coding sequences, which is just one-third of the gene content of the human genome. Thus it is a compact genome; and introns are rare and both intergenic regions and introns are relatively short. Noncoding regions are extremely AT rich (generally over 90% A/T nucleotides), making them easily recognizable from the less A/T-rich coding regions, which comprise 70% of the genome. Despite its compactness, the *Dictyostelium* genome contains a large proportion of low complexity sequence, much of which is present in repeats. In fact, more than 11% of all bases in the genome are part of simple repeats with a bias toward repeats of three to six bases.

D. discoideum encodes 66 ABC transporters, which are generally used to export toxic compounds from the cell. Many of these transporters are expressed during growth to protect the cells from harmful toxic metabolites or to create immunity to toxins produced by other soil dwellers. However, a few ABC transporters, such as TagC, also harbor protease domains. TagC plays a crucial role in cleaving

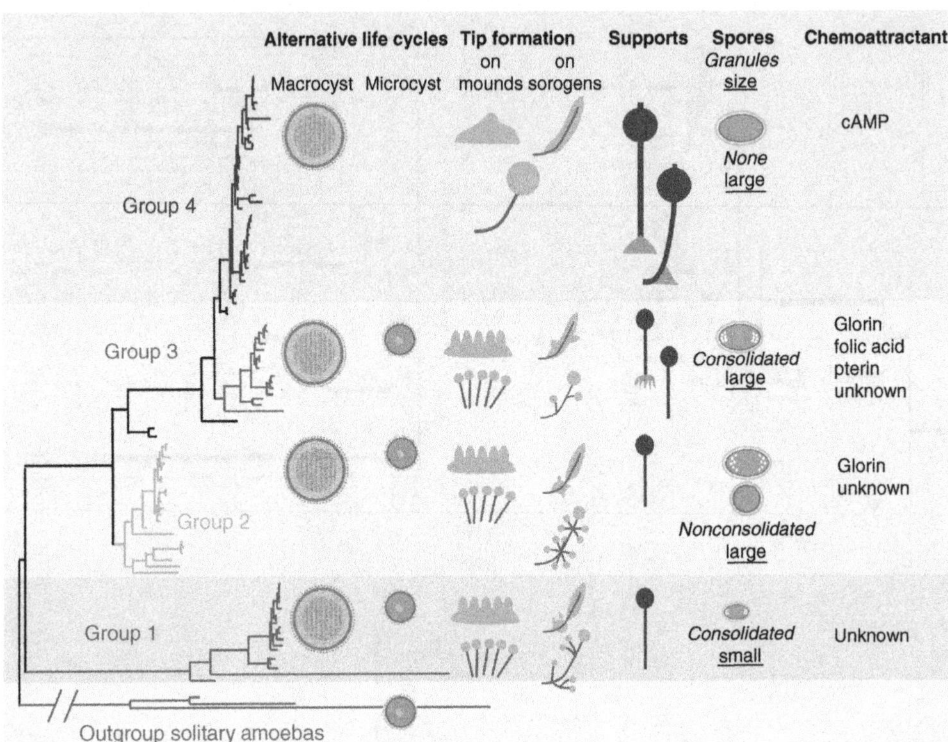

FIGURE 3 Trends in the evolution of morphology. Cartoon representation of the distribution of morphological characters across the molecular phylogeny. Spore size and the position of polar granules in the spores shows strong group-specific trends. Sexual macrocysts are formed by multiple species in each group, but microcysts are no longer formed in group 4. Species in groups 1–3 usually form multiple fruiting bodies from a single aggregate that often carry side branches, while species in group 4 tend to form large solitary and unbranched structures. They also form cellular supporters or basal disks to hold up the stalk. Only a subset of species in group 3 forms crampon-like supports. None of the species in groups 1–3 use cAMP as chemoattractant, but all investigated species in group 4 do.

off a spore differentiation factor 2 (SDF-2) from a larger protein precursor. Other gene families that are highly represented in the *Dictyostelium* genome include actin-binding proteins, cell adhesion proteins, and G-protein-coupled receptors, which include the four cAMP receptors (cARs) 1–4. One other family of proteins that have an unusually high representation in the genome are the polyketide synthases, which are also found in plants, fungi, and bacteria. These enzymes catalyze the condensation of carboxylic acid precursors into polyketide chains, which can be modified further to yield active antibiotics, antifungals, and insecticides. Although this very likely evolved as a defense mechanism against predation, some polyketide-based compounds, such as the DIFs, have adopted roles as developmental signals.

Conversely, *D. discoideum* is unable to synthesize ten amino acids and five vitamins, although it retains all the machinery required for glycolysis and nucleotide metabolism. The amino acid and vitamin biosynthetic enzymes that are absent are also missing from animals. This is a common trait of organisms that ingest other organic matter and do not have to rely on synthesizing all necessary compounds from basic elements. Intriguingly, despite its high number of genes, the number of transcription factors found in *D. discoideum* is considerably lower than in yeast and metazoans; several groups are completely missing such as the basic helix–loop–helix family, the forkhead family, and the steroid receptor transcription factors. However, other transcription factors commonly found in other eukaryotes are present, if somewhat underrepresented.

Experimental Tractability

Culture

All known Dictyostelids can be grown in the laboratory in coculture with bacteria on agar plates. However, *D. discoideum* and several other species can also be grown without bacteria in simple nutrient-rich liquid media. In this way it is possible to generate large cell numbers cheaply and easily, a distinct advantage for many forms of experimentation such as purification of proteins and biologically active compounds.

By removing the food sources and placing the cells on nutrient-free agar, the cells are forced to initiate aggregation and development into fruiting structures. For most species, the timing of the developmental process is very precise and billions of cells proceed through the developmental stages of aggregate, migrating slugs, and culminating fruiting body in synchrony. This greatly facilitates the study of cellular processes at specific developmental stages.

Experimentation

For most of the life cycle, *Dictyostelium* amoebas have no cell wall and can be disrupted by fairly gentle methods that leave the protease-rich lysosomes intact. This allows biochemical or biophysical analysis of enzyme activity, protein–protein interactions, ligand binding, and protein modification in cell lysates, membranes, and subcellular fractions with relative ease. The cells are nevertheless quite robust and withstand high-speed centrifugation, vigorous shaking, pipetting, and other common laboratory procedures without loss of viability.

Dictyostelium cells and multicellular structures are transparent and are therefore fully amenable to all microscopic techniques. As shown below, proteins can be labeled with a variety of fluorescent tags by gene replacement or knock-in of the tag. By using these approaches in combination with modern imaging techniques, *Dictyostelium* has become a paradigm for studying cytoskeletal remodeling in living cells during chemotaxis, cell division, and vesicle trafficking.

In addition to direct protein–protein interactions, translocation of proteins from the cytosol to the plasma membrane and vice versa, or in and out of the nucleus are common themes in the regulation of cell migration or gene expression by extracellular stimuli. The ability to directly visualize protein–protein interactions and protein translocation in living cells is also one of the great strengths of the *Dictyostelium* system.

Gene modification

Protocols for genetic transformation of *D. discoideum* were established in the mid-1980s and have now been expanded into a broad repertoire. Genes can be expressed from constitutive and inducible promoters with a variety of tags for visualization of proteins in living cells, or purification of proteins and protein complexes. Genes can be disrupted or replaced with mutant allelles or gene function can be abrogated by expression of antisense or inverse RNAs.

Because development is separated from growth and cell division, disruption of most developmental genes does not preclude growth, and hence the analysis of the knockout phenotype. Multiple gene deletions can be achieved by consecutive transformations using different selectable markers or by reiterated excision of the same selectable marker by cre–lox recombination.

Classical genetic procedures and linkage analysis in *D. discoideum* remains problematic due to the very slow and infrequent germination of sexual macrocysts. However, *D. discoideum* can also undergo a parasexual cycle whereby two haploid cells can fuse together to form a diploid cell, which contains the chromosomes of both parents contained within a single nucleus. This occurs with a frequency of about 1 per 30000 amoebae. The resultant diploid cells are relatively stable and divide as normal, producing diploid progeny. However, occasionally the diploids lose a copy of each chromosome and revert to a haploid condition with a random mixture of chromosomes from the parental strains. Diploid cells can be selected by using parental strains with complementary markers. Thus, the parasexual cycle of *Dictyostelium* can be used to generate multiple knockout mutants by recombining several existing mutations.

The generation of tagged mutants by a method called restriction enzyme-mediated integration (REMI) has been very successfully used to identify genes that control specific functions in *D. discoideum*. Briefly, cells are electroporated with a linearized plasmid carrying a resistance cassette and a restriction enzyme that generates cuts in the genome with sticky ends that are compatible with the plasmid. The cell's DNA ligases repair the cuts with occasional plasmid insertions. Resistant cells are recovered, grown clonally on bacterial lawns, and inspected for mutant phenotypes. The gene that carries the plasmid insertion can be obtained by digesting, religating, and transforming the genomic DNA into *Escherichia coli*. Only the circularized fragments that contain the plasmid produce antibiotic-resistant *E. coli* clones. The recovered plasmid can then be sequenced to identify the gene carrying the insertion. To confirm that the mutant phenotype is due to the plasmid insertion, wild-type cells are transformed with the rescued plasmid to disrupt the candidate gene by homologous recombination.

THE DEVELOPMENTAL PROGRAM OF *D. discoideum*

Morphogenesis

Aggregation

cAMP production

D. discoideum amoebas find their bacterial prey by following the trail of metabolites that bacteria leave behind. One of these metabolites is folic acid, and the process whereby cells migrate up a chemical concentration gradient is called chemotaxis. Starving amoebas use the same strategy to find each other to form an aggregate. Some members of earlier diverging clades (group 3, Figure 2) such as *Dictyostelium minutum* start to secrete the same compound, folic acid, as they use to find their food. Others, such as *D. discoideum*, have evolved a specialized mechanism for chemoattractant production and secretion. Upon starvation these cells develop the capacity both to detect cAMP and to secrete cAMP pulses, leading to rapid aggregation of the cells (Figure 4).

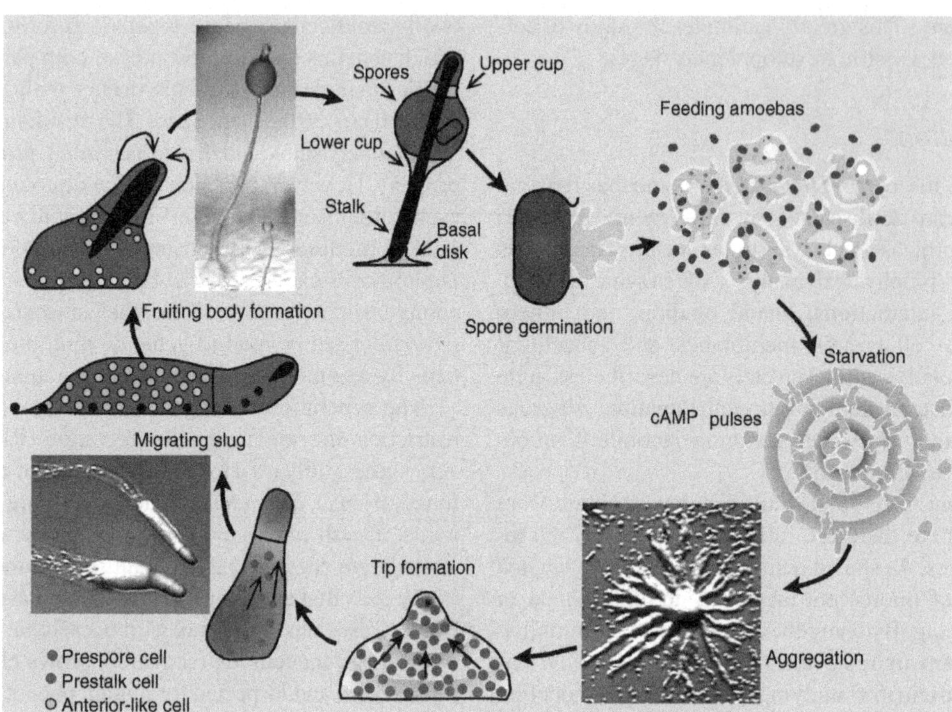

FIGURE 4 The life cycle of *Dictyostelium discoideum*. Solitary amoeba aggregate to form mounds in response to starvation. Within the mounds the cells begin to differentiate into prestalk and prespore cells and the mound morphs into a migratory slug. When suitable conditions for spore dispersal are found, the slug sits down and fruiting body formation commences.

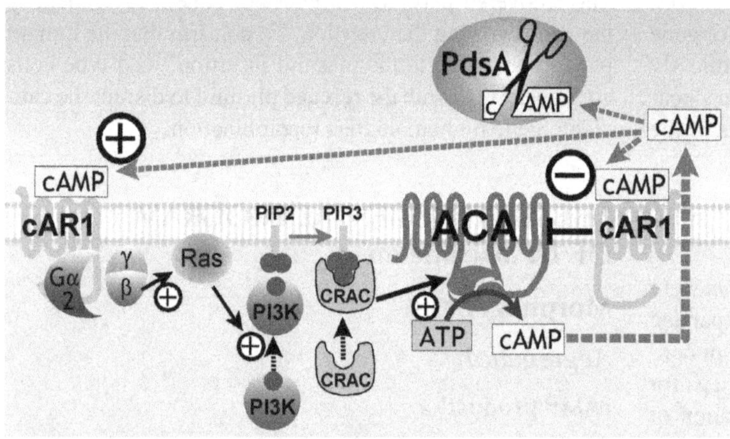

FIGURE 5 Simplified network for production of cAMP pulses. Pulsatile cAMP secretion results from positive and negative feedback loops acting on cAMP production. Amoebas first secrete a little cAMP, which binds to the cAMP receptor cAR1 to activate the adenylate cyclase ACA in an indirect manner that involves phosphorylation of the membrane phospholipid PIP2 to form PIP3. Adenylate cyclase activation terminates because cAMP also causes desensitization of cAR1, which blocks further cAMP accumulation. Hydrolysis of extracellular cAMP by PdsA resets the system for the next pulse.

cAMP is synthesized by the adenylate cyclase ACA, which is homologous to the vertebrate adenylate cyclases, which harbor two sets of six transmembrane domains separated by two adjacent catalytic domains (Figure 5). Enzyme activity is regulated by serpentine receptors that are coupled to heterotrimeric G-proteins. In the case of ACA, the receptors are activated by cAMP, and they exert both positive and negative feedback on cAMP synthesis, resulting in the production of cAMP pulses. This works as follows. Starving cells secrete a small amount of cAMP that acts on cARs to stimulate ACA activity in an indirect manner. This involves production of a specific phospholipid, PIP3, in the plasma membrane that acts as an anchor for the cytosolic regulator of adenylate cyclase (CRAC). CRAC in turn activates ACA. After a small delay, occupied cARs also trigger an inhibitory pathway that blocks further activation of ACA. An extracellular phosphodiesterase A (PdsA) then degrades cAMP, freeing the cARs from cAMP and resensitizing them to respond to cAMP once more.

When cells start to starve, this biochemical network first causes cells to respond to cAMP by secreting a pulse of cAMP, thus relaying the signal to more distant cells. Following this, the cells gain the capacity to spontaneously emit pulses of cAMP. In a field of cells it is always the least nourished cells that start to oscillate first, while the rest propagates the signal through the population, which then starts to move toward the oscillating center.

Chemotaxis

Chemotaxis involves detection of a gradient of chemoattractant and translation of this external gradient into an internal gradient that causes directional movement. Phosphatidyl inositoltrisphosphate (PIP3), the phospholipid mentioned above, plays an important role in this process, although parallel pathways operate to provide additional robustness. The PIP3-mediated pathway is however best resolved. PIP3 is formed when the kinase PI3K phosphorylates phosphatidyl inositolbisphosphate (PIP2). In turn the phosphatase PTEN dephosphorylates PIP3 again to form PIP2 (Figure 6(a)). In resting cells, PTEN is at the membrane and PI3K is in the cytosol. When cAMP binds to cARs, the G-protein G2 dissociates into its α and β–γ subunits. β–γ activates a small G-protein called Ras, which causes PI3K to move toward the membrane and PTEN away from the membrane at the site of the highest cAR activation. PI3K can now phosphorylate PIP2 in the membrane to form PIP3, which then acts as an anchoring site for proteins with pleckstrin homology (PH) domains. Three such proteins, PDK1, PKB, and CRAC, activate pathways that lead to local polymerization of actin and extension of a pseudopod. PTEN remains membrane-localized at the rear of the cell, causing depolymerization of actin (Figure 6(b)).

Slug and fruiting body formation

After cells have aggregated, cAMP pulses continue to be emitted by the aggregation center. The source of the pulses now moves to the top of the aggregate, most likely by moving up an oxygen gradient (Figure 4). The cells on top of the mound also start to secrete cellulose and matrix proteins, which form a sheath around the structure, which eventually extends over the entire aggregate. The cells continue to chemotax toward the cAMP pulses emitted by the tip, but due to the restriction imposed by the sheath the cells are forced to move upward, forming a finger-shaped slug that due to gravity eventually falls over. At this point the slug begins to migrate toward warmth and light, which in nature would be the soil surface. These so-called thermotactic and phototactic responses are important, since localization of the fruiting body at the soil surface facilitates the dispersal of the spores.

While the mound is under construction the cells begin to differentiate into prestalk and prespore cells, which are initially interspersed. Also, here the nutritional status of the cells plays a role in determining the cells spore/stalk fate. The best-fed cells are destined to become spores, while the leaner ones become stalk cells. In nature this inhomogeneity in nutritional status occurs because the cells are at different stages of the cell cycle at the onset of starvation. Those that have just divided and then starve are naturally leaner, while those that divided a while ago have had a chance to build up resources.

About 75% of the cells express genes that will ultimately ensure their survival by causing them to transform into spores. The remaining 25% express genes that are associated with stalk cell differentiation. However, these prestalk cells retain many of the properties of aggregating cells, such as high chemotactive responsiveness and the ability to produce cAMP pulses. As a consequence, the prestalk cells move much more efficiently toward the cAMP pulses secreted by the tip than the prespore cells and thus the cells sort out to form an anterior–posterior prestalk–prespore pattern (Figure 4). While the slug migrates, the prestalk cells are effectively used up as they provide most of the motive force, and they are discarded from the rear of the slug. Meanwhile, the prespore cells at the posterior region dedifferentiate into cells with prestalk properties and these so-called anterior-like cells (ALCs) move forward to replace the lost prestalk cells.

At the onset of fruiting body formation, the slug points its tip upward and the prestalk cells start to construct a

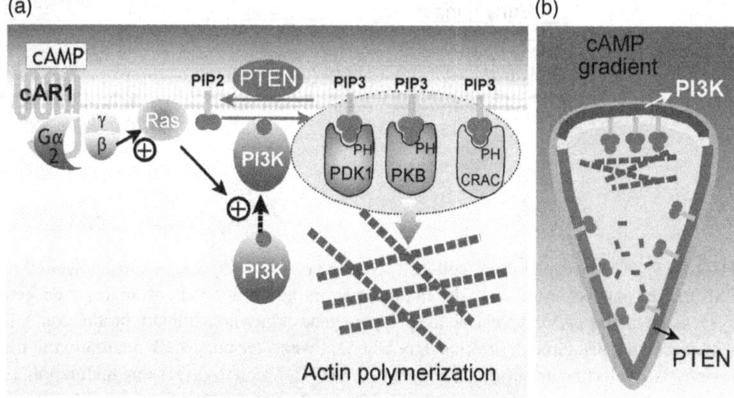

FIGURE 6 How to detect a cAMP gradient and move forward. Amoebas move up a gradient of cAMP by extending a pseudopod in the direction of the highest cAMP concentration and then pulling in their rear end. Pseudopod extension requires polymerization of the major cytoskeletal protein actin at the front of the cell, and depolymerization of actin at the rear. This is dependent on the relative concentrations of the membrane phospholipids PIP2 and PIP3 along the plasma membrane. High PIP3 activates processes that lead to actin polymerization, whereas high PIP2 levels cause depolymerization of actin. PIP3 is formed when PI3 kinase phosphorylates PIP2, and PIP2 is formed when the phosphatase PTEN dephosphorylates PIP3. Binding of cAMP to cARs initiates processes that bring PI3 kinase toward the membrane and move PTEN away, thus favoring production of PIP3 and instructing the cell where to polymerize its actin.

cellulose tube. They move up toward the tip, enter the tube from above, and differentiate into stalk cells. This process involves the construction of a cellulose cell wall, the formation of a central vacuole, and uptake of water to provide rigidity, and ultimately cell death. The prespore cells climb up the stalk and mature into spores, while the ALCs move both up and down through the spore mass to form the supporting structures of basal disk and upper and lower cups. The process of spore maturation involves a swift fusion of prespore vesicles with the plasma membrane forming the first layer of the three-layered spore coat.

Gene Expression and Cell Differentiation

Signals that regulate gene expression during development

Progression through a developmental program is due to regulated expression of genes at different stages, and in different positions in the developing embryo or multicellular structure. The newly synthesized proteins alter the phenotype of the cell, which enables it to proceed to the next developmental stage, and to function according to its specialization in the adult organism. This process of cell differentiation can be linear, that is, the cells change over time, or divergent, that is, a group of cells differentiates into multiple cell types. Although intrinsic factors, such as cell cycle phase in Dictyostelids, can set a bias for cell-type-specific differentiation, virtually all linear and divergent differentiation is controlled by cell–cell communication.

Figure 7summarizes the classes of genes that are expressed during *D. discoideum* development and the signals that control their expression. *Dictyostelium* development is initiated by starvation, but cells also secrete two glycoproteins – prestarvation factor (PSF) and conditioned medium factor (CMF) – that act as quorum sensing factors. Accumulation of these factors to a specific threshold informs the cells that they are present in sufficient numbers to make aggregation a viable option. Many species can individually differentiate into microcysts when this is not the case.

Combined with starvation, the quorum sensing factors downregulate growth genes and induce full expression of the early genes. They also induce low expression of aggregation genes, such as cAR1 and ACA, which allow cells to detect and secrete cAMP. The cells now start to secrete cAMP pulses in the nanomolar concentration range. In addition to triggering chemotaxis and cell aggregation, the cAMP pulses strongly accelerate the expression of aggregation genes, causing cells to aggregate rapidly and efficiently. cAMP pulses also induce competence, that is, components of novel signal transduction pathways for the gene expression events that occur after aggregation.

cAMP pulses continue after aggregation, but ACA is lost from most cells in the rear of the slug. These cells start to express a different adenylate cyclase, ACG, which produces a constant output of cAMP that soon reaches micromolar concentrations. This is the signal that triggers the expression of prespore genes. In turn the prespore cells begin to secrete DIF and DIF-like factors, which inhibit prespore differentiation and cause prespore cells to dedifferentiate into ALCs that eventually end up as stalk cells. However, DIF also induces its degrading enzyme DIF dechlorinase, which keeps DIF levels under control.

Another important signal for development is ammonia (NH_3). The starving cells and particularly the prestalk cell population degrade their own proteins as a resource of

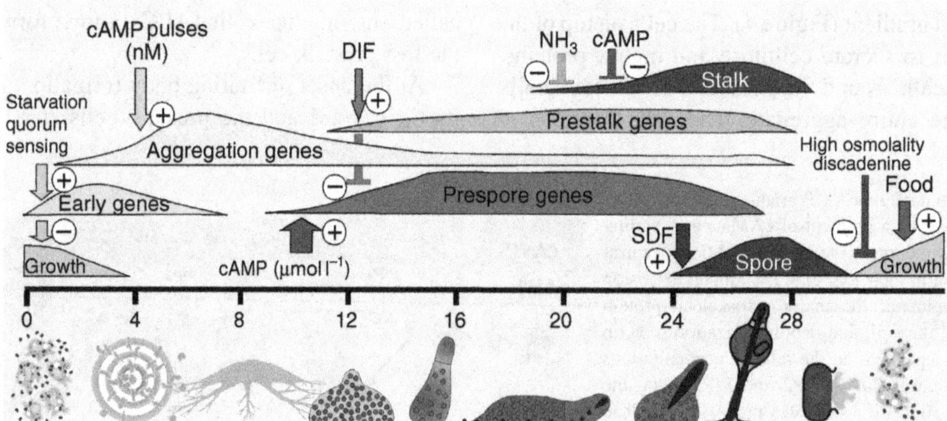

FIGURE 7 Gene regulation by cell–cell signaling during *Dictyostelium discoideum* development. Starvation and the quorum sensing factors CMF and PSF trigger expression of early development genes and low levels of aggregation genes. cAMP pulses further upregulate aggregation genes. Once aggregated higher cAMP levels induce prespore gene expression and DIF production. DIF in turn induces expression of some prestalk genes. Ammonia, which accumulates through protein degradation, prevents terminal stalk maturation in migrating slugs. Rising culminants lose ammonia by gaseous diffusion, stalk cells mature and produce a peptide, SDF-2, which triggers spore maturation. High osmolality and discadenine prevent spore germination in the spore head, but, once dispersed, spores germinate when exposed to food.

energy and metabolites throughout the developmental program. This causes production of large amounts of ammonia, which accumulates in the watery layer through which the slugs move. Ammonia has several roles. First, it prevents the ALCs from moving into the prestalk region. Once the prestalk cells are used up during migration and no longer produce ammonia, the inhibitory effect is relieved and the ALCs move to the front of the slug to take the place of the lost prestalk cells. Second, ammonia prevents the prestalk and prespore cells from maturing into stalk cells and spores as long as the slug is migrating.

However, all this changes during culmination. In nature, incident light causes the slug to point its tip upward and initiate the culmination process. The tip is now exposed to air and loses ammonia by gaseous diffusion. The loss of ammonia allows the prestalk cells to mature into stalk cells. The stalk cells cleave the protein acbA (Figure 8) that is secreted by prespore cells to form a peptide SDF-2 that triggers the maturation of the spores. Together, ammonia and SDF-2 make sure that stalk cells and spores are formed at the right time and position in the fruiting body.

The stalk cells die in the fruiting body, but the spore cells remain viable and propagate the species. To do so, it is vitally important that spores only germinate when the emerging amoebas are surrounded by food. Two factors, ammonium phosphate and a modified adenine called discadenine, are present in the spore head to prevent spore germination in situ. Ammonium phosphate acts by raising the osmolality in the spore head to high levels. High osmolality activates the osmosensor of the adenylate cyclase ACG, causing intracellular cAMP production and inhibition of spore germination.

Processing of developmental signals

In order to be able to act on gene expression, developmental signals need to be detected by receptors. These initiate a chain of events that eventually results in the activation or inactivation of transcription factors that control whether genes will be expressed. Through intensive study many individual components of such signal transduction chains have been identified. However, we do not have complete resolution of any pathway that controls developmental gene expression in *D. discoideum*.

For all its roles, extracellular cAMP is always detected by surface cARs, and G-protein-mediated activation of PI3-kinase is a common component of many pathways. However, some events such as the induction of prespore differentiation by cAMP do not require a G-protein and here the downstream pathway remains obscure. No receptor for DIF has yet been found, but in this case some of the target transcription factors have been identified.

Most remarkable is the fact that many of the other signals that control *Dictyostelium* development, such as PSF, ammonia, high osmolarity, and discadenine, do so by

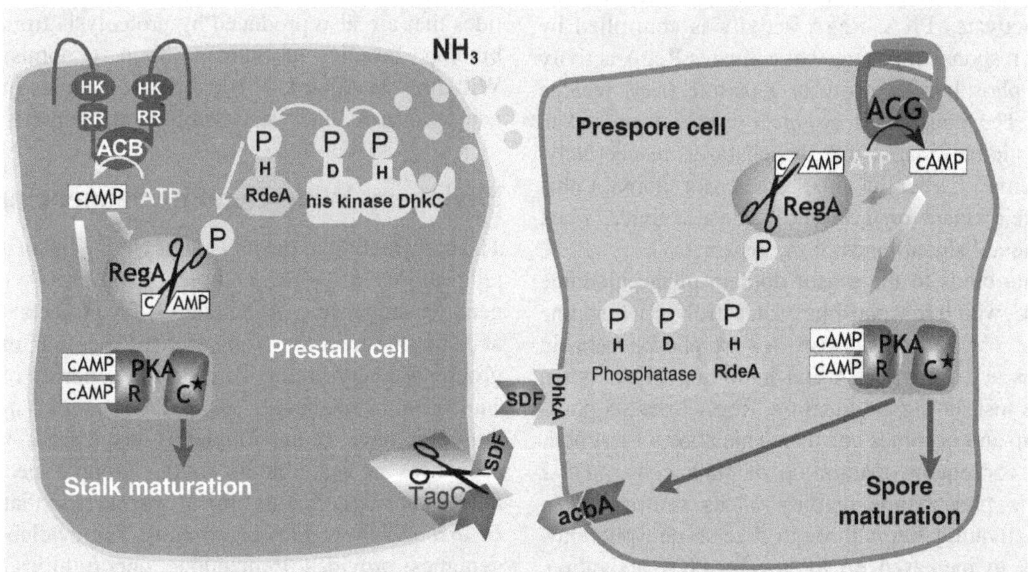

FIGURE 8 External signals regulate spore and stalk maturation by altering intracellular cAMP levels. Terminal spore and stalk maturation require high levels of PKA activation by cAMP. In prestalk cells cAMP is produced by ACB and in prespore cells by ACG. The cAMP phosphodiesterase RegA degrades intracellular cAMP but requires phosphorylation to be active. The phosphoryl group is provided by a sensor histidine kinase that can act as either a kinase or a phosphatase when a ligand binds to the sensor domain. The kinase autophosphorylates a histidine residue and, via a series of histidine to aspartate to histidine transfers, the phosphoryl group is carried over to RegA. Ammonia activates the histidine kinase activity of DhkC, thereby causing activation of RegA and preventing cAMP accumulation. The protease TagC that is expressed on prestalk cells cleaves the protein acbA that is released by prespore cells to yield SDF-2. SDF-2 activates the phosphatase activity of the histidine kinase DhkA. This results in dephosphorylation of RegA and consequent cAMP accumulation and PKA activation in prespore cells.

regulating the activation status of cAMP-dependent protein kinase or PKA. PKA activity is absolutely essential for initiation of development, the maturation of spores and stalk cells, and the control of spore dormancy. PKA is a deeply conserved sensor for cAMP in the eukaryotes. Its cAMP-binding domain is also conserved in prokaryotes, where it is part of the catabolite repressor protein, a cAMP-activated transcription factor. *Dictyostelium* and other protist PKAs consist of a single catalytic subunit (PKA-C) and a single regulatory subunit (PKA-R). Metazoan PKAs have two of each. PKA is activated when cAMP binds to the regulatory subunit, which then dissociates from the catalytic subunit, thus activating the enzyme.

PKA-C and PKA-R levels are low in growing cells and PKA activity is not required for growth. PKA-C is translationally upregulated at the onset of starvation in a process that is most likely activated by PSF. This upregulation is essential for initiation of development. At later stages PKA is regulated by enzymes that control cAMP levels in the cells. Enzymes that upregulate PKA activity are ACG in prespore and spore cells, and a third adenylate cyclase, ACB, which is expressed in prestalk cells. ACB has a similar domain architecture as some cyanobacterial adenylate cyclases, with histidine kinase, receiver, and response regulator domains located in front of the cyclase domain.

However, these adenylate cyclases are not the major focal points that mediate the effect of extracellular signals on PKA activity. That role belongs to an intracellular cAMP phosphodiesterase, RegA, which hydrolyzes cAMP and thereby inactivates PKA. RegA activity is controlled by an attached response regulator that activates RegA activity when it is phosphorylated on an aspartate (asp) residue (Figure 8). The phosphoryl group is provided and taken away by histidine kinases and phosphatases, respectively. These enzymes harbor an attached sensor domain that activates their kinase or their phosphatase activity, when a developmental signal binds to the sensor.

Ammonia binds to the sensor domain of the histidine kinase DhkC, which first phosphorylates itself on a histidine (his) residue. This sets the his–asp–his–asp phosphorelay in motion, thus activating RegA and inhibiting PKA. When ammonia is lost during culmination, RegA loses its phosphoryl group and becomes inactive. This allows PKA activation and consequent maturation of stalk cells. SDF-2 activates the phosphatase activity of its sensor DhkA, thereby inactivating RegA through reverse phosphorelay. This results in increased cAMP levels, PKA activation, and maturation of spores.

A total of 15 sensor-linked histidine kinases/phosphatases are present in the *D. discoideum* genome. Another two, DhkB and DokA, act to control spore germination, and a third DhkK the onset of culmination. The function of the remaining ten is unresolved. Apparently, cAMP signaling has a deeply conserved core function in the Dictyostelids and essentially serves to integrate a range of external stimuli. These stimuli act together to ensure that spores are formed at sites in the soil and positions in the fruiting body that are optimal for dispersal. They also ensure that spore germination only occurs under conditions that are optimal for growth.

RECENT DEVELOPMENTS IN THE *DICTYOSTELIUM* FIELD

The Sporulation Cascade

The fascinating cascade of signals and associated pathways that control the timely maturation of spores was further unraveled by Anjard and coworkers. The previously known components are presented in Figure 8 and consist of AcbA which is secreted by prespore cells and then processed by TagC to form SDF-2, which then acts on DhkA to activate PKA and trigger spore maturation. The new data show that both AcbA secretion and cell surface exposure of the TagC protease domain are triggered by GABA (γ-amino butyric acid) which is secreted by prespore cells in response to another secreted signal, the steroid SDF-3 (Anjard & Loomis, 2006; Anjard, Su, & Loomis, 2009). One of the most remarkable aspects of this cascade is that so many of its components are also found in mammalian neurotransmission. GABA is a well-known neurotransmitter and its *Dictyostelium* $GABA_B$-type receptor, GrlE, is conserved in mammals. SDF-2 is identical to endozepine neuropeptides that are also produced by proteolysis from acylCoA-binding protein in human brain (Loomis, Behrens, Williams, & Anjard, 2010). These examples indicate that neurotransmission is very deeply rooted in protist signaling.

Phylogeny Wide Genome Sequencing

The construction of the molecular phylogeny of all Dictyostelia shortly after the completion of the *D. discoideum* genome sequence sparked interest for projects to sequence at least one genome from each of the four major taxon groups of Dictyostelia. To date, the genomes of the group 1 and group 2 species *D. fasciculatum* and *Polysphondylium pallidum* have been completely sequenced, while draft sequences are available for another group 2 species *Acytostelium subglobosum* and for *D. purpureum* that similar to *D. discoideum* resides in group 4. The availability of the genomes provides tremendous opportunities to retrace how multicellularity evolved in the Dictyostelia.

Evolutionary History of cAMP Signaling

Comparative functional analysis of cAMP signaling genes across the *Dictyostelium* phylogeny provided evidence that the role of intracellular cAMP in induction of spore formation and inhibition of spore germination is evolutionary derived from similar roles in the encystation and excystation of solitary amoebas. The chemoattractant function of cAMP in *D. discoideum* and other group 4 species is evolutionary derived from a universal role of secreted cAMP in organization of fruiting body morphogenesis and induction of prespore gene expression. A double requirement for both secreted cAMP acting on cAMP receptors and intracellular cAMP acting on PKA is what distinguishes induction of sporulation from induction of encystation, which only requires the latter (Kawabe et al., 2009).

CONCLUSIONS

The Dictyostelid social amoebas have fascinated biologists for almost 150 years by their alternation between a protist-like growth phase, animal-like freely moving slugs, and fungal-like fruiting bodies. Combined with their ease of culture and excellent experimental and genetic tractability, this versatility makes them outstanding models to resolve a broad range of questions in molecular, cellular, and developmental biology. With the discovery that many human disease genes are conserved in Dictyostelids, investigation of the molecular basis of human afflictions has recently been added to this repertoire.

Apart from resolving these generally applicable questions, a reconstruction of the evolution of developmental signaling in the social amoebas from likely origins in environmental sensing in their solitary ancestors, combined with an understanding of the ecological factors that shaped this evolution, provides a unique opportunity to understand how and why multicellularity evolved.

This article provides a generalized overview of current knowledge of the biology of the Dictyostelids, summarizing studies of many of our colleagues. More detailed information about their work can be found in the References section.

REFERENCES

Anjard, C., & Loomis, W. F. (2006). GABA induces terminal differentiation of Dictyostelium through a GABA(B) receptor. *Development*, *133*, 2253–2261.

Anjard, C., Su, Y., & Loomis, W. F. (2009). Steroids initiate a signaling cascade that triggers rapid sporulation in Dictyostelium. *Development*, *136*, 803–812.

Kawabe, Y., Morio, T., James, J. L., Prescott, A. R., Tanaka, Y., & Schaap, P. (2009). Activated cAMP receptors switch encystation into sporulation. *Proceedings of the National Academy of Sciences of the United States of America*, *106*, 7089–7094.

Loomis, W. F., Behrens, M. M., Williams, M. E., & Anjard, C. (2010). Pregnenolone sulfate and cortisol induce secretion of acyl CaA binding protein and its conversion into endozepines from astrocytes. *Journal of Biological Chemistry*, *285*(28), 21359–21365.

FURTHER READING

Anjard, C., & Loomis, W. F. (2005). Peptide signaling during terminal differentiation of *Dictyostelium*. *Proceedings of the National Academy of Sciences of the United States of America*, *102*, 7607–7611.

Eichinger, L., Pachebat, J. A., & Gloeckner, G. (2005). The genome of the social amoeba *Dictyostelium discoideum*. *Nature*, *435*, 43–57.

Franca-Koh, J., Kamimura, Y., & Devreotes, P. (2006). Navigating signaling networks: Chemotaxis in *Dictyostelium discoideum*. *Current Opinions in Genetics and Development*, *16*, 333–338.

Insall, R., & Andrew, N. (2007). Chemotaxis in *Dictyostelium*: How to walk straight using parallel pathways. *Current Opinion in Microbiology*, *10*, 578–581.

Kessin, R. H. (2001). *Dictyostelium: Evolution, cell biology and the development of multicellularity*. Cambridge University Press: Cambridge, MA.

Loomis, W. F., & Kuspa, A. (2005). *Dictyostelium genomics*. Horizon Bioscience: Norfolk, UK.

Raper, K. B. (1984). *The dictyostelids*. Princeton University Press: Princeton, NJ.

Saran, S., Meima, M. E., Alvarez-Curto, E., Weening, K. E., Rozen, D. E., & Schaap, P. (2002). cAMP signaling in *Dictyostelium* – Complexity of cAMP synthesis, degradation and detection. *Journal of Muscle Research and Cell Motility*, *23*, 793–802.

Schaap, P. (2007). The evolution of size and pattern in the social amoebas. *BioEssays*, *29*, 635–644.

Schaap, P., Winckler, T., Nelson, M., *et al.* (2006). Molecular phylogeny and evolution of morphology in the social amoebas. *Science*, *314*, 661–663.

Swanson, A. R., Vadell, E. M., & Cavender, J. C. (1999). Global distribution of forest soil dictyostelids. *Journal of Biogeography*, *26*, 133–148.

Thomason, P., Traynor, D., & Kay, R. (1999). Taking the plunge – Terminal differentiation in *Dictyostelium*. *Trends in Genetics*, *15*, 15–19.

Williams, J. G. (2006). Transcriptional regulation of *Dictyostelium* pattern formation. *EMBO Reports*, *7*, 694–698.

Chapter 21

Foraminifera

J. Pawlowski
University of Geneva, Geneva, Switzerland

Chapter Outline

Abbreviations 291
Defining Statement 291
Introduction 291
Cytological and Morphological Characteristics 292
Granuloreticulopodia 292
Nuclei and Other Organelles 292
Test Morphology 293
Life Cycle and Reproduction 294
Ecology 295
Distribution and Abundance 295
Feeding Strategies 295
Symbiosis 295
Collection and Maintenance 297
Morphology-Based Classification 297
Molecular Phylogeny and Diversity 298
Phylogenetic Position 298
Macroevolutionary Relationships 300
Molecular Diversity 308
Evolutionary History and Geological Importance 308
Further Reading 309

ABBREVIATIONS

CCD Carbonate compensation depth
ITS Internally transcribed spacer
LSU rDNA Large subunit rDNA
MOV Motility organization vesicles
SSU rDNA Small subunit rDNA

DEFINING STATEMENT

Foraminifera are common aquatic single-cell eukaryotes, characterized by granuloreticulopodia and an organic, agglutinated, or calcareous test (shell) composed of a single or multiple chambers, although a few are 'naked.' Cells undergo a complex sexual/asexual life cycle, sometimes showing nuclear dimorphism. The group consists of about 5000 modern and 40000 fossil species.

INTRODUCTION

Foraminifera are a highly diverse group of mainly marine protists that are characterized by granulated reticulopodia and usually the presence of an organic, agglutinated, or calcareous test (shell). The foraminiferal test can be composed of a single or multiple chambers divided by septa. The name 'Foraminifera' refers to the internal openings (Latin: foramen, foramina) present in the septa of multichambered tests. The form of the test and its wall composition and structure are traditionally used in the higher-level classification of foraminifera. Recently, the morphology-based taxonomy has been challenged by molecular phylogenetic studies, but a new classification has not been established.

Foraminifera are common in all types of marine environments. They are a numerically important component of benthic meiofauna. A few foraminiferal species are common in marine plankton and some have adapted to freshwater environments. Almost all foraminifera, except the naked ones and those with an organic theca, are well preserved in the fossil record. The tests of planktonic species deposited at the sea bottom form an important layer of marine sediments called Globigerina ooze. Because of their abundance and diversity, the planktonic and benthic foraminifera are the most important group of microfossils, widely used as paleostratigraphic and paleoecological markers. The geological importance of foraminifera contrasts with the relatively limited knowledge of their biology. Very few foraminiferal species have been grown successfully in laboratory cultures and examined cytologically.

CYTOLOGICAL AND MORPHOLOGICAL CHARACTERISTICS

In many respects, foraminifera are unusual protists. Those having multichambered tests look like tiny multicellular organisms and it is not surprising that originally they were described as minute cephalopods. However, despite this particular appearance, the foraminifera are single-cell eukaryotes. Even in those having very complex test architecture, there is a continuous flow of cytoplasm between chambers, through internal openings or canals.

The size of foraminifera is also unusual for protists. Although the mean size of a foraminiferal cell ranges between 0.1 and 0.5 mm, some may reach up to several centimeters. The famous nummulites that form the limestone used to build the pyramids of Gizeh are the best-known example of large calcareous foraminifera. Large species are also common among agglutinated foraminifera living in the deep-sea and polar regions. On the other hand, there also exists a diverse, albeit poorly known, group of foraminifera smaller than 0.1 mm (micro- and nanoforaminifera).

GRANULORETICULOPODIA

The most distinctive structural features of foraminifera are the granuloreticulopodia. They form a unique branching and anastomosing network of granulated filose pseudopodia that allows foraminifera to interact with the environment. All essential functions of foraminiferal cells are accomplished with the help of granuloreticulopodia. They are indispensable for motility, attachment to substrates, and feeding of foraminiferal cells. They also participate in the transport of particles necessary for constructing the test, as well as for reproduction, respiration, and evacuation of metabolic waste products.

In testate foraminifera, reticulopodia extend either through principal or through secondary apertures, or originate from a layer of cytoplasm enveloping the test. Bidirectional movement of granuloreticulopodia is ensured by a dynamic system of microtubules inside the pseudopodia. Microtubules may occur singly or as loosely organized bundles. Like in other eukaryotes, they are formed by molecules of tubulin, which differ, however, in many aspects from those found in other eukaryotes.

A characteristic feature of foraminiferal reticulopodia is the presence of granules moving on the surface and inside the pseudopodia. The identity and function of cytoplasmic granules is not always well understood. The most prominent pseudopodial granules are mitochondria that are essential for aerobic respiration and are probably responsible for transporting metabolic energy within the reticulopodial network. Among other granules whose function has not yet been entirely elucidated, there are 'dense bodies,' clathrin-coated pits and vesicles, 'ellipsoid, fuzzy-coated vesicles,' and 'motility organization vesicles' (MOVs). There are also various types of vacuoles, including phagosomes that contain food particles to be digested and xanthosomes that contain undigested material and metabolic wastes.

NUCLEI AND OTHER ORGANELLES

Some foraminifera possess a single nucleus, which can be very large, up to 100 μm in diameter in *Astrammina rara*. In other species, such as *Reticulomyxa filosa*, there are hundreds of small nuclei loosely dispersed in the cytoplasm (Figure 1). In those foraminifera that undergo an alternation

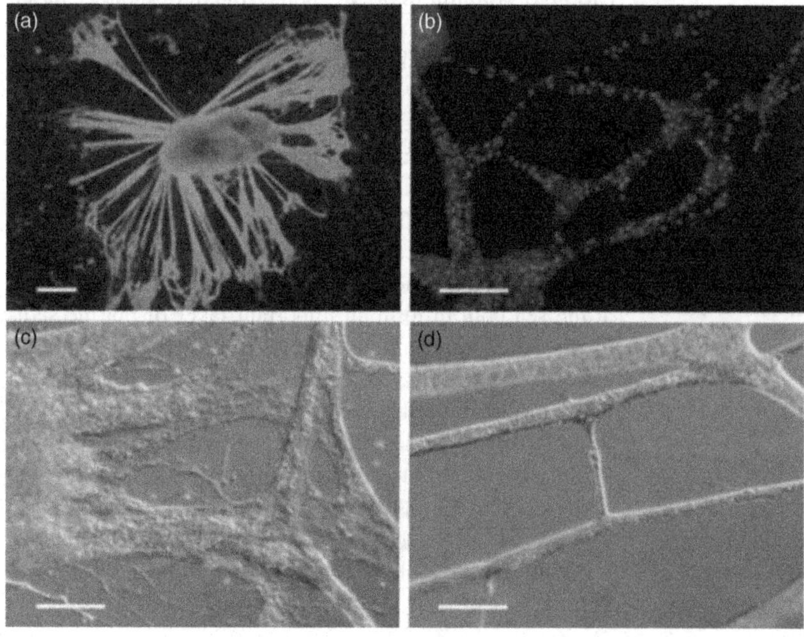

FIGURE 1 *Reticulomyxa filosa* – a naked freshwater foraminifer. (a) An amoeboid cell body, (b) multiple nuclei inside the cell body (DAPI staining), (c) a network of granuloreticupodia, and (d) a detail of anastomosing reticulopodia. Scale bar = 1 mm (a) and 0.05 mm (b–d). *Photo courtesy of Dr. José Fahrni.*

of generation, the number of nuclei is related to generation type. The sexually reproducing gamonts are usually uninuclear, while the asexually reproducing agamonts and schizonts are multinuclear.

Nuclear dimorphism is observed in some foraminifera, and particularly well documented in the family Rotaliellidae. In these heterokaryotic foraminifera, the multiple nuclei of the agamont can be differentiated into somatic and generative types. The somatic nucleus is usually single, larger, and vesicular in form, with many nucleoli. Generative nuclei are more numerous, smaller, and spherical in shape. The generative nuclei are usually found in the central part of the test, while the somatic ones occupy the peripheral chambers. During meiosis, the somatic nucleus disintegrates and only the generative nuclei participate in the formation of gametes. The somatic nucleus seems to have mainly a metabolic function. If it is eliminated experimentally, one of the generative nuclei moves to the peripheral chambers and transforms itself into a somatic one.

Little is known about the DNA content and genome features in foraminifera. In the few species in which chromosomes have been studied, their number ranges from 6 to 24. The DNA content of the nucleus in *A. rara* is estimated to be about 2 ng. This giant nucleus probably contains a large number of genome copies. Although endoreplication has been reported in several foraminiferal species, there is no precise information about their genome size.

In addition to nuclei, the dominant elements in foraminiferal cytoplasm are large digestive vacuoles that are distributed throughout much of the cytoplasm. They are formed outside the test by invagination of pseudopodial membrane around the engulfed prey cells. Digestion occurs typically in intrashell cytoplasm, although there is evidence that it may start earlier in extrashell cytoplasm, at least in some benthic species.

Intrashell and extrashell cytoplasm also contain numerous small mitochondria (1–2 μm) that possess the tubular cristae typical of most protists. They are widely distributed throughout the intrashell cytoplasm, but their density may be higher close to the pores in species living in low-oxygen environments. The intrashell cytoplasm also contains microbodies such as peroxisomes, which serve to degrade toxic compounds and convert metabolic products. In addition, specific organelles such as annulate lamellae and fibrillar bodies have been observed in some planktonic species.

TEST MORPHOLOGY

Despite some 'naked' species, such as *R. filosa* (Figure 1), in which the cell body is covered with a thick glycocalyx, the other foraminifera possess an organic, agglutinated, or calcareous test. The organic test or theca is composed of proteinaceous material, although its precise chemical composition is unknown. The agglutinated test consists of sand grains and other mineral particles embedded into either an organic lining or a calcareous cemented matrix. The calcareous test is composed of secreted calcitic or aragonitic crystals. They can be arranged in one orientation to form a hyaline, transparent wall, consisting of one layer (monolamellar) or two layers separated by an organic lining (bilamellar). Alternatively, they can be randomly oriented and form an opaque 'porcellanous' test. In some foraminifera, the wall is formed of optically monocrystalline calcite, while others are covered with calcareous rod-like structures. Finally, a small group of foraminifera possess a siliceous test.

Test morphogenesis differs in various foraminiferal taxa. The single-chamber species grow by adding new shell material near the aperture or by periodically shedding the test and rebuilding a new one. The growth of multichambered species is accomplished by adding a new chamber. Typically, the process starts by formation of a cyst, within which the granuloreticulopodia build an anlage of the new chamber. Its surface is covered by an inner organic lining and filled by cytoplasm. A layer of calcite is then deposited over the new chamber and the entire test. In some species, the inorganic carbon for calcification is obtained directly from seawater, while the others store large reserves of carbonate in special vesicles prior to calcification.

The form of the test can vary from very simple shapes in unilocular (single-chamber) species to more complex architectures in multilocular (multichambered) foraminifers. The organic-walled foraminifera are usually spherical, ovoid, or elongate. The agglutinated unilocular species may have similar simple shapes or may build complex branching structures. In multilocular species, the form of the test varies depending on chamber arrangement. The test is called uniserial, biserial, or triserial, when the chambers are arranged in single, double, or triple series, respectively. Coiling tests can be either planispiral or trochospiral; the planispiral coil is called evolute when all chambers are visible from both sides or involute when only those of the last whorl are visible. In some foraminifers, the test is composed of only two chambers: the first globular chamber (proloculus) is followed by a second tubular one.

There are different types of openings in the foraminiferal test, which ensure communication between the cell body and the exterior. The majority of testate foraminifera possess a single large aperture, through which the main granoloreticulopodial trunk emerges. This aperture is usually situated at the periphery of the last chamber, but may also be placed near the suture separating the last chamber. When a new chamber is built, this main aperture is transformed into an internal foramen that connects consecutive chambers. Some calcareous species also possess multiple secondary apertures. Moreover, in all calcareous orders, except miliolids, the wall is perforated with tiny openings (pores) which increase exchanges between the cell body and the environment.

The large multilocular calcareous species sometimes develop complex internal architecture. Their chambers can be subdivided into chamberlets. These internal partitions increase the strength of the test, providing better resistance to predation and breakage. They also allow the thinning of the outer walls of the test, which is interpreted as an adaptation for better light penetration in species bearing algal symbionts. Moreover, some foraminifera are characterized by complex canal systems within their test walls. These canals enhance communication between the chambers and the exterior by allowing the extrusion of pseudopodia from any point on the surface of the test. In large foraminifers, the canal system plays an important role in motility, chamber formation, reproduction, and excretion – functions that could be difficult to perform using only a single aperture.

LIFE CYCLE AND REPRODUCTION

The classical life cycle of foraminifera consists of a regular alternation of an asexually reproducing agamont and a sexually reproducing gamont (Figure 2). The gamont is usually uninucleate, while the agamont is multinucleate. The gamont produces amoeboid or flagellated gametes, which fuse to form zygotes. Zygotes develop into diploid agamonts. When the agamonts reach maturity, their nuclei undergo two meiotic divisions. Embryonic gamonts are formed by multiple fission of the agamontic cytoplasm. Because of the intermediary position of meiotic division, which intervenes before multiple fission of the agamont, the gamont phase is haploid while the agamont phase is diploid. This type of alternation of generations is called heterophasic and is known mostly in plants.

In some foraminifera, the classical cycle is modified, as illustrated in Figure 2. More or less regularly, the agamont undergoes mitotic division instead of meiosis and produces another asexually reproducing generation, called schizont or apogamic agamont. The schizont can either undergo meiosis leading to formation of a gamont or it may enter a cycle of successive asexual reproductions during which the new generation of schizonts is produced by schizogony, that is, by multiple fission of a multinucleated parental cytoplasm.

Three types of sexual reproduction are known in foraminifera: gametogamy, gamontogamy, and autogamy. In gametogamy, the biflagellated and motile gametes are released into the surrounding seawater and mating occurs outside the gamontic test. During gamontogamy, also called plastogamy, two or more gamonts join together with their umbilical faces; the flagellated or amoeboid gametes fuse within a limited space formed after a partial dissolution of the umbilical sides of the gamontic tests. In autogamy, which is a form of self-fertilization, the gametes produced by the same gamont fuse inside the gamontic test.

The life cycle of foraminifera is often associated with the concepts of dimorphism and trimorphism of foraminiferal tests, widely used in micropaleontological research. According to the classical concept of foraminiferal dimorphism, the gamont is smaller but has a larger proloculus (initial chamber of the polythalamous test), compared to the agamont, which has a smaller proloculus, but a larger test size than the gamont. The gamont is called megalospheric, while the agamont is called microspheric. The trimorphic concept suggests that in some species the

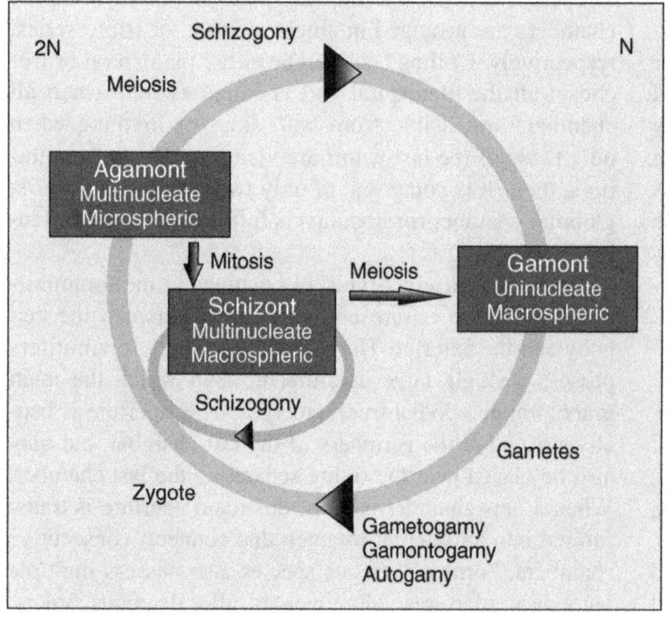

FIGURE 2 Schema of foraminiferal life cycle.

microspheric agamonts produce megalospheric forms that are not gamonts, but asexually reproducing schizonts. Both megalospheric forms are distinguishable by the size of proloculus, which is smaller in agamont as compared to that in a schizont. The differences in proloculus size between megalospheric and microspheric forms are generally explained as a result of the larger size of embryos produced by multiple fission of the cytoplasm compared to the smaller size of zygotes formed by the fusion of gametes. However, the volume of embryos produced by multiple fission of the cytoplasm also depends on the volume of parental cytoplasm and the number of daughter nuclei.

ECOLOGY

Distribution and Abundance

Foraminifera are ubiquitous in all marine habitats. They are widely distributed at all depths and all latitudes, including the extreme environments such as deep-sea trenches and ice-covered polar regions. Some 'naked' foraminifers are present in freshwater settings and one organic-walled species has been described from Australian rain forest soil. However, the diversity of these freshwater and terrestrial foraminifers is poorly known.

Although some foraminifera adopted the planktonic mode of life, most species are benthic. Their distribution is influenced mainly by abiotic factors such as temperature, salinity, oxygen, light, substrate type, turbidity, and nutrients. Different species assemblages are characteristic of different water depths, habitats, and latitudes. Calcareous species dominate in shallow-water and bathyal zones. Their depth distribution is limited by the carbonate compensation depth (CCD). Below this depth, the foraminiferal assemblage is dominated by organic-walled and agglutinated species.

Numerical abundance of foraminifera can be very high. The large calcareous species (soritids, nummulitids, calcarinids, and amphisteginids) are so abundant that they literally pave the seafloor of tropical and semitropical shallow-water habitats. It has been estimated that reef foraminifera annually generate approximately 43 million tons of calcium carbonate; this figure corresponds to about 5% of the global carbonate reef budget. The carbonate production of planktonic foraminifers averages 3.5 g m^{-2} $year^{-1}$. Both planktonic and benthic foraminifers contribute roughly 20% of the global carbonate production.

On the other hand, organic-walled and agglutinated benthic foraminifera are a major component of the total biomass of meiofauna in the deep-sea and high-latitude settings. More than 400 organic-walled allogromiids per 10 cm^2 have been found in the Challenger Deep (Pacific Ocean) at 10 896 m. Macrofaunal-size xenophyophores, shown recently to belong to foraminifera, dominate certain deep-sea benthic communities, with more than 100 specimens occurring per 100 m^2. Highly diverse and abundant assemblages of monothalamous species have been found in the Antarctic shelf waters.

Feeding Strategies

During their long evolutionary history, foraminifera have developed various trophic mechanisms, including grazing, suspension and deposit feeding, carnivory, and parasitism. Those that live within the photic zone commonly feed by grazing on algal cells and bacteria that cover the surface on which the foraminiferal cells move. The species living below the photic zone feed mostly on phytodetritus, exploiting the organic matter and microbiotas associated with it. Many benthic species surround themselves with balls of food (feeding cysts) which they gather with their pseudopodia. Some species develop large arborescent or tubular structures, from which they extend pseudopodia to trap food particles present in the water column. Although foraminifera cannot create water currents and therefore are passive suspension feeders, some of them may use the feeding currents created by invertebrates, on the surface of which they are dwelling.

Some benthic and planktonic foraminifera are carnivorous. They can capture animals as large as 2–3 cm, including small crustaceans and larvae. In laboratory cultures, many planktonic species are commonly fed on copepods and brine shrimp larvae, but it is not known to what extent this feeding strategy is used in nature. The pseudopodia of carnivorous forms are specially adapted for capturing prey. In some of them, for example, *A. rara*, pseudopodia secrete an extracellular matrix material that functions to strengthen pseudopodia and maintain their integrity. They may also have the ability to secrete an adhesive material.

Parasitism in foraminifera has been described in only a few species. The parasitic foraminifera usually live on the shells of other foraminifers or invertebrates. Some of them, for example, *Hyrrokkin sarcophaga*, which penetrates the shell of bivalves, sponges, and stone corals and feeds on their soft tissues, seem to be an obligatory parasite. However, other ectoparasitic foraminifers seem to feed by grazing and by suspension also.

At least some foraminifera are selective feeders. In an experimental study of trophic dynamics, it has been found that of several dozen diatoms and chlorophytes tested as food for foraminifera, only four or five species were eaten in significant amounts. A tiny calcareous species, *Rotaliella elatiana*, has been maintained in laboratory culture for more than 10 years by feeding on the macrophytic alga *Enteromorpha*.

SYMBIOSIS

Many large calcareous benthic foraminifera, as well as some planktonic species, host algal symbionts belonging to the rhodophytes, chlorophytes, dinoflagellates, chrysophytes,

TABLE 1 Foraminiferal hosts and their symbionts

Order	Family	Symbiont class	Symbiont genus
Miliolida	Alveolinidae	Bacillariophyceae	?
	Peneroplidae	Rhodophyceae	*Porphyridium*
	Soritidae/Archaiasinae	Chlorophyceae	*Chlamydomonas*
	Soritidae/Soritinae	Dinophyceae	*Symbiodinium*
Rotaliida	Amphisteginidae	Bacillariophyceae	?
	Calcarinidae	Bacillariophyceae	?
	Nummulitidae	Bacillariophyceae	*Thalassionema*?
Globigerinida	Globigerinidae	Dinophyceae	*Gymnodinium*
		Chrysophyceae	?
	Globorotaliidae	Chrysophyceae	?
	Candeinidae	Chrysophyceae	?

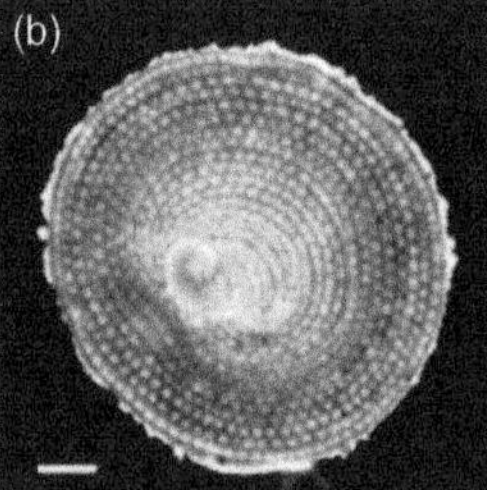

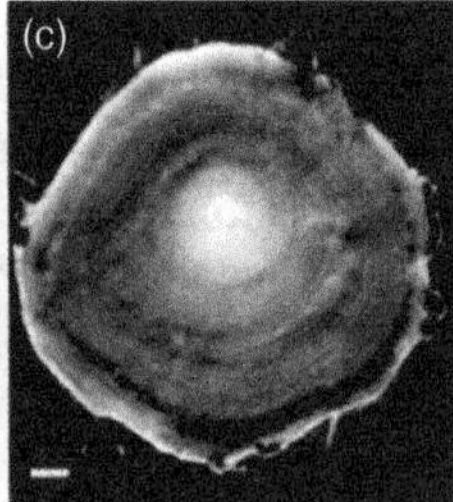

FIGURE 3 Symbiont-bearing soritid foraminifers (a) *Peneroplis planatus*, (b) *Parasorites* sp., and (c) *Marginopora vertebralis* with rhodophytes, chlorophytes, and dinoflagellates, respectively. Scale bar = 0.1 mm.

and diatoms (Table 1). Algal symbiosis appears to have arisen independently in different foraminiferal groups. Among benthic foraminifera, the diatom symbionts have been acquired independently in some miliolid (Alveolinidae) and some rotaliid families. In the evolutionary history of Soritacea, there were consecutive changes of symbionts from rhodophytes in the Peneroplidae to chlorophytes in the Archaiasinae, and *Symbiodinium*-like dinoflagellates in the Soritinae (Figure 3). In planktonic foraminifera, the symbionts are present in most of the tropical and subtropical species. As a general rule, symbioses with chrysophytes appear to be facultative.

The broad range of endosymbionts suggests that foraminifera are particularly good habitats for the establishment and maintenance of algal symbiosis. In fact, the morphological complexity of large foraminifera is often interpreted in terms of adaptation to endosymbiosis. The exposure of algal cells to appropriate light levels is promoted by increasing the size of the foraminiferal test and by its subdivision into chamberlets, which allows the outer walls to be thinner and more transparent. At the same time, the regionalization of foraminiferal tests presumably protects symbionts from host digestive activities.

In spite of the wide diversity of algal symbionts in foraminifera, their symbiotic relationships seem relatively specific. Recent molecular studies show that each of the examined families (Archaiasinae, Soritinae, and Nummulitidae) bears closely related symbionts. All symbionts in Archaiasinae belong to a single-clade sister to *Chlamydomonas* sp., while all symbionts in Nummulitidae group together with the genus *Thalassionema*. The dinoflagellate symbionts found in Soritinae belong to four different types of the *Symbiodinium* complex. Interestingly, only one of these types is present in *Symbiodinium*-bearing corals and other invertebrates living in the same habitat as soritids.

Algal symbiosis in larger foraminifera is often viewed as an adaptation for survival and growth in the extremely oligotrophic tropical and subtropical seas. Foraminiferal hosts are completely dependent on their algal endosymbionts for growth. It has been demonstrated that they will not grow if they are incubated in the dark or when the symbionts have been experimentally removed. It is assumed that algal symbiosis provides foraminiferal hosts with substantial energetic advantages, promotes calcification, and plays a role in removing host metabolites. Yet, the physiological mechanisms involved in foraminiferal endosymbiotic relationships are not well understood, and there is little evidence to support these advantages.

While algal symbiosis characterizes only large benthic and planktonic foraminifera, some small-size benthic species belonging to the families Elphidiidae, Nonionidae, and Rotaliellidae have been shown to sequester and house ingested chloroplasts (kleptoplasts). Ultrastructural studies and photosynthetic pigment analyses suggest that the chloroplasts are of diatom origin, but it is not known whether there is any specific relationship between the chloroplast donors and foraminiferal hosts. The chloroplasts are functional for at least a limited period of time and their half-life varies from 2 to more than 10 weeks depending on the species.

Some foraminifera are also hosts to bacteria. In particular, bacterial symbionts are common in foraminiferal species living in anoxic conditions. Some of them, for example, *Buliminella tenuata*, have numerous rod-shaped bacteria in their cytoplasm. In others, for example, *Nonionella stella*, the bacteria are found inside the test but not intracellularly. Bacterial endobionts and kleptoplasts coexist in a benthic foraminifer *Virgulinella fragilis*, which lives in sulfide-enriched environments. Recently, some foraminifera have been found to be capable of denitrification. The question whether this activity is related to bacterial symbiosis is disputable.

COLLECTION AND MAINTENANCE

Foraminifera are relatively easy to collect and maintain but much more difficult to establish in long-term laboratory culture. To collect foraminifera from coastal epibiotic communities, algae, plants, or small stones should be gently rubbed and washed to dislodge the specimens. Benthic foraminifera from deeper water can be collected by coring and examining the surface layer (1–2 cm) of sediments. In both cases, the collected material can be sieved through coarse (0.5–1 mm) sieves to separate foraminifera from large mineral material, macroalgae, and invertebrates.

Benthic foraminifera can survive for several weeks or months in crude cultures together with other microorganisms collected from the same site. The culture dishes with foraminifera are placed at room temperature, in controlled light with a diurnal cycle, and occasionally fed with algae (diatoms or chlorophytes). To promote their growth, it is recommended to maintain them in Erdschreiber medium composed of seawater enriched with soil extract, sodium phosphate, and sodium nitrate. The culture medium and seawater should be filtered and changed weekly. However, attempts to obtain continuously growing and reproducing monocultures of benthic foraminifera are often unsuccessful. There are less than 30 species for which such cultures have been established and most of them have lasted for a relatively short time. Usually, the cultured foraminifera only reproduce asexually, and few of them are able to complete their life cycle in laboratory conditions. The majority of them are tiny species that reproduce sexually by autogamy or gamontogamy and complete their life cycle within 1–2 weeks. One notable exception is *Heterostegina depressa*, a large benthic foraminifer, for which the complete life cycle has been observed in specimens cultivated in aquaria for several years. Planktonic foraminifera can also be maintained in laboratory culture by feeding on *Artemia* larvae or other animal prey. Some cultured planktonic species undergo gametogenesis. However, their gametes do not fuse in culture conditions and all attempts to continuously reproduce planktonic species have so far been unsuccessful.

Although, none of the foraminiferal species is currently available from culture collections (ATCC, CCAP), the cultures of the freshwater naked species *R. filosa* as well as different strains of *Allogromia* spp. can be obtained from laboratories working on the biology of foraminifera.

MORPHOLOGY-BASED CLASSIFICATION

In traditional morphology-based classification of foraminifera, the high-level taxonomic groups are distinguished according to the composition and structure of the test wall, presence/absence of an internal toothplate, and benthic versus planktonic adaptation. The most recent ordinal classification recognized 16 orders, which can be grouped, according to their wall composition, into five broad classes: (1) the organic-walled (thecate) allogromiids; (2) the agglutinated astrorhizids, lituolids, trochamminids, and textulariids; (3) the calcitic rotaliids, buliminids, miliolids, fusulinids, carterinids, lagenids, and globigerinids; (4) the aragonitic robertinids and involutinids; and (5) the siliceous silicoloculinids (Table 2).

The criteria used to distinguish superfamilies and families are mainly the arrangement of chambers, form of the test, position and form of the main aperture, presence of secondary apertures, and complexity of the interior structures. There are 164 extant families recognized. With 52 families, the rotaliids form the most diverse group, followed by agglutinated Lituolida (28) and calcitic Buliminida (22). The Allogromiida are probably also very diverse but they are much less well known than the others, because their organic-walled tests do not fossilize. The orders Carterinida, Silicoloculinida, and Involutinida are represented by only one family and one genus each, while the order Fusulinida is extinct (Figures 4–6).

TABLE 2 Morphology-based ordinal classification of foraminifera

Order	Wall	Other characteristic features	Modern genera
Allogromiida	Organic	Single-chamber	53
Astrorhizida	Agglutinated	Single-chamber	80
Lituolida	Agglutinated	Planispiral or serial arrangement of chambers	114
Trochamminida	Agglutinated	Test trochospiral	40
Textulariida	Agglutinated with particles cemented by low-Mg calcite	Test uniserial, biserial, triserial, or trochospiral; wall canaliculate	35
Fusulinida	Microgranular, calcitic	Extinct	0
Miliolida	High-Mg calcite, imperforate, porcelaneous	Single-, bi-, or multichambered, chambers arranged in one or several planes	130
Carterinida	Low-Mg calcite, imperforate, with large spicules	Test trochospiral	1
Spirillinida	Low-Mg calcite, optically a single crystal	Test planispiral or trochospiral	8
Lagenida	Low-Mg calcite, monolamellar	Single- or multichambered, with serial or planispiral chamber arrangement	120
Rotaliida	Low-Mg calcite, bilamellar	Low trochospiral, planispiral or annular chamber arrangement	224
Buliminida	Low-Mg calcite, bilamellar	High trochospiral or serial chamber arrangement, aperture usually with internal toothplate	86
Globigerinida	Low-Mg calcite, bilamellar	Planktonic	29
Involutinida	Aragonitic	Bichambered test	1
Robertinida	Aragonitic, perforate	Multichambered	15
Silicoloculinida	Opaline silica, imperforate	Multichambered, miliolid coil	1

Modified from Sen Gupta, B. K. (1999). *Modern foraminifera*. Dordrecht: Kluwer Academic; Debenay, J.-P., Pawlowski, J., & Decrouez, D. (1996). *Les foraminifères actuels*. Paris: Masson.

The taxonomic system for foraminifera is currently undergoing dramatic changes based on molecular phylogenetic data (see below). However, a new classification based on molecular and morphological data is not yet completed. The family-level classification presented here (Table 3) reflects the morphological diversity within each order but may not be congruent with phylogenetic studies.

The number of modern morphospecies is estimated between 5000 and 10000. The great majority of them are described solely on the basis of morphological features. Species are morphologically distinguished by the size and form of the test, number and form of chambers, shape of sutures, and diverse ornamentation structures. Descriptions of very few species refer to cell characteristics, reproductive features, or ecological adaptations. As in many other groups, foraminiferal taxonomy is overloaded with a large number of synonymous names. Moreover, due to the plasticity of tests in some species, which can change morphology depending on environmental factors, it is particularly difficult to distinguish the morphospecies from the ecophenotypes. The use of molecular data has helped resolve some of these problems, but species definition in many foraminifera remains very unclear (see below).

MOLECULAR PHYLOGENY AND DIVERSITY

Recent application of molecular tools to study foraminifera has shed new light on their evolutionary history and systematics. Molecular data have been particularly useful in (1) revealing the phylogenetic position of foraminifera within the eukaryotic tree, (2) revising the phylogenetic relationships between higher-level taxonomic groups, and (3) uncovering numerous cryptic species that were undistinguishable on the basis of classical morphological features.

PHYLOGENETIC POSITION

Traditionally, foraminifera were classified in the subphylum Sarcodina, within the class Granuloreticulosea, together with all other protists having granuloreticulopodia. The

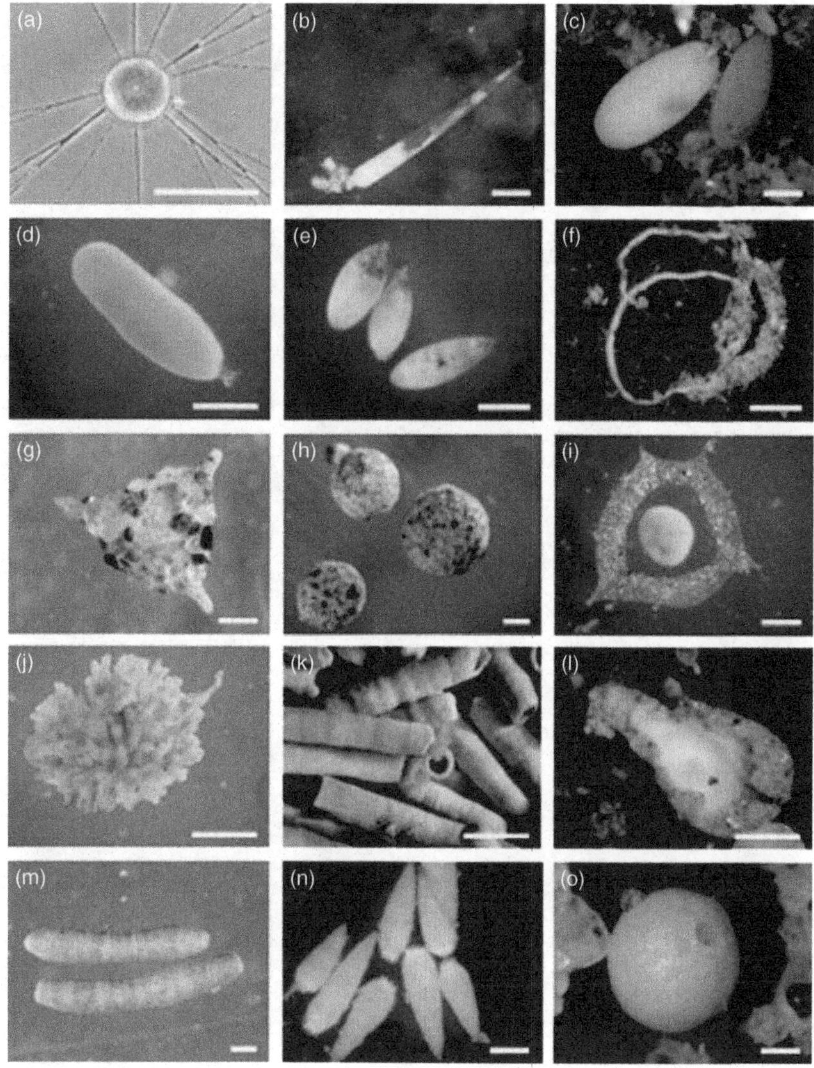

FIGURE 4 Monothalamous foraminifera: (a) *Allogromia* sp., (b) *Micrometula* sp., (c) *Gloiogullmia* sp., (d) undetermined allogromid, (e) *Psammophaga* sp., (f) *Nemogullmia* sp., (g) *Astrammina triangularis,* (h) *Saccammina sphaerica,* (i) *Vanhoeffenella* sp., (j) *Komokia* sp., (k) *Bathysiphon flavidus,* (l) *Crithionina delacai,* (m) *Hippocrepinella hirudinea,* (n) *Hippocrepina indivisa,* and (o) Silver saccaminid. Scale bar = 0.2 mm, except for f, j, k, l = 0.5 mm.

Granuloreticulosea have been divided into naked Athalamae, single-chambered, nontestate Monothalamae, and single- or multichambered, testate Foraminifera. Molecular phylogenetic studies confirmed the monophyly of this class. However, because the 'naked' granuloreticulosean protists have been shown to branch within foraminiferan clade, it has been proposed to include all of them into the phylum of Foraminifera.

In the new classification of eukaryotes, the foraminifera are included in the supergroup of Rhizaria. In addition to foraminifera, the Rhizaria include various amoeboid protists that have been traditionally classified among the Rhizopoda (euglyphid testate ameobae, gromiids, chlorarachniophytes) and Actinopoda (desmothoracid and taxopodid heliozoans and three classes of radiolarians: Phaeodarea, Acantharea, and Polycystinea). They also comprise some groups of taxonomically enigmatic amoeboid and flagellate protists, such as cercomonads and the parasitic haplosporidians and plasmodiophorids.

The majority of Rhizaria are characterized by the presence of reticulate pseudopodia, but any clearly defined morphological or ultrastructural synapomorphy for this supergroup remains to be discovered. At present, the most compelling molecular evidence for the monophyly of Rhizaria is a single or double amino acid insertion at the junction of the polyubiquitin monomers. The monophyly of Rhizaria is also strongly supported by actin and RNA-polymerase phylogenies, as well as by most recent phylogenomic analyses, including more than 100 protein-coding genes. These analyses suggest that Rhizaria are closely related to Stramenopiles and Alveolates, with whom they form the SAR assemblage.

The relationships within Rhizaria are not well resolved. Various flagellated genera, together with the euglyphids, phaeodarians, desmothoracids, chlorarachniiophytes, and

FIGURE 5 Polythalamous foraminifera (a–c) Miliolida, (d) Spirillinida, (f) Lagenida, (g–i) Lituolida, (j) Textulariida, (k, m, and o) Rotaliida, (l and n) Buliminida). (a) *Cornuspira antarctica*, (b) *Pyrgo elongata*, (c) *Triloculinella antarctica*, (d) *Patellina* sp., (e) *Spirillina* sp., (f) *Fissurina* sp., (g) *Reophax hoeglundi*, (h) *Hormosina normani*, (i) *Cyclammina cancellata*, (j) *Clavulina communis*, Rotaliida, (k) *Pullenia subcarinata*, (l) *Trifarina earlandi*, (m) *Globocassidulina biora*, (n) *Brizalina* sp., and (o) *Hyrrokkin sarcophaga*. Scale bar = 0.2 mm.

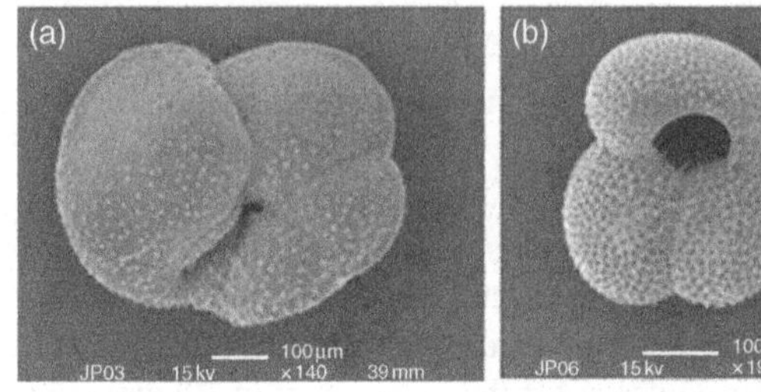

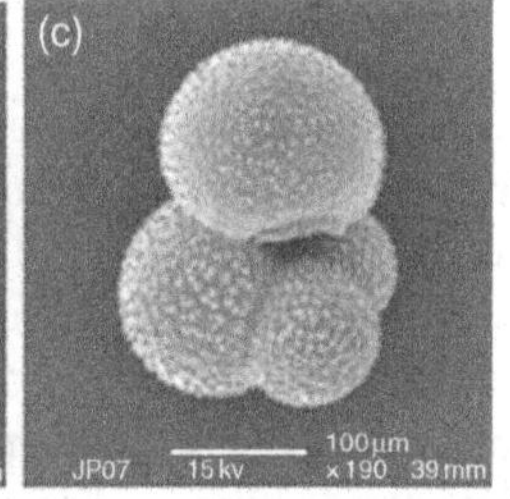

FIGURE 6 Planktonic foraminifera (a) *Globorotalia hirsuta*, (b) *Globigerinoides ruber*, and (c) *Globigerinita glutinata* (SEM views of umbilical side).

ebriids form the clade Cercozoa. Parasitic Phytomyxea and Haplosporidia, as well as Gromiida, Aconchulinida, and Reticulosida, as well as many new environmental lineages branch at the base of this clade. Foraminifera branch with Radiolaria (Acantharea), according to some rDNA-based and recent phylogenomic analyses. The root of Rhizaria is either between the clade Foraminifera+Radiolaria and other Rhizaria, or more probably at Phytomyxea or one of the lineages branching at the base of Cercozoa.

MACROEVOLUTIONARY RELATIONSHIPS

Phylogenetic analyses based on sequences of the 18S rDNA, actin, β-tubulin, and RNA-polymerase genes show a relatively congruent view of foraminiferal macroevolution. In all phylogenetic trees, the monothalamous (single-chamber) species appear as a basal paraphyletic group, from which the polythalamous (multichambered) lineages emerged independently 2–3 times (Figure 7).

TABLE 3 Morphology-based family-level classification of foraminifera

Order Allogromiida – wall organic, test unilocular	
	Family Lagynidae – test small with single or multiple apertures, biflagellate gametes (*Myxotheca, Boderia*)
	Family Allogromiidae – test globular, with single aperture (*Allogromia*)
	Family Shepheardellidae – test tubular, elongate with apertures at both ends (*Nemogullmia, Tinogullmia, Shepheardella*)
Order Astrorhizida – wall agglutinated, test unilocular	
Superfamily Astrorhizacea – test globular, domed, or tubular (branched or unbranched), sometimes irregularly multichambered with incomplete septa	
	Families: Astrorhizidae (*Astrorhiza*), Bathysiphonidae (*Bathysiphon*), Rhabdamminidae (*Rhabdammina, Rhizammina*), Hippocrepinellidae (*Hippocrepinella*), Psammosphaeridae (*Psammosphaera*), Schizamminidae (*Schizammina*), Saccamminidae (*Saccammina*), Hemisphaeramminidae (*Crithionina, Hemisphaerammina*).
Superfamily Komokiacea – test formed of multiple branching tubules, with no distinctive aperture and stercomata accumulated within tubules	
	Families: Komokiidae (*Komokia, Normanina, Septuma*) and Baculellidae (*Baculella, Edgertonia*)
Superfamily Hippocrepinacea – test typically tubular with slightly constricted terminal aperture	
	Families: Hippocrepinidae (*Hyperammina*) and Notodendrodidae (*Notodendrodes*)
Order Lituolida – wall agglutinated, test multichambered	
Superfamily Ammodiscacea – two-chambered test with a globular initial chamber and tubular second chamber	
	Family Ammodiscidae (*Ammodiscus, Glomospira*)
Superfamily Rzehakinacea – test planispiral or with milioline coiling	
	Family Rzehakinidae (*Miliammina*)
Superfamily Hormosinacea – test with uniserial chambers	
	Families: Aschemocellidae (*Aschemonella*), Telamminidae (*Telammina*), Hormosinidae (*Reophax, Hormosina*), Dusenburyinidae (*Dusenburyina*), Thomasinellidae (*Protoschista*)
Superfamily Lituolacea – wall imperforate, test with planispirally coiled chambers	
	Families: Haplophragmoididae (*Cribrostomoides*), Discamminidae (*Ammoscalaria*), Sphaeramminidae (*Sphaerammina*), Lituotubidae (*Lituotuba*), Lituolidae (*Ammobaculites*), Placopsilinidae (*Placopsilina*)
Superfamily Haplophragmiacea – test streptospirally coiled with interiomarginal or areal aperture	
	Families: Ammosphaeroidinidae (*Adercotryma*) and Ammobaculinidae (*Ammobaculinus*)
Superfamily Coscinophragmatacea – test perforate, attached, uniserial, or branching	
	Families: Coscinophragmatidae (*Bdelloidina*) and Haddoniidae (*Haddonia*)

Continued

TABLE 3 Morphology-based family-level classification of foraminifera—cont'd

Superfamily Loftusiacea – test wall with an imperforate outer layer and an alveolar inner layer, planispiral, streptospiral, or trochospiral coil	
	Family Cyclamminidae (*Cyclammina*)
Superfamily Spiroplectamminacea – test planispiral or streptospiral in early part, and biserial or uniserial in later part, or biserial to uniserial or high spiral	
	Families: Spiroplectamminidae (*Spiroplectammina*), Nouriidae (*Nouria*), Pseudobolivinidae (*Pseudobolivina*)
Superfamily Verneuilinacea – test trochospiral, triserial, or biserial to uniserial	
	Families: Verneuilinidae (*Gaudryina*), Prolixoplectidae (*Karrerulina*)
Superfamily Ataxophragmiacea – test trochospiral with serial later part	
	Families: Globotextulariidae (*Liebusella*) and Textulariellidae (*Texturaliella*)
Order Trochamminida – wall agglutinated, multichambered	
	Family Trochamminidae – test with low trochospiral coil, aperture interiomarginal to areal, single, or multiple (*Trochammina, Arenoparella, Jadammina*)
	Family Remaneicidae – test trochospiral, interior partially subdivided by infoldings of the umbilical wall and may have secondary septa (*Remaneica*)
Order Textulariida – wall agglutinated, canaliculate	
	Family Eggerellidae – test trochospirally enrolled in the early stage, later triserial, biserial, or uniserial, aperture basal to areal (*Karreriella, Eggerella*)
	Family Textulariidae – test biserial, later may be reduced to uniserial, aperture interiomarginal to areal (*Textularia, Siphotextularia*)
	Family Valvulinidae – test trochospiral to triserial in early stage, later uniserial, aperture with valcular tooth or flap (*Clavulina, Valvulina*)
Order Miliolida – wall of high-Mg calcite, imperforate, porcellaneous	
	Family Squamulinidae – test unilocular (*Squamulina*)
	Family Cornuspiridae – test two-chambered, proloculus rounded, second chamber tubular, planispirally to streptospirally coiled (*Cornuspira, Meandrospira*)
	Family Hemigordiopsidae – test two-chambered, second chamber streptospirally coiled in early part, later becoming planispiral (*Gordiospira*)
	Family Fischerinidae – test free, enrolled tubular portion divided into a few chambers (*Wiesnerella, Fischerina, Planispirina*)
	Family Nubeculariidae – test free or attached, multichambered, early stage planispiral or irregularly coiled, later may be uncoiled, spreading or branching, wall may have outer agglutinated coating (*Nubeculina, Calcituba, Cornuspiramia*)
	Family Riveroinidae – test planispiral, chambers one-half coil in length, partly subdivided by septa (*Riveroina, Pseudohauerina*)

TABLE 3 Morphology-based family-level classification of foraminifera—cont'd

	Family Ophtalmidiidae – test free, proloculus followed by an undivided coiled second chamber, followed by chambers about one-half coil in length (*Ophtalmina, Spirophtalmidium, Cornuloculina*)
	Family Discospirinidae – test discoid, first chamber rounded, second chamber tubular, later chambers annular, incompletely subdivided (*Discospirina*)
	Family Spiroloculinidae – test with two chambers per whorl, wall may be smooth, striate, or costate (*Adelosina, Spiroloculina*)
	Family Hauerinidae – test free with commonly more than two chambers per whorl, chambers added in 1–5 or more planes of coiling (*Pyrgo, Massilina, Miliolinella, Triloculina, Quinqueloculina*)
	Family Alveolinidae – test large, subcylindrical to fusiform, coiled about elongate axis, numerous chambers divided into chamberlets (*Borelis, Alveolinella*)
	Family Peneroplidae – test planispiral in early stage, later may be uncoiled, flabellulate, or cyclical, chambers with simple interior, not subdivided (*Dendritina, Monalysidium, Spirolina, Peneroplis*)
	Family Soritidae – test large, early stage planispiral, later may be uncoiled, flabelliform, fusiform, or cylindrical, chambers subdivided by interseptal pillars or septula (*Androsina, Archaias, Amphisorus, Marginopora, Sorites*)
Order Carterinida – wall of low-Mg calcite, imperforate with large spicules	
	Family Carterinidae – test trochospiral, attached (*Carterina*)
Order Spirillinida – wall of low-Mg calcite, monocrystalline	
	Family Spirillinidae – test planispiral, bichambered, first chamber globular, second chamber tubular (*Spirillina*)
	Family Patellinidae – test coiled, conical, multichambered, with two chambers in each whorl (*Patellina*)
Order Lagenida – wall of low-Mg calcite, monolamellar	
	Family Lagenidae – test unilocular, aperture terminal, may be produced on a neck, entosolenian tube present or absent (*Lagena, Oolina, Fissurina, Parafissurina*)
	Family Glandulinidae – test uniserial, biserial, or polymorphine, with strongly overlapping chambers, aperture terminal with entosolenian tube (*Glandulina, Phlegeria*)
	Family Nodosariidae – test elongate, uniserial or biserial in early stage, aperture terminal, commonly radiate (*Dentalina, Nodosaria*)
	Family Vaginulinidae – test lenticular, ovate to palmate, planispiral at early stage, aperture terminal, commonly radiate (*Lenticulina, Vaginulina*)
	Family Polymorphinidae – test free or attached, ovate to elongate, multilocular, chambers biserial or spirally arranged, strongly overlapping in early part or irregular, aperture terminal (*Webbinella, Polymorphina, Guttulina*)
Order Buliminida – wall of low-Mg calcite, bilamellar	
Superfamily Bolivinacea – test biserial or uniserial in later part, aperture elongate with a toothplate	
	Family Bolivinidae (*Bolivina*)
Superfamily Loxostomatacea – test biserial, plamate, aperture without toothplate	

Continued

TABLE 3 Morphology-based family-level classification of foraminifera—cont'd

	Family Bolivinellidae (*Bolivinella*)
Superfamily Bolinitacea – test biserial, aperture rounded with toothplate	
	Family Bolivinitidae (*Bolivinita*)
Superfamily Cassidulinacea – test biserial, coiled planispirally or trochospirally, aperture interiomarginal or terminal, with or without toothplate	
	Family Cassidulinidae (*Cassidulina*)
Superfamily Turrilinacea – test high trochospiral, triserial or biserial, or changing to biserial or uniserial in later part, aperture with or without toothplate	
	Family Stainforthiidae (*Stainforthia*)
Superfamily Buliminacea – test high trochospiral, biserial to uniserial in later part, aperture loop-shaped with internal toothplate	
	Family Siphogenerinoididae (*Rectobolivina*), Buliminidae (*Bulimina*), Buliminellidae (*Buliminella*), Uvigerinidae (*Uvigerina*), Reussellidae (*Reusella*), Pavoninidae (*Pavonina*)
Superfamily Fursenkoinacea – test twisted or flat, biserial or triserial, aperture loop-shaped, with internal toothplate	
	Families: Fursenkoinidae (*Fursenkoina*) and Virgulinellidae (*Virgulinella*)
Superfamily Delosinacae – test trochospiral or triserial, biserial in later part, aperture areal, elongate, or in the form of sutural pores	
	Family Delosinidae (*Delosina*)
Superfamily Pleurostomellacea– test triserial or biserial to uniserial, or uniserial throughout, aperture areal, slitlike, or cribrate	
	Family Pleurostomellidae (*Pleurostomella*)
Superfamily Stillostomellacea– test uniserial, aperture terminal with phialine lip and small tooth	
	Family Stilostomellidae (*Stilostomella*)
Superfamily Annulopatellinacea – test low conical, round initial chamber enclosed by second chamber, succeeding uniserial chambers arranged in annuli	
	Family Annulopatellinidae (*Annulopatellina*)
Order Rotaliida – wall of low-Mg calcite, bilamellar	
Superfamily Discorbacea – test low trochospiral, aperture umbilical, interiomarginal, with or without supplementary apertures	
	Families: Bagginidae (*Baggina*), Eponididae (*Eponides*), Helenididae (*Helenina*), Mississippinidae (*Mississippina*), Pegidiidae (*Pegidia*), Discorbidae (*Discorbis*), Rosalinidae (*Rosalina*), Sphaeroidinidae (*Sphaeroidina*)
Superfamily Glabratellacea – test trochospiral, umbilicus depressed with radial striations or granules, aperture umbilical	

TABLE 3 Morphology-based family-level classification of foraminifera—cont'd

	Families: Glabratellidae (*Glabratella*) and Heronallenidae (*Heronallenia*)
Superfamily Siphoninacea – test low trochospiral throughout or in early part, aperture interiomarginal or areal with a lip	
	Family Siphoninidae (*Siphonina*)
Superfamily Discorbinellacea – test low trochospiral or nearly planispiral, aperture interiomarginal, archlike or slitlike	
	Families: Pseudoparrellidae (*Epistominella*) and Discorbinellidae (*Laticarinina*)
Superfamily Planorbulincea – test low trochospiral, with planispiral, serial, or irregular arrangement in later part, aperture interiomarginal with or without secondary apertures, wall coarsely perforate	
	Families: Planulinidae (*Planulina, Hyalinea*), Cibicididae (*Cibicides*), Planorbulinidae (*Planorbulina*), Cymbaloporidae (*Cymbaloporetta*)
Superfamily Acervulinacea – earliest chambers coiled, adult chambers irregularly arranged, aperture present only as mural pores, wall coarsely perforate	
	Families: Acervulinidae (*Acervulina*) and Homotrematidae (*Homotrema, Miniacina*)
Superfamily Asterigerinacea – test trochospiral to nearly planispiral, chambers subdivided by internal partitions, primary aperture interiomarginal or areal, secondary apertures sutural or areal.	
	Families: Epistomariidae (*Nuttalides*), Asterigerinidae (*Asterigerina*), Amphisteginidae (*Amphistegina*).
Superfamily Nonionacea – test planispiral, aperture slitlike or a series of pores	
	Family Nonionidae (*Nonion, Haynesina*)
Superfamily Chilostomellacea – test low trochospiral to planispiral with uncoiled later part, later chambers may envelop earlier ones or may have internal partitions, aperture interiomarginal or terminal	
	Families: Chilostomellidae (*Chilostomella*), Quadrimorphinidae (*Quadrimorphina*), Osangulariidae (*Osangularia*), Oridorsalidae (*Oridorsalis*), Gavelinellidae (*Gyroidina, Hanzawaia*), Karreriidae (*Karreria*)
Superfamily Rotaliacea – test typically low trochospiral or planispiral, with internal canal system, aperture interiomarginal or areal, single or multiple	
	Families: Rotaliidae (*Ammonia, Pararotalia*), Calcarinidae (*Calcarina, Baculogypsina*), Elphididae (*Elphidium*).
Superfamily Nummulitacea – test large, planispiral, or annular, numerous chambers with or without chamberlets, with a complex canal system	
	Family Nummulitidae (*Nummulites, Operculina, Heterostegina*)
Order Globigerinida – wall of low-Mg calcite, bilamellar, planktonic mode of life	
Superfamily Heterohelicacea – test biserial or triserial	
	Families: Guembelitriidae (*Gallitellia*) and Chiloguembelinidae (*Laterostomella*)

Continued

TABLE 3 Morphology-based family-level classification of foraminifera—cont'd

Superfamily Globorotaliacea – test trochospiral or streptospiral, surface nonspinose	
	Families: Globorotaliidae (*Globorotalia*), Pulleniatinidae (*Pulleniatina*), Candeinidae (*Candeina*)
Superfamily Globigerinacea – test trochospiral to planispiral, surface spinose, primary aperture interiomarginale, secondary apertures sutural	
	Families: Globigerinidae (*Globigerina*) and Hastigerinidae (*Hastigerina*)
Order Involutinida – wall aragonitic	
	Family Planispirillinidae – test bichambered, initial chamber enclosed by coiled tubular second chamber, coiling planispiral or trochospiral (*Planispirillina*)
Order Robertinida – wall aragonitic	
	Family Robertinidae – test planispiral to trochospiral, chambers divided by double partition forming small supplementary chambers on one or both sides of the test, aperture consisting commonly of two slitlike openings (*Pseudobulimina, Robertina, Robertinoides*)
	Family Ceratobuliminidae – test trochospirally coiled, chambers with internal partitions, aperture a long narrow slit (*Ceratobulimina, Lamarckina*)
	Family Epistominidae – test biconvex, with low trochospire coiling, internal partition present only in the last chamber (*Hoeglundina*)
Order Silicoloculinida – wall siliceous, imperforate	
	Family Silicoloculinidae – test coiled in various planes (*Miliammellus*)

Modified from Sen Gupta, B. K. (1999). *Modern foraminifera*. Dordrecht: Kluwer Academic; Lee *et al.* (2002). Distinctive features of each superfamily or family as well as representative genera are presented.

The radiation of monothalamous foraminifera comprises all unilocular species having an organic or agglutinated wall that is traditionally belonging to the orders Allogromiida and Astrorhizida. Monothalamids also include 'naked' amoeboid species, such as *Reticulomyxa*, which were previously considered members of the Athalamea, a sister group to Foraminifera. The radiation currently consists of 12 distinct clades. However, this number is constantly increasing with new sequences obtained from new isolates and environmental samples, showing that the diversity of this mainly nonfossilized group was largely overlooked in micropaleontologically oriented foraminiferal research. Recently, several novel lineages of environmental foraminiferal sequences were found in soil samples providing new evidence for ubiquitous character of the group. For the moment, only a few of the monothalamous clades can be defined by structural data. Because some clades consist of both organic-walled allogromiids and agglutinated astrorhizids, the traditional distinction between these two orders is not confirmed by molecular data.

Within the radiation of the monothalamids, the multichambered test was developed at least twice: once, in the lineage leading to the clade grouping most of the agglutinated polythalamous species, the aragonitic robertinids, and calcitic rotaliids, buliminids, and globigerinids; the second time, in the lineage leading to some lituolids, the spirillinids, and the miliolids. The multichamber test may also have appeared independently in the lineage leading to the monolamellar calcitic order Lagenida; the origin of this order, however, remains uncertain due to very limited taxon sampling. The independent position of Lagenida within the radiation of monothalamids, as the only calcareous group not associated with any agglutinated polythalamous lineage, seems artifactual and needs confirmation.

Among the two well-defined polythalamous clades, the first consists of most of the diversity of modern foraminifera. This clade includes all representatives of the agglutinated orders Trochamminida and Textulariida examined so far, and the majority of Lituolida as well as calcitic Rotaliida, Buliminida, Globigerinida, and aragonitic Robertinida. The

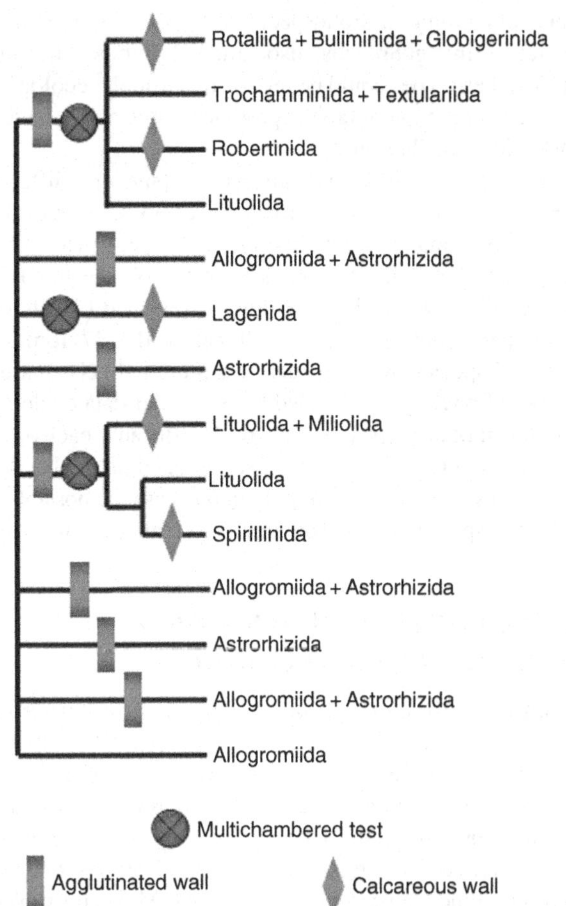

FIGURE 7 Schema of macroevolutionary relationships between the main orders of foraminifera based on molecular data. The transformations from single to multichambered tests, as well as from organic to agglutinated walls and from agglutinated to calcareous walls are indicated.

molecular data do not confirm the monophyly of agglutinated orders. As mentioned previously, they appear rather as a paraphyletic group from which calcareous taxa originated at least twice. Detailed molecular studies show that the distinction between Rotaliida and Buliminida, based principally on the presence/absence of an internal toothplate, is not justified in view of available molecular data. There are also some indications that the planktonic order Globigerinida is polyphyletic, although the exact positions of different planktonic families are difficult to establish due to their unusually rapid evolutionary rates.

The second polythalamous clade includes the representatives of calcareous Spirillinida and Miliolida, as well as some agglutinated Lituolida: the spirally coiled *Ammodiscus* branching as sister group to Spirillinida and the miliolid-like *Miliammina*, branching next to Miliolida. All taxa included in this clade are characterized by the presence of a bi- or multichambered test, typically with an initial globular chamber followed by a coiled tubular second chamber. As the relationship between Spirillinida and Miliolida is based exclusively on small subunit rDNA (SSU rDNA) sequences, which evolve particularly rapidly in Spirillinida and in *Ammodiscus*, the monophyly of this clade and the relationships within it have to be ascertained by multigene phylogenies.

In view of the present molecular data, the transition from single-chamber to multichambered tests seems to be evolutionarily more important than the change in their wall composition and structure. The traditional oversimplified view of the successive evolution of foraminifera from organic-walled to agglutinate and further to calcareous lineages is challenged by molecular evidence for the multiple origins of agglutinated and calcareous walls.

The transformation from organic to agglutinated wall occurred several times in monothalamous lineages. Many of these consist of both organic and agglutinated species and some may even change the nature of their test wall depending on the environmental conditions. As shown in some monothalamous lineages, the process of agglutination is reversible. However, the presence of an agglutinated wall seems to be a prerequisite to the formation of a polythalamous test. Given the available molecular data, the polythalamous agglutinated test evolved at least 3 times independently, probably from an agglutinated monothalamous lineage. Alternatively, its formation could be preceded by the development of a multichambered test in some organic-walled lineages, but the phylogenetic position of the few modern organic-walled polythalamous species is unknown.

The transformation from an agglutinated to a calcareous wall occurred at least 5 times independently (Figure 7). Remarkably, each time a different type of calcareous wall was developed. The three orders with a calcitic bilamellar wall (Rotaliida, Buliminida, and Globigerinida) form one clade. The other calcareous orders, including the aragonitic Robertinida, the calcitic monolamellar Lagenida, the imperforate Miliolida, and the monocrystalline Spirillinida form independent monophyletic groups.

From the perspective of molecular phylogenetic data, the morphology-based ordinal classification of foraminifera (Tables 2 and 3) requires a profound revision. On the one hand, there is definitely no support for the wall-based distinction of monothalamous orders Allogromiida and Astrorhizida. There is also lack of evidence for the monophyly of agglutinated polythalamous orders Lituolida, Textularida, and Trochamminida. On the other hand, the five orders of calcareous foraminifera (Rotaliida, Miliolida, Robertinida, Spirillinida, and Lagenida) appear monophyletic, while two orders (Buliminida and Globigerinida) should be included in the order Rotaliida. Among the remaining four orders, one is extinct (Fusulinida) and three (Carterinida, Silicoloculinida, and Involutinida) are represented by only a few modern species. Further analysis of their sequences probably will not drastically change the molecular view of the macroevolution of foraminifera presented here.

MOLECULAR DIVERSITY

Molecular assessment of foraminiferal diversity is based exclusively on analysis of nuclear ribosomal RNA genes. Most of the studies use the 3′ fragment of SSU rDNA, which contains several rapidly evolving regions specific to foraminifera. Additionally, the 5′ fragment of large subunit rDNA (LSU rDNA) and the ITS (internally transcribed spacer) rDNA region were used to analyze genetic variations within the slowly evolving benthic and planktonic species. Other molecular markers, in particular mitochondrial genes, are not yet available for foraminifera.

The principal advantage of using the nuclear ribosomal genes for foraminiferal diversity studies is the facility of their amplification from single-cell DNA extracts. Moreover, the SSU rDNA gene in foraminifera evolves faster than in other eukaryotes, enabling the distinction of genetic variations at the species level. However, the analyses of foraminiferal ribosomal RNA genes are sometimes problematic because of high intraindividual polymorphism of ribosomal gene copies. This polymorphism is usually below 1% of sequence divergence, but may sometimes exceed this value, making the assessment of the limits between intra- and interspecific variations particularly challenging.

Initial studies of the molecular diversity in foraminifera focused on three taxonomic groups: the monothalamous allogromiids and astrorhizids, the planktonic globigerinids, and some benthic rotaliids. In all these groups, molecular analyses led to the discovery of important cryptic diversity in most of the examined morphospecies.

In monothalamids, the hidden diversity revealed by molecular studies is particularly striking. This generally nonfossilizable group is largely undescribed and many genera are represented by only a single described species. However, the simple morphological features of monothalamids are of limited taxonomic value for species identification. Recently, extensive molecular studies of polar and deep-sea monothalamids indicate the presence of high cryptic diversity at many different taxonomic levels, from higher lineages to the species level. Practically every examined monothalamous genus or morphospecies has been genetically differentiated into several new phylotypes, based on different SSU rDNA sequences, each having a restricted geographic distribution. Given the apparent isolation and genetic differentiation of these phylotypes, we may consider them as representing new species. However, finding the morphological characters that could describe these species will be problematic.

A similar situation occurs in the planktonic foraminifera, where the low number of modern morphospecies (40–50) has been multiplied 4–5 times by molecular analyses. These analyses, based on particularly rapidly evolving partial SSU rDNA sequences, revealed that almost every planktonic morphospecies could be divided into several distinct genotypes, also called phylotypes. In some species, these types could be related to particular ecological conditions or biogeographic provinces, but many have a worldwide distribution.

In benthic rotaliids, the situation is somewhat different because the taxonomy of rotaliid genera is overloaded with synonymous descriptions. For example, phylogenetic analyses of LSU rDNA sequences from 202 specimens of *Ammonia* from 30 localities worldwide, revealed 13 different phylotypes. This number is well below the 37 formally described species of this genus. It is quite unlikely that the number of phylotypes revealed by molecular data could approach that of described morphospecies. Because each of the *Ammonia* phylotypes could be discriminated on the basis of external test characters, it should, in this case, be possible to revise the species description and to validate their names.

EVOLUTIONARY HISTORY AND GEOLOGICAL IMPORTANCE

Foraminifera first appear in the fossil record during the early Cambrian. However, they were most probably an important component of the protist community in the Precambrian. According to the molecular timescale of early foraminiferal evolution, a large radiation of nonfossilized organic-walled and agglutinated monothalamous lineages occurred in the Neoproteozoic (550–1000 Ma). This molecular dating agrees with the interpretation of some Upper Vendian (650 Ma) microfossils as early foraminifers, but it requires confirmation by further genetic studies.

The earliest Cambrian foraminiferal genus *Platysolenites* has the appearance of a large, simple, agglutinated tube resembling the modern foraminiferal genus *Bathysiphon*. Other straight and coiled tubular agglutinated foraminifera have been reported from the Lower and Middle Cambrian and became abundant in the Ordovician. The earliest multichambered agglutinated foraminifera arose in the Devonian and became common in the Carboniferous. During this period, the first calcareous foraminifera also appeared; these belonged to the orders Miliolida and Lagenida, as well as the Fusulinida. The microgranular fusulinids underwent a spectacular diversification in the Permian, but disappeared completely at the end of the Paleozoic.

Following the Permian extinction, several new orders of small benthic foraminifera, including the Rotaliida, Robertinida, Involutinida, and Spirillinida, appeared. The monolamellar Lagenida dominated in the Jurassic, while the Rotallida and Miliolida became most abundant in the Cretaceous. Planktonic foraminifera first appeared in the Middle Jurassic, evolving probably from the benthic family of Oberhauserellidae. After a period of extensive diversification of planktonic lineages in the Middle and Late Cretaceous, most of the planktonic lineages suddenly became extinct at the Cretaceous–Tertiary boundary. This mass extinction was much less marked in shallow benthic foraminiferal faunas and did not occur among deeper benthic assemblages.

The Cenozoic assemblage of benthic foraminifera was dominated by rotaliids and miliolids, while Globigerinidae and Globorotaliidae became the most abundant planktonic families. Deepwater foraminiferal faunas exhibit three major episodes of change: (1) in the late Paleocene, (2) at the Eocene–Oligocene boundary, and (3) in the middle Miocene. The modern neritic and shallow foraminiferal assemblages originated mainly in the Eocene and Miocene. Large benthic foraminifers, including nummulitids, lepidocyclinids, soritids, and alveolinids, were abundant in the shallow and well-illuminated Eocene seas, forming well-known limestone. Many genera of modern smaller benthic and planktonic foraminifera are recorded for the first time in the Miocene strata.

The continuous evolutionary history of foraminifera from the Cambrian to the Recent makes them by far the most important group of microfossils. Owing to their small size and abundance in the fossil record, foraminifera are widely used for the determination of the age of the sedimentary strata and the environmental conditions of the past oceanic environments. Some foraminifera, particularly planktonic lineages, are important biostratigraphic tools used for the development of fine-scale biozonation. Foraminiferal assemblage composition is widely used in paleobathymetry for depth reconstruction, while the characteristic morphological features can be used to deduce the paleohabitats of particular taxa. Variations in past temperatures of surface and bottom waters are commonly inferred from the ^{18}O:^{16}O and Mg:Ca ratios in calcitic shells of planktonic and benthic foraminifera. Long recognized for their economic importance in the oil industry, the foraminifera have today become a major tool in the study of past global climate changes.

FURTHER READING

Boltovskoy, E., & Wright, R. (1976). *Recent foraminifera.* Dr. W. Junk: The Hague.

Bowser, S. S., Habura, A., & Pawlowski, J. (2006). Molecular evolution of foraminifera. In L. Katz & D. Bhattacharya (Eds.), *Genomics and evolution of microbial eukaryotes* (pp. 78–93). Oxford University Press: Oxford.

Culver, S. J. (1993). Foraminifera. In J. H. Lipps (Ed.), *Fossil prokaryotes and protists* (pp. 203–247). Blackwell: Boston, MA.

Debenay, J.-P., Pawlowski, J., & Decrouez, D. (1996). *Les Foraminiféres actuels.* Masson: Paris.

Haynes, J. R. (1981). *Foraminifera.* Wiley: New York.

Hemleben, C., Spindler, M., & Anderson, O. R. (1989). *Modern planktonic foraminifera.* Springer: Berlin.

Lee, J. J., & Anderson, R. O. (1991). *Biology of foraminifera.* Academic Press: London.

Lee, J. J., Pawlowski, J., Debenay, J.-P., *et al.* (2000). Phylum granuloreticulosa. In J. J. Lee, G. F. Leedale & P. Bradbury (Eds.), *An illustrated guide to the protozoa* Vol. 2(pp. 872–951). Allen: Lawrence, KS.

Loeblich, A. R., & Tappan, H. (1987). *Foraminiferal genera and their classification* (Vols. 1–2). Van Nostrand Reinhold: New York.

Pawlowski, J., Holzmann, M., Berney, C., *et al.* (2003). The evolution of early Foraminifera. *Proceedings of the National Academy of Science of the United States of America, 100,* 11494–11498.

Sen Gupta, B. K. (1999). *Modern foraminifera.* Kluwer Academic: Dordrecht.

Travis, J. L., & Bowser, S. S. (1991). The motility of foraminifera. In J. J. Lee & O. R. Anderson (Eds.), *Biology of foraminifera* (pp. 91–156). Academic Press: London.

Chapter 22

Euglenozoa

M.A. Farmer

University of Georgia, Athens, GA, USA

Chapter Outline

Abbreviations	311
Defining Statement	311
Taxonomy	311
Cell Structure	312
Flagellar Structure	312
Mitochondria	314
Nuclei	314
Cytoskeleton	314
Euglenids	314
Pellicle	314
Flagella	315
Chloroplasts	315
Feeding Apparatus	316
Kinetoplastids	317
The Kinetoplast	317
RNA Editing in Kinetoplastids	318
The Glycosome	318
Diplonemids and Other Euglenozoa	319
Evolutionary Relationships	320
Further Reading	321

ABBREVIATIONS

COX1 Cytochrome oxidase subunit 1
gRNA Guide RNA
rRNA Ribosomal RNA
VSG Variant surface glycoprotein

DEFINING STATEMENT

The Euglenozoa consists of three monophyletic groups: Euglenids, Diplonemids, and Kinetoplastids. The article describes the life history, cell biology, and evolutionary relationships of Euglenozoans, primarily focusing on free-living (Euglenids and Bodonids) and parasitic taxa (Trypanosomes and Leishmania). There is also discussion of some lesser known members of the group of uncertain affiliation.

TAXONOMY

Historically, members of the Euglenozoa presented a problem for the classical taxonomic treatment of protozoa and algae in that they consisted of both photosynthetic and nonphotosynthetic genera. The evidence is now overwhelming that three groups of flagellates – the Euglenids, the Kinetoplastids, and the Diplonemids – form a natural, monophyletic assemblage. In addition there are a number of protists that clearly fall within the Euglenozoa but that are not easily assigned to the Euglenid, Diplonemid, or Kinetoplastid clades (Table 1).

As originally defined by Cavalier-Smith, the Euglenozoa are united by the presence of discoidal (paddle-shaped) mitochondrial cristae, nontubular flagellar hairs, paraxial rods, and an unusual form of mitosis in which the nuclear envelope does not break down and the microtubular mitotic spindle forms within the nucleus. Simpson modified this and identified three uniting characters (synapomorphies) shared by all members of the group and presumably present in their last common ancestor. These are (1) a flagellar root pattern with two basal bodies and three asymmetrically arranged microtubular roots (one of which lines a portion of the feeding apparatus), (2) a paraxial rod in one or both of the flagella that has a certain biochemical composition and structure and thus differs from flagellar inclusions found in other groups, and (3) tubular, thick-walled extrusive organelles (trichocysts, mucocysts, etc.). The last character may have been lost or reduced multiple times within Euglenozoa.

The monophyly of the Euglenozoa has been confirmed by numerous studies of gene sequences including those for SSU rRNA genes as well as protein-coding genes. One reason that the euglenids and kinetoplastids were not originally recognized as a monophyletic group is that the

Eukaryotic Microbes.

TABLE 1 Subgroups of Euglenozoa

Euglenozoa (Cavalier-Smith 1981). Eukaryotes, mitochondriate, with heteromorphic (tubular/whorled and parallel) lattice paraxonemal rods, and their descendants
Euglenids (Euglenida, B¨tschli 1884). Euglenozoa with a pellicle of protein strips (fused in many species), and their descendants
Kinetoplastids (Kinetoplastida, Honigberg 1963). Euglenozoa with a kinetoplast (including polykinetoplasts), and their descendants
Diplonemids (Diplonemea, Cavalier-Smith 1993). Currently *Diplonema* and *Rhynchopus* only. Taxonomic and systematic study required
Postgaardi (Fenchel *et al.* 1995)
Calkinsia (Lackey 1960)

Reproduced from Simpson, A. G. B. (1997). The identity and composition of the Euglenozoa. *Archiv Fur Protistenkunde, 148*, 318–328, with permission.

most famous members of each group, *Euglena* (a well-studied photosynthetic euglenid) and *Trypanosoma* (a blood-borne, obligate parasite in the kinetoplastids), are quite strikingly different. It is only upon closer inspection of fine-level cell structure that the important similarities between them become clear.

CELL STRUCTURE

The defining features (morphological apomorphies) of the Euglenozoa include a distinctive flagellar apparatus with three asymmetrically arranged microtubular roots, paraxial rods, and tubular ejectile organelles (mucocysts, trichocysts). The cytoskeleton is unique and striking in that it allows many species to dramatically alter their cell shape and is generally composed of a complete, or nearly complete, corset of microtubules that are longitudinally arranged in either a linear or a helical fashion. Most species have functional mitochondria, although these have been lost or modified on several occasions. When present, these mitochondria generally have a distinctive morphology with cristae in the shape of table tennis paddles often described as being 'discoidal.' Although no euglenid species that possess chloroplasts are capable of consuming prey, many other heterotrophic species are capable of ingesting bacteria or small eukaryotes by way of a feeding apparatus. This apparatus, when present, is supported by microtubule-reinforced rods that extend into the cytoplasm and has a single opening toward the apex of the cell, often adjacent to the flagellar opening through which the flagella emerge from the cell.

Flagellar Structure

Eukaryotic flagella, unlike the evolutionarily unrelated flagella of bacteria, have a complex structure consisting of microtubules and an associated complex of motor and connective proteins collectively known as the axoneme. The axoneme has a ninefold symmetry of microtubular doublets surrounding two central microtubules. The outer nine doublets of the axoneme extend into the cytoplasm where they add a third partial microtubule (triplet) and form an anchor known as the basal body. The basal bodies of the two euglenid flagella are often connected by way of a striated fiber and serve as the anchor site for additional microtubular roots, some of which give rise to cytoskeletal structures and ultimately determine the shape of the cell. Collectively, the basal bodies, connective fibers, and microtubular roots form the 'flagellar apparatus' and the complex arrangement of these structures often determines the manner in which flagella are used by a cell. These characters are relatively conserved among species and allow for their use in defining major groups of eukaryotes. In addition to certain other major morphological features, the components of the flagellar apparatus are some of the most commonly used higher-level diagnostic characters.

The basic Euglenozoan flagellar apparatus consists of two functional basal bodies (i.e., each gives rise to a flagellum), three asymmetrically arranged microtubular roots, and a striated connective fiber that links the two basal bodies to each other (Figure 1). In many euglenozoans, both flagella emerge from the flagellar opening and are heterodynamic, that is, differing from each other in their structure, position, movement, and function. The anteriorly directed or dorsal flagellum typically undulates in front of the cell and is thought to be primarily responsible for locomotion. The posterior flagellum often lies between the cell and the substrate and for this reason is sometimes referred to as the ventral flagellum. In *Euglena* and its close relatives, the ventral flagellum is highly reduced to the point where it is not much more than a stub that never exits the flagellar reservoir (a membrane-bound pocket in the anterior portion of the cell). In species in which one flagellum is highly reduced or the two emergent flagella are homodynamic (similar in structure and function) the striated connective fiber between the two basal bodies may be greatly reduced. Many euglenozoans (e.g., *Peranema*, *Entosiphon*, *Bodo*) move not by swimming through the water but rather by gliding along the substrate. This gliding motility has variously been attributed to both the dorsal and the ventral flagella and the mechanism behind gliding motility is unknown.

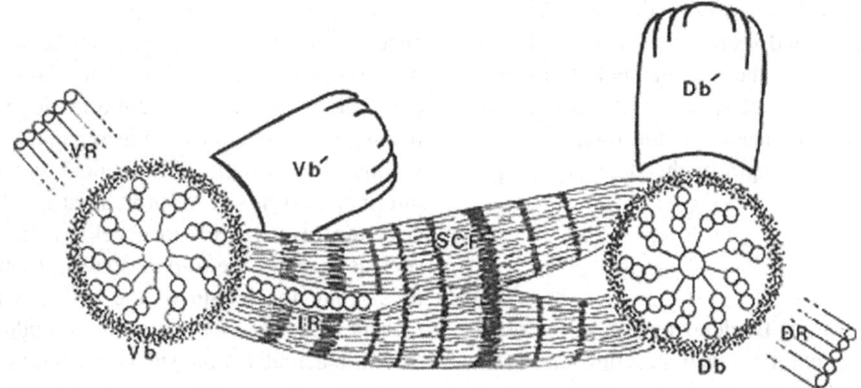

FIGURE 1 Diagrammatic illustration of the interphase flagellar apparatus from the euglenid *Ploeotia vitrea*. The functional basal body of the ventral flagellum (Vb) is associated with ventral (VR) and intermediate (IR) microtubular roots whereas the dorsal basal body (Db) has a single dorsal root (DR). The functional basal bodies are connected by a striated connective fiber (SCF) and have immature basal bodies that do not give rise to flagella in the interphase cell.

One or both of the flagella may be additionally supported by way of a paraxial (synonym: paraflagellar) rod (Figure 2). The paraxial rod consists primarily of two polypeptides whose molecular weights range between 65 and 80 kDa. These proteins share a 60% amino acid identity with one another and are designated as PFR1 and PFR2. Antibodies raised against them react with paraxial rods in both euglenids and kinetoplastids. These PFR polypeptides can be arranged as a paraxial rod in either an amorphous or a highly crystalline array that typically spans the entire length of the flagellum but does not extend into the 'flagellar transition zone,' which is the region in which the axoneme becomes the basal body. Unlike other flagellar elaborations (hairs, scales, etc.) the paraxial rod lies within the flagellar membrane adjacent to the axoneme. The exact role of the paraxial rod is unknown but it is believed to give rigidity to the flagellum. It has been demonstrated that inhibition of the paraxial rod protein gene (PFR1) in the kinetoplastid *Leishmania* prevents normal paraxial rod formation and results in a pronounced loss of motility.

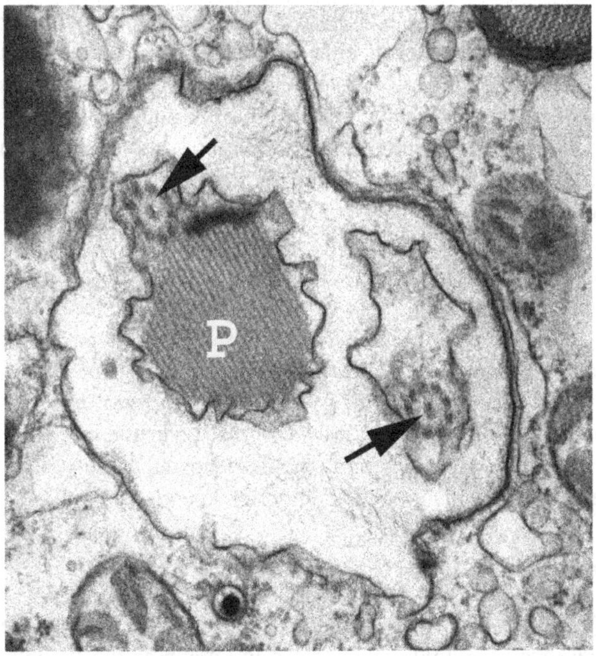

FIGURE 2 Transmission electron micrograph of the cross section through the flagellar reservoir of the euglenid *Dinema sulcatum* showing the axonemes of both flagella (arrows) and the paraxial rod (P) of the ventral flagellum.

In most Euglenozoa the two flagella are dissimilar in terms of both their structure and their function. In many heterotrophic species, the ventral flagellum is more pronounced with a more defined paraxial rod than is the dorsal flagellum. The ventral flagellum usually lies nearly rigid along the ventral surface of the cell while the anteriorly directed dorsal flagellum actively beats or twitches at its end. As mentioned above, in species such as *Euglena*, the ventral flagellum is reduced to a nonemergent stub. In a few species, such as *Eutreptia* and *Eutreptiella*, the two emergent flagella are distinguished from one another only by their relative lengths. In all species the flagella undergo a developmental transformation in which the dorsal flagellum of the parent cell becomes the ventral flagellum of the daughter. Sometimes this requires an extensive reorganization of the flagellum itself with either an increase in the paraxial rod or a drastic reduction in size, as in *Euglena*. Ventral flagella do not undergo further differentiation and are said to be in the 'mature' state whereas the daughter basal bodies give rise to new dorsal flagella in the daughter cells and these must undergo an additional round of cell division before reaching the mature or determinate state.

In many heterotrophic species, the primary mode of cell locomotion is flagellar gliding, not swimming. Often, the anteriorly directed dorsal flagellum lies nearly motionless in front of the cell as it glides along the substrate. In some taxa (e.g., *Peranema* and *Petalomonas*), the tip of the dorsal flagellum is very active and seems to play a role in prey identification and sensing of the environment. The mechanism of flagellar gliding is unknown but it has been suggested that the flagellar hairs or mastigonemes that typically line the exterior length of the flagella may play a role in locomotion.

Mitochondria

The mitochondrion of most Euglenozoa is a single organelle that extends as a multibranched structure throughout the cell. Mitochondrial cristae (infoldings of the inner mitochondrial membrane) are typically discoidal in shape, reminiscent of table tennis paddles. While mitochondrial genomes are well characterized for representatives of nearly all major groups of eukaryotes, the mitochondria of many Euglenozoa remain an enigma. In the kinetoplastids the mitochondrial genome is fragmented and scrambled in such a way that it requires major posttranscriptional modification of the resulting mRNAs, a process known as 'RNA editing,' to create functional gene transcripts (see below). In others, such as the Diplonemids, the mitochondrial genome consists of numerous partially overlapping fragments. The result is that individual genes, such as the gene for cytochrome oxidase subunit 1 (COX1), are fragmented and distributed in pieces, in the case of COX1 among up to seven different minichromosomes. In a few Euglenozoa that are found in anaerobic environments there are no mitochondria with visible cristae although membrane-bound organelles that are thought to be derived from mitochondria are found throughout the cytoplasm.

Nuclei

All known Euglenozoa have a single nucleus per cell. These single nuclei often have chromosomes that appear to be permanently condensed throughout the life cycle and are sometimes visible under a light microscope. In contrast to most eukaryotes, the nucleolus and nuclear envelope remain intact throughout the cell division process. Nuclear division is accomplished by way of a number of intranuclear subspindles, the microtubules of which span the nucleus through pores in the nuclear membrane and attach to kinetochore-like structures on the individual chromosomes.

Cytoskeleton

The cytoskeleton of Euglenozoa is based on a complete, or nearly complete, corset of microtubules that extend along the length of the cell and are arranged in a longitudinal to helical fashion. These microtubules originate from the microtubular root, which is associated with the dorsal flagellar basal body. Many euglenozoans are capable of a wriggling motion that in some euglenids takes on a peristaltic appearance and is often referred to as euglenid 'metaboly.' Metaboly is most often observed in those taxa in which the cytoskeletal microtubules are arranged in a helical fashion but some Diplonemids are also capable of this movement without the benefit of a helical pellicle, the complex microtubule-reinforced protein strips found as the cell covering of euglenids (see below). In other Euglenozoa, such as many of the trypanosomes, the emergent flagellum runs along the length of the cell and is attached to the plasma membrane by means of an 'undulating membrane,' which is actually a physical connection between the outer membrane of the cell and the flagellar membrane. Then, when the flagellum beats, it causes the entire cell to wriggle or bend. The flexible nature of the Euglenozoan cytoskeleton allows for these contortions although the mechanism of cell movement differs significantly between taxa.

EUGLENIDS

Perhaps the earliest recorded report of a microscopic protist comes from Anthony van Leeuwenhoek who in a 1674 letter to a friend described what is now thought to be *Euglena viridis* as being "... green in the middle and before and behind white."

The euglenids are best known by the genus *Euglena*, specifically by the species *Euglena gracilis*, which has been used in countless introductory laboratories as an example of 'both plant and animal.' Of course, this is only true in the superficial sense that *Euglena* can both move and photosynthesize because it is clearly neither a plant nor an animal; nor is it an ancestor of either group. It is unfortunate that *E. gracilis* is used as the 'typical' euglenid for in many ways no other species is more atypical of the group. *E. gracilis* is a photosynthetic autotroph whereas most euglenid genera are either bacteriotrophs/eukaryotrophs, which actively consume prey by means of phagocytosis, or osmotrophs, which meet their nutritional needs by absorbing compounds across the plasma membrane. *Euglena* has a single emergent flagellum, the other flagellum being so reduced that it does not exit the cell, while most euglenids have two emergent flagella. *Euglena* can modify its cell shape by 'euglenoid metaboly' but many euglenids are rigid or nearly so. It is the last feature, however, that gives the euglenids their unique identity among the Euglenozoa, namely, a peculiar arrangement of the cell cortex known as a pellicle.

Pellicle

Euglena and all of the euglenids are characterized by the presence of a unique cell covering known as the pellicle. This pellicle is a complex structure consisting of a proteinaceous layer or 'membrane skeleton' that is underlain by microtubules and covered by the plasma membrane of the

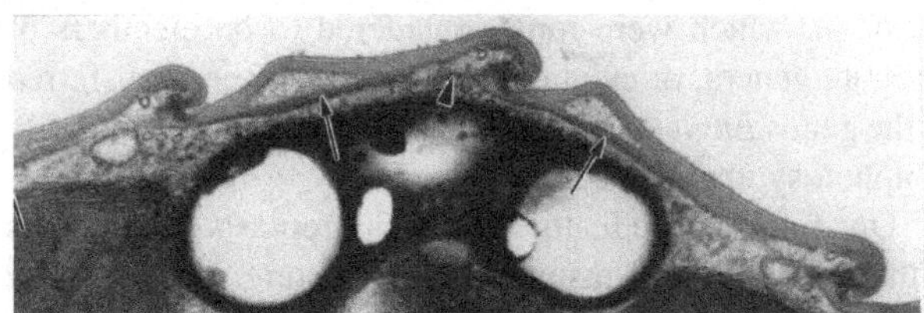

FIGURE 3 Transmission electron micrograph of the cross section of a euglenid pellicle showing the proteinaceous plates (arrows), connective proteins (arrowhead), and arrangement of microtubules beneath overlapping strips.

cell. The protein layer consists of two types (80 and 86 kDa) of articulin proteins that are linked to the plasma membrane by way of a 39 kDa intramembranous protein. This protein layer is then underlain by a series of microtubules that are arranged in a repeating fashion underneath the junction point of adjacent pellicle strips (Figure 3). The pellicle is organized into individual strips that extend along the entire length of the cell, the number and position of which is often diagnostic for various species. In many euglenids the pellicular strips run in a longitudinal fashion along the long axis of the cell while in others, such as *Euglena*, the pellicular strips are organized into a spiral that normally forms a right-handed helix and may encircle the cell one or more times. A sliding movement of one strip relative to another is what allows *Euglena* and other species to change their cell shape. In some taxa the cells can rapidly change from very elongate to nearly round (Figure 4).

Euglenoid metaboly is only known from those taxa that, like *E. gracilis*, have their pellicular strips organized into a helical arrangement; this is true of all phototrophs, eukaryotrophs, and some osmotrophs. However, some species with helically arranged pellicles are still capable of only limited or even no metaboly, having lost the ability to change their cell shape, and are therefore secondarily rigid. Species whose pellicles consist of longitudinally arranged ridges are always rigid and these species are either bacteriotrophs or osmotrophs. The longitudinal pellicle and rigid condition is thought to be ancestral with the helical pellicle and

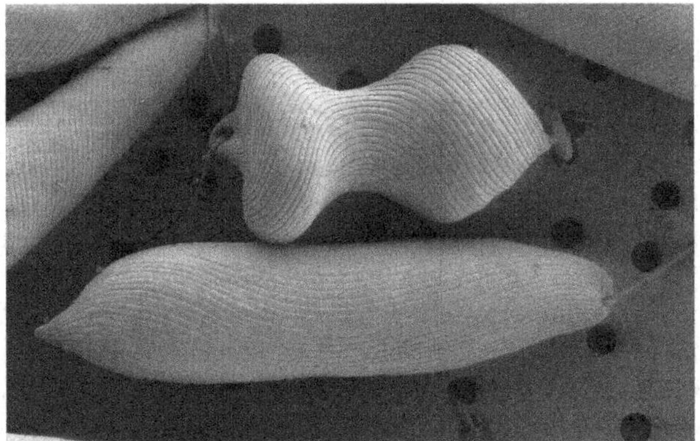

FIGURE 4 Scanning electron micrograph showing the range of shapes (metaboly) found in individual cells of *Euglena gracilis*.

euglenid metaboly arising before the acquisition of chloroplasts. Therefore, all autotrophic taxa either are capable of metaboly or have secondarily lost the ability to change cell shape.

Flagella

One structure sometimes found in association with the flagella of euglenids is the eyespot/flagellar swelling complex. The eyespot (stigma) consists of a number of carotenoid (β-carotene) pigment granules that lie in the cytoplasm adjacent to a particular portion of the flagellar reservoir, which is an invagination that extends from the flagellar opening into the anterior portion of the cell. These red granules do not appear to be membrane-bound and unlike the eyespots of many other photosynthetic protists they are not a part of the chloroplast. In close proximity to the eyespot, but on the emergent dorsal flagellum, there is a prominent flagellar swelling known as the paraflagellar body. The paraflagellar body is made up of a paracrystalline material containing flavins, pterins, and rhodopsin. It is believed that the paraflagellar body is the structure that is actually responsible for light detection and that the pigment granules of the eyespot complex act by shading the flagellar swelling. As the cell rotates through the water, a greater or lesser amount of light will strike the photoreceptive paraflagellar body as the pigmented stigma blocks the light, thus giving the cell the ability to orient either toward or away from the source of light.

Together, the paraflagellar body and the eyespot allow *Euglena* and other phototrophic taxa to exhibit positive phototaxis and maximize photosynthesis. Thus, for the most part a flagellar swelling/eyespot is found only in those euglenids that have chloroplasts or those that have secondarily lost them (see below). This makes sense considering that being able to sense and move into optimal light regimes would be of most use to photosynthetic organisms. However, phototaxis has also been reported for the heterotrophic *Peranema*, and structures resembling the flagellar swelling/eyespot have also been observed in the phagotrophic euglenid *Urceolus*. This suggests the interesting possibility that the eyespot of euglenids may have arisen as an adaptation to prey capture, such as feeding on photosynthetic diatoms that would be drawn to the light, and only subsequently was this organelle found to be useful to those euglenids that acquired a chloroplast. This novel theory may explain the fact that unlike the eyespots of most other photosynthetic protists, the light sensory structure of euglenids is not directly associated with the chloroplast but is instead an integral part of the flagellum.

Chloroplasts

The chloroplasts of photosynthetic euglenids are green in color, contain chlorophylls *a* and *b*, and generally have their thylakoids arranged in groups of three. Thus, they are in many ways very similar to the chloroplasts found in green algae. One important difference is that the chloroplasts of the euglenids have a third membrane (vs. two in the green algae) that surrounds the plastid. The food storage product of euglenids is also quite different from the starch (α-1,4-glucan) found in green algae in that it is a β-1,3-glucan known as paramylon. Based on these and other facts, it has been suggested that the chloroplasts of euglenids are secondarily derived from an endosymbiotic event whereby a phagotrophic euglenid acquired a eukaryotic green alga. Today, all that remains of this symbiosis is the original chloroplast of the green alga contained in the remnants of what was perhaps a phagocytic membrane vesicle. However, this presents a logistical dilemma, namely, if an intact photosynthetic eukaryote was engulfed by a euglenid, there should be four, not three, membranes surrounding the chloroplast. These would correspond to the inner and outer membrane of the chloroplast, the plasma membrane of the green alga, plus the food vacuole membrane of the euglenid.

An interesting possible explanation for the lack of a fourth membrane was put forth by Cavalier-Smith. Euglenids, along with some other protists, are capable of feeding by way of a process known as myxocytosis in which a hole is pierced through the plasma membrane of the prey cell and the cellular contents are sucked out into a food vacuole. This has been observed in the phagotrophic euglenid *Peranema* sucking the cytoplasm out of another euglenid *Lepocinclis*. As other ingested prey organelles such as nuclei and mitochondria are lost or digested, the chloroplast may be retained, and this might ultimately result in a chloroplast with three membranes (two original plastid membranes and one that remains from the food vacuole).

One difficulty in understanding the evolutionary relationships among the euglenids is the fact that many members of the group lack chloroplasts and are therefore colorless. In some cases, such as many of the phagotrophic species, this is apparently an ancestral condition (i.e., these species are descended from ancestors that were never lucky enough to have acquired chloroplasts). In other cases, the colorless condition is the result of species having 'lost' chloroplasts, or at least the photosynthetic capacity of chloroplasts and therefore their pigments. Comparable examples of this phenomenon can be found among colorless epiphytic plants such as the Indian pipe (*Epifagus*). Dramatic evidence for this was found in the colorless euglenid *Astasia longa*, which possesses a membrane-bound organelle found to have a circular 73 kb genome closely related to the chloroplast genome of *Euglena*, albeit substantially reduced in gene content. A closely related colorless euglenid, *Khawkinea*, also lacks photosynthetically functional chloroplasts but these species have retained the flagellar swelling/eyespot complex. Based on nuclear gene sequence and structural similarities, all *Khawkinea* species and *A. longa*, which were long considered to be members of separate genera, or even distinct families, were transferred to the genus *Euglena* and are now recognized as secondarily nonphotosynthetic species of *Euglena*.

One final unique feature of the chloroplasts of euglenids is that they have an unusual genetic structure. In the plastids of virtually all higher plants, algae, and cyanobacteria, the genes that comprise the ribosomal RNA (rRNA) operon are linked together and arranged in an opposite orientation and on opposite strands of the circular genome resulting in an 'inverted repeat.' Not only does *Euglena* not have this inverted repeat structure (it does have three tandem repeats of the rRNA operon but on only one strand) but it also has an unusually high number of introns in its chloroplast genome. To date, over 130 introns have been mapped on the *Euglena* chloroplast genome whereas the chloroplasts of higher land plants typically have fewer than a dozen. Even more amazing is that some *Euglena* plastid introns have introns within them. These self-splicing introns are spliced out in stages; only once the inner intron has been removed can the 'outer' intron assume the correct three-dimensional structure for its self-splicing.

Feeding Apparatus

The idea that *Euglena* and other photosynthetic euglenids are descended from a colorless phagotrophic ancestor is further supported by the fact that many extant euglenids are perfectly capable of ingesting both bacteria and other, sometimes large, eukaryotes. Some euglenids have an extremely well-developed feeding apparatus or 'cytostome' that consists of microtubule-reinforced rods and a set of curved vanes that act as a sphincter in opening and closing the central portion of the cytostome. Interestingly, the ability to ingest eukaryotic prey is apparently restricted to those colorless euglenids that, like *Euglena*, have a helically arranged pellicle. It is possible that the ability to ingest eukaryotic prey is dependent on being able to change cell shape through metaboly, thus explaining why all photosynthetic euglenids, which acquired their plastid by ingestion, have a helically arranged pellicle. Even those heterotrophic species that exhibit myxocytosis are capable of euglenid metaboly.

A large common pocket or 'vestibulum' at the anterior end of the cell contains both the opening of the feeding apparatus and the flagellar opening or 'reservoir.' It is also the only portion of the plasma membrane that is not underlain by a proteinaceous 'skeleton' and thus is believed to be the site of endocytosis. Another small membranous pocket that is lined by microtubules of the flagellar apparatus is designated as the microtubule-reinforced 'MTR' pocket. An apparently homologous structure is found in many heterotrophic euglenids and kinetoplastids where it has a separate opening adjacent to the flagellar opening and serves as

FIGURE 5 An undescribed species of euglenid from anaerobic sediments of the Santa Barbara Basin. Ectosymbiotic bacteria are clustered around the posterior portion of the cell.

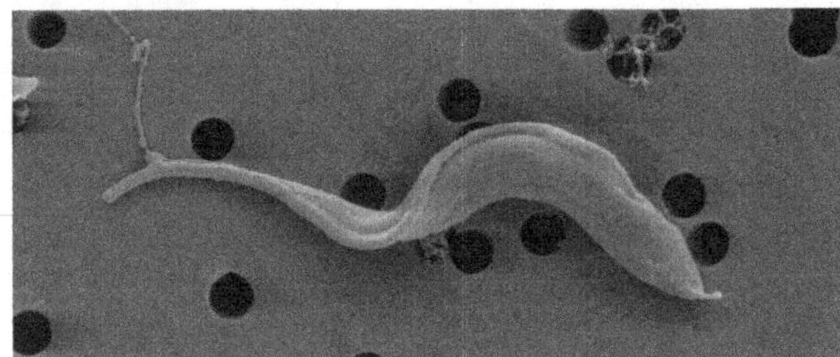

FIGURE 6 Scanning electron micrograph of *Leishmania* showing the attachment of the emergent flagellum along the plasma membrane of the cell.

the primary feeding apparatus for these bacteriotrophs. A much reduced MTR pocket is also still found in phototrophic euglenids and may play a role in additional nutrient uptake. In fact, the presence of an MTR pocket may be an apomorphy for the Euglenozoa, although it has most likely been lost or significantly modified in many taxa.

Habitat

Euglenids have been reported from a very wide variety of habitats ranging from marine sediments to ephemeral 'wagon ruts' or tree hollows. Most typically they are found in nutrient-rich freshwater ponds. In addition to being primary producers and heterotrophs of both bacteria and other eukaryotes, some euglenids are found as symbionts of metazoans. For example, euglenids have been reported from the intestines of poison dart frogs of Central America. These euglenids have three to four flagella per cell but otherwise appear to be closely related to colorless forms such as *Astasia*. It is not known if they are harmful to the host or merely commensal and so, for the time being, they are not referred to as parasites.

Euglenids have also been reported from anaerobic conditions in the sediments off the coast of Santa Barbara, California. Ultrastructural investigation revealed that these taxa appear to lack conventional euglenid mitochondria. Instead, membrane-bound organelles resembling hydrogenosomes were found in the periphery of the cytoplasm. Most strikingly, many of these euglenids harbor ectosymbiotic bacteria attached to their cell surface. These bacteria tend to congregate in areas adjacent to where the putative hydrogenosomes are localized (Figure 5).

KINETOPLASTIDS

The kinetoplastids, of which there are about 600 described species, lack a euglenid-like pellicle although the plasma membrane is supported by an underlying corset of microtubules. Kinetoplastids are united as a group within the Euglenozoa by three features: (1) the kinetoplast, the distinct mitochondrial genome that is found in close association with the flagellar bases; (2) glycosomes, which are microbody-like organelles in which the primary enzymes of the glycolytic pathway are concentrated; and (3) trans-splicing of short RNA leader sequences on to the 5′ end of all mRNAs. The kinetoplastids are divided into two groups: the biflagellate free-living or parasitic bodonids, which typically have retained an MTR-pocket type of feeding apparatus, and the parasitic or endocommensal (living within a host body) trypanosomes. The trypanosomes often have a reduced feeding apparatus and are osmotrophs. They typically have complex life cycles that involve various morphological stages in multiple hosts including an insect host that acts as the transmission vector. Unlike the bodonids in which the two flagella are unattached, the trypanosomes have a distinctive phase in which the flagellum is attached along its length to the cell body to form an undulating membrane (Figure 6). A special set of cytoskeletal microtubules often underlie the region of the undulating membrane and these may define the initiation site of division when the cells begin to undergo cytokinesis.

The Kinetoplast

The defining feature of the kinetoplastids is the kinetoplast, a distinct DNA structure found in their mitochondria. The DNA of the kinetoplast, or kDNA, has received the attention of cell biologists for it is the only known DNA that is in the form of thousands of interlocked circles known as 'catenations.' These are organized as either 25 maxicircles (20–38 kb) or 5000 minicircles (0.46–2.5 kb). Under the electron microscope, this network of interconnected kDNA has a very distinct appearance, not unlike that of the banded chromosomes of dinoflagellates (Figure 7). The kinetoplast may be found in various parts of the mitochondrion but in most species it lies immediately adjacent to the flagellar basal bodies. In fact, this is how the structure obtained its name meaning 'body near the kinetosomes.' When the basal bodies replicate prior to cell division, the kinetoplast also replicates and segregates one mitochondrion to each daughter cell. In some micrographs, there

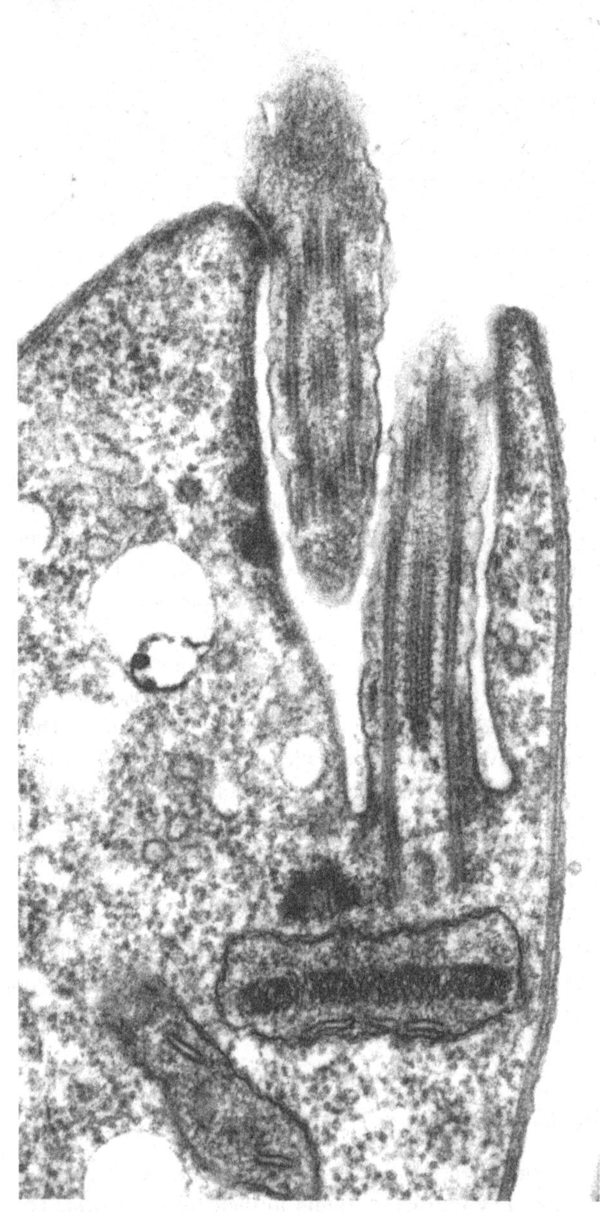

FIGURE 7 Transmission electron micrograph of *Leishmania* showing the kinetoplast adjacent to the flagellar bases.

appears to be a filamentous connection between the basal bodies and the portion of the mitochondrion that contains the kinetoplast.

The functions and modes of replication of the maxicircles and minicircles are quite different. The kDNA maxicircles contain genes that generally resemble the genes of mitochondria from other eukaryotes and code for essential mitochondrial proteins and mitochondrial rRNA. Many parasitic trypanosomes undergo a developmental sequence and at least in one species that has mutations in its maxicircles, this process is affected. Both maxicircles and minicircles are transcribed, but the role of the minicircles appears to be to act as templates for posttranscriptional RNA editing of the genes encoded in the maxicircles (see below). In addition, the minicircles may play a functional role in keeping the maxicircles physically connected. During replication of the kinetoplast (of which there is only one per cell), the minicircles (which make up about 95% of the DNA in the kinetoplast) are released from the linked structure and replicate as free minicircles. The replicated minicircles, utilizing a unique form of topoisomerase, then reconnect to reform a complete concatenated kinetoplast.

RNA Editing in Kinetoplastids

In the 1980s it was discovered that the kDNA genes coding for some essential mitochondrial proteins led to mRNAs that were essentially unusable in that they could not be translated into a functional polypeptide. These genes were essentially not open reading frames. It was recognized that in order for the mRNA to be functional, it required the posttranscriptional addition and deletion of uridine residues. The solution to this enigma came from the discovery of guide RNAs (gRNAs), small transcripts that contained the correct sequence of nucleotides that could bind to and 'repair' the primary RNA transcripts by guiding the insertion or deletion of uridine tracts into the mRNA. The gRNAs were primarily transcribed from the kDNA miniciricles, and by utilizing a complex set of RNA-editing enzymes collectively known as the 'editosome,' the gRNAs complement sections of the original mRNA transcript and ultimately lead to the production of a fully translationally competent mRNA transcript. This unique process of RNA editing seems to have arisen independently in the kinetoplastids; although mRNA editing is now known to occur in other eukaryotes including in mitochondria, chloroplasts, and even human nuclear-encoded genes, it is not known to occur anywhere else to the extent seen in kinetoplastids. The selective pressures that led to kinetoplastid RNA editing remain unknown but the presence of RNA editing in free-living taxa suggests that it did not arise as a result of adaptation to a parasitic lifestyle.

The Glycosome

The bloodstream forms of trypanosomes and leishmanias lack a functional Krebs cycle and they utilize the large amounts of available blood glucose to generate ATP through glycolysis. In these kinetoplastids, the first nine enzymes of glucose and glycerol metabolism are compartmentalized in a peroxisome-related microbody: the glycosome. Under aerobic conditions, glycerol-3-phosphate produced by the glycosome is reoxidized to dihydroxyacetone phosphate by mitochondrial glycerol-3-phosphate oxidase. Since the glycerol-3-phosphate oxidation is not coupled to oxidative phosphorylation in the mitochondrion only a minimal amount of ATP is produced from glycolysis carried out in the cytoplasm. However, under anaerobic conditions glucose metabolism results in equal amounts of pyruvate and glycerol being produced. The compartmentalization achieved in the membrane-bound glycosome is thought to facilitate the production of ATP by isolating the reactions from enzymes in the cytoplasm.

Habitat

The bodonids are recognized as a paraphyletic group in which life styles range from free-living bacteriotrophs (e.g., *Bodo*) to parasites (e.g., *Trypanoplasma*). The insect-borne, blood-inhabiting parasitic kinetoplastids consist of the trypanosomes and leishmanias. As the cells go through their two hosts (mammalian and insect), a number of different morphological changes may take place, resulting in different forms. Briefly, these different forms may be as follows:

- Amastigote – round to oval body with a short flagellum that does not emerge from the pocket.
- Promastigote – with a kinetoplast close to the anterior end and emergent and unattached flagellum.
- Opisthomastigote – similar but with a postnuclear kinetoplast and flagellar pocket that forms a long canal toward the anterior end of the cell.
- Choanomastigote – a body pyriform with the kinetoplast just in front of the nucleus.
- Epimastigote – with a prenuclear kinetoplast and a flagellum that emerges from a pocket part way along the cell body and attached to the body along its anterior portion.
- Trypomastigote – similar to epimastigote but with a kinetoplast and flagellar pocket that are postnuclear.

Many parasitic trypanosomes have variant surface glycoproteins (VSGs) that enable the population of infective cells to constantly change the antigens that they present to the host and thus avoid the ravages of the host immune system.

Leishmaniasis or 'kala-azar' is a widespread disease of humans and other mammals caused by *Leishmania*. The protist is spread between hosts by the female sand fly, which is common throughout the tropics. In the mammalian host, the protists are found as amastigotes in vacuoles of macrophages and lymphoid cells. Here they escape detection and destruction by the normal host lysozymic enzymes. In the insect vector phase of the life cycle, the flagellates transform within the insect host gut into the promastigote form and multiply before being delivered to a new mammalian host. The two species of *Trypanosoma* that are of greatest concern to humans are *Trypanosoma cruzi*, which causes Chagas disease and is endemic to South and Central America and *Trypanosoma brucei*, which is the causative agent of African sleeping sickness. As with *Leishmania*, *Trypanosoma* is spread by biting insects with the 'kissing bug' *Rhodnius* being the vector for *T. cruzi* and the tsetse fly *Glossina* spreading *T. brucei*.

DIPLONEMIDS AND OTHER EUGLENOZOA

In addition to the kinetoplastids and euglenids, there are several other taxa that belong to the Euglenozoa. One of these is *Diplonema* (synonym: *Isonema*), a free-living unicellular protist from marine environments. Like the euglenids, diplonemids have a corset of supporting microtubules under their cell membrane but unlike the euglenids these microtubules are not organized into pellicular strips. Details of the mitotic and flagellar apparatus as well as gene sequences clearly indicate that *Diplonema* and the closely related *Rhynchopus* belong to the Euglenozoa but their exact evolutionary position relative to the euglenids, trypanosomes, and bodonids remains unclear. There are also a number of free-living taxa that are clearly members of the Euglenozoa but are not easily assigned to the euglenids, kinetoplastids, or diplonemids. These include *Postgaardi* (Figure 8) and *Calkinsia* (Figure 9), both of which are found in anaerobic marine environments in association with ectosymbiotic bacteria.

One of the most intriguing euglenozoans is *Petalomonas cantuscygni*, which has a distinctive pellicle composed of longitudinal strips (Figure 10) that should identify *P. cantuscygni* as a euglenid. However, *P. cantuscygni* also has a number of fibrous mitochondrial inclusions in the vicinity of the flagellar apparatus that are very similar to the multiple small and dispersed kinetoplasts found in some bodonids (Figure 11). Thus, in one individual organism the diagnostic characters of both euglenids and kinetoplastids can be found, making *P. cantuscygni* appear to be the ideal Euglenozoan 'missinglink,' spanning the two well-defined clades. There are a number of other species that are currently assigned to either the euglenids or the kinetoplastids that upon closer inspection may not possess characters that would clearly identify them as a member of either group.

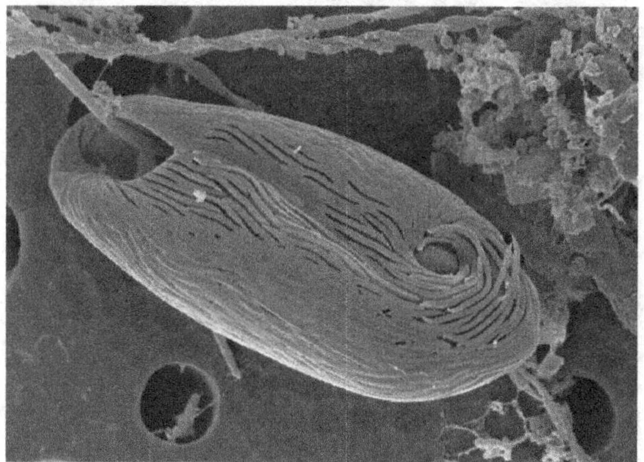

FIGURE 8 *Postgaardi*, another anaerobic Euglenozoan discovered in the Santa Barbara Basin, covered with ectosymbiotic bacteria.

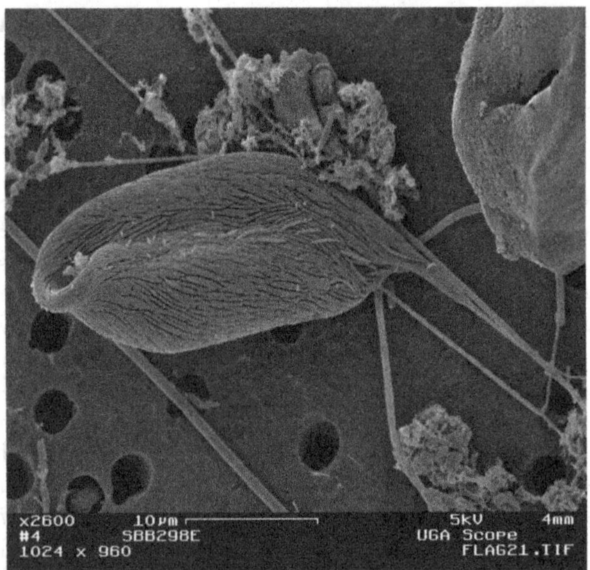

FIGURE 9 *Calkinsia*, an anaerobic Euglenozoan from the Santa Barbara Basin, covered with ectosymbiotic bacteria.

FIGURE 10 Scanning electron micrograph of *Petalomonas cantuscygni*.

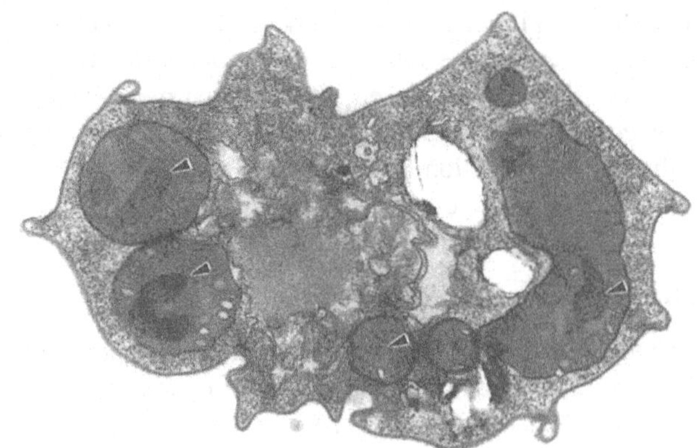

FIGURE 11 Transmission electron micrograph of *Petalomonas cantuscygni* showing euglenid pellicle and kinetoplast-like DNA in mitochondria (arrowheads).

EVOLUTIONARY RELATIONSHIPS

Ultrastructural and gene sequence studies have demonstrated that the trypanosomes are derived from a free-living bodonid-like ancestor. As in the euglenids, the cytoskeletal microtubules of the kinetoplastids are derived from one of the microtubular roots of the flagellar apparatus. In the bodonids, many heterotrophic euglenids, and *P. cantuscygni* one of the three microtubular flagellar roots extends anteriorly toward the tip of the cell and then recurves backward into the cytoplasm where it lines a pocket or invagination of membrane from the cell surface. It has been shown that this microtubule-reinforced MTR pocket acts as the ingestion organelle or cytostome and enables these organisms to ingest bacterial prey. The discovery of a homologous structure in the photosynthetic euglenid *Euglena* strongly supports the hypothesis that the phototrophic euglenids are descended from a heterotrophic ancestor, which is further supported by gene sequence data. In many heterotrophic euglenids and diplonemids, the feeding apparatus is considerably more complex than the simple MTR pocket but these complex cytostomes are believed to have evolved separately in these different lineages. The fact that a reduced MTR pocket is still maintained in photosynthetic euglenids, while any evidence of a more complex feeding apparatus has apparently been lost, suggests that the MTR pocket is no longer involved in bacterial ingestion in these species yet may still play a role in providing needed nutrients to the cell.

All lines of evidence point toward the ancestral Euglenozoan being descended from a bacteriotrophic ancestor that possessed an MTR-pocket cytostome. Analyses of gene sequence data suggest that *P. cantuscygni* lies at or near the base of both the euglenid and kinetoplastid lineages and may serve as a good proxy for the ancestral Euglenozoan.

It has also been proposed that kinetoplastids are descended from a photosynthetic ancestor. The enzyme arginine decarboxylase, which is present in plants but not found in animal cells, is also found in trypanosomes. In fact, comparative sequence analyses of the gene for arginine decarboxylase from the parasite *Leishmania* show that it is most closely related to the homologous sequence from tobacco. Since this enzyme is also present in *T. cruzi* and *T. brucei*, it raises the possibility of treating these diseases with some sort of polyamine inhibitor that would selectively affect the parasite but not the animal host. The fact that this enzyme may also be present in several other protistan groups suggests that it is a very ancient enzyme that has been lost in the animal lineage rather than provides direct evidence for a photosynthetic ancestor of trypanosomes. Likewise, the discovery of genes that code for fructose-1,6-bisphosphatase and glucose-bisphosphatase have been reported from trypanosomes and phylogenetic analysis shows these genes to group with those of photosynthetic organisms.

Despite these anomalies, the preponderance of the evidence suggests that phototrophy arose only once in the Euglenozoa and that this involved a eukaryotrophic euglenid with a flexible pellicle feeding on a green alga. All phototrophic euglenids, including those that have subsequently lost the ability to make a photosynthetically competent chloroplast, are descended from this single common ancestor and the kinetoplastids and diplonemids diverged from the euglenids before the plastid was acquired via secondary symbiosis.

The relationship of the Euglenozoa to other protists remains more uncertain. Certain gene sequences as well as the presence of discoidal mitochondrial cristae suggest that the Euglenozoa may be sister to the heteroloboseans, a large group of heterotrophic amoeboflagellates that also have discoid mitochondrial cristae. Furthermore, there is growing evidence that the Euglenozoa/Heterolobosea clade is part of an even larger clade known as the Excavata. Additional genomic data and their analysis should resolve these questions in the next few years and provide insights into understanding the origins of this unique and important group of organisms.

FURTHER READING

Adl, S. M., Simpson, A. G. B., Farmer, M. A., *et al.* (2005). The new higher level classification of eukaryotes with emphasis on the taxonomy of protists. *Journal of Eukaryotic Microbiology*, *52*, 399–451.

Breglia, S. A., Slamovits, C. H., & Leander, B. S. (2007). Phylogeny of phagotrophic euglenids (Euglenozoa) as inferred from hsp90 gene sequences. *Journal of Eukaryotic Microbiology*, *54*, 86–92.

Cavalier-Smith, T. (1981). Eukaryote kingdoms: Seven or nine? *Biosystems*, *14*, 461–481.

Leander, B. S., Esson, H. J., & Breglia, S. A. (2007). Macroevolution of complex cytoskeletal systems in euglenids. *Bioessays*, *29*, 987–1000.

Michels, P. A. M., Bringaud, F., Herman, M., & Hannaert, V. (2006). Metabolic functions of glycosoes in trypanosomatids. *Biochimica et Biophysica Acta – Molecular Cell Research*, *1763*, 1463–1477.

Milanowski, R., Kosmala, S., Zakrys, B., & Kwiatowski, J. (2006). Phylogeny of photosynthetic euglenophytes based on combined chloroplast and cytoplasmic SSU rDNA sequence analysis. *Journal of Phycology*, *42*, 721–730.

Rodriguez-Ezpeleta, N., Brinkmann, H., Burger, G., *et al.* (2007). Toward resolving the eukaryotic tree: The phylogenetic positions of jakobids and cercozoans. *Current Biology*, *17*, 1420–1425.

Simpson, A. G. B. (1997). The identity and composition of the Euglenozoa. *Archiv Fur Protistenkunde*, *148*, 318–328.

Simpson, A. G. B., Stevens, J. R., & Lukes, J. (2006). The evolution and diversity of kinetoplastid flagellates. *Trends in Parasitology*, *22*, 168–174.

Triemer, R. E., Linton, E., Shin, W., *et al.* (2006). Phylogeny of the euglenales based upon combined SSU and LSU rDNA sequence comparisons and description of discoplastis gen. Nov (Euglenophyta). *Journal of Phycology*, *42*, 731–740.

Chapter 23

Protozoan, Intestinal

K. Pierce and C.D. Huston
University of Vermont College of Medicine, Burlington, VT, USA

Chapter Outline

Abbreviations	**323**	**Recent Developments**	**331**
Defining Statement	**323**	*Cryptosporidium* species	331
Introduction	**323**	*G. intestinalis*	332
The Apicomplexa: *Cryptosporidium* Species	**324**	*E. histolytica*	332
The Microsporidia	**326**	**Conclusion**	**332**
The Flagellates: *Giardia intestinalis*	**327**	**Further Reading**	**332**
The Amoebae: *E. histolytica*	**329**		

ABBREVIATIONS

ALA Amoebic liver abscess
COWP Cysteine-rich oocyst wall protein
Gal/GalNAc Galactose/*N*-acetyl-D-galactosamine
IFN-γ Interferon-γ
MLCK Myosin light chain kinase
VSP Variant surface protein

DEFINING STATEMENT

This chapter reviews pathogenesis of disease due to the major intestinal protozoa that infect humans. Common themes are highlighted, including occurrence of both parasite- and host-induced tissue damage in each case, and the abilities of each parasite to survive environmental conditions, adhere to epithelial cells, and evade the immune response.

INTRODUCTION

Gastrointestinal parasitic protozoa are responsible for a significant amount of morbidity and mortality worldwide. *Cryptosporidium*, *Entamoeba histolytica*, and *Giardia* are the causative agents for the majority of protozoal diarrhea. Estimates of true prevalence are hindered by the technical and financial difficulties associated with diagnosis. Entamoeba is estimated to infect up to 50 million persons worldwide; 100 000 deaths annually are attributed to complications from infection. *Giardia*, *Cryptosporidium*, and *E. histolytica* have been identified in 11%, 8.4%, and 8%, respectively, of children with diarrhea in Bangladesh. In North America, *Giardia* is the parasitic cause of diarrheal illness responsible for 2–5% of cases; similarly *Cryptosporidium* affects up to 2% of those affected with diarrheal illness in the United States. This devastating impact on human health is most evident in the developing world. The burden of illness is greatest among children and adolescents for whom recurrent diarrhea often results in malnutrition and has deleterious effects on growth potential. In developed nations, the majority of severe disease occurs in people living with AIDS, recipients of solid organ transplants, and other immunosuppressed persons. These infections can be severe and life threatening. Limited treatment options and the diversity of clinical disease make understanding the pathophysiology of disease crucial to effective diagnosis and treatment. Additionally, at a time when violent acts of terrorism continue to increase, many of these microorganisms are potential agents of bioterrorism and may pose a significant threat to vulnerable populations, or cause societal disruption through widespread illness in healthy persons as occurred in an outbreak of *Cryptosporidium* in Milwaukee.

In this chapter, we will briefly review the clinical manifestations associated with infection by the medically relevant intestinal protozoa that infect humans (see Table 1) and the pathophysiology of disease caused by each. The emphasis will be on general themes that emerge regarding the host–parasite relationship. The pathogens fall into two broad categories, intracellular and extracellular, and each pathogen is highly adapted to its own unique lifestyle. However, several common requirements for disease causation

TABLE 1 Major protozoan pathogens of the human gastrointestinal tract

Major taxonomic classification		Species
Apicomplexa		*Isospora belli*
		Cryptosporidium species
		Cyclospora cayetanensis
		Toxoplasma gondii
Microsporidia		*Enterocytozoon bieneusi*
		Encephalitozoon intestinalis
Sarcomstigophora		
	Amoebae	*Entamoeba histolytica*
	Flagellates	*Giardia intestinalis*
		Dientamoeba fragilis

are evident, including an ability of the infectious form of each parasite to survive in harsh environmental conditions, an ability to attach to intestinal epithelial cells, and an ability to evade the host immune response. In most cases, the host's immune response is necessary for protection, but also contributes to pathology. Obvious differences also emerge. For example, infection with the intracellular intestinal protozoa is of greater severity in individuals with cell-mediated immune defects, such as those with AIDS, but infection with the extracellular pathogens (e.g., *E. histolytica*) has not been shown to be more common or severe in this population (see Table 2). In fact, much of the tissue damage that occurs during *E. histolytica* infection appears to result from a robust immune response.

TABLE 2 The effect of AIDS on clinical outcome of infection by the intestinal protozoa correlates with whether the parasite is intra- or extracellular

Pathogen	More common and severe in persons with AIDS	Site of infection (intra- vs. extracellular)	Dysentery
Cryptosporidium	Yes	Intracellular	No
Microsporidia	Yes	Intracellular	No
Isospora	Yes	Intracellular	No
Cyclospora	Yes	Intracellular	No
Giardia intestinalis	No	Extracellular	No
Entamoeba histolytica	No	Extracellular	Yes

THE APICOMPLEXA: *CRYPTOSPORIDIUM* SPECIES

Members of the phylum Apicomplexa, *Cyclospora cayetanensis*, *Isospora belli*, and *Cryptosporidium* species, are all spore-forming coccidian protozoa (see Table 1); intracellular parasites reside within the epithelial cells of the host's intestinal tract; all cause dysfunction or abnormalities of absorption/secretion and motility of the intestinal tract of the infected host. All three organisms produce a watery diarrhea, which can be accompanied by constitutional symptoms. In the immunocompromised host, infection with any of these pathogens can be quite severe and protracted, leading to chronic diarrhea and malabsorption. Although they are primarily gastrointestinal parasites, rare cases of disseminated disease have been reported in immunocompromised patients.

Each of these intestinal protozoa is a frequent AIDS-related pathogen; *Cryptosporidium* species and microsporidia (see section 'The Microsporidia') are responsible for the lion's share of AIDS-associated diarrhea (see Tables 2 and 3). In immunocompetent individuals, they also constitute a major cause of traveler's diarrhea, and have been implicated in both community and institutional outbreaks. Despite this, these protozoa were largely unrecognized as human pathogens prior to the onset of the AIDS epidemic. The presence of cell-mediated immune deficiency leads to a marked increase in clinical severity. In many series, cryptosporidia have been recovered from the stools of 10–20% of patients with AIDS-related diarrhea. The prevalence of these organisms and severity of illness are directly related

TABLE 3 Effect of immune defects on susceptibility to infection by different intestinal protozoa

Immune defect	Illness associated with defect	Protozoa
Cell mediated		
Lymphocyte	AIDS, ALL	*Cryptosporidium*
Macrophage	Steroids, immune modulating therapy	*Isospora*
	Organ transplants	*Cyclospora* Microsporidia
Humoral		
Antibody, complement	CLL, antibody or complement deficiency, splenectomy	*Giardia*

to declining CD4 counts, with higher rates noted in patients with CD4 counts < 100. *I. belli* infection is endemic to Africa, Asia, and South America and is less commonly associated with AIDS-related diarrhea in the United States (0.2%). In Haiti, the prevalence of isosporiasis is about 15% in persons living with AIDS. *C. cayetanensis* has been identified worldwide as a cause of diarrheal illness in both immunocompetent and immunocompromised persons.

As noted above, all organisms in this group mature and multiply within the host epithelial cell enterocyte and also produce an infectious particle called an oocyst. The oocysts are excreted into the host stool where infection can be acquired by another host following ingestion of fecally contaminated food or water. Although all three organisms primarily infect the gastroepithelium of the small intestine, their intracellular location is not the same. The morphologies and sizes of infectious particles coupled with location within the host enterocyte allow for diagnosis to be made on the basis of biopsy or aspirate. More specific fecal antigen detection tests are also available. Nevertheless, the similarities in life cycle and biology help to explain why their clinical manifestations are so similar. Additionally, the mechanisms by which these organisms infect their host are similar. Of the intestinal Apicomplexa, *Cryptosporidium* species have been most well studied, and features of its life cycle and pathogenic ability can serve as a model for understanding this group of pathogens.

The oocysts are one of the most conserved features of the intestinal Apicomplexan parasites. Their small size and heartiness make the oocysts quite resistant to standard water-treatment procedures. The oocyst of cryptosporidium, which is resistant to chlorination, is often found in drinking water supplies, awaiting ingestion by a host. The cryptosporidium oocyst contains four sporozoites surrounded by walls made of two electron-dense layers comprised of lipids, carbohydrates, and a family of cysteine-rich oocyst wall proteins (COWPs). The heartiness of the oocyst against environmental pressures is partially due to the COWP family members. Similar gene families are present in other intestinal Apicomplexans as well as *Toxoplasma gondii*, another Apicomplexan parasite, as evidenced by immunogold labeling and electron microscopy.

Products of the gene family COWP are localized to the inner layer of the oocyst walls as well as in wall-forming components found in macrogametes. Repeated cysteine-rich amino acid motifs and the frequent presence of cysteine residues linked by disulfide bonds create a rigid outer wall conferring protection for the cyst until it can find a suitable host. In addition, these cysteine-rich proteins may provide protection from acidic conditions present in the stomach and from digestive enzymes present in the intestinal lumen.

Several factors within the host may stimulate excystation. The oocysts are activated in the stomach and upper intestines. In vitro treatments believed to mimic the normal passage of the oocysts through the host digestive tract, such as exposure to acidic environments, bile salts, and proteases, all enhance excystation of cryptosporidium. Proteolytic enzymes that appear to influence excystation of the oocysts include zinc-binding aminopeptidases and arginine peptidases. These may provide a clue to the significance/importance of host cell factors in influencing/resisting clinical disease. Additionally, trypsin has been observed to increase the motility of sporozoites in experimental models, suggesting that digestive enzymes may play a role in stimulating parasite motility as a precursor to invasion of epithelial cells.

The Apicomplexan parasites are obligate intracellular parasites, and the extracellular forms that exist outside the host cell – the microgametes, merozoites, and sporozoites – are all particularly vulnerable to the acidic environment of the host gastrointestinal tract. Therefore, after excystation of the infectious oocyst within the lumen of the small intestine, the extracellular sporozoites must quickly invade epithelial cells in order to survive. This process involves directed motility, adherence, penetration, and formation of a parasitophorous vacuole, a specialized compartment inside the host cell in which the parasite resides and replicates.

The general process of host cell invasion is highly conserved among the Apicomplexan parasites, and is mediated by an initial adhesion event followed by extension of a unique apical structure called the conoid, and the coordinated secretion of the contents of a group of specialized, secretory organelles collectively known as the apical complex (hence the name Apicomplexan). This complex consists of micronemes, dense granules, rhoptries, and, except for Cryptosporidium, an apicoplast. Secretion of microneme contents following adherence serves to deliver additional adhesins, increasing parasite adherence to the host cell. The rhoptries are club-shaped organelles associated with the endoplasmic reticulum and Golgi complex. In contrast with other Apicomplexa, the sporozoites of Cryptosporidium have only one. The rhoptries are discharged after the micronemes. They open at the apex of the parasite and extend to the site of attachment to the host cell. Rhoptry proteins are required for formation of a moving junction. This is a ring of parasite proteins incorporated into the host cell membrane that excludes host cell membrane proteins from the forming parasitophorous vacuole and likely provides an anchor with which the parasite pulls itself into the host cell. The contents of the dense granules are also discharged during cell invasion, and also help to create the parasitophorous vacuole.

Since the Apicomplexa do not contain pseudopods or flagellae, they employ a unique mechanism of gliding motility, which facilitates both transit from the intestinal lumen to the epithelial cells and penetration of epithelial cells. The sporozoites of Cryptosporidium follow both circular and helical gliding patterns, leaving behind gliding trails rich with surface-associated proteins. Unlike some of the other intestinal parasites, gliding does not require that the parasite changes its shape. The movements of these Apicomplexa are driven by an actin–myosin motor, which is anchored to an inner membrane complex and mediates movement of adhesins on the parasite surface from its apical to its posterior end. Thus, movement of adhesins bound to a host cell pulls the parasite forward analogous to the forward propulsion by a tank tread. This mechanism allows the sporozoite to travel along cell membranes of the intestinal epithelium. In vitro studies using cytochalasins to interfere with actin polymerization have confirmed the presence and requirement of actin for locomotion. DNA sequences from several apicomplexan parasites, furthermore, suggest the existence of evolutionarily different myosins that some investigators feel may be suitable targets for drug development.

For the intestinal Apicomplexa, the first step in host cell invasion is attachment of the sporozoite to the microvillus border of the small intestinal epithelium. The mucinous lining of the gut affords the host a barrier of protection against the acidic environment as well as against microbial invasion. The layer of mucus, however, may assist the cryptosporidium in reaching its goal. Oocyst attachment may be enhanced by mucus production. For the vulnerable sporozoite, this mucus may provide protection and transport assistance to the host cell. Degradation of mucin by secreted proteases then affords the parasite intimate contact with the microvillus border, followed by attachment to epithelial cells. In vitro models have demonstrated that a variety of environmental conditions including parasite numbers, changes in pH, presence of magnesium and calcium, and temperature can all affect sporozoite attachment. Inhibition of attachment by antibodies specific for surface membrane proteins of the sporozoite has demonstrated the importance of protein–protein and protein–glycoconjugate interactions in facilitating attachment. Specifically, galactose/*N*-acetyl-D-galactosamine (Gal/GalNAc)-specific lectins on the surface of the sporozoite bind to protein glycoconjugates located on the apical surface of the host epithelial cells. The importance of host cell surface protein glycoconjugates in Cryptosporidium attachment and invasion has been further elucidated using the enzymes β-galactosidase and *O*-galactosidase to remove carbohydrates from target cells. As noted above, an initial attachment event triggers microneme secretion, and, presumably due to secretion of additional adhesins, microneme secretion is followed by more intimate attachment of the parasite to the host cell.

It is the attachment of motile sporozoites to the gastric epithelium and subsequent invasion of the cell membrane that signal the formation of the parasitophorous vacuole. This unique structure provides protection from the harsh environment of the host gastrointestinal tract. Within this organelle, the merozoites develop both asexually and sexually. Other coccidia also reside within a parasitophorous vacuole, although Cryptosporidium's parasitophorous vacuole is unique in that it is partially formed by elongation and fusion of the microvilli most closely associated with the invading parasite. In vitro data suggest that host cytoskeletal proteins and the water channel aquaporin I are necessary factors for cryptosporidium cell entry. The vacuole itself is surrounded by the parasitophorous membrane, which provides a vehicle for nutrient exchange. In addition, Cryptosporidium uses a feeder organelle membrane that keeps the parasite highly sequestered within the host cell. This ensures that the parasite is protected from the outside environment of the host's gastrointestinal tract as well as assisting with evasion of the host's immune response. Finally, mature meronts exit the host cell, causing cell lysis. Though not studied for cryptosporidium, recent studies on the closely related Apicomplexan parasites *Toxoplasma gondii* and *Plasmodium falciparum* indicated that host calpain proteases are required.

The cytotoxic effects of cryptosporidium may be a combination of direct effects on the host cell and an immune-mediated process. The number of parasites infecting the mucosa influences disease severity, with larger numbers leading to more robust disease. This is especially true in patients who cannot mount an adequate immune response. Direct cytotoxic effects have been demonstrated in both human and animal models. Infection with all three intestinal Apicomplexa can produce profound pathologic effects including villus shortening, crypt hyperplasia, atrophy of microvilli, and an increase in leukocytes in the lamina propria; ulceration is rare. Whether these changes are induced directly by the parasite or a result of the host response to infection remains a subject of investigation.

Murine models have demonstrated that the key mediator of the host immune response to infection with Cryptosporidium is interferon-γ (IFN-γ), which is produced by antigen-stimulated T cells and NK cells. The role of IFN-γ in protective immunity in human disease is less clear. In human experimental models not all infected subjects produce IFN-γ. In addition, this cytokine has not been detected in children with self-limited disease. However, IFN-γ producing dendritic cells, located on the mucosal surface of the intestinal epithelium, may play an important role in activating the lymphoid tissue within the lamina propria, resulting in production of specific intraepithelial lymphocytes and control of infection ensues.

THE MICROSPORIDIA

The phylum of Microspora contains a large number of species. The microsporidia were first recognized as pathogens in silkworms in 1857, but have subsequently been found to be the cause of disease in a wide range of hosts. Over 1000 species of microsporidia are known. These organisms infect host species ranging from protists to humans, and were first associated with human infections in 1924. Until 1984, however, cases of documented human infection were rare. Since the beginning of the AIDS epidemic, human infections with *Enterocytozoon bieneusi* have been increasingly recognized in immunocompromised patients, and it is now clear that microsporidia are the most common cause of chronic diarrhea in individuals with advanced HIV infection/AIDS. In some case series, microsporidia have been identified in as many as 45% of patients. As is the case with the intestinal Apicomplexa, disease frequency and severity is inversely correlated with CD4+ T cell counts.

Seven genera of microsporidia are known to infect humans, but the vast majority of infections are caused by two species: *E. bieneusi* and *Encephalitozoon intestinalis*. Recent sequence analysis of Microsporidia genes suggest that these obligate intracellular parasites are related to fungi. Infectious spores of microsporidia are found throughout the environment, but mainly in ground surface water, municipal water supplies, and sewage treatment facilities. As the name Microspora implies, the infectious spores are quite small, measuring 1–2 μm. Similar to other intestinal protozoa, the hearty oocyst has a thick, resistant wall ensuring survival in the harsh environment outside of the host. Within its life cycle there are two basic morphologic forms: the meront, or proliferative stage, and the spore, which is responsible for transmission.

For many species of microsporidia, the route of transmission of an organism may play a key role in determination of virulence. These parasites may be transmitted either horizontally or vertically. Horizontal transmission occurs as the parasite, in contaminated food or water, is ingested by the host and infects the gut epithelium; infection may spread to other tissues. Viable spores have been recovered from stool, urine, and even respiratory secretions, leading some investigators to conclude that person-to-person transmission does occur. For cases of nonhuman infections, specifically invertebrates, vertical infections are described as transovarial. This describes infection passed from mother to zygote through the cytoplasm of the egg. Vertically transmitted parasites typically have reduced virulence as the host's survival is integral to their success.

Like the intestinal Apicomplexa, all microsporidia have an unusual, highly resistant spore wall. The sporuloplasm contains a nucleus (depending on the genus and species it may have either one or two nuclei), a posterior vacuole, and ribosomes. The electron-dense spore coat is comprised of proteinaceous material. The inner membrane is separated from the external coat by an endospore comprised of chitin. Changes in hydrostatic pressure within the spore provide the trigger for spore discharge. It is the spore, commonly found in surface water, which is the infectious stage of the organisms. Ingestion of the spores leads to initial infection of the host gastrointestinal tract.

The microsporidia use a unique organelle called the polar tube in host cell invasion. This apparatus is attached to the anterior end of the spore, and serves as a bridge through which the sporuloplasm is injected into the host cell cytoplasm. Similar to attachment of the Apicomplexa, lectin–carbohydrate interactions appear to be important for initial attachment of the spore to the host epithelium and for signaling discharge of the polar tube.

Invasion for microsporidia begins with spore discharge, which in itself requires three processes to occur: activation following change in osmotic pressure within the spore, realignment of the polar tube, and injection of the sporuloplasm into the host cell. Activation of the spore requires a variety of conditions, specific to each species. These conditions vary from an acidic pH to an alkaline pH; some species require a dramatic change between the two. Once activated, regardless of mode, the spore experiences a change in osmotic pressure. This pressure change triggers spore discharge, during which the polar filament is extended. This leads to piercing of the host cell membrane, and passage of the sporuloplasm into the host cell then ensues. Many theories exist on how this intricate pressure change occurs within the tiny spore. Changes in the proton gradient, decreased trehalose levels, and the displacement of calcium from the polarplast membrane have all been investigated.

The polar tube provides safe passage for the sporuloplasm as it enters the host cell by bridging the distance between host cytoplasm and parasite. Traditionally, the polar tube is described as piercing the cell. In fact, the exact mechanism of this interaction is not well characterized. Some investigators have postulated that host cell actin mediates this interchange. The polar tube may also enable ingested spores to escape from macrophage phagosomes, allowing the organism to escape one cell and infect an adjacent cell, thereby disseminating the infection. Consistent with this theory, researchers have demonstrated the ability of these spores to germinate within a phagosome.

THE FLAGELLATES: *GIARDIA INTESTINALIS*

A number of flagellates inhabit the small and large bowel lumen. These include *G. intestinalis*, *Dientamoeba fragilis*, *Chilomastix mesnili*, *Trichomonas hominis*, *Enteromonas hominis*, and *Retortamonas intestinalis*. Both *G. intestinalis* and *D. fragilis* have pathogenic potential, but *G. intestinalis* is by far the most medically important.

The clinical manifestations of *Giardia* infection are varied. The spectrum of disease ranges from no symptoms to severe diarrhea, malabsorption, and life-threatening malnutrition. The biology of the parasite and how it causes such severe disease is not well understood. The degree of disease is related to the number of parasites ingested, but as few as ten cysts are needed to cause clinical disease. The immune system of the host also plays a key role in determining the severity of disease. The strain of *Giardia* is also likely to be important. Seven *G. intestinalis* genotypes (called assemblages) have been defined, of which only assemblages A and B are known to infect humans. Assemblage B is more prevalent, and is the only laboratory isolate with confirmed virulence for animals and humans. These studies suggest that virulence correlates with genotype, but molecular epidemiologic studies remain inconclusive. No virulence factors have been identified to account for potential strain differences. However, recent genomic sequencing efforts have demonstrated major differences in genes encoding a family of immunodominant variant surface proteins VSPs (see below).

The adaptability of this organism to survive harsh environments contributes to its virulence. Much like the oocysts of the Apicomplexa, the cysts of *G. intestinalis* are hearty and survive in the ambient environment for months to years. These infectious cysts, dormant in the environment, are transmitted person to person as well as by ingestion of contaminated food or water. Once the cysts are ingested by the host, the low pH of stomach acid, followed by exposure to bile salts, and host intestinal proteases trigger excystation in the small bowel. Trophozoites (Figure 1) emerge from the cyst. Following excystation, the trophozoites use their four pairs of flagellae to swim through the fluid present in the intestinal lumen and infect the small bowel. As is the case with the Apicomplexa, cysts pass into the environment from the host, spreading the infection to others.

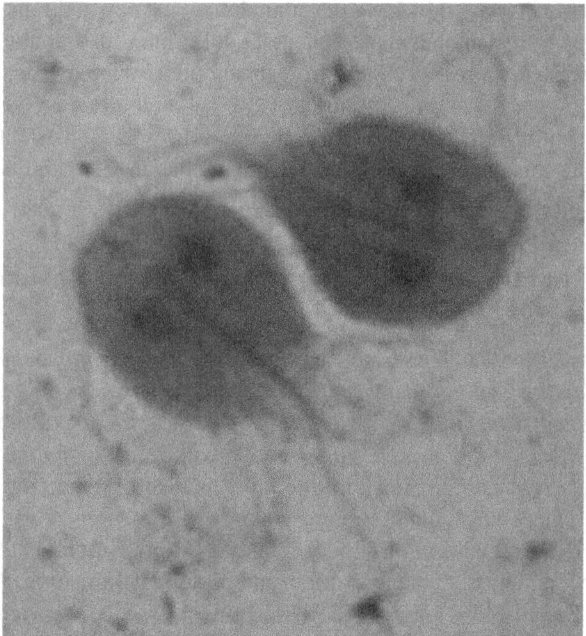

FIGURE 1 Light micrograph of *Giardia intestinalis* trophozoites present in a fecal sample. The trophozoites are characteristically pear-shaped and binucleate. Four pairs of flagellae are located posteriorly. The adhesive disc is present on the ventral surface of the anterior end; by light microscopy it is visible as an area of pallor.

Giardia trophozoites also adhere to epithelial cells, colonizing the small bowel where they may remain for weeks to years. Attachment of the trophozoites to the small bowel is guided by a unique feature, the ventral suction disk. This is a large, rigid structure on the surface of the parasite. The edge of the disk contains a crest, allowing the organism to insert itself between the microvilli of the small intestine. Adherence is believed to be mediated by a clasping or suction-like mechanism, but specific receptor–ligand interactions may also participate. One candidate receptor, taglin (trypsin-activated *Giardia* lectin), has been identified. This lectin on the trophozoite's surface is activated by exposure to trypsin, and, once activated, it binds to protein glycoconjugates on intestinal epithelial cells.

The trophozoites of *Giardia* are neither invasive nor do they secrete any identified toxin. However, attachment of the trophozoite itself may disrupt the integrity of the brush border. This disruption has been studied in both human and murine models and correlates with a decrease in disaccharidases normally present along the brush border. In addition, the immunological response to infection contributes to damage the intestinal architecture. Disruption of the epithelial brush border results in absorptive changes and, therefore, osmotic diarrhea. Disruption of the epithelial barrier is required for the development of disease. This is accomplished via alterations in the tight junction structure, which leads to an increase in intestinal permeability. In vitro models of epithelial barrier function during *Giardia* infections have demonstrated that the trophozoite damages the tight junctional zona occludens, leading to increased permeability of the gut wall and apoptosis of epithelial cells.

Host cell myosin light chain kinase (MLCK) is integral to the disruption of the intestinal cell barrier by *Giardia*, and, therefore, contributes to the clinical manifestations of the disease. MLCK phosphorylates myosin light chains in the gastrointestinal epithelial cells, leading to changes in tight junction structure and subsequent increase in epithelial permeability. Activity of MLCK may be increased in the presence of *Giardia* trophozoites. T cells, activated in response to intestinal infection, initiate a signaling cascade ending in the induction of TNF-α, which also upregulates MLCK activity and increases gastrointestinal permeability. Thus, as is the case for the intracellular intestinal protozoa, an effective immune response is critical to clearance of the infection, but also contributes to the development of abnormal intestinal pathology.

The host immune response to *Giardia* is activated at both the cellular and the humoral levels. The importance of humoral immunity is strongly suggested by the increased incidence and severity of *Giardia* in patients with common variable immunodeficiency, an immune defect characterized in part by agammaglobulinemia (see Table 3). Secretory IgA is involved in local control of the organism. In studies of B cell-deficient mice, researchers found that animals with decreased levels of IgA were unable to resolve their infections. The precise role of antibodies in controlling the infection is still not completely elucidated; however, they may play an important role in decreasing rates of bowel colonization by inhibiting adherence of trophozoites to intestinal epithelial cells. Experimental data from murine models also demonstrate an important role for the cellular immune response in controlling *Giardia* infections. In persons living with HIV/AIDS (and the severe cell-mediated immune defect it imparts), however, the incidence of *Giardia* is no greater than in the general population (see Table 2).

During giardia infection, the parasite's immunodominant surface protein (the VSP) undergoes continuous variation, which is believed to enable evasion of the host's humoral immune response (see Table 4). This phenomenon has been demonstrated in both animal and human experimental models. The VSPs are a family of related proteins, which cover the entire surface of the parasite. They contain repetitive cysteine-rich motifs with hydrophobic tails and zinc-finger motifs, and, as is the case for proteins on the *E. histolytica* surface (see section 'The Amoebae: *E. histolytica*') and the *Cryptosporidium* oocyst, disulfide bonds are believed to confer resistance to intestinal proteases and allow the organism to survive within the host gastrointestinal tract. Genes encoding approximately 190 VSPs are present in the *Giardia* genome. Expression of the VSP genes is tightly regulated; at any given time, only one VSP gene is expressed. The mechanisms that regulate VSP expression remain poorly understood, but recent data suggest a RNA-mediated interference dependent mechanism. The triggers for switching the expressed VSP gene are also unclear. These proteins rapidly vary their surface antigens in vitro; experimental models in humans using *Giardia* clone (GS/H7) demonstrated changes in VSP expression between 2 and 3 weeks after the initial infection.

TABLE 4 Summary of mechanisms used by different parasites for evasion of the host immune response

Parasite	Mechanism
Apicomplexa	Parasitophorous vacuole
Giardia lamblia	Antigenic variation (VSPs)
Entamoeba histolytica	Alteration of MHC II
	Inactivation of complement
	Degradation of IgA IgG antibodies
	Monocyte locomotion inhibitor
	Secreted proteases

It is likely that both immune pressures and environmental pressures result in selection of a dominant clone, so that at any given time only one VSP is expressed in the population. For example, development of a neutralizing antibody response to a given VSP (clonal population) would drive emergence of a new clone expressing an as yet unencountered VSP. The immune selection likely occurs via the host's IgA antibodies selecting against certain VSPs. In murine models looking at immunocompromised and immunocompetent mice and gerbils, investigators found that while the immune-competent controls made successful antibodies to all VSPs, the immunocompromised animals selected for some of the VSPs. Investigators concluded that both positive and negative selective pressures contribute to the emergence of a dominant clone. The nonimmune-mediated mechanisms that factor into selection of VSP changes are likely environmental, such as intestinal proteases, but the precise factors remain to be clearly defined. The ability of *Giardia* to alter its surface coat and adapt to changing pressures within the host (both immune and environmental) likely plays a critical role in the parasite's ability to establish prolonged infections and colonization.

THE AMOEBAE: *E. HISTOLYTICA*

Many *Entamoeba* species infect the human colon. Of these, only one, *E. histolytica*, is pathogenic. *E. histolytica* causes invasive amebiasis, a disease characterized by dysentery and, in a minority of patients, liver abscesses (Figure 2). Infection with *E. histolytica* must be distinguished from infection with *Entamoeba dispar*, a morphologically identical, nonpathogenic species that has been recently recognized as distinct from *E. histolytica* based on genetic and

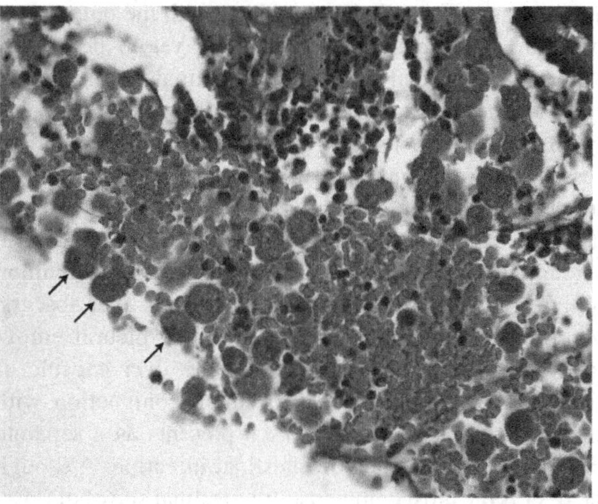

FIGURE 2 Hematoxylin and eosin-stained colonic biopsy from a patient with invasive amebiasis. Numerous *Entamoeba histolytica* trophozoites are visible (arrows), many containing ingested erythrocytes and cellular debris. The colonic epithelium is ulcerated, and inflammatory cells and blood are present in the intestinal lumen.

antigenic differences. The cysts of the nonpathogenic *E. dispar* are often found in men who have sex with men and so-called 'cyst carriers', contributing to diagnostic confusion. These amoebae are all obligate fermenters, and, therefore, their genomes do not encode Krebs cycle enzymes and those necessary for oxidative phosphorylation. Other species within this family that infect humans but are considered nonpathogenic include *Entamoeba moshkovskii*, *Entamoeba hartmanni*, *Entamoeba gingivalis*, *Entamoeba polecki*, *Entamoeba coli*, and *Blastocystis hominis*. Presence of these organisms may represent evidence of fecal oral transmission. In some cases (e.g., *B. hominis*), pathogenic potential and the significance of isolation remains controversial, but most believe that identification of these organisms is not clinically significant except as a marker of potential exposure to pathogenic protozoa.

E. histolytica is found throughout the world, although the greatest burden of disease exists in areas of poor sanitation. Populations at increased risk for amebiasis include institutionalized persons, men who have sex with men, travelers to the developing world, and individuals living in developing nations. Globally, over 50 million cases of symptomatic infections are thought to occur annually. In Dhaka, Bangladesh, approximately 50% of children are infected by 5 years of age. There are two distinct clinical syndromes: amoebic colitis and amoebic liver abscess (ALA). The presentation of these two entities varies geographically. In some areas of the world the predominant manifestation is mostly colitis, whereas in other areas ALA is the classic presentation. These epidemiologic observations suggest that genetic differences in the host and/or parasite strains likely determine the outcome of infection. One molecular epidemiologic study has demonstrated that strain variation correlates with the outcome of infection, but specific differences are yet to be clearly defined. The clinical spectrum of intestinal disease ranges from asymptomatic infection to fulminant colitis and dysentery, leading to toxic mega colon and perforation. Less severe amebiasis occurs in a nondysenteric syndrome, which is often chronic; the clinical syndrome can often be confused with inflammatory bowel disease, a noninfectious disease of the colon. The most common nonintestinal manifestation of the disease is liver abscess, which can be complicated by peritonitis, pleural empyema, and pericarditis if rupture of the liver capsule or diaphragm results. This may occur in conjunction with intestinal disease, but more often presents as a separate entity with no evidence of intestinal infection. Although the liver is the most common extraintestinal site of disease, abscesses of both the lung and the brain are also reported. Immune compromise is not necessarily a risk factor for these unusual presentations as they occur in immunocompetent individuals as well.

The cyst of *E. histolytica*, unlike the trophozoite form, has four nuclei and a wall rich in chitin. This cyst is also highly resistant to desiccation and chlorination. It can survive in the outside environment for many weeks, retaining its infectious potential. Similar to the other intestinal protozoa, pH changes within the host and intestinal proteases are the likely triggers of excystation, though the factors driving excystation and encystation of *E. histolytica* remain poorly defined. Studies using *Entamoeba invadens*, a pathogen of reptiles, have demonstrated an important role for quorum sensing and signaling through a D-Gal/GalNAc-specific amoebic surface lectin (see below) in the encystation process. This same process is likely to be conserved in the human pathogen. Following excystation, the amoeboid trophozoites move through directed chemotaxis, toward bacteria, from which they derive nutrition via phagocytosis, and the intestinal epithelium to which they adhere.

The first step in *E. histolytica* infection is colonization, which is facilitated by the interaction of an amoebic Gal/GalNAc-specific surface lectin with protein glycoconjugates present in intestinal mucus. In addition to playing a critical role in adherence to mucins, attachment via the amoebic Gal/GalNAc-specific lectin is essential for parasite adherence to host epithelial cells and for parasite-induced host cell lysis. Intestinal mucus and its carbohydrate moieties are the host's first layer of defense against invasion by the parasite. The galactose and *N*-acetyl-D-galactosamine residues of the host mucins prevent binding of the amoebic Gal-lectin to the host cell surface, which prevents amoeba-induced contact-dependent cell killing. Binding of the *E. histolytica* trophozoite to the colonic mucosa may induce further production of mucin. The increase in mucin production is a double-edged sword. Although an increase in secretions may offer some protection against the cytotoxic effects of attachment, this will also increase the attachment area for bowel colonization. Noninvasive infection of the bowel is the result of the interaction between the parasite and host mucin cell proteins; thus, overproduction of mucus may favor colonization as opposed to invasive disease. However, as its name implies, when it comes into contact with host cells, *E. histolytica* induces host cell death via secreted pore-forming peptides (see below) and contact-dependent activation of host cell caspases. Apoptosis of host cells is a prominent mechanism of cell death in the cecum of animals infected with *E. histolytica*, and caspase inhibitors reduce the burden of infection. Interestingly, apoptotic cell death appears to be a precursor for amebic phagocytosis of host cells, but the consequences of this for pathogenesis remain unknown.

Secretion of proteolytic enzymes also plays a key role in attachment and invasion by *E. histolytica* trophozoites. Secreted cysteine proteases degrade colonic mucins and extracellular matrix proteins, thereby facilitating access to the epithelial layer and tissue penetration. A direct correlation exists between virulence and *E. histolytica* cysteine proteinase activity. Genes encoding 36 putative cysteine proteinases are present in the *E. histolytica* genome database. A growing body of evidence suggests that these proteases (EhCp1, 2, . . .) are responsible for much of the tissue destruction, as they degrade extracellular matrix proteins including collagen and fibronectin. The *MUC2* gene product, which is the main glycoprotein component of colonic mucus, contains threonine- and proline-rich sequences, which confer resistance to degradation by proteolytic enzymes. However, several areas of the MUC2 mucin are now thought to be susceptible to proteolytic cleavage by proteases. This is especially true of the multiple proteolytic enzymes of *E. histolytica*. Once the EhCPs degrade the colonic mucus, the parasite is able to attach to the host cell via the Gal-lectin.

In addition to a wide armament of proteases, *E. histolytica* possesses a unique peptide known as 'amoebapore.' This 77 amino acid peptide has structural similarity to the pore-forming peptides granulysin and NK-lysin. In vitro experiments have revealed that the purified amoebapore peptide leads to lysis of the host cell. Furthermore, investigators have demonstrated that the peptide is secreted following amoebic contact with mammalian cells. The mechanism of cell death has been likened to that mediated by granulysin and NK-lysin present in cytotoxic lymphocytes and natural killer cells. Further evidence that the amoebapore plays a critical role in pathogenesis comes from animal studies, in which *E. histolytica* strains deficient in amoebapore production have reduced virulence.

The immune response to *E. histolytica* is an excellent example of the contrasting beneficial and deleterious effects of the immune response to intestinal infection. Although the basis of protective immunity remains unclear, it is clear from epidemiologic studies that protective immunity to *E. histolytica* develops following infection of humans. It is equally clear from animal studies that much of the pathology that occurs during acute infection results from the immune response. Attachment of *E. histolytica* trophozoites to colonic epithelial cells triggers release of pre-IL1-β, which is cleaved and activated by the parasite's cysteine proteases. Activated IL1-β is then thought to stimulate production of multiple inflammatory cytokines, including IL-8, IL-6, and COX-2. As is the case for Cryptosporidium, NFκB-dependent signaling plays an essential role in generation of the immune response. Interference with intestinal cell NFκB signaling in animals has been shown to reduce tissue damage following *E. histolytica* infection. However, animals with targeted disruption of NFκB signaling have greater susceptibility to infection. Collectively, these data suggest that NFκB plays a significant role in innate resistance to amebiasis, but also contributes to tissue damage.

E. histolytica has evolved multiple mechanisms to evade the host immune response and cause long-standing infections (see Table 4). The hydrolytic enzymes and proteases secreted by *E. histolytica* not only contribute to penetration of the intestinal mucus and degradation of extracellular matrix proteins, but they also interfere with the host's humoral immune response by degrading complement, IgG, and IgA antibodies. As secretory IgA is part of the host's initial defense against the parasite, loss of this initial defense may facilitate persistent colonic infection. The IgG antibodies are essential in combating infection once it has spread beyond the intestine. Interruption of the complement cascade by inactivation of C3a and C5a may limit recruitment of immune cells to the site of infection. In addition, the amoebic Gal-lectin has homology to CD59, and this lectin inhibits complement-depending killing of amoebae in vitro. The parasite also interferes with the efficacy of macrophages. The respiratory burst present in activated macrophages is inhibited by amoebic trophozoites through a monocyte locomotion inhibitor. A significant reduction in MHC II antigen presentation due to lack of activated macrophages is another effective mechanism of immune evasion, which may reduce development of an effective adaptive immune response.

RECENT DEVELOPMENTS

Cryptosporidium species

While the severity and chronicity of cryptosporidiosis in patients with AIDS emphasizes the importance of cell-mediated immunity in control of cryptosporidium infection, a growing number of studies have highlighted the importance of the innate immune system in protection. A recent epidemiologic study demonstrated an association of serum mannose binding lectin (MBL) deficiency and the MBL2 polymorphism with susceptibility to cryptosporidiosis. No association was found with *Giardia* and *Entamoeba* infections in the same cohort. Human and mouse C1q and MBL are both able to bind *C. parvum* oocysts, and a possible causal effect is supported by mouse studies that demonstrated that MBL-deficient mice have more prolonged shedding of *C. parvum*. It has also become evident that NK cells are an important source of gamma interferon during *C. parvum* infection in mice, further stressing an important role for innate immune mechanisms in protection from *Cryptosporidium* parasites.

Cryptosporidium species remain technically very difficult to study in vitro, but recent progress has been made in understanding the role of host proteins in infection by this obligate intracellular parasite. In addition to host cytoskeletal proteins, the water channel aquaporin I facilitates entry of cryptosporidium into the cell. It appears that this channel, which becomes enriched in the plasma membrane at the site of host cell invasion, facilitates local swelling of the host cell as membrane protrusions enlarge to surround the invading parasite. Similarly, upregulation of antiapoptotic proteins enables persistence of infected cells, and, though not yet examined for *Cryptosporidium* species, host calpain proteases facilitate exit of other Apicomplexan parasites from the cell. It is likely that future studies will identify other host proteins required for efficient cryptosporidium infection.

G. intestinalis

There have been important recent advances in understanding the molecular epidemiology of *G. intestinalis*. Seven genotypes (or assemblages) of *G. intestinalis* have been defined, of which only assemblages A and B have been shown to infect humans. Assemblage B is more prevalent and is the only strain shown to be virulent in human and animal challenge studies, leading to the notion that there may be strain differences in virulence; however, this remains to be definitively demonstrated and no virulence factors have been identified that might account for proposed differences. Recent application of high-throughput DNA sequencing methods has enabled comparison of assemblages A and B on a genome-wide scale. The two genomes were only 77% identical at the nucleotide level, protein coding regions were 78% identical at the amino acid level, 28 genes unique to assemblage B were identified, 3 genes unique to assemblage A were identified, and almost the entire complement of VSPs was different in the two strains. These remarkable differences have led to the proposal that these strains be considered two different species. An additional advance has been the demonstration that VSP gene regulation occurs via an RNA-mediated interference-based mechanism. This finding not only has important implications for understanding *Giardia* biology, but may lead to additional tools to manipulate this parasite in the laboratory.

E. histolytica

It was recently demonstrated using a DNA-based molecular typing method that different *E. histolytica* strains are associated with different outcomes of infection (i.e., asymptomatic infection, dysentery, or amebic liver abscess). This study extended prior work demonstrating human genetic polymorphisms associated with susceptibility to amebiasis and help to partially explain the tremendous variation in outcome of infection with *E. histolytica*, which ranges from an asymptomatic to a fatal infection. Recent in vitro and in vivo animal work have demonstrated an important role for parasite-induced host cell apoptosis in invasive disease, and, interestingly, as a precursor to amebic phagocytosis of dying cells. Finally, significant advances have come in understanding mechanisms of protection from *E. histolytica* infection using a mouse model of cecal infection. These studies have demonstrated an important role for NFκB-dependent signaling in limiting amebic infection. Additional studies using this model have demonstrated protection from intestinal amebiasis following vaccination with a fragment of the major *E. histolytica* adhesion lectin (the Gal/GalNAc lectin). This furthers hopes for development of an effective vaccine to prevent amebiasis.

CONCLUSION

Diarrheal disease due to the intestinal protozoa remains an enormous source of morbidity and mortality globally. Provision of safe freshwater supplies for drinking and cooking, and improved sewage handling could eliminate disease due to these infections in an ideal world. Unfortunately, the political and economic situation in much of the developing world is likely to preclude the necessary improvement to infrastructure for the foreseeable future, making improved methods to prevent these infectious diseases necessary. A better understanding of the steps leading to disease caused by intestinal protozoa is an essential first step in developing such methods. We have attempted to draw attention to several general themes that have emerged from studies on pathogenesis of the intestinal protozoa. Although these parasites have distinctly different niches (e.g., intracellular vs. extracellular, or small intestine vs. colon), they must overcome many of the same pressures in order to successfully infect the host. The common themes include an ability of the infectious particle, be it an oocyst or cyst, to survive harsh environmental conditions outside the host and transit through the acidic stomach environment, an ability to penetrate the intestinal mucous layer and reach the epithelium, an ability to attach, and, in the case of the intracellular parasites, an ability to invade the host cells. The disease caused by each of these intestinal protozoa partly relates to these abilities; partly relates to abilities specific to a given parasite (e.g., direct destruction of small intestinal epithelial cells by cryptosporidium, and the ability of *E. histolytica* to induce host cell death); and partly relates to the inflammatory response each stimulates. A deeper molecular understanding of the mechanisms underlying any of these essential processes may lead to a new method for treatment and/or prevention, the general strategy for which could then possibly be applied to other members of the group.

FURTHER READING

Ali, I. K., Mondal, U., Roy, S., *et al.* (2007). Evidence for a link between parasite genotype and outcome of infection with *Entamoeba histolytica*. *Journal of Clinical Microbiology*, *45*(2), 285–289.

Carmolli, M., Duggal, P., Haque, R., *et al.* (2009). Deficient serum mannose-binding lectin levels and MBL2 polymorphisms increase the risk of single and recurrent Cryptosporidium infections in young children. *Journal of Infectious Diseases*, *200*, 1540–1547.

Chandramohanadas, R., Davis, P. H., Beiting, D. P., *et al.* (2009). Apicomplexan parasites co-opt host calpains to facilitate their escape from infected cells. *Science*, *324*(5928), 794–797.

Deitsch, K. W., Moxon, E. R., & Wellems, T. E. (1997). Shared themes of antigenic variation and virulence in bacterial, protozoal, and fungal infections. *Microbiology and Molecular Biology Reviews, 61*, 281–293.

Diamond, L. S., & Clark, C. G. (1903). A redescription of *Entamoeba histolytica* Schaudinn, 1903 (Amended Walker, 1911) separating it from *Entamoeba dispar* Brumpt, 1925. *The Journal of Eukaryotic Microbiology, 40*, 340–344.

Franzén, O., Jerlström-Hultqvist, J., Castro, E., *et al.* (2009). Draft genome sequencing of *Giardia intestinalis* assemblage B isolate GS: Is human giardiasis caused by two different species? *PLoS Pathogens, 5*, 1–14.

Gillian, F. D., Reiner, D. S., & McCaffrey, J. M. (1996). Cell biology of the primitive eukaryote *Giardia lamblia*. *Annual Review of Microbiology, 50*, 679–705.

Goodgame, R. W. (1996). Understanding intestinal spore-forming protozoa: Cryptosporidia, Microsporidia, Isospora, and Cyclospora. *Annals of Internal Medicine, 124*, 429–441.

Hill, D. R., & Nash, T. E. (1999). Intestinal flagellate and ciliate infections. In R. L. Guerrant, D. H. Walker & P. F. Weller (Eds.), *Tropical infectious diseases: Principles, pathogens and practice* (pp. 703–720). Churchill Livingstone: Philadelphia, PA.

Huang, D. B., & White, A. C. (2006). An updated review on Cryptosporidium and Giardia. *Gastroenterology Clinics of North America, 35*, 291–314.

Huston, C. D. (2004). Parasite and host contributions to the pathogenesis of amebic colitis. *Trends in Parasitology, 20*, 23–26.

Huston, C. D., & Petri, W. A. (2001). Emerging and reemerging intestinal protozoa. *Current Opinion in Gastroenterology, 17*, 17–23.

Loftus, B., Anderson, I., Davies, R., *et al.* (2005). The genome of the protist parasite *Entamoeba histolytica*. *Nature, 24*, 865–868.

Morrison, H. G., McArthur, A. G., Gillin, F. G., *et al.* (2007). Genomic minimalism in the early diverging intestinal parasite *Giardia lamblia*. *Science., 317*, 192–196.

Petri, W. A., Haque, R., & Mann, B. J. (2002). The bittersweet interface of parasite and host: Lectin–carbohydrate interactions during human invasion by the parasite *Entamoeba histolytica*. *Annual Review of Microbiology, 56*, 39–64.

Petri, W. A., Singh, U., & Ravdin, J. I. (1999). Enteric Amebiasis. In R. L. Guerrant, D. H. Walker & P. F. Weller (Eds.), *Tropical infectious diseases: Principles, pathogens and practice* (pp. 685–702). Churchill Livingstone: Philadelphia, PA.

Prucca, C. G., Slavin, I., Quiroga, R., *et al.* (2008). Antigenic variation in *Giardia lamblia* is regulated by RNA interference. *Nature, 456*(2008), 750–754.

Sibley, L. D. (2004). Intracellular parasite invasion strategies. *Science, 304* (2004), 240–253.

Ward, H., & Cevallos, A. M. (1998). Cryptosporidium: Molecular basis of host–parasite interaction. *Advances in Parasitology, 40*, 152–179.

Weiss, L. M. (2005). Microsporidiosis. In G. L. Mandell, J. E. Bennet & R. Dolin (Eds.), *Principles and practice of infectious diseases* (pp. 3237–3249). (6th edn.). Churchill Livingstone: Philadelphia, PA.

Xiao, L., Fayer, R., Ryan, U., & Upton, S. J. (2004). Cryptosporidium taxonomy: Recent advances and implications for public health. *Clinical Microbiology Reviews, 17*, 902–972.

Chapter 24

Leishmania

G.B. Ogden[1] and P.C. Melby[2]

[1]*St. Mary's University and The University of Texas Health Sciences Center at San Antonio, San Antonio, TX, USA*

[2]*South Texas Veterans Health Care System and The University of Texas Health Science Center at San Antonio, San Antonio, TX, USA*

Chapter Outline

Abbreviations 335
Defining Statement 335
Classification and Morphology 335
Life Cycle and Ecology 336
Cellular Biology 337
Plasma Membrane and Surface Molecules 337
Flagellum 338
Other Membrane-Bound Organelles 338
Molecular Biology and Control of Gene Expression 339
Genomic Organization 339
mRNA Processing 339
Control of Gene Expression 340
DNA Transfection and Gene Targeting 340
Pathogenesis and Host Response 340
Epidemiology and Disease 341
Diagnosis, Treatment, and Control 343
Recent Developments 344
Further Reading 344

ABBREVIATIONS

CR Complement receptor
DCL Diffuse cutaneous leishmaniasis
DTH Delayed-type hypersensitivity
ER Endoplasmic reticulum
GPI Glycosylphosphatidylinositol
IFN-γ Interferon-γ
IL Interleukin
LCL Localized cutaneous leishmaniasis
LPG Lipophosphoglycan
MAC Membrane attack complex
MIF Migration inhibitory factor
ML Mucosal leishmaniasis
PFR Paraflagellar rod
PPG Proteophosphoglycan
TNF-α Tumor necrosis factor-α
VL Visceral leishmaniasis

DEFINING STATEMENT

Leishmania are intracellular protozoan parasites of vertebrates that are transmitted by the bite of a sandfly vector. Their fascinating life cycle, cell biology, novel mechanisms of RNA processing, and gene regulation are discussed. *Leishmania* are also an important model of intracellular parasitism and host–parasite immune interactions. The leishmaniases are a diverse set of zoonotic diseases of global importance, including diseases of the skin, mucosa, and a potentially fatal systemic disease; these are discussed along with modern approaches to treatment and control.

CLASSIFICATION AND MORPHOLOGY

Leishmania are eukaryotic protozoan parasites of vertebrates in the family Trypanomastidae (order Kinetoplastida). Characteristic of members of the order Kinetoplastida is the presence of a conspicuous Feulgen stain-positive (i.e., DNA-containing) kinetoplast (see section 'Other Membrane-Bound Organelles'). All members of the family Trypanomastidae are parasitic for vertebrates or invertebrates and undergo morphological changes during transition between stages of their life cycle. Of the many genera in this family, only species of *Leishmania* and *Trypanosoma* are human pathogens. Two subgenera of *Leishmania*, *L.* (*Leishmania*) and *L.* (*Viannia*) are recognized. Conservatively, at least 14 *Leishmania* species are pathogenic for mammals, of which nine are recognized parasites of humans.

Leishmania exist as morphologically distinct forms. In mammals, amastigotes are an obligate intracellular parasite of mononuclear phagocytes. The elongated motile promastigote form is found in female sandflies (genus *Phlebotomus* in the Old World and *Lutzomyia* and *Psychodopygus*

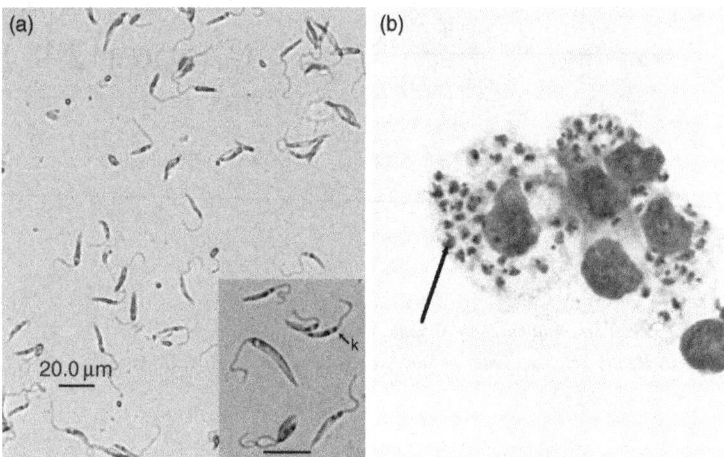

FIGURE 1 Morphological forms of *Leishmania*. In (a) (including insert), the elongated forms of cultured *L. donovani* promastigotes are shown (stained with giemsa, bar in insert = 10 μm). An arrow points to a kinetoplast (k) in the insert. (b) Shows the intracellular *L. donovani* amastigotes infecting hamster spleen cells after staining with giemsa. Each of the darkly stained bodies within the cell's cytoplasm (arrow) is an amastigote. Photo of promastigotes by GBO, amastigote photo kindly provided by Yaneth Osorio.

in the New World), which are the only known vector for *Leishmania*. Promastigotes possess a single nucleus and are variable in length (15–25 μm) and shape (ellipsoid to slender). The most prominent features of stained promastigotes are the nucleus, the kinetoplast, and the flagellum (Figure 1); the kinetoplast and the origin of the flagellum define the anterior region of the parasite (see section 'Flagellum'). Amastigotes are round to oval in shape with a 2–10 μm diameter. This stage is aflagellar (i.e., the flagellum does not extend past the cell boundary), but the kinetoplast and nucleus remain visible in stained amastigotes.

LIFE CYCLE AND ECOLOGY

Leishmania spp. have a complex life cycle (Figure 2) with two morphologically distinct forms, the amastigote and the promastigote. The female phlebotomine sandfly vector becomes infected during a bloodmeal when the fly repeatedly bites an infected vertebrate host, creating a pocket of pooled blood. Upon drinking the bloodmeal, the fly ingests amastigotes, and within the bloodmeal itself, the amastigotes transform into the promastigote form. These early stage promastigotes, termed procyclics, actively replicate within a chitin-containing gut lining (the peritrophic membrane) secreted by fly midgut epithelial cells (the members of the *Viannia* subgenus replicate in the hindgut). About 3 days after being ingested by the fly, procyclic forms escape the peritrophic membrane and travel to the anterior portion of the midgut, where they attach to epithelial cells. This marks the early stage of metacyclogenesis, where procyclics transform into the nonreplicating, infectious, metacyclic promastigotes. Metacyclics are thinner and faster moving than procyclic forms, but more importantly, about 5 days after the bloodmeal, biochemical changes occur in the surface coat (see section 'Lipophosphoglycan'), which cause the release of metacyclics from midgut epithelial cells. The newly released, and now highly infectious, metacyclic promastigotes are free to migrate anteriorly through

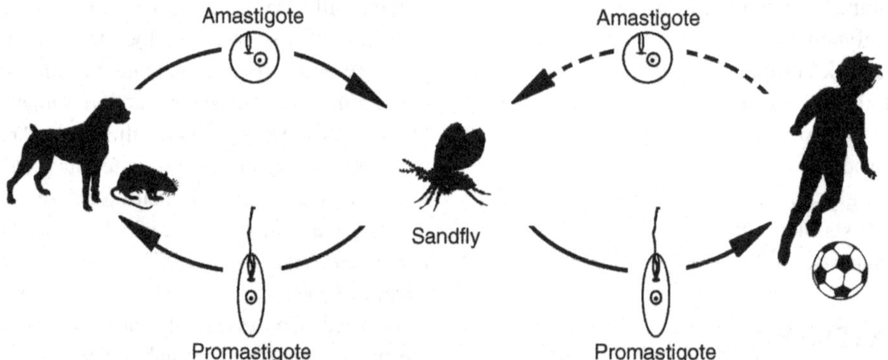

FIGURE 2 Life cycle of *Leishmania*. *Leishmania* have two morphologically distinct forms, the amastigote and the promastigote. In this schematic diagram, transmission of the infectious promastigotes is shown to occur from the infected female phlebotomine sandfly vector to the two main reservoir animals (i.e., dogs and rodents). The infectious metacyclic promastigotes then infect mononuclear phagocytes in the animal reservoir and within the phagolysosome transform into nonmotile amastigotes (the amastigotes replicate within the phagolysosome and eventually lyse the host cell). The cycle is completed when a sandfly ingests amastigotes from the reservoir during a bloodmeal. As is typical in zoonotic transmission, humans (frequently children) are incidentally infected; however, as indicated by the broken arrow, humans rarely serve as reservoirs for the parasite (see full description of life cycle in 'Pathogenesis and Host Response').

the foregut, eventually reaching the fly buccal cavity and mouth parts. From the mouth parts, these infectious promastigotes are dislodged into the wound created during the next bloodmeal. Upon contaminating the wound of the vertebrate host, promastigotes are phagocytosed by tissue phagocytic cells (neutrophils and macrophages) responding to the damaged tissue. Within the macrophage phagolysosome, the promastigotes rapidly transform into nonmotile amastigotes. The amastigotes replicate within the acidic and hostile environment of the phagolysosome, eventually lysing the macrophage and releasing amastigotes to infect other macrophages. The cycle continues when a sandfly ingests free amastigotes or amastigote-infected macrophages during a bloodmeal.

The transmission of *Leishmania* in most endemic areas is maintained through a zoonotic cycle during which humans are only incidentally infected. In general, the species that cause cutaneous disease are maintained in rodent or other small mammal reservoirs, whereas the domestic dog is the usual reservoir for the species that cause visceral disease. In most instances, multiple mammalian species are found to be infected with a given parasite, but usually there is a principal reservoir and the other animals are incidental hosts that do not play a major role in the transmission cycle. The transmission between reservoir and sandfly is highly adapted to the specific ecology of the endemic region. For instance, the reservoirs for *L. major* in the Middle East are desert rodents and the sandfly vectors cohabit the rodent burrows where they are protected from the extreme heat and low humidity of the desert environment. In the New World, the zoonotic cycle of parasites of the *Viannia* subgenus typically involves small mammals and sandflies that inhabit dense tropical and subtropical forests. Human infections occur when their activities bring them in contact with the zoonotic cycle. Anthroponotic transmission, where humans are the presumed reservoir, occurs with *L. tropica* in some urban areas of the Middle East and with *L. donovani* in India.

CELLULAR BIOLOGY

Plasma Membrane and Surface Molecules

Plasma membrane

Leishmania possess a bileaflet plasma membrane similar to the membranes of metazoan cells (i.e., multicelled eukaryotes). A membrane indistinguishable in appearance from the plasma membrane itself also surrounds the flagellum, and during development of promastigotes these membranes undergo biochemical changes, which modulate parasite attachment to host tissues (see the following section). The plasma membrane that lines the flagellar pocket actively acquires nutrients by pinocytosis, and pinocytotic vesicles have been observed to travel to the cytoplasm, where they fuse with lysosomes. The flagellar pocket is also active in secretion by exocytosis.

Lipophosphoglycan

The entire surface of promastigotes, including the flagellum, is covered by a cell coat or glycocalyx composed predominantly of lipophosphoglycan (LPG), as well as lesser amounts of gp63 (see the following section), proteophosphoglycans (PPGs), and other molecules. LPG is anchored to the parasite via a glycosylphosphatidylinositol (GPI) anchor, as are most other surface molecules (including gp63 and PPGs). This type of lipid anchor is unusual in animal cells. In addition to the GPI anchor, LPG has three other domains: a glycan core, repeats of a saccharide–phosphate region, and an oligosaccharide cap. The glycan core and lipid anchor are conserved among all species of *Leishmania*, whereas the carbohydrates in the repeat region and the cap are highly variable in composition. Several other *Leishmania* surface and secreted proteins contain repeating units of phosphoglycan and GPI-anchor domains, which has complicated the analysis of LPG's contribution to virulence. However, the creation of mutants unable to synthesize LPG, by targeted gene disruption, has helped to establish LPG's role in *L. major*'s virulence. LPG is thought to contribute to *L. major*'s virulence in key ways – providing resistance to oxidative damage in the phagolysosome and protecting the parasite from the effects of human serum complement (LPG may not be needed for complement resistance in mice). Infective *L. donovani* or *L. major* metacyclic promastigotes possess a thickened cell coat because of an increase in the number of phosphorylated saccharides in the LPG. The LPG molecules in metacyclic promastigotes of *L. major* are almost double the length of those found in procyclics, long enough to prevent the membrane attack complex (MAC) of the complement system from reaching the plasma membrane and lysing the parasite. Noninfective procyclic promastigotes have a thinner LPG coat and are readily lysed by host complement; LPG is almost absent from amastigotes. Interestingly, *L. mexicana* LPG mutants behave differently, as they show little or no reduction in virulence in macrophages or in mice, demonstrating that *Leishmania* virulence factors may be species specific.

Early in the development of *L. major* promastigotes in the sandfly, the terminal sugars of the LPG cap are predominantly galactose residues, but during metacyclogenesis these sugars are replaced by arabinopyranose. Because of these changes, procyclic promastigotes can attach to receptors on the epithelial layer of the sandfly midgut, allowing developing promastigotes to remain in the gut during digestion and excretion of the bloodmeal. Once the developmental transformation to the infectious metacyclic stage has taken place, changes in LPG composition enable the parasites to be released for passage to the sandfly's mouth parts to infect the next host. Similar events occur in other species of *Leishmania*, although the specific changes in LPG composition may differ. Certain species of sandfly are competent hosts for particular species of *Leishmania*. Variations in LPG structure, which accompany varying ability to attach to sandfly epithelial cells, seem to determine, at least in part, what species of sandfly are suitable hosts for a particular species of *Leishmania*.

Attachment and entry of promastigotes into macrophages may require the interaction of several parasite surface molecules with macrophage receptors. Identification of the specific molecules required for uptake remains to be determined. LPG plays a role in the uptake of complement-opsonized parasites into macrophages; however, because LPG mutants are efficiently endocytosed by macrophages, LPG is not required for uptake (see the following section). The macrophage CR1 and CR3 surface complement receptors, which bind C3b and inactivated C3b (iC3b), respectively, mediate adhesion of the promastigotes to macrophages. Attachment via complement receptors does not trigger the oxidative burst during phagocytosis (the oxidative burst in the phagolysosome is required for efficient intracellular killing of *Leishmania*). Consistent with these observations, in vitro experiments with *L. major* have demonstrated the increased intracellular survival of complement-fixed metacyclic promastigotes in macrophages.

Glycoprotein 63

The major protein found on the surface of promastigotes in all pathogenic species of *Leishmania* is a 63-kDa phophotidylinositol-linked glycoprotein designated gp63 (gp63 synthesis is downregulated in amastigotes). Biochemical and molecular analysis has identified gp63 as a type of zinc-dependent metalloprotease. Like LPG, gp63 undergoes changes during the development of promastigotes, and an increase in the surface expression of gp63 during metacyclogenesis has been documented. Surface exposure of gp63, however, is largely blocked by the elongated metacyclic LPG. Gp63 may play a secondary role in the binding of *Leishmania* to the macrophage surface and subsequent entry by receptor-mediated endocytosis. As C3b is converted to iC3b by gp63, it has been suggested that an accumulation of iC3b on the parasite's surface would promote phagocytosis by CR3; as mentioned earlier, this would prevent a triggering of a more destructive oxidative burst during phagocytosis.

Flagellum

The flagellum originates from the anterior of the cell, arising from an invaginated region termed the flagellar pocket, and pulls, rather than pushes, the promastigote. In addition to serving as a means of locomotion, electron microscopy reveals that the *Leishmania* flagellum sometimes functions as a tether, attaching the developing promastigotes to the tissues lining the midgut of the sandfly vector (see section 'Lipophosphoglycan'). In the mammalian host, the functional role of the flagellum is curtailed, as the flagellum in amastigotes does not extend beyond the flagellar pocket. The flagellum has the typical 9+2 axoneme structure of metazoan cells, possessing a cytoskeletal core or axoneme of nine microtubule doublets surrounding two unattached inner microtubules. In addition, there is a complex cytoskeletal structure running parallel to the axoneme, termed the paraflagellar rod (PFR). The PFR is unique to kinetoplastids, euglenoids, and some dinoflagellates. Little is known about the role of the PFR, but *Leishmania* with mutations in PFR proteins have reduced swimming velocity. The basal bodies at the base of the flagellum are structurally similar to centrioles found in animal cells.

Nucleus

The *Leishmania* nucleus is spherical or slightly ovoid in shape, with a diameter of approximately 1.5–2.5 μm. It is unusual in that the nuclear membrane remains intact during nuclear division. During division, the nucleus elongates and then constricts, splitting the nuclear membrane into two. Spindle microtubules are believed to elongate and push the dividing nucleus apart, but how the genetic material is partitioned is unknown. The nuclear membrane itself is typical in appearance, possessing a double unit membrane studded with 65–100 nm diameter nuclear pores. The nucleus contains chromatin, but the chromosomes do not condense or become visible during nuclear division.

Other Membrane-Bound Organelles

Mitochondrion and kinetoplast

A single mitochondrion is present and is functional in both the insect and mammalian forms of *Leishmania* (i.e., cyanide-sensitive respiration is found in both forms of the parasite). Within the mitochondrion is the kinetoplast, one of the defining morphological features of members of the order Kinetoplastida. This structure lies at the base of the flagellum, close to the basal bodies from which the flagellum arises, and is visible by light microscopy. The kinetoplast is a large rod-shaped assemblage of kinetoplast DNA (kDNA) located within the mitochondrion. The kDNA itself is unusual, as spreads of purified kDNA visualized under the electron microscope appear predominantly as thousands of 'minicircles' and a few 'maxicircles' catenated (i.e., interlocked) into a large DNA network. This contrasts sharply with the mitochondrial DNA found in metazoan cells, which typically exists as a single circle of DNA.

Moreover, maxicircles encode the usual mitochondrial genes but pre-mRNA transcripts for these genes must be extensively edited (by uridine insertion and deletion) to generate a translatable form of mRNA. The 'RNA editing' process is dependent upon guide RNAs (gRNAs) acting as templates for editing, and these gRNAs are encoded by the minicircles (see section 'RNA editing').

Endoplasmic reticulum and Golgi apparatus

The endoplasmic reticulum (ER) is similar in appearance to that found in metazoan cells. It is contiguous with the outer nuclear envelope and may be rough (containing attached ribosomes) or smooth (lacking attached ribosomes) in appearance. Closely associated with the ER is the Golgi apparatus, which appears as flattened membranous structures frequently stacked at the base of the flagellar pocket (in addition to being a site of endocytosis, the flagellar pocket is active in exocytosis).

MOLECULAR BIOLOGY AND CONTROL OF GENE EXPRESSION

Genomic Organization

Recently, the genomes of several trypanomastid protozoa (including *L. major*, *L. infantum*, and *L. V. braziliensis*) have been sequenced and analyzed. *Leishmania* spp. have 35 or 36 linear chromosomes, with each genome measuring about 3.0–3.8 × 10^7 bp, which is approximately eight times the size of the genome of the bacterium *Escherichia coli*. The estimates for the number of putative protein-encoding genes found are very similar for the three species of *Leishmania* ranging from about 8200 to 8400 genes. Many of these genes are repeated multiple times in tandem, forming large gene clusters. About 50% of the genes identified have no known function; however, 3–4% of these genes encode leucine-rich repeat sequences that are thought to bind to complement receptor CR3. *Leishmania* spp. appear to be asexual; sexual crosses between these organisms have not been successfully performed under laboratory conditions.

mRNA Processing

RNA editing

Two of the most uncommon types of messenger RNA (mRNA) processing, namely, RNA editing and trans-splicing, were first discovered in studies on trypanomastid protozoa. One of these, RNA editing, has changed our understanding of how gene expression may be controlled in eukaryotes. The kDNA maxicircles mentioned earlier encode genes for ribosomal RNA, and certain proteins necessary for oxidative phosphorylation and electron transport in the mitochondrion. The study of many of these genes revealed that the mature mRNA sequences are different from the genomic DNA sequence from which they are transcribed. Moreover, many of the genomic sequences do not code for a functional protein. These startling and controversial observations were reconciled when it was discovered that a posttranscriptional process, now termed RNA editing, resulted in the site-specific deletion or addition of uridine (U) residues in the mRNA, which created functional initiation sites for protein translation and/or created an open reading frame for protein translation. RNA editing, in effect, repairs 'defects' in certain protein-coding regions in the DNA at the RNA level, enabling the expression of a functional protein from the processed mRNA. gRNAs are known to be critically important to this process. The gRNAs carry the template for proper mRNA editing and may also contribute U residues to the editing process. In *L. tarentolae*, a lizard parasite, about 20 different classes of gRNAs are encoded in both mini- and maxicircles. What is lacking in the study of RNA editing, however, is an explanation for the reliance of the trypanomastids on this unusual form of gene regulation.

Trans-splicing

It is believed that within any species of trypanomastid protozoa all mature mRNAs possess an identical 5′ exon sequence. This 5′ sequence, termed the 'spliced leader' or 'mini-exon,' is encoded by a gene separate from the genes corresponding to the mRNA transcripts. The mechanism for linking the mini-exon to the mRNA, termed trans-splicing, was first discovered in *Trypanosoma brucei* and is now known to also occur in about 10–15% of nematode mRNAs and in some flatworm mRNAs. (Recent findings indicate that trans-splicing occurs in a chordate, the tunicate *Ciona intestinalis*, opening the possibility for future discovery in vertebrate species as well.) The biochemistry of trans-splicing is similar to conventional cis-splicing where one exon is fused to a second downstream exon by two sequential transesterification reactions, resulting in the covalent joining of the two exons and the excision of the intervening or 'intron' sequence (in the form of a partially circular molecule termed a 'lariat structure') from the precursor mRNA. Although both mechanisms use small nuclear RNA–protein complexes (i.e., snRNPs) in catalyzing this reaction, the crucial difference between cis- and trans-splicing is that the two exons linked in trans-splicing are not encoded on the same pre-mRNA molecule, but rather are transcribed from separate genes. As a consequence of splicing two separate transcripts, the trans-splicing reaction releases a branched or 'Y-shaped' intron instead of a lariat. In all *Leishmania* spp. studied, the spliced leader RNA is transcribed from clusters of tandemly repeated genes (ranging from about 90 to 200 copies) on the genome. In *L. enriettii*, an 85 nt spliced leader RNA is transcribed from a 440 bp long gene. At the 5′ end of this transcript is the 35 nt mini-exon sequence found at the 5′ end of all mature *L. enriettii* mRNAs.

Control of Gene Expression

For most eukaryotes and bacteria, gene expression is controlled primarily by regulating the initiation of transcription. In this paradigm, transcriptional promoters on the genome largely dictate where, and how often, initiation of transcription takes place. RNA polymerase I and RNA polymerase III promoters have been described. No classic RNA polymerase II promoter sequence containing a TATA box and other typical polI promoter elements has been identified in any trypomastid protozoa, including *Leishmania*. In *L. tarentolae*, however, a TATA-less promoter that drives the expression of SL-RNA has been characterized. This promoter has two well-described elements at −60 (−67 to −58) and −30 (−41 to −31). TATA-less promoters are common in constitutively expressed (unregulated) genes in mammalian cells. Overall, transcription of protein-encoding mRNAs in *Leishmania* appears to be non-stringent, as promastigotes transfected with plasmids containing the gene for neomycin resistance (but containing no *Leishmania* sequences) were shown to actively transcribe the neomycin gene. All that was required was the incorporation of a signal for trans-splicing in the neomycin gene sequence. So far, other evidence for a true *Leishmania* promoter has been merely suggestive or inconclusive. Histone modifications influencing chromatin structure are likely to play a role in transcription initiation, as well. Regulation of protein-encoding genes in *Leishmania* and other kinetoplastids, however, seems to depend largely on post-transcriptional mechanisms. This may be a consequence, again, of a peculiarity of trypanomastid molecular biology. Unlike most eukaryotes studied, trypanomastids produce polycistronic mRNAs (transcribed from genes clustered in tandem on the genome, as mentioned earlier). Translation of these transcripts, however, requires their conversion to monocistrons by trans-splicing at the 5′ end and polyadenylation at the 3′ end. The polyadenylation of an upstream gene is functionally coupled to trans-splicing of a downstream gene. Formation of mature mRNA may then be controlled by regulating splicing and/or polyadenylation. However, it should be noted that the tandemly repeated genes observed in the genomes of trypanomastids may enable the cell to increase the expression of a protein (by increasing gene copy number) without modulating the initiation of transcription. The stability of leishmanial mRNA is also important, because the rate of mRNA degradation of a heat-shock protein hsp83 is dependent on 5′ and 3′ untranslated regions. In addition, gp63 transcripts have been shown to have a much shorter half-life in logarithmic phase cells, compared to transcripts in stationary phase cells.

DNA Transfection and Gene Targeting

A major contribution to the molecular analysis of *Leishmania* was the development of functional plasmid expression vectors in 1990. Two groups used an *E. coli* plasmid in combination with the intergenic region of the α-tubulin gene or the flanking region of the *dhfr-ts* gene to construct a *Leishmania* plasmid. More recently, other plasmids have been made: several selectable markers are available for stable transfection (e.g., neomycin, hygromycin, bleomycin, and puromycin) and a number of phenotypic markers are available for transient assays (e.g., β-galactosidase, luciferase, and β-glucuronidase). Negative selection is also possible using the thymidine kinase gene. Important questions about the pathogenesis of *Leishmania* have been answered using these plasmids. As mentioned earlier, transfection of gp63-variant strains with gp63-expressing plasmids has helped to elucidate the role of gp63 in infection. Similarly, targeted mutations in LPG have contributed to our understanding of this molecule's role in virulence. Moreover, homologous recombination between a plasmid and the *Leishmania* genome has proved to be a powerful tool in studying the function of particular genes in these parasites. Researchers may now specifically target and eliminate (knockout) the activity of a gene or add a gene to a species' genome. Although gene knockout experiments are potentially powerful in studying *Leishmania*, difficulties may arise, as many *Leishmania* genes are present in tandem repeats on the chromosome. Transgenic strains of *Leishmania* have also been created (e.g., red fluorescent protein expressing *L. Mexicana*).

PATHOGENESIS AND HOST RESPONSE

The *Leishmania* parasite has multiple ways in which it adapts to, and survives within, both the invertebrate and vertebrate hosts. Within the sandfly, promastigotes undergo metacyclogenesis to the infective stage (see section 'Life Cycle and Ecology'). When the sandfly takes a bloodmeal, the infective promastigotes are deposited in the host skin, which initiates the infection. As noted earlier (see section 'Cellular Biology'), the parasite's surface plays an important role in the entry and survival of *Leishmania* in the mammalian host by conferring complement resistance and by facilitating entry into the macrophage without triggering a respiratory burst.

The nature of the host immune response plays a critical role in the evolution and outcome of infection. There is extensive evidence from experimental animal models that cellular immune mechanisms mediate resistance to *Leishmania* infection. Human studies have confirmed in a general sense the findings of the experimental animal studies. The immunological mechanisms related to resistance and susceptibility are summarized in Figure 3. Resistance to leishmanial infection is associated with the capacity of

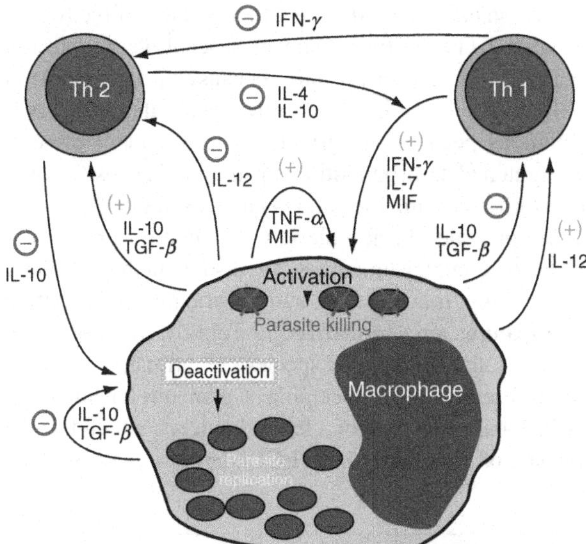

FIGURE 3 Immunopathogenesis of *Leishmania* infection. The balance between the Th1 and Th2 cell response and the effect on macrophage activation and parasite killing is shown schematically. Inhibitory effects (−) and stimulatory effects (+) are noted. When Th1 cells are activated by *Leishmania* antigens, IFN-γ is produced, which downregulates the Th2 response and promotes macrophage activation. IL-7 and MIF may play a less prominent role in macrophage activation and parasite killing. TNF-α, produced by macrophages, synergizes with IFN-γ to induce macrophage activation and parasite killing. If *Leishmania* antigens induce a Th2 response with IL-4 and/or IL-10 production, the Th1 response is inhibited and the macrophage is deactivated, making it permissive for parasite replication. The infected macrophage itself may also produce IL-10 and TGF-β, which promote parasite survival by deactivating the macrophage.

$CD4^+$ T cells (Th1 subset) to generate interferon-γ (IFN-γ) in response to the parasite. Tumor necrosis factor-α (TNF-α) may contribute to protective responses by augmenting IFN-γ-mediated macrophage activation. Interleukin (IL)-12, which stimulates T cells and NK cells to produce IFN-γ, promotes Th1 subset expansion, suppresses Th2 cells (IL-4 and IL-10 producers), and plays a critical role in the generation of a protective immune response. Other cytokines, such as IL-7 and macrophage migration inhibitory factor (MIF), have been shown to activate macrophages to eliminate intracellular *Leishmania*, but their role is less significant when compared to that of IFN-γ. The killing of intracellular *Leishmania* is mediated by the generation of reactive oxygen and nitrogen intermediates. The latter mechanism is dominant in some rodent models of leishmaniasis, but it has a less prominent role in killing of *Leishmania* by human macrophages.

Susceptibility to leishmanial infection and progression of disease is associated with the production of IL-4 and IL-10 by Th2 cells, and IL-10 and TGF-β by infected macrophages (see Figure 3). IL-4 is a cross-regulatory cytokine that inhibits the expansion of the Th1 subset of T cells. IL-10 inhibits the production of cytokines by Th1 cells by suppression of accessory cell function and IL-12 production.

Infected macrophages have a diminished capacity to initiate and respond to an inflammatory response, thus providing a safe haven for the intracellular parasite. IL-10 and TGF-β, which are potent inhibitors of the antimicrobial activity of macrophages, are produced by T cells or infected macrophages at the site of infection. The expression of IL-10 mediates the persistence of parasites in the infected tissue. In addition to the production of deactivating cytokines, infection of the macrophage, especially in the presence of the type 2 cytokines (IL-4 and IL-13), can lead to an alternative form of activation characterized by blunted IFN-γ-mediated activation, low NO production, high arginase activity, increased production of prostaglandin E_2, and a gene expression profile that includes increased expression of phagocytic receptors, IL-1 receptor antagonist, and certain chemokines. Infected macrophages may also inhibit antigen presentation to effector T cells and there may be expansion of regulatory T cell populations. Both $CD4^+CD25^+Foxp3^+$ regulatory T cells and IL-10-producing $CD4^+CD25^-Foxp3^-$ T cells have been shown to promote chronic disease.

After inoculation of the *Leishmania* spp. by the sandfly vector, large numbers of neutrophils (polymorphonuclear leukocytes) followed by T lymphocytes, dendritic cells, and macrophages, migrate into the site of infection. Host factors (genetic background, concomitant disease, nutritional status), parasite factors (virulence, size of the inoculum) and possibly, vector-specific factors (sandfly genotype, immunomodulatory salivary constituents) influence the subsequent development and evolution of disease. Patients who develop localized cutaneous leishmaniasis (LCL, see following text) generally exhibit strong protective antileishmanial T cell (Th1) responses and spontaneously heal within 3–6 months. Patients with mucosal leishmaniasis (ML, see section 'Pathogenesis and Host Response') exhibit a hyperresponsive cellular immune reaction, which may contribute to the prominent tissue destruction seen in this form of the disease (see section 'Pathogenesis and Host Response'). Patients with active disseminated disease (diffuse cutaneous leishmaniasis (DCL) or visceral leishmaniasis (VL)) demonstrate minimal or absent *Leishmania*-specific cellular immune responses.

EPIDEMIOLOGY AND DISEASE

In most instances, when *Leishmania* are transmitted to humans, the infection is controlled by the host response and overt disease does not develop. Field studies in endemic areas reveal people who have no history of disease often have evidence of infection by virtue of a positive leishmanin (delayed-type hypersensitivity, DTH) skin test. In most endemic areas, subclinical infection (no overt disease) occurs more frequently than active disease. Epidemiological studies indicate that individuals with previous active cutaneous leishmaniasis or subclinical infection are usually immune to a subsequent clinical infection. Within endemic areas the prevalence of DTH skin test positivity (related to previous subclinical or clinical infection) increases with age and the incidence of clinical disease decreases with age, indicating that immunity is acquired in the population over time.

The leishmaniases (clinical manifestations of disease) are estimated to affect 10–50 million people in endemic tropical and subtropical regions on all continents except Australia and Antarctica. Over the past decade, an increased number of cases have been reported throughout the world, either as new sporadic cases in endemic regions or new epidemic foci. Severe epidemics with more than 100000 deaths from VL have occurred in India and Sudan. The disease has emerged in new foci because of the following factors: (1) the movement of susceptible people into existing endemic areas, usually because of agro-industrial development, (2) an increase in the vector and/or reservoir populations because of agriculture development projects, (3) heightened anthroponotic transmission because of rapid urbanization in some foci, and (4) increased sandfly density because of scaled-back malaria vector control programs.

The different leishmaniases are distinct in their etiology, epidemiology, transmission, and geographical distribution. These features are summarized in Table 1. With only rare exceptions, the *Leishmania* species that primarily cause cutaneous disease do not visceralize. Cutaneous leishmaniasis may be localized to one or a few skin ulcers, which localize at the site of the sandfly bite (LCL), or in rare instances the

TABLE 1 Major species of *Leishmania* and their geographic distribution

Parasite	Disease form	Reservoirs	Vectors	Distribution
Old World Leishmaniasis				
L. (*L.*) *major*	LCL	Desert rodents (*Psammomys, Meriones, Gerbillus*)	*Phlebotomus papatasi*	North Africa, the Middle East, Central Asia, and the Indian subcontinent
L. (*L.*) *tropica*	LCL	Humans, rock hyraxes, and unknown animals	*P. sergenti*	North Africa, the Middle East, Central Asia, and the Indian subcontinent
L. (*L.*) *aethiopica*	LCL, DCL	Hyraxes	*P. pedifer, P. longipes*	Ethiopian highlands, Kenya
L. (*L.*) *infantum*	VL, LCL	Domestic dog, wild canines	*P. perniciosus, P. ariasi, P. tobbi, P. langeroni*	Mediterranean basin, Middle East, and Central Asia
L. (*L.*) *donovani*	VL	Humans	*P. argentipes, P. orientalis, P. martini*	Kenya, Sudan, India, Pakistan, and China
New World Leishmaniasis				
L. (*L.*) *mexicana*	LCL, DCL	Forest rodents	*Lutzomyia olmeca olmeca, Lu. cruciata*	Southern Texas through Mexico and northern Central America
L. (*L.*) *amazonensis*	LCL, DCL	Forest spiny rats	*Lu. faviscutellata*	South America in the Amazon basin and northward
L. (*L.*) *pifanoi*	LCL, DCL	Probably rodents	Unknown	Venezuela
L. (*L.*) *garnhami*	LCL	Unknown	*Lu. youngi*	Venezuela
L. (*L.*) *venezuelensis*	LCL	Unknown	*Lu. olmeca bicolor*	Venezuela
L. (*V.*) *braziliensis*	LCL, ML	Forest rodents, opossums, sloths, domestic dogs, donkeys	*Psychodopygus wellcomei* and others	South America from the northern highlands of Argentina and northward to Central America
L. (*V.*) *panamensis*	LCL, ML	Sloths	*Lu. trapidoi, Lu. ylephiletor*	Panama, Costa Rica, Colombia
L. (*V.*) *guyanensis*	LCL	Sloths, lesser anteater	*Lu. umbratilis*	Guyana, Surinam, northern Amazon Basin
L. (*V.*) *peruviana*	LCL	Unknown	*Lu. peruensis*	Peru, Argentinian highlands
L. (*L.*) *chagasi*	VL, LCL	Domestic dogs and foxes	*Lu. longipalpis*	Mexico (rare) through Central and South America

parasite may disseminate to involve large areas of skin (DCL). Mucosal leishmanisasis (ML), an uncommon complication of LCL, is a destructive, disfiguring infection of the nasal and oropharyngeal mucosa. VL is the most serious form because there is progressive parasitism of the visceral reticuloendothelial organs (liver, spleen, and bone marrow).

LCL (also called oriental sore) can affect individuals of any age but children are the primary victims in many endemic regions. It typically presents as one or a few raised or ulcerated lesions located on skin that is exposed to the sandfly bite, for example, the face and extremities; rare cases of more than a 100 lesions have occurred. The lesions develop at the site of the sandfly bite and increase in size and often ulcerate over the course of several weeks to months. The lesions remain localized to the skin and the patient generally has no systemic symptoms. Lesions caused by *L. major* and *L. mexicana* usually heal spontaneously after 3–6 months leaving a depressed scar. Lesions on the ear pinna caused by *L. mexicana*, called Chiclero's ulcer because they were common in chicle harvesters in Mexico and Central America, often follow a chronic, destructive course. In general, lesions caused by *L.* (*Viannia*) spp. tend to be larger and more chronic.

DCL is a rare form of leishmaniasis caused by organisms of the *L. mexicana* complex in the New World and *L. aethiopica* in the Old World. DCL manifests as large nonulcerating skin lesions, which often involve large areas of skin and may resemble the lesions of leprosy. The face and extremities are most commonly involved. Dissemination of the parasite from the initial lesion usually takes place over several years. It is thought that an immunological defect underlies this severe form of cutaneous leishmaniasis.

ML (also called espundia) is an uncommon but serious manifestation of leishmanial infection resulting from the spread of the parasite from the skin to the nasal or oropharyngeal mucosa because of a cutaneous infection. ML is usually caused by parasites in the *L.* (*Viannia*) complex. ML occurs in less than 5% of individuals who have, or have had, skin lesions caused by *L. (V.) braziliensis*. Patients with ML most commonly have nasal mucosal involvement, which results in symptoms of nasal congestion, discharge, and recurrent nose bleeds. In severe cases over the course of several years, there is marked soft tissue, cartilage, and even bone destruction, which lead to visible deformity of the nose or mouth, perforation of the nasal septum, and even obstruction of the airways.

VL (also called kala azar) typically affects children less than 5 years of age in the New World (*L. chagasi*) and Mediterranean region (*L. infantum*), and older children and young adults in Africa and Asia (*L. donovani*). After inoculation of the organism into the skin by the sandfly, the child may have a completely asymptomatic infection (no clinical signs of disease), or 2–9 months later show signs of active VL. The classic clinical features of VL include high fever, enlargement of the liver and spleen, and severe cachexia (weight loss and muscle wasting). There is loss of bone marrow function, so the leukocyte and erythrocyte counts fall. Malnutrition is a risk factor for the development and more rapid evolution of active VL. Without the institution of drug therapy death occurs in more than 90% of patients.

VL has been increasingly recognized as an opportunistic infection associated with HIV infection. Most cases have occurred in southern Europe and Brazil, but there is potential for many more cases as the endemic regions for HIV and VL converge. Disease may result from reactivation of a long-standing subclinical infection.

DIAGNOSIS, TREATMENT, AND CONTROL

A definitive diagnosis of leishmaniasis rests on the microscopic detection of amastigotes in tissue specimens, or isolation of the organism by culture (performed only in specialized laboratories). A positive culture enables speciation of the parasite (usually by isoenzyme analysis by a reference laboratory). Serological tests are useful for the diagnosis of VL because of the very high level of antileishmanial antibodies, but not for ML or LCL where the antibody response is minimal.

Specific antileishmanial therapy is not routinely needed for uncomplicated LCL caused by strains that have a high rate of self-healing (*L. major*, *L. mexicana*). Lesions that are extensive or located where a scar would result in disability or disfigurement should be treated. Cutaneous lesions suspected or known to be caused by members of the *Viannia* subgenus (New World) should be treated because of the low rate of spontaneous healing and the potential risk of developing mucosal disease. Likewise, patients with lesions caused by *L. tropica* (Old World), which are typically chronic and nonhealing, should be treated. All patients with VL or ML should receive therapy.

The pentavalent antimony compounds (sodium stibogluconate and meglumine antimoniate) have been the mainstay of antileishmanial chemotherapy for more than 40 years. These drugs are difficult to administer and have moderate toxicity. Using the recommended regimen, cure rates of 80–100% for LCL, 50–70% for ML, and 80–100% for VL can be expected. Failure of antimony therapy is more common in some parts of the world, and is particularly problematic in young children. Relapses are common in patients who do not have an effective antileishmanial cellular immune response, such as patients who have DCL or are coinfected with HIV. These patients often require multiple courses of therapy.

Several alternative therapies have been used in the treatment of the leishmaniases, including the antifungal agent amphotericin B and its related lipid-associated compounds, parenteral paromomysin (aminosidine) and pentamidine.

Recombinant human IFN-γ has been successfully used as an adjunct to antimony therapy in treatment of refractory cases of ML and VL. Oral miltefosine has been used with success in treating VL in India, but significant resistance may be a problem in the future. The oral antifungal drug ketoconazole has been effective in treating some patients with LCL.

Leishmaniasis is best prevented by avoiding exposure to sandflies and use of insect repellents. Where transmission is occurring near homes, community-based insecticide spraying has had some success in reducing human infections. Control or elimination of infected reservoir hosts (e.g., seropositive domestic dogs) has had limited success. Where anthroponotic transmission is thought to occur, the early recognition and treatment of cases is essential. Because humans acquire long-standing immunity following infection with *Leishmania*, prevention of the disease through vaccination is a possible approach. Recent immunization studies involving experimental animal models are promising. Vaccination of humans, or domestic dogs to prevent transmission of VL, may have a role in the prevention of leishmaniases in the future.

RECENT DEVELOPMENTS

Leishmania must infect and replicate in the macrophages of the mammalian host. However, a new appreciation for the role neutrophils play in the early stages and progression of a cutaneous infection is being developed. Recent studies have employed transgenic parasites and mice, with *L. major* expressing red fluorescent protein and the neutrophils of mice expressing green fluorescent protein. The behavior of fluorescent neutrophils and parasites over the course of infection can now be monitored in vivo and in real time, using state-of-the-art fluorescent imaging techniques.

The short-lived neutrophils are known to arrive within 30 min at the site of a skin injury created by a needle stick or the bite of a sandfly vector. As actively phagocytic cells possessing powerful reactive oxygen metabolites, lytic enzymes, and other mechanisms for killing phagocytosed microbes, neutrophils occupy an important place in the host's arsenal of innate immune defenses – or so it has long been thought. Recent work studying cutaneous leishmaniasis suggests a very different picture. Compared to a wound resulting from a needle stick, neutrophils persist a significantly longer time at the site of a sandfly bite to the skin (with either infected or uninfected flies, suggesting salivary constituents cause this effect). Neutrophils have also been shown to contribute to infection: in experiments where neutrophils were depleted from infected mice, the parasite burden and progression of cutaneous disease was impaired. Infected neutrophils undergo apoptosis (programmed cell death) at the infection site, and contact with these apoptotic cells can impair the ability of macrophages to kill ingested parasites. Neutrophils may also impair the response of *L. major*-specific adaptive immune cells to the site of infection. These observations suggest that neutrophils responding to the natural route of infection, a sandfly bite to the skin, may play an important role in contributing to the progression of disease. They do not completely explain the role neutrophils play in this process, however.

A recent 'Trojan Horse' model proposed that macrophages actively clearing apoptotic neutrophils at the infection site became 'silently' infected when they phagocytosed parasites contained within these neutrophils. A subsequent study employing real-time imaging of fluorescent *L. major* and neutrophils in mice, however, found no evidence to support this model and suggested alternative explanations. One possibility was that parasites better adapted for intracellular survival (due to having survived within neutrophils) escape from apoptotic neutrophils and subsequently infect macrophages (and these macrophages may have been impaired by contact with the apoptotic neutrophils, as mentioned above). In this way neutrophils actually promote the course of cutaneous infection. The authors of this and another study have suggested that the role of neutrophils in promoting infection may help to explain their observation that a vaccine that protects animals from cutaneous disease caused by *Leishmania* delivered by needle stick but fails to protect animals infected by the bite of infected sandflies. The possible role of neutrophils in promoting cutaneous leishmaniasis is intriguing and warrants additional study.

FURTHER READING

Berman, J. (2005). Recent developments in leishmaniasis: Epidemiology, diagnosis, and treatment. *Current Infectious Disease Reports*, *7*(1), 33–38.

Desjeux, P. (2004). Leishmaniasis: Current situation and new perspectives. *Comparative Immunology, Microbiology and Infectious Diseases*, *27*, 305–318.

Dye, C., & Williams, B. G. (1993). Malnutrition, age, and the risk of parasitic disease: Visceral leishmaniasis revisited. *Proceedings of the Royal Society B*, *254*, 33–39.

El-On, J. (2009). Current status and perspectives of the immunotherapy of leishmaniasis. *The Israel Medical Association Journal*, *11*(10), 623–628.

El-Sayed, N. M., Myler, P. J., Blandin, G., *et al.* (2005). Comparative genomics of trypanosomatid parasitic protozoa. *Science*, *309*(5733), 404–409.

Grevelink, S. A., & Lerner, E. A. (1996). Leishmaniasis. *Journal of the American Academy of Dermatology*, *3*(4), 257.

Kreier, J. P., & Baker, J. R. (1987). *Parasitic protozoa.* Allen & Unwin: Boston, MA.

Martınez-Calvillo, S., Vizuet-de-Rueda, J. C., Florencio-Mart´inez, L. E., *et al.* (2010). Gene expression in trypanosomatid parasites. *Journal of Biomedicine and Biotechnology*, *2010*, e525241.

Molyneux, D. H., & Ashford, R. W. (1983). *The biology of trypanosoma and leishmania, parasites of man and domestic animals.* Taylor & Francis: London.

Murray, H. W., Berman, J. D., Davies, C. R., & Saravia, N. G. (2005). Advances in leishmaniasis. *The Lancet*, *366*, 1561–1577.

Okwor, I., & Uzonna, J. (2009). Vaccines and vaccination strategies against human cutaneous leishmaniasis. *Human Vaccines*, *5*(5), 291–301.

Olivier, M., Gregory, D. J., & Forget, G. (2005). Subversion mechanisms by which *Leishmania* parasites can escape the host immune response: A signaling point of view. *Clinical Microbiology Reviews*, *18*, 293–305.

Peacock, C. S., Seeger, K., Harris, D., *et al.* (2007). Comparative genomic analysis of three *Leishmania* species that cause diverse human disease. *Nature Genetics*, *39*(7), 839–847.

Peters, W., & Killick-Kendrick, R. (1987). *The leishmaniases in biology and medicine*. Vols. 1 and 2, Academic Press: London.

Peters, N. C., Kimblin, N., Secundino, N., *et al.* (2009). Vector transmission of leishmania abrogates vaccine-induced protective immunity. *PLoS Pathogens*, *5*(6), e1000484.

Peters, N., & Sacks, D. (2006). Immune privilege in sites of chronic infection: *Leishmania* and regulatory T cells. *Immunological Reviews*, *213*, 159–179.

Peters, N. C., & Sacks, D. L. (2009). The impact of vector-mediated neutrophil recruitment on cutaneous leishmaniasis. *Cellular Microbiology*, *11*(9), 1290–1296.

Ritter, U., Frischknecht, F., & van Zandbergen, G. (2009). Are neutrophils important host cells for *Leishmania* parasites? *Trends in Parasitology*, *25*(11), 505–510.

Sacks, D. L. (1992). The structure and function of the surface lipophosphoglycan on different developmental stages of *Leishmania* promastigotes. *Infectious Agents and Disease*, *1*, 200–206.

Sacks, D., & Anderson, C. (2004). Re-examination of the immunosuppressive mechanisms mediating non-cure of *Leishmania* infection in mice. *Immunological Reviews*, *201*, 225–238.

Sacks, D., & Kamhawi, S. (2001). Molecular aspects of parasite-vector and vector-host interactions in leishmaniasis. *Annual Review of Microbiology*, *55*, 453–483.

Scott, P., Artis, D., Uzonna, J., & Zaph, C. (2004). The development of effector and memory T cells in cutaneous leishmaniasis: The implications for vaccine development. *Immunological Reviews*, *201*, 318–338.

Sharma, U., & Singh, S. (2009). Insect vectors of *Leishmania*: Distribution, physiology and their control. *Journal of Vector Borne Diseases*, *45* (12), 255–272.

Späth, G. F. (2003). The role(s) of lipophosphoglycan (LPG) in the establishment of *Leishmania* major infections in mammalian hosts. *Proceedings of the National Academy of Sciences of the United States of America*, *100*(16), 9536–9541.

Chapter 25

Oomycetes (Water Mold, Plant Pathogenic)

S. Kamoun
The Sainsbury Laboratory, Norwich, UK

Chapter Outline

Abbreviations	**347**
Defining Statement	**347**
Impact on Agriculture and Environment	**347**
Biology	**348**
Evolutionary History	348
General Biological Features	349
Unique Biological Features	350
Genome Structure	350
Pathology	**350**
Infection Cycle	350
Adhesion, Penetration, and Colonization of Host Tissue	350
Induction of Defense Responses and Disease-Like Symptoms	351
Inhibition of Host Enzymes	351
Effectors	352
Conclusions	**352**
Further Reading	**353**

ABBREVIATIONS

Avr Pathogen avirulence
endoPGs Endopolygalacturonases
GFP Green fluorescent protein
GIPs Glucanase inhibitor proteins
HGT Horizontal gene transfer
HR Hypersensitive response
HT Host translocation
NLPs Nep1-like proteins
PAMPs Pathogen-associated molecular patterns
R Plant disease resistance

DEFINING STATEMENT

Plant pathogenic oomycetes cause devastating diseases on several crops and ornamental and native plants and continue to have a significant impact on agriculture and the environment. This chapter reviews our current understanding of the biology and pathology of plant pathogenic oomycetes.

IMPACT ON AGRICULTURE AND ENVIRONMENT

The more than 500 species of oomycetes, commonly known as water molds, white rusts, or downy mildews, are essentially saprophytic but include pathogens of plants, insects, crustaceans, fish, vertebrate animals, and various microorganisms. Plant pathogenic oomycetes cause devastating diseases on several crop, ornamental, and native plants and are classified in several genera (the major genera are listed in Table 1). The most notorious plant pathogenic oomycetes are the ~60 species of the genus *Phytophthora*, arguably the most devastating pathogens of dicotyledonous plants. *Phytophthora* cause enormous economic damage to important crop species, such as potato, tomato, pepper, soybean, and alfalfa, as well as environmental damage to natural ecosystems. Virtually every dicot plant is affected by one or more species of *Phytophthora*, and several monocot species are infected as well.

The most notable pathogenic oomycete is *Phytophthora infestans*, the Irish potato famine pathogen. This species causes late blight, a ravaging disease of potato and tomato. Introduction of this pathogen to Europe in the mid-nineteenth century resulted in the potato blight famine and the death and displacement of millions of people. Today, *P. infestans* remains a devastating pathogen causing up to $5 billion losses in potato production worldwide. The appearance of highly aggressive and fungicide-insensitive strains in North America and Europe in the 1990s resulted in a new wave of severe and destructive potato and tomato late blight epidemics and has enhanced the impact of this disease on potato production. In developing countries, late blight affects

Eukaryotic Microbes.

TABLE 1 Major genera of plant pathogenic oomycetes

Genus	Host(s)	Description
Albugo	Plants	Obligate pathogens of plants
Aphanomyces	Animals and plants	Mostly pathogens of aquatic animals, but includes pathogens of plants
Bremia	Plants	Obligate pathogens of plants
Hyaloperonospora	Plants	Obligate pathogens of plants
Peronospora	Plants	Obligate pathogens of plants
Phytophthora	Plants	Most extensively studied genus with more than 60 species that cause some of the most severe diseases of dicot plants
Plasmopara	Plants	Obligate pathogens of plants
Pseudoperonospora	Plants	Obligate pathogens of plants
Pythium	Plants, animals, and microbes	Mostly plant pathogens, but includes mycoparasites, and at leastone species that infects humans

subsistence potato production. For instance, in 2003, potato production was nearly eliminated in Papua New Guinea, one of the few countries in the world that was previously free of the disease. Remarkably, the disease spread through the entire country within 2 months of first incidence. Disturbing reports predict that potato late blight will continue to cause food shortages and hunger in several parts of the world.

There are many other economically important *Phytophthora* diseases. These include root rot of soybean caused by *Phytophthora sojae*, rot and blight of several vegetable crops caused by *Phytophthora capsici*, black pod of cocoa caused by *Phytophthora palmivora* and *Phytophthora megakarya*, a recurring threat to worldwide chocolate production, dieback and related root rot diseases caused by *Phytophthora cinnamomi* on crops and native plant communities, and sudden oak death caused by the recently discovered *Phytophthora ramorum*. *P. cinnamomi* has had a tremendously destructive effect on the native flora of Australia. In recent years, sudden oak death has decimated oak tree populations along the Pacific coast of the United States and might be expanding to other hosts, such as redwoods, and to other regions in North America. At least one species, *Phytophthora brassicae* (previously known as *Phytophthora porri*), infects the model plant *Arabidopsis thaliana*, and this pathosystem was proposed to be easier to dissect at the genetic level than the agronomically important *Phytophthora* diseases.

Important oomycete plant pathogens also occur outside the genus *Phytophthora* (Table 1). Among these, a phylogenetically diverse group of obligate pathogens are represented by several genera. Among these, species of the genus *Plasmopara* are best represented by *Plasmopara viticola*, the agent of downy mildew of grapevine and a major problem of this crop. Many other downy mildew genera, such as *Bremia*, *Peronospora*, and *Hyaloperonospora*, cause serious diseases on a variety of plant species. Species of the genus *Albugo* are also obligate parasites and cause the white rust disease on several crop plants.

The genus *Pythium* forms another group of important pathogens, which includes more than a hundred species. *Pythium* species are abundantly present in water and soil habitats and cause a diversity of plant diseases, mainly in the root tissue. *Pythium* infections are usually limited to the meristematic tips, epidermis, cortex of roots, and fruits; but occasionally, severe *Pythium* infections occur when the pathogen moves deeper into the plant tissue and reaches the vascular system. Some *Pythium* species, such as *Pythium oligandrum*, are essentially beneficial and reduce infections caused by the more severe pathogenic microbes. This can occur directly through antagonistic effects or mycoparasitism or indirectly by induction of defense responses in plants.

Most modern research focuses on innovative approaches for the management of oomycete diseases, including the use of plant breeding, genetic engineering, and genomic technologies. Nonetheless, management of oomycete diseases is typically expensive and challenging.

BIOLOGY

Evolutionary History

Traditionally and essentially because of their filamentous growth habit, oomycetes have been classified as fungi. However, modern molecular and biochemical analyses as well as morphological features suggest that oomycetes share little taxonomic affinity to filamentous fungi, but are more closely related to brown algae and diatoms in a group known as stramenopiles (or heterokonts). This position is supported by molecular phylogenies based on ribosomal RNA sequences, compiled amino acid data for mitochondrial proteins, and protein-encoding chromosomal genes. The oomycetes also display a number of biochemical and morphological characteristics that distinguish them from the fungi and confirm their affinity to brown algae and other heterokonts. The cell walls of oomycetes are composed mainly of glucans and cellulose and, unlike fungal cell walls, contain little or no chitin. The zoospores display two flagella with an ultrastructure similar to the flagella of the motile spores of heterokont algae. The oomycetes also contain the energy storage chemical mycolaminarin, a molecule that is also found in kelps and diatoms.

As stramenopiles, the oomycetes share an evolutionary history with photosynthetic organisms and appear to have evolved from benign phototrophic ancestors. In fact, modern phylogenetic analyses provide robust support for the chromalveolates, a supergroup of algae and protists that brings together the stramenopiles (oomycetes, diatoms, and brown algae) and the alveolates (apicomplexans, ciliates, and dinoflagellates). The chromalveolate lineage is postulated to have been derived from a common ancestor that acquired a chloroplast endosymbiont from a red alga. The chromalveolate hypothesis predicts that nonphotosynthetic lineages like oomycetes have lost chloroplasts throughout their evolution. The recent completion of the genome sequences of *P. sojae* and *P. ramorum* provided strong support for this hypothesis and revealed that the endosymbiont left a genomic footprint through the transfer of chloroplast genes to the host nucleus. Phylogenetic analyses indicated that numerous *Phytophthora* genes for biosynthetic enzymes group in two distinct phylogenetic branches: one with affinity to proteobacteria indicative of a mitochondrial origin, and the other with affinity to cyanobacteria suggestive of a plastid origin.

The oomycetes are more closely related to the apicomplexans, the alveolate taxon that includes *Plasmodium* and *Toxoplasma* parasites, than to any other eukaryotic parasite, including major groups like fungi and trypanosomids. Striking similarities in pathogenicity mechanisms between oomycetes and apicomplexans have been noted with regard to host translocation (HT) signals in virulence (effector) proteins, secretion of protease inhibitors, and attachment to host tissues. These shared mechanisms of pathogenesis between oomycetes and apicomplexans may be a reflection of their common evolutionary history.

The peculiar phylogenetic affinities of oomycetes are also reflected in their distant relation to the true fungi despite apparent similarities, such as filamentous growth habit, heterotrophic lifestyle, and specialized infection structures. Surprisingly, comparative analyses of *Phytophthora* and fungal genome sequences identified a number of related genes that group together in phylogenetic analyses. Some authors explain these findings by multiple horizontal gene transfer (HGT) events that enabled the oomycetes to acquire fungal genes. Alternatively, differential gene loss might be the most parsimonious explanation for the observed biased phylogenetic distribution, particularly if the candidate HGT genes happen to occur in a wider range of eukaryotes. Nonetheless, the observed overlap among plant pathogenic filamentous microbes in genes encoding enzymes involved in cell wall hydrolysis, virulence, and osmotrophy points to a complex evolutionary history for the oomycetes, which resulted in a distinctive gene content.

The subdivison of oomycetes into taxonomic classes continues to be under debate. Typically, four classes of oomycetes are identified. These include Saprolegniales, Leptomitales, Lagenidales, and Peronosporales. Some authors elevated the plant pathogenic genera *Phytophthora* and *Pythium* to a separate class, named Pythiales. Recent molecular phylogenetic studies using ribosomal and mitochondrial sequences have started to unravel the evolutionary relationships between the different classes of oomycetes. The Peronosporales/Pythiales, which comprise the majority of plant pathogenic genera, form an ancient monophyletic group, suggesting that acquisition of plant pathogenicity probably occurred early in the evolution of this lineage. Most of the saprophytic and animal pathogenic species are restricted to the other classes. *Aphanomyces*, a genus with strong affinity to the Saprolegniales, includes both animal- and plant-pathogenic species.

General Biological Features

The oomycetes primarily inhabit aquatic and moist soil habitats. They are often very abundant and can be easily cultured from both freshwater and saltwater ecosystems, as well as from a variety of agricultural or natural soils. Several species are mainly terrestrial, including obligate biotrophic pathogens of plants that depend on air currents to disperse their spores.

The basic somatic structure of a majority of oomycete species is an extending fungus-like thread, the hypha, which grows into a branched network of filaments, the mycelium. Oomycetes are known as coenocytic, that is, their mycelium lacks septa or cross walls that divide the hypha, except to separate it from the reproductive organs. Both asexual and sexual reproductive structures occur. The primary asexual reproductive organ is the sporangium that differentiates at the tip of a vegetative hypha to produce and release motile zoospores with two flagella. The zoospores can germinate directly or indirectly to produce a vegetative mycelium, or can differentiate into secondary zoospores. Sexuality commonly occurs in oomycetes. Sexual reproduction involves the interaction of a male antheridia with a female oogonia through a fertilization tube that allows the male nuclei to migrate into the oogonium. Some oomycetes are self-fertile or homothallic, whereas others are self-sterile or heterothallic and require that strains with different mating types come into contact to achieve sexual reproduction. The sexual spores are the oospores, which can survive desiccation and starvation over long periods. Under favorable environmental conditions, the oospores germinate to form vegetative mycelium or to release zoospores. Oospores are also the structures that gave the oomycetes their name of egg fungi. Oomycetes are diploid in the dominant vegetative phase, with meiosis occurring only during gametogenesis.

Unique Biological Features

The oomycetes have a number of remarkable biological features that distinguish them from several other eukaryotic microorganisms. For example, the cell walls of oomycetes are mainly composed of β-1,3-glucan polymers and cellulose, and unlike fungal cell walls, contain little chitin. Nonetheless, chitin synthase genes are widely distributed among oomycete species, and the chitin synthase inhibitor polyoxin D caused significant reduction in growth in *Saprolegnia*, indicating that chitin is a minor but important component of the oomycete cell wall. Other singular features include the energy storage carbohydrate mycolaminarin, a β-1,3-glucan that is also found in kelps and diatoms; diploidy at the vegetative stage; and complex life, infection, and sexual cycles. Within the oomycetes, *Phytophthora* are both sterol and thiamine auxotrophs, and typically require exogenous sources of β-hydroxy sterols for sporulation and thiamine for growth.

Genome Structure

Genome sequence drafts of five plant pathogenic oomycete species have been completed and more are under way. The five species that were sequenced include four *Phytophthora* species, namely, *P. capsici*, *P. infestans*, *P. ramorum*, and *P. sojae*, and the downy mildew *Hyaloperonospora parasitica*, a pathogen of *Arabidopsis thaliana* that figures prominently in research on this model plant. The genomes vary extensively in size ranging from 65 Mbp for *P. capsici* and *P. ramorum* to 240 Mbp for *P. infestans*. Gene content is estimated to vary from 15 000 to more than 22 000 genes per genome. More than 50% of the sequenced genomes consist of repetitive DNA.

PATHOLOGY

Infection Cycle

Plant-associated oomycetes may be facultatively or obligately pathogenic. Many plant pathogenic oomycetes form specialized infection structures such as appressoria (penetration structures) and haustoria (feeding structures). Similar structures are also produced by fungal plant pathogens. The infection cycle of many oomycetes includes a biotrophic phase in which the pathogen requires living host cells. Pathogens like *P. infestans* adopt a two-step infection style typical of hemibiotrophs. An early biotrophic phase is followed by extensive necrosis of host tissue, resulting in colonization and sporulation. Infection events and host responses following *P. infestans* attack are well understood at the cellular level. Infection generally starts when motile zoospores that swim on the leaf surface encyst and germinate. Occasionally, sporangia can also initiate infections. Germ tubes form an appressorium and then a penetration peg, which pierces the cuticle and penetrates an epidermal cell to form an infection vesicle. Branching hyphae with narrow, digit-like haustoria expand from the site of penetration to neighboring cells through the intercellular space. Later on, infected tissue necrotizes and the mycelium develops sporangiophores, which emerge through the stomata to produce numerous asexual spores called sporangia. Pathogen dispersal usually occurs through the sporangia which release zoospores under cool and humid conditions. Resistance, whether displayed by host or by nonhost plants, is frequently associated with the hypersensitive response (HR), a cell death defense response of plants.

For an oomycete pathogen-like *P. infestans* to successfully infect and colonize its hosts, a series of pathogenic processes are necessary. These include adhesion to plant surfaces, as well as penetration and colonization of host tissue. The molecular basis of these processes remains, in large part, poorly understood. However, the gene products that facilitate the infection process and mediate pathogenicity are being discovered at an unprecedented rate. Pathogenesis involves the secretion of proteins and other molecules by *P. infestans*. Some of these participate in helping the pathogen attach to plant surfaces, while others help in breaking down physical barriers, such as plant membranes or cell walls, to infection. Other molecules influence the physiology of the host by suppressing or inducing host defense responses.

Adhesion, Penetration, and Colonization of Host Tissue

In *Phytophthora*, infection generally starts when motile zoospores released from sporangia reach a leaf or root surface, encyst, and germinate. Adhesion of cysts to plant surfaces occurs rapidly following zoospore encystment. The cysts germinate, and the germ tubes swell to form appressoria or appressoria-like structures that facilitate adhesion and penetration of plant surfaces. In root-infecting species, penetration can occur in between cells without the aid of an appressoria.

Little is known about the developmental processes that lead to appressoria formation. It was proposed that appressoria formation may result from the difficulty that germlings experience while attempting to penetrate plant surfaces. Some members of a small gene family, *Car*, encode extracellular mucin-like proteins and are upregulated in germinating cysts and appressoria shortly before penetration of the plant tissue and could function in adhesion. Recently, CBEL, a cellulose-binding protein of *Phytophthora parasitica*, was shown to be essential in adhesion to cellulosic substrates. *P. parasitica* strains silenced for the *CBEL* gene were impaired in their ability to attach to cellophane membranes, but remained able to infect tobacco plants. CBEL may also play a role in signaling, since the purified protein induces defense responses in tobacco plants.

Penetration and colonization of host tissue involves the secretion of a range of degradative enzymes that break down physical barriers to infection. A number of degradative enzymes such as cutinases, proteases, endo- and exoglucanases, and chitinases have been identified. A handful of *Phytophthora* genes encoding degradative enzymes have been characterized in detail, including phospholipases, a β-glucosidase/xylosidase, exo-1,3-β-glucanases, an endo-1,3-β-glucanase, and endopolygalacturonases (endoPGs). The endoPG family is remarkable in many respects. In *P. cinnamomi*, endoPGs form a major family with at least 19 members. Birth-and-death evolution, reticulate evolution, and diversifying selection were detected in *P. cinnamomi* endoPGs and may have contributed to the evolution of this structurally diverse and complex family. Phylogenetic analyses indicated that *Phytophthora* endoPGs are more similar to fungal endoPGs than to their plant and bacterial counterparts. Similar phylogenetic affinity to fungal sequences was also observed for exo-1,3-β-glucanases and an endo-1,3-β-glucanase from *P. infestans*. The phylogenies of these enzymes are unexpected and contrast with phylogenies obtained using ribosomal sequences or compiled protein sequences from mitochondrial and housekeeping chromosomal genes. These exceptional phylogenies possibly reflect convergent evolution through which phylogenetically distinct enzymes evolved to share significant similarity, perhaps by targeting similar substrates. On the other hand, HGT events in which plant pathogenic oomycetes acquired endoPG and glucanase genes from fungi may have taken place.

Induction of Defense Responses and Disease-Like Symptoms

Several *Phytophthora*-secreted proteins are known to induce a variety of cellular defense responses in plants. Some of these proteins induce defense responses in both susceptible and resistant plants and are referred to as general elicitors. Based on similarity to self and nonself recognition models of the animal innate immune system, these general elicitors have been likened to pathogen-associated molecular patterns (PAMPs), which are surface-derived molecules that induce the expression of defense response genes and the production of antimicrobial compounds in host cells. Oomycete PAMPs include glucan and various other cell wall components.

Phytophthora species produce 10 kDa extracellular proteins, known as elicitins, which induce the HR and other biochemical changes associated with defense responses in tobacco species. Tobacco is resistant to most *Phytophthora* species, and recognition of elicitins is thought to be one component of this resistance. Direct evidence for this model was obtained using *P. infestans* strains engineered to be deficient in the elicitin INF1 by gene silencing. These strains induce disease lesions on the wild tobacco species *Nicotiana benthamiana*, suggesting that the elicitin INF1 conditions avirulence to this plant species. In *P. parasitica*, some strains are known to naturally infect tobacco and cause the black shank disease. These strains evade the defense surveillance system of tobacco by either not producing elicitins or exhibiting downregulation of elicitin genes *in planta*.

Nep1-like proteins (NLPs) are ~25 kDa proteins that are widely distributed in bacteria, fungi, and oomycetes, particularly in plant-associated species. The canonical 24 kDa necrosis- and ethylene-inducing protein (Nep1) was originally purified from culture filtrates of the fungus *Fusarium oxysporum* f. sp. *erythroxyli*. NLPs have subsequently been described in species as diverse as *Bacillus*, *Erwinia*, *Verticillium*, *Pythium*, and *Phytophthora*. Despite their diverse phylogenetic distribution, NLPs share a high degree of sequence similarity, and several members of the family have the remarkable ability to induce cell death in as many as 20 dicotyledonous plants. The wide phylogenetic conservation and broad-spectrum activity of NLPs distinguish them from the majority of cell death elicitors and suggest that the necrosis-inducing activity is functionally important. NLPs are thought to function as toxins that facilitate colonization of host tissue during the necrotrophic phase of growth.

Inhibition of Host Enzymes

Oomycete-secreted proteins that suppress host defense responses have been described in several pathosystems. Suppression of host defenses can occur through the production of inhibitory proteins that target host enzymes. Genes encoding secreted proteins that inhibit soybean endo-β-1,3-glucanase have been cloned from *P. sojae*. These proteins, termed glucanase inhibitor proteins (GIPs), share significant structural similarity to the trypsin class of serine proteases, but bear mutated catalytic residues and are proteolytically nonfunctional as a consequence. GIPs are thought to function as counterdefensive molecules that inhibit the degradation of β-1,3/1,6-glucans in the pathogen cell wall and/or the release of defense-eliciting oligosaccharides by host endo-β-1,3-glucanases.

Protease inhibitors are another class of secreted inhibitory proteins. EPI1 and EPI10 are multidomain-secreted serine protease inhibitors of the Kazal family (MEROPS family I1) that are thought to function in counterdefense. They inhibit and interact with the PR protein P69B, a subtilisin-like serine protease of tomato that is thought to function in defense. The *epi1*, *epi10*, and *P69B* genes are concurrently expressed and upregulated during infection of tomato by *P. infestans*. The mechanism by which inhibition of P69B by EPI1 and EPI10 affects the late blight disease is not yet elucidated.

Cystatin-like cysteine protease inhibitors EPIC1 and EPIC2 (InterPro IPR000010, MEROPS family I25) form another class of secreted protease inhibitors of *P. infestans*. Recent findings suggest that these two inhibitors target an apoplastic papain-like cysteine protease of tomato, suggesting multifaceted inhibitions of tomato proteases by secreted effectors of *P. infestans*.

Effectors

Secreted oomycete proteins that alter plant responses have been defined as effectors and are currently the focus of extensive research activities. In susceptible plants, effectors, encoded by pathogen virulence genes, promote infection by suppressing defense responses, enhancing susceptibility, or inducing disease symptoms. Alternatively, in resistant plants, effectors, encoded by pathogen avirulence (*Avr*) genes, are recognized by the products of plant disease resistance (*R*) genes, resulting in the HR and effective defense responses.

In the oomycetes, two classes of effectors target distinct sites in the host plant: Apoplastic effectors, such as the glucanase and protease inhibitors described earlier, are secreted into the plant extracellular space, while cytoplasmic effectors are translocated inside the plant cell, where they target different subcellular compartments. One class of cytoplasmic effectors, the so-called RXLR effectors, has been the subject of much research in recent years. These effectors, first identified as avirulence proteins in *H. parasitica*, *P. infestans*, and *P. sojae*, carry a conserved motif, termed RXLR (arginine, any amino acid, leucine, arginine), that is located downstream of the signal peptide and has been implicated in HT. Remarkably, the RXLR motif is similar in sequence and position to the plasmodial HT/Pexel motif that functions in delivery of parasite proteins into the red blood cells of mammalian hosts. A ~30 amino acid region encompassing the RXLR motif of *P. infestans* RXLR proteins AVR3a and PH001D5 mediates the export of the green fluorescent protein (GFP) from the *Plasmodium falciparum* parasite to the host red blood cells, suggesting that the RXLR and HT/Pexel domains are functionally interchangeable. This and related findings led to the view that oomycete RXLR effectors are modular proteins with two major functional domains. While the N-terminal domain encompassing the signal peptide and RXLR leader functions in secretion and targeting, the remaining C-terminal region carries on the effector activity and operates inside the plant cells.

The genome sequences of plant pathogenic oomycetes enabled genome-wide cataloging of RXLR effectors using computational approaches. The RXLR effector secretomes turned out to be more complex than expected, consisting of hundreds of candidate effectors. Tyler and colleagues reported 350 RXLR effectors each in the genomes of *P. ramorum* and *P. sojae* using iterated similarity searches. Analyses performed using combinations of motif and hidden Markov model searches uncovered at least 50 candidates in the downy mildew *H. parasitica* and more than 200 each in *P. capsici*, *P. infestans*, *P. ramorum*, and *P. sojae*.

Comparative analyses of the RXLR effectors indicate that they are undergoing birth-and-death evolution resulting in divergent sets of effectors in the three species. Patterns of accelerated rates of gene loss and duplication at some RXLR effector loci were reported for *P. sojae* and *P. ramorum*. In another example, *Avr3a* of *P. infestans* and *ATR1* of *H. parasitica* occur in conserved syntenic chromosomal regions but are highly divergent. Rapid evolutionary rates in these effector genes may reflect evolutionary adaptations to host plants.

The remarkably large numbers of candidate RXLR effectors identified in the genome-wide analyses suggest that oomycetes extensively modulate host processes during infection. The next challenge is to unravel the virulence activities of these effectors to understand how they perturb plant processes to increase the reproductive success of the pathogen. For instance, a possible virulence function was ascribed to *P. infestans* RXLR effector Avr3a, which is able to suppress the hypersensitive cell death induced by another *P. infestans* protein, INF1 elicitin.

CONCLUSIONS

Oomycetes evolved the ability to infect plants independently of other eukaryotic microbes and have probably developed unique mechanisms of pathogenicity. With the availability of a respectable molecular toolbox and a multitude of gene sequences, significant progress has been made in understanding the molecular basis of infection by oomycetes. However, most of our knowledge remains limited to economically important species in the *Phytophthora* genus, and little is known about infection by other plant- or animal-pathogenic oomycetes. Future research will exploit emerging information about *Phytophthora* genetics to ask pertinent questions about oomycete pathology and evolution. The increased availability of genome sequences will offer unique opportunities to address these questions and perform comparative genomics among pathogenic oomycetes and between oomycetes and other eukaryotic microbes. Such studies will improve our overall understanding of the evolution of parasitic and pathogenic lifestyles in eukaryotes.

FURTHER READING

Agrios, G. N. (2004). *Plant pathology*. Academic Press: San Diego, CA.

Erwin, D. C., & Ribeiro, O. K. (1996). *Phytophthora diseases worldwide*. APS Press: St. Paul, MN.

Fry, W. (2008). Phytophthora infestans: The plant (and R gene) destroyer. *Molecular Plant Pathology*, *9*(3), 385–402.

Haldar, K., Kamoun, S., Hiller, L. N., Bhattacharjee, S., & van Ooij, C. (2006). Common infection strategies of pathogenic eukaryotes. *Nature Reviews Microbiology*, *4*, 922–931.

Kamoun, S. (2006). A catalogue of the effector secretome of plant pathogenic oomycetes. *Annual Review of Phytopathology*, *44*, 41–60.

Kamoun, S., & Smart, C. D. (2005). Late blight of potato and tomato in the genomics era. *Plant Disease*, *89*, 692–699.

Margulis, L., & Schwartz, K. V. (2000). *Five kingdoms: An illustrated guide to the phyla of life on Earth*. W.H. Freeman: New York.

Raffaele, S., Farrer, R. A., Cano, L. M., Studholme, D. J., MacLean, D., Thines, M., *et al.* (2010). Genome evolution following host jumps in the Irish potato famine pathogen lineage. *Science*, *330*(6010), 1540–1543.

Schornack, S., Huitema, E., Cano, L. M., Bozkurt, T. O., Oliva, R., Van Damme, M., *et al.* (2009). Ten things to know about oomycete effectors. *Molecular Plant Pathology*, *10*(6), 795–803.

Thines, M., & Kamoun, S. (2010). Oomycete-plant coevolution: recent advances and future prospects. *Current Opinion in Plant Biology*, *13*(4), 427–433.

Tyler, B. M., Tripathy, S., Zhang, X., *et al.* (2006). Phytophthora genome sequences uncover evolutionary origins and mechanisms of pathogenesis. *Science*, *313*, 1261–1266.

Chapter 26

Picoeukaryotes

R. Massana

Institut de Ciències del Mar, Barcelona, Catalonia, Spain

Chapter Outline

Abbreviations 355
Defining Statement 355
Introduction
What Are Marine Picoeukaryotes? 355
Method-Driven History of Marine Picoeukaryotes 356
Biology of Cultured Marine Picoeukaryotes 358
Cultured Strains 358
Cellular Organization 359
Physiological Parameters 360
The Implications of Being Small 360
Picoeukaryotes in the Marine Environment 361
Bulk Abundance and Distribution 361
Ecological Role of PP 362
Ecological Role of HP 362
Molecular Tools to Study Picoeukaryote Ecology 363
Elusive View of In Situ Diversity by Nonmolecular Tools 363
Cloning and Sequencing Environmental Genes 363
Beyond Clone Libraries: FISH and Fingerprinting Techniques 364
In Situ Phylogenetic Diversity 365
Overview of 18S rDNA Libraries 365
Relatively Well-Known Groups 365
Marine Alveolates and Marine Stramenopiles 367
Novel High-Rank Phylogenetic Groups 367
Biogeography 367
The Genomic Era 368
Genome Projects on Cultured Picoeukaryotes 368
Environmental Genomics or Metagenomics 368
Recent Developments 368
Concluding Remarks 370
Further Reading 370
Relevant Websites 371

ABBREVIATIONS

ARISA Automated ribosomal interspacer analysis
BAC Bacterial artificial chromosomes
ciPCR Culture-independent PCR
DGGE Denaturing gradient gel electrophoresis
DOM Dissolved organic matter
FISH Fluorescent in situ hybridization
HNF Heterotrophic nanoflagellates
HP Heterotrophic picoeukaryotes
HPLC High-performance liquid chromatography
MALV Marine alveolates
MAST Marine stramenopiles
PNF Phototrophic nanoflagellates
PP Phototrophic picoeukaryotes
TEM Transmission electron microscopy
T-RFLP Terminal-restriction fragment length polymorphism

DEFINING STATEMENT

Microorganisms play fundamental roles in marine ecosystems, sustaining food webs and driving biogeochemical cycles. Unicellular eukaryotes smaller than 2–3 μm (picoeukaryotes) are recognized as important members of microbial assemblages in terms of both biomass and activity. They show a large functional and phylogenetic diversity and include many poorly or entirely uncharacterized taxa.

INTRODUCTION

What Are Marine Picoeukaryotes?

Marine picoeukaryotes are a heterogeneous assemblage of very small eukaryotic organisms. Although some examples of cultured picoeukaryotes exist, most have only been detected in the last 5–6 years using culture-independent sampling techniques and therefore have not been characterized in any detail. Nonetheless, we now know that picoeukaryotes are ubiquitous throughout the marine environment, occupying a wide variety of habitats. This includes distinct layers of the water column, each with its own dominant biogeochemical regime, marine sediments, and unique ecosystems such as hydrothermal vents.

Salinity is the main parameter that differentiates marine and freshwater habitats, and these habitats are for the most part populated by very different species. This article will focus on planktonic marine picoeukaryotes, those that live suspended in seawater, and especially those living in the upper water column where photosynthesis occurs. Since most primary production in marine systems is due to photosynthesis by planktonic microorganisms (cells smaller than 200 μm), the microbial component plays a key role in marine food webs. Among these organisms, picoeukaryotes, many of which have only recently been discovered, are qualitatively and quantitatively important. Marine primary production accounts for roughly half of the Earth's production, indicating that the oceans and their microorganisms are crucial in sequestering inorganic carbon from the atmosphere and potentially in mitigating global change.

Picoeukaryotes have a typical eukaryotic cell structure in a miniaturized state. This includes the presence of a nucleus, an endomembrane system (endoplasmic reticulum, Golgi body, and vesicles), mitochondria, and in the case of photosynthetic picoeukaryotes, a chloroplast. Nonetheless, these cells are extremely small, the diameter ranging from 0.8 μm in the case of *Ostreococcus tauri*, the smallest known eukaryote, to an upper range of 2–3 μm. Due to this small size, picoeukaryotes are largely indistinguishable by light microscopy, the usual method for studying eukaryotic microbial diversity. Thus, very few picoeukaryotes have been isolated and characterized.

In 1978 a scheme for classification of marine organisms according to size was delineated largely based on sieving technology. Microorganisms were operationally split into three categories: picoplankton (0.2–2 μm in cell diameter), nanoplankton (2–20 μm), and microplankton (20–200 μm). Initially, the picoplankton was thought to be almost exclusively made up of prokaryotes and the nanoplankton mostly of small single-celled eukaryotes. However, the existence and abundance of protists within the picoplankton size class was soon recognized. Today, the term picoeukaryotes is often used a bit loosely to include protists with a size up to 3 μm. Direct inspections of marine protist assemblages indicate that the 2 μm limit often falls in the middle of the size spectra and that a more coherent group is delimited using a 3 μm upper boundary.

Picoeukaryotes thus defined (protists smaller than 3 μm) are abundant in the marine plankton. They include diverse phototrophic and heterotrophic cells, and they play crucial roles as primary producers, bacterial grazers, and parasites. In recent years their diversity, abundance, and widespread distribution has begun to be recognized and they are attracting more attention.

Method-Driven History of Marine Picoeukaryotes

Microorganisms are invisible to the unaided human eye, so the history of their study is inevitably linked to the development of new methods for their observation and characterization (Figure 1). The existence of very small cells in the marine plankton was known from the beginning of the twentieth century by phytoplanktologists that inspected concentrated samples of seawater by light microscopy. The first cultured picoeukaryote, the pigmented flagellate *Micromonas pusilla* (formerly, *Chromulina pusilla*), was described in 1952. Cultured picoeukaryotes provided the basis for many early microscopic and physiological studies, leading to the description of new species. The easy cultivability of *M. pusilla* allowed the initial estimations of its abundance by the serial dilution method. This showed that it is widely distributed in the marine environment and can reach abundances as high as 10^4 cells ml^{-1}. Despite this particular culturing success with *M. pusilla*, microbial ecologists soon became aware that the dominant microorganisms were often not easily cultured, so direct inspections of natural samples and culture-independent approaches were still much needed.

Electron microscopy, which had been used on cultured material since the 1950s, started to be applied to inspect natural protist assemblages during the 1970s. Transmission electron microscopy (TEM), which allows the inspection of intact specimens, revealed conspicuous features of nanoplanktonic protists. However, these studies rarely targeted picoeukaryotes. The latter were first detected by TEM in thin sections of centrifuged natural samples in 1982. *M. pusilla* and an unknown prasinophyte (later identified as *Bathycoccus prasinos*) were seen in many coastal and oceanic samples, sometimes being relatively abundant. These observations also provided the first clue on mortality of picoeukaryotes, since cells often appeared infected with large viruses. Soon after this study, it was shown that a large fraction of marine primary production in offshore regions is due to picophytoplankton (cyanobacteria and picoeukaryotes).

The first accurate counts of marine picoplankton were obtained by epifluorescence microscopy in the early 1980s. This was based on the fluorescence (natural or stain induced) emitted by cells retained quantitatively on the surface of flat polycarbonate filters. Bacterial numbers revealed by this approach were orders of magnitude higher than expected from previous cultivation-dependent techniques. Protists of different sizes were also evident at abundances

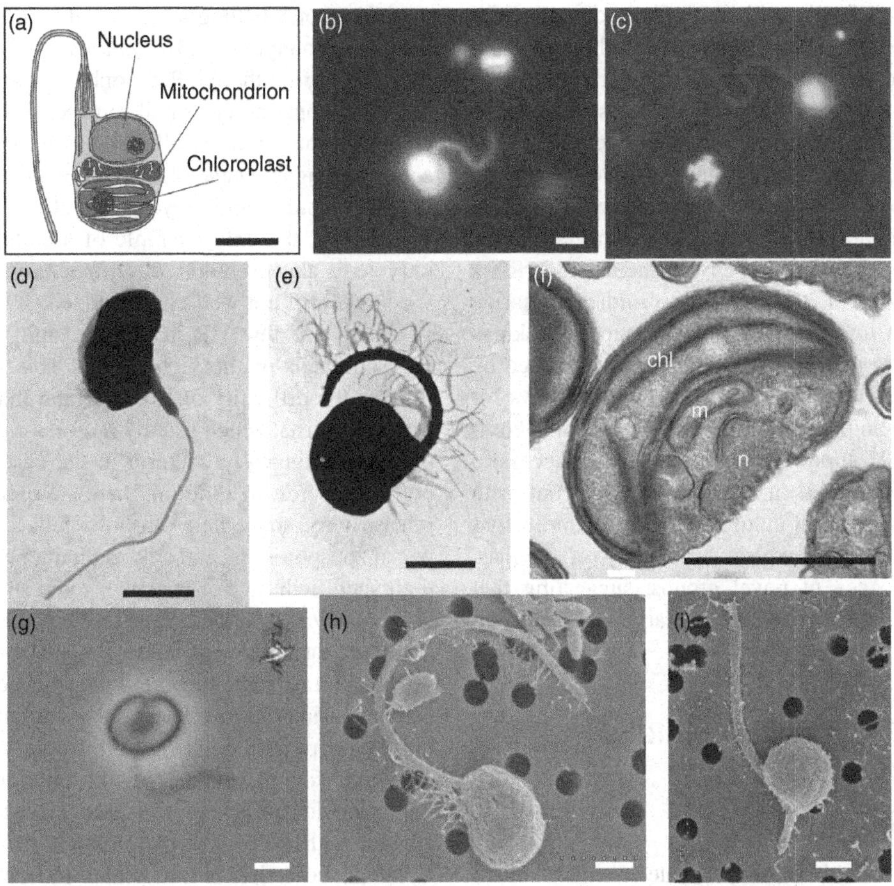

FIGURE 1 Examples of marine picoeukaryotes. (a) Drawing of *Micromonas pusilla. Reproduced from Slapeta, J., Lopez-Garcia, P., & Moreira, D. (2006). Global dispersal and ancient cryptic species: The smallest marine eukaryotes.* Molecular Biology and Evolution, 23, *23–29.* (b) Unidentified flagellate seen by epifluorescence under UV radiation after DAPI staining. (c) Unidentified phototrophic (left) and heterotrophic (right) flagellates seen by epifluorescence under blue light. Photo courtesy by Dolors Vaqué. (d) Stained whole mount of *M. pusilla. Reproduced from Guillou, L., Eikrem, W., Chrétiennot-Dinet, M.-J., et al. (2004). Diversity of picoplanktonic prasinophytes assessed by direct nuclear SSU rDNA sequencing of environmental samples and novel isolates retrieved from oceanic and coastal marine ecosystems.* Protist, 155, *193–214.* (e) Stained whole mount of *Symbiomonas scintillans. Reproduced from Guillou, L., Chrétiennot-Dinet, M.-J., Moon-van der Staay, S. Y., Boulben, S., & Vaulot, D. (1999).* Symbimonas scintillans *gen. and sp. nov. and* Picophagus flagellatus *gen et sp. nov. (Heterokonta): Two new heterotrophic flagellates with picoplanktonic size.* Protist, 150, *383–398.* (f) Thin section through *Ostreococcus tauri* with chloroplast (chl), mitochondrion (m), and nucleus (n). *Reproduced from Guillou, L., Eikrem, W., Chrétiennot-Dinet, M.-J., et al. (2004). Diversity of picoplanktonic prasinophytes assessed by direct nuclear SSU rDNA sequencing of environmental samples and novel isolates retrieved from oceanic and coastal marine ecosystems.* Protist, 155, *193–214.* (g) *Pelagomonas calceolata* by phase contrast. *Reproduced from http://starcentral.mbl.edu/microscopeportal.php. (h, i) Unidentified flagellates by scanning electron microscopy (SEM). Scale bar = 1 μm.*

in the thousands of cells per milliliter. In addition, phototrophic or heterotrophic protists, which play fundamentally different ecological roles, could be counted separately based on the presence or absence of chlorophyll autofluorescence. Epifluorescence microscopy still remains the method of choice (albeit time-consuming) for counting heterotrophic picoeukaryotes (HP). Phototrophic picoeukaryotes (PP), on the other hand, can be more easily counted by flow cytometry, a technique imported from biomedicine based on the laser detection of single cells flowing through a small aperture. Flow cytometry was first applied to marine ecology in the 1980s, and was instrumental in the discovery of the most abundant phototroph on Earth, the cyanobacterium *Prochlorococcus*. This tool has been extensively applied to describe the global distribution of marine PP.

The twenty-first century started with a fair knowledge of the global abundance and distribution of marine picoeukaryotes in the sea and a sizeable collection of characterized cultures. However, there was a remarkable lack of knowledge about which species dominate natural assemblages, and whether or not the cultured strains represented relevant ecological models. This was because the techniques used for identification (culturing, electron microscopy) were not quantitative, and conversely the techniques used for quantification (epifluorescence microscopy, flow cytometry) did not permit species identification.

This changed radically with the advent of molecular tools, particularly culture-independent sampling of ribosomal RNA sequences (phylotypes) from selected environments (see section 'Molecular Tools to Study Picoeukaryote Ecology'). These provided exciting results for marine bacteria and archaea during the 1990s, revealing whole new phyla and possibly even kingdoms of organisms. The first studies on the in situ diversity of marine picoeukaryotes by cloning and sequencing environmental 18S rRNA genes were published in 2001. Similar to studies of marine prokaryotes, these revealed new major groups of eukaryotes. Later, specific phylogenetic groups were targeted by fluorescent in situ hybridization (FISH), which allows for directly observing and quantifying natural cells in the environment. The FISH approach was particularly successful for organismal lineages seen in clone libraries but with no known close relative in culture. The use of molecular tools has revealed an unexpected diversity of protists, including the presence of novel groups, suggesting that marine picoeukaryotes represent a large reservoir of unexplored biodiversity.

BIOLOGY OF CULTURED MARINE PICOEUKARYOTES

Cultured Strains

Marine picoeukaryotes have been isolated using standard methods, such as mineral media and light for phototrophic cells and rice (or yeast extract) enrichment media for heterotrophic cells, often inoculated with 2–3 μm filtered seawater. To describe a new species, isolates must be characterized by a set of complementary techniques. Optical microscopy reveals the cell shape and motility pattern. Electron microscopy uncovers the cell ultrastructure: architecture of mitochondria, chloroplast, and flagellar apparatus, number, size, and ornamentation of flagella, and presence of external structures. Molecular markers, mostly 18S rDNA but also other genes, allow for phylogenetic placement of new isolates. Biochemical markers, such as pigment analysis by high-performance liquid chromatography (HPLC), storage products, or fatty acid profiles also provide specific information. A subset of these techniques may be used to assign new isolates to a given species, but molecular markers are used most commonly because they are faster and easier to gather. Once formally described, strains are deposited in culture collections, such as the Provasoli–Guillard National Center for Culture of Marine Phytoplankton in the United States (http://ccmp.bigelow.org/) or the Roscoff Culture Collection in France (http://www.sb-roscoff.fr/).

Since the first culture of *M. pusilla* was isolated in 1952, there has been some success in isolating additional picoeukaryotes (Table 1). Most are phototrophic and belong to the green algae and the stramenopiles. Cultured green algal picoeukaryotes belong mostly to the order Mamiellales in the Prasinophyceae (*Micromonas*, *Ostreococcus*, and *Bathycoccus*), although Pedinophyceae and Trebouxiophyceae also contain very small representatives. Stramenopile picoeukaryote cultures all belong to novel algal classes, such as Pelagophyceae (described in 1993), Bolidophyceae (in 1999), and Pinguiophyceae (in 2002). Apart from these two groups, the only example of a cultured picophytoeukaryote is the cryptophyte *Hillea marina*, although this species is still not well characterized.

There are few HP in culture, and all belong to the chrysomonads or the bicosoecids. This scarcity probably reflects the difficulty of isolating and maintaining heterotrophic protists, which usually requires culturing with other organisms, typically bacteria in the case of phagotrophic picoeukaryotes. In addition, there are parasitic protists that release very small heterotrophic cells as free-living dispersal zoospores and can only be maintained in culture with their specific host. For instance, some strains of the alveolates *Amoebophrya* and *Parvilucifera*, parasites of marine dinoflagellates, have zoospores as small as 3 μm.

In addition, some larger cultured species have a minimal cell dimension ≤ 3 μm. These are not strictly picoeukaryotes but would pass through the 3 μm pore size prefilter used to select for picoeukaryotes in environmental surveys. Phototrophic species of this size category belong to all classes with picoeukaryotes shown in Table 1 (except Bolidophyceae that only contains picoeukaryotes) as well as Prymnesiophyceae and some additional stramenopiles (e.g., Bacillariophyceae, Dictyochopyceae, and Eustigmatophyceae). Some of these, such as the Bacillariophyceae (diatoms) and the Prymnesiophyceae, include very important marine phytoplankters. Additional heterotrophic species span the breadth of the eukaryotic tree including cercomonads (supergroup Rhizaria), kinetoplastids and jakobids (supergroup Excavata), choanoflagellates (supergroup Opisthokonts), and apusomonads (unclear affiliation).

Some picoeukaryotic species have been named based on direct observations of natural samples, enrichments, or temporary cultures. Examples are the heterotrophic flagellate *Pseudobodo minimum* (a bicosoecid 2.0 μm in size) and the green alga *Chlorella minima* (1.5–3 μm). Morphological descriptions are sometimes accompanied by ultrastructural characters obtained by electron microscopy, such as for the cercozoan *Phagomyxa odontellae* (a diatom parasite with zoospores of $3–4 \times 2–3$ μm) or the Parmales. The latter is an intriguing algal group, commonly observed in the sea, which includes the picoplankton species *Tetraparma pelagica* (2.2×2.8 μm). Although tentatively classified within the class Chrysophyceae, the phylogenetic position of Parmales is still unknown, and it is likely that molecular analyses, once they are possible, will reveal that they deserve a new class rank, as has occurred with other algal stramenopile lineages.

TABLE 1 Examples of marine picoeukaryotes in culture, including the cell size (minimal and maximal length), trophic mode, and where the culture is deposited

Taxonomic group	Species	Size (μm)	Trophic	Culture collection
Archaeplastida				
Pedinophyceae	*Marsupiomonas pelliculata*	3.0–3.0	P	PCC
	Resultor micron	1.5–2.5	P	SCCAP
Prasinophyceae	*Bathycoccus prasinos*	1.5–2.5	P	RCC, CCMP
	Dolichomastix lepidota	2.5–2.5	P	
	Micromonas pusilla	1.0–3.0	P	RCC, CCMP
	Ostreococcus tauri	0.8–1.1	P	RCC, CCMP
	Picocystis salinarum	2.0–3.0	P	CCMP
Trebouxiophyceae	*Chlorella* sp.	2.0–3.0	P	RCC
	Picochlorum eukaryotum	3.0–3.0	P	RCC, CCMP
Stramenopiles				
Bicosoecida	*Caecitellus pseudoparvulus*	2.0–3.0	H	RCC
	Cafeteria roenbergensis	3.0–3.0	H	RCC
	Symbiomonas scintillans	1.2–1.5	H	RCC, CCMP
Bolidophyceae	*Bolidomonas pacifica*	1.0–1.7	P	RCC, CCMP
Chrysophyceae	*Picophagus flagellatus*	1.4–2.5	H	RCC, CCMP
Pelagophyceae	*Aureococcus anophagefferens*	1.5–2.0	P	RCC, CCMP
	Pelagomonas calceolata	2.0–3.0	P	RCC, CCMP
Pinguiophyceae	*Pinguiochrysis pyriformis*	1.0–3.0	P	MBIC
CCTH				
Cryptophyceae	*Hillea marina*	2.0–2.5	P	

P, phototrophic; H, heterotrophic; PCC, Plymouth Culture Collection; SCCAP, Scandinavian Culture Centre for Algae and Protozoa; RCC, Roscoff Culture Collection; CCMP, Provasoli–Guillard National Center for Culture of Marine Phytoplankton; MBIC, Marine Biotechnology Institute Culture Collection.

Cellular Organization

Picoeukaryotes are miniaturized unicellular organisms that nonetheless retain all typical eukaryotic subcellular structures. They mostly divide asexually, a common feature in many protists, and a complete life cycle with the plausible presence of a sexual phase is totally unknown. The algal class with most cultured picoeukaryote species is the Prasinophyceae. Picoprasinophyte cells have a single chloroplast (often with a starch granule) with typical prasinophyte pigments (chlorophyll *b* and prasinoxanthin), one mitochondrion, and one Golgi body. The three most common cultured genera illustrate the variability within the class. *B. prasinos* lacks flagella and is covered by spider web-like organic scales. *M. pusilla* is naked, has a single flagellum with a short wide base and long thin distal end and a characteristic swimming behavior. *O. tauri* is coccoid, nonmotile, and naked. For each of these, strains with indistinguishable ultrastructural features have been isolated from distant geographic sites. For *B. prasinos* these strains appear to be genetically similar, but there is a clear genetic structure among *M. pusilla* (at least five clades) and *O. tauri* (at least four clades) strains. It has been proposed that these clades can be viewed as ecotypes with specific adaptations to their environments (see later). The remaining picoeukaryotic green algal species show similar minimal cell structure but have different flagellar architecture, pigment signatures, and swimming behavior.

The second group with a significant number of cultured picoeukaryote strains is the stramenopiles. This is a vast and extremely diverse clade of autotrophic and heterotrophic taxa, most of which have two unequal flagella, the longer being covered by tripartite hairs that reverse the thrust from the flagellum. Photosynthetic stramenopiles have chlorophyll *c* as the main accessory pigment and the chloroplast is surrounded by an endoplasmid reticulum continuous with the outer nuclear membrane. Picostramenopiles have a simplified cell structure, with a single mitochondrion, one chloroplast, and one Golgi body. *Pelagomonas calceolata* is covered by a thin organic theca and has only one flagellum with two rows of bipartite hairs, and even lacks the basal body of the second flagellum. Its main carotenoid is 19′-butanoxylofucoxanthin. *Bolidomonas pacifica* is naked and has two unequal flagella, the longer with tubular hairs similar to those of *P. calceolata*. It can swim vigorously, up to 1.5 mm per second, and contains fucoxanthin as the main carotenoid, like diatoms. *Pinguiochrysis pyriformis* is coccoid, naked, lacks flagella, and produces large amounts of polyunsaturated ω-3 fatty acids. HP within the stramenopiles also have a simplified cell structure, with a nucleus, a single Golgi body, one to two mitochondria, and no chloroplast. *Symbiomonas scintillans* is naked and has a single flagellum (and only one basal body) with two rows of tripartite tubular hairs. It contains several endosymbiotic bacteria located close to the nucleus. *Picophagus flagellatus* is naked, has two unequal flagella, the longer with two rows of tripartite hairs, and swims energetically.

Physiological Parameters

Cultured picoeukaryotes provide necessary material not only for ultrastructural, biochemical, and molecular studies, but also for defining physiological properties. Some physiological parameters deal with how fast unicellular organisms use environmental resources, such as inorganic nutrients and light for phototrophs and prey for phagotrophs. For instance, the relationship between ingestion rate and prey availability in a phagotrophic protist (functional response) depends on prey concentration: at low levels, the ingestion rate increases linearly with prey concentration; at medium levels, the rate still increases but not linearly; and at high levels, the rate reaches a maximum. This relationship can be described by several hyperbolic models. The most commonly used is analogous to the Michaelis–Menten equation for enzyme kinetics, which is based on the maximal ingestion rate (U_m) and the prey concentration allowing half U_m (K_m). Both parameters are characteristic of a given species and have interesting ecological implications: U_m gives the upper limit of a species grazing capacity and K_m roughly indicates the prey concentration at which the species is adapted to live. The uptake of inorganic nutrients by a phototroph can be described by the same equation. The hyperbolic relationship between light and photosynthetic activity, on the other hand, uses a different model to incorporate the inhibitory effect of high irradiances.

The differential use of resources translates into different growth rates. In the case of phagotrophic protists, a similar relationship can be found between growth rate and prey availability, known as numerical response, again modeled by the Michaelis–Menten equation. A crucial parameter for phagotrophs is growth efficiency, the fraction of the food ingested that is converted to biomass. This relates ingestion rates and growth rates and has strong implications for respiration and nutrient remineralization. For phototrophic protists, growth rate is often the activity parameter measured to follow their relationship with inorganic nutrients and light. The plasticity with which protists change their reproductive rates according to the available resources is remarkable.

This simplistic view of resource–activity relationships becomes more complex when taking into account the properties of given resources, providing an additional set of more realistic species-specific physiological parameters. In photosynthetic protists, the quality of light and the chemical state of inorganic nutrients can be important. Accessory pigments can tune the cell to a given region of the light spectrum, whereas membrane transporters, genetically codified, determine the nutrient state that can be used. Heterotrophic protists can choose prey depending on size, phylogenetic composition, surface properties, and motility behavior. Finally, there are other species-specific responses related to environment parameters, such as temperature. Altogether, physiological parameters for a given species may explain its competitive advantage and success in the environment. Conceptual models illustrating how variations in these parameters may induce similar species (or ecotypes of the same species) to occupy different ecological niches are shown in Figure 2.

O. tauri represents a good example of ecotype differentiation. Twelve *O. tauri* strains are ultrastructurally indistinguishable but form distinct genetic clades using rDNA sequences. One clade is formed by strains isolated from the bottom of the photic zone (~100 m deep), which grow well at low irradiances but are inhibited at irradiances typical of surface waters. The clade formed by strains isolated from surface waters, on the other hand, represents a high-light-adapted ecotype only growing at surface irradiances. The low-light ecotypes possess additional photosynthetic pigments absent from the high-light ecotypes. These different ecotypes, which together are able to exploit a wide range of light levels, might explain the success of *O. tauri* throughout the photic zone. Ecotype differentiation has also been observed in *M. pusilla*, where strains of one clade seem to be adapted to live in polar waters. It is plausible that ecotype diversity is a widespread phenomenon in the microbial world.

The Implications of Being Small

Cell size is the single trait that most influences the physiological and ecological properties of a given organism. Smaller cells, by virtue of their higher surface-to-volume

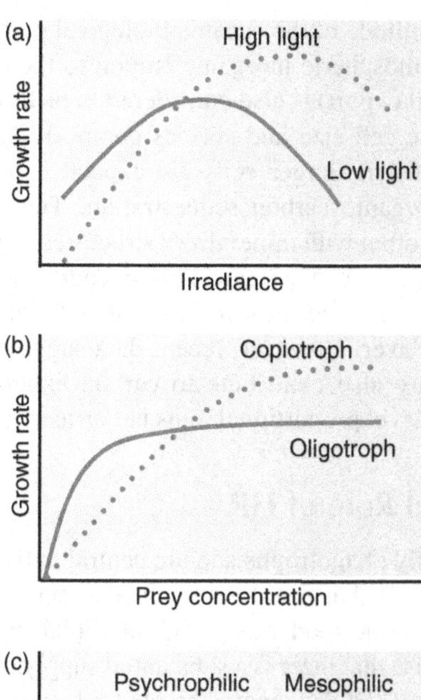

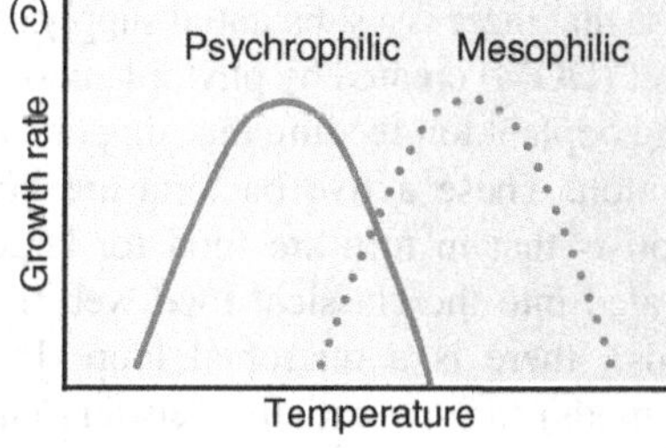

FIGURE 2 Conceptual models illustrating how physiological responses induce ecological adaptation. (a) The low-light adapted phototrophic ecotype grows better at low irradiances, and the high-light adapted ecotype at high irradiances. (b) The oligotroph phagotropic ecotype grows better at low prey concentration and the copiotroph at high levels. (c) The psychrophilic ecotype grows better at cold temperatures and the mesophilic ecotype at warmer temperatures.

ratio as compared to larger cells, are generally more efficient in resource acquisition and therefore may have higher specific metabolic rates. Very crudely, physiological rates are inversely proportional to body length, the so-called allometric relationship. Thus, picoeukaryotes would be the protists with the highest growth rates and better adapted to oligotrophic conditions. Picoeukaryotes live in an environment with low Reynolds numbers, where their motility is dominated by viscous forces and inertial forces are negligible. This implies that all movements have to be active and that cells do not sink passively. Finally, from an ecological perspective, cell size is the best indicator of the level an organism occupies in the trophic food web. Although there are many exceptions, phagotrophs eat organisms smaller than themselves with a general predator-to-prey ratio of 10:1 (length). So, picoeukaryotes would always be near the base of food webs, and their biomass would arrive at macroscopic trophic levels only after several trophic transfers.

PICOEUKARYOTES IN THE MARINE ENVIRONMENT

Bulk Abundance and Distribution

Very small protists are found in essentially all seawater samples inspected by epifluorescence microscopy. Since these protists were considered nanoplanktonic and many have flagella, they are routinely referred to as phototrophic nanoflagellates (PNF) if they contain chlorophyll, and heterotrophic nanoflagellates (HNF) if they are colorless. However, few cells are larger than 5 μm and many are ≤2 μm. A recent effort in protist counting and sizing by epifluorescence in contrasting marine systems reveals that 84% of phototrophic protists (between 75% and 91% in different systems), and 76% of heterotrophic protists (between 64% and 84%) are 3 μm or smaller. So, most PNF and HNF cells would qualify as picoeukaryotes in all marine habitats studied. It should be noted that, due to the absence of flagella in some picophytoeukaryotes, the terms PP and HP are probably more appropriate when referring to the epifluorescence counts of small protists.

There is an extensive database on epifluorescence counts of marine PP and HP. These operationally defined groups are ubiquitous throughout the photic zone at concentrations of thousands of cells per milliliter. PP are generally the most abundant picoeukaryotes and show a large variability, with cell counts typically increasing with the trophic state of the sample. Conversely, HP are several times less abundant than PP but vary only moderately across systems, generally less than one order of magnitude, and are often correlated with bacterial abundance. In coastal systems, typical ranges for PP are $1.1–8.5 \times 10^3$ cells ml^{-1}, with episodic peaks well above 10^4 cells ml^{-1}, whereas HP concentrations typically vary between 0.6 and 3.1×10^3 cells ml^{-1}. Cells are less abundant in offshore, more oligotrophic systems: typical PP concentration ranges from 1.0 to 3.3×10^3 cells ml^{-1} and HP ranges from 0.6 to 1.5×10^3 cells ml^{-1}. For example, in offshore Indian Ocean samples, average cell counts were 1.7×10^3 cells ml^{-1} for PP and 0.45×10^3 cells ml^{-1} for HP.

Understanding of the patterns of PP distribution has been considerably expanded by the semiautomatic counts provided by flow cytometry. Inspection of marine samples by this technique reveals an assemblage of photosynthetic picoeukaryotes, generally at thousands of cells per milliliter, yielding counts consistent with those of PP counted manually by epifluorescence. Flow cytometry is routinely used in oceanographic cruises and monitoring programs and allows direct comparisons of picophytoplankton groups. Results indicate that picoeukaryotes generally covary with *Synechococcus* and dominate in coastal waters, whereas these protists are less abundant offshore where *Prochlorococcus* dominates. The ubiquity, abundance, and constancy of picoeukaryotes suggest they must be important players in the photic zone and that their growth and mortality rates are relatively tightly coupled.

Photosynthesis does not occur below the photic zone, in the mesopelagic (200–1000 m deep) and bathypelagic (1000–3000 m) regions; so this extensive biome is largely devoid of photosynthetic protists. Nevertheless, bacterial production can occur in these deep waters based on sedimenting organic matter from the photic zone and on chemolithoautotrophic bacteria gathering their energy from reduced inorganic compounds. However, both bacterial abundance and production is still 1–2 order of magnitude lower here than at the surface, and the numbers of heterotrophic picoeukaryotes as seen by epifluorescence microscopy are also lower, ranging between 10 and 100 cells ml^{-1}.

Ecological Role of PP

It has long been known that microorganisms are responsible for most of the marine primary production. However, prior to 1983 we were unaware of the importance of picoplankton in this crucial process. Picophytoplankton generally dominates primary production in offshore, oligotrophic systems, and can also be important in coastal systems on a seasonal basis. For instance, picoplankton averaged 71% of photosynthetic biomass and 56% of primary production during four Atlantic Ocean cruises (50° N–50° S) crossing coastal, upwelling, and central oceanic regions. Picocyanobacteria, specifically *Synechococcus* in nutrient-rich and *Prochlorococcus* in oligotrophic regions, can reach abundances of up to 10^5 cells ml^{-1}, so they were initially thought more important in supporting food webs than picoeukaryotes. However, albeit less abundant, picoeukaryotes are larger (their biovolume can be 100 times that of picocyanobacteria), so they can contribute significantly to biomass and primary production. For example, a study in the central North Atlantic indicated that while only 10% of the surface picoplankton were eukaryotes, these contributed 61% of the biomass and 68% of the primary production.

The size spectrum of marine primary producers strongly influences the food web complexity and has crucial implications for transfer efficiency to higher trophic levels. Large cells such as diatoms and large dinoflagellates can be readily consumed by copepods and directly sustain fish larvae. However, primary production based on very small cells has to pass through several trophic steps (flagellates, ciliates) and a significant fraction of this organic matter is respired at these lower levels. Thus, even though picoeukaryotes appear to be optimal food for small predators, systems largely reliant on their production are less efficient in sustaining higher trophic levels. Other mechanisms, such as viral lysis, also potentially diminish trophic efficiency by causing direct remineralization of picoeukaryotic biomass. However, the role of viruses as mortality agents in picoeukaryotes (and other algae) remains largely unexplored.

The magnitude of the ocean's biological pump (sequestration of atmospheric inorganic carbon to the deep ocean by biological export) is also considered to be heavily influenced by the cell size and species composition of phytoplankton. That is, larger cells are expected to contribute more to inorganic carbon sequestration. This is because larger cells, often with mineralized structures, can sediment directly (alone or in aggregates) or be readily consumed by copepods that excrete fecal pellets that sink out of the surface mixed layer. However, recent data suggest that picoplankton may also contribute to carbon export from the surface at a level proportional to its net primary production.

Ecological Role of HP

HP are mostly phagotrophs and are central in the microbial loop, a concept that has driven a fundamental revision of models of marine food webs. The microbial loop is based on the premise that there is a substantial supply of dissolved organic matter (DOM) created by phytoplankton exudation or inefficient zooplankton feeding that supports active bacterial production. These active bacteria are grazed upon by small protists that in turn are food for larger grazers. So, incorporated into the classical food web (phytoplankton, copepods), there is a microbial loop (DOM, small grazers, copepods) that potentially transfers energy from dissolved pools to higher trophic levels. The reality is closer to a web than a loop, since small grazers also consume picophytoplankton, often the dominant producers. Regardless, the main grazers of bacteria and picophytoplankton are picoeukaryotes, as seen by microscopic inspections of protists with ingested bacteria and by size fractionation experiments showing that most bacterial grazing occurs in the fraction below 3 μm.

Grazing by HP represents an important mortality factor for bacterial assemblages over large oceanic scales. Other mortality agents like viruses may contribute in coastal systems but seem less important in oligotrophic regions. Mixotrophic protists, cells capable of both phagotrophy and photosynthesis, can contribute to half of bacterial mortality in coastal systems, but their relevance in offshore systems is not well defined. Bacterial grazing, by either heterotrophic or mixotrophic protists, is the first of a multistep food chain, during which bacterial production is mostly respired and inorganic nutrients bound to bacterial biomass (often enriched in P and N) are released. Thus, the main ecological roles of phagotrophic picoeukaryotes are controlling bacterial (and other picoplankton) abundances and remineralizing inorganic nutrients in the photic zone, allowing sustained primary production in oligotrophic systems.

HP may have other nutritional modes besides phagotrophy. Strictly osmotrophic protists are probably unable to compete with heterotrophic bacteria, which are likely more efficient in using the diluted and refractory marine DOM.

Nevertheless, the existence of phagotrophic protists that can supplement their diet with DOM is plausible and ecologically relevant. Parasitism has long been known to occur in the sea, and there are many descriptions of parasites infecting protists (e.g., diatoms, dinoflagellates) and metazoans (e.g., copepods, crabs). These parasites always have a free-living dispersal form (zoospore), colorless and often very small, which would be considered as HP. Parasitism is fundamentally different from phagotrophy because the parasite is smaller than the host, has a limited host range, and does not always kill its host. Given the high diversity of microbial assemblages and the dilute environment in which they live, parasitism was not considered as a major process in the marine plankton. However, recent environmental molecular surveys (see section 'In Situ Phylogenetic Diversity') reveal many sequences of putatively parasitic protists retrieved from the picoplankton, even at the more oligotrophic stations studied. The ecological relevance of parasitism in coastal and offshore microbial assemblages is an open field for future research.

MOLECULAR TOOLS TO STUDY PICOEUKARYOTE ECOLOGY

Elusive View of In Situ Diversity by Nonmolecular Tools

Despite the ecological importance of picoeukaryotes as primary producers, bacterial grazers, and parasites, it is remarkable how little we knew about in situ diversity before the application of molecular tools. This is because picoeukaryotes cannot be identified by the techniques that provide accurate cell counts, such as epifluorescence microscopy or flow cytometry. Specific ultrastructural features can often be revealed by electron microscopy, but this cannot be applied routinely to all cells of an assemblage, although sometimes it has served to identify dominant species, such as *Micromonas* and *Bathycoccus*.

Culturing is an excellent approach to obtain biological models (see section 'Biology of Cultured Marine Picoeukaryotes'). However, many cells do not grow easily in the lab, so culturing provides a severely biased view of microbial diversity. In a few cases, there are dominant populations in the sea that are also easily cultured, so they can be counted by diluting and culturing, yielding a most probable number. An excellent example is the prasinophyte *M. pusilla*, which is widely distributed in the sea in numbers ranging from 10^2 to 10^5 cells ml^{-1}, being more abundant in coastal areas but also found in the oligotrophic open sea.

Pigment analysis by HPLC yields information on the composition of marine phytoplankton, since each algal group has a specific pigment signature, and has often been used to compare samples during oceanographic cruises. When applied to samples prefiltered through 3 μm, it can provide a general view of picophytoplankton composition. Since the pigment profile derives from a complex assemblage, algorithms have been developed to infer the contribution of different algal classes to total chorophyll *a* (the only pigment common to all phytoplankters). However, translating pigment profiles to diversity depends on the pigments ratios generated in cultured species, which may not totally represent natural populations. Even, in the best scenarios, HPLC pigment analysis only provides identification to an algal class, not to lower ranks such as genus or species.

Cloning and Sequencing Environmental Genes

Molecular tools were introduced in marine microbial ecology relatively recently. The most widely used gene for these studies is the SSU rDNA (16S rDNA in prokaryotes, 18S rDNA in eukaryotes), which codes for the small subunit ribosomal RNA. This gene has the distinct advantages that codifies for the same function in all organisms, it has both highly conserved and variable regions, and its product (rRNA) is present in high abundance in all living cells. SSU rDNA has been used to delimit the three domains of life and to classify organisms within a given class, genus, and species. The basis of molecular protist ecology is identifying cells in situ by directly retrieving their rDNA. This is achieved by extracting DNA from marine picoplankton assemblages, amplifying the 18S rRNA genes by PCR, cloning and sequencing the PCR products, and comparing the sequences with SSU rDNA databases.

The first culture-independent PCR (ciPCR) studies on marine picoeukaryotes appeared in 2001. These showed that picoeukaryotes are extremely diverse, so the indistinguishable cells seen by epifluorescence hide phylogenetically different organisms. In fact, environmental picoeukaryote sequences are scattered throughout the eukaryotic tree of life (Figure 3). Furthermore, while some of these sequences match well-known species, others form phylogenetic clades (sets of related sequences) that represent novel and unexplored diversity. Studies from widely separated sites often show similar phylogenetic clades, suggesting that few biogeographic barriers exist in the marine realm and similar protists thrive when conditions are similar. Besides 18S rDNA, other genes have been used for similar purposes, such as the chloroplast genes 16S rDNA, *rbcL*, and *psbA*. These target only the phytoplankton, and complement and expand the results obtained with 18S rDNA.

The cloning and sequencing approach is critical to unveil novel microbial diversity but is limited in its ability to reveal the true community composition. First, samples are obtained by sequential filtration (through 3 and 0.2 μm pore size filters), and small cells can break during the process and be lost from the picoplanktonic sample. Organisms larger than 3 μm, on the other hand, can break

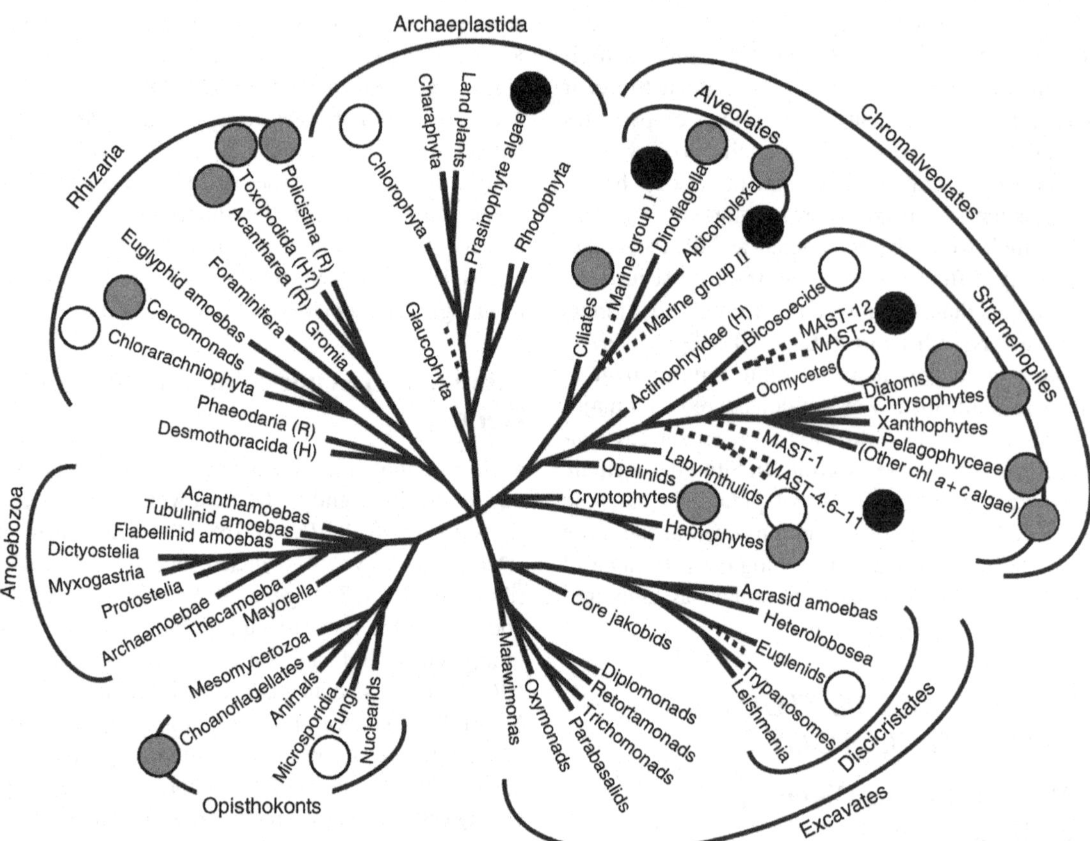

FIGURE 3 Representation of eukaryotic phylogeny displaying all lineages in a few supergroups (modified from Baldauf, S. L. (2003). The deep roots of eukaryotes. *Science 300*, 1703–1706). The supergroup CCTH (cryptomonads, centrohelids, telonemids, and haptophytes) has been proposed recently after phylogenomic analysis (Burki, F., Inagaki, Y., Bråte, J., et al. (2009). Large-scale phylogenomic analyses reveal that two enigmatic protist lineages, Telonemia and Centroheliozoa, are related to photosynthetic chromalveolates. *Genome Biology and Evolution 231*, 213–218). Groups found in picoplankton 18S rDNA libraries are marked with a dot.

into small fragments or squeeze through pores and be collected as picoplankton. This surely explains the finding of metazoan and ciliate sequences in clone libraries of picoeukaryotes. Second, some species can be more resistant to DNA extraction than others, depending on their cell structure and outer layers. Third, the required PCR step may bias the original gene abundance, both by primer mismatches and by preferential amplification in some groups. Finally, the rDNA copy number varies widely between eukaryotic species, from one to several thousands, and this obviously influences clonal representation. Thus, 18S rDNA clone libraries are fundamental and informative but do not completely describe in situ protist diversity.

Beyond Clone Libraries: FISH and Fingerprinting Techniques

After environmental sequencing has identified dominant phylogenetic groups, their abundance and distribution in the sea can be assessed by FISH, a technique that elegantly complements the rRNA approach. FISH also reveals critical features of novel phylogenetic clades such as cell size, thus confirming whether they are picoeukaryotes. During the FISH process, fixed microbial cells immobilized on a filter are hybridized with a taxon-specific 18S rDNA probe labeled with a fluorochrome. So, target cells fluoresce under epifluorescence microscopy. A crucial step in FISH is probe design, the selection of a short DNA sequence (around 20 nucleotides) specific for a given group, be it a species, genus, class, or domain. For probes targeting a novel clade, no culture is available in which to optimize hybridization conditions, so instead a natural sample where these sequences have been detected can be used. Once the FISH protocol is optimized, it is relatively easy to apply, albeit time-consuming. FISH has been successfully used for three groups of marine picoeukaryotes: prasinophytes, stramenopiles, and picobiliphytes. It has provided data on their cell size and abundance, thus giving unambiguous estimates of their contribution to the bulk picoeukaryote assemblage. Presently, the FISH approach is mostly limited by the probes available, so major efforts in probe design and optimization are required.

Fingerprinting techniques are popular in microbial ecology because they allow a fast comparison of the community composition among samples. They provide a characteristic fingerprint for each sample, typically a banding pattern where each band represents a specific taxa. The most commonly used techniques are denaturing gradient gel electrophoresis (DGGE), terminal-restriction fragment length polymorphism (T-RFLP), and automated ribosomal interspacer analysis (ARISA). All start with a PCR amplification of rDNA genes from the assemblage in question using general primers. Individual sequences in this mixture are then separated by electrophoresis based on their melting domains (DGGE), the location of their first restriction site (T-RFLP), or the size of their ITS region (ARISA). Fingerprinting techniques are used to reveal spatiotemporal patterns of picoeukaryotic assemblages in the sea. Picoeukaryotic assemblages appear to have a relatively wide distribution, with large oceanic areas with similar composition, but they also show significant changes with depth, along onshore–offshore gradients and across frontal regions. Fingerprinting techniques are also very useful to follow microbial dynamics during manipulative experiments.

IN SITU PHYLOGENETIC DIVERSITY

Overview of 18S rDNA Libraries

The phylogenetic diversity contained within marine picoeukaryotes is astonishing, as only one eukaryotic supergroup is not represented in 18S rDNA clone libraries (Figure 3). Some of the retrieved sequences are very similar to cultured picoeukaryotes (Figure 4(a)). Others clearly affiliate within a given group but are distantly related to its cultured representatives, so likely represent new species or genera (Figure 4(b)). Other sequences show a profoundly deep distance from all cultured organisms and can only be unambiguously affiliated to a supergroup, such as the marine alveolates (MALV) and the marine stramenopiles or MAST (Figure 4(c)). Finally, in a few cases the retrieved sequences form a novel high-rank taxonomic entity that cannot be assigned to any given supergroup, for example, the picobiliphytes.

A summary of the clonal representation in 18S rDNA libraries published so far from surface samples is shown in Table 2. These libraries derive from Atlantic, Pacific, Indian, and Southern oceans and Mediterranean and North seas (35 libraries and 2175 clones). The better-represented phylogenetic groups are alveolates (42.5% of clones), stramenopiles (22.8%), and prasinophytes (15.0%). Some groups are more abundant in coastal libraries (ciliates, MALV, cryptophyta, prasinophyta, and cercozoans), whereas others are more abundant in offshore libraries (MAST, pelagophytes, prymnesiophytes, and radiolarians). The different levels of phylogenetic novelty shown in Figure 4 are discussed below.

Relatively Well-Known Groups

Prasinophytes show the best correspondence between molecular and culturing approaches. They are well represented in both approaches, and environmental sequences are often identical to that of cultured cells, particularly from *M. pusilla*, *O. tauri*, and *B. prasinos*. Using FISH probing, *M. pusilla* has been found to be very abundant in coastal systems (up to 10^5 cells ml^{-1}), supporting previous views from culturing, electron microscopy, and pigment analysis. It appears that a substantial fraction of picoeukaryotes in coastal systems are prasinophytes (50–90%), and these account for a lower percentage of cells (>20%) in offshore samples. Other picophytoplankton groups with cultured species are poorly represented in clone libraries: only 1% of the clones affiliate with pelagophytes, 0.5% with bolidophytes, and the pinguiophytes have never been detected. The pelagophyte sequences are nearly identical to *P. calceolata* (Figure 4(a)). This species appears to have a wide distribution, but its true abundance remains uncertain, since the attempted FISH probing

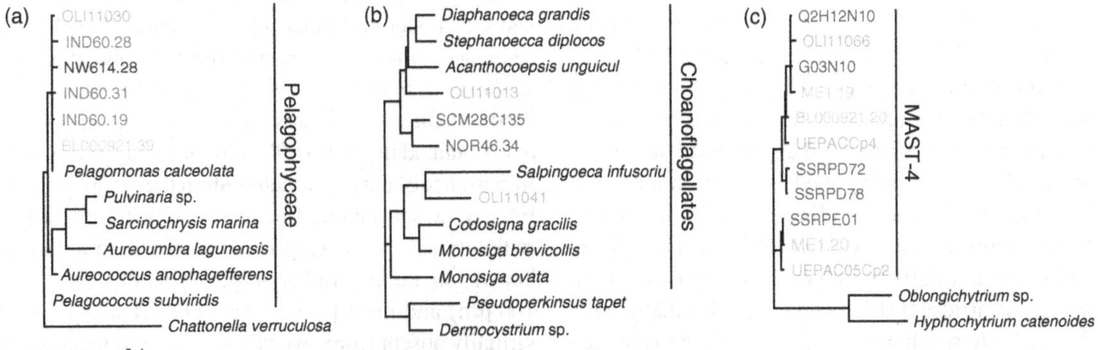

FIGURE 4 Pelagophytes (a), choanoflagellates (b), and MAST-4 (c) phylogenetic trees with 18S rDNA sequences from the marine picoplankton and closest cultured representatives. Environmental sequences are colored depending on their origin: Atlantic (red), Pacific (green), Indian (gray), Arctic (blue), and Mediterranean (yellow).

TABLE 2 Clonal representation of phylogenetic groups in 18S rDNA libraries prepared from surface picoplankton from coastal (23 libraries and 1349 clones from Mediterranean, North Sea, English Channel, and Pacific Coast) and offshore (12 libraries and 826 clones from the Mediterranean Sea, Antarctica, and Indian, Pacific, and Atlantic oceans) systems

		Total (%)	Coast (%)	Offshore (%)	Tendency
Alveolates	Ciliates	4.9	7.6	2.7	Coastal
	Dinoflagellates	5.3	6.1	4.7	
	MALV	32.4	35.8	29.3	Coastal
Stramenopiles	Chrysomonads	2.7	2.2	3.1	
	Diatoms	2.5	0.6	4.0	
	Dictyochales	1.1	1.5	0.9	
	MASTs	13.4	9.0	17.0	Open
	Pelagophytes	1.0	0.1	1.8	Open
Archaeplastida	Prasinophytes	15.0	19.3	11.6	Coastal
Rhizaria	Cercozoans	2.6	4.8	0.9	Coastal
	Radiolarians	5.6	0.3	9.8	Open
CCTH	Cryptophytes	2.4	3.6	1.4	Coastal
	Prymnesiophytes	4.5	1.2	7.2	Open
	Picobiliphytes	0.9	1.2	0.7	
Opisthokonts	Choanoflagellates	1.0	0.9	1.0	
Remaining groups		4.7	5.8	3.9	

still gives unclear results. Nevertheless, HPLC pigment analysis suggests a significant contribution of pelagophytes in open ocean systems.

Prymnesiophytes are important marine algae that seem to contribute significantly to the picoplankton according to HPLC analysis. Also, many cultured prymnesiophytes have the smallest cell dimension ≤ 3 μm. Nevertheless, their contribution to clone libraries is moderate (4.5%), and the environmental sequences often form novel clades. FISH probing shows that prymnesiophytes are rather abundant in coastal and offshore samples (more than their clonal abundance would suggest), reaching up to 30% of the picoeukaryotic cells. Other algal groups detected in clone libraries are the dinoflagellates (5.3%), important marine protists with phototrophic and heterotrophic species. Since the smallest dinoflagellates observed are around 5 μm, their molecular signal has been interpreted as filtration artifacts, although the presence of picodinoflagellates cannot be excluded. Similar concerns apply to cryptophytes (2.4% of clones) and diatoms (2.5% of clones), although the latter have very small centric and slim pennate species that would pass the 3 μm prefilter. A few sequences are close to the dictyochophyte *Florenciella parvula*. Finally, some sequences affiliate distantly to the chlorarachniophytes, but it cannot be determined if they represent phototrophic protists.

Libraries show a large diversity of putatively heterotrophic protists at low clonal abundance. Some of these groups contain cultured heterotrophic pico- and nanoflagellates: apusomonads, bicosoecids, cercozoans, chrysomonads, choanoflagellates, katablepharids, and *Telonema*. Other groups detected in marine surveys are known to contain osmotrophic or parasitic forms, such as apicomplexa, fungi, labyrinthulids, oomycetes, and pirsonids. A significant number of sequences relate to ciliates (4.9%), important components of the nano- and microzooplankton but at least 10 μm in size, so their presence is most likely a filtration artifact. The chrysomonads also show a moderate clonal representation (2.7%) and contain sequences close to well-known heterotrophic flagellates, such as *Paraphysomonas*, as well as novel clades. The trophic mode of these novel chrysomonads is uncertain, but recent data obtained in plastid rDNA libraries suggest an unexpected importance of chrysomonads within the picophytoeukaryotes. Finally, a substantial fraction of clones affiliate with the radiolarians. This is surprising, since the radiolarian species known so far are rather large (typically between 20 and 100 μm) and most possess mineralized skeletons. They are virtually absent from coastal systems, and reach a significant clonal abundance in the surface of the open sea (9.8%). Moreover, the fraction of radiolarian sequences increases with depth in the water column, and can reach 20–30% of clones at the bottom of the photic zone and below. Marine radiolarian sequences are diverse and form several clades related to acanthareans and polycystinea. The existence of these diverse radiolarian sequences from the picoplankton is a current enigma.

Marine Alveolates and Marine Stramenopiles

Environmental surveys have revealed novel clades that affiliate to a given eukaryotic supergroup but without a clear affiliation to any of its members. Among these, the MALV and MAST clades are particularly interesting because they appear in virtually all marine surveys at high clonal abundance. MALV are divided into five main groups, of which MALV-I (five clades) and MALV-II (16 clades) are the most widely represented. The first described MALV-II species was *Amoebophrya*, a dinoflagellate parasite, and additional parasite sequences have been recently published within both groups. So, it now appears that the whole MALV assemblage might consist of parasites of marine organisms. The specific interaction with different hosts could explain their high level of genetic diversity. This opens new avenues for exploration of the role of parasitism as a trophic mechanism in the open sea.

MASTs form more than ten clades at the basal part of the stramenopile tree, where all protists are heterotrophic, free-living flagellates, parasites, osmotrophs, or commensals. This suggests these MASTs are heterotrophs, which has been confirmed by FISH for several clades. On average, MAST sequences account for 13.4% of picoeukaryotic clones, and most of this signal is explained by clades MAST-1, -3, -4, and -7. These are colorless protists, with a size from 2 to 8 μm, able to grow in the dark and to ingest bacteria (Figure 5). MASTs are widely distributed and account for a significant fraction of heterotrophic flagellates. For example, the very small MAST-4 protist (2–3 μm) is found in all samples except the polar ones, averages 130 cells ml^{-1}, and accounts for 9% of heterotrophic flagellates. This shows that still-uncultured groups can be dominant in marine picoeukaryote assemblages and that model cultured organisms might have limited use for understanding how marine ecosystems work.

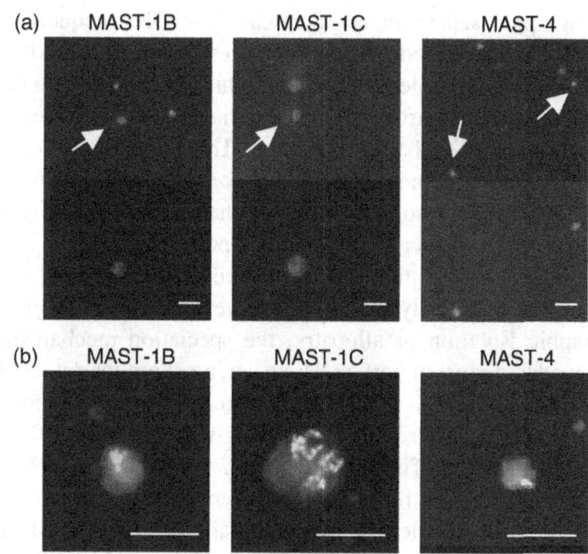

FIGURE 5 Epifluorescence images of MAST-1B, -1C, and -4 cells. (a) Same microscopic field observed by UV (upper panels: DAPI-stained blue cells) and green light (lower panels: fluorescent in situ hybridization (FISH)-stained orange cells). MAST cells in the upper panels are indicated with a white arrow. (b) Combination of three images in several MAST cells: DAPI staining (blue nucleus), FISH (orange cytoplasm), and ingested FLBs (yellow spots). Scale bar = 5 μm.

Novel High-Rank Phylogenetic Groups

Early molecular surveys claimed the discovery of novel groups at the highest taxonomic rank, in an effort to highlight the potential of this approach. However, it was soon recognized that some of these novel groups were artifacts due to the presence of undetected chimeras, misplacement of fast-evolving lineages, and incomplete representation of cultured strains. Nevertheless, it is clear that some novel high-rank groups are real and form robust and deep phylogenetic clades that cannot be assigned to any known eukaryotic supergroup. These are generally found at low clonal abundance, so probably are not very important ecologically. Instead, their interest resides in their evolutionary novelty. Perhaps the best example is the picobiliphytes, a novel phytoplanktonic class that probably has phycobilin-containing plastids and can be locally abundant. However, most of these novel potentially high-rank groups await formal characterization.

Biogeography

There is considerable debate about the extent of diversity and biogeographical distribution of microorganisms. It has been argued that given the huge population sizes and the potential for distant dispersal, most microbial species must be cosmopolitan and the total number of species must be relatively low. This is what is observed using the morphospecies concept to identify protists. However, this concept cannot account for cryptic species, the level of which may not be insignificant among protists, given that strains from the same morphospecies can belong to genetically, reproductively, and ecologically isolated groups. Thus, the use of molecular tools to compare assemblages from distant systems provides a more systematic and objective assessment of the extent of protist biogeography and diversity.

Clone libraries from widely separated picoeukaryote assemblages yield similar phylogenetic groups, suggesting that their overall community structure is comparable on a worldwide scale. Further, there appears to be no particular geographic separation as identical 18S rDNA sequences have been retrieved from distant oceans (Figure 4). Altogether, we find little sign of geographic barriers for picoeukaryotes in the marine environment, and most groups seem to be roughly globally distributed. This supports the cosmopolitan view mentioned above for the marine habitat, although finer resolution markers than 18S rDNA might be needed to detect locally adapted populations.

Global distribution has been used as an argument for low protist diversity as it should prevent speciation by geographic isolation or allopatry, the speciation mechanism considered most important. However, most groups detected in environmental surveys show a large 18S rDNA sequence variability, thus strongly disagreeing with the view of low protist diversity. The significance of this large intragroup rDNA diversity is presently not understood but can be relevant, since genetic distances as small as 0.01 (equivalent to 99% similarity) in the 18S rDNA might imply millions of years of evolutionary divergence. It is also not clear how this large phylogenetic diversity of protists translates into functional diversity in the marine environment or how this diversity is generated and maintained. One possibility is to view the seemingly homogeneous pelagic habitat as a continuum of environmental niches. Thus, there is a large genetic diversity of marine protists, which may have a global distribution, but the ecological implications of this large diversity remain to be investigated.

THE GENOMIC ERA

Genome Projects on Cultured Picoeukaryotes

Currently there are 129 complete published genome projects of eukaryotes. These include a few free-living marine protists such as the diatoms *Thalassiosira pseudonana* and *Phaeodactylum tricornutum*, the prasinophytes *Ostreococcus tauri, O. lucimarinus*, and *Micromonas pusilla*, the green algae *Chlamydomonas reinhardtii*, the choanoflagellate *Monosiga brevicollis*, and the ciliates *Paramecium tetraurelia* and *Tetrahymena thermophila*. In addition, there are 1326 ongoing eukaryotic genome or total mRNA (EST) sequencing projects, including relevant marine strains affiliating to the dinoflagellates, chrysophytes, pelagophytes, and haptophytes.

The first marine picoeukaryote genome published was *O. tauri*. This prasinophyte has a 12.56 Mb haploid nuclear genome organized in 20 chromosomes. The genome is highly compacted, mainly due to the reduction in size of intergenic regions and the low copy numbers of most genes. The 8166 identified genes include all basic cell functions, such as photosynthesis, central metabolism, and cell–environment interaction. Genes encoding enzymes for C_4-photosynthesis have been identified, which may help the cells adapt to the limiting CO_2 concentrations of phytoplankton blooms. There are genes for transport and assimilation of nitrate, ammonium, and urea, and the larger number of ammonium transporters as compared to nitrate transporters indicates that *O. tauri* could be a strong competitor for ammonium, which is uncommon in eukaryotic algae. *O. tauri* also has two chromosomes that differ structurally, being biased toward a lower GC content and a larger number of transposable elements. The first seems to have an alien origin, while the second could represent a sexual chromosome, preventing recombination with similar but not identical strains. Genes related to meiosis have been identified and are apparently functional, suggesting that this protist that usually reproduces asexually may also possess a sexual phase never observed. Thus, the genome of *O. tauri* follows prediction of compaction that might be driven by its specific lifestyle and ecology.

Environmental Genomics or Metagenomics

A striking advance in microbial ecology has been the gathering of gene content of natural communities, an approach that is not culture dependent nor hypothesis driven. There are two main strategies depending on the size of the DNA being cloned. Large DNA fragments (40–200 kb) are cloned in bacterial artificial chromosomes (BAC) or fosmid vectors, and clones with these large inserts are screened before being completely sequenced. This may provide complete operon information, and in the case of prokaryotes may also link phylogenetic and functional markers. Alternatively, small DNA fragments (3–5 kb) may be cloned by routine methods and randomly sequenced at high throughput, which is known as shotgun sequencing. This provides a huge amount of genetic data, but the short sequences obtained are difficult to assemble due to the large diversity of natural assemblages.

Metagenomic approaches have been particularly useful for marine bacteria and archaea, to identify previously unknown metabolic pathways, such as novel uses of light and reduced compounds to generate energy, and putatively novel enzymes, antibiotics, and signaling molecules. This approach is still not routinely applied to picoeukaryotes, which are prefiltered from the samples. However, these will be included in more recent projects such as the Global Ocean Sampling Expedition, which has provided so far 6.3 billion base pairs of sequencing information from 41 distant marine sites.

RECENT DEVELOPMENTS

The study of marine picoeukaryotes is a field in expansion that is attracting new scientists, mostly moved by the possibilities given by molecular tools to investigate novel ecological, evolutionary, and phylogenetic aspects of these minute cells. Recent results have confirmed previous observations regarding their large genetic and functional diversity and the presence of novel groups. In addition, the most innovative contributions have derived from recent technical advances like NGS (next generation sequencing) technologies that offer an unprecedented sequencing capacity, and flow cytometry routines that allow cell sorting before molecular or physiological analysis.

A few genomes of cultured marine free-living protists have been published during 2008 and 2009, including three additional prasinophytes (*Ostreococcus lucimarinus* and two *Micromonas* strains), one diatom (*Phaeodactylum tricornutum*), and one choanoflagellate (*Monosiga brevicollis*). Each genome has been analyzed under a different evolutionary story, covering gene organization, speciation patterns, or the origin of multicellularity. These studies are expected to reveal the gene basis of the ecological success of these minute cells. At present, genomic projects are limited by the availability of relevant ecological models into culture, since it is well known that a large fraction of the in situ diversity is not represented in culture. This occurs at all phylogenetic scales, but it is more critical for the high-rank taxa such as the uncultured MAST or the picobiliphytes. The culturing gap is difficult to bridge, since it requires dedicated laboratory work and the design of innovative culturing strategies. The recent culture of *Triparma* sp. (cf. *Triparma laevis*) exemplifies how original attempts can promote the growth of novel diversity. The parmales is an algal group formed by silica-covered cells that were described by electron microscopy. *Triparma* sp. has been brought into culture by using fluorescent precursors that target the deposition of silica during cell growth. Based on its 18S rDNA sequence, it affiliates with the bolidophytes, the sister group of diatoms, and not to the chrysophytes as was proposed. Additional ecologically relevant cultures will surely solve other scientific conundrums.

Flow cytometry has become a standard technique in oceanography for counting viruses, bacteria, cyanobacteria, and phototrophic picoeukaryotes. A recent study explains how to count heterotrophic flagellates (HF) as well. The protocol is based on DNA staining of microbial cells and separating HF from heterotrophic bacteria by their larger cell size and higher DNA content. This seems to work better in oligotrophic conditions, where there are very few large bacteria that can interfere HF counts. The full potential of flow cytometry is attained when the counting routine is accompanied with cell-sorting capabilities. Cell populations with a particular flow cytometric signature have been sorted to identify their contribution to a particular process or to study their diversity by classical molecular tools. Thus, sorted pigmented protists are contributing to half of oceanic bacterivory, highlighting the importance of mixotrophy in aquatic habitats. Molecular analyses of sorted pigmented picoeukaryotes have expanded their diversity, with novel lineages of prasinophytes, chrysophytes, and haptophytes being identified. Finally, the capacity to isolate individual cells by flow cytometry is opening new possibilities for single cell analyses of microorganisms. These single cells can be used as inoculum to start pure cultures, or as template for whole genomic amplification, which can be then used in genomic projects. This single cell approach has been successfully applied to marine bacteria but not yet to picoeukaryotes.

Thirty-eight papers have been published so far studying the diversity of marine protists by clone libraries of environmental rDNA genes, 11 of these during 2008 and 2009. These papers confirm and expand previous results and provide a better coverage of the distribution of specific lineages. In addition, useful complementary data has been obtained with slight modifications of the general approach, such as the analysis of flow cytometry-sorted populations (as mentioned above), the use of group-specific primers (typically expanding the diversity of the group), or the targeting of other genes than rDNA. Comparing the diversity obtained with a standard rDNA library and a library constructed after environmental ribosomes has revealed that some groups with high clonal abundance in the rDNA analysis (MALV, radiolarians) are little represented in the rRNA approach. So, cells from these two groups could be less active, could have a disproportionately high rDNA copy number or could be more represented in detrital DNA. The in situ diversity of marine protists has also been approached using the high sequencing capability provided by NGS technologies like 454 pyrosequencing of short rDNA amplicons. Several studies have already been published and more are in the pipeline. It is expected that NGS will allow to asses the true extent of phylogenetic diversity and provide exhaustive datasets for comparative purposes. These analyses indicate a large contribution of rare taxa, although at present it is difficult to fully discriminate new diversity from sequencing errors. A more ambitious use of the NGS technology is to obtain a full inventory of all genes in natural assemblages. Thus, the GOS metagenomic dataset has been explored for the presence of several anchor eukaryotic genes, and the diversity displayed after the detected 18S rDNA genes is strikingly similar to that observed in standard clone libraries. This culture- and PCR-independent approach can potentially link diversity markers with putative functions, specially if large pieces of DNA are cloned before sequencing. Metagenomic studies on marine microorganisms, including picoeukaryotes, are expected to burst in the near future.

The natural complement of sequencing surveys is the FISH technique that labels cells belonging to a given rDNA clade, which can be formed by environmental sequences only. When combined with functional experiments and natural observations, FISH allows linking novel lineages to specific ecological roles. As culturing, this technique is time-consuming and can only target a single taxa at a time, and at present there is plenty of room for more probe design to address critical questions. Combining FISH with bacterivory experiments has revealed important functional differences between several uncultured MAST lineages, which seem to have different grazing rates and consume different bacteria. Also, it has been recently demonstrated that several MALV lineages are parasites of specific dinoflagellates that are controlling the host abundances and dynamics in a coastal marine system.

In summary, the study of marine picoeukaryotes is a very active discipline that is benefiting by methodological improvements that allow opening new windows to unexplored fields. Challenges for the future is to get in culture the most abundant species, FISH probing to study abundances and biogeochemical implications of the novel groups, and to obtain a full inventory of in situ phylogenetic diversity to study biogeographical patterns and understand the evolutionary an ecological factors driving the large diversity observed.

CONCLUDING REMARKS

Picoeukaryotes are very small organisms (≤ 3 μm in the maximal dimension) that are morphologically and structurally simple but with all components required for an independent life. It is fascinating to find all organelles necessary for cell metabolism, growth, and reproduction assembled in such a small package. The few model cultures available so far provide useful tools for ecophysiological and genomic studies. In the marine environment, picoeukaryotes are ubiquitous, account for a significant share of planktonic biomass, and play key ecological roles as primary producers, bacterial predators, and parasites of marine life. The true diversity of marine picoeukaryotes has been recently unveiled by the use of molecular tools, which have revealed a very high diversity within this assemblage and the presence of many novel eukaryotes uncharacterized and uncultured. This increase in diversity occurs at almost all possible phylogenetic scales: high-rank novel groups, novel clades within supergroups, and putatively new orders, families, genera, and species within known lineages. The challenge is to retrieve in culture these novel organisms and to determine their ecological role. The implication of this large and novel eukaryotic diversity for biodiversity surveys and ecosystem functioning opens new avenues for future research.

FURTHER READING

Amaral-Zettler, L. A., McCliment, E. A., Ducklow, H. W., & Huse, S. M. (2009). A method for studying protistan diversity using massively parallel sequencing of V9 hypervariable regions of small-subunit ribosomal RNA genes. *PLoS ONE*, *4*, e6372.

Azam, F., Fenchel, T., Field, J. G., Gray, J. S., Meyer-Reil, L. A., & Thingstad, F. (1983). The ecological role of water-column microbes in the sea. *Marine Ecology Progress Series*, *10*, 257–263.

Baldauf, S. L. (2003). The deep roots of eukaryotes. *Science*, *300*, 1703–1706.

Bowler, C., Allen, A. E., & Badger, J. H. (2008). The *Phaeodactylum* genome reveals the evolutionary history of diatom genomes. *Nature*, *456*, 239–244.

Burki, F., Inagaki, Y., Bråte, J., *et al.* (2009). Large-scale phylogenomic analyses reveal that two enigmatic protist lineages, Telonemia and Centroheliozoa, are related to photosynthetic chromalveolates. *Genome Biology and Evolution*, *231*, 213–218.

Chambouvet, A., Morin, P., Marie, D., & Guillou, L. (2008). Control of toxic marine dinoflagellate blooms by serial parasitic killers. *Science*, *322*, 1254–1257.

Derelle, E., Ferraz, C., Rombauts, S., *et al.* (2006). Genome analysis of the smallest free-living eukaryote *Ostreococcus tauri* unveils many unique features. *Proceedings of the National Academy of Sciences of the United States of America*, *103*, 11647–11652.

Hughes Martiny, J. B., Bohannan, B. J. M., Brown, J. H., *et al.* (2006). Microbial biogeography: Putting microorganisms on the map. *Nature Reviews Microbiology*, *4*, 102–112.

Ichinomiya, M., Yoshikawa, S., Kamiya, M., Ohki, K., Takaichi, S., & Kuwata, A. (2010). Isolation and characterization of Parmales (Heterokonta/Heterokontophyta/Sramenopiles) from the Oyashio region, Western North Pacific. *Journal of Phycology*, *47*(1), 144–151. doi:10.1111/j.1529-8817.2010.00926.x.

Johnson, P. W., & Sieburth, J. Mc.N. (1982). *In situ* morphology and occurrence of eucaryotic phototrophs of bacterial size in the picoplankton of estuarine and oceanic waters. *Journal of Phycology*, *18*, 318–327.

Jürgens, K., & Massana, R. (2008). Protistan grazing on marine bacterioplankton. In D. L. Kirchman (Ed.), *Microbial ecology of the oceans* (pp. 383–441). (2nd ed.). Wiley: Hoboken, NJ.

King, N., Westbrook, M. J., Young, S. L., *et al.* (2008). The genome of the choanoflagellate *Monosiga brevicollis* and the origin of metazoans. *Nature*, *451*, 783–788.

Li, W. K. W. (1994). Primary production of prochlorophytes, cyanobacteria, and eucaryotic ultraphytoplankton: Measurements from flow cytometric sorting. *Limnology and Oceanography*, *39*, 169–175.

Massana, R., Unrein, F., Rodríguez-Martínez, R., Forn, I., Lefort, T., Pinhassi, J., *et al.* (2009). Grazing rates and functional diversity of uncultured heterotrophic flagellates. *The ISME Journal*, *3*, 588–596.

Moon-van der Staay, S. Y., De Wachter, R., & Vaulot, D. (2001). Oceanic 18S rDNA sequences from picoplankton reveal unsuspected eukaryotic diversity. *Nature*, *409*, 607–610.

Not, F., del Campo, J., Balagué, V., de Vargas, C., & Massana, R. (2009). New insights into the diversity of marine picoeukaryotes. *PLoS ONE*, *4*, e7143.

Not, F., Valentin, K., Romari, K., *et al.* (2007). Picobiliphytes: A marine picoplanktonic algal group with unknown affinities to other eukaryotes. *Science*, *315*, 252–254.

Piganeau, G., Desdevises, Y., Derelle, E., & Moreau, H. (2008). Picoeukaryotic sequences in the Sargasso Sea metagenome. *Genome Biology*, *9*, R5.

Richardson, T. L., & Jackson, G. A. (2007). Small phytoplankton and carbon export from the surface ocean. *Science*, *315*, 838–840.

Rodríguez, F., Derelle, E., Guillou, L., Le Gall, F., Vaulot, D., & Moreau, H. (2005). Ecotype diversity in the marine picoeukaryote *Ostreococcus* (Chlorophyta, Prasinophyceae). *Environmental Microbiology*, *7*, 853–859.

Shi, X. L., Marie, D., Jardillier, L., Scanlan, D. J., & Vaulot, D. (2009). Groups without cultured representatives dominate eukaryotic picophytoplankton in the oligotrophic South East Pacific ocean. *PLoS ONE*, *4*, e7657.

Vaulot, D., Eikrem, W., Viprey, M., & Moreau, H. (2008). The diversity of small eukaryotic phytoplankton ($\leq$3 μm) in marine ecosystems. *FEMS Microbiology Reviews*, *32*, 795–820.

Viprey, M., Guillou, L., Ferréol, M., & Vaulot, D. (2008). Wide genetic diversity of picoplanktonic green algae (Chloroplastida) in the Mediterranean Sea uncovered by a phylum-biased PCR approach. *Environmental Microbiology*, *10*, 1804–1822.

Worden, A. Z., Lee, J.-H., Mock, T., *et al.* (2009). Green evolution and dynamic adaptations revealed by genomes of the marine picoeukaryotes *Micromonas*. *Science*, *324*, 268–272.

Worden, A. Z., & Not, F. (2008). Ecology and diversity of picoeukaryotes. In D. L. Kirchman (Ed.), *Microbial ecology of the oceans* (pp. 159–205). (2nd ed.). John Wiley & Sons, Inc.: Hoboken, NJ.

Woyke, T., Xie, G., Copeland, A., *et al.* (2009). Assembling the marine metagenome, one cell at a time. *PLoS ONE*, *4*, e5299.

Zubkov, M. V., & Tarran, G. A. (2008). High bacterivory by the smallest phytoplankton in the North Atlantic Ocean. *Nature*, *455*, 224–227.

RELEVANT WEBSITES

http://www.icm.csic.es/bio/projects/icmicrobis/bbmo/ – Blanes Bay Microbial Observatory (BBMO).

http://keydnatools.com/ – KeyDNATools.

http://www.arb-silva.de/ – SILVA.

http://www.genomesonline.org/ – Genome Projects Home (GOLD).

http://www.tolweb.org/tree/ – Tree of Life Web Project.

http://www.sb-roscoff.fr/. – Station Biologique de Roscoff.

http://ccmp.bigelow.org. – The Provasoli–Guillard National Center for Culture of Marine Phytoplankton, CCMP.

Chapter 27

Stramenopiles

H.S. Yoon[1], R.A. Andersen[1], S.M. Boo[2] and D. Bhattacharya[3]

[1]*Bigelow Laboratory for Ocean Sciences, West Boothbay Harbor, ME, USA*

[2]*Chungnam National University, Daejeon, Republic of Korea*

[3]*University of Iowa, Iowa City, IA, USA*

Chapter Outline

Abbreviations 373
Defining Statement 373
Evolutionary History of the Stramenopiles 373
Origin of the Stramenopiles 374
Fossil Record and Divergence Times for Stramenopiles 374
Diversity of the Stramenopiles 376
The Stramenopile Plastid 377
Cell Covering 379
Flagella 379
Phylogeny and Classification of the Stramenopiles 380
Colorless Stramenopiles 380
Photosynthetic Stramenopiles 381
Phaeophyceae 383
Conclusion 383
Further Reading 383
Relevant Website 384

ABBREVIATIONS

CER Chloroplast endoplasmic reticulum
ER Endoplasmic reticulum
PER Periplastidal endoplasmic reticulum

DEFINING STATEMENT

The stramenopiles are a monophyletic group of eukaryotes that possess tripartite hairs along one flagellum. Twenty-one classes are recognized including five nonphotosynthetic groups. Here, we review knowledge of the diversity and evolutionary history of the stramenopiles, which originated in the late Mesoproterozoic and arose to great ecological and economic importance.

EVOLUTIONARY HISTORY OF THE STRAMENOPILES

The stramenopiles (Latin, *stramen* – straw + *pila* – hairs) are a distinct, highly diverse, and yet, clearly monophyletic group of eukaryotes, all of whose swimming cells possess two different types of flagella (Figure 1). A long anteriorly directed flagellum has two rows of tripartite hairs (i.e., a basal section, a long tubular shaft, and terminal fibrils). The posteriorly directed flagellum is generally short and lacks these hairs. Because the two flagella act differently (i.e., beating pattern), the term heterokont was originally used for this lineage. Subsequently, the term Heterokontae was used over 100 years ago to name some algae now placed in the Xanthophyceae and Raphidophyceae. Approximately 40 years ago, the term Heterokontophyta was introduced for what is today's algal stramenopiles (stramenochromes). The terms chromophyte, Chromista, and ochrophyte are also sometimes used for these taxa, although the former two terms denote additional taxa not confirmed to be directly related to stramenopiles (haptophytes and cryptophytes) and the latter term only denotes a subset of species. Although the term stramenopiles is relatively new (1989), it has gained wide acceptance and is the most commonly used moniker to denote this lineage.

The stramenopiles comprise more than 100 000 species, including very diverse life forms from single cells to large plasmodia to complex multicellular thalli. The best known members of the group are the colorless oomycetes (aquatic 'fungi'), diatoms, chrysophyte algae, and giant kelp seaweeds. Photosynthetic stramenopiles are the predominant eukaryotes in most aquatic environments, and they play an important role in ecosystems as major primary producers. The colorless groups include major plant pathogens for cultivated crops (oomycetes) and several groups of

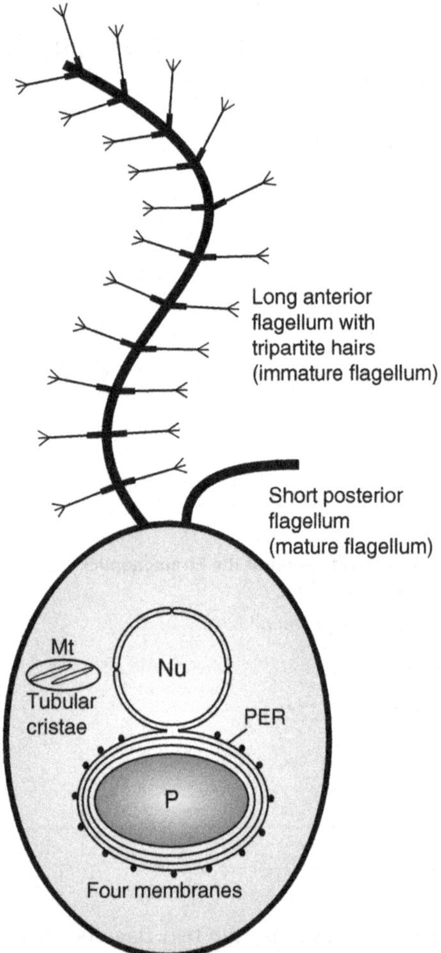

FIGURE 1 Simplified stramenopile cell. Stramenopiles possess two flagella: one, the long anterior flagellum, has two rows of tripartite hairs, each consisting of a basal section, a longer tubular shaft, and terminal fibrils; and a short and smooth posterior flagellum. The stramenopile plastid (P) is surrounded by four membranes that are located in the periplastidal endoplasmic reticulum (PER) lumen. Tubular cristae are found in the mitochondria (Mt).

flagellate or amoeboid cells that are important in the microbial food web. Molecular evolutionary studies suggest that some, if not all, of the colorless stramenopiles are derived from the photosynthetic stramenopiles by loss of their plastids.

Origin of the Stramenopiles

It is now widely accepted that the plastid of stramenopiles originated from a red alga through secondary endosymbiosis (cf. chapter 'Secondary Endosymbiosis'). That is, a nonphotosynthetic protist engulfed a red alga but rather than digesting it as prey, the alga was maintained permanently. Over many generations, this red alga became a reduced and enslaved photosynthetic organelle; that is, it converted its new host from an animal-like organism into a plant-like organism. Evidence suggests that this event occurred early during the evolution of the chromalveolates, which include not only the stramenopiles, but also the cryptophytes, haptophytes, and alveolates (ciliates, apicomplexa, and dinoflagellates). Thus the chromalveolate hypothesis unites all chlorophyll *c*-containing algae into a major taxon that also includes a large number of colorless relatives. This hypothesis has received considerable support from molecular phylogenetic studies of nuclear-encoded plastid-target genes and recently, from large multiprotein datasets. More recent phylogenetic studies suggest that chromalveolates may also include the supergroup Rhizaria, or even that Rhizaria may be embedded within what would then be referred to as the 'former chromalveolates' (Figure 2).

Fossil Record and Divergence Times for Stramenopiles

Stramenopiles have an old and extensive fossil record. The earliest known fossil that has been assigned to the stramenopiles is *Palaeovaucheria*, which was found in the upper Mesoproterozoic Lakhanda Formation (~1000 Ma) in eastern Siberia. This fossil was regarded as closely related to the modern xanthophycean genus *Vaucheria*. Another xanthophycean algal fossil, *Jacutianema*, was found in the early Neoproterozoic Svanbergfjellet Formation (~750 Ma), Spitsbergen of Arctic Norwegian island, where the earliest known filamentous green algal fossil (*Proterocladus*) was also identified. These two *Vaucheria*-like fossils, however, may also be interpreted as filamentous green algae due to their simple morphology. Microfossils have also been reported from the upper Tindir Group (~740 Ma) in northwestern Canada. On the basis of scale morphologies, the latter were first postulated to be chrysophyte algae but later were identified as scaled amoebae belonging to the Rhizaria. Other more compelling eukaryotic fossil remains were found from the late Neoproterozoic Doushantuo Formation (~600 Ma), South China – *Miaohephyton* and *Konglingiphyton*. These were characterized as being regularly dichotomous, multicellular thalli, with apical and intercalary growth and conceptacles, suggesting an affinity with brown algae. It is noteworthy that there is a very well-preserved fucoid-like brown algal fossil, *Thalassocystis*, from the Silurian of Michigan. The *Thalassocystis* fossils are branched, and each branch terminates in an inflated structure that closely resembles the bladders of modern fucoid algae. There are still uncertainties as to the affinity of some stramenopiles fossils, but it is likely that the multicellular stramenopiles diverged prior to the late Neoproterozoic era.

It is sometimes possible to estimate evolutionary dates using a method referred to as a molecular clock analysis. This requires a reliable fossil record and an accurate

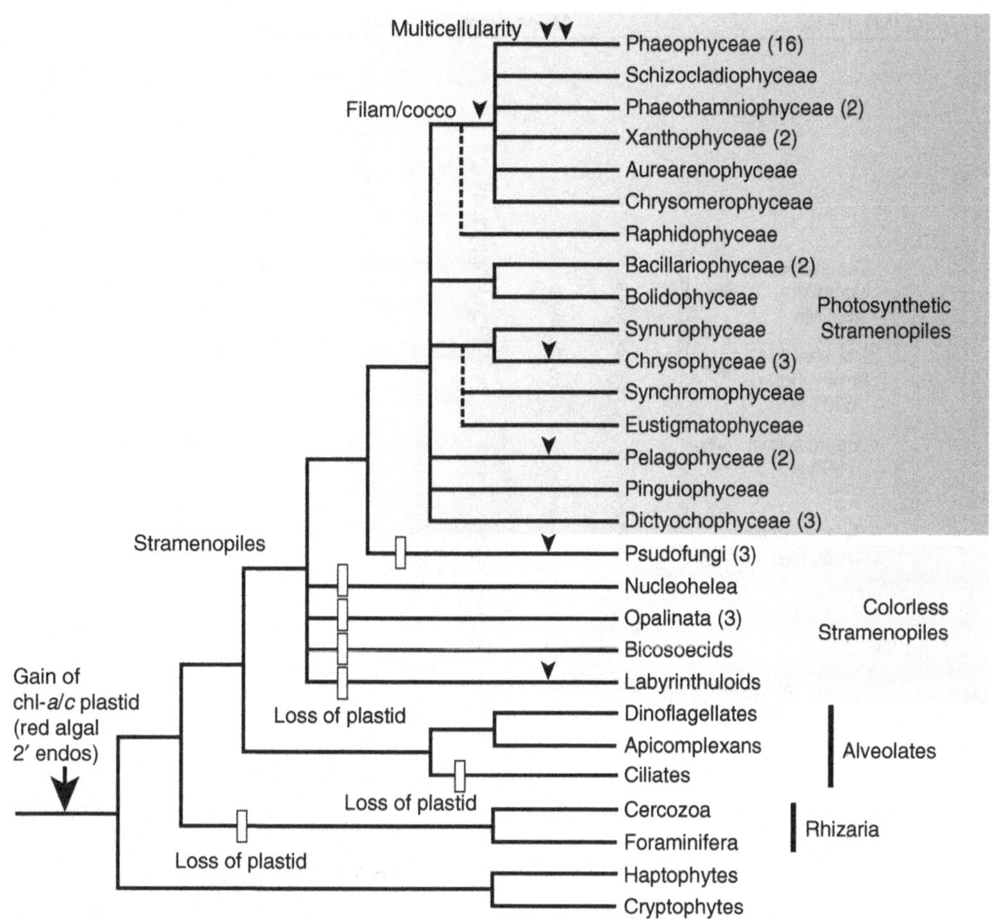

FIGURE 2 Evolutionary relationships among stramenopiles. An ancient red algal secondary endosymbiosis is thought to have occurred in a unique common ancestor of the chromalveolate + Rhizaria clade. This cell would then have given rise to the diverse chlorophyll *c*-containing protist groups including stramenopiles. Several plastid losses (marked as empty rectangles) occurred in the tree. The alveolates is the most closely related lineage to stramenopiles. The stramenopiles are composed of 21 classes or class-level of lineages including five colorless groups. The colorless pseudofungi and photosynthetic stramenopiles are sister groups of each other. The Phaeophyceae shows true multicellularity (double arrowhead), whereas some lineages have simple filamentous or coccoid forms (Filam/Cocco). Adapted from Adl *et al.* (2005); Andersen, R. A. (2004). Biology and systematics of heterokont and haptophyte algae. *American Journal of Botany, 91*, 1508–1522; Cavalier-Smith, T., & Chao, E.-Y. (2006). Phylogeny and megasystematics of phagotrophic heterokonts (Kingdom Chromista). *Journal of Molecular Evolution, 62*, 388–420; Daugbjerg, N., & Guillou, L. (2001). Phylogenetic analyses of Bolidophyceae (Heterokontophyta) using rbcL gene sequences support their sister group relationship to diatoms. *Phycologia, 40*, 153–161; Hackett, J. D., Yoon, H. S., Li, S., Reyes-Prieto, A., Rümmele, S. E., & Bhattacharya, D. (2007). Phylogenomic analysis supports the monophyly of cryptophytes and haptophytes and the association of 'Rhizaria' with chromalveolates. *Molecular Biology and Evolution, 24*, 1702–1713; Horn *et al.* (2007); Kai, A., Yoshii, Y., Nakayama, T., & Inouye, I. (2008). Aurearenophyceae classis nova, a new class of Heterokontophyta based on a new marine unicellular alga *Aurearena cruciata* gen. et sp. nov. inhabiting sandy beaches. *Protist, 159*, 435–457.

molecular phylogeny for the group in question, in which case this method can provide a powerful approach to estimate divergence times for ancient evolutionary groups. In fact, the molecular clock method may provide, in some cases, the only possible approach because of the limited fossil data that are available for many ancient groups. However, modeling DNA sequence evolution is the most error prone in such instances due to the accumulation of superimposed mutations in molecular sequences. Using various corrections for these possible sources of error, a recent molecular clock study provided a hypothetical molecular timeline for the origin of photosynthetic eukaryotes using relaxed clock methods with several fossil constraints on a multigene phylogeny (Figure 3). This study estimated the date for the red algal secondary endosymbiosis that gave rise to the stramenopile plastid at ~1261–1305 Ma. This is after the split of the Cyanidiales from other red algae (~1350–1416 Ma), that is, considerably after the origin of red algae. This study also placed the date for the primary eukaryotic plastid endosymbiosis (involving a cyanobacterium) at ~1531–1602 Ma. The stramenopiles and haptophytes appear to have split ~1025–1077 Ma after the cryptophyte's divergence (~1172–1219 Ma) with each of these lineages radiating early in the Neoproterozoic (e.g., 791 Ma for haptophytes, 712 Ma for stramenopiles, and 720 Ma for cryptophytes). These estimates are still quite preliminary as not all

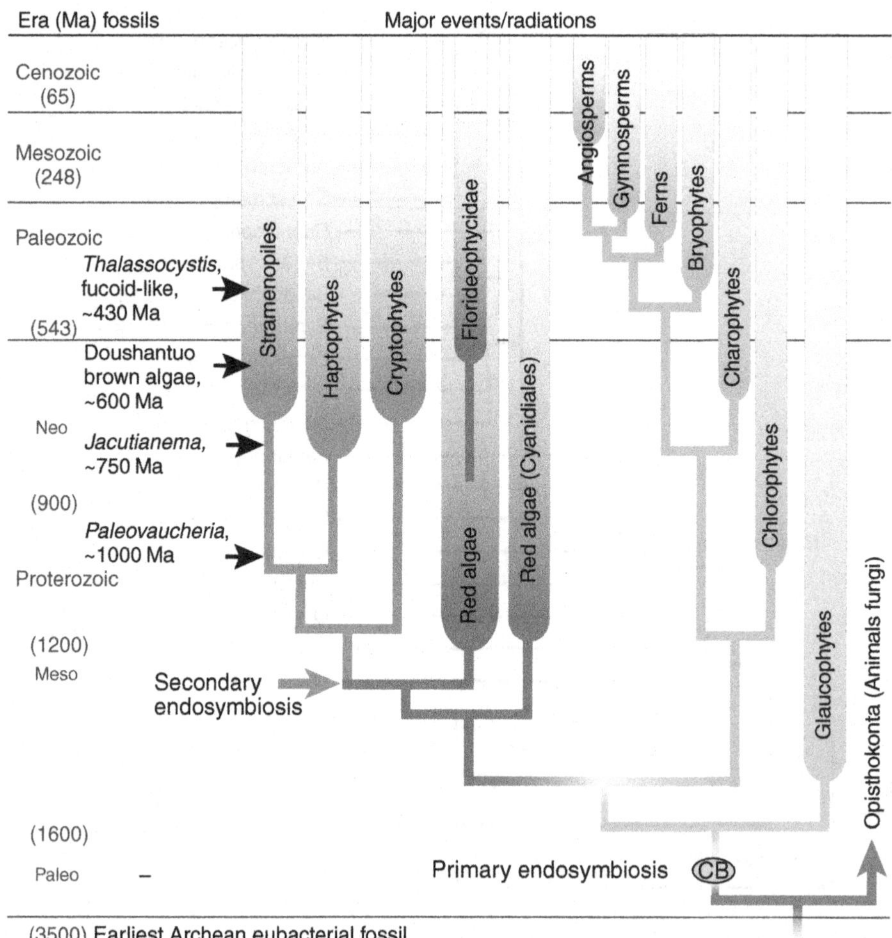

FIGURE 3 Molecular timeline for the origin of photosynthetic eukaryotes. The red, green, and glaucophyte algae diverged after primary endosymbiosis. Through secondary endosymbiosis, an ancestor of cryptophyte, haptophyte, and stramenopiles acquired its plastid from a red alga. The existence of Proterozoic fossils that are identified as stramenopiles are generally consistent with the divergence time of stramenopiles shown here. Modified from Yoon, H. S., Hackett, J. D., Ciniglia, C., Pinto, G., & Bhattacharya, D. (2004). A timeline for the origin of photosynthetic eukaryotes. *Molecular Biology and Evolution*, *21*, 809–818, with permission from Oxford University Press, Oxford, UK and adapted from Butterfield (2004), Allison and Hilgert (1986), and Xiao (1998) for fossil records.

major lineages could be included (e.g., the dinoflagellates), and the addition of earlier diverging taxa could potentially push these dates further back into the Neoproterozoic. Also some parts of the molecular tree were not well resolved. The tree used plastid genes, and the relationship between these and host divergence times is still not clear. Nonetheless, this study suggests a late Mesoproterozoic origin of photosynthetic eukaryotes and a Neoproterozoic diversification of the major eukaryotic lineages. This estimate is consistent with the fossil record of eukaryotes based on biomarker data, such as steranes (2715–2600 Ma) from the Marra Mamba and Maddina Formations in Australia; *Tawuia* and *Chuaria* (1800–1900 Ma) from the Changzhougon Formation in China; acritarchs (1400–1500 Ma) from the Roper Group of Australia; and a reliable red algal *Bangiomorpha* fossil (1198 Ma) from the Hunting Formation in arctic Canada. Thus, this molecular clock suggests the origin of stramenopiles to be around 1000 Ma followed by a major radiation at ~700 Ma.

DIVERSITY OF THE STRAMENOPILES

The stramenopiles are one of the most diverse eukaryotic groups. Twenty-one class-level divisions are currently recognized, including five that are not photosynthetic. The stramenopiles encompass over 100 000 species, although some authors suggest there may be over 1 million species for the diatoms alone. (The diatoms are the 'insects' of the algal world, i.e., there are many species, yet their morphological diversity is constrained.) Regardless of the exact number of species that have been formally described, most scientists agree that there are many more undescribed species, probably including entire new classes (cf. chapter 'Picoeukaryotes'). Transmission electron microscopy and molecular systematic studies have resulted in the description of 13 new class-level lineages of stramenopiles during the past several decades. That is, earlier classifications based on light microscopy did not recognize the details that became apparent when the internal cell structures were examined and genetic comparisons were made. We now know that the stramenopiles also include species previously classified as distinct kingdoms of algae, fungi, and protozoa 60 years ago.

Stramenopiles thrive in many environments, from freshwater to seawater, on soils, in animals, within snow banks, as parasites on plants, and even growing as disease organisms within the guts of humans. They also have diverse life forms from typical autotrophic algae or heterotrophic flagellates and amoebae to mixotrophic organisms that are capable of both autotrophy (photosynthesis) and heterotrophy (herbivory, carnivory). The size and the shape of these organisms vary from tiny, microscopic single cells (two thousandths of a millimeter in diameter) to giant kelp seaweeds that often reach up to 60 m in length (Figure 4). Unicellularity is the most widely occurring life form, but colonies of cells are common, and the true multicellular seaweeds produce a complex thallus with rootlike, stemlike, and leaflike structures.

The Stramenopile Plastid

The plastid in photosynthetic stramenopiles is structurally distinctive in that it possesses four outer membranes, the outermost of which corresponds to the rough endoplasmic reticulum (ER) (Figure 1). This chloroplast endoplasmic reticulum (CER) or periplastidal endoplasmic reticulum (PER) is continuous with the outer nuclear membrane. The existence of four membranes appears to be a result of the secondary endosymbiosis that gave rise to the stramenopile plastid. Thus, the inner three membranes are retained from the original red algal symbiont – the first, second, and third membrane corresponding to the cyanobacterial membrane, the primary host vacuolar membrane, and the primary host cytoplasmic membrane, respectively (for further information, see chapter 'Secondary Endosymbiosis').

Secondary acquisition of plastids is complicated by the fact that over 95% of the genes encoding plastid proteins are located in the host nucleus. This is a consequence of endosymbiotic gene transfer, where roughly 500–1000

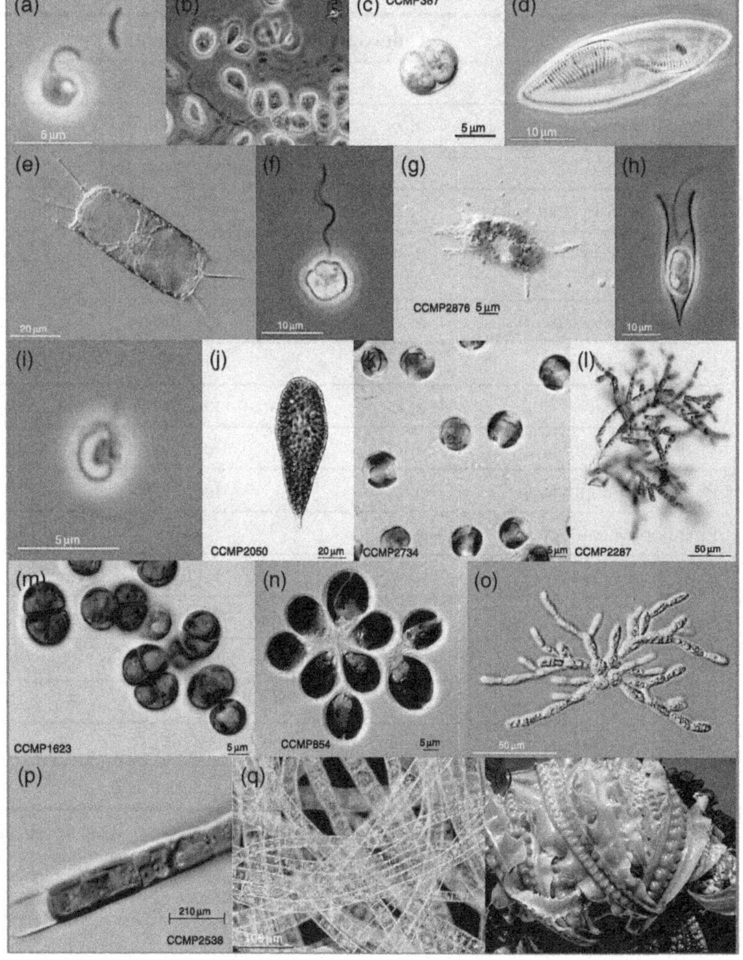

FIGURE 4 Examples of different stramenopile types. (a) *Symbiomonas* (Bicoeceae); (b) *Labyrinthula* (labyrinthulids); (c) *Eustigmatos* (Eustigmatophyceae); (d) *Entomoneis* (Bacillariophyceae – pinnate diatom); (e) *Odontella* (Bacillariophyceae – Centric diatom); (f) *Pseudopedinella* (Dictyochophyceae); (g) *Synchroma* (Synchromophyceae); (h), *Dinobryon* (Chrysophyceae); (i) *Bolidomonas* (Bolidophyceae); (j) *Chattonella* (Raphidophyceae); (k) *Chrysoreinhardia* (Pelagophyceae); (l) *Schizocladia* (Schizocladophyceae); (m) *Glossomastix* (Pinguiophyceae); (n) *Synura* (Synurophyceae); (o) *Phaeothamnion* (Phaeothamniophyceae); (p) *Tribonema* (Xanthophyceae); (q) *Pylaiella* (Phaeophyceae); (r) *Laminaria* (Phaeophyceae). All images are from the National Center for Culture Collection of Marine Phytoplankton of USA (CCMP), except (b) from Microscope and (r) from south coast of Korea taken by SMB.

formerly plastid-encoded genes have, over time, been relocated to the host nucleus. The nuclear-encoded plastid-target proteins have complex N-terminal extensions (transit peptides) to direct their posttranslational import into the plastid. Thus, secondary endosymbiosis required not only the acquisition of a plastid, but also the acquisition of hundreds of nuclear genes required to service and reproduce these plastids. This is further complicated in the new (secondary) host by the requirement that the protein products of these genes must now be passaged through four plastid membranes. This is accomplished by the ribosomes attached to the outermost membranes of the PER lumen. Thus targeting into the PER membrane is achieved with a signal peptide that has an additional hydrophobic region. This bipartite leader sequence subsequently targets proteins across the inner two plastid membranes using a downstream transit peptide.

Stramenopile plastids have lamellae, each formed by three stacked or apressed thylakoids. In almost all species (excluding Eustigmatophyceae), a girdle lamella surrounds the other lamellae. Whereas the regular lamellae are more or less platelike, the girdle lamella is spherodial, completely enclosing the other lamellae. The photosynthetic product is a β-1,3-linked glucan that is stored in a vacuole in the cytosol. Photosynthetic pigments vary in different lineages and are summarized in Table 1. The major photosynthetic pigments are chlorophylls a, c_1, c_2, c_3, and carotenoids (e.g., diatoxanthin, fucoxanthin, violaxanthin). The distribution of pigments is too variable to identify any clear evolutionary trend.

An eyespot is located within the plastid in most cases, except in the Eustigmatophyceae where it is located outside of the plastid. This eyespot is associated with the mature (generally shorter) flagellum, and together these

TABLE 1 Plastid pigmentation and cell coverings of the stramenopiles

Class	Chlorophyll	Carotenoid	Cell covering
Aurearenophyceae	a	fuc, dia, vio	Cell walls
Bacillariophyceae	a, $c_{1,2}$ (c_3)	fuc, hex, dia	Siliceous frustules
Bolidophyceae	a, c_{1-3}	fuc, hex, dia	Naked flagellates
Chrysomerophyceae	a, $c_{1,2}$	fuc, vio	Cell walls
Chrysophyceae	a, $c_{1,2}$	fuc, vio	Cell walls, organic loricas, organic or silica scale cases, gelatinous coverings, naked cells
Dictyochophyceae	a, $c_{1,2}$	fuc, dia	Silica skeletons, organic scales, or naked cells
Eustigmatophyceae	a	vio, vau	Cell walls
Pelagophyceae	a, $c_{1,2}$	fuc, (hex), but, dia	Cell walls, thecae, gelatinous coverings, naked cells
Phaeophyceae	a, $c_{1,2}$	fuc, vio	Cellulosic cell walls
Phaeothamniophyceae	a, $c_{1,2}$	fuc, dia, het	Cell walls
Pinguiophyceae	a, $c_{1,2}$	fuc, vio	Mineralized loricas, gelatinous coverings, or naked cells
Raphidophyceae-FW	a, $c_{1,2}$	fuc±, dia, het, vau	Naked cells
Raphidophyceae-Mar	a, $c_{1,2}$	fuc±, vio	Naked cells
Schizocladophyceae	a, c (type ?)	fuc, (?)	Cell walls without cellulose
Synchromophyceae	a, c_2	fuc, vio	Lorica
Synurophyceae	a, c_1	fuc, vio	Silica scales
Xanthophyceae	a, $c_{1,2}$	vio, het, vau	Cell walls

fuc, fucoxanthin; hex, 19′-hexanoyloxyfucoxanthin; but, 19′-butanoyloxyfucoxanthin; dia, pigments of the diatoxanthin and diadinoxanthin cycle; vio, pigments of the violathanin, antheraxanthin, zeaxanthin cycle; het, heteroxanthin; vau, vaucherioxanthin; ±, present or absent; (?), unknown.
Modified from Andersen, R. A. (2004). Biology and systematics of heterokont and haptophyte algae. *American Journal of Botany, 91*, 1508–1522; Horn *et al.* (2007); Kai, A., Yoshii, Y., Nakayama, T., & Inouye, I. (2008). Aurearenophyceae classis nova, a new class of Heterokontophyta based on a new marine unicellular alga *Aurearena cruciata* gen. et sp. nov. inhabiting sandy beaches. *Protist, 159*, 435–457.

are referred to as the photoreceptor apparatus. These structures detect the direction of the light source, which is used in turn to change the direction of cell movement (i.e., phototaxis).

There are five major colorless groups of stramenopiles, and, assuming the ancestral stramenopile was photosynthetic, this means that plastid loss has occurred, possibly several times independently. Since transfers of plastid genes to the nucleus are common in plastid-containing cells, some of these transferred plastid genes may remain in the genomes of taxa that have secondarily lost the plastid. There are two possible examples of this in colorless stramenopiles. Glutamine synthetase, an essential gene in glutamine biosynthesis and ammonium assimilation, has three gene family members (GSI, GSII, and GSIII). A recent study indicated that the GSII gene from two oomycetes (colorless stramenopiles) shows a strongly supported monophyletic relationship with the GSII genes of photosynthetic diatoms and green and red algae. This suggests that the GSII gene of oomycetes evolved through endosymbiotic gene transfer from a red algal plastid, presumably early in stramenopile evolution. Similarly, a 6-phosphogluconate dehydrogenase gene (*gnd*), in members of the oomycete genus *Phytophthora*, shows a cyanobacterial origin and groups together with homologues found in photosynthetic stramenopiles and red and green algae. Thus, while GSII and *gnd* are not directly involved in photosynthesis, their presence strongly suggests that 'footprints' of a red algal endosymbiosis are still detectable in the colorless oomycetes. This hypothesis was recently substantiated with the finding of at least 30 genes of putative cyanobacterial and algal (i.e., endosymbiotic) origin in the completely sequenced nuclear genomes of two oomycete species (*Phytophthora* spp.).

In addition to the exclusively heterotrophic stramenopiles (e.g., oomycetes, thraustochytrids), there are some colorless stramenopiles reported among the photosynthetic lineages. These are most certainly due to the secondary loss or reduction of the ancestral plastid. For example, *Pteridomonas danica* and *Ciliophrys infusionum* are colorless species of the Dictyochophyceae, a class that consists largely of photosynthetic members. These two species, however, retain an unpigmented plastid (leucoplast) that still encodes the *rbc*L gene. This indicates that the loss of photosynthetic ability occurred relatively recently, and apparently also independently in these two species. A remnant plastid is also found in some chrysophycean algae (e.g., *Spumella*, *Paraphysomonas*, and *Anthophysa*) as well as in certain diatom species.

Cell Covering

The cell coverings found among stramenopiles are very diverse (Table 1). The cell wall is the most common covering, but there are several types. Probably the most striking are the diatoms, which have cell walls (frustules) made of opaline glass, identical to that found in windowpanes. Organic cell walls are found in the brown algae, oomycetes, yellow-green algae, and certain species in many other classes. Oomycetes generally have a cellulosic wall, but some species have chitinous walls and others have walls based on β-1,3-linked glucans. Meanwhile, brown algae have a cellulosic wall that is impregnated with alginate and silicate, but Schizocladiophyceae have a cellulosic wall without alginate. For other classes, the composition of walls is still not known (e.g., Pelagophyceae and Phaeothamniophyceae).

Scales are the second most common cell covering. Silica scales are found on all Synurophyceae and some Chrysophyceae. Organic scales cover the cells of some Chrysophyceae, some Dictyochophyceae, some labyrinthulids, and organic scales cover the flagella in the Synurophyceae. Loricas, wall-like coverings with at least one opening, are common among the Bicosoeciophyceae and Chrysophyceae. Loricas may be composed of an intertwining network of cellulose-like fibrils, or of numerous organic scales or sometimes the lorica is impregnated with mineral deposits (e.g., iron, manganese). Naked cells are also widespread among some of the stramenopile groups (e.g., Bolidophyceae, Chrysophyceae, and labyrinthulids), whereas other naked cells are surrounded by only a gelatinous matrix. The silicoflagellates have a silica skeleton, which is neither a wall nor a scale.

Flagella

The stramenopiles usually have two flagella (Figure 1), a long anteriorly directed flagellum (immature flagellum) that bears two rows of tripartite hairs (mastigonemes), and a (usually) short posteriorly directed flagellum (mature flagellum). The flagellar hairs on the flagellum reverse the swimming direction, that is, the same flagellar action causes the cell to be pushed backward when the hairs are lacking, but the cell is pulled forward when the hairs are present. A small number of stramenopiles have only one flagellum, specifically diatom sperm cells, *Pelagomonas* species (Pelagophyceae), some species of *Mallomonas* (Synurophyceae), and the zoospores of Hyphochtridio-mycetes. Nonetheless, this single flagellum in all cases still has the tripartite hairs of the distinctive stramenopile type. The flagellum axoneme typically has the 9 + 2 microtubular arrangement, but the diatom sperm flagellum is an exception because it lacks the central pair of microtubules (9 + 0 arrangement). Stramenopile flagella are inserted subapically or laterally and are usually supported by four microtubule roots that are arranged in a distinctive pattern. However, a number of groups have a reduced flagellar apparatus with few or no microtubular roots (e.g., diatom sperm, Pelagomonadales, and Dictyotales). Other features

TABLE 2 Features of the flagellar apparatus of the stramenopiles

Class	Tri-H	# flagella	TH	Flagella beat	Lateral hairs	Mt roots	Green flagellum	Paraflagella rod	Striated root
Bacillariophyceae	+	1,0	0	Pulls	−	0	−	?	−
Bolidophyceae	+	2	0	Pulls	−	0	?	−	−
Chrysomerophyceae	+	2	6↑	Pulls	−	4	?	−	?
Chrysophyceae	+	2	4–6↑	Pulls	+	2–4	+	−	+
Dictyochophyceae	+	1	0–2↓	Pulls	−	0	−	+	−
Eustigmatophyceae	+	2,0	6↑	Pulls	−	4	−	−	+
Pelagophyceae	±	2,1,0	0–2↓	Pulls	−	0	−	+	−
Phaeophyceae	+	2	0	Pulls	−	4	+	−	−
Phaeothamniophyceae	+	2	6↑	Pulls	−	4	+	−	+
Pinguiophyceae	±	2,1,0	0,2↓	Pulls, ?	−	3,4	±	−	+
Raphidophyceae	+	2	0	Pulls	−	?	−	−	?
Schizocladophyceae	+	2	6↑	Pulls	−	?	+	−	?
Synurophyceae	+	2,1	6–9↑	Pulls	+	2	+	−	+
Xanthophyceae	+	2,0	2 × 6↑	Pulls	−	4	+	−	+

Tri-H, tripartite tubular hairs; TH, transitional helix; mt roots, microtubular roots. Arrows (↑; ↓) indicate the position (distal, proximal) of the transitional helix with respect to the major transitional plate. +, present; −, absent; ±, present or absent; ?, unknown.
Adapted from Andersen, R. A. (2004). Biology and systematics of heterokont and haptophyte algae. *American Journal of Botany, 91*, 1508–1522.

of the flagellar apparatus include a green autofluorescent substance in many organisms, the presence or absence of a transitional helix, and the occasional presence of a paraflagellar rod (Table 2).

PHYLOGENY AND CLASSIFICATION OF THE STRAMENOPILES

Stramenopiles constitute at least 21 major classes. Five colorless lineages, which are presumed to have lost their plastid secondarily, are positioned at the base of the group (Figure 2). Phylogenetic relationships among these classes are still mostly unresolved (Figure 2), probably due to the insufficient phylogenetic information from current molecular phylogenies, which are based on single gene datasets (i.e., SSU or *rbc*L). Multigene studies are likely to solve this problem but these are currently extremely limited in taxon sampling within stramenopiles, often only including a single class. Beginning in 2007, the National Science Foundation, USA, funded a tree of life (ATOL) project on the 'Algal Heterokont Tree of Life' (http://ccmp.bigelow.org/), which will generate sequences for 7 genes from 300 stramenopiles species and plastid genomes from representatives of 30 genera. As with similar multigene studies of major eukaryotic taxa, it is expected that phylogenetic analyses of these data will resolve most if not all class-level relationships among stramenopiles at the end of this 5-year project.

Colorless Stramenopiles

Five colorless stramenopile class-level lineages are recognized. These appear to diverge at the base of the clade, although relationships among these lineages are not yet resolved. Nonetheless, a few tentative conclusions can be drawn at this time. The Opalinata, which includes the Opalinea, Proteromonadea, and Blastocystea, shows a monophyletic relationship with the Nucleohelea (Figure 5). The genus *Opalina* is unusual because it occurs only in the cloaca of frogs and other amphibians. The labyrinthulids (Figure 4 (b)) form a monophyletic group, whereas relationships of the Bicoecea are unresolved. *Labyrinthula* spp. typically grow on coastal seaweeds, where they form networks of slime along which the cells glide. The genus *Bicosoeca* (Figure 4(a)) is made up of naked or loricate flagellates. In these species, the mature flagellum is used to anchor the cell to a substrate, and the immature (hairy) flagellum draws bacteria to the anterior of the cell, where these prey are ingested by phagocytosis. In addition to the 21 recognized classes, a number of class-level stramenopile assemblages have recently been identified from environmental surveys.

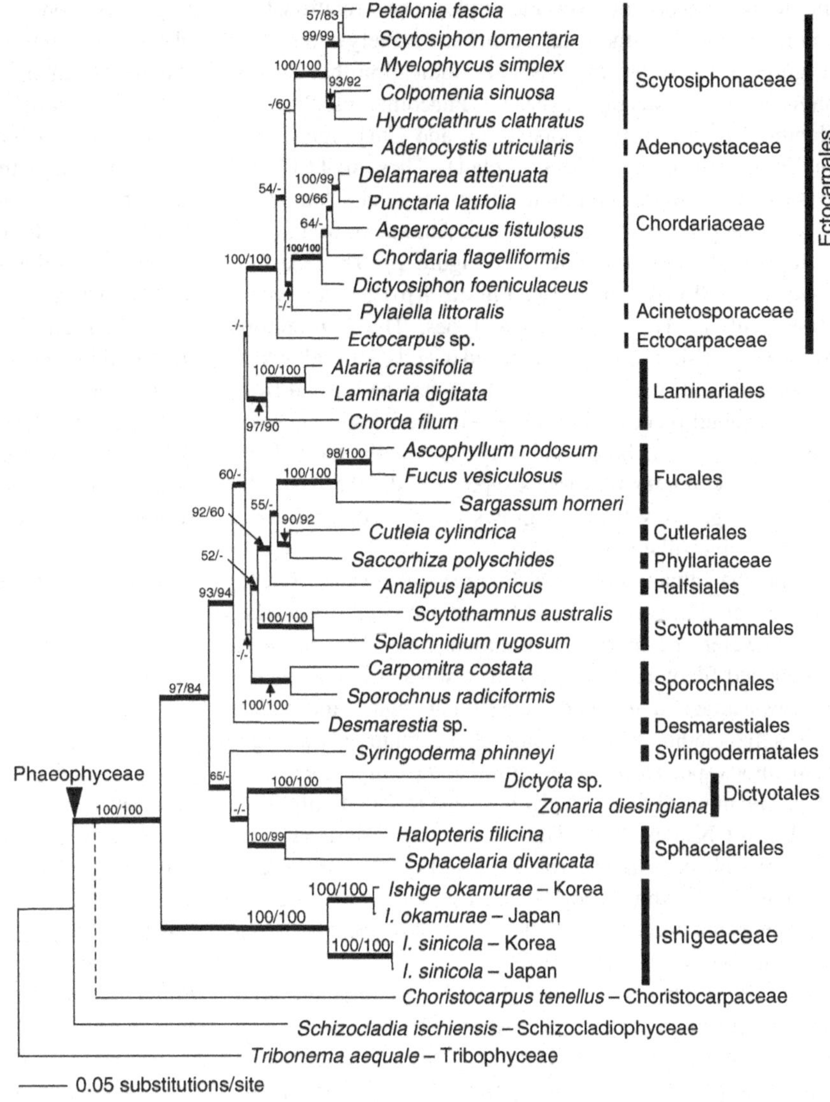

FIGURE 5 Maximum likelihood (ML) tree of the Phaeophyceae using a concatenated *rbc*L, *psa*A, and *psb*A dataset. The bootstrap values shown above the branches are from maximum likelihood/maximum parsimony methods and dashes indicate <50% bootstrap support. The thick branches indicate Bayesian posterior probabilities >0.9. The dashed line for the Christocarpaceae was based on a *rbc*L phylogeny in the same study. Modified from Cho, G. Y., Lee, S. H., & Boo, S. M. (2004). A new brown algal order, Ishigeales (Phaeophyceae), established on the basis of plastid protein-coding rbcL, psaA, and psbA region comparisons. *Journal of Phycology, 40*, 921–936.

Since the organisms corresponding to these sequences have yet to be identified and isolated, they are still known only from partial 18S rRNA sequences (phylotypes). Therefore, these environmental samples are generally depicted as independent lineages at the base of the clade. Among other things, it is not even unknown if these sequences are from pigmented (photosynthetic) or unpigmented organisms.

An informal grouping, the pseudofungi appears to be the sister group to the photosynthetic stramenopiles. The monophyly of pseudofungi is uncertain, but it appears to include the oomycetes, the hyphochytrids, and the Bigyromonadea (certain zooflagellates like *Developiella*). The Oomycetes (water molds), which used to be classified as fungi, include some of the most important known plant pathogens. For example, *Phytophthora infestans* causes late blight of potato, which led to the Great Irish potato famine. Other *Phytophthora* species are the main causes of some of the most important diseases of plants, such as dieback, sudden oak death, rhododendron root rot, and ink disease in the American chestnut tree. Various species of *Pythium* cause diseases such as seed rot and 'damping off' in young plant seedlings. A freshwater mold, *Saprolegnia*, causes tail rot on fish such as salmon and trout. Therefore, these colorless stramenopiles are important both in economic and in scientific terms. For these reasons, genome sequencing projects are completed or underway for three *Phytophthora* species (*P. infestans* T30-4, *Phytophthora ramorum*, *Phytophthora sojae*).

Photosynthetic Stramenopiles

Sixteen photosynthetic classes of stramenopiles (including two recently described classes, the Synchromophyceae and the Aurearenophyceae) are currently known. As with the nonphotosynthetic taxa, the phylogenetic relationships among the photosynthetic classes are still largely unresolved (Figure 2). However, there is some indication that Phaeophyceae, Schizocladiophyceae, Phaeothamniophyceae, Xanthophyceae, Aurearenophyceae, and Chrysomerophyceae may form a clade (PSPXAC clade). These six PSPXAC classes also share common morphological characters, such as the presence of coccoid, filamentous, or true multicellular thalli (see arrowhead in Figure 2). The Xanthophyceae (yellow-green algae; Figure 4(p)) are a diverse class, with many morphological types. The filamentous forms may be siphonous or composed of cells with cell walls. These cell walls may be entire or they may be formed from two H-shaped pieces. The single cells are mostly coccoid, but capsoid, amoeboid, and flagellate cells have been described. Colonies of various types are also known. The siphonous genus *Vaucheria* grows on wet soils where it forms a velvet-like mat.

The Phaeothamniophyceae (Figure 4(o)) are also diverse, with unicellular, colonial, and filamentous species occurring in freshwater and marine environments. *Phaeothamnion* is a branched filament that occurs in freshwater ponds, where it grows attached to pondweeds, other algae, and rocks. The Schizocladiophyceae (Figure 4(l)) contain only a single filamentous species, which is similar to the Chrysomerophyceae. The unusual *Schizocladia* has been reported only from the Bay of Naples, Italy. The recently characterized class Aurearenophyceae includes only coccoid species. *Aurearena* is known only from sandy beaches in Japan; the coccoid cell is unusual because inside the cell wall resides the living cell with two fully formed flagella.

The Phaeophyceae (brown algae; Figure 4(q) and 4(r)) are filamentous or parenchymatous in form. Some members of the class have true multicellularity with different tissue types and different organ-like structures including a holdfast, stipe, and blade that are structurally analogous to the root, stem, and leaf of land plants. In addition, these phaeophytes possess complex male and female reproductive structures, phloem sieve cell-like tissue for conductance, and plasmodesmata for communication between cells. This is the only group among the stramenopiles in which true multicellarity has arisen (see double arrowhead in Figure 2), although filaments and colonies of multiple cells are found in many different stramenopile groups. The Raphidophyceae (Figure 4(j)) may also belong to this lineage. This class has marine members that are brown in color but freshwater species that are bright green. Species of the marine genus *Heterosigma* have caused massive salmon mortalities when blooms of the alga occur near salmon farms.

A second major clade within the photosynthetic stramenopiles is the strongly supported CSSE clade, including the Chrysophyceae, Synurophyceae, Synchromophyceae, and Eustigmatophyceae. The Chrysophyceae (golden-brown algae; Figure 4(h)) are an old taxonomic class, which originally contained numerous organisms that are now placed in other classes. Most members are freshwater flagellates, including single cells, colonies, loricate cells, and scaled cells. *Uroglenopsis* may grow abundantly in reservoirs, and blooms of this alga cause taste and odor problems when the reservoir water is used as a human drinking water source. The Synurophyceae (silica-scaled algae; Figure 4(n)) are mostly freshwater single-celled or colonial flagellates that bear bilaterally symmetrical silica scales. *Synura* and *Mallomonas* are common in unpolluted freshwater, and *Synura* may cause taste and odor problems in drinking water. The Synchromophyceae (Figure 4(g)) contain one described species, which is an unusual marine amoeba with 20 or more chloroplasts per cell. *Synchroma* is known only from the Canary Islands where it grows on basaltic rock in tide pools. The Eustigmatophyceae (Figure 4(c)) are a class of almost exclusively coccoid microalgae that occur in a variety of habitats, including freshwater, seawater, and soils. The genus *Nannochloropsis* is used widely as a food for shellfish, and more recently it has been grown as a source of lipids in efforts to produce biodiesel.

The phylogenetic affinities of the remaining classes of photosynthetic stramenopiles, Pelagophyceae, Pinguiophyceae, and Dictyochophyceae, remain uncertain. The Pelagophyceae (Figure 4(k)) are predominately marine microalgae, and *Pelagomonas* and *Pelagococcus* are important components in the open-ocean phytoplankton (open oceans cover 66% of the earth's surface). The Pinguiophyceae (Figure 4(m)) are a marine microalgal group that is famous for its high content of omega-3 fatty acids. For example, *Pinguiococcus* forms large vacuoles full of eicosapentaenoic acid, which is known to provide human health benefits by reducing heart disease, breast cancer, and rheumatism. The Dictyochophyceae (Figure 4(f)) are flagellate and amoeboid algae found in oceans and freshwater lakes and ponds. The group is named after the silicoflagellates, which are flagellate cells that each occupy an elaborate siliceous skeleton or basket. Conversely, the Bacillariophyceae and Bolidophyceae form a strong monophyletic group in all phylogenetic analyses. The Bolidophyceae (Figure 4(i)) are a small group of naked biflagellate cells that occur in the open oceans, especially in tropical waters. The Bacillariophyceae (diatoms; Figure 4(d) and 4(e)) are the most species-rich classes of stramenopiles, and they comprise more than 200 genera and at least 100 000 species.

Diatoms are characterized by a silica frustule with two parts fitting one within the other like a glass box. This structure encases the cell through most of its life cycle. Traditionally, diatom species are classified based on the shape of the frustule into the centric diatoms (radially symmetrical; Figure 4(e)) and pennate diatoms (bilaterally symmetrical; Figure 4(d)), although this classification is now rejected by molecular phylogenetic studies. Because of these elaborate and beautiful silica frustules, diatoms have been known and studied for a long time and are popular in applied fields, such as architectural science and design. Diatoms thrive in almost all environments from freshwater and oceans to wet soils. Therefore, they are important primary producers in all water bodies, contributing about 45% of total eukaryotic marine primary production. Diatoms also play an important role in the global silicate cycle, and silicate sediments are frequently important in the geological record. Diatoms are also used in water quality assessment, as different species are characteristic of different trophic conditions. Because of their ecological importance, genome sequencing projects have been completed for two diatom species, *Phaeodactylum tricornutum* and *Thalassiosira pseudonana*, and a number of others are under way (e.g., *Pseudo-nitzschia*).

Phaeophyceae

The Phaeophyceae (brown algae) include 16 orders with approximately 285 genera and about 1800 species. Morphologies range from simple microscopic filaments to giant kelps that may reach 60 m in length. The large multicellular Phaeophyceae, or giant kelps, play important roles in coastal marine environments. These are truly multicellular organisms, consisting of a variety of differentiated multicellular structures including a holdfast, stipe, and blade(s). The latter use specialized conductive cells (called trumpet hyphae based on their shape) for translocation of metabolites. Giant kelps produce a large biomass with high growth rates and form marine forests that serve as habitats for a diversity of organisms. Many brown algal species are also important as human food sources, and brown algal-based industries have existed for thousands of years in Asia. Aquaculture of kelp species is very popular in Korea and Japan with a significant economic impact. Bioremediation using kelp cultivation is also under development.

Early studies of brown algae morphological traits (i.e., thallus organization, mode of growth, and type of sexual reproduction and life history), led to the suggestion that the simple filamentous Ectocarpales were the earliest diverging group, whereas the morphologically complex Fucales and Durvilleales were the most derived. This 'simple is primitive' concept was challenged by molecular phylogenetic studies. Phylogenies based on rDNA and/or *rbc*L sequences of brown algae suggested that reevaluation of the traditional classification system was needed. For example, the concept of the Ectocarpales was changed to include the Chordariales, Dictyosiphonales, Punctariales, and Scytosiphonales, while the Durvilleales was synonymized with the Fucales. The Laminariales now contain primitive kelps such as Pseudochordaceae and Akkeshiphycaceae, and it excludes the Phyllariaceae.

A recent, detailed multigene study using a fairly taxonomically broad sampling of species (Figure 5) provides an overall picture of brown algal phylogeny. Although much remains to be resolved, it is clear that the Choristocarpaceae are the earliest diverging lineage, followed by a newly established order, Ishigeales, which is in turn followed by a clade, including Sphacelariales, Dictyotales, and Syringodermatales. The remaining taxa (e.g., Fucales, Laminariales) cluster together in a monophyletic 'crown' group, within which the branching orders remain unresolved. Interestingly, the simple filamentous Ectocarpales are not located at the base but rather at the tip of the phylogeny, suggesting that ectocarpalean morphology represents a secondarily derived simplification. Taken together, contemporary taxa belonging to the basal group of brown algae occur mostly in tropical to warm water, whereas most brown algae that exhibit the greatest diversity in terms of species and morphology occur in cold waters.

An extreme level of conflict between morphological and molecular concepts for genera is shown in *Colpomenia*, *Petalonia*, and *Scytosiphon*. These taxa are found in most intertidal areas from tropical to temperate regions, and cryptic species and misapplied names are common. Because convergent evolution is a common theme in brown algae, robust multigene phylogenies are greatly needed in order to reevaluate brown algal diversity and the evolution of their ultrastructural and life history traits.

CONCLUSION

The stramenopiles, which include diatoms, brown algae, and oomycetes, are a well-known and important group of aquatic organisms and plant pathogens. Major advances in understanding their origin and evolution have occurred with the development of advanced molecular and ultrastructural techniques. However, many questions remain, including many of the deep phylogenetic relationships within the group. Multigene analyses from a broad taxon sampling are essential in our efforts to unravel the evolutionary history of this group, and this is the aim of the Heterokont Tree of Life project. In addition, several genome projects, including sequences from the oomycetes *P. sojae* and *P. ramorum*, the diatoms *T. pseudonana* and *P. tricornutum*, and the brown alga *Ectocarpus* sp., are contributing significant new knowledge to our understanding of the stramenopiles.

FURTHER READING

Andersen, R. A. (2004). Biology and systematics of heterokont and haptophyte algae. *American Journal of Botany*, *91*, 1508–1522.

Adl, S. M., Simpson, A. G., Farmer, M. A., *et al.* (2005). The new higher level classification of eukaryotes with emphasis on the taxonomy of protists. *J Eukaryot Microbiol*, *52*(5), 399–451.

Cavalier-Smith, T., & Chao, E.-Y. (2006). Phylogeny and megasystematics of phagotrophic heterokonts (Kingdom Chromista). *Journal of Molecular Evolution*, *62*, 388–420.

Cho, G. Y., Lee, S. H., & Boo, S. M. (2004). A new brown algal order, Ishigeales (Phaeophyceae), established on the basis of plastid protein-coding rbcL, psaA, and psbA region comparisons. *Journal of Phycology*, *40*, 921–936.

Daugbjerg, N., & Guillou, L. (2001). Phylogenetic analyses of Bolidophyceae (Heterokontophyta) using *rbcL* gene sequences support their sister group relationship to diatoms. *Phycologia*, *40*, 153–161.

Graham, L. E., & Wilcox, L. W. (2000). *Algae*. Prentice Hall: Upper Saddle River, NJ.

Hackett, J. D., Yoon, H. S., Li, S., Reyes-Prieto, A., Rümmele, S. E., & Bhattacharya, D. (2007). Phylogenomic analysis supports the monophyly of cryptophytes and haptophytes and the association of 'Rhizaria' with chromalveolates. *Molecular Biology and Evolution*, *24*, 1702–1713.

Kai, A., Yoshii, Y., Nakayama, T., & Inouye, I. (2008). Aurearenophyceae classis nova, a new class of Heterokontophyta based on a new marine unicellular alga *Aurearena cruciata* gen. et sp. nov. inhabiting sandy beaches. *Protist*, *159*, 435–457.

Qudot-Le Secq, M. P., Loiseaux-de, Goër S., Stam, W. T., & Olsen, J. L. (2006). Complete mitochondrial genomes of the three brown algae (Heterokont: Phaeophyceae) *Dictyota dichotoma, Fucus vesiculosus*, and *Desmarestia viridis*. *Current Genetics*, *49*, 47–58.

Van den Hoek, C., Mann, D. G., & Jahns, H. M. (1995). *Algae: An introduction to phycology*. Cambridge University Press: Cambridge, MA.

Yoon, H. S., Hackett, J. D., Ciniglia, C., Pinto, G., & Bhattacharya, D. (2004). A timeline for the origin of photosynthetic eukaryotes. *Molecular Biology and Evolution*, *21*, 809–818.

RELEVANT WEBSITE

http://ccmp.bigelow.org/ – National Center for Culture of Marine Phytoplankton.

Chapter 28

Toxoplasmosis

J.C. Boothroyd
Stanford University School of Medicine, Stanford, CA, USA

Chapter Outline

Abbreviations 385
Defining Statement 385
Introduction 385
Classification 386
Life Cycle 386
Clinical Aspects and Public Health 387
Symptoms 387
Diagnosis 387
Treatment 387
Public Health 388
Population Biology 388
Major Genotypes 388
Strain-Specific Virulence 389
Molecular Biology and Genetics 389
Genome and Gene Expression 389
Genetics 389
Molecular Genetic Tools Available for the Study of *Toxoplasma* 390
Cell Biology 390
Organelles 390
The Lytic Cycle 391
Host Immune Response 393
Nature of the Host Response 393
Genetics of Host Susceptibility 394
Immunization Studies 394
Effect on Behavior 394
Prospects for Future Improvements in Controlling Toxoplasmosis 394
Public Health 394
Vaccination 394
Chemotherapy 395
Further Reading 395
Relevant Website 396

ABBREVIATIONS

DHFR Dihydrofolate reductase
GAG Glycosaminoglycan
HXGPRTase Hypoxanthine/xanthine/guanine phosphoribosyl transferase
IL12 Interleukin 12
IPN Intraparasitophorous vacuolar network
PVM Parasitophorous vacuole membrane
RFLP Restriction-fragment-length-polymorphism
STAT Signal transducers and activators of transcription
TRAP Thrombospondin-related anonymous protein
UPRTase Uracil phosphoribosyl transferase

DEFINING STATEMENT

Toxoplasma is an obligate, intracellular parasite that causes a range of symptoms in its exceptionally broad host range.

INTRODUCTION

Toxoplasmosis is caused by the protozoan *Toxoplasma gondii*, an obligate, intracellular parasite whose definitive host is the family Felidae (cats). *T. gondii* has an exceptionally broad range of intermediate hosts, including humans, in which the asexual cycle can occur resulting in serious disease. It is found throughout the world and, in terms of the number and range of hosts infected, *Toxoplasma* may be the most common protozoan parasite of warm-blooded animals.

The genus *Toxoplasma* falls within the phylum Apicomplexa and is a close relative of the nonhuman pathogens, *Neospora* and *Sarcocystis*. It is more distantly related to *Eimeria*, the causative agent of coccidiosis in birds and more distant still to *Plasmodium* and *Cryptosporidium*, the etiologic agents of malaria and cryptosporidiosis, respectively. At the still more distant level, *Toxoplasma* is also related to free-living dynoflagellates but the evolutionary timescales involved are enormous (e.g., humans and flies are relatively close cousins by comparison).

Eukaryotic Microbes.

Infection of healthy humans by *Toxoplasma* is usually of little consequence but human disease will often occur in either of two scenarios: in the child of a woman infected for the first time during pregnancy and in individuals who are immunocompromised. Occasionally, however, serious eye infections can occur even in people who are otherwise healthy. This unusual occurrence may be the result of an unusually large inoculum, the particular genetic makeup of the person infected or, perhaps, the strain of parasite responsible for the infection.

In animals, toxoplasmosis is also generally self-limiting with the most serious consequences (abortion) being from acute infection during pregnancy. This is a particularly serious problem in the raising of sheep.

The relative ease of handling this haploid parasite in the laboratory and the wide range of natural hosts have made *T. gondii* a popular model for experimental study of intracellular parasitism. Through such studies, much has been learned about the way in which this common parasite interacts with its host, about its population biology, and about the molecular details behind its obligate intracellular life style.

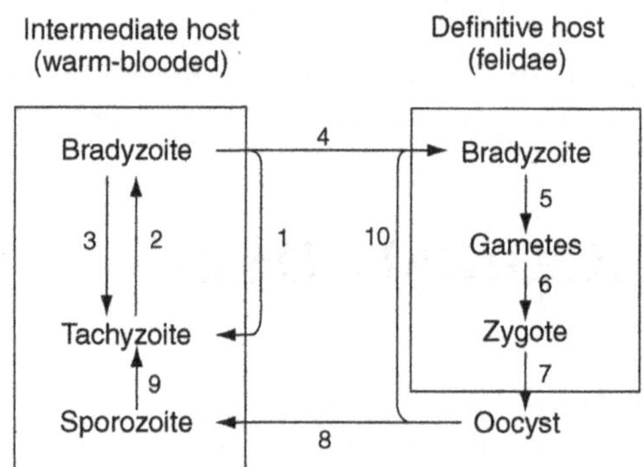

FIGURE 1 The life cycles of *Toxoplasma gondii*. The asexual cycle can occur in a large number of warm-blooded animals and is shown on the left. It involves an equilibrium between the rapidly dividing tachyzoite and the more slowly dividing bradyzoite (arrows 2 and 3). Transmission in the asexual cycle is through the ingestion (carnivorism) of meat and other tissue containing the infectious encysted bradyzoites (arrow 1; e.g., rat to pig or pig to human). Scavenging allows the parasite to cycle back down the food chain (e.g., pig to rat). The sexual cycle is shown on the right and involves schizogony, gametogenesis, and fertilization (arrows 5, 6, and 7) in the gut epithelium of felines. Cat-to-cat transmission through the sexual cycle is theoretically possible by ingestion of oocysts in fecal contamination (arrow 10). Crossover between the two cycles is represented by arrows 4 and 8, but the extent of this in nature is not known.

CLASSIFICATION

T. gondii is a protozoan parasite within the phylum Apicomplexa, class Conoidasida, order Eucoccidiorida, suborder Eimeriorina, family Sarcocystidae. The genus name derives from the Greek word *toxon* meaning bow and *plasma* meaning form and refers to the bow-shaped tachyzoite. The species name derives from the name of the North African rodent from which the parasite was first isolated in 1908 by Nicolle and Manceaux. It was simultaneously described by Splendore in Brazilian rabbits.

LIFE CYCLE

The complete life cycle of *T. gondii* was not established until about 1970 when four groups independently identified the cat as the definitive host, meaning this is the host in which *Toxoplasma* enters into its sexual cycle. Figure 1 depicts the complete life cycle as consisting of two, potentially independent cycles, one asexual and the other sexual. This has been done in order to emphasize that the parasite appears fully capable of propagating itself through either cycle. The degree to which these two cycles intermix (arrows 6 and 7) is not known. In practical terms, this uncertainty means that it is not clear what fraction of infection in the intermediate host (e.g., humans) is a result of ingesting meat-containing tissue cysts (arrow 3) versus accidental ingestion of oocysts in the environment (arrow 7).

The tachyzoite stage is much the best studied, mainly because it is easy to grow this form in vitro. The other asexual stage, the slow-growing encysted bradyzoite is poorly understood. Fortunately, however, protocols have recently been developed that can cause tachyzoites to efficiently differentiate to the bradyzoite stage in vitro. These protocols rely on a number of stresses (certain drugs, heat, pH) that stimulate the differentiation pathway. Although the differentiation in vitro may not be complete, the availability of these protocols has opened up the way for more detailed studies of this critically important stage, which previously could only be obtained in small amounts from the tissue of infected animals.

The two asexual stages are similar in overall morphology with mostly the same complement of organelles. The major differences are in the presence of many amylopectin granules in the bradyzoite and a cyst wall that encloses the entire complement of bradyzoites in any given vacuole. The metabolic differences are not clear, although the existence of stage-specific isoforms of some glycolytic enzymes (e.g., enolase and lactate dehydrogenase) suggests the possibility of a fundamental difference in sugar metabolism. Such differences might be related to the presence of the amylopectin granules and/or the polysaccharide-containing cyst wall.

The sexual stages have only been described morphologically in the gut of infected cats (it has not yet been possible to grow any of these stages in vitro). The process of gametogenesis, which occurs in the epithelium of the small intestine, conforms to classic coccidian rules including schizogony where a multinucleated state is reached followed by segmentation to yield multiple uninucleated merozoites. Mating between micro- and macrogametes has never been directly observed but is presumed to occur within infected epithelial cells. The process results in a zygote, which then encysts yielding an immature oocyst. This is released into the environment where, after ~2 days of exposure to air, it develops into a mature oocyst comprising two sporocysts, each with four sporozoites. The genetic relationship of the eight sporozoites within a single oocyst has not yet been determined (i.e., when the three divisions occur to take the oocyst from one to eight organisms and which of those divisions are mitotic vs. meiotic).

CLINICAL ASPECTS AND PUBLIC HEALTH

Symptoms

Toxoplasma causes disease in humans primarily under two circumstances: the developing fetus of a woman who acquires her first infection during pregnancy and persons who are immunocompromised. The disease that results varies from mild to severe and sometimes fatal. Note, however, that in congenital infections the disease may not appear until many years later, when the child is in the second decade of life or older.

The prevalence of seropositive adults ranges from ~5 to 15% in the United States to as high as 60–85% in France and many developing countries. Historically, severe disease has not been common because of the low number of persons who fall within the two disease-susceptible groups mentioned above: the children of acutely infected pregnant women and people who are immunocompromised. The AIDS epidemic obviously produced a huge increase in the number of people in the immunocompromised category. Initial results indicated that, if left untreated, about one-fourth of AIDS patients who harbored a chronic infection of *Toxoplasma* (most cases of toxoplasmosis in AIDS patients are thought to be due to reactivation of historic, quiescent infections) were liable to develop the potentially fatal condition of toxoplasmic encephalitis. Within the United States, this proportion has dropped in the last decade, probably as a result of three factors. (1) Improved anti-HIV therapies (with a resulting improvement in general immune competence). (2) Prophylactic treatment with trimethoprim/sulfamethoxazole for another opportunistic infection, *Pneumocystis*: such treatment may have the added benefit of controlling *Toxoplasma* infection, if present. (3) Deliberate prophylaxis for *Toxoplasma* in patients who are seropositive and whose CD4 T cell counts are low. Data on toxoplasmosis in AIDS patients in the developing world are scant but it appears that toxoplasmic encephalitis may be a significant problem in the AIDS populations of those countries just as it was in the United States prior to the aggressive treatment protocols that are in place today.

Congenital infection rates vary by country (in proportion to the seropositivity of the adult population); in the United States, there are no exact numbers but current estimates are that ~500–3000 congenitally infected children are born each year. The probability of transmission from an acutely infected mother to the developing fetus varies with the stage of the pregnancy: acute infections in the first trimester are the least likely to be transmitted, whereas third trimester infections are the most likely to be transmitted. As for many congenital diseases, however, the outcome is worse when the infection occurs earlier in the pregnancy: first trimester infections can lead to severe neurological disease (blindness and retardation) or even death, whereas third trimester infections are more mild and may even yield no symptoms at birth. This is not to say that no disease will occur in this latter group: it appears that the parasite can lie essentially dormant and erupt in early adulthood leading to various symptoms, most notably sudden blindness in one eye due to severe retinal destruction. Drug treatment of acutely infected pregnant women can reduce the probability of congenital transmission but, as discussed below, such treatment is not without significant risk to the fetus.

Diagnosis

Serology is the most common method for diagnosis. Acute infection in pregnant women is usually diagnosed from a high IgM or IgA antibody titer and/or a rise in IgG in successive serological samples taken at multiweek intervals during the pregnancy. In countries like France where the incidence is high, strict monitoring is in place: women are tested for anti-*Toxoplasma* titers at the time of marriage and those who are seronegative and later become pregnant are monitored repeatedly for signs of seroconversion (equating to an acute infection). If an acute infection is detected, it is now possible to monitor for transplacental transmission to the fetus (which is the real concern) by amniocentesis and polymerase chain reaction to detect the parasite (or, rather, its DNA) directly.

Treatment

Current treatment of acute infection in nonpregnant adults relies on the synergistic action of pyrimethamine plus sulfadiazine, which combine to block folate metabolism through inhibition of dihydrofolate reductase (DHFR) and dihydropteroate synthase, respectively. Unfortunately, even this powerful combination has little effect on the chronic form of the parasite (the bradyzoite), and thus treatment does not completely eliminate the infection. As a result, the drug therapy must be maintained for long periods to prevent recrudescence of the infection. This is not a simple option as sulfadiazine is poorly tolerated by many patients and its use often must be stopped. Other therapies, while showing some promise, have yet to prove as effective as pyrimethamine plus sulfadiazine (e.g., pyrimethamine plus clindamycin and the newer drug, atovaquone), although better formulations and treatment protocols could dramatically improve their efficacy.

Because of drug toxicity, treatment of pregnant women is always a clinically complicated decision. In countries like France, which have the most experience of managing such patients, acute infection in the mother is treated with spiromycin although the benefit of such treatment is not firmly established. If transmission to the fetus has been demonstrated, then the more potent but more toxic combination of pyrimethamine plus sulfadiazine is used because the risk of serious disease outweighs the possible harmful effects of the drugs. Therapeutic abortion is also sometimes chosen.

Public Health

In the absence of a vaccine and given the shortcomings of the drugs mentioned above, the best approach to controlling human toxoplasmosis is avoidance. This is easier said than done, however, as evidenced by the fact that infection with this parasite is so prevalent worldwide.

Meat-eaters are especially at risk from undercooked lamb and pork, although the latter may not be such a problem because of centuries of knowledge about severe disease that can result from pork and the resulting common knowledge that it should always be well cooked. Lamb, on the other hand, is frequently eaten extremely rare in many cultures. (Note that beef, while also often eaten rare or raw, is not such a problem because for unknown reasons, bovines do not harbor substantial numbers of tissue cysts in their muscle or other tissue.) It is thus critical that persons at risk, for example, persons who are immunocompromised and seronegative women who might become pregnant, avoid eating meat that is not thoroughly cooked. Fortunately, freezing also kills tissue cysts and thus this simple preventative measure can also be used to reduce the risk of infection.

Toxoplasma's unusual life cycle means that simple vegetarianism will not enable someone to avoid infection. The oocysts shed by cats are extremely stable in the environment and cats are particularly fond of defecating in loose soil such as that found in vegetable gardens. It is thus fairly simple for vegetables to be contaminated with minute amounts of oocysts and so at-risk persons should take special care to eat only cooked or thoroughly washed fresh vegetables.

Finally, the stability of the oocysts also means that they can be ingested through daily activities as innocuous as gardening or playing in a dusty field or sandbox. Obviously, cleaning a pet cat's litter tray is not appropriate for at-risk persons although because oocysts take ~48 h to become infectious after shedding, regular daily removal of the feces can reduce the risk of infection. Even better, keeping cats strictly indoors and feeding them only on tinned or dried foods can virtually eliminate the chance of their acquiring an infection.

One of the key deficiencies in our understanding of toxoplasmosis is what fraction of human infection comes from ingestion of tissue cysts in undercooked meat products versus oocysts in environmental contamination. This point is discussed further below.

POPULATION BIOLOGY

Major Genotypes

In recent years, the population biology of *T. gondii* has come under intense scrutiny through studies of isoenyzmes, restriction-fragment-length-polymorphisms (RFLP), microsatellite analysis, and direct sequencing of specific loci. Out of these studies a very clear, consistent, but surprising picture has emerged. First, despite the existence of the well-described sexual cycle in cats, the parasite appears to be reproducing largely clonally. Worldwide, there appear to be just 11 population clades. In fact, looking at samples from humans and domestic animals/livestock in Europe and North America, reveals the vast majority of strains from these regions come under one of just three genotypes (referred to as 'Type' I, II, or III, respectively). Because *T. gondii* is haploid, this does not necessarily mean that the sexual cycle is not occurring. Instead, it could mean that the mating that does occur is almost entirely between individuals of the same genotype ('self'). This is not unreasonable, given that a single haploid tachyzoite can ultimately (via an infected mouse) give rise to a full sexual cycle in the cat, including genetically identical micro- and macrogametes, and thus the parasite does not have fixed mating types or sexes. Moreover, it is known that once a cat is infected, oocysts appear in the feces for about 1–2 weeks but thereafter, the cat is relatively immune to further infection and oocyst shedding decreases precipitously. Whether a cat infected with strain A and subsequently some weeks later with strain B will yield A/B recombinant progeny has not been directly examined but all available data suggest the answer will be no. These factors alone could account for the rarity of recombinants.

There is a second possible explanation for the clonal propagation: cats may play little if any role in the transmission of some strains. For example, it has been noted that Type I strains grown in the laboratory are difficult to pass in cats. As cats are the only known host in which the sexual cycle occurs, a loss of this function would require exclusively asexual expansion with transmission through carnivorism/scavenging or, perhaps, vertically in some host species.

The above could explain why most strains appear to be expanding clonally but why do most strains infecting domestic animals in North America and Europe come under only three genotypes? For one of the most common protozoan parasites on earth (at least in terms of warm-blooded animals), it is hard to imagine that a major 'bottle-neck' in the population has ever occurred; that is, it seems doubtful that virtually all individuals in the *Toxoplasma* population were eliminated and only a few survivors went on to reestablish the species with their specific (chance) genotype. Instead, it appears that the three 'types' that have arisen are remarkably successful recombinants that have been selected for their ability to thrive in today's environment and that there is also a small but important population of recombinant parasites that collectively possess tremendous genetic diversity. Support for this hypothesis comes from the fact that when strains are isolated from remote regions and unusual species (e.g., South American wild monkeys or North American sea otters), quite different genotypes can be found. These latter strains are presumably representative of the total diversity to be found when more unusual hosts and regions are eventually sampled and carefully analyzed.

Strain-Specific Virulence

The three major types of this parasite are not just minor or random polymorphic states. Indeed, one of the first clues that the population biology of *T. gondii* was unusual that the virulence of multiple different strains, when measured as the dose that resulted in a lethal infection in mice (LD_{100}) was either a single parasite or more than a thousand without a substantial middle ground. We now know that in susceptible mice, Type I parasites are highly virulent organisms whereas Types II and III are relatively avirulent.

Recently, crosses between the three types have been used to identify some of the genes responsible for the differences in virulence. The two genes identified so far have been found to encode one or other of two protein kinases that are released from the secretory organelles called rhoptries (see below). These proteins, ROP16 and ROP18, are ultimately found in the nucleus of the host cell or on the vacuolar membrane that encloses the parasites, respectively. As yet, the proteins that ROP18 phosphorylates are not known – they could be host or parasite proteins – and the mechanism by which it impacts virulence is a complete mystery. For ROP16, the picture is much clearer: different alleles of the *ROP16* gene encode a tyrosine kinase that, depending on the allele, phosphorylates STAT3 and STAT6 to a greater or lesser extent. STATs are 'signal transducers and activators of transcription' and are key to turning the immune response up or down. Hence, differences in STAT3/6 activation will have a profound effect on how the host handles a *Toxoplasma* infection and, hence, virulence.

As to the reasons why the strains differ in virulence, there has been recent speculation that each strain evolved in association with a specific intermediate host (e.g., mice or sparrows) and is optimized for infection/transmission in that host species. When changes in the environment present such a strain with the opportunity to infect a new host species, it may do so but with a very different outcome from infection in its natural host. It seems unlikely, for example, that the Type I strain could have evolved as a natural infection of mice, given its extreme virulence in that species. Instead, it may have evolved with birds as its natural host and only recently been given the opportunity to infect mice. A corollary of this is that the serious consequences that result from such infections are not the result of selection because this combination of parasite strain and host species is a dead end (just as all infections with *Toxoplasma* are in humans).

There is very little information on how differences in mouse virulence correlate with disease outcome in humans and livestock. The reason is that the knowledge that there are a limited number of strains is still relatively new and methods to distinguish them have been tedious and/or only recently applied. Nevertheless, some studies have been done and the data suggest that Type I strains may be more virulent in otherwise healthy humans as well. In the absence of information on what strains are present in asymptomatic infections, however, and on what is the relative frequency of different types in the various animal reservoirs of human infection, it is impossible to draw any definitive conclusions about relative virulence. Determining whether strain type is an important factor in the severity of human infection is an important future goal of *Toxoplasma* researchers.

MOLECULAR BIOLOGY AND GENETICS

Genome and Gene Expression

Except for the diploid zygote, all stages of *Toxoplasma* are haploid with a genome of $\sim 6.5 \times 10^7$ bp per nucleus. Using a combination of physical and genetic means, it has been shown that *Toxoplasma* has 14 chromosomes in its nucleus. The genomic sequences for Type I, II, and III strains have recently been essentially completed and are available at www.toxodb.org/toxo/. This is an extraordinary advance and has opened up many new areas of investigation. Unlike some other parasitic protozoa, gene expression in *Toxoplasma* appears to be similar to that seen in other model eukaryotes such as yeast or humans. For better or worse, this has allowed investigators to focus on the coding function of the genes rather than any subtleties about how they are regulated.

In addition to the nuclear DNA, two other genomes are present in *Toxoplasma*: the ~ 35 kb genome of the plastid-like structure known as the apicoplast (see below) and as yet incompletely characterized mitochondrial DNA.

Genetics

Because of its haploid nuclear genome, it is relatively easy to select phenotypic mutants in *Toxoplasma*. A large number of such mutants have been described including lines that are temperature-sensitive for growth, deficient in development, disrupted in cell cycle, or resistant to one or more drugs. Mutant libraries have been made using chemicals or random insertion of DNA sequences carrying selectable markers. The latter has also been done with plasmids carrying a unique sequence that serves as a 'signature' or bar code allowing pools of mutants to be analyzed as a group.

The recent determination of the genome sequences for Types I, II, and III means that an extremely high-resolution genetic map now exists. This will greatly facilitate mapping of F1 progeny generated by crossing any two of these strains. To date, this has been done with crosses between Types II and III and Types I and III. The fact that the crosses must be done in cats, however, makes these sorts of experiments far from trivial.

Molecular Genetic Tools Available for the Study of *Toxoplasma*

In the past decade or so, there has been a substantial effort invested into the development of molecular genetic techniques for the study and manipulation of *Toxoplasma*. The large number of such techniques now available makes *Toxoplasma* one of the most easily studied and 'engineered' of any protozoan parasite.

Initially, transient expression of suitably engineered DNA plasmids was achieved using electroporation. There are now several selectable markers reported for stable transformation of *Toxoplasma*: bacterial chloramphenicol acetyl transferase (conferring resistance to chloramphenicol); pyrimethamine-resistant DHFR; bacterial tryptophan synthase (*trpB*, conferring tryptophan prototrophy); bacterial *ble*, conferring bleomycin and phleomycin resistance; bacterial β-galactosidase (*lacZ*); jellyfish green fluorescent protein and, in the appropriate null backgrounds, *Toxoplasma* hypoxanthine/xanthine/guanine phosphoribosyl transferase (HXGPRTase); uracil phosphoribosyl transferase (UPRTase); and the major tachyzoite surface antigen *SAG1*. With the exception of the last, all have been successfully used to introduce exogenous or engineered genes into the parasite for expression.

This large collection of selectable markers has made reverse genetics a staple of the *Toxoplasma* field. Many genes have been specifically disrupted through targeted insertion/deletion using one or other of the selectable markers listed above. This has allowed the function of many loci to be investigated. Of course, essential genes in a haploid organism cannot be deleted, by definition, unless a back-up copy is provided. This has been done using the popular tetracycline-regulated expression system that allows the backup copy to be specifically turned on or off by adding/subtracting the drug.

Episomal vectors have also been developed, which should allow for simpler recovery of introduced DNA, multicopy suppression, and tests for essentiality using negative selectable markers on the episome. This approach has yet to be exploited but direct complementation using plasmid and cosmid libraries has and these have opened up the field by allowing genes that are mutated in random mutagenesis protocols to be identified.

CELL BIOLOGY

Organelles

As a dedicated, intracellular parasite, *Toxoplasma* has many specialized organelles dedicated to this lifestyle. Some of these are shown in Figure 2 and each will now be briefly described.

Apicoplast

This is a spherical, plastid-like organelle just anterior to the nucleus. Its importance to *Toxoplasma* and related parasites is evident from the fact that it is found in Apicomplexan parasites ranging from *Toxoplasma* to *Eimeria* and *Plasmodium*. Its existence has been known for decades, going by various names (e.g., Golgi adjunct, and spherical body). It is delimited by four membranes and contains its own genome whose sequence revealed its relatedness to plastids of algae. The data further argue that it was acquired by the ancestor of *Toxoplasma* (and other Apicomplexa) through a secondary endosymbiosis of an algal-like eukaryote. A thorough study of the proteins encoded in the nucleus that ultimately are targeted to the apicoplast argue that it has an important role in fatty acid metabolism among other

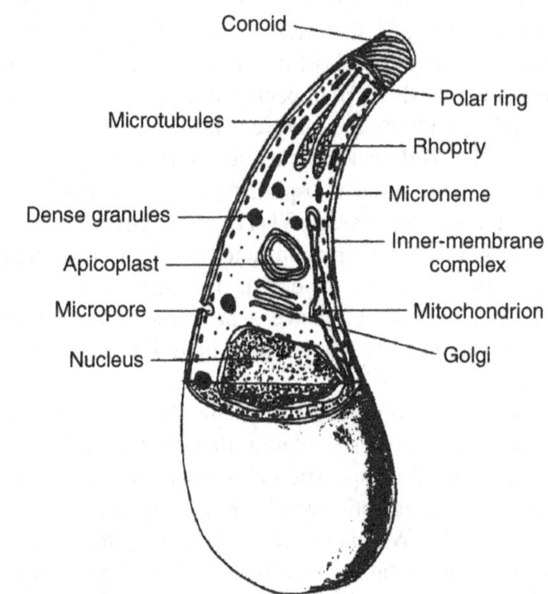

FIGURE 2 Ultrastructure and morphology of the tachyzoite. This schematic shows a cut-away of the tachyzoite stage of the parasite. The overall morphology is very similar for the bradyzoite and sporozoite. Note that the microtubules gently spiral down the body, starting from the anterior polar ring and ending about two-thirds along the length of the parasite. This is why they have an oval appearance in the cross-section shown.

functions. As mentioned below, its conservation throughout the Apicomplexa and the complete lack of a similar organelle in the animal hosts of *Toxoplasma* make it an extremely attractive drug target. Already, one drug that had been empirically determined to be very effective against Apicomplexan parasites, clindamycin, has been shown to act through interfering with translation in the apicoplast. It is likely that this organelle will provide rich pickings for future drug-discovery programs.

Dense granules

These are small, spherical, secretory organelles distributed throughout tachyzoites, bradyzoites, and sporozoites. They release their contents upon invasion into a host cell and are believed to help modify the vacuole to make it better suited for growth of the parasites within. One of the most abundant constituents of these granules is a potent NTPase that may have a role in scavenging nucleosides from the host cell although this has yet to be directly demonstrated. At least one dense granule protein, GRA7, has been reported to be present on the surface of the infected host cell although its function there is unknown.

Inner-membrane complex

This is a set of collapsed vesicles almost completely enveloping the parasite and closely apposed to the parasite's plasma membrane. Together with the latter, they make the parasite surface appear as if it consists of a triple layer of membranes. It is one of the first organelles to appear de novo during endodyogeny and behaves as if it were the nucleating point for formation of the two daughter parasites. The inner-membrane complex likely plays many roles, including an anchor for some of the machinery involved in 'gliding motility' of the parasite.

Micronemes

These are numerous, thin, anterior organelles that secrete their contents onto the parasite's surface at the start of the invasion process. Their components appear key to motility and invasion. One microneme protein, MIC2, has been directly implicated as a way that the parasite attaches to an exterior substrate like a host cell and then pulls itself in. Another, AMA1, has been found at the so-called 'moving junction' that forms as the parasite enters a host cell (see the section on 'Invasion,' below). This moving junction is a ring-shaped entity that represents the point of contact between the parasite and host cell surfaces as the parasite invades into the nascent parasitophorous vacuole. The moving junction also comprises several proteins from another secretory compartment, the rhoptries, as discussed next.

Rhoptries

These are club-shaped, apical organelles that release their contents early in the invasion process. The functions of their protein contents are just beginning to be understood and it is looking like they play several crucial roles in the parasite's interaction with the host. First, as just mentioned, the rhoptries release proteins that collaborate with AMA1 to form the moving junction. Interestingly, these rhoptry proteins all emanate from the more apical, neck-like region of the organelles, rather than the bulbous base. On the basis of this Rhoptry Neck origin, they have been designated RON proteins. How these proteins interact with AMA1 to form the moving junction and how this structure maintains an interaction between the parasite and host surfaces are mysteries yet to be solved.

Many rhoptry bulb proteins (which are designated ROPs) are released into the host cell where they are found associated with the parasitophorous vacuole membrane (PVM). At least some of these are exposed to the host cell cytosol. It has been speculated that others might span the PVM, perhaps acting as pores for transport of nutrients out of the host cytosol into the vacuole. Such an activity has been described but the responsible proteins have yet to be identified.

Finally, some ROP proteins (ROP16 and a protein phosphatase of the 2c class) have been shown to somehow end up in the host cell nucleus. As discussed above, ROP16 appears to have key functions in the host cell; the function of the protein phosphatase has proven more elusive because even a complete deletion of the gene has little impact on *Toxoplasma's* ability to invade and co-opt a host cell in vitro or cause serious disease in the infected animal. Chances are these negative results simply reflect the fact that the experiments were done with only one strain, a very limited number of specific host cell types and in only one host species (laboratory mice).

Micropore

This small, single invagination anterior to the nucleus has an apparent clathrin coat and is presumed but not yet demonstrated to be involved in endocytosis.

The Lytic Cycle

Toxoplasma is an obligate intracellular parasite. As such, and given that it causes lysis of the host cell after going through several rounds of replication, the classic virology term, 'lytic cycle' is perfectly suited to describing the various stages of intracellular growth. The cycle begins with attachment followed by invasion, replication (including vacuolar modification), and egress (Figure 3).

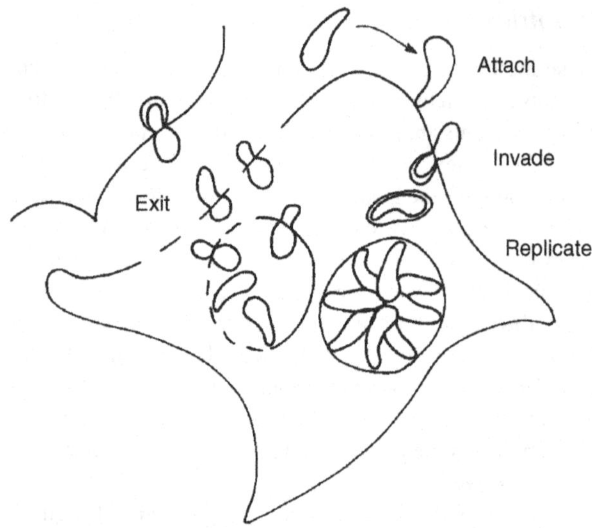

FIGURE 3 The lytic cycle of asexual reproduction. In this cartoon, the stages of attachment, invasion, replication, and egress are shown. Note that the host cell may have multiple vacuoles. Egress is shown here as involving both membrane destruction and active egress through semi-intact membranes. This is the least well-studied part of the lytic cycle.

Attachment

Taking all the data together, the most likely model for attachment involves at least two steps. First, a resident surface molecule, perhaps one or more of the glycolipid-anchored surface antigens, may interact with as yet unidentified host ligands. This part of the model is based on studies showing that antibodies to the major surface antigen, P30 or SAG1, can block binding of parasites to host cells (fixed or live) and thus prevent invasion. The putative ligand on the host cell side is likely to be something that is nearly ubiquitous as *Toxoplasma* is remarkable for its ability to invade almost any vertebrate cell it encounters, in vitro, at least.

Following such initial interactions, a signal is sent which results in release of additional molecules involved in invasion, including the micronemal MIC2 and AMA1 proteins, mentioned above. MIC2 is a homolog of the better studied 'thrombospondin-related anonymous protein' (TRAP) molecule of *Plasmodium* sp. TRAP and MIC2 have domains whose sequence suggests an ability to bind glycosaminoglycans (GAGs). GAGs and other proteoglycans have been implicated as a key ligand on the host side. AMA1 was also first described in studies on Plasmodium; its ligand is, as yet, completely unknown in any system.

Electrophysiological studies have provided further insight into this initial attachment stage. They show that there is a brief, transient spike of conductance across the host cell plasma membrane suggesting that there is a profound perturbation of the integrity of this membrane associated with the initial attachment. The significance and role of this event are not yet known, although it is tempting to speculate that it somehow is connected to the process of introducing proteins into the host cell; the perturbation seems too brief to reflect the introduction itself but it could be related to the process that does this.

Invasion

Invasion is an active process involving many parasite functions and is accompanied by the sequential discharge of several distinct intracellular compartments including the micronemes, rhoptries, and dense granules. The contents of these compartments serve many functions. Among these, it is presumed that some number of molecules facilitate the invasion process itself, perhaps providing the link between the host cell surface and the parasite's actin/myosin-based motors that provide the driving force of invasion. Thus, molecules such as MIC2 or AMA1 might enable the parasite to literally pull itself into the host cell, creating the moving junction described above and a nascent parasitophorous vacuole as it proceeds. Physiology studies have clearly demonstrated that the PVM is largely derived from the host cell's plasma membrane.

Replication

The first challenge to the intracellular parasite is to create the necessary niche for optimal growth. For Toxoplasma, this requires modifying the vacuole created on invasion. The PVM appears largely devoid of host markers suggesting that during its formation, most resident proteins are removed from the invaginating host plasma membrane through some sort of molecular sieve that must exist at the moving circular junction that migrates down the parasite surface creating the PVM in its wake. Very quickly, however, parasite proteins, particularly the ROPs, begin to be detectable on the PVM. Few host proteins are ever detected, suggesting that no fusion with lysosomes or other compartments of the host endocytic pathway occurs. This is consistent with ultrastructure studies on infected cells and the fact that there is no acidification of the vacuole.

The major features of the PVM are as follows:

1. Pores may arise in the PVM that are permeable to molecules of up to about 1300 Da without regard to charge. These structures, which have yet to be seen and are of unknown composition, are presumed necessary to allow the parasite to feed freely on host cell nutrients. They also result in the lack of a physiological barrier between the vacuolar space and the host cell cytosol. Thus aspects like pH, osmolarity, concentration of specific ions, and so on, should be similar if not identical between these two compartments.
2. An extensive intraparasitophorous vacuolar network (IPN) forms inside the PV. This is a loose meshwork of tubules of unknown composition and function.

The IPN has been reported to be partly continuous with the PVM suggesting it may be used to increase the effective surface area of the PVM, although there are no data to show that this is physiologically the case. The IPN is rich with dense granule proteins which may play an active role in its generation.

3. Host cell organelles such as mitochondria and endoplasmic reticulum are recruited around the PVM with extraordinary efficiency and uniformity. Presumably, the parasite proteins that are associated with the PVM are involved.

Through this extraordinary vacuole, therefore, the parasite is able to create the ideal environment for its own growth, which then occurs with considerable efficiency (e.g., a doubling time of as little as 6 h for some strains under ideal conditions in vitro). Division is not by simple binary fission, however. Instead, the parasite undergoes a complex process known as endodyogeny (Figure 4). This term was chosen to convey the fact that two daughter parasites develop within the mother cell, dividing up the organelles between them or creating sets of entirely new ones. Midway through endodyogeny, some organelles that are made de novo are found as three distinct sets – one new set for each of the daughter cells and one old set still in the mother cell. The inner-membrane complex and the rhoptries are examples of this situation. Most organelles, however, are only ever present with a complement of at most two, for example, the nucleus, the mitochondrion, the Golgi apparatus, and the plastid, all of which divide conventionally and flow into the two daughter cells as they develop.

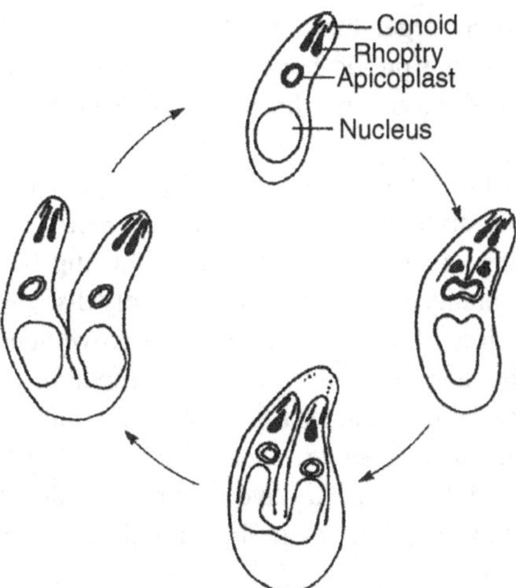

FIGURE 4 Endodyogeny during asexual reproduction. Endodyogeny involves the creation of two daughter cells within a mother cell. As the two daughters develop from the anterior end backward, the apicoplast and the nucleus divide, and new rhoptries, and inner-membrane and apical complexes, appear. Eventually, the mother cell's organelles, including her plasma membrane, are fully co-opted by the daughters, except for the mother's inner-membrane complex, which mysteriously disappears.

Egress

The final stage of the lytic cycle, egress, is also the least studied. The intuitively appealing notion that the parasite's replication simply fills the host cell to the bursting point is argued against by the following data:

1. Video microscopy clearly shows that the parasites become highly motile just prior to egress suggesting a signal is delivered that initiates the exit process.
2. Egress can be artificially stimulated by various treatments including adding calcium ionophores such as A23187 to the medium, or strong reducing agents, such as dithiothreitol. This effect can be achieved as soon as 30 min after invasion (i.e., when there is only a single parasite per vacuole).
3. A parasite homolog to perforin is required for efficient egress.

The simplest explanation is that in the natural cycle, as the parasite numbers grow, a critical concentration of a key molecule in the vacuolar space is reached, which then sends a signal to initiate the cascade of events leading to egress. This molecule would likely be freely diffusible (through the pores described above), thus explaining why the total number of parasites within a given cell is important rather than the number within a given vacuole. Hence, an infected cell harboring four vacuoles each with eight parasites will be lysed at roughly the same time as one infected with a single vacuole containing ~32 parasites.

HOST IMMUNE RESPONSE

Nature of the Host Response

As would be expected for an intracellular parasite of nucleated cells, the major effector mechanism for host immunity is cell-mediated; humoral immunity (antibodies) can confer some protection but this is apparently less important than the cellular response. Based on the use of antibodies that selectively deplete different T cell subsets and/or mice that are deficient in these cells, it is clear that CD8+ cytotoxic T cells are among the most important effectors. Their action is augmented by a potent Th1-type response facilitated by CD4+ T cells as evidenced by the effect of injecting infected animals with various cytokines or antibodies to such cytokines: for example, injection of anti-interferon-γ causes the immune response to be depressed and the cerebral infection is correspondingly more severe. Cellular components of the innate immune response, for example, natural killer cells, also play an important role in controlling the acute stages of the infection.

What tips the immune response in one direction or another is not completely clear but certain parasite molecules do seem to be efficiently recognized by the innate immune system. This can result in the rapid production by neutrophils and macrophages of the potent cytokine, interleukin 12 (IL12). In turn, IL12 can lead to the generation of interferon-γ and a potent inflammatory response characterized by the generation of helper T cells of the Th1 flavor. In some cases, this host immune response can itself produce a severe pathology with lethal consequences in, for example, certain strains of inbred mice (C57Black/6).

Despite a generally robust immune response that effectively controls the acute (tachyzoite) stage of the infection, most animals and humans appear unable to clear the infection completely. Instead, tissue cysts containing bradyzoites persist for the life of the host in many tissues but most notably the brain, perhaps due to the very different nature of the immune response in this organ.

Genetics of Host Susceptibility

There is some information to indicate that in mice, at least, there is a genetic component that influences the outcome of infection. As already mentioned, C57Black/6 mice are especially prone to an overexuberant immune response to oral infections by Type II parasites. Other experiments have implicated the L_d allele within the MHC cluster of genes as key to disease outcome. This is presumably affecting the presentation of a specific key antigen, which has recently been identified as GRA6, a secreted dense granule protein.

There are also data to suggest that MHC plays a role in human infection, at least in the context of toxoplasmosis in AIDS patients, but the effect is relatively small and the basis of the effect is unknown.

Immunization Studies

Immunization with crude or purified antigens (e.g., SAG1) or with attenuated strains (e.g., the temperature-sensitive ts4 mutant or strains that have been engineered to be auxotrophs) can elicit a protective response. Efforts have also been made to use a mutant strain that is unable to complete the sexual cycle in cats. It generates a good immunity in the cat while not producing oocysts and would thus be relatively safe. Neither of these approaches, however, have yet been developed for commercial use.

A commercial vaccine does exist based on the S48 strain, which was serially passed 3000 times through mice. This strain will produce an infection in sheep, which, once cleared, results in an immunity that reduces the problems associated with subsequent infection during pregnancy of the ewes (i.e., abortion). The strain apparently is completely cleared from the infected animals with no tissue cysts developing. Thus vaccination with this strain should not represent a problem if the sheep is subsequently slaughtered for consumption as no transmission to humans should be possible. Likewise, death of the vaccinated sheep in the field should not result in the strain entering the food chain through being ingested by scavengers.

EFFECT ON BEHAVIOR

Experiments in rodents have clearly established that a chronic infection, which includes development of tissue cysts within the brain, can result in a marked change in behavior. These changes include increased activity that may make the animals more likely prey for cats because felines are attracted by motion of the prey. Even more intriguingly, several laboratories have reported that infected rats lose their innate fear of cats as demonstrated by their willingness to go where cats have urinated. This too would have obvious evolutionary benefit to the parasite in terms of increasing the probability of transmission. Unfortunately, it is almost impossible to test experimentally whether this is the real reason for these phenomena but such is an enticing possibility.

The notion that human behavior might be affected in people harboring a chronic infection has not been carefully examined but there are some epidemiological hints that this might be the case. As with most human studies, however, it is extremely difficult to tease out confounding variables and the data in support of this possibility are not yet compelling.

PROSPECTS FOR FUTURE IMPROVEMENTS IN CONTROLLING TOXOPLASMOSIS

Public Health

Educating the public about this disease and ways to avoid it continue to be crucial goals. The clarity of the message would be hugely helped if we better understood the relative contribution to human infection of the two different routes of transmission (undercooked meat containing bradyzoites vs. accidental ingestion of oocysts contaminating drinking water, garden vegetables, hands, etc.). Such information would allow the public health message and control measures to be focused on whichever route proves to be the more serious risk.

Information about which *Toxoplasma* strains cause the most serious disease would also be a huge help in the clinical management of the patient (e.g., in balancing between the use of potentially toxic drugs and the likely disease outcome if the infection is left untreated). Such improvements will depend on the development of diagnostic methods that distinguish between strains and source of the infection. Serological tests that distinguish between people infected with the different 'types' of *Toxoplasma* have begun to be developed but are not yet at the point where they are clinically useful.

Vaccination

Although there can be little doubt that a human vaccine could be developed, the relative rarity of serious human disease (especially compared to the number of people infected) and the fact that such disease is largely restricted to pregnancy or AIDS make testing and use of such a vaccine extremely difficult. Nevertheless, vaccines could have a tremendously important role to play because we are dealing with a zoonosis: *Toxoplasma* is not passed from human to human but instead is acquired from animals. Thus successful vaccination of these animals would be very effective in preventing human infection. The fact that *Toxoplasma* infection in sheep is a major cause of abortion in these animals adds a commercial incentive for the farmer, which provides further motivation for development of an animal vaccine. This could take the form of a recombinant subunit vaccine based on heterologous expression of the genes for one or more of Toxoplasma's major antigens. Alternatively, the relative ease with which the parasite can be genetically manipulated makes possible the creation of a specifically engineered, attenuated mutant that would give rise to a high level of immunity without being pathogenic or transmissible from one host to another. Such a strain might be an excellent replacement for the S48 strain that comprises the current sheep vaccine because the genetic basis of its inability to differentiate into tissue cysts is not known.

Chemotherapy

Existing drugs are far from benign and/or are limited in their efficacy, especially against the chronic bradyzoite stage. The recent explosion in our understanding of the parasite's metabolism and cell biology, combined with concomitant improvements in the area of drug design, bode extremely well for the development of safer and more effective drugs. The fact that such drugs might also work on related Apicomplexan parasites such as *Eimeria* (the cause of chicken coccidiosis) and *Plasmodium* (the cause of human malaria) provides further incentive for efforts in this direction.

FURTHER READING

Besteiro, S., Michelin, A., Poncet, J., Dubremetz, J. F., & Lebrun, M. (2009). Export of a Toxoplasma gondii rhoptry neck protein complex at the host cell membrane to form the moving junction during invasion. *PLoS Pathogens*, *5*, e1000309.

Blanchard, N., Gonzalez, F., Schaeffer, M., Joncker, N. T., Cheng, T., *et al.* (2008). Immunodominant, protective response to the parasite Toxoplasma gondii requires antigen processing in the endoplasmic reticulum. *Nature Immunology*, *9*, 937–944.

Boyle, J. P., Rajasekar, B., Saeij, J. P., *et al.* (2006). Just one cross appears capable of dramatically altering the population biology of a eukaryotic pathogen like *Toxoplasma gondii*. *Proceedings of the National Academy of Sciences of the United States of America*, *103*, 10514–10519.

Buzoni-Gatel, D., & Werts, C. (2006). *Toxoplasma gondii* and subversion of the immune system. *Trends in Parasitology*, *22*, 448–452.

Carruthers, V., & Boothroyd, J. C. (2007). Pulling together: An integrated model of *Toxoplasma* cell invasion. *Current Opinion in Microbiology*, *10*, 83–89.

Denkers, E. Y., & Butcher, B. A. (2005). Sabotage and exploitation in macrophages parasitized by intracellular protozoans. *Trends in Parasitology*, *21*, 35–41.

Dubey, J. P., Lindsay, D. S., & Speer, C. A. (1998). Structures of *Toxoplasma gondii* tachyzoites, bradyzoites, and sporozoites and biology and development of tissue cysts. *Clinical Microbiology Reviews*, *11*, 267–299.

Gajria, B., Bahl, A., Brestelli, J., Dommer, J., Fischer, S., *et al.* (2008). ToxoDB: An integrated Toxoplasma gondii database resource. *Nucleic Acids Research*, *36*, D553–D556.

Hakansson, S., Charron, A. J., & Sibley, L. D. (2001). Toxoplasma evacuoles: A two-step process of secretion and fusion forms the parasitophorous vacuole. *The EMBO Journal*, *20*, 3132–3144.

Holland, G. N., & Lewis, K. G. (2002). An update on current practices in the management of ocular toxoplasmosis. *American Journal of Ophthalmology*, *134*, 102–114.

Kasper, L., Courret, N., Darche, S., *et al.* (2004). *Toxoplasma gondii* and mucosal immunity. *International Journal for Parasitology*, *34*, 401–409.

Khan, A., Taylor, S., Su, C., *et al.* (2005). Composite genome map and recombination parameters derived from three archetypal lineages of *Toxoplasma gondii*. *Nucleic Acids Research*, *33*, 2980–2992.

Lehmann, T., Marcet, P. L., Graham, D. H., Dahl, E. R., & Dubey, J. P. (2006). Globalization and the population structure of *Toxoplasma gondii*. *Proceedings of the National Academy of Sciences of the United States of America*, *103*, 11423–11428.

Montoya, J. G., & Liesenfeld, O. (2004). Toxoplasmosis. *Lancet*, *363*, 1965–1976.

Nishi, M., Hu, K., Murray, J. M., & Roos, D. S. (2008). Organellar dynamics during the cell cycle of Toxoplasma gondii. *Journal of Cell Science*, *121*, 1559–1568.

Remington, J. S., Thulliez, P., & Montoya, J. G. (2004). Recent developments for diagnosis of toxoplasmosis. *Journal of Clinical Microbiology*, *42*, 941–945.

Roos, D. S. (2005). Genetics. Themes and variations in apicomplexan parasite biology. *Science*, *309*, 72–73.

Saeij, J. P., Boyle, J. P., Coller, S., *et al.* (2006). Polymorphic secreted kinases are key virulence factors in toxoplasmosis. *Science*, *314*, 1780–1783.

Saeij, J. P., Coller, S., Boyle, J. P., Jerome, M. E., White, M. W., & Boothroyd, J. C. (2007). Toxoplasma co-opts host gene expression by injection of a polymorphic kinase homologue. *Nature*, *445*, 324–327.

Sher, A., Collazzo, C., Scanga, C., Jankovic, D., Yap, G., & Aliberti, J. (2003). Induction and regulation of IL-12-dependent host resistance to *Toxoplasma gondii*. *Immunologic Research*, *27*, 521–528.

Sibley, L. D., & Ajioka, J. W. (2008). Population Structure of Toxoplasma gondii: Clonal expansion driven by infrequent recombination and selective sweeps. *Annual Review of Microbiology*, *62*, 329–351.

Straub, K. W., Cheng, S. J., Sohn, C. S., & Bradley, P. J. (2009). Novel components of the Apicomplexan moving junction reveal conserved and coccidia-restricted elements. *Cellular Microbiology*, *11*, 590–603.

Su, C., Evans, D., Cole, R. H., Kissinger, J. C., Ajioka, J. W., & Sibley, L. D. (2003). Recent expansion of *Toxoplasma* through enhanced oral transmission. *Science*, *299*, 414–416.

Taylor, S., Barragan, A., Su, C., *et al.* (2006). A secreted serine–threonine kinase determines virulence in the eukaryotic pathogen *Toxoplasma gondii*. *Science*, *314*, 1776–1780.

Yamamoto, M., Standley, D. M., Takashima, S., Saiga, H., Okuyama, M., *et al.* (2009). A single polymorphic amino acid on Toxoplasma gondii kinase ROP16 determines the direct and strain-specific activation of Stat3. *The Journal of Experimental Medicine*, *206*, 2747–2760.

RELEVANT WEBSITE

www.toxodb.org/toxo/. – ToxoDB.

Chapter 29

Trypanosomes

L.V. Kirchhoff[1], C.J. Bacchi[2], F.S. Machado[3], H.D. Weiss[4], H. Huang[4], S. Mukherjee[4], L.M. Weiss[4] and H.B. Tanowitz[4]

[1]*University of Iowa, Iowa City, IA, USA*

[2]*Pace University, New York, NY, USA*

[3]*Federal University of Minas Gerais, Belo Horizonte, Minas Gerais, Brazil*

[4]*Albert Einstein College of Medicine, Bronx, NY, USA*

Chapter Outline

Abbreviations 397
Defining Statement 397
Trypanosoma Cruzi 397
Transmission 398
Organism and Life Cycle 398
Epizootiology and Epidemiology 400
Clinical Manifestations 401
Pathogenesis 402
Diagnosis 403
Treatment 403
Prevention 404
Trypanosoma Brucei 404
Organism and Life Cycle 404
Epidemiology 405
Pathology and Pathogenesis 406
Clinical Manifestations 407
Diagnosis 407
Treatment 407
Prophylaxis and Prevention 409
Further Reading 409

ABBREVIATIONS

CNS Central nervous system
DFMO Difluoromethylornithine
ELISA Enzyme-linked immunosorbent assay
ES Expression site
ESAG ES-associated gene
Hsp70 Heat shock protein 70
ODC Ornithine decarboxylase
PARP Procyclic acid-rich protein
PCR Polymerase chain reaction
RBBB Right bundle branch block
VSG Variant surface glycoprotein

DEFINING STATEMENT

The genus *Trypanosoma* consists of approximately 20 species. However, only *Trypanosoma cruzi*, the cause of American trypanosomiasis, and *Trypanosoma brucei gambiense* and *T. b. rhodesiense*, the causes of African trypanosomiasis, have been linked to human disease. In Africa, there are a number of *Trypanosoma* species, including *T. b. rhodesiense*, which cause trypanosomiasis in wild and domesticated animals and are of immense economic importance.

TRYPANOSOMA CRUZI

Trypanosoma cruzi is the protozoan parasite that causes American trypanosomiasis or Chagas' disease. It is enzootic and endemic in all Latin American countries outside the Caribbean, including Mexico, and it is an important cause of chronic heart and digestive diseases in these nations. Studies of autopsied Native American mummies indicate that Chagas' disease in humans dates back to the pre-Colombian era, and given the widespread distribution of mammalian reservoirs (e.g., opossums, armadillos, and many dozens of others), it is reasonable to assume that humans have been infected since they first invaded enzootic regions roughly 14000 years ago. Seroepidemiologic and clinical studies indicate that approximately 8 million persons are chronically infected with *T. cruzi* in endemic countries and roughly 20000 die annually as a result of the infection. Several autochthonous cases of *T. cruzi* infection have been reported in the United States since 1955, and small numbers of infected blood donors who have not visited endemic countries have been identified since serologic screening of the US blood supply began in early 2007. Naturally infected dogs as well as a wide variety of wild mammals have also been described in the southern and southwestern United States, and infected insect vectors have been found in these areas as well.

Transmission

In endemic areas, transmission is most common among the rural poor whose dwellings become infested with infected vectors due to the proximity of sylvatic reservoirs. Vector-borne transmission of *T. cruzi* is contaminative, since the infective parasites are in the excreta of the insects. During a blood meal, insect feces containing metacyclic trypomastigotes by chance find their way onto mucosal surfaces, the conjunctivas, or breaks in the skin. There they enter susceptible cells of the new host and establish infection. Control programs directed at decreasing contact with insect vectors through housing improvement, insecticide spraying, and education have met with considerable success in many endemic areas. The transmission of *T. cruzi* by blood transfusion has historically been the second most important mode of spread of the parasite. This was a major public health problem in many endemic regions, but its importance has been reduced markedly in recent years through widespread implementation of serologic screening of donated blood. Several cases of transfusion-associated transmission of *T. cruzi* have been reported in the United States and Canada, resulting from the transfusion of blood donated by asymptomatic *T. cruzi*-infected immigrants from Latin America. Other modes of transmission include congenital transmission, organ transplantation, laboratory accidents, ingestion of contaminated food or drink, and breastfeeding.

Organism and Life Cycle

T. cruzi has a complex life cycle in which four morphologically and biochemically distinct forms exist (Figure 1). In mammalian hosts, there are two forms: extracellular nondividing trypomastigotes (blood forms) (Figures 2 and 3) and intracellular dividing amastigotes (Figures 4 and 5). During a blood meal, insect vectors ingest blood forms, which then transform into epimastigotes in the midgut. After 3–4 weeks, infective but nondividing metacyclic trypomastigotes are present in the hindgut of the vector. These forms are then deposited with the excreta of the vector during subsequent blood meals. Transmission to a new mammalian host occurs when these infective parasites contaminate susceptible tissues. The infective trypomastigotes parasitize cells by direct penetration or through phagocytosis, and then transform into amastigotes. The mechanisms by which

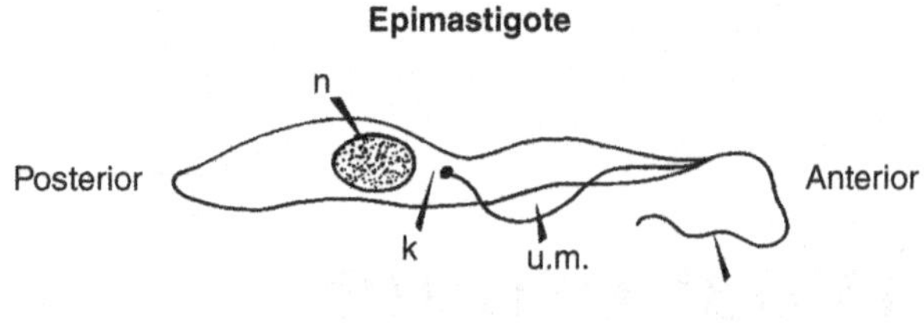

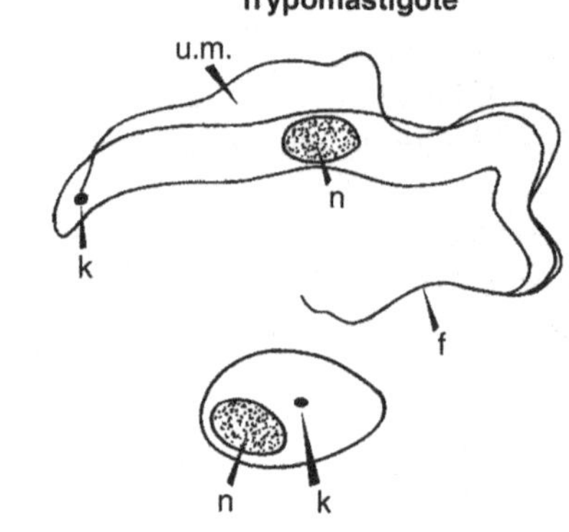

FIGURE 1 Life-stage forms of *Trypanosoma cruzi*. Note the kinetoplasts; see Figure 5 for the ultrastructure of this organelle. The trypomastigotes measure 15–20 μm in length, the epimastigotes measure 20 μm length, and the amastigotes measure 3–4 μm in diameter.

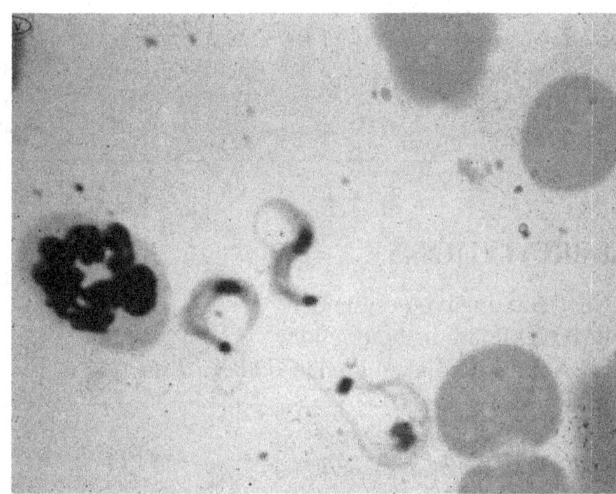

FIGURE 2 Trypomastigotes of *Trypanosoma cruzi* in the bloodstream. Note the 'C' shape. Courtesy of the American Society of Tropical Medicine and Hygiene/Zaiman 'A Presentation of Pictorial Parasites.'

the trypomastigotes gain entry into host cells are not entirely understood. Studies have implicated the enzyme *trans*-sialidase in this process. Penetrin, another parasite molecule, binds to heparin sulfate receptors on nonphagocytic host cells, thereby promoting adhesion and subsequent parasite invasion. Several host cell receptors that aid in the entry of the parasite into cells have been described. In the parasitophorous vacuole, amastigotes synthesize a hemolysin that lyses the vacuolar membrane, thus allowing the

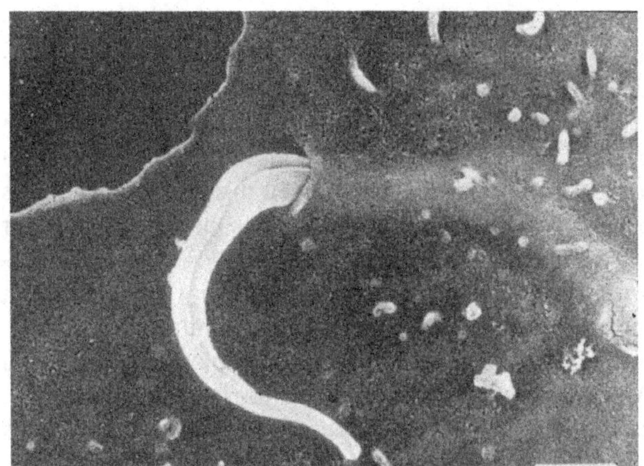

FIGURE 3 Scanning electron micrograph of a trypomastigote of *Trypanosoma cruzi* invading a mammalian host cell.

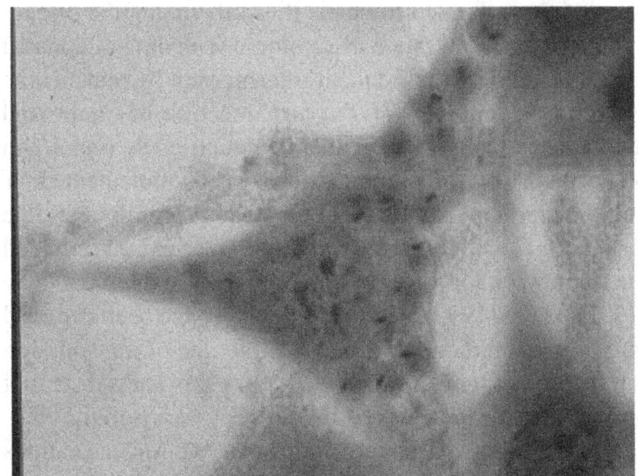

FIGURE 4 Giemsa-stained endothelial cells infected with *Trypanosoma cruzi*. Note the numerous amastigotes.

parasite to escape the cytocidal mechanisms of the cell. Once in the cytoplasm of the host cells, amastigotes (Figures 4 and 5) undergo binary fission and as sizable numbers accumulate, the cell is overwhelmed and transformation to trypomastigotes occurs. Although any nucleated mammalian cell can be invaded by this parasite, cells of the reticuloendothelial, nervous, and muscle systems appear to be favored. In addition, recent evidence suggests that adipose tissue and adipocytes are readily invaded by trypomastigotes. The parasitized cells eventually rupture, releasing amastigotes and trypomastigotes that infect adjacent cells and, importantly, trypomastigotes disseminate through the lymphatics and the bloodstream, ultimately infecting a wide variety of tissues, including cardiac and skeletal muscles, as well as cells of the reticuloendothelial and nervous systems. Both cell necrosis and apoptosis have been reported to be associated with *T. cruzi* infection. The chromosomal complement of *T. cruzi* is approximately 50 Mb or roughly ten times the amount of DNA found in

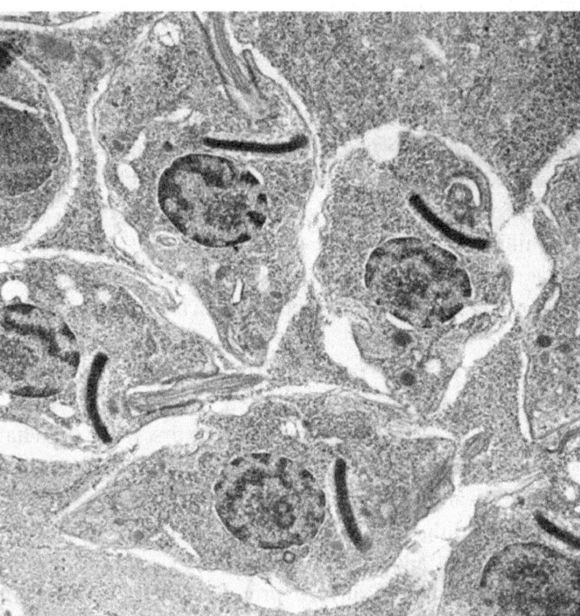

FIGURE 5 Transmission electron micrograph of parasitized myoblast cell line. Note the amastigotes in the cytoplasm and their prominent disk-shaped kinetoplasts.

Escherichia coli. Pulse-field gel electrophoresis reveals at least 20, and as many as 40, discrete chromosomal bands, ranging in size from 200 to 2000 kb. *T. cruzi* molecular karyotypes are somewhat strain-dependent (i.e., the reference strain CL Brener contains approximately 60–80 Mb DNA). Among characterized strains, there appears to be only one that contains minichromosomes, in sharp contrast to *Trypanosoma brucei*. The genome of *T. cruzi* appears to be diploid in all stages. Homologous chromosomes, however, can differ markedly in size, and chromosomes containing conserved linkage groups can vary extensively in size among isolates. *T. cruzi* exhibits significant polymorphism at a number of characterized alleles, far in excess of that seen in higher eukaryotes. It appears that multiplication of *T. cruzi* is exclusively, or nearly exclusively, clonal without any evidence of a sexual stage. Beginning in the 1980s, numerous genes of *T. cruzi* were cloned and characterized, and the complete DNA sequencing of the CL Brener strain of *T. cruzi* was recently completed. Many *T. cruzi* genes are arranged in tandem arrays of nearly identical copies, as is the case with GAPDH, histone H1, and heat shock protein 70 (Hsp70). Importantly, *T. cruzi* genes lack introns and all mature mRNAs have a highly conserved 5′ cap (spliced leader) that is added to pre-mRNAs by trans-splicing. *T. cruzi* mRNAs contain poly-A tails, but there is no highly conserved poly-A addition site consensus sequence, as is the case with other eukaryotes. The classification of *T. cruzi* strains has been accomplished by a variety of methods, including isoenzyme differentiation (zymodemes), restriction endonuclease digestion of kinetoplast DNA (schizodemes), and sequencing of noncoding segments of rRNA genes. There is considerable correlation among the groupings determined by the various approaches. These observations are consistent with the concept that *T. cruzi* consists of many clonal lineages with little sexual recombination. Moreover, rRNA sequence data also suggest that the clonal lineages come under two main branches (*T. cruzi* I and *T. cruzi* II), with a very ancient split between them. Assuming a constancy of the molecular clock, the divergence appears to have occurred contemporaneously with the divergence of reptiles and amphibians, which predated the first mammals.

Analyses of the *T. cruzi* CL Brener strain have revealed that the diploid genome contains a predicted 22570 proteins, of which 12570 represent allelic pairs. More than half of the genome consists of repeated sequences, such as retrotransposons and genes for large families of surface molecules, including the *trans*-sialidase gp85 superfamily, mucins, gp63s, and a large novel family (>1300 copies) of mucin-associated surface protein genes. Analyses of the *T. cruzi*, *T. brucei*, and *Leishmania major* genomes imply differences from other eukaryotes with respect to DNA repair and initiation of replication and reflect their unusual mitochondrial DNA, which may offer targets for intervention. Proteomic analysis of the four life cycle stages of *T. cruzi* revealed that peptide mapping to 2784 proteins in 1168 protein groups from the annotated *T. cruzi* genome was identified across all stages. The different parasite stages use distinct energy sources depending on the host in which they are found – histidine for stages in the insect vectors and fatty acids by trypomastigotes and amastigotes in the mammalian hosts.

Epizootiology and Epidemiology

T. cruzi is found only in the southern United States, Mexico, and the nations of Central and South America. There is no Chagas' disease in the Caribbean Islands. Infection with this parasite is a zoonosis and the involvement of humans in the life cycle essentially has nothing to do with the perpetuation of the parasite in nature. The parasite primarily infects both wild and domestic mammals as well as insects. The triatomine insect vectors of *T. cruzi* are distributed from the southern United States to central Argentina. Transmission of the parasite typically occurs through infected vectors and nonhuman mammalian hosts in hollow logs, palm trees, burrows, and other animal shelters. Large numbers of insect vectors can also be found near houses in piles of roof tiles, wood, and vegetation. *T. cruzi* has been found in more than 100 species of domestic and wild mammals, in a range similar to that of its insect vectors. Opossums, nonhuman primates, wood rats, armadillos, raccoons, dogs, and cats all are typical hosts, but *T. cruzi* is not a problem in livestock.

Nontypical hosts, such as polar bears, can become infected when held in zoos in areas in which *T. cruzi* is enzootic. This lack of species specificity, combined with the fact that infected mammalian hosts have lifelong parasitemias, results in an enormous domestic and sylvatic reservoir in enzootic areas. Humans have become part of the cycle of *T. cruzi* transmission as farmers and ranchers open new lands in enzootic areas. When this development takes place, the vectors, known variously as the 'kissing bugs' or 'assassin bugs' (e.g., *Rhodnius prolixus*, *Triatoma infestans*, *Panstrongylus megistus*, and many other species), invade the nooks and crannies of the primitive wood, mud-walled, and stone houses that are typical of rural Latin America. In this manner, the vectors become domiciliary, establishing a cycle of transmission involving humans and peri-domestic mammals that is independent of the sylvatic cycle. Historically, Chagas' disease has been an infection in poor people living in rural environments. However, over the decades, an enormous number of *T. cruzi*-infected people have migrated to cities, thus urbanizing the disease and, prior to the initiation of serologic screening of donated blood, resulting in its frequent transmission by transfusion.

The epidemiology of *T. cruzi* infection has improved markedly in many of the endemic countries as vector and blood bank control programs have been implemented. A major international vector eradication program in the 'Southern Cone' countries of South America has provided the framework for much of this progress. To date, Uruguay (1997), Chile (1999), and Brazil (2006) have been declared free of transmission. In Argentina, the rate of transmission is a fraction of what it was 15–20 years ago, and substantial progress has been achieved in Paraguay and Bolivia. Similar programs have been implemented in the Andean nations and in Central America. The obstacles hindering the elimination of *T. cruzi* transmission to humans are economic and political, and no technological breakthroughs are necessary for overall control of the problem.

Approximately 70% of people who harbor *T. cruzi* chronically never develop associated cardiac or gastrointestinal symptoms. In the past, the high frequency of sudden death among young adults in some areas was attributed to rhythm disturbances due to chronic Chagas' disease and in one highly endemic area of Brazil, chagasic cardiac disease was found to be the most frequent cause of death in young adults. There is considerable geographic variation in the relative prevalence of cardiac and gastrointestinal disease in patients with chronic *T. cruzi* infection. It is not known if this variable clinical expression is caused primarily by parasite strain or host factors. In this regard, studies have implicated host genetic factors as important variables in host response to infection.

The number of people in the United States with chronic *T. cruzi* infections has increased substantially in recent decades. Current estimates put the number of immigrants from Chagas-endemic areas living in the United States at 13 million. Roughly, 80000–120000 of these persons are thought to be chronically infected with *T. cruzi*. Their presence creates a risk of transfusion-associated transmission of *T. cruzi*, and seven such cases have been reported in the United States and Canada. Almost all the reported cases occurred in immunocompromized patients in whom the diagnosis of *T. cruzi* was made because the transfusion recipients became severely ill. This suggests that additional cases have occurred in immunocompetent patients and have gone unnoticed because they were more able to control the acute infection. However, in the United States, this risk has been essentially eliminated by serologic screening of donated blood, which started in January 2007. At the time of this writing (September 2008), 85–90% of the 15.2 million units donated annually are being screened for Chagas', and the confirmed prevalence among donors is about 1 in 29000. Moreover, five instances of transmission of *T. cruzi* through organ transplantation have been reported in the United States.

Clinical Manifestations

Acute Chagas' disease

Seroepidemiologic studies in endemic areas indicate that most seropositive patients have no recollection of having had acute Chagas' disease and do not carry a specific diagnosis. This results from the generally mild nature of acute Chagas' disease, as well as from the fact that the persons most at risk for acquiring *T. cruzi* infection have little access to medical care. Some infected persons may develop severe symptoms after an incubation period of 7–14 days. This has been documented with vector-borne infection as well as infections acquired by blood transfusions or laboratory accidents. Signs and symptoms include fever, chills, nausea, vomiting, rash, diarrhea, and meningeal irritation. A chagoma (a raised inflammatory lesion at the site of parasite entry), Romana's sign (unilateral periorbital edema), conjunctivitis, lymphadenopathy, and hepatosplenomegaly have all been described in patients with acute Chagas' disease. Laboratory abnormalities include anemia, thrombocytopenia, and elevated liver and cardiac enzymes. The diagnosis of acute Chagas' disease is primarily parasitologic, as motile trypomastigotes can often be found in blood and cerebrospinal fluid. Serologic tests for parasite-specific IgG are often negative during this stage, and assays for IgM have not been standardized and are not widely available. During the acute phase of the illness, asynchronous cycles of parasite multiplication, cell destruction, and infection of new cells occur. Myocarditis and cardiomegaly, sometimes associated with congestive heart failure, are present in some patients and may be more common than was previously appreciated. The appearance of arrhythmias, heart block, or congestive heart failure in the setting of acute *T. cruzi* infection is a poor prognostic indicator. It is not known if the severity of acute Chagas' disease is related to the likelihood of development of the cardiac and gastrointestinal manifestations of chronic symptomatic Chagas' disease years later. A small percentage of acutely infected patients, often children, die of acute myocarditis or meningoencephalitis. As antibodies appear, the parasitemia wanes, and almost all infected persons resolve the acute phase of the illness in 2–4 months, but remain infected for life. These patients then enter the indeterminate phase of the illness, which is characterized by a lack of symptoms, easily detectable antibodies to a variety of *T. cruzi* antigens, and low parasitemias. Congenital Chagas' disease is another example of acute infection. It occurs in about 5% of infants born to mothers with chronic *T. cruzi* infections. These babies are most often asymptomatic but can present with fever, jaundice, and occasionally seizures, and may be difficult to distinguish from infants with other congenital infections such as toxoplasmosis. Children with congenital disease may also have cardiac and gastrointestinal manifestations.

Chronic cardiopathy

Chronic cardiac Chagas' disease may present insidiously as congestive heart failure or abruptly with arrhythmias and/or thromboembolic events. Dilated congestive cardiomyopathy is an important manifestation of chronic Chagas' disease that typically occurs years or even decades after acute infection. Apical aneurysm of the left ventricle is one of the hallmarks of this disease. Chronic chagasic heart disease is associated with myonecrosis and myocytolysis. Contraction band necrosis occurs after transient hypoperfusion followed by reperfusion, as occurs with local spasm of the coronary microvasculature. Focal and diffuse areas of myocellular hypertrophy are observed with or without inflammatory infiltrates, and fibrosis replacing previously damaged myocardial tissue is evident (Figure 6). The destruction of conduction tissue results in atrioventricular and intraventricular conduction abnormalities. In areas

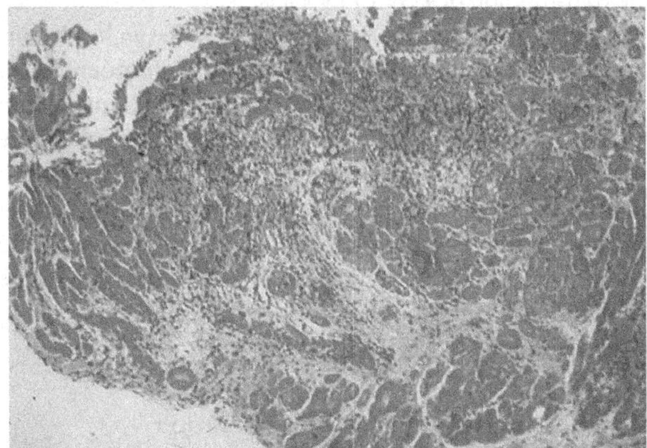

FIGURE 6 Endomyocardial biopsy from a patient with chronic Chagas' disease. Note the inflammation and fibrosis. Courtesy of Dr. Alain C. Borczuk, North Shore University Hospital, Manhasset, NY.

where the disease is endemic, the presence of right bundle branch block (RBBB), associated with an anterior fascicular block, is highly suggestive of chagasic cardiopathy. Conduction defects may necessitate the placement of a pacemaker.

Chronic gastrointestinal manifestations

Chagas' disease is associated with the loss of neurons of the myenteric plexus. Lesions of the autonomic nervous system produce functional disorders in hollow visceral muscular organs. Initial damage to the nervous system occurs during acute *T. cruzi* infection and may be the consequence of the increase in local inflammatory mediators. Further neuronal loss occurs slowly during the chronic phase of the infection. The development of clinically manifested chronic gastrointestinal disease may take several decades, and the determinants of this progression are unknown.

The colon is frequently affected in chronic Chagas' disease. Although the entire colon may be enlarged, dilatation is often confined to the sigmoid colon. Constipation is a common complaint. Symptomatic patients may have disorders of motility that antedate by years the radiological diagnosis of megacolon. In patients with chagasic megacolon, there is a reduction in the number of ganglia throughout the colon. The esophagus is also frequently affected in patients with long-standing *T. cruzi* infection. Patients with megaesophagus may have less than 5% of the normal number of ganglia and this loss of neurons is uniform along the length of the organ, including the grossly normal abdominal portion of the esophagus. Typical symptoms of megaesophagus include dysphagia, odynophagia, chest pain, cough, and regurgitation. Mega-gall bladder and mega-ureter also have been reported in patients with chronic *T. cruzi* infections.

Other clinical syndromes

Transfusion-associated T. cruzi infection

The transmission of *T. cruzi* by transfusion of blood donated by asymptomatic persons who harbor the parasite has historically been a major public health problem in the countries in which Chagas' disease is endemic. This mode of transmission of *T. cruzi* has been essentially eliminated in most of the endemic nations as mandatory programs for screening donated blood have been implemented. As noted earlier, several transfusion-associated instances of *T. cruzi* transmission have been reported in the United States as a consequence of the immigration of infected persons, and widespread screening of donated blood began here in early 2007.

Immunosuppression and transplantation

The incidence of reactivation of *T. cruzi* infection in immunosuppressed people is not known. In view of the possibility of acute reactivation of *T. cruzi*, patients who are at risk of being infected and for whom immunosuppressive treatment is being planned, either as primary or as posttransplantation therapy, should be tested serologically and, if positive, should be monitored closely while immunosuppressed for evidence of reactivation. Several dozen HIV-infected patients with reactivation of chronic *T. cruzi* infections have been described. Interestingly, many of these patients developed cerebral lesions radiographically similar to those observed in AIDS patients with cerebral toxoplasmosis. As with the latter, cerebral lesions do not occur in immunocompetent patients chronically infected with *T. cruzi*. Finally, more than 100 patients with end-stage Chagas' heart disease have undergone cardiac transplantation in Brazil, and the procedure has been performed in a couple of dozen *T. cruzi*-infected people in the United States. Because of the intense postoperative immunosuppression caused in the early series of patients transplanted in Brazil, life-threatening reactivation of *T. cruzi* infection occurred in some of these patients. However, lower doses of immunosuppressive drugs have been used for many years now and reactivation of *T. cruzi* infection has become much less of a problem. The overall survival of chagasic patients receiving cardiac transplants is better than the total group of patients transplanted for other reasons.

Pathogenesis

Animal models have been used extensively to evaluate the pathogenesis of chagasic heart disease. For example, alterations in cardiac choline acetyltransferase, acetylcholine, norepinephrine, and β-adrenergic adenylate cyclase complex in association with chronic *T. cruzi* infection have been described in experimental animals. The expression of cytokines and NOS2 in the myocardium has been suggested as a possible cause of myocardial dysfunction. Morphological changes during acute *T. cruzi* infection may also be important in the pathogenesis of the myocardial lesions. For example, reduced numbers of autonomic ganglia are associated with clinically manifest cardiac as well as gastrointestinal Chagas' disease, and as noted, this may be a direct result of destructive processes that are part of the early phase of the illness. The immunology of *T. cruzi* infection has been studied extensively in animal models and in humans, and a large portion of these efforts has been focused on a possible role for autoimmunity in the pathogenesis of chagasic lesions. There is evidence from studies in mice that both humoral and cell-mediated components of the immune system are important in host resistance, as is genetic background. CD8+ T cells are important in the pathogenesis of *T. cruzi* infection. The possible roles of cytokines in the pathogenesis of *T. cruzi* infections have also been the subject of intense investigation. Cytokines are elevated during acute murine infection and may contribute to the pathogenesis of the disease. Increased levels of

IL-1β, TNF-α, and IL-6, expressed in infected endothelial cells, result in leukocyte recruitment, coagulation, and smooth muscle cell proliferation. The role of cytokines and nitric oxide in the killing of the parasite has received considerable attention as well. Nitric oxide, IFN-γ, TNF-α, and IL-12 all appear to be involved in intracellular killing. Recently, the 21-amino acid peptide endothelin-1 and the eicosanoid thromboxane A_2 have also been implicated in the pathogenesis of *T. cruzi* infection. Trypanosomes contain proteases, gelatinases, and collagenases capable of degrading native type I collagen, heat-denatured type I collagen (gelatin), and native type IV collagen. Proteolytic activities against laminin and fibronectin are also present. These enzymes may play important roles in the degradation of extracellular matrix and the subsequent tissue invasion by *T. cruzi*. It has been proposed that the degradation of the collagen matrix, evident in acute murine Chagas' disease, may result in chronic pathology such as apical thinning of the left ventricle.

Diagnosis

The diagnosis of acute *T. cruzi* infection is generally made by the detection of parasites. Active trypomastigotes can frequently be seen by microscopic examination of fresh anticoagulated blood or buffy coat, and organisms can often be seen in Giemsa-stained thin and thick blood smears as well. If the organisms cannot be detected by these approaches, one may inoculate blood specimens or buffy coat into specialized liquid medium or intraperitoneally into mice. The disadvantages of these approaches are their lack of sensitivity and the fact that the parasites are usually not seen in positive cultures or infected mice for several weeks. Assays based on polymerase chain reaction (PCR) have been described, and the results obtained suggest that this approach may be the most sensitive method for detecting acute *T. cruzi* infections. When acute Chagas' disease is suspected in an immunocompromized patient and these methods fail to demonstrate the presence of parasites, additional tissue specimens should be examined. These patients can pose a difficult diagnostic problem because they may present with fulminant clinical disease and low parasitemias that cannot be readily detected. Surprisingly, parasites can sometimes be seen in atypical sites, such as pericardial fluid, bone marrow, brain, skin, and lymph nodes, and thus, these tissues should also be investigated when indicated.

The diagnosis of chronic Chagas' disease is generally based on detecting specific antibodies that bind to *T. cruzi* antigens. Several dozen serological assays based on a variety of technologies are used in Latin America for detecting antibodies, such as the indirect immunofluorescence test and the enzyme-linked immunosorbent assay (ELISA). These and other conventional serologic assays are used widely for clinical diagnosis and for screening donated blood, as well as in epidemiological studies. False-negative and false-positive reactions, however, have been a persistent problem with many of these assays. Presently, only two tests available in the United States have been approved by the FDA for clinical testing. The first is a lysate-based ELISA (Hemagen Chagas Kit, Hemagen Diagnostics, Inc., Columbia, MD) and the other is an ELISA based on recombinant antigens (Chagatest Elisa Recombinante; Laboratorios Wiener, Rosario, Argentina). Screening of the US blood supply is currently being done with a lysate-based ELISA (Ortho *T. cruzi* ELISA Test System; Ortho-Clinical Diagnostics, Raritan, NJ), but this test has not been cleared for clinical use. An automated blood screening assay based on four chimeric recombinant antigens is being developed (PRISM Chagas Assay; Abbott Laboratories, Abbott Park, IL) and a test based on a blot format is being developed with the same antigens for confirmatory testing (Abbott Chagas Immunoblot Assay).

An immunoprecipitation assay based on iodinated *T. cruzi* proteins (RIPA), developed by one of the authors (LVK), has been shown to be highly specific as well as sensitive when used in clinically and geographically diverse groups of infected people. The RIPA currently is being used as the confirmatory assay to test all donor samples that are positive in the Ortho screening assay, and it is also available for clinical testing.

The detection of chronic infection by testing for parasite antigens in blood and urine has been studied, but this approach has not achieved results comparable to those obtained by serologic methods. PCR-based assays for detecting chronic *T. cruzi* infections have been studied extensively, but their usefulness has not been established definitively. It would appear that this approach is particularly suited for the task of detecting the low number of parasites circulating in the blood of chronically infected patients, but sampling issues and the possibility that parasitemias may in fact be intermittent might limit the sensitivity of the assays. Moreover, false-positive results have been an issue in some laboratories. The most likely niche for PCR-based assays is in diagnosing congenital *T. cruzi* infections in the days and weeks after birth when serological tests are made useless by the presence of maternal anti-*T. cruzi* antibodies.

Treatment

The treatment of *T. cruzi* is unsatisfactory. Nifurtimox (Lampit, Bayer 2502) and benznidazole (Rochagan, Roche 7–1051), the two drugs available for treating this infection, lack efficacy, must be taken for extended periods, and may cause severe side effects. Both drugs reduce the severity of acute Chagas' disease, and it is thought that about 70% of acute patients treated with a full course of either nifurtimox or benznidazole are cured parasitologically. This cure rate decreases as a function of the time patients have been infected and may be less than 10% in persons who have harbored the parasite for many years. There are no convincing data from properly controlled trials that treatment with either nifurtimox or benznidazole is beneficial in persons with long-standing infections. Chagas' experts in both Brazil and Argentina currently recommend specific treatment only for patients with acute and congenital *T. cruzi* infections and for chronically infected children. Therapy for adults assumed to have long-standing infections is not recommended, regardless of clinical status. A large trial designed to address the efficacy of benznidazole is under way (the BENEFIT Multicentric Trial).

Allopurinol and several antifungal azoles have been shown to have some anti-*T. cruzi* activity in in vitro experiments and in animal studies, but none has a level of activity that would warrant its use in place of nifurtimox or benznidazole.

Surgery is often necessary for gastrointestinal megasyndromes. As noted, persons with severe Chagas' heart disease may benefit from heart transplantation. Stem cell transplantation currently is being evaluated in patients with severe heart failure due to *T. cruzi* infection.

Prevention

No vaccine or chemoprophylactic drug is available for preventing the transmission of *T. cruzi* in any context. The elimination of domiciliary vectors by spraying residual insecticides, improving housing conditions, and educating populations at risk are the linchpins of the widely successful programs for elimination of *T. cruzi* transmission to humans in the endemic countries. Serologic screening of blood donors is practiced throughout most of the endemic range, and as a consequence, transmission of *T. cruzi* by transfusion has largely been eliminated. The risk of acquiring *T. cruzi* infection faced by short-term travelers to endemic regions is extremely low, as only three such instances have been reported. Nonetheless, persons traveling to rural areas in endemic countries should avoid sleeping in primitive dwellings and should use an insect repellent. Laboratorians who work with parasites or infected vectors should use appropriate protective equipment.

TRYPANOSOMA BRUCEI

Human African trypanosomiasis, or 'sleeping sickness,' is caused by two subspecies of hemoflagellate protozoa: *T. b. gambiense* and *T. b. rhodesiense*. The former causes the West African or gambiense form and the latter the East African or rhodesiense form. Although these subspecies cause similar diseases, the West African form usually evolves over months to years and ends fatally if it is not treated. The East African form usually kills its host in weeks to months. These diseases exist in Africa wherever the various species of tsetse flies belonging to the genus *Glossina* are found.

Organism and Life Cycle

T. b. gambiense and *T. b. rhodesiense* are pleomorphic flagellates 15–30 μm in length and 1.5–3.5 μm in breadth. The two subspecies are morphologically indistinguishable. There are two forms of trypomastigotes that circulate in the bloodstream: long slender organisms that are capable of dividing and short stumpy forms that are thought to be nondividing parasites that are infective for tsetse flies. There are no intracellular forms. At various stages of the disease, trypomastigotes may be found in peripheral blood, lymphatics, lymph nodes, cerebrospinal fluid, and neural tissue (Figure 7). Other than humans, there is no important reservoir host for *T. b. gambiense*, whereas *T. b. rhodesiense* is primarily a parasite of wild game animals. In the tsetse fly, trypomastigote forms ingested with a blood meal settle in the posterior midgut, where they multiply by binary fission for approximately 7–10 days and then migrate anteriorly to the foregut, where they remain for 2–3 weeks. Finally, they enter the salivary glands, continue to replicate, and after several cycles of division, transform into infective metacyclic trypomastigote forms. These organisms are inoculated the next time a mammalian host is bitten, and once in a human host, trypomastigotes multiply by binary fission in the blood, lymph, and other extracellular spaces. The central nervous system (CNS) is eventually invaded and multiplication continues there as well.

The haploid genome size of *T. brucei* spp. is approximately 35 Mb, although there is up to 14% variation in isolates of the same subspecies and up to 29% between the two subspecies. There is a minimum of seven resolvable chromosome pairs on pulse-field gel electrophoresis in the size range of 1.1–6 Mb. Homologous chromosomes, when probed in Southern blots, can differ in size by up to 20%. In addition to the large chromosome pairs, *T. brucei* contains approximately 100 linear minichromosomes ranging in size from 50 to 150 kb. Minichromosomes contain

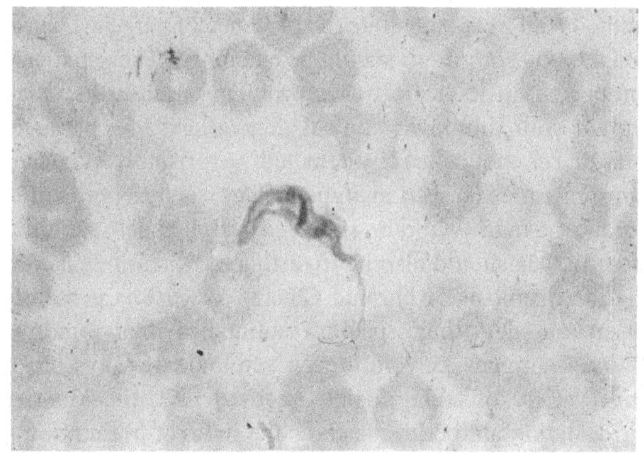

FIGURE 7 Trypomastigote of *Trypanosoma brucei gambiense* in the bloodstream. Courtesy of the American Society of Tropical Medicine and Hygiene/Zaiman 'A Presentation of Pictorial Parasites.'

tandem arrays of 177 bp repeats as well as transcriptionally silent copies of variant surface glycoprotein (*VSG*) genes proximal to their telomeres. *T. brucei* contains approximately 1000 genes capable of coding for *VSG* genes, which are switched at a rate of 102–106 switches per generation. This process serves as the main mechanism of immune evasion for *T. brucei*. Only one *VSG* expression site (ES) is active at any given time. There are 15–20 ESs per genome, all at subtelomeric locations. In addition to the *VSG* genes, several upstream genes, called ES-associated genes (ESAGs), are also transcribed. Three types of DNA rearrangements are associated with ES switching – duplicative transposition, telomere exchange, and telomere conversion. Transposition involves the 1.6 kb *VSG*, in addition to a 1.5 kb proximal sequence. Thus, a minimum of 8% of the genome is devoted to *VSG* coding and flanking sequences. As mentioned previously, a large fraction of the 25% of the genome found in the minichromosomes also contributes to *VSG* diversity. The modified DNA base 'J' (β-D-glucosyl-hydroxymethyl-uracil), unique to organisms in the order Trypanosomatidae, replaces up to 1% of thymidines, and its frequency is higher in repetitive DNA adjacent to transcriptionally silent telomeres, including the $(GGGTTA)_n$ telomeric hexamer.

T. brucei genes are transcribed as large polycistronic units that then undergo trans-splicing so that all mature mRNAs contain the same 39 nt sliced leader at their 5′ ends. In contrast, with the exception of an 11 nt intron in a tRNA-tyr, *T. brucei* genes contain no introns and, hence, do not undergo cis-splicing. The spliced leader is also notable for the presence of a 7-methyl-guanosine 5′ cap, a promoter-like region consisting of four methylated nucleotides at the 5′ end, similar to that found upstream from ES regions and from genes that encode procyclic stage-specific coat protein (procyclic acid-rich protein (PARP), also called procyclin). Transcription from the PARP and VSG promoters, like that from the trypanosome rRNA promoter, is α-amanitin-resistant, suggesting transcription by RNA polymerase I. Stage-specific gene expression is also influenced by 3′ untranslated portions of mRNA, mediated through changes in mRNA stability and in efficiency of mRNA maturation. The sequence influencing stage specificity of *VSG* mRNA abundance is localized to a region 97 nt upstream from the polyadenylation site. Retroposon-like elements are also scattered throughout the genomes of these parasites. The best studied is a 5 kb sequence, designated *ingi* (Swahili for many), that is similar to reverse transcriptase genes in other organisms. There are approximately 400 copies of *ingi*, making up to 5% of the *T. brucei* genome.

Another interesting feature of trypanosome genetics, which is shared with other members of the class Kinetoplastida, is the phenomenon known as RNA editing. The kinetoplast DNA is organized as an interlocking and supercoiled network of approximately 50 maxicircle DNA molecules (20–30 kb) and many thousands of minicircle DNA molecules (1.0 kb in *T. brucei* and 1.6 kb in *T. cruzi*). Maxicircle DNA encodes about a dozen mitochondrial proteins. The maxicircles and minicircles both encode small (50–100 nt) guide RNAs that serve as templates for the insertion, and less frequently deletion, of uridines in the primary RNA transcripts of the maxicircle mitochondrial genes. In some cases, nearly 50% of the mature mRNAs consist of uracils inserted posttranscriptionally by the editing process. Studies comparing homologous nuclear genes between *T. brucei* and *T. cruzi* have demonstrated a large evolutionary divergence in codon use. A comparison of the nuclear small and large subunit rRNA gene sequences yields genetic distances comparable to those between plants and animals.

Epidemiology

Sleeping sickness (*T. b. gambiense* and *T. b. rhodesiense*) and veterinary trypanosomiasis caused by *T. brucei* subgroup parasites continue to be responsible for much human suffering and economic loss (Table 1). These agents are endemic and enzootic in an area of sub-Saharan Africa covering 10×10^6 km^2. Approximately 50 million people are at risk of becoming infected with these parasites and tens of thousands of new cases of human African trypanosomiasis occur each year. Exact numbers are not available because the acquisition of reliable health statistics is difficult in the developing countries where sleeping sickness is endemic. Human African trypanosomiasis has undergone resurgence, and in recent years, major epidemics have occurred in the Central African Republic, Ivory Coast, Chad, Sudan, and several other endemic countries. Losses of cattle due to trypanosomiasis have had an enormous economic impact in many regions. Human African trypanosomiasis is restricted to those areas south of the Sahara in which the annual rainfall exceeds 500 mm (i.e., 20 in.) because the larval stages of the tsetse fly are vulnerable to desiccation. Thus, the gambiense form occurs in the western portion of tropical Africa and focal incursions eastward all the way over to Lake Victoria and into southern Sudan. The rhodesiense form is found in the southeastern portion of Africa, north of South Africa. There is some overlap in the endemic ranges of these two forms of the disease. Humans are the only important reservoir of *T. b. gambiense*. The cycle of gambiense disease is maintained only where there is a close relationship between humans and tsetse flies that belong to the *Glossina palpalis* group and preferentially feed on human blood. In contrast to gambiense disease, the cycle of *T. b. rhodesiense* is maintained in wild mammals, and humans are only incidental hosts. The vectors of *T. b. rhodesiense*, which belong to the *Glossina morsitans* group, inhabit the relatively dry eastern African

TABLE 1 African animal trypanosomiasis

Parasite	Host	Occurrence
Trypanosoma vivax	Cattle	Common, mild
	Equines	Rare, mild
	Sheep and goats	Rare, severe
	Camels	Rare, mild
	Dogs	Not reported
	Pigs	Not reported
Trypanosoma bruceibrucei	Cattle	Common, but most cattletolerate well
	Equines	Common, severe
	Sheep and goats	Rare
	Camels	Common, severe
	Dogs	Common, severe
	Pigs	Not reported
Trypanosomacongolense	Cattle	Common, severe
	Equines	Rare, mild disease
	Sheep and goats	Rare
	Camels	Not reported
	Dogs	Not reported
Trypanosoma evansi	Cattle	Mild disease
	Equines	Common, severe
	Sheep and goats	Not reported
	Camels	Common, severe

Animal trypanosomiasis is of great economic importance in Africa and is associated with fever, anemia, thrombocytopenia, wasting, and death. These manifestations are mediated, in part, by cytokines.

savannas and preferentially feed on wild ungulates that are relatively trypanotolerant and maintain infective parasitemias for long periods. Nonetheless, there have been instances in which East African trypanosomiasis has reached epidemic proportions. Human cases of rhodesiense trypanosomiasis typically occur in young adult men who have occupations in which they are exposed to tsetse flies. In epidemics, however, all age groups are infected, and mechanical transmission is thought to occur in this context. Tourists from nonendemic areas may acquire this infection. This occurs via the blood-filled proboscis of a fly, which may be interrupted while taking a blood meal from an infected individual. When the fly bites an uninfected host, within 2–3 h, the blood from the infected host becomes a parasite-bearing inoculum. In contrast to *T. cruzi*, congenital transmission of African trypanosomes is extremely rare. Laboratory-acquired transmission has been reported.

Pathology and Pathogenesis

Metacyclic infective trypomastigote forms are inoculated into the skin by the tsetse fly and multiply there. A characteristic hard and sometimes painful chancre is formed. By about day 10, long slender forms are found in the bloodstream and lymphatics, and for the next several days their numbers increase exponentially. Soon thereafter, the organisms nearly disappear from the bloodstream, only to reappear later. The interval between waves of parasitemia may vary from 1 to 2 weeks, with clinical symptoms accompanying each bout of parasitemia. Each successive wave of organisms represents a new crop of parasites expressing a *VSG* not previously expressed in that host, and it is through this process of sequential antigenic variation that the parasites stay one step ahead of the host's specific antibody responses. In response to infection, a marked early humoral antibody response, consisting predominantly of IgM, is seen regularly. These macroglobulins not only consist of antitrypanosomal antibodies directed against parasite surface antigens, but also include a variety of other antibodies such as heterophile and rheumatoid factor. As a result of polyclonal B cell activation, there also are many antibodies produced to a wide variety of antigens, including brain-specific autoantibodies. In addition, antibodies directed against myelin basic protein, gangliosides, and cerebrosides have been found in experimental models. Circulating immune complexes have been reported regularly and these may be responsible for the glomerulonephritis often accompanying acute and chronic disease. Cell-mediated immunity is also important in this disease, and nitric oxide may be important in the depression of T-cell responsiveness and generalized immunosuppression.

Lymphadenopathy usually involves the posterior cervical, submaxillary, supraclavicular, and mesenteric lymph nodes. The advanced disease, often called stage II disease, involves the CNS. The microscopic examination of lymphatic tissues usually reveals generalized hyperplasia with diffuse proliferation of lymphocytes. Initially, affected lymph nodes are markedly hemorrhagic and contain a large number of trypomastigotes; later, the nodes may become small and fibrotic. A progressive, chronic leptomeningitis develops in stage II disease. The brain becomes edematous and there is prominent perivascular cuffing by glial cells, lymphocytes, and plasma cells. Morula cells, reactive astrocytes, and hyperplasia of microglial cells all have been described in brain specimens from patients with CNS disease, and demyelination occurs in chronic cases. Organisms can be found in the brain tissue near vessels and may also be present in the cerebrospinal fluid. There is a striking lymphocytosis in the cerebrospinal fluid and most of these lymphocytes are B cells. Glomerulonephritis, myocarditis, pericardial effusion, pulmonary edema, and hypoplastic bone marrow with an associated anemia may develop in some patients. The pathogenesis of the neuropsychiatric manifestations is poorly understood. Experimental models suggest that a variety of factors may be responsible (e.g., deposition of immune complexes and aberrant levels of brain neurotransmitters, prostaglandins, and cytokines).

Clinical Manifestations

Clinical manifestations of infection with *T. b. gambiense* and *T. b. rhodesiense* are similar, except that the rhodesiense disease usually runs a much more fulminant course. Untreated patients with the latter typically die in weeks to months, whereas persons infected with *T. b. gambiense* may live for years and have long periods without symptoms. In both forms, the trypanosomal chancre may be evident several days after the bite of an infected tsetse fly. Within a week or two, the lesion becomes a large, red, painful nodule that may reach 5–10 cm in diameter. This nodule, which is reported to be more common in people of European descent, subsides spontaneously in a few weeks. In rhodesiense trypanosomiasis, the incubation period of the systemic disease is typically 2–3 weeks, whereas with the gambiense infection, the first symptoms may be noted many weeks or months after the acquisition of the infection. The early systemic disease is characterized by intermittent fevers, chills, headache, and generalized lymphadenopathy. In gambiense disease, the nodes in the posterior cervical triangle may become enlarged (i.e., Winterbottom sign) and, when present, strongly suggest the diagnosis in a patient with exposure to tsetse flies. Delayed deep hyperesthesia may occur over the tibia and moderate hepatosplenomegaly may be noted. Anemia and thrombocytopenia are frequent as well. Intermittent fevers may last for months to years with gambiense infection. In Europeans, a circinate erythematous rash or erythema multiforme is sometimes present. Untreated patients eventually develop signs of CNS invasion. Severe headaches, loss of nocturnal sleep, and a feeling of impending doom are typical. Following this, there may be progressive mental deterioration, with patients becoming incapable of caring for themselves. Tremors, especially of the tongue, hands, or feet, as well as generalized or focal convulsions may occur. Almost any neurologic and psychiatric manifestation can be seen with progressive mental deterioration until patients finally lapse into coma and die of intercurrent infections.

Diagnosis

A definitive diagnosis of African trypanosomiasis is made by detecting trypanosomes, and this can be done by looking for parasites in fresh and stained specimens of blood, bone marrow, lymph node aspirates, or, in late disease, the cerebrospinal fluid. If parasites are not seen in these specimens obtained from a patient whose history and clinical findings suggest African trypanosomiasis, efforts should be made to concentrate the organisms in blood. This can be done most simply by using commercially available quantitative buffy coat analysis tubes. The parasites are separated from the blood components by centrifugation in these acridine orange-coated tubes and are easily seen under light microscopy because of the stain. Alternatively, buffy coat obtained by centrifugation of 10–15 ml of anticoagulated blood can be examined microscopically as a wet preparation and after Giemsa staining. All of these specimens can also be inoculated into specialized liquid culture medium. Finally, a highly sensitive method for detecting *T. b. rhodesiense* is the inoculation of small volumes of specimens obtained from the patient into rodents. Patent parasitemias usually develop within a week or two in animals injected with specimens from people infected with *T. b. rhodesiense*. Unfortunately, host specificity precludes the use of this approach for diagnosing *T. b. gambiense*.

Several serologic assays are available to aid in the diagnosis of sleeping sickness, but the variable accuracy of these tests mandate that treatment decisions still be based on detection of the parasite. These assays are useful, nonetheless, in epidemiologic surveys. Simple direct agglutination tests for trypanosomes performed on cards (CATT and TrypTect CIATT) are available commercially. A role for PCR-based assays for detecting African trypanosomes has not been defined. In advanced untreated sleeping sickness, the IgM level in the cerebrospinal fluid is often elevated, but it has no relationship to the presence of trypanosomes in the cerebrospinal fluid. After successful treatment, the IgM level declines gradually, disappearing after approximately 1 year and, thus, a persistently elevated level or an abrupt rise in the IgM months after treatment may indicate a relapse. IgM levels, however, should not be used as the sole method of diagnosis or prognosis.

Treatment

Remarkably, the chemotherapy of these diseases, both human and veterinary, has lagged behind that of other tropical diseases. The main chemotherapeutic agents for human trypanosomiasis are pentamidine and suramin for the early-stage disease and melarsoprol (Mel B, Arsobal) for the late-stage (CNS) disease. Difluoromethylornithine (DFMO, Ornidyl) is the only new useful addition to this list since the early 1950s. With the exception of suramin, resistance to the established agents is growing, and the toxicity of these drugs continues to be a problem.

Pentamidine is a water-soluble aromatic diamidine that has been in use since the 1930s. It is effective against early-stage *T. b. gambiense* infection, but is less effective against

T. b. rhodesiense infection, and is ineffective against late-stage disease. African trypanosomes have a nucleoside transporter (adenine/adenosine: P2) that takes up pentamidine, resulting in the concentration of the agent at levels many times that in plasma. Many studies have focused on the mechanism of pentamidine action; however, none appears to have conclusively defined the target. It is known to bind to the minor groove of kinetoplast (i.e., mitochondrial) DNA and to promote the cleavage of kinetoplast minicircle DNA, eventually leading to the development of dyskinetoplastic cells. Despite its effects on kinetoplast DNA, pentamidine has no effect on nuclear DNA, and dyskinetoplastic forms can persist in the bloodstream of mammals. Pentamidine has also been found to be a reversible inhibitor of *S*-adenosylmethionine (AdoMet) decarboxylase, an enzyme in the polyamine biosynthetic pathway. Although ki values were in the 200 mmol l^{-1} range, we now know that this internal concentration is achievable via uptake through the P2 nucleoside–pentamidine transporter. Other targets studied previously in trypanosomes include the inhibition of glycolysis and lipid metabolism, as well as the effects on amino acid transport and ion exchange. The facts that pentamidine does not kill trypanosomes outright and bloodstream forms persist after treatment argue for a sustained effect more consistent with interference with nucleic acid metabolism.

Diminazene aceturate (Berenil) is an aromatic diamidine developed as treatment for bovine trypanosomiasis; however, its apparent low incidence of adverse reactions and significant therapeutic activity has led some physicians in endemic countries to use it extensively. It is effective against early-stage *T. b. gambiense* and *T. b. rhodesiense.* Diminazene has also been used in combination with melarsoprol for late-stage disease. Mechanistically, like pentamidine, diminazene has also been linked to kinetoplast DNA binding at the minor groove and cleavage of minicircle DNA. Like pentamidine, diminazene may also interfere with RNA editing and trans-splicing. Diminazene is also a more effective and noncompetitive inhibitor of AdoMet decarboxylase in trypanosomes, resulting in the reduction of spermidine content and elevating putrescine in the parasite. As with pentamidine, diminazene uptake occurs via the P2 nucleoside transporter, which allows significant accumulation from the external environment. Although diminazene has been used for many years on thousands of sleeping sickness patients, there is little published material on its toxicity. This may in part be due to physicians who are unwilling to document human studies with an agent licensed for veterinary use. However, personal accounts of those using diminazene in humans indicate that it is well tolerated.

Suramin is a sulfonated naphthylamine that has been used successfully against early-stage sleeping sickness caused chiefly by *T. b. rhodesiense.* It was first used in 1922, developed from the closely related azo dyes trypan red and trypan blue. Suramin has an extremely long half-life in humans, 44–54 days, as a result of avid binding to serum proteins. Suramin binds to many plasma proteins including LDL, to which trypanosomes avidly bind and endocytose as a result of specific membrane receptors. LDL is a prime source of sterols for bloodstream trypanosomes. The uptake of suramin as a protein complex results in an internal concentration of 100 mmol l^{-1}. Suramin has been shown to inhibit all of the glycolytic enzymes in *Trypanosoma brucei brucei* in the range of 10–100 mmol l^{-1}, which in most cases is severalfold lower than for the corresponding mammalian enzyme. This specificity for trypanosomal enzymes was attributed to higher (i.e., more basic) isoelectric points of the parasite enzyme than the mammalian enzymes, allowing the negatively charged suramin to bind preferentially to the parasite enzymes. In practice, because most trypanosome glycolytic enzymes are contained in a membrane-bound cytosolic organelle, the glycosome, it is not likely that rapid massive binding occurs. This would rapidly induce lysis in bloodstream forms that depend on glycolysis as the sole energy-generating source. Rather, animals that are heavily infected with trypanosomes and given suramin show a slow decrease in parasite numbers, indicating that enzyme inhibition occurs slowly. Suramin may be affecting newly synthesized enzyme molecules in the cytosol before they are imported into the glycosome. Beyond the inhibition of glycolytic enzymes, suramin has also been found to affect thymidine kinase and dihydrofolate reductase. It is likely that suramin's action may be attributable to the inhibition of several of these enzymes.

Melarsoprol (Mel B, Arsobal) is an arsenical resulting from the efforts of Ernst Freidheim in the late 1940s. His initial compound, melarsen oxide, *p*-(4,6-diamino-*s*-triazinyl-2-*yl*) aminophenylarsenoxide, was complexed with dimercaptopropanol (British Anti-Lewisite) to form a less-toxic complex, melarsoprol. Until 1990, this was the only agent available for curing the late-stage (CNS) disease in both East African and West African infections. Toxicity is an important concern with melarsoprol. This takes the form of reactive arsenical-induced encephalopathy, which is often followed by pulmonary edema and death within 48 h in more than half the cases. Although the mechanism of melarsoprol action has been extensively studied, it still remains unclear. Parasites exposed to low levels (1–10 mmol l^{-1}) rapidly lyse. Because the bloodstream forms are intensely glycolytic, any interruption of glycolysis should produce this effect. Thus, a series of reports has detailed melarsoprol inhibition of trypanosome pyruvate kinase (ki, 100 mmol l^{-1}), phosphofructokinase (ki, >1 mmol l^{-1}), and fructose-2,6-bisphosphate (ki, 2 mmol l^{-1}). It is likely that the rapid inhibition of fructose-2,6-bisphosphate production is a key factor in halting glycolysis through the downregulation of pyruvate kinase. Other studies have indicated that melarsoprol and melarsen oxide formed adducts with trypanothione (*N*1,*N*8-bisglutathionyl spermidine), a metabolite unique to trypanosomes and believed to be responsible for the redox balance of the cell and detoxification of peroxides. The melarsen–trypanothionine adduct (Mel T) inhibits trypanothione reductase, which has been attributed to the mode of action. However, melarsoprol and related arsenicals may also bind to other sulfhydryl-containing agents in the cell, including dihydrolipoate and the closely adjacent cysteine residues of many proteins. Like pentamidine and diminazene, melarsoprol uptake into African trypanosomes has been attributed to entry through the P2 purine nucleoside transporter; thus, significant levels can be concentrated in the cell from a low external (plasma) concentration. Although most laboratory-generated melarsoprol-resistant strains have lost or modified the P2 transporter, clinical isolates appear to have retained uptake capacity.

Eflornithine DFMO (DL-*a*-difluoromethylornithine, Ornidyl) is the most recently developed agent for late-stage *T. b. gambiense* sleeping sickness. After initial testing in model infections, DFMO was studied extensively in human trials. The standard treatment regimen resulting from the trials indicate that DFMO is $>95\%$ active when given intravenously. DFMO cured children, adults, and patients with melarsoprol-refractory strains, and patients with late-stage disease. The short plasma half-life of DFMO necessitates continuous infusion when given intravenously. The most frequent toxic reaction was reversible bone marrow suppression, which was alleviated upon reduction of the doses. The major drawbacks with respect to DFMO are its cost, the duration of treatment, and its availability. DFMO rapidly and irreversibly binds to the catalytic site (cysteine 360 in mouse ornithine decarboxylase (ODC)), inactivating it. In culture, it blocks division of bloodstream trypanosomes, but it is not trypanocidal. In laboratory infections, DFMO cures when administered continuously in the drinking water as a 2% solution. Within 48 h of administration, DFMO reduces putrescine levels to zero and reduces spermidine levels by $\sim 75\%$. Trypanothione levels are also significantly reduced. As noted, DFMO is not trypanocidal and depends on a functional immune system to rid the host of nondividing forms. Morphologically, trypanosomes with multiple kinetoplasts and nuclei are common, as are forms resembling 'stumpy' blood forms. DFMO is curative for laboratory infections of *T. b. brucei* and *T. b. gambiense*, but not to all strains of *T. b. rhodesiense*. The reason for this selectivity is not known, although it is not due to the uptake of DFMO, because it enters by passive diffusion, not transport. Levels of AdoMet are highly elevated in susceptible strains, but less so in refractory isolates. The elevated levels of AdoMet are due to an AdoMet synthase insensitive to its product. DFMO treatment leads to intracellular concentrations of ~ 5 mmol l^{-1}, a ~ 50-fold increase over untreated parasites. Trypanosome ODC is missing the c-terminal PEST sequence in both procyclic and bloodstream trypanosomes, and this seems to be the major reason for the stability of the trypanosome enzyme. The remainder of the ODC molecule has $>60\%$ sequence identity with the mammalian enzyme, including a cysteine 360 residue at the demonstrated DFMO-binding site for the mammalian enzyme. Beyond this, trypanosomes lack a polyamine oxidase, which in mammalian cells converts spermine to the biologically active spermidine. Trypanosomes are also limited in their ability to transport putrescine and spermidine.

The usefulness of nifurtimox in combination with DFMO or melarsoprol as treatment for human African trypanosomiasis has been assessed in several recent trials especially with the emergence of resistance to DFMO. Outcomes obtained to date suggest that the addition of nifurtimox to adjusted doses of either of these two drugs results in increased efficacy, while reducing overall toxicity.

Prophylaxis and Prevention

Trypanosomes cause complex public health and epizootic problems in many developing countries in Africa. Control programs concentrating on the eradication of vectors and drug treatment of infected people and animals have been in operation in some areas for decades. Considerable progress has been made in a number of regions, but the lack of agreement on the best approach to solving the problem of African trypanosomiasis, combined with a paucity of resources, stands in the way of effective control. Individuals can reduce their risk of becoming infected with trypanosomes by avoiding tsetse fly-infested areas, by wearing clothing that reduces the biting of the flies, and by using insect repellants. Chemoprophylaxis with suramin or pentamidine can be effective, but it is not clear which populations should use this as a preventive measure. No vaccine is available to prevent the transmission of the parasites.

FURTHER READING

Ashton, A. W., Mukherjee, S., Nagajyothi, F. N. U., *et al.* (2007). Thromboxane A_2 is a key regulator of pathogenesis during *Trypanosoma cruzi*infection. *The Journal of Experimental Medicine*, *204*, 929–940.

Bacal, F., Silva, C. P., Pires, P. V., Mangini, S., Fiorelli, A. I., Stolf, N. G., *et al.* (2010). Transplantation for Chagas' disease: an overview of immunosuppression and reactivation in the last two decades. *Clinical Transplantation*, *24*, E29–E34.

Bern, C., Montgomery, S. P., Herwaldt, B. L., *et al.* (2007). Evaluation and treatment of Chagas' disease in the United States: A systematic review. *The Journal of the American Medical Association*, *298*, 2171–2181.

Bisser, S., N'Siesi, F. X., Lejon, V., *et al.* (2007). Equivalence trial of melarsoprol and nifurtimox monotherapy and combination therapy for the treatment of second-stage *Trypanosoma brucei gambiense* sleeping sickness. *The Journal of Infectious Diseases*, *195*, 322–329.

Brun, R., Blum, J., Chappuis, F., & Burri, C. (2010). Human African trypanosomiasis. *Lancet*, *375*(9709), 148–159.

Carod-Artal, F. J., & Gascon, J. (2010). Chagas disease and stroke. *Lancet Neurology*, *9*, 533–542.

Centers for Disease Control and Prevention. (2006). Chagas disease after organ transplantation – Los Angeles, California, 2006. *Morbidity and Mortality Weekly Report*, *55*, 798–800.

Centers for Disease Control and Prevention. (2007). Blood donor screening for Chagas disease – United States, 2006–2007. *Morbidity and Mortality Weekly Report*, *56*, 141–143.

Checchi, F., Piola, P., Ayikoru, H., Thomas, F., Legros, D., & Priotto, G. (2007). Nifurtimox plus eflornithine for late-stage sleeping sickness in Uganda: A case series. *PLoS Neglected Tropical Diseases*, *1*, e64.

Cheng, K. Y., Chang, C. D., Salbilla, V. A., *et al.* (2007). Immunoblot assay using recombinant antigens as a supplemental test to confirm antibodies to *Trypanosoma cruzi*. *Clinical and Vaccine Immunology*, *14*, 355–361.

Combs, T. P., Nagajyothi, F. N. U., Mukherjee, S., *et al.* (2005). The adipocyte as an important target cell for *Trypanosoma cruzi* infection. *The Journal of Biological Chemistry*, *280*, 24085–24094.

del Puerto, R., Nishizawa, J. E., Kikuchi, M., Iihoshi, N., Roca, Y., Avilas, C., *et al.* (2010). Lineage analysis of circulating *Trypanosoma cruzi* parasites and their association with clinical forms of Chagas disease in Bolivia. *PLoS Neglected Tropical Diseases*, *4*, e687.

El-Sayed, N. M., Myler, P. J., Blandin, G., *et al.* (2005). Comparative genomics of trypanosomatid parasitic protozoa. *Science*, *309*, 404–409.

Gascon, J., Bern, C., & Pinazo, M. J. (2010). Chagas disease in Spain, the United States and other non-endemic countries. *Acta Tropica*, *115*, 22–27.

Kirchhoff, L. V., Paredes, P., Lomeli-Guerrero, A., *et al.* (2006). Transfusion-associated Chagas' disease (American trypanosomiasis) in Mexico: Implications for transfusion medicine in the United States. *Transfusion*, *46*, 298–304.

Kirchhoff, L. V., & Pearson, R. D. (2007). The emergence of Chagas' disease in the United States and Canada. *Current Infectious Disease Reports*, *9*, 347–350.

Machado, F. S., Tanowitz, H. B., & Teixeira, M. M. (2010). New drugs for neglected infectious diseases: Chagas' disease. *British Journal of Pharmacology*, *160*, 258–259.

Molyneux, D., Ndung'u, J., & Maudlin, I. (2010). Controlling sleeping sickness – "When will they ever learn?". *PLoS Neglected Tropical Diseases*, *4*, e609.

Picozzi, K., Carrington, M., & Welburn, S. C. (2008). A multiplex PCR that discriminates between *Trypanosoma brucei brucei* and zoonotic *T. b. rhodesiense*. *Experimental Parasitology*, *118*, 41–46.

Priotto, G., Kasparian, S., Mutombo, W., Ngouama, D., Ghorashian, S., Arnold, U., *et al.* (2009). Kande V (2009) Nifurtimox-eflornithine combination therapy for second-stage African *Trypanosoma brucei gambiense* trypanosomiasis: A multicentre, randomised, phase III, non-inferiority trial. *Lancet*, *374*(9683), 56–64.

Priotto, G., Pinoges, L., Badi Fursa, I., *et al.* (2008). Safety and effectiveness of first line elornithine for *Trypanosoma brucei gambiense* sleeping sickness in Sudan: Cohort study. British *Journal of Medicine*, *336*, 705–708.

Rassi, A., Jr., Rassi, A., Little, W. C., *et al.* (2006). Development and validation of a risk score for predicting death in Chagas' heart disease. *The New England Jounal of Medicine*, *355*, 799–808.

Rassi, A., Jr., Rassi, A., & Marin-Neto, J. A. (2010). Chagas disease. *Lancet*, *375*(9723), 1388–1402.

Sartori, A. M., Ibrahim, K. Y., Nunes Westphalen, E. V., *et al.* (2007). Manifestations of Chagas disease (American trypanosomiasis) in patients with HIV/AIDS. *Annals of Tropical Medicine and Parasitology*, *101*, 31–50.

Schofield, C. J., Jannin, J., & Salvatella, R. (2006). The future of Chagas disease control. *Trends in Parasitology*, *22*, 583–588.

Chapter 30

Sleeping Sickness

S.C. Welburn[1], K. Picozzi[1], I. Maudlin[1] and P.P. Simarro[2]

[1]*Centre for Infectious Diseases, The University of Edinburgh, 1 Summerhall Square, Edinburgh, UK*

[2]*World Health Organization, Control of Neglected Tropical Diseases, Innovative and Intensified Disease Management, Geneva, Switzerland*

Chapter Outline

Abbreviations	**411**
Defining Statement	**411**
Background	**411**
Epidemiology of Sleeping Sickness	**412**
Distribution	412
Reservoirs of Disease	413
Transmission Cycles	414
Causes of Epidemics	**415**
Diagnosis	**416**
Clinical Signs	416
Diagnostic Tests	416
Treatment	**417**
Disease Burden	**418**
Economic Impact	418
Sleeping Sickness Control	**419**
Controlling Gambian Sleeping Sickness	419
Controlling Rhodesian Sleeping Sickness	419
Future Prospects for Controlling Sleeping Sickness	**420**
A Neglected Disease	420
Treatment	422
Diagnostics	422
Vector Control	423
The Future: Control or Eliminate?	**423**
Gambian Sleeping Sickness	423
Rhodesian Sleeping Sickness	423
Further Reading	**425**
Relevant Websites	**426**

ABBREVIATIONS

CATT Card agglutination test for trypanosomiasis
DALY Disability-adjusted life year
FIND Foundation for Innovative New Diagnostics
GPS Global positioning systems
HAT Human African trypanosomiasis
PATTEC Pan African Tsetse and Trypanosomosis Eradication Campaign
PCR Polymerase chain reaction
SP Synthetic pyrethroids
SRA Serum resistance-associated gene
VATs Variable series of antigenic types

DEFINING STATEMENT

Both forms of sleeping sickness, acute and chronic, are fatal if left untreated. Drugs currently available are toxic and drug resistance is of increasing concern. Sleeping sickness is defined as a neglected tropical disease, a status compounded by severe underreporting. The burden of sleeping sickness is much greater than might be expected from its relative incidence. Public–private partnerships are addressing the urgent need for new nontoxic drugs, diagnostics, and innovative and sustainable control measures.

BACKGROUND

Sleeping sickness is used to describe two quite distinct diseases caused by different subspecies of *Trypanosoma brucei*: *T. b. gambiense* and *T. b. rhodesiense* are both human-infective; the third subspecies *T. brucei brucei* is morphologically indistinguishable from the others, infects a range of mammalian species both domestic and wild, but is not human-infective. *T. b. gambiense* and *T. b. rhodesiense* are invariably fatal in humans if left untreated; in contrast, *T. b. brucei* is pathogenic to some mammalian hosts (e.g., horses, dogs, and exotic breeds of cattle) and not to others (e.g., indigenous African cattle breeds). Molecular taxonomic studies have shown that *T. b. gambiense* has diverged from both *T. b. brucei* and *T. b. rhodesiense*, while a difference in expression of a single gene (SRA – serum resistance-associated gene) distinguishes *T. b. rhodesiense* from *T. b. brucei* and *T. b. gambiense*.

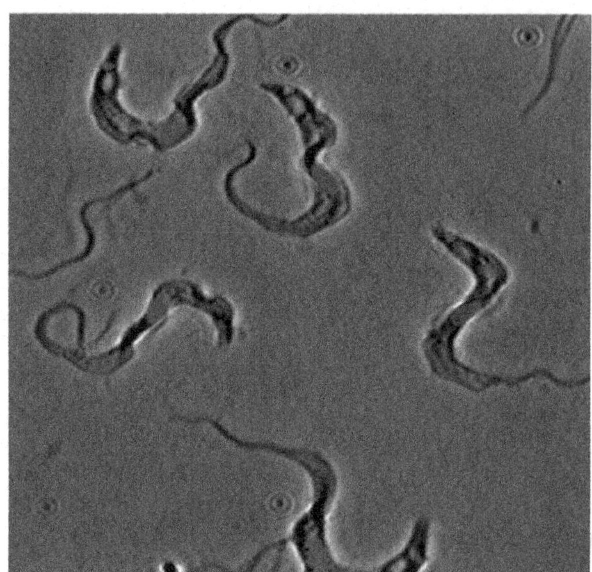

FIGURE 1 Bloodstream form of *Trypanosoma brucei* with kinetoplast and nuclear DNA stained blue. *Photograph provided by Prof. K. Matthews, University of Edinburgh.*

These two diseases have distinct clinical pictures and their discrete geographical distribution across Africa serves to emphasize these differences; understanding these biological, clinical, and epidemiological differences has been crucial in devising effective methods of disease control and in defining health policy in relation to sleeping sickness.

Sleeping sickness infections begin when in the course of obtaining a further blood meal, a previously infected tsetse fly injects saliva into a human host together with trypanosome parasites (Figure 1). These protozoan parasites divide in the host's tissue and vasculature and this burst of population growth is met by a specific immune response by the host. Trypanosomes have however evolved a sophisticated strategy – antigenic variation – for surviving this immune response; a clonal subset of the population survives by expressing a different protective antigenic coat all over the parasite surface, which is unrecognized by the current host antibody response. Intriguingly, this antigenic variation does not simply arise as a response to the host attack but is spontaneous and rapid, giving rise to successive waves of parasites that present a variable series of antigenic types (VATs) to the host. Antigenic variation also explains why a vaccine has not become available to deal with trypanosomiasis, and, as a consequence, control of sleeping sickness is still dependent on a variety of technologies designed to deal with the vector as well as the parasite.

EPIDEMIOLOGY OF SLEEPING SICKNESS

Distribution

Sleeping sickness is also known as human African trypanosomiasis (HAT) to distinguish it from another trypanosome infection in humans, Chagas' disease, that is found in the Americas and is caused by a distantly related trypanosome, *Trypanosoma cruzi*. Sleeping sickness is confined to sub-Saharan Africa, its distribution strictly limited by the distribution of its vectors, flies of the genus *Glossina* referred to as tsetse flies, which are not found outside Africa. *T. b. gambiense* and *T. b. rhodesiense* are also referred to as Gambian and Rhodesian sleeping sickness, respectively, following their first recorded isolation in the early years of the twentieth century; these common names have remained apposite to this day with *T. b. gambiense* confined to the west of Africa and *T. b. rhodesiense* to the east.

Sleeping sickness is not distributed randomly or evenly across sub-Saharan Africa but is found in pockets of infection or foci (Figure 2), which may vary in size but whose centers have not changed much over recorded history (i.e., since the colonial period in Africa). It is believed that these foci of infection are of ancient origin, reflecting the distribution of reservoirs of trypanosome infection and of their tsetse vector populations. Sleeping sickness foci are however dynamic, expanding and contracting over time, reflecting changes in tsetse distribution and human activity, as well as in response to control activities. Uganda provides good evidence of the antiquity of such foci with a *T. b. rhodesiense* focus in the southeast and a *T. b. gambiense* focus in the northwest of the country; these foci have been isolated for at least the last 100 years for which there are records. Uganda remains the only state in Africa that hosts both types of sleeping sickness.

The distribution of these two diseases is neatly delineated by the Rift Valley running north to south through Africa (see Figure 2), raising intriguing questions about the evolution and distribution of the disease. Given the present-day knowledge about the evolution of hominids in the Rift Valley, would sleeping sickness have affected this evolutionary process? The climate of East Africa became drier between about 5 and 2.5 Ma, and it is thought that this may have been the catalyst that forced our ancestors to adapt to a savannah environment as the forests dwindled. This change of habitat would have forced the apes and early hominids to forage in open country, bringing them into contact with parasites different from those found in the forests, including trypanosomes circulating in the reservoir of savannah-adapted game animals. Apes, like humans, are partially adapted to *T. b. gambiense*, developing chronic infections; *T. b. rhodesiense* is an acute disease in apes and monkeys as well as in humans and, we can assume, would have rapidly killed early hominids. The only strategy early hominids could adopt in these circumstances would have been avoidance of tsetse-infested areas. This may have limited human population size within Africa, which grew dramatically following movement 'out of Africa.'

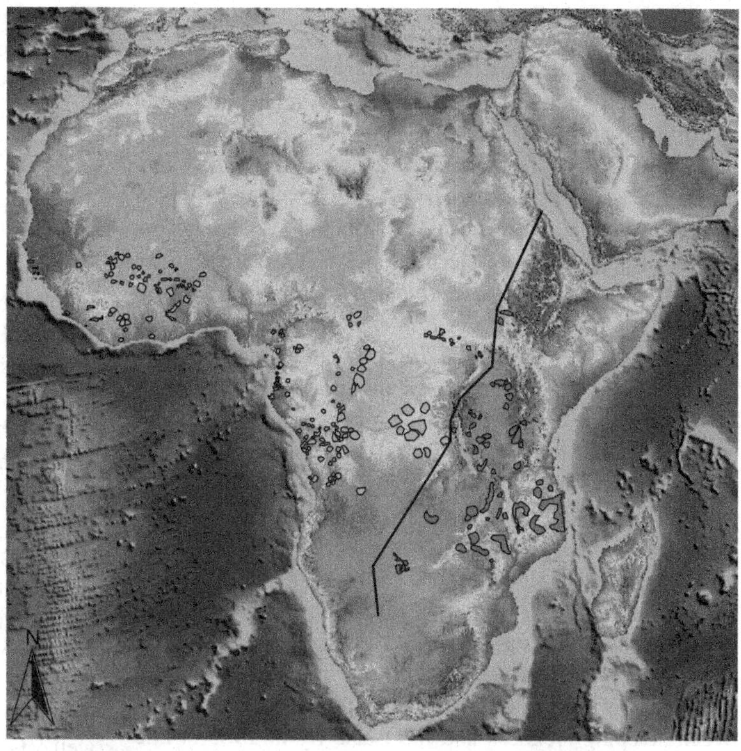

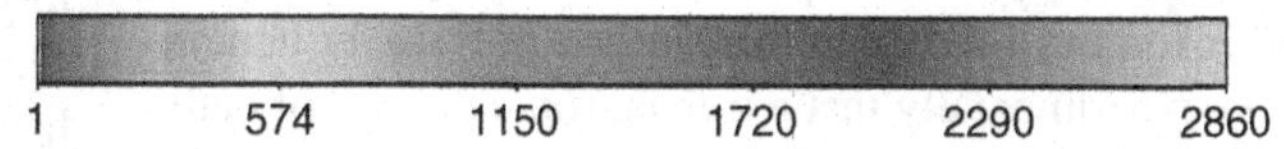

FIGURE 2 The geographical separation of *Trypanosoma b. gambiense* (Tbg, yellow) and *T. b. rhodesiense* (Tbr, red) foci follows closely the western edge of the African Rift Valley. *Reproduced from Welburn, S. C., Févre, E. M., Coleman, P. G., Odiit, M., & Maudlin, I. (2001). Sleeping sickness: A tale of two diseases.* Trends in Parasitology, 17, *19–24, with permission from Elsevier.*

Reservoirs of Disease

For a disease to continue to exist there must be some reservoir of infection but we must be clear how we define 'reservoir'; the following definition is helpful when considering the epidemiology of sleeping sickness:

> *[O]ne or more epidemiologically connected populations or environments in which the pathogen can be permanently maintained and from which infection is transmitted to the defined target population. Existence of a reservoir is confirmed when infection within the target population cannot be sustained after all transmission between target and non-target populations has been eliminated.*
>
> Haydon et al., 2002

The critical issue in this definition of a reservoir is the persistence of infection in the reservoir, which can only be determined through longitudinal studies. Such studies were carried out for over 20 years (1930s–1950s) in Tinde, Tanzania, to show that animals were the reservoir for *T. b. rhodesiense*. In the Tinde experiment, trypanosomes isolated from humans were serially transmitted by tsetse from sheep to sheep, and, most importantly, remained human-infective – the latter being tested by inoculation into human volunteers from time to time. However, this experiment on its own did not prove that wild animals (as opposed to laboratory animals) were the reservoir of Rhodesian sleeping sickness; that this was indeed the case was demonstrated unequivocally in the 1950s by taking blood from an infected bushbuck and injecting it into human volunteers. It later became clear that domestic animals could also act as reservoirs of *T. b. rhodesiense* – again volunteers were infected with blood from Kenyan cattle. In present-day Uganda, cattle form the major reservoir of Rhodesian sleeping sickness, profoundly affecting the control strategy.

Unlike Rhodesian sleeping sickness, the case for an animal reservoir for Gambian sleeping sickness is far from clear. Recent molecular studies in West Africa using PCR techniques have detected *T. b. gambiense* DNA in domestic pigs and in a range of wild animals including primates; this has been taken as evidence of an animal reservoir for Gambian sleeping sickness. However, there have been no equivalents of the Tinde experiment demonstrating persistence of Gambian sleeping sickness in the animal reservoir and animal-to-human transmission. Why such critical transmission studies were not carried out in the colonial period in West Africa is unclear, when, as we have seen from East Africa, ethical restrictions on such experimentation were not in place at that time. We can only conclude that wildlife and domestic livestock should only be regarded as potential reservoir hosts for Gambian sleeping sickness; this conclusion has critically affected control strategies for Gambian sleeping sickness.

Transmission Cycles

There are 33 extant species and subspecies of *Glossina*, which are grouped, by their ecology and relatedness, into riverine, savannah, or forest types. All tsetse species, both males and females, feed exclusively on blood and are capable of disease transmission. Historically, it was widely held that specific tsetse–trypanosome combinations were necessary for transmission of sleeping sickness; riverine tsetse species such as *Glossina palpalis palpalis* were thought to be the exclusive vectors of *T. b. gambiense*, and savannah species such as *Glossina pallidipes* were thought to transmit *T. b. rhodesiense*. This very basic misconception, which affected thinking around the epidemiology and control of sleeping sickness for many years, arose simply from the chronology in which the diseases were identified. Sleeping sickness had long been recognized in West Africa by those involved in the slave trade but it was not until 1901 that the infectious agent of sleeping sickness was isolated from a patient and named *T. gambiense*, reflecting its geographical origins. David Livingstone observed that the tsetse fly was "a perfect pest, injecting a poison like tiny scorpions" but it was Sir David Bruce who made the important link between sleeping sickness and the tsetse fly in 1903, while studying the very serious epidemic of sleeping sickness that had broken out in Uganda. Bruce had previously demonstrated experimentally that a trypanosome disease of cattle in South Africa was spread between animals by tsetse flies and so it is no great surprise that he concluded from his work in Uganda that (1) "sleeping sickness is caused by ... a species of trypanosome"; (2) "this species is probably that described by Dutton from the West Coast of Africa and called by him *Trypanosoma gambiense*"; and (3) "trypanosomes are transmitted by *Glossina palpalis*, and it alone." Bruce has since been shown to be correct about the identity of the vector of disease in Uganda but the special relationship between riverine tsetse and *T. b. gambiense* transmission is no longer accepted.

The human disease caused by *T. b. rhodesiense* was first recognized as a distinct entity in 1910 in what was then Rhodesia (now Zambia). It was assumed at the time that Rhodesian sleeping sickness was a zoonosis, which added to the proposition that there was an affinity between vector and host. It was inferred that flies found in savannah areas with abundant wildlife (i.e., flies of the morsitans group such as *G. pallidipes*) were the exclusive vectors of Rhodesian sleeping sickness whereas riverine flies (i.e., flies of the palpalis group) were the vectors of the disease in West Africa. It followed from these preconceptions that the transmission cycle for *T. b. rhodesiense* would be described as game–fly–human while for *T. b. gambiense* the cycle was simply human–fly–human. The game–fly–human cycle definition did not however survive the demonstration that domestic animals were involved in the cycle of Rhodesian sleeping sickness; Figure 3 illustrates current opinion of transmission cycles for sleeping sickness.

We now know that, unlike other vector-borne diseases such as malaria, there are no specific vector–parasite combinations required for transmission of sleeping sickness. However, some tsetse species have been shown in the laboratory to be superior vectors to others. Within tsetse species, it is well established that most flies within a population are refractory to trypanosome infection and that newly emerged flies (described as teneral) are more readily infected than flies that have previously taken a blood meal. Selection experiments have shown that susceptibility to trypanosome infection is a maternally inherited character in tsetse. There is an evolutionary downside for flies developing mature salivary gland infections as this significantly shortens the lifespan of infected flies. Male tsetse are far more likely to produce mature salivary gland infections than female tsetse but this is probably balanced from the trypanosome survival viewpoint by the fact that female tsetse outlive males.

Further complicating the transmission picture are the apparent feeding preferences of different tsetse species. For example, *G. pallidipes* has been shown to be strongly

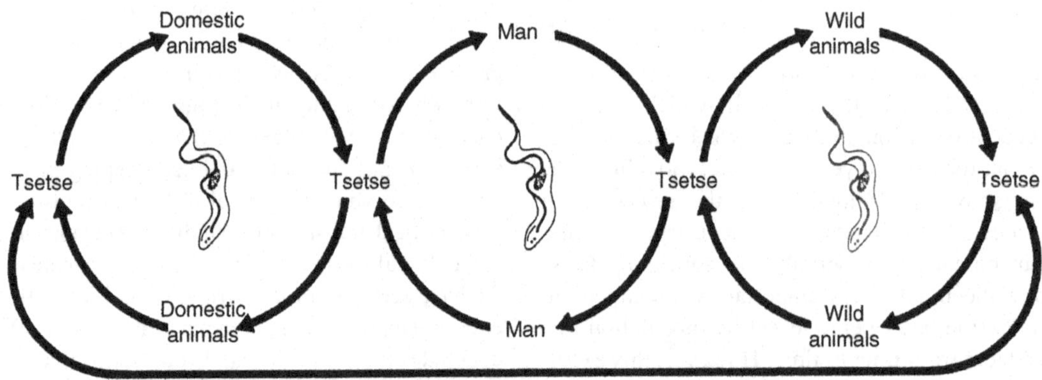

FIGURE 3 Transmission cycles of sleeping sickness. The epidemiology of the two forms of sleeping sickness is highly dependent on the reservoir of disease. Transmission of *Trypanosoma b. gambiense* is largely a simple man–fly–man cycle while *T. b. rhodesiense* has a reservoir in domestic and wild animals.

attracted to odors from oxen and partially repelled by odors from humans. Such variables can affect the epidemiology of sleeping sickness and have impacted on the development of control strategies involving insect attractants or baits.

There is a further complication to the trypanosome life cycle in the fly that is particularly significant in relation to sleeping sickness transmission. It has been shown in the laboratory that trypanosomes are capable of some form of genetic recombination in the tsetse fly. Much of the literature suggests there is a sophisticated and formal sexual phase in the trypanosome life cycle resulting in Hardy–Weinberg genetic equilibria. It follows that any *T. brucei* could transform into *T. b. rhodesiense* simply by mating with a trypanosome carrying the SRA gene. It would also follow that the SRA gene would be found wherever *T. brucei* occurred (i.e., across the fly belts of Africa), maintained by selection as a balanced polymorphism with the wild-type gene. It is difficult to reconcile these laboratory results with data from the field showing that the human-infective trypanosomes of Africa are clonal in nature and apparently do not undergo sexual recombination. Moreover, the SRA gene is not distributed across populations of *T. brucei* from the field as would be expected from a balanced polymorphism. There is recent evidence from *T. cruzi* and *T. brucei* to suggest that recombination is a random process in trypanosomes involving some primitive form of genetic exchange, often resulting in trypanosomes with different levels of ploidy. This may not be a natural process but rather a consequence of flies being infected in the laboratory with massive doses of parasites, in contrast to the field situation where flies are likely to take in very few trypanosomes at a feed. We may conclude that it is unlikely that human infectivity has spread across Africa as a by-product of mating between infective and nonhuman-infective members of *T. brucei*. The existence of ancient islands of sleeping sickness in an ocean of nonhuman-infective trypanosomes tends to support this conclusion.

CAUSES OF EPIDEMICS

Initial attempts to understand the epidemiology of sleeping sickness were profoundly influenced by events that took place following the scramble for territory in Africa by European powers in the late nineteenth century. Huge numbers of deaths attributable to sleeping sickness were reported from the Congo Free State and Uganda. The British had made enormous investments in East Africa and the loss of workforce due to sleeping sickness alarmed investors. An estimated 300000 Ugandans died around the shores of Lake Victoria (Figure 4) and, in the absence of curative drugs, it was decided in 1908 to evacuate the lakeshore region and isolate infected people in sleeping sickness camps, which in the absence of any cure became death camps. In the Congo Free State infected people were also placed in sleeping sickness camps, which became, as in Uganda, death camps; in 1901 alone half a million people were estimated to have been killed by the disease in the Congo Free State. Hundreds of thousands of cases of sleeping sickness were diagnosed between 1900 and 1940 in this region and, given the diagnostics then available, these are probably gross underestimates. Deaths from sleeping sickness in this period probably ran into millions across the continent, reflecting a situation comparable to the current HIV/AIDS pandemic in Africa. Support for this view comes from recent studies in Uganda showing that for every reported death caused by sleeping sickness, 12 cases go undetected, even with the availability of far more sophisticated diagnostic aids.

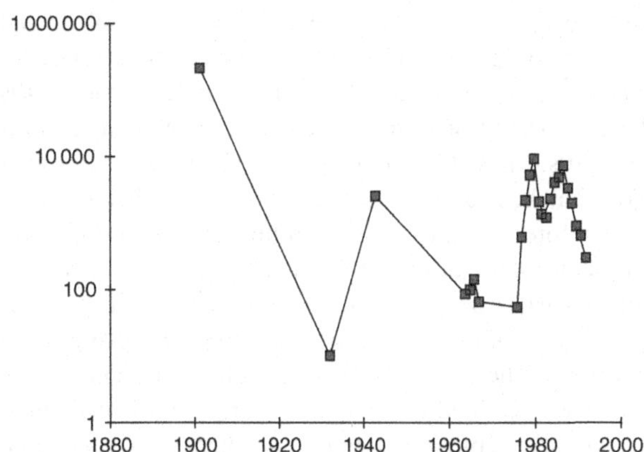

FIGURE 4 Epidemics of sleeping sickness in Uganda. Incidence of new cases of sleeping sickness in Busoga, Uganda, during the twentieth century showing three major epidemics. Sleeping sickness is a fatal disease if left untreated and as there were no drugs available at the time of the first recorded epidemic, which peaked in 1901, the figure reflects estimates of at least 300000 deaths.

The interventions taken to halt these epidemics would nowadays be considered inhuman, but reflected prevailing epidemiological dogma: movement of infected people spread the disease and therefore to interrupt transmission such movements must be stopped. Epidemiologists reasoned that these epidemics had their origins in the large-scale movements of people encouraged by the so-called Pax Britannica that followed colonization. This peace bonus was known to have inhibited internicene strife, promoting freedom of movement and the development of agriculture in previously uncultivated areas where tsetse flies thrived. This newfound freedom was thought to have increased human–fly–human transmission and the risk of epidemics. The apparent spread of Gambian sleeping sickness across Africa in the wake of human migrations was narrated in detail by contemporary epidemiologists: starting on the coast of West Africa in the 1890s and spreading across the Congo and throughout East Africa, giving rise to the 1901 epidemic in Uganda. Similarly, Rhodesian sleeping sickness was thought to have spread by the migration of people from Zambia through Tanzania to Kenya and Uganda, giving rise to an epidemic of Rhodesian sleeping sickness in Uganda in the 1940s. The Pax theory of transmission went unchallenged until the 1970s when John Ford proposed that the people of Africa had been at peace with each other and their environment before the European scramble for Africa. Ford saw European colonization, rather than peace, stimulating ecological warfare as a result of the introduction of new intensive systems of agriculture for cash crops. Ford suggested that prior to the European conquest, sleeping sickness was naturally contained in ancient endemic foci from which, occasionally, epidemics emerged when their ecological balance was upset.

There is now molecular evidence to show that migrant workers did not bring *T. b. rhodesiense* to Uganda from Rhodesia but that this parasite is endemic in and native to Uganda. Indeed, there is now good evidence to show that this parasite was the likely cause of the epidemic in Uganda in 1901, which Bruce mistakenly took to be Gambian sleeping sickness. Epidemics continue to arise from these ancient foci, triggered by various forms of societal disruption, upsetting the natural ecological balance, as suggested by Ford. The three epidemics seen in southeast Uganda in the twentieth century (see Figure 4) would have resulted from such social disorder, a process that we have seen repeated in the epidemics of sleeping sickness that have sprung up across Africa in the wake of civil unrest in Sudan, Uganda, Democratic Republic of the Congo, and Angola.

While Ford dismissed the notion that sleeping sickness spread about Africa like an influenza epidemic, he was convinced that a significant consequence of the induced ecological disorder was the expansion of tsetse populations, which would drive epidemics. Particular emphasis was placed on the effects of the rinderpest epizootic (between 1896 and 1898) on tsetse populations and hence the distribution of trypanosomiasis. This epizootic caused the death of practically all cattle and wildlife with consequent replacement of grasslands by bush; these conditions were thought to promote the spread of tsetse populations throughout East Africa between 1900 and 1960 and hence the epidemics of trypanosomiasis.

This apparent spread of tsetse following the rinderpest epidemic was however most likely due to the inadequate techniques then available for measuring and mapping tsetse populations. The serious sleeping sickness epidemics of the 1990s in both West and East Africa have been shown to be unrelated to the spread of tsetse populations. Failures in the medical services consequent on civil disruption are the likely cause of recent outbreaks in quiescent foci of Gambian sleeping sickness. A combination of molecular diagnostics and global positioning systems (GPS) has shown that recent outbreaks of Rhodesian sleeping sickness in Uganda resulted from movements of the animal reservoir (particularly restocking of cattle populations) following civil unrest.

DIAGNOSIS

Clinical Signs

A person infected with either *T. b. rhodesiense* or *T. b. gambiense* is said to have sleeping sickness but there the similarity ends. Clinically, the two diseases are quite different: *T. b. rhodesiense* is an acute infection, usually fatal within a few months if left untreated, the condition of the patient deteriorating rapidly as the parasite moves from stage 1 (early stage, hemolymphatic) to stage 2 (late stage, meningoencephalitic). *T. b. gambiense* is usually a chronic infection, often with a long period of infection with few or very common and nonalarming symptoms; people living in endemic areas are used to fevers, headaches, and other symptoms linked with other prevalent parasitic diseases. This period is followed by a chronic meningoencephalitic condition during the late stage before it is fatal. Slave shippers in West Africa rejected those with swelling of the posterior cervical lymph nodes – this is still referred to as Winterbottom's sign after the physician who described the disease in the twentieth century.

The first sign of infection may be a chancre or inflammation of the skin produced as an immune response at the site of the bite of the infected tsetse fly. Parasites then multiply in the blood and lymph, eventually invading the central nervous system causing inflammation of the brain tissue (meningoencephalitis). In the absence of treatment, death usually occurs within 6–8 months of infection with *T. b. rhodesiense*. As *T. b. gambiense* is a chronic infection it has been difficult to determine with any degree of accuracy how long a patient may have been infected. However, recent studies of a large group of over 6000 patients in Uganda and Sudan followed during Médecins Sans Frontières control operations have provided much better estimates for the duration of the two stages: mean duration of stage 1 = 526 days and mean duration of stage 2 = 2500 days, giving a mean of nearly 3 years in total, in the absence of treatment.

Diagnostic Tests

Despite recent developments in molecular diagnostics, the decision made by the physician, whether or not to treat a patient, is still based on the direct observation of parasites by microscopy for the simple reason that the drugs commonly used to treat stage 2 sleeping sickness are highly toxic and the decision to treat cannot be taken lightly. Concentration techniques have been developed to improve the sensitivity of microscopy, for example, microhematocrit centrifugation and miniature anion exchange centrifugation technique for examination of blood samples. The Gambian form is diagnosed in more that 40% of cases by examination of lymph gland aspirate using microscopy. To determine whether the disease has entered stage 2, cerebrospinal fluid must be examined for the presence of trypanosomes under the microscope. There are also concentration techniques using centrifugation to improve sensitivity. In the absence of observable parasites the cerebrospinal fluid sample is screened for leukocytes (count of more than 5 cells μl^{-1} is the cutoff point) and/or increased protein concentrations (>37 mg per 100 ml).

The most widely used immunological test for *T. b. gambiense* screening is the card agglutination test for trypanosomiasis (CATT), which detects antibodies to a dominant VAT of *T. b. gambiense*. The CATT test has the benefit of cheapness and simplicity of use; again, clinicians demand direct observation of parasites to proceed with treatment. The CATT test does not recognize *T. b. rhodesiense* infection and screening of Rhodesian sleeping sickness still largely relies on clinical signs whereas diagnosis relies on direct detection of parasites in blood or cerebrospinal fluid by microscopy.

Molecular markers that can be used to detect and differentiate *T. b. rhodesiense* and *T. b. gambiense* by using the polymerase chain reaction (PCR) are now available. The SRA gene provides a genetic marker for Rhodesian sleeping sickness and PCR identification of this gene is a very specific test. As it is impossible to distinguish *T. b. rhodesiense* from nonhuman-infective *T. b. brucei* by microscopy, it was not possible to detect the animals that were carrying human-infective *T. b. rhodesiense* without recourse to human volunteers, who were used as guinea pigs in the past. The discovery of the SRA gene has made it possible to identify animals carrying *T. b. rhodesiense* and thus to measure the risk posed to human health by infected livestock.

Various genes from the trypanosomes' antigenic repertoire have also been used to identify *T. b. gambiense* by PCR but because of sequence variation such tests may not detect parasite populations carrying slightly different antigenic variants.

Research is moving rapidly in this area with novel diagnostics such as loop-mediated isothermal amplification appearing regularly. While such developments are welcome, their use is largely confined to research projects where they can contribute to our understanding of the epidemiology of sleeping sickness. In reality, rural clinics continue to rely on microscopy and, where available, the CATT test for diagnosis; even these simple procedures may fail in the absence of basic inputs such as reagents, refrigeration, power supply, and perhaps, most importantly, trained technical staff.

TREATMENT

Chemotherapy of sleeping sickness has changed little over time, limited by the paucity of new drugs coming on stream. Early stage Gambian sleeping sickness is usually treated with pentamidine while stage 1 Rhodesian sleeping sickness is treated with suramin (pentamidine is less effective against stage 1 in this parasite). Unfortunately, there have been few therapeutic advances in the treatment of stage 2 sleeping sickness, both forms of which are still treated with melarsoprol, a toxic arsenical that results in approximately 5% fatalities due to reactive arsenical encephalopathy. Eflornithine, developed originally as a cancer treatment, was the first new drug to be approved for the treatment of sleeping sickness for 50 years and offered much promise when introduced in the 1980s. Eflornithine has never achieved widespread use because it is difficult to administer under field conditions (requiring intravenous infusion every 6 h for 14 days), nor can it be used to treat *T. b. rhodesiense*.

The effectiveness of melarsoprol treatment for *T. b. gambiense* has diminished recently with treatment failures reaching as high as 20–25% of those treated in certain foci (in northwest Uganda, Southern Sudan, Angola, and the Democratic Republic of Congo). Treatment failures for *T. b. rhodesiense* are much less of a problem. The emergence of resistance to melarsoprol is hardly surprising given that the drug has been in widespread use since its discovery in 1949. The underlying biochemical mechanisms for patient relapses in recent *T. b. gambiense* epidemics are however still not clear and may involve a combination of factors: classical drug resistance mutations (e.g., P2 adenosine transporters in the trypanosome membrane); the drug (e.g., correct administration); and the host (differences between individuals in pharmacokinetics and coinfections).

Eflornithine is now the only licensed treatment recommended for melarsoprol-resistant cases of *T. b. gambiense*. In the absence of new therapies, current efforts focus on reevaluating older drugs and combinations of drugs to determine possible combined or synergistic effects and to find ways to reduce toxicity. For example, nifurtimox, which has been used successfully in the treatment of American trypanosomiasis, is being studied in combination with eflornithine. Treatment schedules with melarsoprol have been subjected to trials and it has been shown that a 10-day course with melarsoprol given once a day is as effective as older protocols in which the drug was given over prolonged periods, interspersed with recovery periods. The 10-day schedule has economic as well as practical advantages over the very lengthy standard schedules. This concise treatment with melarsoprol has only been validated for *T. b. gambiense*; a clinical trial is ongoing for *T. b. rhodesiense*.

Search for new drugs to treat sleeping sickness is a pressing issue not only because of the rise in treatment failures with melarsoprol but also for the pragmatic reason that control programs for *T. b. gambiense* still rely on active case finding and treatment; treatment failure will lead to breakdown in disease control.

DISEASE BURDEN

The World Health Organization regularly collates and updates information gathered by countries in the affected region on the incidence of new cases of sleeping sickness. Of the 36 countries in sub-Saharan Africa endemic for sleeping sickness, 24 are at present affected by *T. b. gambiense*. Of these 24 countries, three reported more than 1000 new cases per year (Angola, Democratic Republic of the Congo, and Sudan); for the period 1997–2006, four reported more than 100 but less than 1000 cases per year (Chad, CAR, Congo, Uganda), with a further six countries having less than 100 cases per year. Therefore, there are currently seven countries in which Gambian sleeping sickness may be described as highly endemic.

Thirteen countries are endemic for Rhodesian sleeping sickness in East, Central, and Southern Africa with only two countries, Uganda and Tanzania, accounting for most of the new cases, each reporting between 100 and 1000 cases per year.

Current sleeping sickness incidence data are available through the World Health Organization Weekly Epidemiological Record at http://www.who.int/en.

Prior to 1997, the situation was very different with the World Health Organization regularly expressing concern about the rising incidence of sleeping sickness in the endemic region (Figure 5). This steep rise in incidence had followed a period of steady decline of the disease from 1930s onward as a result of intensive efforts to control the disease by colonial administrations across Africa. The postcolonial era of the 1960s saw funding for control programs dry up and, more significantly, the emergence of a period of social upheaval within the endemic countries of Africa. The reductions in disease incidence observed between 1997 and 2006 followed resolution of the humanitarian crises resulting from conflict in many of the endemic countries; surveillance and health care programs in the affected regions were restored and incidence of new cases has duly fallen.

Economic Impact

Trypanosomiases have been and remain a serious constraint to economic development in sub-Saharan Africa, impacting on the health of the people as well as on their domestic livestock. The impact of a human disease is expressed in DALY (disability-adjusted life year), which is a time-based measurement unit for estimation of the health burden caused by different diseases designed to facilitate economic comparisons across all locations and cultures.

According to recent World Health Organization/World Bank estimates, the total DALYs lost due to sleeping sickness is 1.53 million; this compares to 40.9 million for malaria, 9.27 million for tuberculosis, and 1.33 million for schistosomiasis. Among the human infectious and parasitic diseases in Africa, sleeping sickness is ranked 9th out of 25

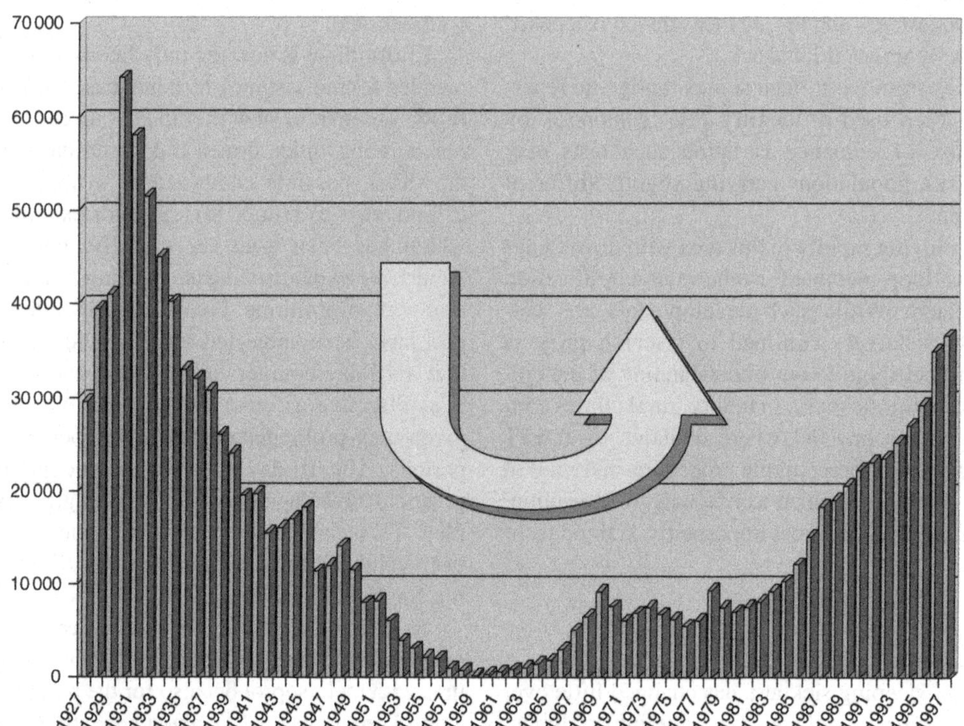

FIGURE 5 New cases of sleeping sickness reported from Africa between 1927 and 1997. Figure taken from Simarro, P. P., Jannin, J., & Cattand, P. (2008). Eliminating human African trypanosomiasis: Where do we stand and what comes next? *PLoS Medicine*, *5*(2), e55. © 2008 Simarro *et al.*

for mortality and 13th out of 25 for DALYs. However, unlike malaria to which people can develop a natural immunity, sleeping sickness is always fatal if left untreated. The costs of sleeping sickness treatment are high because of the drug regimen involved and the need for treatment in hospital.

While global statistics on sleeping sickness incidence are regularly collated from country submissions, these data are inevitably subject to rounding errors and significant but unavoidable underreporting. Few attempts have been made to quantify in detail the economic impact of sleeping sickness, as this demands the study of remote rural areas. A recent examination of a Rhodesian sleeping sickness outbreak in Serere, Uganda, has highlighted the difficulties involved in measuring the real impact of this disease by revealing that underreporting accounts for an astonishing 93% of the DALY estimate. This study has also shown that sleeping sickness cases occupy more patient admission time than all other infectious diseases excepting malaria. In other words, the burden of sleeping sickness is much greater than might be expected from its relative incidence. The total costs (including drugs and hospitalization) of treating a Rhodesian sleeping sickness case in Uganda are estimated as US$147 per patient, a not insignificant sum in the Uganda setting where the per capita total Government expenditure estimated to be available for healthcare in 2008 is US$19.

A similarly detailed case study of the economic impact of Gambian sleeping sickness on a rural community near Kinshasa in the Democratic Republic of Congo has shown that sleeping sickness-related costs for each household are equivalent to 5 months' income for the affected household. The total number of sleeping sickness-related DALYs in this small community was estimated as 2145, and interventions to control sleeping sickness averted 1408 DALYs; the cost per DALY averted was calculated to be US$17. This study concludes that because sleeping sickness has such a serious economic impact on households and controlling the disease is highly cost-effective, health priorities should take into account the impact of sleeping sickness on families in Africa, rather than deferring to global statistics for disease burden.

SLEEPING SICKNESS CONTROL

To control an epidemic, disease transmission must be stopped. Epidemiologists use the basic reproductive number, R_0, to measure the potential for the spread of a disease. R_0 is defined as the expected number of secondary infectious cases generated by an average infectious case in an entirely susceptible population. If R_0 is greater than 1, each patient will on average infect more than one individual, and thus an epidemic will be propagated. If interventions can reduce the reproductive number to less than 1, the epidemic will be under control.

For a disease to continue to exist there must be some reservoir of infection and it follows that to eliminate a disease, control measures must be directed at this reservoir if practicable; if controlling the reservoir is not achievable then decisions must be taken to agree what is an acceptable level of disease control. The two forms of sleeping sickness present different reservoir targets and these differences are reflected in the very different strategies that have been adopted by control programs to stop transmission.

Controlling Gambian Sleeping Sickness

Active case finding by mobile teams was a technique developed by the French scientist Eugene Jamot in West Africa, who saw that passive screening would not solve the problem of Gambian sleeping sickness epidemics. Jamot proposed that infected people should not be moved but should be treated wherever found by specialized mobile medical teams. In the 1920s and 1930s, hundreds of thousands of cases of sleeping sickness in francophone West Africa were diagnosed and treated in this way in the field with tryparsamide, an organo-arsenic compound.

Active case finding by mobile teams is still the principal means for controlling Gambian sleeping sickness, proceeding on the premise that the reservoir of the disease is largely human. While there is continued debate about the involvement of an animal reservoir, the success, over many years, of such control programs based on active case finding suggests that humans form the main reservoir for *T. b. gambiense*.

There are logistical problems with active case finding as sleeping sickness exclusively affects the rural poor, often in remote areas of strife-torn countries with health systems under pressure; low attendance rates at screening are common. Despite these drawbacks, this approach remains effective in reducing the incidence of Gambian sleeping sickness.

Controlling Rhodesian Sleeping Sickness

T. b. rhodesiense control demands an entirely different approach given that there is a recognized animal reservoir sustaining the parasite. Awareness of this reservoir in wild animals – based on the evidence of a single infected bushbuck – unfortunately led to the adoption by colonial regimes of game culling over vast swathes of Zimbabwe, Zambia, Mozambique, Botswana, and Uganda. The aim of these culling activities was primarily to control animal rather than human trypanosomiasis but the ever-present threat of sleeping sickness provided useful cover for such actions. In present-day Africa, game animal populations are much reduced in size and often confined to reserves but the threat to human health now comes mainly from domestic livestock that are symptomless carriers of *T. b. rhodesiense*. Recent studies of endemic foci in Uganda using molecular markers for the SRA gene have revealed just how important this domestic reservoir can be, with up to 40% of cattle carrying *T. b. rhodesiense*.

Given the overriding influence of the animal reservoir in *T. b. rhodesiense* transmission, control programs for this disease have ineluctably concluded that control could only be achieved by reducing vector numbers. For vector-borne diseases in general, and Rhodesian sleeping sickness in particular, the logic of vector control as a means of disease control is unarguable, if we ignore economics – which in the developing world cannot and should not be ignored.

As with all vector-borne diseases, the most significant variable in transmission is the longevity of the vector – first set out in the classical Macdonald–Ross model for malaria transmission. For Rhodesian sleeping sickness, it has been shown that R_0 is very sensitive to changes in the chance of a tsetse fly surviving through 1 day – reducing the chances of tsetse survival will reduce the reproductive number to less than 1 and hence control the epidemic. Modeling studies support this conclusion (Figure 6), showing that when there is an animal reservoir, tackling the tsetse vector is the best bet control option.

Prior to the introduction of insecticides, tsetse control was based on (1) the destruction of the environment thought to be conducive to tsetse survival – the ruthless clearing of bush to remove tsetse habitat and/or (2) culling of game animals providing blood meals for tsetse; of course removing all trees and shrubs would in any case remove the wild animal population so these strategies were not independent. While these severe measures were more or less effective in controlling tsetse populations they were abandoned when insecticides became cheap and plentiful in the 1940s as insecticides provide the most cost-effective way to reduce the chances of a fly surviving through 1 day. Tsetse control using insecticides became a significant part of government expenditure in the anglophone countries of West, East, and South Africa. Some of these control programs were carried out on a vast scale starting with the relatively primitive technology of ground spraying insecticides (used to treat large areas of Northern Nigeria) with ever more sophisticated insecticide techniques being introduced (e.g., aerial spraying) together with the integrated use of the sterile insect technique, which involves the release of laboratory-reared sterile males.

The enormous attendant costs of these large-scale tsetse control programs meant they became unsustainable if the goal of tsetse elimination was not reached. There are examples of effective elimination of tsetse but unsurprisingly success has often been based on the prior geographical isolation of insect populations or populations already on the ecological limits of tsetse distribution, for example, the sterile insect project on Unguja Island, Zanzibar, and the aerial spraying of the Okavango Delta in Botswana. If isolation cannot be assured then reinvasion is a constant threat to any nonisolated population of tsetse and then questions of economic sustainability come to the fore. Such operations are in any case beyond the means of the most affected countries of sub-Saharan Africa and indiscriminate use of insecticides is no longer considered environmentally acceptable.

As a result of changing economic circumstances in the countries of sub-Saharan Africa, large-scale tsetse control operations have largely given way to local interventions reliant on community participation or on the efforts of individual farmers. Facilitating this tactical change has been the development of a variety of innovative tools designed to trap and kill tsetse flies (Figure 7). Much effort has gone into trap design, to simplify manufacturing and reduce costs, which led to the deployment of targets – simply pieces of cloth impregnated with insecticides – that have proved effective on quite large scales, even in savannah environments. This bait technology became far more effective when it was shown that tsetse could be attracted over quite large distances by using cheap chemical attractants that mimicked host animal odors – odors that research had shown tsetse naturally use for host finding. Much cheaper traps offered the possibility of a sustainable approach to tsetse control, not reliant on government or external finance but based on the shared aims of a community. Working together, it was hoped that communities could afford to maintain sufficient traps around their villages to stop disease transmission. Unfortunately, community-based trapping schemes have rarely proved sustainable in practice for a variety of socioeconomic reasons centering on the fair use of a public good.

As baits for tsetse were developed to mimic both odor and visual cues of host animals, the natural progression was to treat host animals directly with insecticide; this became practical with the introduction of long-lasting formulations of synthetic pyrethroids (SP) designed to be poured on or sprayed on cattle. The costs of the original pour-on insecticide formulations proved prohibitive for smallholder farmers in Africa. However, further research on tsetse behavior revealed that it was not necessary to treat the whole animal with SP but only the parts of cattle on which tsetse preferentially feed – the legs and the belly. Costs of insecticide could then be substantially cut; this restricted application technology has brought tsetse control within the reach of poor farmers in Africa at a cost of around US$1 per animal per year. Live-bait techniques are of course only effective in controlling tsetse populations in settings where there is a sufficient density of treated livestock.

FUTURE PROSPECTS FOR CONTROLLING SLEEPING SICKNESS

A Neglected Disease

The necessary recent focus of health policy makers on HIV/AIDS, tuberculosis, and malaria, as well as on emerging or reemerging diseases, has had the unintended consequence

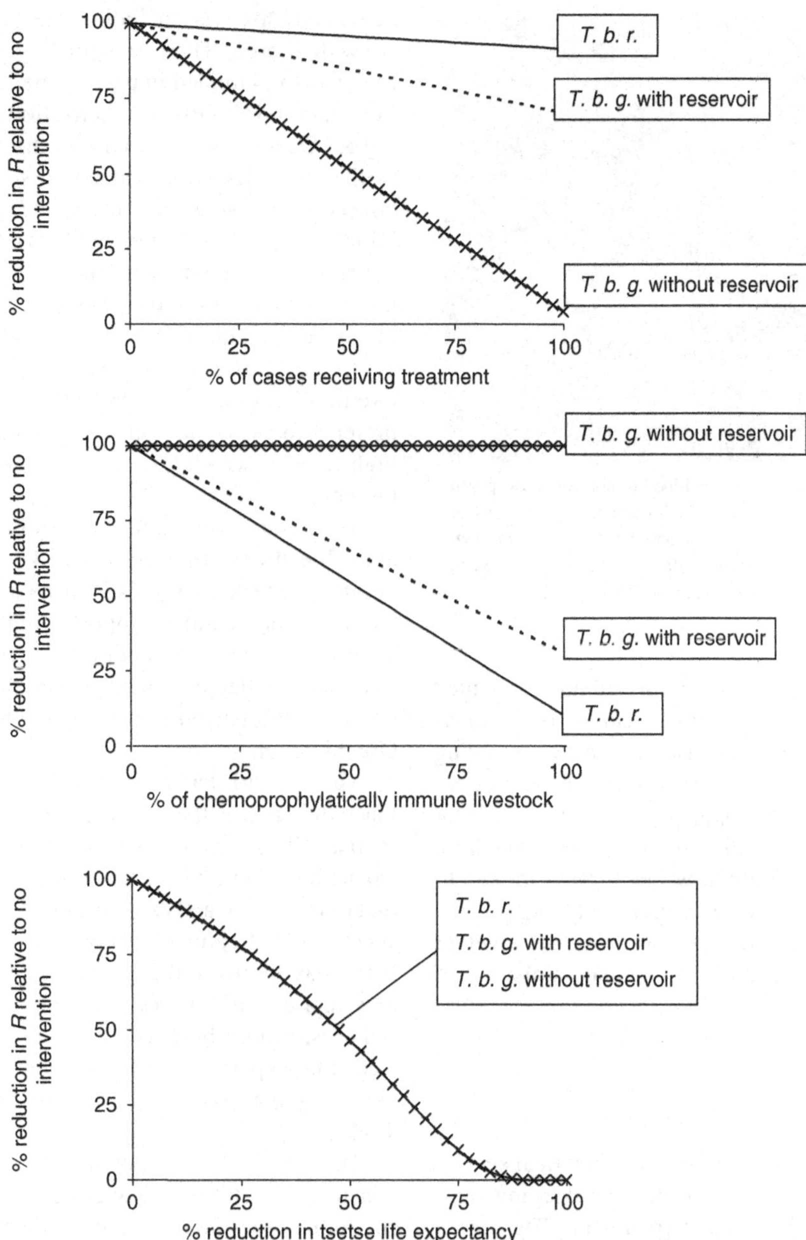

FIGURE 6 The animal reservoir and strategies for sleeping sickness control. Differences in the epidemiology of *T. b. gambiense* and *T. b. rhodesiense* have profound effects on the relative effectiveness of sleeping sickness interventions. The effectiveness of different interventions – early curative treatment of humans, chemoprophylatic treatment of animals, and vector control – has been modeled mathematically – to examine how sleeping sickness transmission will vary between the two species under different assumptions of the existence of an animal reservoir. The outcomes of this modeling are shown here. In the absence of treatment, the duration of infection of Rhodesian sleeping sickness is much shorter than the duration of untreated Gambian infections. In the model, the basic reproduction number of infection, R_0 (i.e., the number of secondary infections arising from one infectious host in a population of fully susceptible hosts), is linearly related to the duration of infectiousness, such that R_0 increases with increasing length of infectiousness. The effectiveness of early treatment of human cases (a), chemoprophylaxis of domestic livestock reservoir (b), and vector control (c) are expressed as a percentage reduction in the effective reproductive number, R, of infection relative to no control. For each intervention, the results are shown for *T. b. rhodesiense* infections assuming a livestock reservoir, and *T. b. gambiense* infections with and without a livestock reservoir. Early case treatment is most effective against Gambian infections if there is no animal reservoir (a). The presence of an animal reservoir reduces the relative effectiveness more for Rhodesian than for Gambian sleeping sickness because the duration of infectiousness is greater in the latter, so transmission in humans accounts for a greater proportion of the total R_0. Chemoprophylaxis of animals is most effective for *T. b. rhodesiense* infections, as the animal reservoir is a more important component of the overall R_0 (b). In the absence of a livestock reservoir for Gambian sleeping sickness, chemoprophylaxis of the livestock population has no effect. Vector control, shown here as increasing tsetse mortality, is equally effective for all three scenarios, regardless of duration of infection or existence of a livestock reservoir (c). *Reproduced from Welburn, S. C., Fèvre, E. M., Coleman, P. G., Odiit, M., & Maudlin, I. (2001). Sleeping sickness: A tale of two diseases.* Trends in Parasitology, 17, *19–24. Permission to use this illustration and text was granted by Elsevier.*

FIGURE 7 Simple tools have been developed to control tsetse populations. Shown here is the 'Lancien' trap, which was used to monitor tsetse populations during the sleeping sickness epidemic in Uganda in the 1980s (see Figure 4). These traps are cheap to manufacture and, when treated with insecticide, can also be effective control tools. *Photograph provided by I. Maudlin.*

of diminishing funding for control of the so-called neglected tropical diseases. Sleeping sickness is acknowledged as one of these neglected diseases – major disabling conditions that are among the most common chronic infections of the world's poorest people. It is now recognized that the (most) neglected tropical diseases have been given little attention and a number of initiatives have been formed to tackle the problems of neglected diseases involving private, public, and international organizations working together with pharmaceutical partners and national ministries of health.

Treatment

It is recognized that a safe, effective, and practical stage 2 drug (preferably stage 1 + 2) is needed for sleeping sickness as well as a simpler stage 1 treatment. The search for novel drugs to deal with the problems of resistance to currently available treatments is clearly a priority. Recent work has highlighted the importance of combined therapies in the treatment of sleeping sickness; drugs such as nifurtimox and the experimental treatment fexinidazole have both been seen to result in the rapid acquisition of resistance and as such should not be considered as a stand alone treatment. The recent announcement that WHO has approved the use of the combination of nifurtimox and eflornithine to treat chronic Gambian sleeping sickness, caused by *Trypanosoma brucei gambiense*, is a welcome step in the seemingly interminable process of searching for less toxic drugs to treat this devastating disease.

In 2003, the Drugs for Neglected Diseases initiative was set up to address global problems for treatment of neglected diseases including sleeping sickness (http://www.dndi.org/). The Drugs for Neglected Diseases initiative, which is funded in part by the Bill & Melinda Gates Foundation, is a virtual organization that integrates existing R&D capacity, especially in the developing world, as well as provides additional expertise as required. The discovery of a potent and selective *N*-myristoyltransferase inhibitor, currently designated DDD85646, that could cure experimental infections in mice with an oral dose administered over 4 days may well represent an opportunity for the development of a safe oral treatment against this neglected disease. Trypanosomes appear to be hypersensitive to inhibition of this enzyme, whose function enables proteins to associate with the cell membrane, due to the high rates at which the protective VSG coat is recycled by these parasites.

To ensure the continued supply of the currently available drugs for sleeping sickness treatment, collaboration between Aventis Pharma and the World Health Organization that also supports drug delivery in endemic countries was established in 2001. As a result, drugs for treatment of sleeping sickness are now available free of charge with distribution organized by the World Health Organization.

Recent work has highlighted the importance of combined therapies in the treatment of sleeping sickness; drugs such as nifurtimox and the experimental treatment fexinidazole have both been seen to result in the rapid acquisition of resistance and as such should not be considered as a stand alone treatment. The recent announcement that WHO has approved the use of the combination of nifurtimox and eflornithine to treat chronic Gambian sleeping sickness, caused by *Trypanosoma brucei gambiense*, is a welcome step in the seemingly interminable process of searching for less toxic drugs to treat this devastating disease.

The discovery of a potent and selective *N*-myristoyltransferase inhibitor, currently designated DDD85646, that could cure experimental infections in mice with an oral dose administered over 4 days may well represent an opportunity for the development of a safe oral treatment against this neglected disease. Trypanosomes appear to be hypersensitive to inhibition of this enzyme, whose function enables proteins to associate with the cell membrane, due to the high rates at which the protective VSG coat is recycled by these parasites.

Diagnostics

The Foundation for Innovative New Diagnostics (FIND – launched on 22 May 2003 at the World Health Assembly in Geneva) was established to develop affordable and novel diagnostic tests for diseases, including sleeping sickness, in high-burden countries. FIND (http://www.finddiagnostics.org/)

is a public–private partnership between the World Health Organization, the diagnostics industry, and other organizations supported by the Bill & Melinda Gates Foundation. FIND aims to improve trypanosome separation from blood and CSF, serological and molecular diagnostic tests, and, most importantly, 1 or 2 staging of the disease that determines the course of treatment.

Vector Control

The African Union adopted a resolution in 2000 to embark on a Pan African Tsetse and Trypanosomosis Eradication Campaign (PATTEC) for the whole of sub-Saharan Africa. Implementation of this policy is under way with six countries having initiated the first phase of a PATTEC project with financial support from the African Development Bank. PATTEC was established in general to deal with trypanosomiasis in Africa rather than to tackle human sleeping sickness.

A public–private partnership among cattle owners, a pharmaceutical company, and veterinarians has been set up specifically to control the northward spread of Rhodesian sleeping sickness in Uganda. The Stamp Out Sleeping Sickness campaign (http://www.sleepingsickness.org/) has received support from the pharmaceutical company CEVA Santé Animale and the investment group, IK Investment Partners, together with research inputs from the World Health Organization, the Wellcome Trust, and the UK Government's Department for International Development. The Stamp Out Sleeping Sickness campaign uses insecticide-treated cattle combined with trypanocidal drugs to attack vector and reservoir of the disease simultaneously to prevent convergence of the *T. b. rhodesiense* and *T. b. gambiense* sleeping sickness foci in Uganda (Figure 8).

THE FUTURE: CONTROL OR ELIMINATE?

Gambian Sleeping Sickness

Gambian sleeping sickness is unique among infectious diseases in the use of population screening as a control tool and this makes considerable demands on the workforce and budgets of endemic countries. However, given a period of political stability in the affected region and with it a steady input of trained and dedicated medical staff, there is every reason to suppose that active case finding will remain the intervention of choice and could effectively control the incidence of new cases of Gambian sleeping sickness. Recent increases in treatment failure with Gambian sleeping sickness cases suggest that, in the absence of new drugs, elimination of this disease will prove difficult. If active case finding were to be combined with tsetse control, then it may be possible to eliminate Gambian sleeping sickness from some foci provided they are not subject to reinvasion by tsetse. This is a critical proviso but the alternative strategy – area-wide elimination of tsetse from the endemic region – is an enormous and prohibitively expensive objective.

As the World Health Organization has affirmed, the challenge now is to achieve and sustain cost-effective surveillance and control for Gambian sleeping sickness. The World Health Organization advises this will be best implemented by specialized teams of trained staff working in harmony with local primary health care systems.

Rhodesian Sleeping Sickness

We have only recently become aware of the sheer scale of the domestic animal reservoir of Rhodesian sleeping sickness; given this information, control will inevitably involve efforts to reduce the risk presented by this reservoir of disease. As in the case of Gambian sleeping sickness, elimination of Rhodesian sleeping sickness will be very difficult to achieve unless tsetse control is considered a practical and economic proposition in affected countries (see Figure 6). The Stamp Out Sleeping Sickness Public Private Partnership is now successfully, and cost-effectively, controlling the northwards spread of sleeping sickness in eight districts of Uganda, fuelled by cattle movements in Uganda. The campaign has prevented convergence of the two forms of sleeping sickness in this country by a combination of animal-based chemotherapy and farmer-based restricted application of insecticides to cattle to interrupt disease transmission.

A recent study looking at the prevalence of *T. brucei* s.l. infections within 13 wild African lion prides across the Greater Serengeti Ecosystem, observed a distinct peak and decrease in age prevalence of *T. brucei* s.l. infections. High natural exposure and increasing infection resulted in the clearance of most infections of *T. brucei* by 3–5 years of age and elimination of all human infective *T. b. rhodesiense* parasites from the lions by 6 years. Frequent challenge should accelerate onset and extent of immunity, driven by an exposure dependent increase in cross-immunity following infections with more genetically diverse species. Such partial protection may be sufficient to protect animals from harbouring human infective *T. b. rhodesiense*. The theory of acquired immunity may prove the stimulus to view existing data from a different angle, and ultimately provide the empirical basis for successful vaccine development and control of human infections by clearing these parasites from animal reservoirs.

Unlike most neglected tropical diseases, particularly helminth infections, sleeping sickness in humans cannot be controlled simply by mass drug administration. The simplest conceptual approach to control remains the removal of the tsetse fly. While a declaration by African

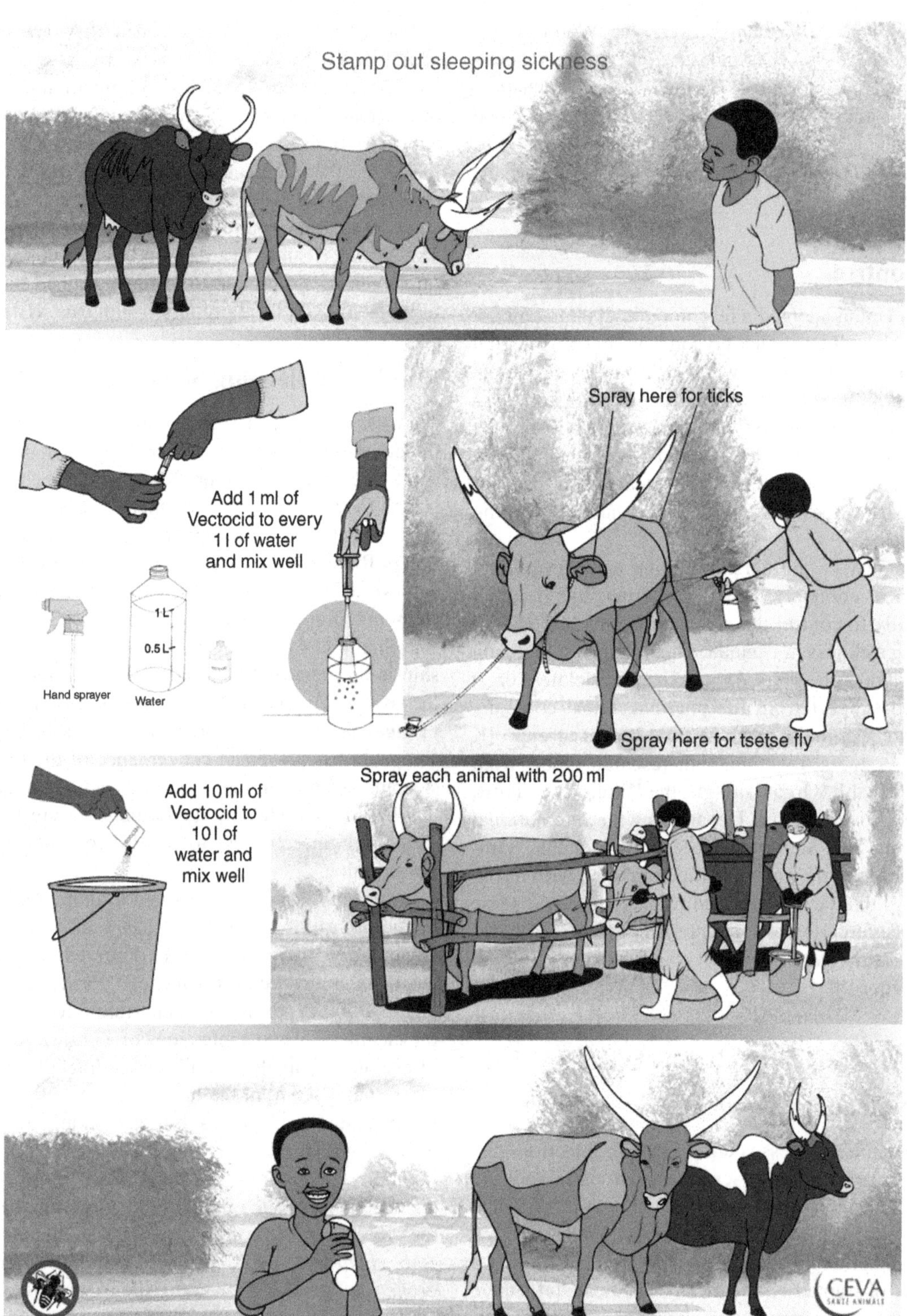

FIGURE 8 Poster from the Stamp Out Sleeping Sickness campaign uses restricted application technology to control Rhodesian sleeping sickness in Uganda. Cattle are sprayed only on the parts on which tsetse preferentially feed – the legs and the belly. Stamp Out Sleeping Sickness (http://www.sleepingsickness.org/) is a public–private partnership set up and jointly managed by the Universities of Edinburgh (UK) and Makerere (Uganda) together with the international veterinary laboratory CEVA Sante Animale (headquartered in France) and further supported by the pan European private equity firm IK Investment Partners.

heads of State in 2000 called for the eradication of tsetse from Africa, the universal desirability of this aim has been questioned:

> *[W]hether tsetse should be controlled or eradicated and, if so, how this should be achieved, have been debated with varying intensities during the last 100 years. No universally accepted answers have been found, largely because the perceived benefit of eradicating tsetse depends on the extent to which the flies either upset, or appear to protect, the distinctive type of land use beloved by each lobby, and because the people and institutions that do recommend attacking the fly often prefer the particular control technique that they themselves have developed or refined, or are best able to afford.*
>
> Hargrove, 2003

Besides these technical considerations there are compelling economic arguments against large-scale tsetse control campaigns, which may divert scarce resources, reducing those available for the multitude of more pressing needs and priorities in rural development in Africa. Today we remain still barely able to manage sleeping sickness even during the comfort afforded by the present interepidemic period. The huge rise in philanthro-capitalist investments that has been welcome in the past decade needs to translate into practical solutions for rural peoples to manage this devastating disease. Investments that we have seen in genetics and genomics may reap rewards in years to come, but in the meantime, funds must be provided to sustain effective and practical control strategies.

The Stamp Out Sleeping Sickness Public Private Partnership is now successfully, and cost-effectively, controlling the northwards spread of sleeping sickness in eight districts of Uganda, fuelled by cattle movements in Uganda. The campaign has prevented convergence of the two forms of sleeping sickness in this country by a combination of animal-based chemotherapy and farmer-based restricted application of insecticides to cattle to interrupt disease transmission.

A recent study looking at the prevalence of *T. brucei* s.l. infections within 13 wild African lion prides across the Greater Serengeti Ecosystem, observed a distinct peak and decrease in age prevalence of *T. brucei* s.l. infections. High natural exposure and increasing infection resulted in the clearance of most infections of *T. brucei* by 3–5 years of age and elimination of all human infective *T. b. rhodesiense* parasites from the lions by 6 years. Frequent challenge should accelerate onset and extent of immunity, driven by an exposure dependent increase in cross-immunity following infections with more genetically diverse species. Such partial protection may be sufficient to protect animals from harboring human infective *T. b. rhodesiense*. The theory of acquired immunity may prove the stimulus to view existing data from a different angle, and ultimately provide the empirical basis for successful vaccine development and control of human infections by clearing these parasites from animal reservoirs.

Today we remain still barely able to manage sleeping sickness even during the comfort afforded by the present interepidemic period. The huge rise in philanthro-capitalist investments that has been welcome in the past decade needs to translate into practical solutions for rural peoples to manage this devastating disease. Investments that we have seen in genetics and genomics may reap rewards in years to come, but in the meantime, funds must be provided to sustain effective and practical control strategies.

FURTHER READING

Batchelor, N. A., Atkinson, P. M., Gething, P. W., Picozzi, K., Fèvre, E. M., Kakembo, A. S., *et al.* (2009). Spatial predictions of Rhodesian human African trypanosomiasis (sleeping sickness) prevalence in Kaberamaido and Dokolo, two newly affected districts of Uganda. *PLoS Neglected Tropical Diseases, 3*, e563.

Chappuis, F., Loutan, L., Simarro, P., Lejon, V., & Büscher, P. (2005). Options for field diagnosis of human African trypanosomiasis. *Clinical Microbiology Reviews, 18*, 133–146.

Checchi, F., Filipe, J. A. N., Haydon, D. T., Chandramohan, D., & Chappuis, F. (2008). Estimates of the duration of the early and late stage of gambiense sleeping sickness. *BMC Infectious Diseases, 8*, 16.

Delespauxde, V., & de Koning, H. P. (2007). Drugs and drug resistance in African trypanosomiasis. *Drug Resistance Updates, 10*, 30–50.

Fevre, E. M., Odiit, M., Coleman, P. G., Woolhouse, M. E., & Welburn, S. C. (2008). Estimating the burden of rhodesiense sleeping sickness during an outbreak in Serere, eastern Uganda. *BMC Public Health, 8*, 96.

Fevre, E. M., Picozzi, K., Jannin, J., Welburn, S. C., & Maudlin, I. (2006). Human African trypanosomiasis: Epidemiology and control. *Advances in Parasitology, 61*, 167–221.

Ford, J. (1971). *The role of the trypanosomiases in African ecology: A study of the tsetse fly problem.* Clarendon Press: Oxford, UK.

Frearson, J. A., Brand, S., McElroy, S. P., Cleghorn, L. A., Smid, O., Stojanovski, L., *et al.* (2010). N-myristoyltransferase inhibitors as new leads to treat sleeping sickness. *Nature, 464*, 728–732.

Hargrove, J. W. (2003). *Tsetse eradication: Sufficiency, necessity and desirability.* (Research Report, DFID Animal Health Programme, Centre for Tropical Veterinary Medicine). Edinburgh: University of Edinburgh. http://www.sun.ac.za/.

Kennedy, P. G. (2006). Diagnostic and neuropathogenesis issues in human African trypanosomiasis. *International Journal of Parasitology, 36*, 505–512.

Lyons, M. (1992). *The colonial disease: A social history of sleeping sickness in northern Zaire, 1900–1940.* Cambridge University Press: Cambridge, UK.

Maudlin, I. (2006). African trypanosomiasis. *Annals of Tropical Medicine and Parasitology, 100*, 679–701.

Maudlin, I., Eisler, M. C., & Welburn, S. C. (2009). Neglected and endemic zoonoses. *Phil. Trans. R. Soc. B, 364*, 2777–2787.

Molyneux, D., Ndung'u, J., & Maudlin, I. (2010). Controlling sleeping sickness – 'When will they ever learn?'. *PLoS Neglected Tropical Diseases*, *4*, e609.

Odiit, M., Coleman, P. G., Liu, W. C., *et al.* (2005). Quantifying the level of under-detection of *Trypanosoma brucei rhodesiense* sleeping sickness cases. *Tropical Medicine and International Health*, *10*, 840–849.

Shaw, A. P. M. (2004). The economics of African trypanosomiasis. In I. Maudlin, P. H. Holmes & M. A. Miles (Eds.), *The trypanosomiases* (pp. 369–402). CABI International: Wallingford, UK.

Simarro, P. P., Jannin, J., & Cattand, P. (2008). Eliminating human African trypanosomiasis: Where do we stand and what comes next. *PLoS Medicine*, *5*, e55.

Torr, S. J., Maudlin, I., & Vale, G. A. (2007). Less is more: Restricted application of insecticide to cattle to improve the cost and efficacy of tsetse control. *Medical and Veterinary Entomology*, *21*, 53–64.

Welburn, S. C., Coleman, P. G., Maudlin, I., Fevre, E. M., Odiit, M., & Eisler, M. C. (2006). Crisis, what crisis? Control of Rhodesian sleeping sickness. *Trends in Parasitology*, *22*, 123–128.

Welburn, S., Picozzi, K., Coleman, P. G., & Packer, C. (2008). Patterns in age-seroprevalence consistent with acquired immunity against Trypanosoma brucei in Serengeti lions. *PLoS Neglected Tropical Diseases*, *2*(12), e347.

RELEVANT WEBSITES

http://www.dndi.org/ – Drugs for Neglected Diseases *initiative*.

http://www.finddiagnostics.org/ – Foundation for Innovative New Diagnostics.

http://www.sleepingsickness.org/ – Stamp Out Sleeping Sickness.

http://www.who.int/en/ – World Health Organization.

Chapter 31

Secondary Endosymbiosis

J.M. Archibald
Dalhousie University, Halifax, NS, Canada

Chapter Outline

Abbreviations 427
Defining Statement 427
The Origin of Eukaryotic Photosynthesis 427
Primary Endosymbiosis 427
Endosymbiotic Gene Transfer and Plastid Protein Import 429
Secondary Endosymbiosis 430
Diversity of Secondary Plastid-Containing Algae 430
Gene Transfer and Protein Import in Secondary Plastids 431
Number of Secondary Endosymbiotic Events 431
Nucleomorphs and Their Genomes 433
Further Reading 434

ABBREVIATIONS

ER Endoplasmic reticulum
EST Expressed sequence tag
GAPDH Glyceraldehyde-3-phosphate dehydrogenase
TIC Translocons of the inner chloroplast membrane
TOC Translocons of the outer chloroplast membrane

DEFINING STATEMENT

Secondary endosymbiosis is the uptake and retention of a eukaryotic alga by a nonphotosynthetic host, a process that has given rise to a significant fraction of algal biodiversity. This chapter summarizes current knowledge of the pattern and process of secondary endosymbiosis from the perspective of cell biology, genetics, and evolution.

THE ORIGIN OF EUKARYOTIC PHOTOSYNTHESIS

Primary Endosymbiosis

Endosymbiosis has had a profound impact on the evolution and diversification of eukaryotes. Mitochondria and plastids, the energy-generating organelles of modern-day eukaryotes, evolved from free-living prokaryotes that were taken up by eukaryotic hosts and transformed into permanent subcellular compartments. In 'modern' cells, these organelles now function as the sites of respiration and oxygenic photosynthesis, respectively. Unlike the endosymbiotic origin of mitochondria, which appears to have occurred in the common ancestor of all known eukaryotes, the endosymbiosis that gave rise to plastids occurred after the deepest divergences in eukaryotic evolution had taken place. This pivotal event paved the way for the evolution of a diverse array of algal lineages and for the spread of plastids between unrelated groups of eukaryotes by 'secondary' (i.e., eukaryote–eukaryote) endosymbiosis.

Photosynthesis is widespread in bacteria, but oxygenic photosynthesis occurs only in the cyanobacteria, a lineage of unicellular and colony-forming prokaryotes once referred to as blue-green algae. Among bacteria, only cyanobacteria possess both photosystems I and II. This allows them to use light energy to split water and produce ATP, NADPH, and glucose, liberating oxygen in the process. These products, together with carbon dioxide, are then converted to carbohydrate via the Calvin cycle. The fundamental similarities of cyanobacterial photosynthesis and that of plastids, which also possess photosystems I and II, argue strongly that eukaryotes derived their plastids from a harnessed cyanobacterium, in what is generally referred to as a 'primary' endosymbiotic event (Figure 1(a)).

The evidence supporting a cyanobacterial ancestry for plastids is overwhelmingly strong and is based on a large amount of biochemical, ultrastructural, and molecular phylogenetic data. As discussed below, three eukaryotic lineages – the green algae (and land plants), red algae, and glaucophyte algae – possess primary plastids (Table 1), indicating that they arose directly from this landmark event.

One of the hallmark features of plastids is their membranes. Most bacteria have two membranes, with a layer

Eukaryotic Microbes.

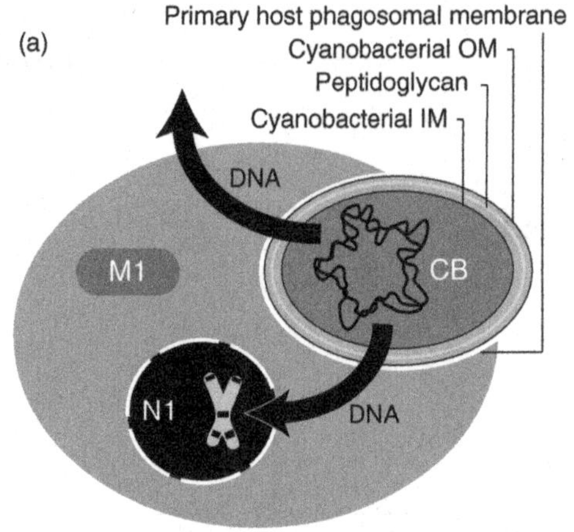

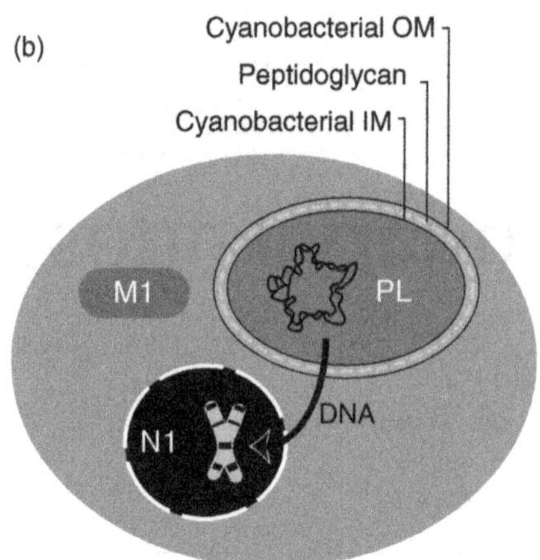

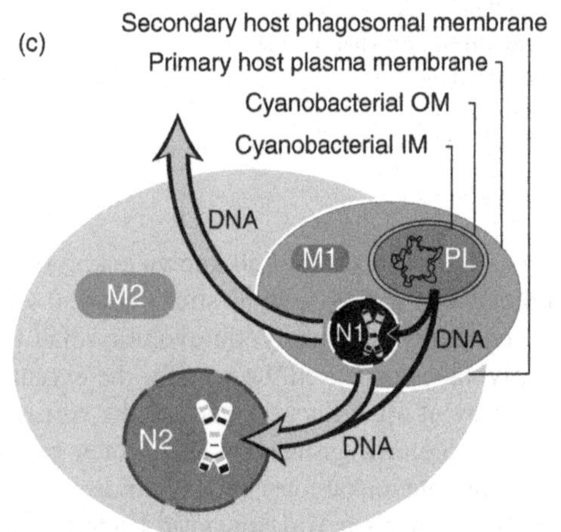

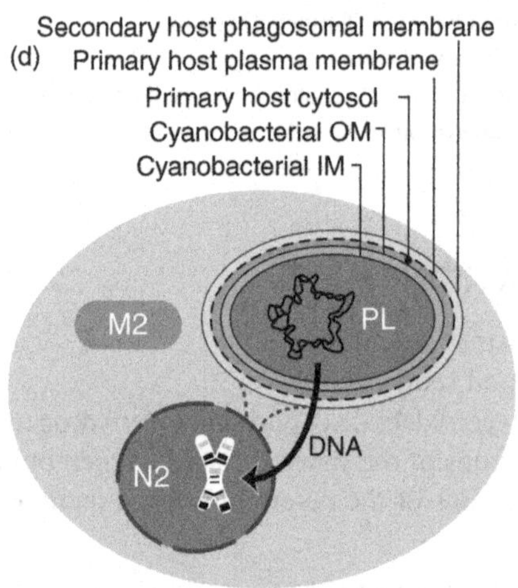

FIGURE 1 Plastid evolution by primary and secondary endosymbiosis. (a) Diagram depicting the process of primary endosymbiosis in which a non-photosynthetic eukaryote engulfs a cyanobacterium. The process involves extensive loss of DNA from the cyanobacterial genome as well as transfer of cyanobacterial genes to the nuclear genome of the host eukaryote. (b) Primary plastid-containing cell with two membranes surrounding its plastid, both of which are cyanobacterial in nature. The plastids of glaucophyte algae retain a layer of peptidoglycan between the two membranes. (c) Diagram shows the process of secondary endosymbiosis, whereby a primary plastid-containing alga is taken up by a heterotrophic eukaryote. Secondary endosymbiosis involves the large-scale movement of cyanobacterial and eukaryotic DNA from the primary host nucleus (N1) to the secondary host nucleus (N2), as well as DNA loss. DNA transfers from the plastid to the primary nucleus or directly to the secondary host nucleus are also possible. (d) Secondary plastid-containing eukaryote with three or four membranes surrounding the plastid (the primary host plasma membrane is believed to have been lost in euglenids and some dinoflagellates). The space between the inner and outer pairs of membranes corresponds to the remnant cytosol of the primary plastid-containing alga and, in cryptophytes and chlorarachniophytes, still harbors the primary host nucleus. The outermost plastid membrane in cryptophytes, haptophytes, and stramenopiles is contiguous with the nuclear envelope and endoplasmic reticulum (ER). Movement of DNA involving the mitochondrion has been omitted for simplicity. Abbreviations: CB, cyanobacterium; PL, plastid; OM, outer membrane; IM, inner membrane; M, mitochondrion; N, nucleus.

of peptidoglycan in between. Based on what is understood about the process of phagocytosis in eukaryotes, one would expect plastids to be surrounded by three membranes, two from the cyanobacterial endosymbiont and a third, outermost membrane derived from the phagosomal membrane of the heterotrophic eukaryote (Figure 1(a)). However, primary plastids possess only two membranes, both of which appear to be cyanobacterial in origin (the plastids of glaucophyte algae also possess a peptidoglycan layer, suggesting that they might represent an early divergence from the main line of plastid decent) (Figure 1(b)). One possible explanation is that the engulfed cyanobacterium 'escaped' the confines of the host cell phagocytic vacuole and took up residence in its cytoplasm. Regardless, it is reasonable to assume that the photosynthetic abilities of the endosymbiont were of great benefit to the

TABLE 1 Distribution of primary and secondary plastids and their basic characteristics

Organism	Putative origin	Cellular location	Membranes	Pigmentation
Glaucophytes	1°	Cytosol	2[a]	Chl *a* Phycobiliproteins
Green algae + land plants	1°	Cytosol	2	Chl *a*+*b*
Euglenids	2° (green)	Cytosol	3	Chl *a*+*b*
Chlorarachniophytes[b]	2° (green)	Cytosol	4	Chl *a*+*b*
Red algae	1°	Cytosol	2	Chl *a* Phycobiliproteins
Cryptophytes[b]	2° (red)	RER lumen[c]	4	Chl *a*+*c* Phycobiliproteins Alloxanthin
Haptophytes	2° (red)	RER lumen[c]	4	Chl *a*+*c*
Stramenopiles	2° (red)	RER lumen[c]	4	Chl *a*+*c*
Dinoflagellates[d]	2° (red)	Cytosol[e]	2–3[d]	Chl *a*+*c* Peridinin
Apicomplexans	2° (red)	Cytosol[e]	4[f]	None (nonphotosynthetic)
Chromera velia[g]	2° (red)	?	4	Chl *a*

Chl, chlorophyll; RER, rough endoplasmic reticulum.
[a]*Glaucophyte plastids possess a layer of peptidoglycan between the inner and outer membranes.*
[b]*The cryptophytes and chlorarachniophytes are unusual in that the nucleus of their red and green algal endosymbionts persists in highly degenerate form called a nucleomorph. In both lineages, the nucleomorph is located in the space between the inner and outer pairs of plastid membranes, which is derived from the remnant cytosol of the primary algal host cell and sometimes referred to as the periplastid space.*
[c]*In cryptophytes, haptophytes, and most stramenopiles (heterokonts), the outermost plastid membrane is continuous with the endomembrane system of the host cell. The plastid thus physically resides within the lumen of the rough endoplasmic reticulum (RER), an arrangement sometimes referred to as the chloroplast endoplasmic reticulum.*
[d]*Dinoflagellates are exceptionally diverse in terms of the type of plastid they possess. Among photosynthetic dinoflagellates, the chlorophyll* a+c *and peridinin-pigmented plastid are most common. Plastid membrane number varies depending on plastid type (see main text).*
[e]*The plastids in these lineages are not physically connected to the host cell endoplasmic reticulum, although they are surrounded by additional membranes and thus enveloped by a region of endomembrane lumen of unknown origin.*
[f]*The general consensus is that four membranes surround the apicomplexan plastid, though some studies suggest that this number can vary.*
[g]Chromera velia *is a newly discovered photosynthetic organism with a four-membrane plastid whose genes most closely ally it with apicomplexans and peridinin plastid-containing dinoflagellates.*

eukaryotic host and provided a strong selective advantage to those cells that happened to retain their endosymbionts for progressively longer periods of time. The two cells became increasingly dependent on one another and, eventually, the cyanobacterial endosymbiont evolved into a fully integrated component of the host.

Endosymbiotic Gene Transfer and Plastid Protein Import

While the genomes of modern-day cyanobacteria range from ~2 to 10 Mbp in size and encode >1000 genes, the largest known plastid genomes are only ~0.2 Mbp and possess at most ~250 genes (e.g., the red alga *Porphyra*). Roughly half of this massive genome reduction involved the elimination of genes no longer essential for intracellular life. The remainder involved the transfer of genes from the endosymbiont to the host nuclear genome, a process known as endosymbiotic gene transfer (Figure 1(a)). Based on the similarities in gene content of plastid genomes of red, green, and glaucophyte algae, much of the plastid-to-nucleus gene transfer probably occurred early in the evolution of photosynthetic eukaryotes, in a common ancestor shared by all primary plastid-containing algae. However, this has been an ongoing process throughout the evolution of photosynthetic eukaryotes and probably still occurs in at least some lineages. Indeed, examination of plant nuclear genomes reveals that they are littered with numerous and often large (albeit presumably nonfunctional) fragments of plastid DNA.

An early requirement for the functional transfer of endosymbiotic genes to the host nucleus was the establishment of a protein-targeting apparatus capable of directing the products of these nuclear-encoded genes back to their organelle of origin. Genomic and proteomic analyses of primary plastid-containing organisms have revealed that ~1000 proteins function in the chloroplast. Most of these are now nucleus-encoded and targeted to the organelle posttranslationally. This targeting is accomplished by an N-terminal 'transit peptide' extension of ~20–150 amino acids present on nearly all plastid-targeted proteins. These extensions are poorly conserved at the primary sequence level but possess a net positive charge and a hydrophobic N-terminus.

Transit peptides are recognized by receptors in the plastid membrane and the preprotein is then pulled into the plastid by a pair of translocation machineries. These are referred to as TIC and TOC, for translocons of the inner and outer chloroplast membranes, respectively. Once inside the plastid, the transit peptide is cleaved from the preprotein to yield the mature protein product. Genomic comparisons have revealed that the plastid import machineries of red and green algae are predominantly cyanobacterial in nature and quite similar to one another, suggesting that they were established in their common ancestor (little is known about protein targeting to the plastids of glaucophytes, though it is likely to be similar). While the TIC–TOC-based import pathway appears to be the predominant system for plastid protein import, it should be noted that protein targeting in the absence of canonical N-terminal signals is also possible. Exactly how this occurs is at present unclear.

While the nuclear genomes of plants and algae harbor hundreds of genes encoding proteins that service the plastid, a significant fraction of the cyanobacteria-derived genes residing in the nuclear genomes of modern-day photosynthetic eukaryotes appear to have nothing to do with photosynthesis. For example, while ~1700 of the ~9300 genes in the nuclear genome of the flowering plant *Arabidopsis thaliana* appear to be of cyanobacterial origin, less than 50% of these appear to encode proteins that are targeted to the plastid. The exact numbers are still being debated and analyses of other primary plastid-containing eukaryotes, such as the green alga *Chlamydomonas reinhardtii*, the red alga *Cyanidioschyzon merolae*, and the glaucophyte *Cyanophora paradoxa*, have yielded slightly lower estimates. Nevertheless, the cyanobacterial progenitor of the plastid clearly had a major impact on the nuclear genome of the ancestral eukaryotic phototroph, donating hundreds of genes whose products took on roles in a wide range of cellular processes unrelated to photosynthesis.

SECONDARY ENDOSYMBIOSIS

Canonical plastids appear to have evolved only once in the history of life. That is, all eukaryotic plastids appear to trace to a single endosymbiotic event. However, primary plastid-containing organisms account for only a fraction of the known diversity of photosynthetic eukaryotes. A wealth of data clearly indicates that on multiple occasions plastids have spread from one eukaryote to another by a process known as 'secondary endosymbiosis.' These events gave rise to a diverse set of environmentally, economically, and medically significant algal lineages. Secondary endosymbiosis (Figure 1(c)) occurs when a primary plastid-containing alga is taken up by a nonphotosynthetic eukaryote and transformed into a 'secondary' or 'complex' plastid (Figure 1(d)).

The process of secondary endosymbiosis is complicated by the fact that the bulk of plastid-targeted proteins are encoded in the primary host nuclear genome. Thus secondary endosymbiosis is not just the acquisition of a plastid by a new host. In order for a secondary plastid to be stably maintained and inherited, hundreds of nuclear genes must be transferred from the primary host nucleus to the secondary host nucleus. Comparative genomic analyses have shed light on the genetic processes that accompany this transformation. One of the most important advances has been the discovery that the nucleus of the algal endosymbiont still exists in some secondary plastid-bearing lineages (albeit in a miniaturized form). Among other things, this proves beyond all doubt that secondary endosymbiosis of plastids has occurred on multiple occasions.

Diversity of Secondary Plastid-Containing Algae

At the most fundamental level, secondary plastids are of two basic types: those derived from green algal endosymbionts and those that evolved from red algae. Two evolutionarily distinct lineages, the euglenids and chlorarachniophytes, possess 'green' secondary plastids (Table 1). The euglenids are a relatively well-studied group of marine and freshwater algae, which include the laboratory specimen *Euglena*. In contrast, the chlorarachniophytes are a rare and poorly understood group of amoeboflagellate algae that together with the cryptophytes are unique in that they still possess the nucleus of their algal endosymbiont, referred to as a 'nucleomorph.' Four membranes surround the chlorarachniophyte plastid, whereas the euglenid plastid has three membranes (Table 1).

The evidence supporting a green algal ancestry for the plastids of both chlorarachniophytes and euglenids comes primarily from their shared possession of chlorophyll *b* (as do green algae and land plants) and from molecular phylogenies inferred from plastid-encoded genes. In the case of chlorarachniophytes, this has also been confirmed by analysis of nucleomorph genes. While the euglenid and chlorarachniophyte plastids are clearly of green algal origin, molecular phylogenies have thus far not revealed the closest free-living relatives of their endosymbionts within the green algal/land plant radiation.

A much large number of eukaryotic algae harbor plastids believed to be of red algal origin. Perhaps best known are the stramenopiles (or heterokonts), which include unicellular forms such as diatoms and multicellular groups such as the giant kelps. Stramenopile plastids are surrounded by four membranes and contain chlorophyll $a+c$

and the accessory pigment fucoxanthin (Table 1). The haptophytes are another well-known and ecologically significant algal lineage with 'red' secondary plastids. Other groups include the cryptophytes, a relatively abundant group of marine and freshwater algae that possess a nucleomorph.

Red algae-derived plastids are also found in alveolates such as the dinoflagellates, notorious bloom formers that are also well known for their ability to 'steal' the plastids of other secondary plastid-containing algae, as well as the apicomplexans, a secondarily nonphotosynthetic group of parasites that include the malarial parasite *Plasmodium.* The most recently discovered red secondary plastid-containing lineage is *Chromera velia,* which appears to be specifically related to apicomplexans and dinoflagellates. *C. velia* is a unicellular phototroph with a four-membrane, chlorophyll *a*-containing plastid (Table 1). As will be discussed below, the evolutionary history of red secondary plastids is complex and controversial. There are currently many unanswered questions regarding when these plastids were acquired and how they are related to one another.

Gene Transfer and Protein Import in Secondary Plastids

As is the case for primary endosymbiosis, endosymbiotic gene transfer and gene loss play an important role in the establishment of secondary plastids. The plastid genomes of a variety of secondary plastid-containing algae have been sequenced and, as expected, their coding capacity is similar to that of the red or green algal plastids from which they evolved. This suggests that at the time of engulfment, most of the cyanobacteria-derived genes were already located in the nuclear genome of the primary alga. In secondary endosymbiosis, these genes must again be transferred en masse, this time from the primary host nucleus to that of the secondary host (Figure 1(c)). Genes can also be transferred directly from the plastid to the secondary host nuclear genome.

In most secondary plastid-containing algae the process of gene transfer has gone to completion and the endosymbiont nucleus has disappeared. Even in cryptophytes and chlorarachniophytes, where a nucleomorph persists, its coding capacity is extremely limited, indicating that most of the genes present in the primary algal nucleus have relocated to the secondary host nucleus. A unique aspect of secondary endosymbiosis is the potential for transfer of eukaryotic genes from the primary algal nucleus to the secondary host nucleus. The functions of many of these genes presumably overlap with those already present in the nuclear genome of the secondary host, and they would be prone to degradation and loss. However, they also have the potential to acquire novel functions in their new genomic and cellular context or to replace their secondary host nuclear counterparts. The frequency and significance of 'eukaryote–eukaryote' endosymbiotic gene transfer is currently poorly understood.

Unlike primary plastids, which are surrounded by two membranes, most secondary plastids are characterized by the presence of three or four membranes (Table 1), the outermost of which is thought to be derived from the phagosomal membrane of the secondary host (Figures 1(c) and 1(d)). This fact has important ramifications for the process of protein targeting. Thus, a critical stage in the evolution of a secondary plastid is the establishment of a second import pathway – in addition to the transit peptide-based system already used by the engulfed algal cell – capable of targeting the products of plastid-derived genes now encoded in the new host nuclear genome to their compartment of origin. Such proteins are characterized by the presence of a bipartite N-terminal extension consisting of a signal peptide as well as a transit peptide. The exact nature of the targeting process varies depending on the membrane topology of the plastid in question.

In cryptophytes, haptophytes, and stramenopiles, whose plastids are surrounded by four membranes, the outermost membrane is continuous with the nuclear envelope and endoplasmic reticulum (ER) (Figure 1(d)), and plastid preproteins are first inserted cotranslationally into the ER lumen. The signal peptide is cleaved by signal peptidase and the transit peptide is then used to divert the preprotein from the secondary host cell's normal secretion pathway and direct it across the remaining plastid membranes. In organisms whose plastids do not physically reside within the host cell's endomembrane system, such as chlorarachniophytes and euglenids (Table 1), bipartite N-terminal extensions are also used, with plastid proteins presumably being shuttled from the ER lumen to the plastid by vesicular transport. In some cases, host nucleus-encoded proteins must be targeted to the residual cytosol of the engulfed algal cell (the periplastidal space) but not to the plastid itself. In these instances, specific sequence characteristics of the N-terminal end of the transit peptide appear to determine whether or not the protein will be translocated across the remaining plastid membranes.

Number of Secondary Endosymbiotic Events

Determining the number of endosymbioses that have spawned the known diversity of secondary plastid-containing algae has proven challenging. Given that secondary plastids are either of green or red algal origin, a minimum of two endosymbioses must be inferred. However, as many as seven independent events, involving multiple green and red algal endosymbionts, have also been proposed. This discrepancy is based on different interpretations of the evolutionary significance of nonphotosynthetic relatives of the major secondary plastid-containing lineages described above. For example, in addition to possessing multiple photosynthetic groups, the stramenopiles also include many nonphotosynthetic (and plastid-lacking) lineages such as the oomycete plant pathogens. In the case of dinoflagellates, only ~50% of known species are actually photosynthetic, and the most basal lineage of cryptophytes, the genus *Goniomonas*, is a heterotrophic group that does not appear to harbor a plastid. In addition, the chlorarachniophytes and euglenids are evolutionarily deeply embedded within lineages of predominantly nonphotosynthetic groups (Rhizaria and Euglenozoa, respectively).

This 'patchy' distribution can thus be explained by invoking either a separate, recent, endosymbiotic event for each secondary plastid-containing lineage, or fewer, more ancient, secondary endosymbioses coupled with extensive plastid loss. Distinguishing between these two possibilities relies on our ability to accurately infer whether or not the phylogenies of plastid and host nuclear genes agree with one another and are consistent with all other available data, such as plastid pigmentation, ultrastructure, and biochemistry. Significant advances have been made in this area, though many uncertainties remain and the topic is actively debated in the primary literature.

Regarding the origin(s) of photosynthesis in chlorarachniophytes and euglenids, current data are most consistent with the idea that their green algal plastids are the product of two separate secondary endosymbioses. Single-locus phylogenies of plastid and nuclear genes typically provide little information about the evolutionary placement of these two lineages relative to one another, but the picture emerging from analyses of concatenated plastid- and nucleus-encoded proteins is that two evolutionarily distinct green algal endosymbionts were taken up independently by the nonphotosynthetic ancestors of the two groups.

Discerning the evolutionary history of photosynthesis in lineages harboring red algae-derived plastids is somewhat more complicated. At one extreme, the 'chromalveolate' hypothesis posits a single, ancient secondary endosymbiosis in an ancestor of cryptophytes, haptophytes and stramenopiles (chromists), and dinoflagellates, and apicomplexans and ciliates (alveolates). This demands secondary plastid loss in all plastid-lacking lineages specifically related to these groups (e.g., ciliates, oomycetes). Alternatively, a separate secondary endosymbiosis could have given rise to the plastids in each of these groups, eliminating the need to invoke plastid loss.

A variety of data have been brought to bear on the question of congruence between host cell and plastid phylogenies, with mixed results. Plastid gene phylogenies typically do not strongly favor either hypothesis, although such analyses are often hampered by limited taxonomic sampling and artifacts associated with the highly divergent sequences that are found in some of these plastid genomes. Interesting evidence consistent with the chromalveolate hypothesis has come from nuclear-encoded plastid-targeted proteins such as glyceraldehyde-3-phosphate dehydrogenase (GAPDH). Eukaryotic phototrophs typically possess two nucleus-encoded GAPDH isoforms, one noncyanobacterial protein functioning in the cytosol, and a second cyanobacteria-derived protein targeted to the plastid. In chromists, dinoflagellates, and alveolates, the plastid-targeted GAPDH protein is not cyanobacterial in origin, but is instead the product of a duplicated version of the cytosol-localized protein. This phenomenon, referred to as 'endosymbiotic gene replacement,' is potentially an important source of rare genomic characters that can shed light on deep evolutionary relationships, although it is often difficult to rule out the possibility that such genes have evolved in a nonvertical fashion (i.e., have been horizontally transferred between unrelated species).

With the increasing availability of complete genome sequences and large expressed sequence tag (EST) data sets, it is now possible to analyze large data sets of concatenated protein sequences in an effort to better resolve the interrelationships among chromalveolate taxa. To date, such analyses have provided support for a close specific relationship between some, but not all, chromalveolate lineages. Specifically, cryptophytes and haptophytes appear to be related to one another. However, there is little evidence that this pair is directly related to the stramenopiles, despite the fact that this would be predicted based on their plastid gene sequences and the membrane topology of their plastids. Instead, stramenopiles often branch as the closest relatives of the alveolates and, unexpectedly, the Rhizaria, to which the green algal plastid-containing chlorarachniophytes belong. This branching pattern is inconsistent with the chromalveolate hypothesis as originally conceived.

Inferring the minimum number of endosymbioses necessary to explain the diversity of secondary plastid-bearing eukaryotes is desirable from the perspective of understanding the processes of gene transfer and protein import. As summarized above, secondary endosymbiosis is very complex at the genetic and biochemical level: each event requires (1) the relocation of hundreds of plastid-derived genes from the primary host nucleus to the secondary host nucleus; (2) the acquisition of a functional signal peptide by each of these genes/proteins; and (3) the adaptation of the existing signal peptide secretion system such that plastid proteins containing bipartite N-terminal leader sequences can be targeted to the plastid.

Conversely, plastid loss would also seem to be difficult. Critics of the chromalveolate hypothesis point out that, in addition to photosynthesis, the plastids of plants and algae are the site of essential cellular processes such as the biosynthesis of isoprenoids and fatty acids, which explains why nonphotosynthetic organisms such as the malaria parasite *Plasmodium* retain a vestigial form of the organelle (the apicoplast). However, evidence for outright plastid loss in the apicomplexan *Cryptosporidium* and in the stramenopile *Phytophthora* has come from the presence of cyanobacteria-derived genes of apparent plastid origin in the nuclear genomes of these organisms. In addition, organisms previously thought to be plastid lacking, such as the dinoflagellate-like genus *Perkinsus*, have been found to harbor a remnant plastid. Finally, the discovery of organisms such as *C. velia*, a unicellular phototroph that appears to be the closest known relative of the apicoplexan parasites, has the potential to fill in some of the existing gaps between secondary plastid-containing lineages, such that plastid gain/loss scenarios can be better examined. In sum, there is a steady stream of new molecular and ultrastructural data to account for when considering the evolutionary history of red secondary plastids.

NUCLEOMORPHS AND THEIR GENOMES

As mentioned above, the cryptophytes and chlorarachniophytes are of special interest with respect to the study of secondary endosymbiosis. This is because they are the only two lineages in which the nucleus of the engulfed algal cell still exists inside the new host. These remnant nuclei (nucleomorphs) lie within the residual cytosolic compartment of the alga, which is located between the inner and the outer plastid membrane pairs (Figure 1(d)). The nucleomorph genomes of cryptophytes and chlorarachniophytes have been completely sequenced and are surprisingly similar in basic structure and composition. This is intriguing given that they evolved independently of one another (the cryptophyte plastid is derived from a red alga, while the chlorarachniophyte plastid has a green algal ancestry). Analysis of nucleomorph genome sequences has provided fascinating insights into the consequences of nuclear genome reduction and compaction.

The first nucleomorph genome to be analyzed in detail was that of the model cryptophyte *Guillardia theta*. The complete *G. theta* genome comprises three small chromosomes totaling a mere 551 kbp. Each of the three chromosomes is capped with unusual telomeres $(([AG_7]AAG_6A)_{11})$ and near-identical subtelomeric ribosomal DNA (rDNA) repeats. As is often the case with genomes of endosymbiotic origin, the *G. theta* genome is compositionally biased ($\sim$75% AT in single-copy regions, $\sim$55% in the repeats) and its gene density is high ($\sim$1 kbp/gene). Of 465 protein-coding genes, only 30 encode plastid-targeted proteins, counter to initial speculation that nucleomorph genomes would be enriched in genes related to photosynthesis and plastid-related processes. Instead, the bulk of the *G. theta* genome encodes proteins involved in core eukaryotic processes such as transcription, translation, and protein folding/degradation. These proteins presumably function to maintain the nucleomorph genome, express its 'housekeeping' genes, and enable it to produce the few proteins that it continues to target to the plastid. The *G. theta* gene repertoire is in fact far from complete and it is already missing numerous genes for essential cellular processes such as DNA replication as well as important structural proteins (e.g., actin). The 'missing' nucleomorph proteins are presumably encoded in the *G. theta* host cell nucleus, translated on cytosolic ribosomes, and transported to the endosymbiont using the import machinery described above.

The second cryptophyte nucleomorph genome to be sequenced was that of *Hemiselmis andersenii*, an organism quite distantly related to *G. theta*. At 572 kbp, the genome is similar in size and coding capacity to that of *G. theta* but possesses a number of interesting differences. The *H. andersenii* nucleomorph telomeres are $(GA_{17})_{4-7}$, very different from those of *G. theta*, and only three of the six chromosome ends possess intact rDNA operons, the other three harboring stand-alone 5S rDNA loci. Most unexpected is the complete absence of spliceosomal introns and genes for splicing RNA and proteins in the *H. andersenii* genome. This contrasts with the *G. theta* genome, which possesses 17 small introns and encodes a variety of proteins that make up the spliceosome, including the splicing 'platform' prp8. The *H. andersenii* genome represents the only known instance of complete intron loss in a nuclear genome, presumably as a result of intense evolutionary pressure to reduce genome size.

The nucleomorph genome of the chlorarachniophyte *Bigelowiella natans* is 373 kbp in size, significantly smaller than that of *G. theta*, *H. andersenii*, or, indeed, any known cryptophyte nucleomorph genome. Like the cryptophyte nucleomorphs, the *B. natans* genome consists of three small, AT-rich chromosomes, each with subtelomeric rDNA operons. Remarkably, the nucleomorph genomes of cryptophytes and chlorarachniophytes have converged upon the same karyotype and basic chromosome structure despite being derived from different algal endosymbionts, each of which presumably possessed dozens of nuclear chromosomes. The significance of this convergence is unknown. The *B. natans* nucleomorph encodes 293 protein genes, only 17 of which are involved in photosynthesis. Unlike the situation in cryptophytes, introns are abundant in the nucleomorph genes of *B. natans*, with an average of $\sim$3 introns/gene. The introns themselves are, however, greatly reduced in size, being between 18 and 21 bp in size, the smallest spliceosomal introns known. Overall, the functional distribution of nucleomorph genes in *B. natans* is similar to that seen in *G. theta* and *H. andersenii*, with the exception of genes involved in RNA metabolism, for which *B. natans* is somewhat enriched. Like in cryptophytes, the *B. natans* host nuclear genome has presumably acquired a substantial number of endosymbiont nuclear genes, whose protein products are posttranslationally targeted to the endosymbiont cytoplasm.

One of the most fundamental, and unanswered, questions regarding nucleomorph evolution is why they exist at all. By definition the process of secondary endosymbiosis includes a 'transition period' during which the nucleus of the algal endosymbiont resides within its residual cytosolic compartment and continues to service its plastid. However, other than cryptophytes and chlorarachniophytes, the nucleomorphs of all other secondary plastid-containing lineages have completely disappeared, having transferred all of their essential genes to the secondary host nucleus. There are two main possibilities for the persistence of nucleomorphs. First, it is possible that not enough time has elapsed for nucleomorph-to-host-nucleus gene transfer to go to completion and the nucleomorph genomes (and nucleomorphs) of cryptophytes and chlorarachniophytes will eventually disappear. Alternatively, there may be barriers to the successful transfer of the remaining genes in the cryptophyte and chlorarachniophyte nucleomorph genomes, as has been proposed to explain the persistence of mitochondrial and plastid genomes.

There are currently not enough data to determine which of these two possibilities is most likely, though the comparison of the gene content of the nucleomorph genomes of the cryptophyte *G. theta* and the chlorarachniophyte *B. natans* is consistent with the former scenario. There is very little overlap in the suite of plastid proteins encoded in the *G. theta* and *B. natans* genomes, suggesting that nucleomorph to host nucleus gene transfer could essentially be a random process. Although the nucleomorph genome of the cryptophyte *H. andersenii* encodes the identical set of 30 plastid proteins found in *G. theta*, it is difficult to assess the significance of this observation, given that the two genomes are the product of the same secondary endosymbiotic event.

Complete genome sequences from additional cryptophytes and chlorarachniophytes should make it possible to better discern patterns of gene content variation within and between the nucleomorphs of both lineages and lead to a more comprehensive understanding of underlying forces and the possible reason(s) for nucleomorph persistence. In particular, exploration of nucleomorph genomes that are significantly smaller and larger than those so far examined will be particularly informative. Such analyses will make it possible to identify and study specific instances of nucleomorph to host nucleus gene transfer, which may lead to a deeper understanding of the general process of gene transfer among unrelated organisms. In addition, these data will help elucidate the cellular processes taking place in the remnant cytoplasmic compartments of cryptophyte and chlorarachniophyte algae, which are models of genome compaction and distillation of essential cellular processes. The ultimate goal will be to gain a better understanding of the process of secondary endosymbiosis and its role in the evolution of eukaryotic photosynthesis.

FURTHER READING

Archibald, J. M. (2007). Nucleomorph genomes: Structure, function, origin and evolution. *BioEssays*, *29*, 392–402.

Bhattacharya, D., Yoon, H. S., & Hackett, J. D. (2003). Photosynthetic eukaryotes unite: Endosymbiosis connects the dots. *BioEssays*, *26*, 50–60.

Cavalier-Smith, T. (1999). Principles of protein and lipid targeting in secondary symbiogenesis: Euglenoid, dinoflagellate, and sporozoan plastid origins and the eukaryote family tree. *Journal of Eukaryotic Microbiology*, *46*, 347–366.

Gould, S. B., Waller, R. F., & McFadden, G. I. (2008). Plastid evolution. *Annual Review of Plant Biology*, *59*, 491–517.

Graham, L. E., & Wilcox, L. W. (2000). *Algae*. Prentice-Hall: Upper Saddle River, NJ.

Keeling, P. J., Burger, G., Durnford, D. G., *et al.* (2005). The tree of eukaryotes. *Trends in Ecology and Evolution*, *20*, 670–676.

Lane, C. E., & Archibald, J. M. (2008). The eukaryotic tree of life: Endosymbiosis takes its TOL. *Trends in Ecology and Evolution*, *23*, 268–275.

Larkum, A. W., Lockhart, P. J., & Howe, C. J. (2007). Shopping for plastids. *Trends in Plant Science*, *12*, 189–195.

Martin, W., Rujan, T., Richly, E., *et al.* (2002). Evolutionary analysis of arabidopsis, cyanobacterial, and chloroplast genomes reveals plastid phylogeny and thousands of cyanobacterial genes in the nucleus. *Proceedings of the National Academy of Sciences of the United States of America*, *99*, 12246–12251.

McFadden, G. I. (1999). Plastids and protein targeting. *Journal of Eukaryotic Microbiology*, *46*, 339–346.

Palmer, J. D. (2003). The symbiotic birth and spread of plastids: How many times and whodunit? *Journal of Phycology*, *39*, 4–11.

Timmis, J. N., Ayliffe, M. A., Huang, C. Y., & Martin, W. (2004). Endosymbiotic gene transfer: Organelle genomes forge eukaryotic chromosomes. *Nature Reviews Genetics*, *5*, 123–135.

Chapter 32

Algal Blooms

P. Assmy and V. Smetacek
Alfred Wegener Institute for Polar and Marine Research, Bremerhaven, Germany

Chapter Outline

Abbreviations 435
Defining Statement 435
Introduction 435
Physical Environment of Blooms 436
Chemical Environment of Blooms 437
Major Contributors to Algal Blooms 438
Cyanobacteria 438
Haptophytes (Prymnesiophytes) 440
Dinoflagellates 441
Other Groups 442
Pathogens and Grazers of Algal Blooms 442
Recurrent and Unusual Algal Blooms 443
Spring Blooms 443
Autumn Blooms 445
Blooms in Upwelling Regions of Low Latitudes 445
Miscellaneous Algal Blooms 446
Harmful Algal Blooms 446
Iron-Fertilized Blooms 447
Future Research Avenues 448
Recent Developments 448
Marine Genomics 448
Abandoning Sverdrup's Critical Depth Hypothesis 449
The Role of Grazing in Suppressing Nondiatom Blooms and Recycling Iron in Pelagic Ecosystems 449
Further Reading 450
Relevant Websites 450

ABBREVIATIONS

DMS Dimethylsulfide
DMSP Dimethylsulfoniopropionate
HABs Harmful algal blooms
HNLC High-nutrient, low-chlorophyll
PCD Programmed cell death
PSP Paralytic shellfish poisoning
SML Surface mixed layer

DEFINING STATEMENT

Dense aggregations of phytoplankton cells of one or more species are loosely referred to as algal blooms. They play a central role in the ecology and biogeochemistry of all water bodies from ponds to the ocean, but a mechanistic understanding of the factors leading to the rise and fall of blooms is still lacking.

INTRODUCTION

'Algal bloom' is a term of convenience applied to an outbreak of phytoplankton cells well above the average for a given region or water body. Blooms show up as peaks in the annual cycle of phytoplankton biomass and chlorophyll concentrations. They are ephemeral phenomena that arise when growth rates of one or more species of the phytoplankton assemblage exceed their mortality rates. As a result, cells of these species accumulate in the water column until their growth is checked by resource depletion, generally a nutrient such as phosphorus, reactive nitrogen, or iron. The magnitude of the bloom peak is determined by the nutrient concentrations prior to outbreak of the bloom. Since all phytoplankton contain chlorophyll *a*, which is easily assessed by a range of methods, the size of a bloom is conveniently expressed in units of chlorophyll. In productive water masses with high-nutrient availability, such as eutrophic lakes and coastal seas, chlorophyll concentrations at bloom peaks can reach >20 mg Chl m^{-3}, but over much of the ocean tenfold lower values are also commonly referred to as blooms because their concentrations again are 5- to 10-fold higher than the concentrations prevailing there over much of the year.

Phytoplankton blooms develop in the surface mixed layer (SML), so the total biomass or standing stock of a bloom is expressed as the concentration (cell numbers or biomass m^{-3}) multiplied by the depth of the mixed layer (cell numbers or biomass m^{-2}). The latter varies from less

Eukaryotic Microbes.

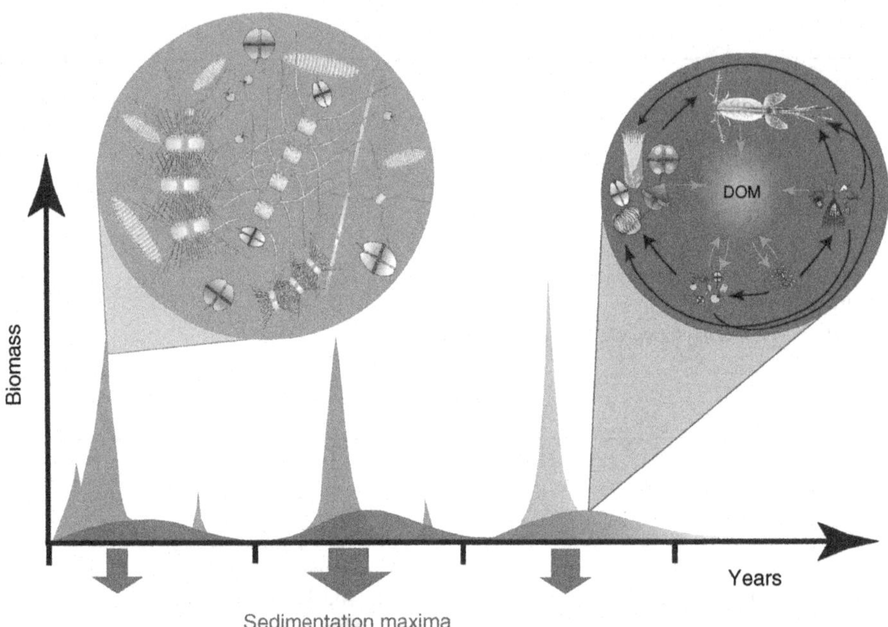

FIGURE 1 A schematic representation of the relationship between seasonality of the microbial network (regenerating system) and superimposed blooms. The latter are followed by sedimentation pulses. DOM, dissolved organic matter. *Modified by Christine Klaas from Smetacek, V., Scharek, R., & Nöthig, E. M. (1990). Seasonal and regional variation in the pelagial and its relationship to the life history cycle of krill. In K. R. Kerry, & G. Hempel (Eds.),* Antarctic ecosystems: Ecological change and conservation *(1st ed., pp. 103–114). Berlin, Heidelberg: Springer. © Christine Klaas.*

than a few meters in lakes to tens of meters in the ocean, so standing stocks of blooms in both regions are broadly similar because the higher concentrations of algae generally found in shallower mixed layers are compensated by the greater depth of the surface layer where concentrations are lower. It is the standing stock of the bloom that determines how much carbon is fixed per unit area and hence how much food is available to the larger organisms of higher trophic levels and to the microbes and animals (benthos) of the underlying sediments. It follows that the highest standing stocks (largest blooms) develop in SMLs of intermediate depths.

Algal blooms are short-lived phenomena superimposed on the ubiquitous microbial food web, which is the characteristic pelagic ecosystem prevailing in the surface layer of most of the oceans and large lakes (Figure 1). The system maintains itself by recycling regenerated nutrients (ammonia, phosphate, and iron) between phytoplankton, comprising $<2\ \mu m$ solitary cyanobacteria and small ($<10\ \mu m$) autotrophic flagellates, and heterotrophs, comprising bacteria, their nanoflagellate grazers, and larger protozoa in particular ciliates and dinoflagellates that feed mainly on the eukaryotic components of the food web. They in turn are grazed by copepods and other suspension-feeding zooplankton. Algal biomass is generally <20 mg Chl m^{-2}, that is, concentrations of <2 mg Chl m^{-3} and <0.5 mg Chl m^{-3} in coastal and open-ocean regimes, or eutrophic and oligotrophic lakes, respectively. Heterotrophic biomass is generally several fold that of the autotrophs. The opposite situation prevails during blooms: additional or new nutrients, generally in the form of nitrate, are introduced by admixture of nutrient-rich water, which disrupts the quasi-steady state of the microbial food web. The microbial community also profits from the new nutrients, but the bulk of the biomass increase is due to larger phytoplankton inaccessible to smaller grazers. These are the bloom-forming species.

In the following we first deal with the so-called bottom-up factors that influence bloom formation – the physical and chemical properties of the environment. After an overview of the major genera and species contributing to blooms, we deal with the top-down factors – pathogens, parasitoids, and predators – that kill phytoplankton cells. Since the term parasitoid is not very commonly used, we define it as an organism that kills the host as a result of parasitism. The various types of algal blooms – recurrent and sporadic – are discussed followed by brief accounts of harmful algal blooms (HABs) and artificial blooms induced by iron fertilization in the open ocean. We conclude with some avenues that need to be explored to further our mechanistic understanding of the rise and fall of algal blooms.

Physical Environment of Blooms

The depth of the SML is generally determined by wind mixing, which in turn is a function of wind force and fetch, which is why the SML is shallowest in small, protected lakes and deepest in the open ocean. The density difference between surface and deeper layers constrains the depth of mixing because the greater the gradient the more energy required to mix across it. Generally, temperature determines density, so the SML is at its shallowest during the summer and in low latitudes. The steep density gradient between the warm SML and the underlying colder water is known as the thermocline. Exceptions are found in estuaries, brackish seas, and at the sea-ice edge during melting where the vertical salinity gradient can have a greater effect than temperature in constraining depth of the SML. Here, the boundary between the fresher SML and the saltier underlying water is known as the halocline. Pycnocline is the collective term for a density gradient whether caused by temperature, salinity, or both.

The depth and rate of vertical mixing of the SML is of critical importance in bloom formation because it determines the light climate experienced by the phytoplankton suspended in it. Light intensity decreases exponentially with depth because light is absorbed by water and converted into heat. Suspended particles and some classes of dissolved organic matter such as humic acids (gelbstoff) also absorb light or scatter it, contributing to attenuation of light intensity with depth. In the clear blue water of the open ocean, living phytoplankton are the major light-absorbing particles because detritus plays a minor role and most protozoa and zooplankton are transparent to avoid visual predators. At the other extreme, in the green waters of turbid, shallow lakes and tidally mixed shallow seas, suspended particles (detritus and lithogenic particles) scatter and absorb a far greater percentage ($>90\%$) of incoming light than the phytoplankton. So with increasing depth and turbidity of the SML, the light climate experienced by the phytoplankton population as a whole declines.

Net growth of an algal cell occurs only when carbon fixed by photosynthesis exceeds remineralization by respiration. The light intensity at which the two opposing processes balance each other is termed as the compensation light intensity and is in the order of 0.1% of incoming radiation at the surface. Its location in the water column sets the lower boundary of the euphotic zone within which net growth occurs. The depth of the euphotic zone can range from decimeters to meters at low solar angles (winter of high latitudes) and in highly turbid water, to >50 m in the open ocean. The relationship between the depths of the euphotic zone and the SML determines the rate of bloom growth. If the SML is much deeper, then loss processes outweigh gains by photosynthesis, and phytoplankton biomass cannot increase. In the reverse case, given the presence of adequate nutrients, growth can be sufficiently rapid to attain bloom proportions.

However, the relationship between SML and euphotic zone depths on bloom development is rendered complex by the biology of plankton organisms. Thus, phytoplankton can adjust their photosynthetic machinery to work efficiently at low light levels (shade adaptation) of a deepwater column, but the cells suffer damage at high light intensities prevailing closer to the surface if transported there by vertical mixing. A day or two of full sunshine in otherwise cloudy weather can have the same negative effect on shade-adapted cells. Clearly, the species with the highest growth rates will have achieved the optimal trade-off respective to the inherently fluctuating physical environment. On the other hand, loss rates due to pathogens and grazers also vary widely and are generally of much greater importance than respiratory losses of the phytoplankton.

Satellite observations of chlorophyll distribution indicate that the occurrence of algal blooms is related to nutrient rather than light fields. Light limitation of growth rate is restricted to the few winter months at high latitudes. It should be mentioned that the daily photon flux in midsummer at the poles is equivalent to that in the tropics because low solar angle is compensated by the much greater day length. Temperature also has a secondary effect on algal growth rates because of cold adaptation. Unlike land plants that are exposed to widely fluctuating temperatures in the transition phase from dormant to growth seasons, aquatic plants can adapt to a narrow temperature range. Thus, maximum division rates close to the freezing point can be as high as 1 per day, which is more than half that attained in the tropics. Indeed among the highest rates of primary production measured anywhere (<10 g C m^{-2}) were recorded in dense blooms developing in the nutrient-rich, shallow SML of the Bering Sea inflow to the Arctic Ocean.

Chemical Environment of Blooms

A prerequisite for the growth of blooms is the availability of essential elements of which the macronutrients phosphorus and reactive nitrogen and the micronutrient iron generally limit buildup of blooms. Phytoplankton grow by taking up dissolved nutrients and incorporating them into biomass. With the exception of lipids, organic matter is denser than water, so particularization eventually results in loss of nutrients from the SML due to sinking particles. Nutrients are supplied by mixing of deep, nutrient-rich water to the SML. In high and midlatitudes, this is invariably due to cooling and breakdown of thermal stratification by convective mixing during autumn and winter. In the tropics, upwelling of deeper water due to large-scale lateral advection of the surface layer is the major source of nutrients, although runoff from land is also locally important. The passage of storms also mixes nutrients into the SML.

Bloom biomass is limited by the concentration of the nutrient in shortest supply. There is now general consensus that it is phosphate in lakes and some oligotrophic seas (e.g., the Mediterranean), nitrate in iron-rich coastal waters, and iron in the open ocean with the exception of most of the North Atlantic. The concentrations of phosphate and iron are limited by their geochemical properties. Thus, maximum winter concentrations of ~ 1 μmol PO_4 l^{-1} seem to be the rule in coastal regions where surface water is in interaction with oxygenated sediments. The corresponding iron concentrations are <1 nmol Fe l^{-1}, which are well in excess of the P:Fe ratio required by phytoplankton. In contrast, dissolved molecular nitrogen is abundant and can be fixed to reactive forms by cyanobacteria, which results in maintenance of N:P ratios commensurate with the demands of phytoplankton. Because of the extreme insolubility of ferric hydroxide (rust), iron is selectively lost from the SML and becomes the limiting nutrient in land-remote regions. The subarctic North Pacific Ocean and the entire Southern Ocean harbor perennially high concentrations of nitrate and phosphate because of iron limitation of phytoplankton growth. The same is true of the equatorial Pacific. Phytoplankton growth is not continuously iron-limited in the high latitude and equatorial Atlantic presumably due to iron input by dust emanating from the adjacent arid regions.

As unicellular algae are too small to experience shear stress caused by the kinetic energy of water, they do not require investment in supporting structures such as carbohydrates to maintain shape and position as do all larger metaphytes. They also do not store reserves such as starch or lipids, which, in sessile plants, enable spurt growth at the beginning of the growth season as a means to overgrow competitors for space and light. As a result, the C:N:P ratio of aquatic unicellular organisms resembles that of living cytoplasm and varies <20% around the global average of 106:16:1. This is known as the Redfield ratio after the scientist who first discovered its relative constancy not only in suspended matter but also in dissolved nutrients in the deep ocean. It should be pointed out that variation in the Redfield ratio within species as a result of adaptation to the growth environment can be as great as between the algal classes.

MAJOR CONTRIBUTORS TO ALGAL BLOOMS

Of the several thousand phytoplankton species described so far, <5% have been reported to contribute significantly to algal blooms in lakes and oceans, which implies that the ability to form blooms is not an indication of evolutionary success. Many species are widely distributed and regularly contribute to blooms throughout their range, implying that adaptation to a specific set of environmental conditions, such as nutrient ratios or light conditions, is not a prerequisite. Bloom-forming species differ widely from one another in shape, size, and behavior, and – because they belong to different phylogenetic groups – also differ in their pigment composition and biochemistry. Not surprisingly, their blooms also have very different impacts on the ecosystem and biogeochemistry. Their degree of recurrence also varies. Some species bloom every year in the same water body and are an integral part of the seasonal cycle. Others reach bloom proportions only in some years but always in the same season. Yet others form blooms seemingly haphazardly. It is thus not possible to formulate generalizations about bloom-forming species other than that their size (cell diameter, chain length, and colony diameter) tends to be larger than 10 μm. Since smaller cells tend to have higher growth rates, the ability to grow fast per se cannot be regarded as a prerequisite for bloom formation. Apparently, the ability to withstand attack by pathogens and herbivores by means of larger size and other defenses is hence of greater importance in enabling accumulation of cells. Here, we refer to species differentiated under the microscope, that is, on the basis of their morphology (Figure 2). Genetic studies are revealing ‘cryptic’ species within morphologically similar cell walls. In the following we deal with the bloom-forming species according to taxonomic groupings with some brief notes on their ecology and impact where appropriate.

Cyanobacteria

This group, formerly known as blue-green algae, is prokaryotes and the progenitors of the chloroplasts of all eukaryotes. They possess essentially the same photosynthetic machinery as the algae but, unlike all algae, some species of cyanobacteria are able to fix molecular nitrogen and channel it into the ecosystem. Another unique property of the larger species is their ability to secrete gas bubbles within the cell, which renders them positively buoyant and results in formation of conspicuous scum on the surface. Since nitrogen fixing is an energy-intensive process, it is hypothesized that the gas bubbles enable them to stay at the surface.

Cyanobacteria tend to be rare when fixed nitrogen in the form of nitrate or recycled ammonia is available, but increase their numbers when nitrogen becomes limiting for algal growth but other nutrients, such as phosphate and iron, are still available. This situation invariably arises in the summer months when the N:P ratio, normally at 16:1, reaches its annual minimum. In oxygenated waters characteristic of most of the ocean as well as oligotrophic deep lakes this happens because P is recycled more rapidly than N, hence selectively lost via sinking particles. In shallow eutrophic lakes with suboxic deep water, nitrate is oxidized to N_2 by denitrifying bacteria, which further lowers the N:P ratio. The lowest N:P ratios (<10:1) are found in water bodies overlying anoxic sediments that mobilize not only P but also Fe. It follows that the highest concentrations of cyanobacteria occur in highly eutrophic lakes and ponds.

The most common bloom-forming genera in freshwater, but also brackish seas such as the Baltic, are dealt with next. The ubiquitous genus *Microcystis* builds spherical colonies, but all the other genera occur either as solitary cells (such as the bacteria-sized *Synechococcus*) or as chains (*Nodularia*, *Aphanizomenon*, and *Anabaena*). The latter generally form aggregates in the shape of bundles or globular clumps of

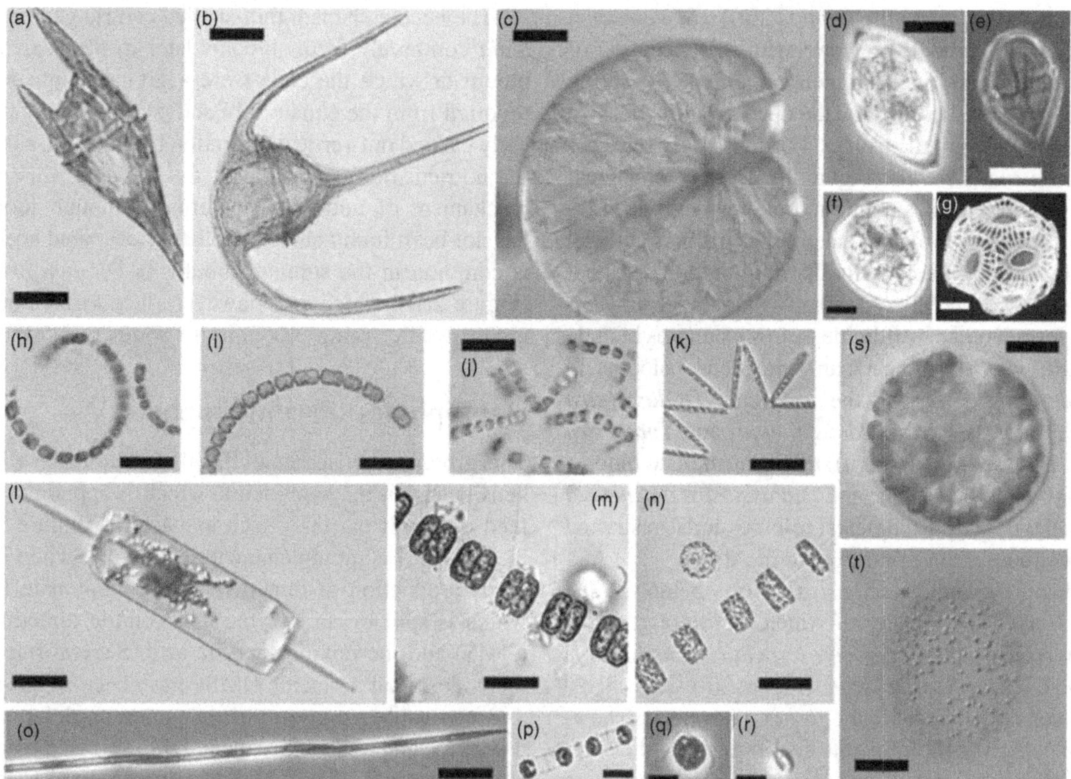

FIGURE 2 Pictures of prominent bloom-forming species. (a) *Ceratium furca*, (b) *Ceratium tripos*, (c) *Noctiluca scintillans*, (d) *Heterocapsa triquetra*, (e) *Scrippsiella trochoidea*, (f) *Prorocentrum minimum*, (g) *Emiliania huxleyi*, (h) *Chaetoceros debilis*, (i) *Chaetoceros curvisetus*, (j) *Chaetoceros socialis*, (k) *Thalassionema nitzschioides*, (l) *Ditylum brightwelli*, (m) *Thalassiosira nordenskioeldii*, (n) *Thalassiosira rotula*, (o) *Pseudo-nitzschia lineola*, (p) *Skeletonema costatum*, (q) *Chrysochromulina polylepis*, (r) *Phaeocystis antarctica* solitary cell, (s) *P. antarctica* compact small colony, and (t) *P. antarctica* large colony. Scale bars = 2 μm (g), 5 μm (f), 10 μm (d, e, m, and p–r), 15 μm (a), 20 μm (l, n, o, and s), 30 μm (b), 50 μm (h, j, and k), 60 μm (i), 100 μm (t), 200 μm (c). Light micrographs (o), (r), (s), and (t) have been kindly provided by Marina Montresor. Scanning electron micrograph (g) was taken by Philipp Assmy. All other light micrographs were taken from the open access repository for plankton-related information PLANKTON*-NET (URL: http://planktonnet.awi.de) of which (a), (b), (h–k), (m), and (p) were kindly provided by Mona Hoppenrath; (d) and (f) by Regina Hansen; (l) and (n) by Tanya Morozova; (c) by Susanna Knotz; (q) by http://www.algaebase.org/index.lasso; and (e) by Alexandra Kraberg.

chains. These species have specialized cells known as heterocysts in which nitrogen is fixed. Cyanobacteria are generally well defended and some species, particularly *Microcystis*, *Anabaena*, and *Nodularia*, sometimes secrete potent toxins. Blooms of these species can pose a serious problem in reservoirs and lakes. The only genus that regularly forms blooms in the ocean is *Trichodesmium*, which grows in sheathed chains (trichomes) that form large aggregates that often color the surface of many warm ocean regions a reddish-brown. Toxins are apparently not produced, but this genus is not relished by grazers.

Diatoms

Most algal blooms in oceans and lakes are dominated by a broad range of species belonging to this group (Figure 2(h)–2(p)). Diatoms have an obligate need for silicic acid because the cell is encased by silica shells comprising two half-boxes that fit tightly into one another. The shell surfaces are dotted with minute pores that only allow dissolved substances to enter into the cells. The fact that, in contrast to other algal groups, only few instances of viral infection of diatoms have been reported so far suggests that diatoms are well defended against pathogens possibly due to the silica shells. Since silica shell formation is an energetically cheap process, diatoms tend to have high growth rates. They can also regulate buoyancy despite the ballast effects of the shell. Shape and size of bloom-forming diatoms vary widely from long-needle-shaped cells up to few millimeters long to small cylindrical cells in long chains. Some species have long silica spines or chitin threads that protrude outward from the cells. Bloom-forming species are scattered over all diatom lineages (raphid and araphid pennates and various families of centrics) and many genera.

The most prominent recurrent feature of the seasonal plankton cycle in temperate and boreal systems is the spring bloom, which is generally dominated by comparatively few species of unrelated genera of diatoms. Typical examples of recurrent bloomers are species of the cosmopolitan diatom genera *Skeletonema*, *Thalassiosira*, and *Chaetoceros*

(Figure 2(h)–2(j), 2(m), 2(n), and 2(p)). *Skeletonema* blooms occur throughout the world with the exception of the polar oceans. They are prominent not only in the spring blooms of the brackish Baltic and the open Atlantic, but also in upwelling blooms of the tropics. Formerly only one species, *Skeletonema costatum* (Figure 2(p)), was recognized, but recent detailed studies on the morphology and genetics of this species have so far revealed eight distinct entities, which differ in their geographic distribution and seasonal occurrence. A wide variety of species within the genus *Thalassiosira* form blooms worldwide. A few examples include *Thalassiosira antarctica* that can form dense blooms in layers underlying sea ice in the Antarctic; *Thalassiosira weissflogii*, *Thalassiosira nordenskioeldii*, and *Thalassiosira rotula* (Figure 2(m) and 2(n)) that contribute to blooms in temperate and boreal areas; and *Thalassiosira partheneia* that builds large (> 1 cm diameter) colonies and is restricted to the upwelling region of northwestern Africa.

The bloom-forming *Chaetoceros* species belong to the subgenus *Hyalochaete*, many of which are bipolar. Three prominent species are *Chaetoceros socialis*, *Chaetoceros debilis*, and *Chaetoceros curvisetus* (Figure 2(h)–2(j)), all of-which are cosmopolitan, with the former two species showing a more polar-to-cold temperate distribution and the latter species a warm temperate distribution. *C. socialis* grows in spherical colonies that bear superficial resemblance to those of the haptophyte *Phaeocystis*. The species not only dominates in open-ocean, meltwater-associated blooms in the Antarctic and Arctic, but also in blooms of upwelling areas. Most bloom-forming *Chaetoceros* species are capable of converting vegetative cells into thick-walled resting spores that overwinter on the sediment surface in shallow environments, inside the sea ice in the seasonal sea-ice zone, and in the pycnocline in the open ocean. The widespread and consistently high accumulation rates of *Chaetoceros* resting spores in the Southern Ocean throughout the last glacial and their restricted occurrence during the Holocene provide compelling evidence for substantially higher productivity and organic carbon export into the deep ocean during the last glacial and highlight the potential of resting spores as biological proxies for paleoceanography.

The cosmopolitan needle-shaped genus *Pseudo-nitzschia* comprises many species that are difficult to differentiate under the light microscope, but some of the species can contribute substantially to both open-ocean and coastal blooms. Detailed studies on *Pseudo-nitzschia delicatissima* and *Pseudo-nitzschia pseudodelicatissima*, two morphologically distinct entities, have shown reproductive isolation among sympatric cryptic species in the Gulf of Naples. The factors selecting for these subtle differences are as yet unknown but likely represent either reproductive isolation or adaptations at the physiological but not the morphological level such as defenses against specific pathogens or predators.

The needle-shaped, thin-shelled centric genus *Rhizosolenia* commonly forms blooms later in the year. A dense bloom in which the cells were aggregated into mats was reported from the equatorial Pacific where the mats apparently carried out vertical migration between the surface and the nutrient-rich deeper layer. Surprisingly, this effective mechanism of nutrient acquisition, although looked for, has not been found since. Another widespread species that is common in the summer months is *Ditylum brightwelli* (Figure 2(l)), which has, however, once formed an almost monospecific spring bloom in the southern North Sea.

Haptophytes (Prymnesiophytes)

This group comprises small flagellates equipped with a prehensile organ, the haptonema, which is apparently used to feed on small particles such as bacteria (Figure 2(g) and 2(q)–2(t)). A common characteristic of this group is the copious production of dimethylsulfoniopropionate (DMSP), which is split by an enzyme into volatile dimethylsulfide (DMS) and the noxious acrylic acid. Several functions of the widespread molecule DMSP have been suggested such as a compatible solute in osmoregulation, defense against predators, and as an antioxidant. DMS released from haptophyte blooms is a major global source of sulfur to the atmosphere where it is oxidized to hygroscopic sulfate, which acts as a cloud condensation nucleus. Large amounts of sulfur in the atmosphere result in smaller droplets, that is, whiter clouds that scatter sunlight back into space, thus contributing to cooling. The global significance of haptophyte blooms in cooling the planet is under debate.

The calcareous coccolithophores belong to this group and are second in importance to the diatoms in their contribution to extensive oceanic blooms. These normally arise in the aftermath of the spring diatom bloom and, because of the high reflectance of their calcareous plates, are visible from space. The species *Emiliania huxleyi* (Figure 2(g)) is the most prominent member of this group and forms blooms in both coastal and open-ocean regions. Coccolithophores exhibit highly complex life cycles in which haploid, diploid, and polyploid stages, some bearing different types of coccoliths, alternate with one another. The adaptive significance of this life cycle is not understood. Blooms of *E. huxleyi* are regularly terminated by the spread of viral infection within the population. Interestingly, despite its numerical dominance in the surface layer, *E. huxleyi* contributes only a minor fraction of the total calcareous material accumulating in the deep North Atlantic. Less abundant but more heavily calcified coccolithophores like *Calcidiscus leptoporus* contribute the bulk of accumulated calcareous sediments.

Another group of haptophytes involved in extensive blooms in both temperate and polar waters are species of the genus *Phaeocystis*. At least six species have been identified worldwide based on morphological and genetic characteristics, but only three species – *Phaeocystis pouchetii* (Arctic), *Phaeocystis globosa* (temperate and tropical waters), and *Phaeocystis antarctica* (Antarctic, Figure 2 (r)–2(t)) – dominate blooms. These are also the only species thus far observed to form colonial stages comprising many hundreds of cells. The others occur only as the solitary nanoflagellate stage. The flagellate stage suffers much higher mortality by both viral infection and protozoan grazing than the colonial one, indicating that colony formation is a defense strategy. The fact that colonies in their early stages are often found attached to the spines of diatoms (e.g., *Chaetoceros*, *Corethron*) also suggests that small colonies are more vulnerable and find protection from protozoan grazing on these large diatoms (Figure 3). In the North Sea, *P. globosa* tends to attain dominance in the later spring, in the aftermath of the diatom bloom following Si exhaustion. Extensive blooms dominated by *P. antarctica* are a regular feature of the Ross Sea but, for unknown reasons, are more sporadic elsewhere around Antarctica. On rare occasions dense blooms of the solitary flagellate genera *Chrysochromulina* and *Prymnesium* occur in some regions. Some species are highly toxic and result in mass mortality of other organisms.

Dinoflagellates

This is an ancient, heterogeneous group comprising autotrophic, heterotrophic, and mixotrophic modes of nutrition across a wide range of shapes and sizes (Figure 2(a)–2(f)). Interestingly, the morphology of photosynthetic and predatory forms does not differ, signifying that their shapes and sizes have not evolved to maximize resource acquisition. Like the coccolithophorids, small-celled dinoflagellates ($\sim$10 μm) encased in cellulose armor (e.g., *Prorocentrum*, *Heterocapsa*, *Scrippsiella*) (Figure 2(d)–2(f)) commonly form blooms in the aftermath of the spring diatom bloom in coastal and shelf areas. Larger, often unarmored forms such as *Gymnodinium* form sporadic blooms not only in

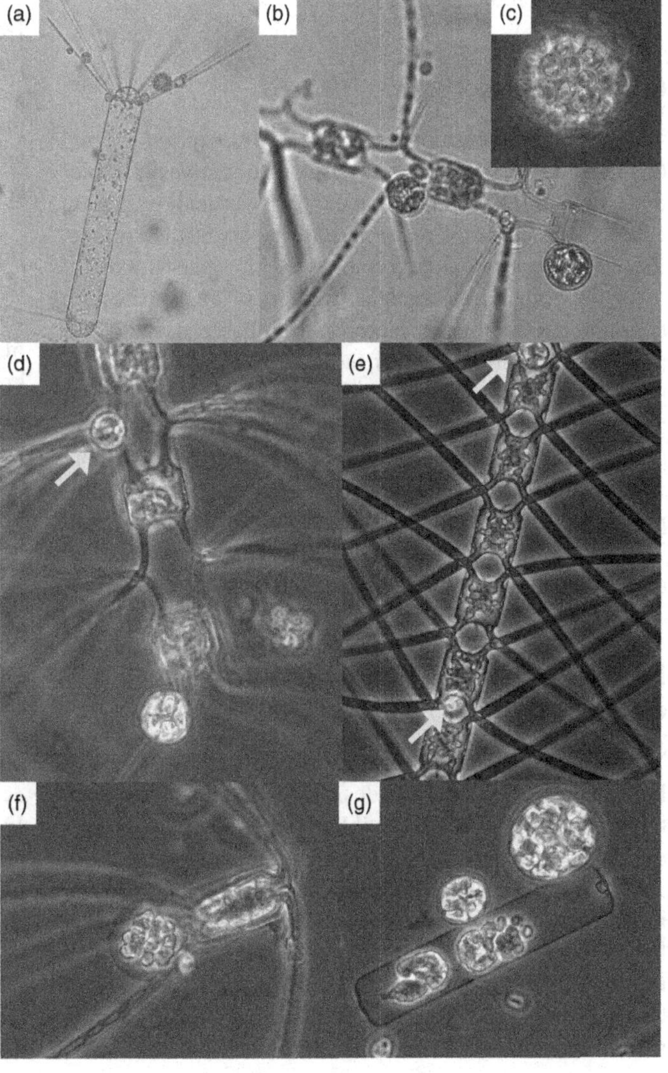

FIGURE 3 Colonies of *Phaeocystis antarctica* attached to different diatom species (from field material): (a) spines of *Corethron pennatum*; (b) setae and cell wall of *Chaetoceros dichaeta*; (c) close-up of colony; (d) *Chaetoceros dichaeta*; (e) *Chaetoceros atlanticus*; (f) *Chaetoceros peruvianus*; (g) *Guinardia cylindrus*. Light micrographs by Philipp Assmy.

the summer months but also in upwelling regions. Spectacular blooms of the heterotrophic bioluminescent *Noctiluca scintillans* (Figure 2(c)), colored by pigments from their algal food, cause discoloration of large stretches of surface waters mainly in sheltered bays or estuaries. Because of their motility and comparatively large size, some dinoflagellates have been observed to migrate vertically to take up nutrients below a shallow SML. Many species also form dense layers at density discontinuities, which reach the surface along hydrographic fronts where they are highly conspicuous. The standing stocks of such blooms are low, despite their dense concentrations because they are restricted to thin layers.

The largest dinoflagellate blooms in both lakes and the ocean are formed by species of the large, armored genus *Ceratium* during the autumn. For unknown reasons, grazing pressure on this genus is low, suggesting that survival rather than fast growth rates is the major reason for bloom build up over the late summer. Although there are many species, only a few cosmopolitan ones (e.g., *Ceratium tripos*, *Ceratium furca*, *Ceratium fusus*) (Figure 2(a) and 2(b)) form blooms throughout their range.

Other Groups

The autotrophic, cosmopolitan ciliate *Mesodinium rubrum* (also known as *Myrionecta rubra*) is an active swimmer that, like dinoflagellates, carries out vertical migration and sometimes forms dense reddish layers at the surface. Its cells are packed with cryptophyte chloroplasts that still carry the cryptophyte nucleus. Blooms of this species are common and have been reported from habitats as diverse as tropical upwelling regions, the inner Baltic and the Arctic shelf. Large blooms of this ciliate tend to be sporadic but patchy blooms occur regularly in the North and Baltic seas in early summer. Other groups reported to contribute significantly to algal blooms are the euglenophytes, cryptophytes, and raphidophytes; their blooms are only of local importance. In lakes, various species of chlorophytes and chrysophytes form blooms under eutrophic conditions.

PATHOGENS AND GRAZERS OF ALGAL BLOOMS

The important role of viruses in the sea has only come to light over the past two decades. It is now known that bacterial populations are susceptible to viral infection and that their density is partly regulated by this factor. Less is known about viral infection of algae but coccolithophorid blooms in mesocosms have been reported to have been terminated by mass infection. So far, only few viruses have been found to infect diatoms, and the pathogens of other groups have barely been studied. This also applies to pathogenic bacteria that are known to occur but whose influence is unknown.

A wide range of grazers is present in the plankton: puncturing and ingesting protists that prey on individual cells and chains, suspension-feeding copepods that select and ingest particles individually, and filter-feeding larger metazooplankton such as euphausiids (krill) that collect particles *en masse* and ingest them indiscriminately. Complexity is introduced by the tendency of copepods to selectively feed on protistan grazers of phytoplankton, in particular ciliates and heterotrophic dinoflagellates, which results in a trophic cascade favoring accumulation of algal cells.

Protistan grazers have growth rates equivalent to their phytoplankton prey and many have evolved techniques to cope with prey sizes as large as or even larger than themselves. Thus, small flagellates belonging to various protistan lineages can attach themselves to large diatoms and insert feeding tubes through pores in the frustule or between the girdle bands. The cell multiplies while feeding on the diatom plasma, releasing large numbers of flagellated cells that can infect other diatom cells. These specialized grazers are known as parasitoids, because of their small size relative to their prey. They are generally species-specific in prey selection and have the potential to decimate prey populations during blooms. However, this has rarely been observed presumably because they are kept in check by grazing of larger protists, in particular ciliates that are specialized grazers of small flagellates, although many ciliate species can ingest cells as large as themselves.

Dinoflagellates have evolved the broadest range of feeding techniques: some ingest prey cells up to their own size, others pierce the cell wall and suck out the contents with a feeding appendage known as the peduncle. Several genera of armored dinoflagellates are able to extrude a feeding veil known as the pallium, which can envelope an entire chain of diatoms many times the size of the predator and digest it outside the cell wall (Figure 4). The pallium is retracted after digestion is completed. The entire process of prey capture and pallium retraction lasts for 10–20 min and the pallium-feeding dinoflagellates can divide faster than their algal prey. Again, decimation of algal blooms by dinoflagellates is rarely reported, although they have the potential to do so. In all likelihood, the ubiquitous small copepods selectively feed on large heterotrophic protists, which releases grazing pressure on the algae, particularly diatoms.

In contrast to unicellular herbivores, metazooplankton, in particular copepods and euphausiids have complex life cycles involving a range of larval stages that take weeks or months to complete, depending on temperature in opportunistic species, or innate life cycle traits in species with pronounced seasonality. The latter generally have larval stages that overwinter at depth and ascend to the surface in spring. It is widely believed that blooms arise because copepod populations are unable to match their appearance at the surface with the timing of bloom initiation and subsequently gear their population size to the growing bloom.

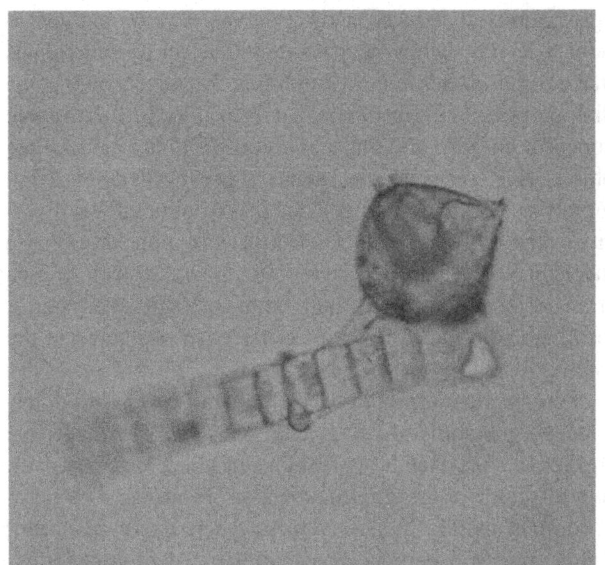

FIGURE 4 A pallium-feeding dinoflagellate of the genus *Protoperidinium* capturing a pennate chain-forming diatom. Light micrograph by Christine Klaas.

However, this neither explains why other grazers, in particular unicellular herbivores, are unable to suppress blooms, nor does it explain why copepods, if they are indeed so inefficient at utilizing potential food sources, nevertheless manage to maintain the bulk of zooplankton biomass in oceans and oligotrophic lakes.

It has long been appreciated that, unlike terrestrial vegetation, the bulk of phytoplankton cells are grazed within the growth environment. So, given the diversity of phytoplankton grazers and their potential to overgraze their food, it is indeed surprising that blooms develop at all. However, it should be remembered that only a few of the species present contribute significantly to algal blooms. The other species either do not respond with maximized growth rates to the advent of favorable conditions or are grazed faster than the ones making the bloom. The conclusion is that understanding bloom dynamics requires basic knowledge of the ecophysiology and species-specific regulatory mechanisms inherent to the bloom formers as well as a deeper understanding of the complex interactions and potential trophic cascades within pelagic food webs. Given the great variability inherent to complex biological systems, it is indeed surprising that there is any degree of recurrence in the annual cycles of species composition and biomass in lakes and the sea.

RECURRENT AND UNUSUAL ALGAL BLOOMS

Algal blooms can be broadly differentiated into seasonally recurrent and unusual stages in the annual cycles of plankton of a given water body or region. The former tend to be initiated by seasonal change in the physical regime. At higher latitudes, this is due to warming and shallowing of the SML by thermal stratification in the spring, and its breakdown by convective mixing during the autumn. The former process improves the light supply of the cells trapped in the SML and the latter reintroduces nutrients lost by sinking from it. Spring and autumn blooms, respectively, are the result of these physical processes (Figure 1). Blooms are also a regular feature of the upwelling regions of lower latitudes. Blooms that are not recurrent features of the annual cycle or that only occur in some regions are dealt with separately. In particular long-term data series on the annual cycles of phytoplankton gathered from coastal sites all around the world have provided valuable information on many of the recurrent and unusual aspects of bloom phenomena outlined in the following paragraphs.

Spring Blooms

Algal blooms, almost always dominated by diatoms, are regular events in the annual cycles of all lakes and coastal oceans that arise when deep mixing during the winter is stopped by the advent of surface stratification during spring (Figure 1). The timing of the spring bloom is influenced as much by the hydrography of the region as its latitude (light intensity). Thus, spring blooms along the melting ice edge off Greenland occur well before the bloom on the Norwegian shelf 1000 km ($\sim 10°$ latitude) further south. This fact was first noticed in the 1930s and attributed to earlier improvement of the light climate by shallow stratification due to meltwater in the Arctic as compared to the slower establishment of thermal stratification in the stormy northern North Sea. Similarly, blooms start about a month earlier in protected bays and fjords than in the adjacent, open shelf. Turbidity of the water also plays an important role in spring bloom initiation, which occurs much later in the shallow, highly turbid water overlying the tidal mud flats of the German Bight in the southern North Sea than in the clearer water of the tide-free Baltic Sea on the other side of the Danish Peninsula.

The dynamics and composition of the spring bloom varies widely from region to region and also from year to year in the same region, depending on weather conditions, phytoplankton seeding stocks, and prevailing grazing pressure. Thus, when bloom initiation coincides with prolonged calm, sunny weather the bloom peak is reached within 2–3 weeks and is often dominated by only one or two diatom species with a high initial seeding stock. At the other extreme, bouts of stormy weather alternating with calmer periods at intervals of a few days can prolong the life time of the bloom by many weeks. In such cases, different species dominate biomass across a succession of smaller peaks until nutrient exhaustion is reached and the recycling community established.

The fate of bloom biomass varies depending on its dynamics. In the first case, algal biomass is much higher than that of the heterotrophs and the bloom is terminated by mass sinking of ungrazed cells within a few days resulting in a nutrient impoverished SML. Although the termination phase of blooms is less well studied than the longer growth phase, it is likely that morphology and behavior of the species involved influence the depth and extent to which bloom biomass reaches the sea floor. Thus, diatom species that form resting spores tend to sink out quantitatively and the long chains of the spiny genus *Chaetoceros* clump into aggregates, aided by sticky mucilage that entrains other cells in the large, rapidly sinking flocs. Such boom-and-bust-type blooms transfer a large amount of organic matter to the benthos and, when they overlie deeper water, play a significant role in sequestering carbon in the deepwater column and sediments.

Blooms prolonged by fluctuating favorable and unfavorable weather conditions should, theoretically, enable grazer populations to keep pace with the slower accumulation rate of biomass. As a result, more organic matter should be channeled to the zooplankton and higher pelagic trophic levels, leaving a smaller percentage of bloom biomass for export to deeper water and the sediments. However, food web relationships in the plankton during prolonged blooms are not well documented and are likely to be more complex than the theoretical case portrayed earlier.

Consensus is emerging that diatoms are ubiquitous in spring blooms not only because they have high growth rates but also because they are better defended than algae from other groups and hence do not represent an ideal food source. This is at odds with the long-held view of diatoms representing the pastures of the sea. The first and most obvious line of defense constitutes the silica shell that provides mechanical protection against a range of crustacean predators, wards off protistan grazers through long and spiny cell appendages, and restricts access to the encased cell for piercing predators, parasites, and pathogens. Although copepods have evolved silica-edged mandibles to crack diatom shells, laboratory experiments indicate and field observations suggest that they prefer protozoa over diatom food. Why this is so is still under debate. One line of explanation suggests that diatoms are nutritionally less valuable than protozoa; another explanation is that diatoms produce substances noxious to copepods, either directly or by reducing their reproductive efficiency. Thus, it has been shown that some diatom species contain enzymes that cleave structural lipids into toxic aldehydes when the shells are crushed. Such wound-induced herbivore deterrents are commonplace in terrestrial plants, but systematic investigation of this aspect of planktonic food webs has commenced only recently. Diatom aldehydes are teratogens, that is, the feeding adults are not affected, but egg viability is reduced and the larvae develop malformations. Clearly, such a mechanism will select for individuals that eat food other than just diatoms. However, only some diatoms possess the enzymes that form aldehydes, and some copepod species are more susceptible to its effect than others. So, this is by no means a universal defense but merely one of the weapons in an arsenal of unknown diversity. Again, such species-specific variation in defense mechanisms among herbivores and plants is well known from terrestrial, benthic, and limnic pelagic ecosystems, but this evolutionary arms race has been neglected in the marine plankton.

Perhaps not surprisingly, the genera that tend to dominate spring blooms – *Skeletonema* and *Thalassiosira* – have species that produce large amounts of aldehydes; however, they do not always dominate bloom biomass. The other prominent genus *Chaetoceros* has been shown to harbor lipids toxic to copepods. These three morphologically dissimilar genera dominate spring blooms worldwide, from the coasts to the open ocean. This is not to say that one of them will invariably be present at every spring bloom but that they are more likely to contribute significantly to spring bloom biomass everywhere than other genera. Their contribution varies widely both spatially and temporally, possibly with implications for the fate of bloom biomass.

The dominant spring bloomers in lakes also tend to be diatoms whereby the centric genera *Melosira* and *Cyclotella* and the pennate species *Asterionella formosa* often dominate. Indeed, the latter species has always dominated the spring bloom of Lake Windermere over the past 60 years of continuous observation. In other lakes, its presence fluctuates annually as a result of infection by a parasitoid. Summing up, there is still a lot we do not understand about the dynamics and fate of spring blooms, particularly at the species level.

A pronounced spring bloom, generally dominated by diatoms, was long considered an invariable feature of high and midlatitude seasonal cycles, but recent analyses of long-term data indicate otherwise. Thus, a winter–spring bloom of *Skeletonema* used to occur every year over decades in Narragansett Bay, Rhode Island, United States, but is absent since about a decade. Blooms now occur later in the year and have different characteristics to the spring bloom. A similar situation has been reported from San Francisco Bay and could be related to the spread of an invasive filter-feeding bivalve that prevented the bloom from developing. The situation changed when a basin-scale shift in the climate regime led to more input of shelf water to the Bay, which introduced fish predators that checked benthic filter feeders from grazing down the phytoplankton. These examples indicate the complex interactions between bottom-up and top-down factors in shaping the annual cycles of pelagic ecosystems. Long-term data are normally recorded adjacent to research institutes and it is likely that similar variation occurs in deeper, offshore water.

Autumn Blooms

Autumn blooms arise when enhanced vertical mixing due to stormy weather and cooling deepens the summer thermocline, thereby introducing new nutrients to the SML (Figure 1). The biomass and species composition of autumn blooms vary regionally more than those of the spring and they do not occur in all high- and midlatitude oceans. However, much less effort has been invested in their study as compared to spring blooms. In many regions, the autumn bloom is dominated by diatoms that, in the Kiel Bight, can reach their peak as late as mid-November. The species involved are similar to the spring bloom, but genetic studies have revealed that the spring *Skeletonema* species is different to the autumn one. Although the characteristic spring genera are present in autumn diatom blooms, the species assemblage tends to be more diverse and includes dinoflagellates and larger ciliates. The latter is probably due to the low number of copepods. Where studied, the bulk of bloom biomass sinks out of the water column.

In regions where dinoflagellates dominate late summer plankton, an early autumn bloom comprising species belonging to this assemblage develops in the initial stages of thermocline erosion. Most prominent are species of the armored genus *Ceratium*, in particular, the cosmopolitan species *C. tripos*, *C. fusus*, and *C. furca*. These species are apparently unpalatable, although their defense mechanisms, other than their cellulose armor plating, are not known. They have comparatively low growth rates but because their populations accumulate over the summer months, a few divisions over several weeks in the nutrient-rich, autumn water column suffices to build up large biomasses that, in some regions, can rival those of the spring bloom. Growth of their bloom is generally accompanied by declining zooplankton populations, which can be interpreted as a sign of success in the evolutionary arms race.

Ceratium blooms regularly occur in many, but not all, coastal environments from the tropics to the Arctic but they are absent around Antarctica. They are apparently terminated by mass cell death and disintegration in the water column followed by sinking of phytodetritus, but the sinking rates and the depth to which they sink are not known. Recurrent large *Ceratium* blooms regularly lead to anoxic events in the Laholm Bay of southern Sweden. A massive *C. tripos* bloom, which for unknown reasons survived through the winter and sank the following spring, caused widespread anoxia and collapse of the commercial scallop fishery in New York Bight in 1976. This was a one-time event. The same can be said of a massive *C. furca* bloom that caused extensive anoxia in the late summer of the North Sea in 1981, but has not occurred since.

Blooms in Upwelling Regions of Low Latitudes

Upwelling of nutrient-rich, cold, deep water occurs seasonally or for longer periods along the western continental margins. The intensity of upwelling varies regionally and over periodic cycles of several years of which the El Nino oscillation is the best known. Dense blooms of diatoms develop in regions with intense upwelling but, because of the high sun angle and rapid warming of the newly replenished SML, algal growth rates are much higher and the bloom peak is reached faster than in spring blooms. Nevertheless, as mentioned in the section on diatoms, much the same diatom genera typical of the midlatitude spring blooms – *Skeletonema*, *Thalassiosira*, *Chaetoceros* – are prominent contributors to biomass, although *Phaeocystis* blooms have only been reported from the Arabian Sea. Most diatom species of these genera are different but a few of the prominent species are the same. Thus, genetic analyses indicate that the same species *S. costatum* that makes dense blooms off the Indian coast during monsoon upwelling also contributes to North Sea blooms. The *Phaeocystis* species – *P. globosa* – is also the same. Given the vast differences in light climate and nutrient regimes (Si:N:P concentrations and ratios), it is surprising that the same cosmopolitan species can outgrow locally evolved competitors (species present only in one or the other region), casting doubt on the role of resource competition (bottom-up factors) in selecting species dominance in blooms. There is considerable regional and interannual variability in bloom magnitude and composition depending on the intensity of upwelling. Thus, the depth from which the upwelling water emanates can favor either diatom or dinoflagellate blooms. Extensive blooms of *M. rubrum* have also been reported off Peru.

The structure and response of the food web is generally similar to that of spring blooms and again, much the same genera of protozoa and copepods are represented. However, in some upwelling regions, the local sardine-like fishes have developed fine-meshed gill rakers to graze directly on the diatoms, thus short-circuiting the food chain. Upwelling regions harbor large stocks of pelagic fishes indicating that a significant proportion of primary production is retained in the surface pelagic system. However, in regions where upwelling is prolonged over many months, the sinking flux to deeper layers and the underlying sea floor can be substantial. Thus, suboxic conditions can develop in deeper layers over extensive regions. These zones play a crucial role in oceanic nutrient cycles because denitrifying bacteria reduce nitrate to N_2, thus lowering the N:P ratio. Nitrous oxide (N_2O), which is 300 times more potent as a greenhouse gas than CO_2, is also formed. In the underlying anoxic sediments methane and hydrogen sulfide are produced, which reach such high concentrations off the Namibian coast that methane bubbles, together with H_2S gas, reach the surface and cause mass death of the local benthos.

H_2S is oxidized to elemental sulfur in surface water and the resulting reflective particles can be seen in satellite images. Since global warming steepens the temperature gradient between land and sea, winds are expected to be stronger and upwelling more intense in the future. This will result in greater input of methane and N_2O to the atmosphere further exacerbating global warming.

Miscellaneous Algal Blooms

Blooms that do not fall under the aforementioned categories are grouped here. Thus, blooms caused by vagaries of the weather acting on hydrography are generally singular and local events. For instance, the passage of storms over stratified, nutrient-poor waters of summer months causes deep mixing and upward transport of nutrients, often resulting in a bloom. Generally, the standing stock of these blooms is lower than in seasonal ones, but where motile forms that concentrate their populations along hydrographic gradients (nutricline) are involved, algal densities in thin layers can be very high. Tilting of the pycnocline along hydrographic fronts introduces this layer to the surface resulting in conspicuous stretches of discolored waters generally referred to as red tides. The spatial extent of the discolored patches or streaks ranges from meters in protected bays to many kilometers in the open sea, where they appear as meandering structures in satellite images.

In many regions characterized by surface fronts between water masses with different properties, such as at the entrance of the English Channel, frontal blooms are regular features. However, because they tend to be dominated by a single species, their composition from year to year can vary more than in the case of seasonal blooms. Most often dinoflagellates dominate frontal blooms but in estuaries cryptophyte and haptophyte blooms have also been reported. Because of their local nature, these blooms do not have much impact on their environment and tend to go unnoticed. However, when the species involved in the bloom are toxic or cause harm to the environment, they are grouped under harmful algal blooms (HABs). Not surprisingly, these blooms are among the best-studied phenomena in pelagic ecosystems and are dealt with separately next.

HARMFUL ALGAL BLOOMS

The term HAB is in common usage to denote proliferations of algal species that have a detrimental effect on human use of lakes and the sea, from drinking water reservoirs and recreation to aquaculture. The spread of artificial fertilizers in agriculture in the second half of the last century led to increased nutrient input to lakes and estuaries often resulting in algal blooms. Because both tourism and aquaculture have intensified and spread over the past decades the incidence of HABs in coastal waters has accordingly increased. Thus, they can be regarded as the aquatic equivalent of terrestrial weeds, which by definition are plants that interfere with human usage of the region. There are no 'weeds' in pristine ecosystems. It follows that the number of HABs increases with the intensity and diversification of usage. Furthermore, they can be controlled, or at least kept in check, by adopting practices that minimize their proliferation, analogous to the situation in agriculture or gardening. Thus, algal blooms caused by nutrient runoff from fertilized fields, referred to as eutrophication and particularly prominent in lakes, estuaries, and coastal regions, are being brought under control in Europe and North America by improving agricultural practices and installing sewage treatment plants. Another successful measure has been aeration of deep water in shallow, eutrophied lakes. The proliferation of cyanobacteria during the summer months has been checked as a result.

Since HABs are grouped according to their effect on the environment, their dynamics and species composition vary widely. Here, it is impossible to provide an exhaustive review of HABs, so we shall restrict ourselves to an overview of the major types illustrated with some examples. In many regions, they are a regular feature of the annual cycle, in others they develop only in some years. The species composition is crucial in determining the degree of harm caused by the bloom. The harm is inflicted either by toxins in the algal species or by massive buildup of biomass resulting in oxygen depletion in deeper layers.

The occurrence of toxic algal blooms was first brought to the attention of the European public by the notorious bloom of a little-known, small haptophyte flagellate, *Chrysochromulina polylepis*, in May 1988. In addition to killing off most marine life in contact with the bloom, this species caused extensive fish kills in aquaculture farms along the Swedish and Norwegian coasts. It built up biomass as a dense layer in the pycnocline of the Belt Sea, which moved north with outflowing Baltic water and reached the surface along the Swedish and Norwegian coasts. The bloom of this toxic species was associated with increasing eutrophication at the time but the absence of a similar bloom in subsequent years despite similar weather conditions to those of May 1988 cast doubt on the explanation of a single cause-and-effect relationship. The same is true of other singular blooms that caused widespread anoxia such as in New York Bight in 1976 and in the North Sea in 1981.

The dinoflagellates have the greatest number of harmful species some of which produce potent toxins. Many of the species are responsible for fish kills as well as human poisoning via shellfish. Saxitoxin was the first algal toxin to be described; it blocks the sodium channels of neurons causing muscular paralysis and, at high dosage, death. The condition is termed paralytic shellfish poisoning (PSP). Some related genera of dinoflagellates, particularly various species of *Alexandrium*, cause PSP. The function of the toxins is not known. Not only mussel farms but also wild mussel beds are now closely monitored for toxins, but in the past human deaths due to PSP occurred regularly. A number of other toxic molecules have since been identified. Thus toxins produced by the dinoflagellate *Karenia brevis* escape to the atmosphere and cause respiratory problems in humans. Blooms of this species have been reported for the Florida coast of the Gulf of Mexico where they also cause massive fish kills.

Diarrhetic shellfish poisoning is caused by toxins present in some species of the common dinoflagellate genera *Prorocentrum* and *Dinophysis*. Although not fatal, the toxins can cause severe discomfort to the digestive system. The *Dinophysis* toxin ocadaic acid is potent enough to cause closure of clam and mussel maricultures at cell concentrations >1000 cells l^{-1}. *Pseudo-nitzschia* is the only diatom genus known to produce a potent toxin, in this case domoic acid, which bioaccumulates in shellfish and has caused human deaths by amnesic shellfish poisoning. Seals and birds are also affected, particularly along the California coast. Victor, do you want to add some more information here from your excellent review on toxic Pseudo-nitzschiaspecies?

Blooms that cause anoxia or unsightly accumulations along beaches are also grouped under HABs. Thus, the two exceptionally dense and extensive blooms of *Ceratium* species dealt with under autumn blooms fall under the former category. Blooms of the small-celled *Aureococcus anophagefferens* appeared suddenly along the northeast coast of the United States in the 1980s and spread southward in subsequent years reaching the Gulf of Mexico. The blooms produced unsightly brownish colored water along the coast known as brown tides. The blooms have since subsided. Copious amounts of mucus produced by the large, centric diatom *Coscinodiscus wailesii* that invaded the North Sea from the North Pacific during the late 1970s caused damage to the fisheries by clogging and tearing fishing nets. However, although this species is now established in the North Sea, blooms comparable to those of the 1970s have not since been reported. The slime blooms of the Adriatic appear in some years and are also a nuisance for tourism and fisheries. Since phytoplankton densities in the slime accumulations are low, they cannot be attributed to blooms. Where mucous accumulations settle out on the sea floor, the underlying benthos is asphyxiated.

IRON-FERTILIZED BLOOMS

Large expanses of the open ocean have perennially high macronutrient concentrations (nitrate and phosphate) but low phytoplankton biomass: the phytoplankton of these high-nutrient, low-chlorophyll (HNLC) ocean regions – the subarctic and equatorial Pacific and the entire Southern Ocean – have been shown by in situ iron fertilization experiments to be limited by iron availability. All but one of the ten experiments carried out in the HNLC regions induced diatom blooms. The smaller phytoplankton of the microbial food web were also stimulated by iron addition but their biomass failed to increase, indicating that the increase in growth rates was compensated by a corresponding increase in mortality rates presumably due to pathogens and herbivores. Only the mortality rates of the diatoms were sufficiently low to enable accumulation of biomass. The maximum biomass concentration of 23 mg Chl m^{-3} was reached in an experiment (SEEDS I) carried out in a 10 m SML, whereas the highest standing stock of 280 mg Chl m^{-2} developed in a 100 m deep SML in the Southern Ocean.

The species composition of the blooming diatoms varied widely and depended on the seeding stock present at the time of fertilization. In some experiments many different species accumulated biomass in about equal proportions, others were dominated by one or a few species with exceptionally high accumulation rates. In the experiment carried out in HNLC waters off Japan, which yielded the highest chlorophyll concentrations (SEEDS I) a single species – the cosmopolitan neritic species *C. debilis* – contributed >90% to total biomass at the time of nutrient exhaustion reached within 10 days. A subsequent experiment carried out under the same conditions (SEEDS II) failed to induce a bloom although it did elicit a strong physiological response of the phytoplankton assemblage. The grazer community was apparently not exceptionally large, but *C. debilis* was absent. Southern Ocean experiments also yielded different species compositions: in one case the cosmopolitan pennate species *Pseudo-nitzschia lineola* contributed 25% to diatom biomass with a rising tendency when the experiment had to be ended after 3 weeks. The same ubiquitous species was also present in other experiments but did not accumulate cells faster than the other species. The puzzling results demonstrate the need to carry out more in situ experiments because, unlike laboratory and mesocosm experiments, the complexity of factors enabling bloom formation can only be unraveled under natural physical and biological conditions, in particular an intact mortality environment comprising the natural gamut of pathogens and zooplankton.

The iron fertilization experiments were carried out to test the iron hypothesis, proposed by John Martin, that increased iron transport via dust to the Southern Ocean during the dry glacials led to phytoplankton blooms that sank out of the surface layer thereby transporting carbon from the atmosphere to the deep sea. Martin hypothesized that this carbon sink contributed substantially to the lowering of atmospheric CO_2 levels during glacial as compared to interglacial periods. The four experiments carried out so far in the Southern Ocean have supported the first condition of Martin's hypothesis, that iron addition elicits massive diatom blooms. However, the fate of bloom biomass remains under debate because most experiments were not designed to track particles sinking from the bloom. If Martin's hypothesis is correct and artificial iron fertilization elicits the same response as an outfall of dust, then simulating the glacial ocean by large-scale ocean fertilization should help in sequestering some of the anthropogenic CO_2 now accumulating at an alarming rate. The maximum amount that could be sequestered annually is less than one-third of the accumulation rate in the atmosphere but it is too large to be ignored in the upcoming struggle to mitigate the adverse effects of global warming. The widespread belief that iron fertilization 'will not work' is premature.

FUTURE RESEARCH AVENUES

The phylogenetic diversity of bloom-forming species has interesting implications for our understanding of the evolutionary ecology of marine plankton. Indeed, the fact that the vast majority of phytoplankton species do not contribute significantly to blooms indicates that the density of individuals required for a species population to survive and evolve, that is, to maintain fitness in the marine plankton, is two orders of magnitude below that achieved by bloom-forming species. Clearly, the $>5\%$ of the species reported to form blooms have not outcompeted the $>90\%$ background species over evolutionary time scales, implying that the life cycle of bloom-forming species is one among many viable life cycle strategies in the phytoplankton. The phenology of bloom-forming species, that is, their mode of gearing to seasonality and hydrography of their environment together with their specific defense strategies, needs to be addressed if we are to make headway in understanding the driving forces shaping algal blooms.

Given their importance for food webs and biogeochemical cycles, the fate of blooms requires more dedicated study. Thus, mortality is not just caused by an external agent, that is, a pathogen or grazer. New results show that autocatalytic cell death or apoptosis, analogous to programmed cell death (PCD) in multicellular organisms, does occur in unicellular algae and has already been reported from cyanobacteria, green algae, coccolithophores, dinoflagellates, and diatoms. PCD seems to play a crucial role in the coordinated collapse of phytoplankton blooms but the mechanisms are still not fully understood. The PCD pathway involves the expression and biochemical coordination of a specialized cellular machinery that includes receptors, adaptors, signal kinases, proteases, and nuclear factors. Within this cellular machinery, a specific class of intracellular cysteinyl aspartate-specific proteases (i.e., caspases) is of particular interest. These caspases play a ubiquitous role in both initiation and execution of PCD through the cleavage of various essential proteins in response to proapoptotic signals. This cellular self-destruction process in phytoplankton is triggered by specific environmental stresses that range from cell age, nutrient limitation, high light and/or UV exposure, and oxidative stress.

The collapse of the annual bloom of *Peridinium gatunense* in Lake Kinneret, Israel, is, for example, triggered by CO_2 limitation followed by oxidative stress that initiates a PCD-like cascade. Interestingly, a signaling molecule, a protease, excreted by senescing *P. gatunense* cells triggers synchronized cell death of the whole population by sensitizing younger cells to oxidative stress. Other studies have shown PCD in the cyanobacterium *Trichodesmium* spp. and the diatom *Thalassiosira pseudonana* in response to nutrient starvation (e.g., phosphorous and iron limitation). Furthermore, it has been shown that lytic viral infection and autocatalytic PCD in the coccolithophore *E. huxleyi* work in tandem. Common morphological observation of cells displaying signs of PCD includes DNA fragmentation, degradation of cell organelles, and cytoplasmatic shrinking, while the plasma membrane remains intact. Depriving viruses of their host populations and parasitoids and predators of their food could explain the evolutionary fitness logic of PCD in the pelagic realm. These examples illustrate some of the avenues that need to be explored to further our mechanistic understanding of algal blooms.

RECENT DEVELOPMENTS

Here, we provide an overview of some recent progress in the field of algal research since the publication of our original chapter on algal blooms.

Marine Genomics

The advent of high-throughput sequencing techniques has enabled biological oceanographers to study complete genomes of model species from both prokaryotic and eukaryotic algal lineages. Whole genome sequences are already available for representatives of most algal lineages being constantly amended by new genomes. The complete genomes of two diatom species, the centric diatom *Thalassiosira pseudonana* and the pennate diatom *Phaeodactylum tricornutum*, have shed new light on the evolutionary origins of diatoms and their ecological success. It was originally proposed that diatoms originated through a secondary endosymbiosis in which a red alga became incorporated by a heterotrophic eukaryote and took on the role of the diatom's plastid. Although red algal genes have been found in both the *T. pseudonana* and the *P. tricornutum* genomes providing evidence for a red algal endosymbiont recent findings indicate a predominance of green algal genes suggesting that the green algal endosymbiont preceded the red alga. These findings cast doubt on the assumption that the red lineages succeeded in the ocean whereas the green lineages dominated on land and indicates that diatoms were derived from a serial secondary endosymbiosis and not a single event. Diatoms not only inherited genes from their different endosymbionts and the exosymbiont but also a large number of bacterial genes through horizontal gene transfer. This unique combination of genes has permitted novel metabolic pathways in diatoms, for example, carbon concentrating mechanisms (CCMs), urea cycle, genes encoding the iron storage protein ferritin, and contributed to their predominance in the ocean. Whole genomes of individual species are now being supplemented by metagenomic and metatranscriptomic approaches that aim at circumscribing the genomic variation of natural communities in case of the former and the gene expression profile of environmental mRNA in case of the latter. The combination of genomic approaches with conventional oceanography will improve our understanding of the factors determining species dominance and succession in phytoplankton blooms and thus help to understand how different algal classes will respond to future climate change.

Abandoning Sverdrup's Critical Depth Hypothesis

For half a century Sverdrup's critical depth hypothesis has served biological oceanographers as a model to explain the vernal phytoplankton bloom in temperate and polar latitudes. It posits that blooms can only occur when the surface mixed layer shoals beyond a critical depth defined by the point where phytoplankton growth exceeds losses through respiration, sinking, grazing, viral infection, parasitism, and horizontal and vertical dilution. A multiyear satellite record from the North Atlantic has now challenged the traditional view that blooms are caused by enhanced growth rates in response to improved light climate, rising temperatures and increased stratification. The satellite record shows that bloom initiation coincides with maximum mixed layer depth in winter despite the fact that phytoplankton instantaneous growth rates (μ) are minimal in winter. This apparent discrepancy can now be resolved in the dilution-recoupling hypothesis coined by M.J. Behrenfeld. During deep winter mixing phytoplankton cells and their grazers are diluted thus lowering encounter rates between predator and prey and decoupling net population growth rates (r) from grazing pressure. This illustrates the major flaw in Sverdrup's model, namely that loss rates are constant over time. Recoupling of phytoplankton growth (μ) and loss rates during spring stratification due to entrainment of grazers within the shallow mixed layer compensates for the higher μ attained under more favorable light conditions in spring. Therefore, peak net population growth rates (r) are as likely to occur in midwinter as in spring and along the seasonal cycle r is generally inversely related to μ. These new findings highlight the tight coupling of phytoplankton growth (μ) and loss rates and the superior role of grazing mortality over improvement of the light environment for phytoplankton bloom dynamics.

The Role of Grazing in Suppressing Nondiatom Blooms and Recycling Iron in Pelagic Ecosystems

The most recent Indo-German iron fertilization experiment LOHAFEX (loha is the hindi word for iron) was conducted in severely silicon-depleted waters of the productive southwest Atlantic sector of the Antarctic Circumpolar Current (ACC). Diatom increase was thus limited by the lack of silicon and the other large algal species, that form extensive blooms in coastal waters, were heavily grazed by the large population of zooplankton, chiefly copepods. The bloom was hence composed of naked, motile algae, known as flagellates, smaller (<5 μm) than the size range accessible to copepod grazing. Nevertheless these flagellates exhibited only a muted biomass increase in response to iron addition because they were efficiently kept in check by their protistan grazers, chiefly ciliates, that were themselves channeled to higher trophic levels (copepods) illustrating the efficiency of trophic cascading effects in flagellate dominated phytoplankton communities. The results from LOHAFEX have thus shown that only diatoms are able to escape the grazer gauntlet exerted by the larger zooplankton stocks on an areal basis in oceanic as compared to coastal waters and that despite high growth rates biomass build-up of nondiatom phytoplankton is top-down controlled. This result is supported by observations from previous iron fertilization experiments performed in silicon-replete waters that were dominated by diatoms and provides a mechanistic explanation for the observation that massive phytoplankton blooms in the open ocean are almost exclusively due to diatoms provided that silicon is not limiting. Another striking aspect of the LOHAFEX experiment were the low loss rates of carbon by particles sinking from the surface layer indicating that iron fertilization under these conditions does not lead to significantly more carbon being sequestered in the ocean. The low carbon loss rates despite high copepod grazing pressure and concomitant fecal pellet production rates illustrate the high recycling efficiency of this grazer dominated system and its potential to retain essential nutrients within the productive surface layer. Indeed elevated iron concentrations were measured in copepod fecal pellets during LOHAFEX providing a potential source of regenerated iron for continued algal growth. These findings are being supported by measurements of iron content in baleen whale feces that is approximately 10 million times that of Antarctic sea water, suggesting that defecation of baleen whales is another way of natural iron fertilization apart from continental sources and that prewhaling populations of whales must have recycled more iron in surface waters thereby stimulating primary productivity through this positive feedback loop. The prominent role of higher trophic levels, in particular the megafauna, for ecosystem functioning was long known for terrestrial ecosystems and is only now emerging for the marine realm. Unfortunately industrial whaling and fishing has nearly depleted all of the marine apex predators rendering the assessment of their ecosystem wide impact close to impossible. Allowing whale and fish populations to recover will likely help to restore the ecosystem services they provide and contribute to an overall improvement of ocean productivity.

ACKNOWLEDGMENT

Philipp Assmy was supported by the Bremen International Graduate School for Marine Sciences (GLOMAR) that is funded by the German Research Foundation (DFG) within the frame of the Excellence Initiative by the German federal and state governments to promote science and research at German universities.

FURTHER READING

Behrenfeld, M. J. (2010). Abandoning Sverdrup's critical depth hypothesis on phytoplankton blooms. *Ecology*, *91*, 977–989.

Bowler, C., Vardi, A., & Allen, A. E. (2010). Oceanographic and biogeochemical insights from diatom genomes. *Annual Review of Marine Science*, *2*, 333–365.

Boyd, P. W., Jickells, T., Law, C. S., *et al.* (2007). Mesoscale iron enrichment experiments 1993–2005: Synthesis and future directions. *Science*, *315*, 612–617.

Cembella, A. D. (2003). Chemical ecology of eukaryotic microalgae in marine ecosystems. *Phycologia*, *42*, 420–447.

Cloern, J. E. (1996). Phytoplankton bloom dynamics in coastal ecosystems: A review with some general lessons from sustained investigations of San Francisco Bay, California. *Reviews of Geophysics*, *34*, 127–168.

Falkowski, P. G., Katz, M. E., Knoll, A. H., *et al.* (2004). The evolution of modern eukaryotic phytoplankton. *Science*, *305*, 354–360.

Franklin, D. J., Brussaard, C. P. D., & Berges, J. A. (2006). What is the role and nature of programmed cell death in phytoplankton ecology? *European Journal of Phycology*, *41*, 1–14.

Hamm, C., & Smetacek, V. (2007). Armor: Why, when, and how? In P. Falkowski & A. Knoll (Eds.), *The evolution of aquatic photoautotrophs* (pp. 311–332). Elsevier: Amsterdam.

Margalef, R. (1978). Life-forms of phytoplankton as survival alternatives in an unstable environment. *Oceanologica Acta*, *1*, 493–509.

Martin, J. H. (1990). Glacial-interglacial CO_2 change: The iron hypothesis. *Paleoceanography*, *5*, 1–13.

Nicol, S., Bowie, A., Jarman, S., *et al.* (2010). Southern Ocean iron fertilization by baleen whales and Antarctic krill. *Fish and Fisheries*, *11*, 203–209.

Raven, J. A., & Waite, A. M. (2004). The evolution of silicification in diatoms: Inescapable sinking and sinking as escape? *The New Phytologist*, *162*, 45–61.

Richardson, K. (1997). Harmful or exceptional phytoplankton blooms in the marine ecosystem. In J. H. S. Blaxter & A. J. Southward (Eds.), *Advances in marine biology* (pp. 301–385). Academic Press, Inc.: San Diego, London, New York, Boston, Sydney, Tokyo, Toronto Vol. 31.

Smayda, T. J. (2004). What is a bloom? A commentary. *Limnology and Oceanography*, *42*, 1132–1136.

Smetacek, V., & Cleorn, J. E. (2008). On phytoplankton trends. *Science*, *319*, 1346–1348.

Smetacek, V., Assmy, P., & Henjes, J. (2004). The role of grazing in structuring Southern Ocean pelagic ecosystems and biogeochemical cycles. *Antarctic Science*, *16*, 541–558.

Smetacek, V., Scharek, R., & Nöthig, E. M. (1990). Seasonal and regional variation in the pelagial and its relationship to the life history cycle of krill. In K. R. Kerry & G. Hempel (Eds.), *Antarctic ecosystems: Ecological change and conservation* (pp. 103–114). (1st ed.). Springer: Berlin, Heidelberg.

Tillmann, U. (2004). Interactions between planktonic microalgae and protozoan grazers. *Journal of Phycology*, *51*, 156–168.

RELEVANT WEBSITES

http://www.algaebase.org/index.lasso.– AlgaeBASE.
http://planktonnet.awi.de.– Plankton Net.

Chapter 33

Food Webs, Microbial

E.B. Sherr and B.F. Sherr
Oregon State University, Corvallis, OR, USA

Chapter Outline

Abbreviations	451
Defining Statement	451
Introduction	451
Understanding Microbial Food Webs	452
Components and Pathways	452
Microbes in Aquatic Food Webs	454
Heterotrophic Prokaryotes	454
Autotrophic Prokaryotes	454
Autotrophic Eukaryotes	455
Heterotrophic Eukaryotes	455
Marine Pelagic Habitats	456
Benthic Habitats	457
Role of Microbial Food Webs in Biogeochemical Cycling	458
Food Resource for Metazoans	460
Modeling Microbial Food Webs	460
Chemical Interactions Between Microbes	462
Spatial Structure of Microbial Food Webs	463
Recent Developments Biogeography of Form and Function in Microbial Food Webs	464
Further Reading	465

ABBREVIATIONS

CFB *Cytophaga–Flavobacteria–Bacteriodes*
DMS Dimethyl sulfide
GGE Gross growth efficiency
NPZ Nutrient–phytoplankton–zooplankton
TEP Transparent exopolymer particles

DEFINING STATEMENT

In natural ecosystems, microbial food webs consist of predator–prey interactions of unicellular prokaryotes and eukaryotes. In this chapter, we focus on the structure and ecological and biogeochemical importance of microbial food webs in aquatic ecosystems, and particularly in the oceans.

INTRODUCTION

In his book *The Ecological Theater and the Evolutionary Play*, limnologist G. Evelyn Hutchinson proposed that the environment provides the stage for the drama of evolution of species. If so, microbes are the stage hands, ceaselessly building and remolding the set. For a long time in Earth's history, microbes were the only actors as well, carrying out in microscale the script of production, predation, and dissolution; in other words, microbial food webs. Microbial food webs are similar in some ways to the familiar lynx and hare predator–prey interactions of nature shows. In other ways, they are not. A notable difference is that microbial food webs have an enormously important role in decomposing the plant carbon and all of the other components feces, dead bodies, and so forth, of the macroscopic world.

Most microbes are single prokaryotic or eukaryotic cells, although some form either filamentous chains or colonies of single cells, and fungi produce multicellular fruiting bodies. Microbes form multispecies communities, and thus food webs, throughout the biosphere, including some habitats where multicellular life cannot exist. In natural systems, a large proportion of prokaryotic, or bacterial, cells present in an environment may be relatively inactive, or dormant, but able to start growing when conditions are favorable. Unicellular eukaryotes, or protists, may also form resting stages. Ecologists refer to the total complement of microbes, both active and inactive, in a habitat as the microbial assemblage. The microbial community refers to the subset of microbes that are actively growing and metabolizing at any one time.

Microbial food webs are organized into trophic levels, or compartments, depending on their function. Primary producers make up the first level, or bottom, of a food web. Decomposing organisms, heterotrophic bacteria, and fungi grow on nonliving organic matter. Phagotrophic protists can feed on single or multiple compartments of a food web, depending on the consumers' size and feeding capability. The combined activities of microbial communities result in large-scale cycling of bioactive elements – carbon, nitrogen, phosphorus, sulfur, and trace metals – in ecosystems.

UNDERSTANDING MICROBIAL FOOD WEBS

The concept of systems ecology was crucial to understanding the role of microbes in ecosystems. The text *Fundamentals of Ecology* by Eugene and Howard Odum, first published in 1953, established the systems approach, which focused on ecosystem function rather than on specific populations, and followed the flows of elements such as carbon or nitrogen, or of energy, either solar energy or energy from the respiration of organic compounds, through food web compartments. Initial formulations were linear food chains, from primary producers, which captured solar energy and produced organic matter from carbon dioxide and other inorganic compounds, to primary and secondary consumers; for example, grass to antelopes to lions. Decomposing organisms, heterotrophic bacteria and fungi, were known to be important components of ecosystems, but did not comfortably fit into such a food chain. Early studies of the roles of heterotrophic microbes in ecosystems focused on the rates of respiration of organic carbon to carbon dioxide and on regeneration of bioactive elements from organic compounds into inorganic compounds such as ammonium and phosphate, which algae and higher plants could use for growth.

The notion of trophic interactions between different groups of microbes was developed during research on marine ecosystems during the 1970s and 1980s. Microbial ecologists found that microbial plankton were responsible for the bulk of respiration in the sea. They also highlighted the role of small flagellated protists as consumers of bacteria in seawater. Bacterivorous flagellates kept bacterial stocks in check and at the same time regenerated much of the nitrogen and phosphorus accumulated in bacterial cells. Subsequent work showed that the bacterivorous flagellates were in turn grazed by larger-sized protists, establishing a microbial food chain that resulted in virtually all of the organic carbon used by bacteria being recycled back to carbon dioxide and inorganic nutrient compounds. Similar microbial food chains, from bacteria to flagellates to larger protists, were found in freshwater ecosystems and in benthic habitats in marine and freshwater environments.

Further research on microbes in aquatic ecosystems showed that this microbial food chain, termed the microbial loop, was too simplistic. Both small flagellates and larger protists also fed on autotrophic cells, including photosynthetic bacteria, and on algae of all sizes. Mixotrophic phytoflagellates could ingest bacteria. Viruses infected and lysed both bacterial and algal cells, causing a short circuit of the microbial loop. A more sophisticated view of a complex microbial food web that included autotrophic, mixotrophic, and heterotrophic microbes and formed the basic food resource for metazoans such as copepods and larvae of pelagic and benthic animals emerged. Microbial food webs in terrestrial systems are simpler in that primary production is carried out by large multicellular plants, and microbes are limited to either decomposition of plant material or feeding on decomposing microbes. In this chapter, we will focus on the microbial food webs of aquatic ecosystems, and particularly of the ocean.

COMPONENTS AND PATHWAYS

Aquatic microbial food webs consist of producer and consumer compartments. Examples of microbial producers and consumers in aquatic food webs visualized via epifluorescence microscopy are shown in Figure 1. Epifluorescence microscopy is a method used by microbial ecologists to inspect cells that are either autofluorescent by virtue of their pigments, such as chlorophyll or phycobilins or made fluorescent by staining with dyes that bind to organic compounds in microbial cells and fluoresce at selective wavelengths. In Figure 1, autofluorescent cells fluoresce red, while heterotrophic prokaryotes and phagotrophic protists fluoresce blue due to added DAPI stain.

Microbial species also include a large size range (Figure 2). Most planktonic bacteria are about half a micron in size. The largest-sized phytoplankton and protists are 100–200 μm in length. Cell size is important in microbial food webs since most, although not all, phagotrophic protists feed on organisms smaller than themselves. The smallest microbial cells are less than 2 μm in diameter. This size category, which includes most aquatic prokaryotes, heterotrophic and autotrophic bacteria and archaea, and the smallest eukaryotic cells, is termed picoplankton. Nanoplankton, cells 2–20 μm in size, includes most species of flagellates, autotrophic, heterotrophic, and mixotrophic, along with some smaller-sized nonflagellated green algae and diatoms and the smallest species of dinoflagellates and ciliates. Microplankton, cells and chains of cells 20–200-μm long, covers the larger-sized phytoplankton, mainly single cells and chains of diatoms and larger species of photosynthetic dinoflagellates, and the larger-sized phagotrophic protists, ciliates, and heterotrophic dinoflagellates. Phagotrophic protists in the plankton greater than about 20 μm are termed microzooplankton and are major consumers, or grazers, of phytoplankton in marine and freshwater systems. Viruses that occur in all aquatic systems and are less than 0.2 μm, or 200 nm, are categorized as femptoplankton.

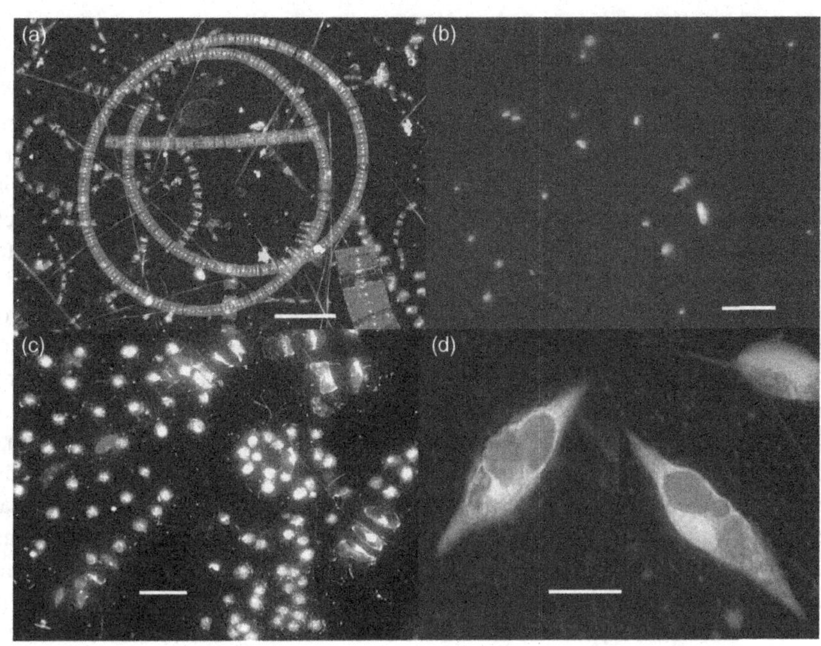

FIGURE 1 Examples of autotrophic and heterotrophic microbes in microbial food webs, visualized by epifluorescence microscopy with DAPI staining and sized using an image analysis system consisting of a Cooke Sensicam QE CCD camera with Image Pro Plus software mated to an Olympus BX61 Microscope with a universal fluorescence filter set. Red color indicates autofluorescence of photosynthetic pigments, blue color indicates DAPI staining of nonfluorescent cells and cytoplasm; the brightest blue staining occurs in the nucleus of the cells. (a) A mixed species bloom of diatoms, major algal producers in aquatic ecosystems, in the western Arctic Ocean, scale bar = 50 μm. (b) Planktonic prokaryotes stained with DAPI in water collected in a Georgia salt marsh estuary, scale bar = 2 μm. (c) Bacterivorous nanoflagellates, probably choanoflagellates, in a decaying diatom bloom in the western Arctic Ocean, scale bar = 10 μm. (d) Two heterotrophic gyrodinium-type dinoflagellates with red-fluorescent food vacuoles full of picoplankton-sized autotrophic cyanobacteria and picoeukaryotes from the western Pacific Ocean off the coast of Oregon, USA, scale bar = 20 μm.

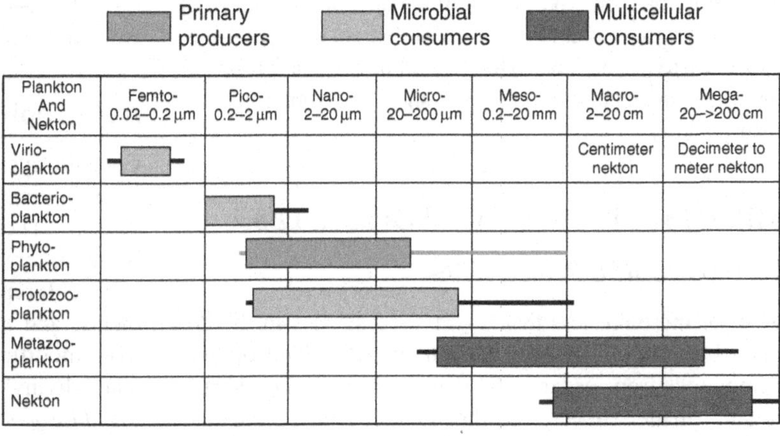

FIGURE 2 Distribution of different taxonomic-trophic compartments of plankton in a spectrum of size fractions, with a comparison of size ranges of zooplankton and nekton. Solid rectangles denote size of most organisms in each size group, bars denote approximate minimum/maximum size range of group. 30% gray bars, heterotrophic microbes; 40% gray bar, autotrophic microbes (phytoplankton); 70% gray bars, animals. Figure is updated from a figure in Sieburth, J. Mc. N., Smetacek, V., & Lenz, J. (1978). Pelagic ecosystem structure: Heterotrophic compartments of the plankton and their relationship to plankton size fractions. *Limnology and Oceanography*, *23*, 1256–1263.

At the bottom level of microbial food webs are the primary producers, usually photosynthetic bacteria and algae. However, in hypoxic habitats in the water column and the benthos, primary producers may be chemosynthetic bacteria that are able to gain energy by the oxidation of reduced chemicals such as sulfide, reduced iron, or methane. Such bacteria form the base of food webs at hot springs on land and at hydrothermal vents and methane seeps on the seafloor.

Trophic level in aquatic food webs is not easily segregated according to the size of the microorganism, since individual species of both heterotrophic and autotrophic microbes occur across the entire range of microbial size categories. In the open ocean and in large lakes, less than 2-μm-sized coccoid-shaped cyanobacteria can be important primary producers. In coastal and shallow water systems, massive blooms of microplankton-sized algae, either diatoms or dinoflagellates, can occur. To make matters more complicated, some species of eukaryotic microbes in all size ranges are mixotrophic. Species of autotrophic flagellates, or phytoflagellates, in both picoplankton and nanoplankton size ranges are capable of ingesting heterotrophic bacteria and even small phototrophic cells. It is likely that most photosynthetic dinoflagellates are also phagotrophic, preying on both autotrophic and heterotrophic protists. Some ciliates temporarily hold chloroplasts from their algal prey just below the cell membrane and use sugars produced by the chloroplasts for supplemental nutrition. One species of marine ciliate, *Myrionecta rubra*, has taken this form of mixotrophy to the extreme. Its captured chloroplasts have retained DNA from the original cryptophyte algal prey and are capable of division along with the host ciliate. *Myrionecta*, unlike other ciliates, is thus primarily autotrophic and can even form blooms in the ocean, although the ciliate is still capable of phagotrophy. There are other types of symbiotic mixotrophy in which one type of microbe lives in or on, and contributes to the metabolism of, another microbe. Examples include nitrogen-fixing cyanobacteria that live in diatom and heterotrophic dinoflagellate host cells, helping them survive in nitrogen-poor environments, and chemosynthetic sulfur-oxidizing bacteria that live on the cell membranes of benthic ciliates, which essentially farm the chemosynthetic bacteria as their main food resource.

Although trophic interactions between microbes in aquatic food webs are more complicated than trophic interactions in macroscopic food webs, the cast of characters in microbial food webs is less so. Most water column, or pelagic, food webs in both the ocean and lakes have a consistent assortment of major groups of microbes with similar trophic roles. Taxonomic groups of microbes appear to be much more uniformly distributed in the ocean, and in lakes, compared to the distribution of species of, for example, copepods and fish, in aquatic systems. Of course, this apparent uniformity may only be a result of the lack of knowledge of genetic differences among strains of distinct microbial species in different habitats. The general taxonomic groups of microbes in aquatic food webs, and their functions, are listed below.

MICROBES IN AQUATIC FOOD WEBS

Heterotrophic Prokaryotes

Most of these are less than 1-μm-sized species in the domain bacteria and live by assimilating dissolved organic compounds from water or by degrading nonliving detrital organic matter. Species in the domain archaea are also present everywhere in the sea and in freshwater habitats. There are four groups of archaea in the marine pelagic environment; the most abundant of these are Marine Group 1 archaea in the Crenarchaea, the same subdomain as sulfur-oxidizing archaea living in hot springs or in hydrothermal vents. The other marine archaea (Marine Groups 2, 3, and 4) are in the Euryarchaeota and include methanogenic and halophytic prokaryotes. Very little is known about the modes of metabolism of most of the marine archaea. Some Marine Group 1 archaea have been found to assimilate amino acids. However, archaeal cells are most abundant in the sea at depths of 200–4000 m, where there is little organic carbon. It has been established that the cold-temperature Crenarchaeota present in the ocean can gain energy for growth by oxidizing ammonium. A member of the Marine Group 1 archaea isolated from a marine aquarium tank has been shown to grow chemoautotrophically in culture by oxidizing ammonium and assimilating carbon dioxide.

Microscopic and flow cytometric methods used to enumerate heterotrophic prokaryotes in aquatic systems are based on fluorescent staining of cells, which does not distinguish between species of bacteria and archaea. For this reason, the general term bacteria, which has long been used by microbial ecologists to mean heterotrophic prokaryotes, is assumed to include cells in both domains. Bacterioplankton refers to heterotrophic prokaryotic cells suspended in seawater or freshwater.

Most strains of aquatic bacteria are in the bacterial phylum Proteobacteria. In the sea, marine bacteria are typically strains of α-Proteobacteria, which includes the most abundant open ocean ribotype, the SAR-ll clade, and of δ-Proteobacteria, which includes fast-growing opportunistic strains of the genera *Pseudomonas* and *Vibrio*. In freshwater habitats and in some coastal and estuarine environments, strains of β-Proteobacteria are abundant. In eutrophic and benthic habitats, heterotrophic bacteria grow on surfaces, forming colonies and biofilms that are an important food resource for small invertebrates. Bacteria adapted to attach to and grow on surfaces are phylogenetically diverse, and commonly include ribotypes in the *Cytophaga–Flavobacteria–Bacteriodes* (CFB) group as well as in other groups of bacteria.

Autotrophic Prokaryotes

Coccoid cyanobacteria, photosynthetic bacteria less than 2 μm in diameter, are ubiquitous in marine and freshwater systems. There are two major groups of these picocyanobacteria: orange-fluorescing *Synechococcus* spp., which have chlorophyll *a* and phycobiliprotein accessory pigments, and red-fluorescing *Prochlorococcus* spp., which have modified chlorophyll pigments, divinyl chlorophyll *a* and divinyl chlorophyll *b* as the main accessory pigment. *Prochlorococcus* spp. are smaller in size than *Synechococcus* spp., and are typically abundant in open ocean habitats, while *Synechoccocus* spp. are most abundant in nearshore to outer continental shelf waters. *Synechococcus* spp. are also abundant in the water column of lakes, while *Prochlorococcus* spp. are predominantly marine. Filamentous cyanobacteria are common in polluted freshwaters and hot springs. In the ocean, the filamentous cyanobacteria *Trichodesmium* spp. form blooms in subtropical regions and are globally important nitrogen fixers. Nonoxygen-producing, bacteriochlorophyll-containing bacteria, which require a source of reduced compounds to grow, can be significant primary producers in lakes and marine systems that have subsurface anoxic water masses rich in sulfide. In addition, many strains of heterotrophic bacteria living in oxic aquatic habitats have been found to contain either bacteriochlorophyll or bacteriorhodopsin pigments, which may be used to generate extra ATP by using energy harvested from light.

In benthic habitats, chemosynthetic bacteria can support both microbial and macroscopic food webs. These autotrophs include free-living single-celled and filamentous strains of sulfur-oxidizing bacteria, such as species of *Beggiatoa* and *Thiospirillum*, and symbiotic sulfur-oxidizing bacteria living in or on both single-celled and multicellular eukaryotes. Examples are the sulfur-oxidizing bacteria that grow on the benthic ciliate *Zoothamnium niveum*, which lives at the oxic–anoxic interface in sandy marine sediments and provides its bacterial crop with both sulfide from below and oxygen from above, and the symbiotic sulfur-oxidizing bacteria that grow in special organs of gutless hydrothermal vent tube worms, providing the worms with all of their nutrition. At methane seeps on the seafloor, methane-oxidizing bacteria grow in the sediments and in the gill tissues of seep mussels, providing most of the primary production for these ecosystems.

Autotrophic Eukaryotes

Also termed algae, single-celled photosynthetic eukaryotes are the most significant primary producers both in the sea and in lakes. Algal cells are diverse both in size and in taxonomic diversity. The smallest algal cells are 0.8-μm-diameter marine *Ostreococcus* spp. and 1–2-μm-diameter *Micromonas* spp, both abundant in the open ocean. There is a great diversity of algal species in the nanoplankton size range. Most of these are golden brown-pigmented, flagellated chrysophytes and prymnesiophytes, orange-pigmented cryptophytes, and green-pigmented prasinophytes, although nonflagellated chlorophytes and diatoms also occur in this size range. Algae larger than 20 μm are less abundant than smaller-sized phytoplankton, but at times form dense blooms in coastal waters or in lakes. Bloom-forming algae greater than 20 μm are typically diatoms or autotrophic dinoflagellates. Many flagellated algae, including chloroplast-bearing nanoflagellates and dinoflagellates, can ingest other microbial cells. Some species of algae capable of ingesting bacteria cannot grow in the absence of prey.

Heterotrophic Eukaryotes

These microbes were known as protozoa, and researchers often still use that term. However, many species of heterotrophic eukaryotes are close kin to photosynthetic species. The word Protozoa, which means first animal life in Greek, is thus not an appropriate label for these microbes, and we prefer to use the term heterotrophic protist. Heterotrophic protists, which do not have chloroplasts, are as ubiquitous and as diverse as autotrophic protists, the algae. There are some protist lineages: the bodonids, the choanoflagellates, and the kinetoplastids, that do not have any chloroplast-containing species, and are strictly phagotrophic. The choanoflagellates are of particular interest to molecular geneticists as they are the group of single-celled protists most closely related to multicellular animals. Phagotrophic protists have the potential to be major predators in microbial food webs because they are in the same general size range as their microbial prey, bacteria, algae, and other heterotrophic protists (Figure 2), and because protist growth rates are on the same temporal scale, hours to days, as those of their prey. The high rate of metabolism of these small, unicellular predators also facilitates carbon and energy flux through ecosystems.

The smallest heterotrophic protists are 1–2-μm-sized flagellated species, which occur in several protist groups, including the chrysophytes and bodonids. Heterotrophic protists 2–20 μm in size, mainly nanoflagellates (e.g., Figure 1(c)), are very diverse taxonomically and are major consumers of picoplankton and smaller-sized nanoplankton cells in aquatic systems. Some species of ciliates and heterotrophic dinoflagellates are also less than 20 μm. Phagotrophic protists larger than 20 μm are predominately ciliates and nonchloroplast-containing dinoflagellates. Examples of these protists are shown in Figure 3. This size class of phagotrophic protist, termed microzooplankton, is abundant in the sea and in lakes, and these protists are major consumers of phytoplankton and of heterotrophic nanoflagellates. Planktonic ciliates are mainly spherical or conical spirotrichs, cells with cilia grouped around an oral end. One subgroup of spirotrichous ciliates, the tintinnids, build species-specific houses, or loricae, in which they live, serving as a protective shelter (Figure 3(b)). Heterotrophic dinoflagellates may have rigid cellulosic plating, or armor, which gives them a distinctive shape (e.g., the cell in Figure 3(d)) and makes it difficult, though not impossible, for them to ingest prey cells directly. Many armored dinoflagellates instead feed externally by extruding a hollow tube into, or a pseudopodial veil around, their algal prey. The dinoflagellate injects digestive enzymes into the prey cells and sucks the digested prey cytoplasm back into itself by these feeding structures. Most species of heterotrophic dinoflagellate are nonarmored, and have an elastic cell membrane that allows them to ingest algal prey up to a size equal to, or even greater than, the dinoflagellate, often greatly distending the dinoflagellate cell in the process (Figures 3(e) and 3(f)).

In benthic habitats, where the fluid environment interacts with surfaces such as grains of sand, clay particles, or organic detritus, the protist community is dominated by surface-feeding hymenostome ciliates similar to *Paramecium* spp., whose main food is bacteria. Some of these are very long and thin, adapted to move through narrow spaces between sediment particles, and some have stiff ventral cilia that allow them to scrape bacterial biofilms off particle surfaces. Amoebae and amoeboid flagellates are also common in benthic environments and on detrital particles in the water column. Phagotrophic protists are

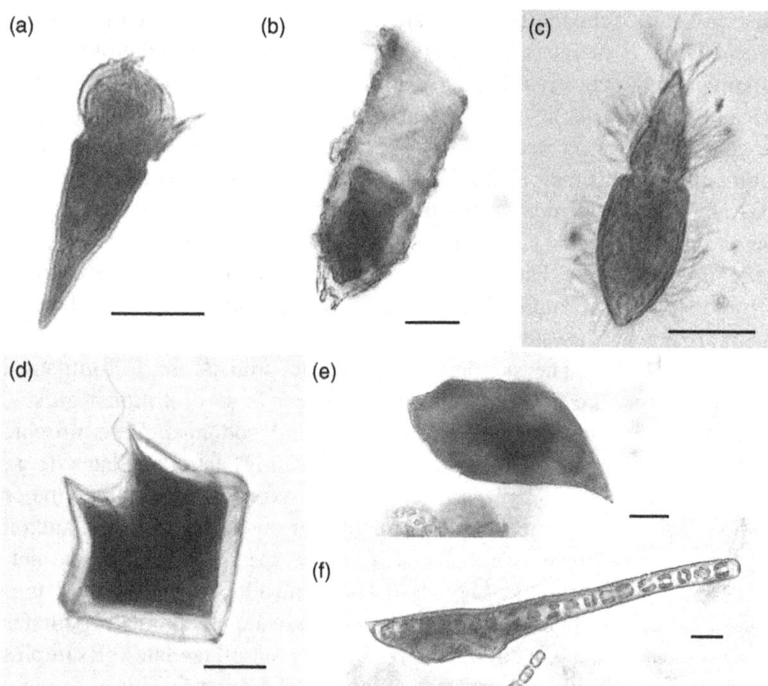

FIGURE 3 Examples of aquatic phagotrophic protists in the 20–200 μm, or microzooplankton, size class. Cells were preserved and stained with iodine-based acid Lugol solution, which gives them a brown color, visualized via light microscopy, and sized and photographed using an image analysis system consisting of a Cooke Sensicam QE CCD camera with Image Pro Plus software mated to an Olympus BX61 Microscope. (a) *Strombidium* sp. ciliate, consumer of nanoplankton-sized microbial prey from the western Arctic Ocean; (b) *Tintinnopsis* sp. tintinnid ciliate in an aggregated lorica, consumer of nanoplankton-sized prey from the western Arctic Ocean; (c) Haptorid ciliate, species unknown, a predatory ciliate that feeds on cells as large as itself from the western Arctic Ocean; (d) *Protoperidinium* sp. armored heterotrophic dinoflagellate that feeds on large cells including diatoms using an extracellular pseudopodial feeding veil from the Oregon upwelling system of the western Pacific Ocean; note flagellum trailing behind the cell; (e) *Gyrodinium* sp. nonarmored heterotrophic dinoflagellate associated with a diatom bloom in the western Arctic Ocean; (f) *Gyrodinium* sp. nonarmored heterotrophic dinoflagellate distended from an ingested chain of the diatom *Thalassiosira* sp., associated with a diatom bloom in the Oregon upwelling system. All scale bars = 20 μm.

important components of microbial food webs in extreme habitats such as sea ice, solar salterns with salinities as high as 150 ppt, and the deep ocean.

MARINE PELAGIC HABITATS

The oceans, which cover more than two-thirds of the earth's surface, provide large and variable habitats for microbial food webs. In the euphotic zone, the upper lighted depths of the sea, approximately from the surface to 50–200 m depending on the region and season, photosynthetic microbes, both prokaryotic and eukaryotic, provide a continuous source of organic carbon and prey cells for heterotrophic bacteria and protists. At subeuphotic depths, where there is insufficient sunlight for photosynthesis, heterotrophic microbes live on organic carbon sinking down from the euphotic zone. At these depths, the microbial food web regenerates most of the sinking organic carbon back to carbon dioxide and inorganic nutrients.

The largest oceanic habitats are the slowly rotating gyres at the center of each of the great ocean basins. These subtropical open ocean regions have warm surface waters and low amounts of inorganic nutrients for phytoplankton growth. They are oligotrophic or nutrient limited. Phytoplankton cells are small-sized, dominated by cyanobacteria and small algal species, and the total biomass of primary producers is low. Microbial food webs are complex, with small heterotrophic protists and mixotrophic flagellated algae ingesting heterotrophic bacteria and cyanobacteria, slightly larger protists feeding on nanoalgae and on bacterivorous flagellates, and the largest protists feeding on algal cells and on heterotrophic protists. Most of the primary production is respired in this multicompartment microbial food web, with little remaining for macroscopic marine life.

At the edge of continents, the oceans cover continental shelves of varying depths and widths. Continental shelf waters are dynamic, and occur in subtropical, temperate, and polar regions. Most shelf systems are mesotrophic, with higher biomass of phytoplankton and of other microbes compared to the oligotrophic gyres. Upwelling or injection of nutrient-rich subsurface water frequently occurs at the edge of the shelf, resulting in eutrophic conditions with high phytoplankton biomass and production. Some regions, such as the narrow continental shelves of the Pacific Northwest of the United States, Northwest Africa, and Peru and Chile in South America, have seasonal or persistent upwelling of nutrient-rich seawater extending to the coast, which results in massive phytoplankton blooms, usually of species of diatoms. Phytoplankton blooms also occur each spring in temperate and polar regions due to increase in day length from winter to summer. Much of the organic matter produced in these mass phytoplankton blooms sinks down to subsurface depths or to the sediment. Some of the bloom production is utilized by heterotrophic bacteria and by microzooplankton-sized protists capable of preying on large algal cells and chains of cells (e.g., the dinoflagellate in Figure 3(f)).

At the margins of the oceans are shallow nearshore and estuarine habitats influenced by inputs of freshwater from rivers or land runoff and by interactions between the water column and the sediments. Depending on their geographic location and local conditions, nearshore and estuarine systems may be oligotrophic, mesotrophic, or eutrophic. Agriculture and other human activity has resulted in a large increase in the amount of plant nutrients such as nitrate and phosphate carried by rivers and by land runoff to nearshore marine systems. As a result of enhanced phytoplankton growth due to these nutrients, microbial food webs in some nearshore regions respire enough organic matter to deplete the water of oxygen, resulting in hypoxic or anoxic dead zones. A classic example is the large and growing zone of low oxygen water off the Mississippi River in the Gulf of Mexico. Hypoxic and anoxic marine habitats also occur naturally beneath persistent upwelling regions and in some enclosed fjords and basins, for example, the Black Sea.

Most of the pelagic habitat of the ocean is subsurface, below 200 m. Only a small fraction of primary production in the euphotic zone sinks out to depths below 200 m. Microbes that live below that depth are much less abundant than in the euphotic zone and must adapt to highly stressful conditions of low food, cold temperatures, and high pressure. Both bacteria and archaea, as well as heterotrophic flagellates, occur at depth in the sea. Cold-temperature Crenarchaeota are relatively more abundant compared to bacteria in this habitat. Currently, not much is known about how deep ocean bacteria survive, or about deep ocean microbial food webs.

BENTHIC HABITATS

Microbial food webs in aquatic sediments are shaped by the characteristics of benthic habitats. In sediments, microscale structure results in large changes in environmental conditions, for example, redox potential, oxygen and nutrient concentrations, over millimeter to centimeter spatial scales. Particles, both inorganic and organic, are a dominant component of the benthic environment; thus, interactions with particles, for example, attachment to, and grazing on, surfaces, are of major importance for microbes living in and on sediments. Benthic ecosystems are mainly heterotrophic; the microbial food web is based on input of organic matter that settles to the bottom as a result of primary production carried out in the water column. Exceptions are hydrothermal vents and methane seeps, where chemosynthetic sulfur-oxidizing bacteria or methanotrophic bacteria form the basis of local food webs. Where the input of organic matter to sediments is high relative to the supply of oxygen, suboxic/anoxic habitats dominate in the benthos. Depth zones in the sediment reflect sequential utilization of compounds as electron acceptors with depth: oxygen, nitrate, sulfate, and carbon dioxide. Food webs in anoxic environments are shorter compared to food webs in oxic environments due to lower growth efficiencies of anaerobic microbes.

The fine-scale habitat differentiation in benthic habitats, in terms of water chemistry, oxidizing/reducing conditions, and sediment texture, yields a high diversity of potential niches for microbes. Since particle surfaces predominate in the benthos, microbial biofilms are prevalent on both organic detrital particles and inorganic grains of sand or clay. Microbial exopolymers, high molecular weight polysaccharide or mucopolysaccharide secretions, are copiously made by benthic bacteria and microalgae such as diatoms. Exopolymers create a microenvironment around a microbial cell, buffering it from rapid environmental changes in pH, salinity, dessication, or nutrient regimes.

The depth to which oxygen is present in sediments depends both on sediment composition: loose sandy sediments allow oxygen to penetrate to a greater depth compared to compact clay sediments, and on the amount of organic matter reaching the sediments: the more organic matter, the greater the rate of oxygen utilization. In coastal waters, usually only the sediment–water interface and the upper few millimeters or centimeters is near oxygen saturation. Oxygen is supplied mainly by diffusion from the overlying water; however, in shallow sediments, some oxygen may be provided by microalgal photosynthesis, and in all sediments, bioturbation by invertebrates results in local oxidizing zones in the top few centimeters, or deeper, in the sediment.

In marine unperturbed sediments there is a standard sequence of redox zones, compounds used as electron acceptors, and associated metabolic processes of microbes with depth (Figure 4). In the upper layer of the sediment

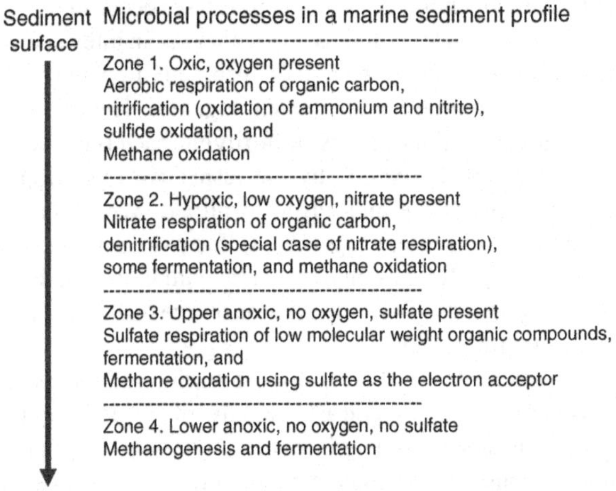

Sediment surface

Microbial processes in a marine sediment profile

Zone 1. Oxic, oxygen present
Aerobic respiration of organic carbon,
nitrification (oxidation of ammonium and nitrite),
sulfide oxidation, and
Methane oxidation

Zone 2. Hypoxic, low oxygen, nitrate present
Nitrate respiration of organic carbon,
denitrification (special case of nitrate respiration),
some fermentation, and methane oxidation

Zone 3. Upper anoxic, no oxygen, sulfate present
Sulfate respiration of low molecular weight organic compounds,
fermentation, and
Methane oxidation using sulfate as the electron acceptor

Zone 4. Lower anoxic, no oxygen, no sulfate
Methanogenesis and fermentation

FIGURE 4 Sequence of redox zones and associated microbial processes with depth in an idealized sediment profile. Moving down from the sediment surface the sequence is as follows: (Zone 1) oxic zone: high oxygen concentration at the sediment surface; aerobic respiration, sulfide oxidation, nitrification, and methane oxidation; (Zone 2) hypoxic zone: low oxygen and measurable nitrate concentration; anaerobic nitrate-based respiration, denitrification, some fermentation, and methane oxidation; (Zone 3) upper anoxic zone: no oxygen but sulfate present, sulfate respiration, sulfide formation, fermentation, and methane oxidation using sulfate as the electron acceptor; (Zone 4) lower anoxic zone: no oxygen or sulfate, methanogenesis using carbon dioxide as the electron acceptor and fermentation.

in contact with overlying waters, respiration of oxygen by prokaryotes and protists occurs. Where overlying water is rich in nitrate, anaerobic nitrate respiration to nitrite by prokaryotes and some protists, or denitrification to nitrogen gas by denitrifying bacteria, dominates when oxygen concentration is depleted. Deeper in the sediment, both oxygen and nitrate are exhausted, but the interstitial seawater is still rich in sulfate. In this zone, sulfate-respiring bacteria grow on hydrogen and fatty acids produced by anaerobic fermenting microbes. The end product of sulfate respiration is sulfide, which builds up in anoxic marine habitats, producing a characteristic rotten egg smell. Still deeper, sulfate is depleted and microbial metabolism is mainly based on methanogenesis by archaea and fermentation by bacteria and protists.

In the anoxic zones of both marine sediments and oxygen-depleted water masses, sulfate-respiring bacteria and methanogenic archaea compete for the metabolites of fermenting bacteria: hydrogen and low molecular weight organic compounds, particularly acetate. Sulfate-reducing bacteria are better competitors and can grow at lower hydrogen concentrations than can methanogens. Thus in marine anoxic habitats, sulfate reducers outcompete methanogens in zones where there are significant concentrations of sulfate. Because sulfate respirers and methanogens can utilize only low molecular weight organic substrates, fermenting microbes are primarily responsible for degradation of particulate detritus and high molecular weight organic compounds in anaerobic sediments.

Two major chemoautotrophic processes occur in marine sediments and water columns where there is an interface between oxic and anoxic habitats. Nitrification occurs when oxygen and ammonium are present together. The two phylogenetically distinct groups of nitrifying bacteria produce energy for carbon fixation by the respective oxidation of ammonium to nitrite and nitrite to nitrate. When oxygen and sulfide are present together, sulfur-oxidizing bacteria produce energy for carbon fixation by oxidation of reduced sulfur compounds. This is the major source of fixed carbon at hydrothermal vents.

The microbial food web of most marine sediments and anoxic water masses consists mainly of detrital organic matter consumed by heterotrophic prokaryotes, which in turn are consumed by heterotrophic protists. However, in coastal and intertidal marine sediments where there is sufficient light, benthic algae are a component of food webs. Ciliates are abundant in benthic habitats and consume both bacteria and algae; the roles of heterotrophic flagellates and amoebae in sediments are less well known.

Anaerobic metabolism is inherently less energetically efficient than is aerobic metabolism. Fermenting organisms typically grow with a gross growth efficiency (GGE) of substrate use of 10%, while aerobic microbes can transform 40% of assimilated organic carbon into biomass. Any process that serves to enhance the low growth efficiencies of anaerobic organisms would give a competitive edge to such organisms. Anaerobic ciliates are characteristic of both marine and freshwater anoxic habitats. These protists generate energy by fermentation of organic compounds obtained by ingesting other microbes, primarily bacteria. In these ciliates, the fermentative processes resulting in oxidation of pyruvate and production of hydrogen occur in unique organelles, hydrogenosomes. Hydrogenosomes appear to be modified mitochondria that have lost the electron transport system. Many species of anaerobic ciliates are full of endosymbiotic prokaryotes. When excited by blue light, the cells fluoresce blue-green, a characteristic of methanogenic archaea. The observation of methane generation in these protists, along with molecular genetic analysis of the endosymbionts, has confirmed that they are in fact methanogens. The cytoplasm of the fermenting ciliate is a microhabitat with high abundance of hydrogen, acetate, and carbon dioxide – waste products of the host and substrates for methanogens. This is of particular significance in marine anoxic habitats, where high concentrations of seawater sulfate foster the growth of anaerobic sulfate-respiring bacteria, which outcompete nonsymbiotic methanogens for available hydrogen and fatty acids. In turn, the endosymbiotic methanogens make the metabolism of the ciliate more efficient by decreasing fermentation waste products in the cell and by serving as a food resource for the ciliate. This unique microbial collaboration is a classic case of syntropy, literally feeding together, in which two organisms grow in a mutually beneficial, intimate association.

ROLE OF MICROBIAL FOOD WEBS IN BIOGEOCHEMICAL CYCLING

Microbes can be viewed as the chemical engineers of the biosphere. Biogeochemical cycles of carbon, nitrogen, and sulfur cannot occur without specific metabolic capabilities of various groups of microorganisms. For many of the cycles of bioactive elements, interactions of both prokaryotic and eukaryotic species in microbial food webs are required for completion of the pathways of the elements. Conversion of organic carbon, organic nitrogen, and organic phosphorus into inorganic compounds, namely, carbon dioxide, ammonium, and phosphate, is facilitated by consumption of prokaryotic and eukaryotic prey cells by phagotrophic protists.

The dominant degradation pathway of organic matter produced by autotrophic microbes, that is, primary production, in aquatic ecosytems is assimilation and respiration by heterotrophic prokaryotes, both bacteria and archaea. Prokaryotes utilize organic matter in many forms and steps in aquatic food webs: in the water column as dissolved organic matter released by growing algae, as nonliving particulate organic matter, or organic detritus, produced during decaying phytoplankton blooms and as waste products of protist and metazoan consumers, and in sediments from the settling of organic particles.

Part of the primary production assimilated by prokaryotes is regenerated via cellular catabolism back to inorganic compounds, carbon dioxide, ammonium, and phosphate, and part is converted into cell biomass. The proportion of organic carbon assimilated by microbes that is used in anabolic processes to produce more cell biomass is termed the GGE (Figure 5). For prokaryotic microbes, the GGE is simply the amount of cell biomass produced as a fraction of the total amount of organic carbon assimilated. The rest of the assimilated carbon is respired to carbon dioxide. This growth efficiency is sometimes termed bacterial growth efficiency (BGE). Although the theoretical maximum BGE is 67%, in nature the community BGE is much lower, and generally ranges between 10 and 40%. A major problem with ascertaining the actual BGE of the community of growing cells in the natural environment is that a variable portion of cells in the prokaryotic assemblage are dead or dormant, which makes it difficult to scale assimilation and growth rates measured for the total assemblage to just the active community.

The relative fractions of assimilated organic nitrogen and phosphate that are released as inorganic compounds by prokaryotic metabolism depend on the elemental ratios of carbon, nitrogen, and phosphate in the organic matter on which the microbial cells are growing. Elemental C:N:P ratios of phytoplankton are variable and depend on factors such as the availability of inorganic nitrogen and phosphorus in the environment, species composition, and growth state of the phytoplankton. The classic C:N:P atom ratio of phytoplankton in the sea is 106C:16N:1P, and the ratio between carbon and nitrogen is 6–7:1. This is known as the Redfield ratio, after the oceanographer Alfred C. Redfield who proposed it as an explanation of why the general elemental ratio between nitrate and phosphate in the sea was about 16:1. However, bacterial cells have a higher requirement for both nitrogen and phosphorus, and a C:N ratio of 4–5:1. Prokaryotic cells also have a higher biomass-specific concentration of iron and other trace metals compared to eukaryotic cells. Thus prokaryotic cells tend to sequester nitrogen and phosphorus and metal ions, which are only released back to the environment by predation or by viral lysis.

Ingestion and digestion of other microbial cells by phagotrophic protists is of vital importance in complete regeneration of the elements fixed by phytoplankton into organic compounds back to inorganic compounds that can be reutilized by autotrophs for further primary production. In both marine and freshwater systems, phagotrophic protists, both flagellates and ciliates, are major consumers of prokaryotic cells. Phagotrophic protists also are major consumers of algal cells, even the large-sized phytoplankton characteristic of mass blooms. The carbon-based growth efficiency of phagotrophic protists is about 40% of ingested prey biomass. Protists release undigested components of ingested prey as dissolved and particulate organic matter, as well as metabolic waste products as dissolved inorganic compounds such as ammonium and phosphate. Thus protistan grazing provides organic and inorganic substrates for further growth of their prey, both heterotrophic bacteria and autotrophic cells. Protists have much higher biomass-specific rates of nutrient excretion than do larger-sized zooplankton, and they regenerate nutrient elements bound up both in bacteria and in phytoplankton. Thus protist consumption of microbial cells is a major process in regeneration of nitrogen and phosphorus compounds in aquatic systems.

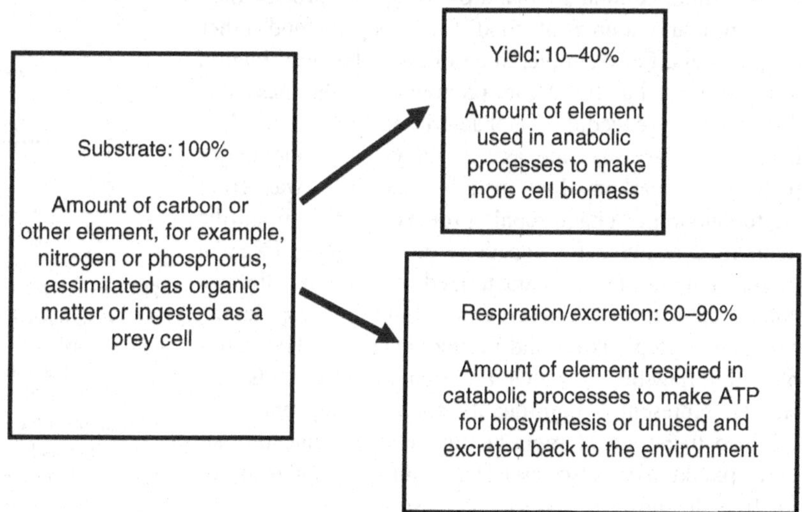

FIGURE 5 Diagram showing gross growth efficiency as cell biomass yield as a proportion of either total substrate or prey biomass assimilated or as a portion of the sum of yield plus amount of ingested food either respired or excreted.

The capacity of many species of autotrophic flagellates to phagocytize gives these phytoplankters an advantage in the acquisition of nutrients in a chemically dilute environment. Mixotrophic algae ingest bacteria and eukaryotic prey to gain both organic substrates and inorganic nutrients. In oceanic systems in which iron is a limiting micronutrient, consumption of iron-rich bacterial cells is an adaptive strategy for phagotrophic algae. Bacterivorous flagellates may also experience iron limitation, and thus ingestion of prokaryotic prey with high iron concentrations can be important to heterotrophic as well as to autotrophic protists.

FOOD RESOURCE FOR METAZOANS

Microbial production forms the base of aquatic food webs. A variable, and at times large, part of the production consumed by aquatic animals is direct consumption of algae. Prokaryotic biomass is a food resource for some animals, for example, rotifers and cladoceran zooplankton such as *Daphnia* spp. in brackish coastal systems and in lakes, and deposit feeding worms in benthic habitats. However, phagotrophic protists play a significant role in channeling microbial, both prokaryotic and algal, production at the base of the food web to higher trophic levels. In addition, phagotrophic protists consume other heterotrophic protists. Species of heterotrophic dinoflagellates and ciliates have been shown in culture to readily ingest heterotrophic flagellates as well as phytoplankton prey. This trophic link in aquatic microbial food webs, although in theory quite important, has, to date, received surprisingly little attention.

There has been debate about the quantitative significance of trophic transfers involving protists; but there is no doubt that heterotrophic protists represent food for a variety of other consumers. The largest body of studies on this subject deals with ciliates and heterotrophic dinoflagellates as food for mesozooplankton. In regions of the ocean where most phytoplankton are less than 5 μm, which is too small for most multicellular zooplankton to capture, protists may be a primary source of food for copepods and other zooplankters. For example, in an oligotrophic atoll lagoon in the tropical Pacific Ocean, phytoplankton biomass was dominated by coccoid cyanobacteria and algal cells less than 3 μm. Grazing rate assays showed that the major pathway of carbon flow in this food web was from phytoplankton to phagotrophic protists, which formed the main food resource for copepods in the lagoon. Even in mesotrophic systems characterized by diatom blooms, phagotrophic protists can serve as an important trophic link between phytoplankton and mesozooplankton. Heterotrophic dinoflagellates, which are rich in fatty acids and sterols, represent a high-quality food for copepods and enhance their rate of reproduction. Phagotrophic protists in the plankton can also serve as a significant food resource for filter-feeding benthos such as oysters.

MODELING MICROBIAL FOOD WEBS

Conceptual, or box, models, such as the ones shown in Figures 6 and 7, and simulation models, which put the flows between the boxes of conceptual models on a mathematical basis, are a standard approach to understanding how the various components of ecosystems function interactively. The first quantitative models of pelagic food webs, dating from the 1940s, were simple nutrient–phytoplankton–zooplankton (NPZ) simulations based on transfers of nitrogen between an inorganic nutrient (nitrate plus ammonium) compartment and phytoplankton and zooplankton compartments. Phytoplankton production was dependent on nutrient availability, and zooplankton consumption of phytoplankton and regeneration of phytoplankton nitrogen back into inorganic nitrogen as ammonium depended on phytoplankton production. The microbial food web, including both heterotrophic prokaryotes and protists, was either ignored or put into an extra organic detritus compartment in the model to account for nitrogen regeneration from decaying phytoplankton cells or zooplankton fecal pellets.

After subsequent research findings proved that heterotrophic microbes played significant and central roles in aquatic food webs, microbes began to be formally included in model diagrams and simulations. The first formulation was to add a decomposing microbial food chain composed of detrital organic matter, heterotrophic prokaryotes, bacterivorous flagellates, and larger protists that fed on the flagellates to the standard, or classic, food chain of phytoplankton to zooplankton to larger consumers (Figure 6). This microbial food chain was termed the microbial loop, and its role in the overall food web was to respire a large

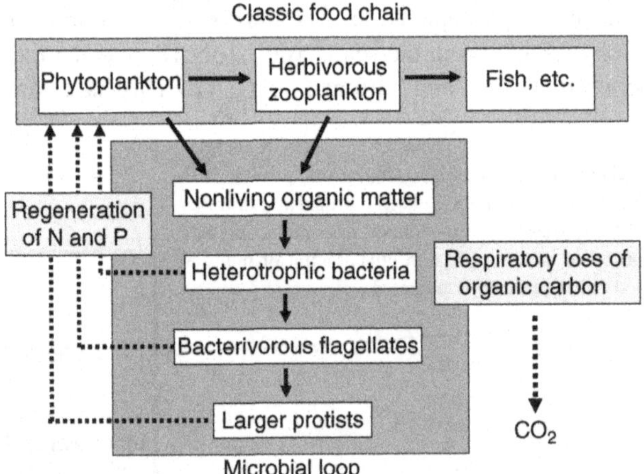

FIGURE 6 Initial conceptualization of the place of microbes in aquatic food webs based on the box model diagram of the microbial loop concept. In this conceptualization, a linear microbial food chain of heterotrophic bacteria to heterotrophic protists to larger protists is added on to the classic phytoplankton to zooplankton to higher consumers food chain. The role of the microbial loop is viewed in this concept as mainly a sink for primary production and a major pathway of regeneration of inorganic nutrients. Redrawn from Figure 1 of Ducklow, H. W. (1983). Production and fate of bacteria in the oceans. *BioScience*, *33*, 494–501.

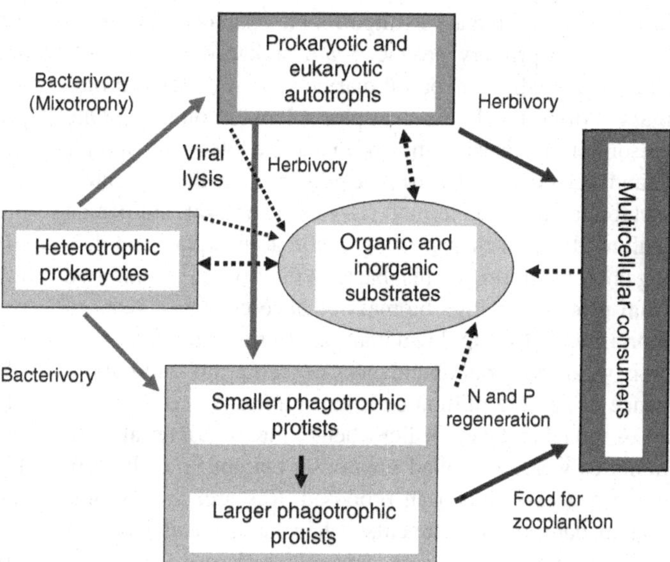

FIGURE 7 Current conceptual model of the major compartments and roles of the microbial food web in aquatic ecosystems. In this conceptualization, the microbial loop is embedded in a larger microbial food web that includes additional pathways of mixotrophy by phytoflagellates, consumption of phytoplankton by phagotrophic protists, viral lysis of both bacteria and phytoplankton, and phagotrophic protists as an important food resource, along with phytoplankton, for multicellular zooplankton.

fraction, 50% or more, of overall phytoplankton production into carbon dioxide and to regenerate nitrogen and phosphorus nutrients for further primary production.

Further work showed that the heterotrophic microbial food chain was embedded in a much larger, more complex microbial food web in which protists of all sizes fed on autotrophic prey and in which viral lysis of prokaryotes and algal cells at times acted to short circuit the microbial loop (Figure 7). Recent simulation models of pelagic food webs have explicitly included microzooplankton as consumers of bacteria and phytoplankton, and as food for mesozooplankton. The proportion of phytoplankton carbon that flows through a multistep microbial food web versus a shorter phytoplankton–mesozooplankton food chain has implications for the capacity of marine ecosystems to sequester organic carbon or to efficiently produce fish biomass. Two theoretical scenarios have been proposed in which pelagic systems characterized by an active microbial food web will export less organic carbon compared to systems in which activity of heterotrophic bacteria and protists is relatively low. The first scenario is termed a microphagous food web, dominated by phagotrophic protists consuming prokaryotic and small algal cells, and the second scenario a macrophagous food web in which large-sized algae such as diatoms are consumed by copepods. To a large extent the factor that determines the degree to which pelagic food webs are microphagous versus macrophagous is the proportion of plankton biomass that consists of heterotrophic bacteria and phytoplankton cells less than about 5 μm.

In an empirical test of the theory, it was found that in the St. Laurence River estuary in Canada export flux from the water column was the same during both the spring diatom bloom with a small microbial food web and the postbloom with a larger, more dynamic microbial food web. However, the nature of the sinking material was different. During the spring bloom, the vertical flux was primarily in the form of organic aggregates consisting primarily of sedimenting phytoplankton cells; during the postbloom period, the major flux was in the form of fecal pellets from omnivorous copepods feeding on heterotrophic protists. Obviously, food web structure alone is not always a good predictor of the quality or quantity of sinking organic carbon. One must also include trophic flux studies coupled with hydrodynamic measurements across time and space. It is also important that studies on microbial food webs include data on all the major components of a pelagic ecosystem, not just the microbes. Since the ocean is the largest reservoir of inorganic carbon that freely exchanges with the atmosphere, understanding the influence of microbial food webs on the ability of the ocean to store atmospheric carbon dioxide is an important research theme for biological oceanographers.

In modeling microbial food webs whether microbial stocks and rates of biomass production are controlled by bottom-up or by top-down processes is of critical importance. Bottom-up processes are characterized by availability of requirements for growth, for example, inorganic nutrients and light for phytoplankton and quantity and quality of organic substrates for heterotrophic prokaryotes. Top-down processes are mortality processes, mainly due to predation but also including viral lysis, as well as cell death due to unfavorable environmental conditions. An enduring question for aquatic microbial ecologists is why bacterial abundances range over only about 1.5 orders of magnitude in the euphotic zone of the ocean, from several hundred thousand cells per milliliter to one or two million cells per milliliter; while phytoplankton biomass, as measured by concentration of the main photosynthetic pigment chlorophyll *a*, varies over 3 orders of magnitude, from 0.05 μg chl *a* l^{-1} in oligotrophic open ocean gyres to 30–50 μg chl *a* l^{-1} in coastal phytoplankton blooms. This results in bacterioplankton biomass being equal to or even greater than phytoplankton biomass in oligotrophic open ocean systems where most primary producers are prokaryotes or small-sized algae, while in coastal marine systems, bacterial biomass is often much less than phytoplankton biomass. One reason for the disparity may be that at the lower end of their abundance range, heterotrophic prokaryote cells enter into a metabolic condition termed starvation survival and persist without dying or disappearing, while at the upper end during phytoplankton blooms, bacteria may either be inhibited from growing to high biomass or have enhanced rates of mortality, that is top-down controls, from grazing and viral lysis when bacterial abundances exceed a threshold abundance of about a million cells per milliliter.

A model that can predict whether marine bacterial communities were controlled mainly by bottom-up or by top-down processes has been proposed. Researchers assumed that in natural environments, when bacterial abundances were close to the carrying capacity of the local environment, bacterial growth rates would be high and variable, and mainly limited by the availability of organic substrate, or bottom-up-controlled. However, when bacterial abundances were far from the carrying capacity of the local environment, growth rates would tend to be lower, and bacterial growth was likely to be limited by mortality processes, or top-down-controlled. Thus, if there is a negative relationship between bacterial growth rate and abundance, it should indicate that the bacterial abundances are near the carrying capacity of the local environment, and that the bacterial community is limited by substrate availability. However, if bacterial abundance and growth rate are not related, it suggests that bacteria are top-down-controlled, with a small range of possible growth rates. This idea was tested by comparing environmental data on bacterioplankton abundance and assemblage growth rates collected from various open ocean habitats. The empirical data sets in fact showed a negative relationship between bacterial growth rate and bacterial abundance in eutrophic regions, indicating bottom-up or resource control, but no relationship between bacterioplankton abundance and growth rate in oligotrophic regions, suggesting top-down or mortality control. Experiments were also carried out in the oligotrophic systems in which water was screened through 0.8-μ m-pore-sized filters to remove heterotrophic protists, the main source of mortality. In most cases the bacterial growth rate in screened water increased compared to growth in the presence of phagotrophic protists, confirming top-down regulation.

The main processes controlling the abundances and growth rates of phagotrophic protists in aquatic ecosystems are less well understood. Comprehensive data sets on protist abundance and in situ growth rates are lacking. Mortality processes are likely to be important in determining the abundances and biomass of populations of protists.

Top-down control of heterotrophic protists in aquatic food webs results in trophic cascades in which enhanced mortality of a trophic group of heterotrophic protists, for example, nanoflagellates that preferentially consume bacteria and less than 5-μm-sized phytoplankton, increases the growth rate of the prey. Factors that set up tropic cascades include variations in intrinsic growth rates of different classes of grazing protists and dependence of protist growth rate on prey abundance. In natural systems, phagotrophic protists are typically food-limited, and they are poised to rapidly increase their growth rate when they encounter higher prey abundance. Another factor is how the available prey in a microbial food web is partitioned among various groups of grazing protists, particularly by differences in selection of prey by size, as different populations of protist grazers in a system tend to have different prey size preferences. Finally, copepods and other zooplankton can exert a strong top-down control on protists larger than about 10 μm.

Trophic cascade effects have been studied by manipulation experiments in which marine or freshwater samples are treated to exclude predators, and effects of the manipulation on growth of various groups of microbes in a food web are followed over time. As an example, an experiment that demonstrated trophic cascades in a microbial food web was carried out in the northern Baltic Sea. To remove various groups of phagotrophic protists, separate volumes of seawater were filtered to yield four size fractions: a less than 0.8-μm fraction that only included heterotrophic bacteria and autotrophic picoplankton, a less than 5-μm fraction that included heterotrophic bacteria, small-sized phytoplankton, and small-sized heterotrophic flagellates, a less than 10-μm fraction that also included 5–10-μm phytoplankton and flagellates, and a less than 90-μm fraction that included most components of the microbial food web but excluded metazoan grazers. The fractionated water samples were incubated in situ in dialysis bags with a molecular weight cutoff of 12–14 kDa, which allowed dissolved nutrients and organic substrates to pass in and out of the incubation bags. The development of the plankton community in the various size-fractioned water samples was followed over 8 days. The results showed that both picoplankton and nanoflagellates were top-down-controlled. Removal of all protists via the 0.8-μm filtration or inclusion of larger-sized protists in the less than 90-μm fraction increased the net growth rates of heterotrophic bacteria and picoplankton, while removal of larger protists by 5- or 10-μm filtration led to enhanced growth rates of bacterivorous nanoflagellates and decreased growth rates of picoplankon. The experiment suggested omnivory among phagotrophic protists, which led to trophic cascades in the food web.

CHEMICAL INTERACTIONS BETWEEN MICROBES

Microbial cells, including prokaryotes, algae, and heterotrophic protists, have patterns of behavioral response to environmental cues that affect food web interactions. Many of these cues are specific chemicals dissolved in the aqueous medium or on surfaces of other cells or nonliving particles. Motile bacterial cells can sense gradients of chemical compounds in their environment and respond by moving up or down the gradient. A large portion of prokaryotic strains cultured from seawater are motile, and it is postulated that this motility allows bacterial cells to move toward microsites of high substrate concentration, for example, phytoplankton cells or rich organic particles. The process of chemoreception is based on bacteria having specific recognition sites on their cell membranes for a particular substance. *Escherichia coli*, for instance, has at least 20 chemoreceptor sites that allow the bacterium to respond positively, that is, be attracted to, or negatively, that is, be repelled by, various chemicals. Generally, aerobic bacteria move toward higher concentration of small molecular weight organic substrates such as amino acids and sugars and toward higher oxygen concentration. Aerobic bacteria move away from metabolic poisons, for example, sulfide and heavy metal cations.

Bacteria can sense changes in chemical concentration over time as they swim through concentration gradients. Motile bacteria respond to changes in their environment by changing their swimming behavior or direction. For bacteria having two or more flagella, the normal swimming behavior is straight-line swimming runs with the flagella operating synchronously and occasional tumbles with the flagella flailing in opposite directions, which allows the cell to orient to a new straight-line direction. An increase in concentration of an attractant results in a shift to straight-line swimming alone, while a decrease in concentration of an attractant results in dramatic increase in tumbling, leading to frequent changes of direction to facilitate the search for a more favorable concentration of substrate. Conversely, increasing the concentration of a repellent chemical results in increased tumbling, and decreasing the repellent concentration leads to straight-line swimming. A bacterium that has only a single flagellum can change in rotational direction of the flagellum, so in this case the cell only moves in forward or reverse direction, and does not tumble.

An example of bacterial behavior based on chemosensory response can be observed in the aggregation of bacteria under a coverslip: strains of bacteria that prefer a high oxygen tension will congregate at the edges of the slip, while microaerobic bacteria will remain in the middle. Motile marine bacteria have been shown to aggregate around organic-rich particles and to track the path of a phytoflagellate, presumably by chemosensory response to organic compounds emanating from the particle or algal cell. This ability allows motile bacteria to aggregate around detrital particles or algal cells, forming locally elevated prey concentrations for heterotrophic protists.

Chemosensory response is thus also important in the feeding behavior of phagotrophic protists. Marine bacterivorous flagellates exhibit chemoattraction to amino acids and to bacterial cells, as has been demonstrated. Marine ciliates whose preferred prey is algae exhibit strong positive chemosensory response to some species of algae, and neutral or negative response to other prey species. Species of heterotrophic dinoflagellates have also been found to move toward algal cells and algal cell lysate, and to preferentially prey on some strains of algae. A modeling exercise designed to estimate the advantage conferred on a phagotrophic protist that is able to chemically detect prey cells showed that this capacity conserves energy used to search for food, and would confer the greatest advantage when prey are scarce.

Bacterial and algal cells may deter protist predation directly by producing chemicals that are either toxic or unpalatable to phagotrophic protists. The pigment violacein, a bacterial secondary metabolite, inhibited predation by nanoflagellates on bacteria containing the pigment. Ingestion of just one to three violacein-containing bacterial cells was found to cause rapid death of the flagellate cells. The heterotrophic dinoflagellate *Oxyrrhis marina* was shown to have a much lower feeding rate on one particular strain of the coccolithophorid algae *Emiliania huxleyi* compared to a second strain of *E. huxleyi*. In the presence of the dinoflagellate, the strain of *E. huxleyi* that the dinoflagellate avoided produced a concentration of dimethyl sulfide (DMS) and acrylic acid that was an order of magnitude higher than in the second strain of *E. huxleyi*. DMS was found to inhibit predation of algal prey by several other species of phagotrophic protists, and appears to be a chemical deterrent to protist feeding. The deterrent mechanism is not known.

The biochemical mechanism for chemosensory response in both prokaryotic and eukaryotic microbes is based on chemoreceptor sites associated with the cell membrane. The chemical compound being sensed binds to a specific chemoreceptor site, triggering signal transduction, the release of secondary messenger compounds within the cell, which results in change in motility, for example, direction and/or speed, or in ingestion of a prey cell by a protist. Such cellular processes are obviously important for survival and growth of aquatic bacteria in terms of locating microsites of higher substrate concentration and for aquatic protists in determining prey location, selection, capture success, and rates of ingestion. Understanding the extent to which cell surface chemoreceptor binding and consequent signal transduction pathways operate in bacterial and protist chemosensory behavior is vital to a predictive understanding of the structure and function of marine food webs.

SPATIAL STRUCTURE OF MICROBIAL FOOD WEBS

Initially, the environment encountered by aquatic microorganisms was perceived as being largely fluid, relatively homogeneous, and governed by diffusive processes; and it was assumed that microbes suspended in water were more or less uniformly distributed. Research on how chemosensory behavior of motile microbes can result in microbial aggregations, combined with observations of patchy distribution of phytoplankton, detrital organic matter, and high molecular weight colloidal dissolved organic substances, has led to the understanding that microbial food webs in fact occur in a nonuniform structured habitat. Phytoplankton and bacteria release polymeric material directly, other polymeric material is produced during feeding and egestion by protists and zooplankton. Biopolymer gels form when polymer chains are hydrated and cross-linked or aggregated, resulting in a three-dimensional network. These polymeric microgels then aggregate to form gel-like sheets, strings, and webs that provide surfaces, form barriers against diffusion, and furnish refuges against predators. Detrital particles and microbial cells are interspersed in these gel webs. The interactions of bacteria with the organic matter continuum from dissolved organic compounds to large particles, and the behavioral response of microbes to the patchy distribution of these particles, create microscale features, hotspots of microbial activity, and food web interactions, with distinctive natures and intensities of biogeochemical transformations.

The larger organic particles formed in this way are termed marine snow because these aggregated particles, which are usually greater than 0.5 mm in diameter, strongly resemble snowflakes when seen in the water. The basic glue holding marine snow particles together is thought to consist of fibrillar, long-chain polysaccharide polymers termed transparent exopolymer particles (TEP). Marine snow particles are initially colonized by heterotrophic prokaryotes that produce extracellular enzymes that hydrolyze the organic matrix of the particle into low molecular weight dissolved organic compounds, for example, monomers or small polymers of sugars and amino acids. The colonizing bacteria produce more of these organic substrates than can be assimilated; thus, a plume of organic material is released from the particle that attracts motile bacteria. Higher abundances of bacteria in turn attract bacterivorous flagellates, which may then attract larger protists and zooplankton. The organic particle and its plume may thus become the focus of a complex food web. Marine snow particles can sink at speeds greater than 100 m in a day, allowing them to travel from the surface to the subsurface ocean within a matter of days. As they sink, the organic material in the particles is continually degraded by microbes. Any organic material that reaches the sea floor is either consumed by benthic organisms or incorporated into sediments. Sinking of marine snow particles is important to the ocean's capacity to sequester atmospheric carbon dioxide.

Biofilms, which, as mentioned previously, form on surfaces, are similar to suspended organic particles in that they are essentially an organic polymer gel with living microorganisms embedded in it. Biofilms form on most abiotic surfaces in aquatic systems and are sites of enhanced microbial activity. The polysaccharides produced by surface-growing microbes act to cement sediment particles together and represent a food resource for benthic animals. The microbial species colonizing the biofilm and the organic composition of the matrix largely determine the physical properties of the biofilm. Once attached to a surface, bacteria begin to produce extracellular polysaccharides. The amount of biopolymer produced can exceed the mass of the bacterial cell by a factor of 100 or more. The gel is mostly water, but has properties that influence the transport of materials at the surface of the biofilm. Changes in transport rates can create unique niches within the biofilm for the proliferation of a variety of microbial species. Bacterial extracellular polysaccharides also tend to absorb cations and organic molecules from the overlying water. The reduced diffusion rates of substrates within a biofilm serve to create localized concentration gradients and the possibility of well-defined spatial relationships between individual bacterial cells. These gradients may result in suboxic and even anoxic regions in the interior of these structures, contributing to the coexistence and metabolic interaction of both aerobic and anaerobic microbes.

RECENT DEVELOPMENTS BIOGEOGRAPHY OF FORM AND FUNCTION IN MICROBIAL FOOD WEBS

Major recent discoveries about patterns and processes in microbial food webs stem from sophisticated molecular genetic and cell biology approaches to understanding the biogeography of microbial community composition and metabolic function. Metagenomics, analysis of DNA sequences of microbial genomes present in an environmental sample, has revealed astoundingly high species or strain diversity in communities of prokaryotic and eukaryotic microbes. Genetic fingerprinting methods have demonstrated that communities of marine bacteria have richer diversity in warm tropical waters compared to cold polar waters, a pattern similar to that found for phytoplankton and zooplankton. On a finer scale, these genetic methods have shown that at a coastal ocean station, community structure of planktonic bacteria can shift dramatically as environmental conditions vary, in patch sizes of several kilometers and over timescales of months. Metabolic capabilities also vary. New data have confirmed that the Crenarchaea which dominate marine microbial communities at depths greater than 200 m can be autotrophic nitrifyers, gaining energy for carbon fixation from ammonium oxidation. Analysis of the complete genome of the archaeal nitrifyer *Nitrosopumilus maritimus*, which grows autotrophically in culture, has found structurally different enzymes and pathways for ammonium oxidation and carbon fixation compared to those of nitrifying bacteria. Metagenomic studies of marine bacterial communities have identified abundant genes coding for proteorhodopsin, a light-harvesting pigment that can serve as an energy-generating proton pump.

Proteomics, the study of what proteins are expressed by individual species and communities of microbes, links genetic potential to in situ biochemical functions. A detailed analysis was made of the proteins expressed in bacterial cell membranes in surface waters of the South Atlantic Ocean. The dominant proteins were TonB-dependent transporters, which require a trans-membrane proton gradient to move substrates across the cell membrane. Diverse rhodopsins were also detected. A close association of these transporter molecules with rhodopsin proton pumps in bacterial membranes suggests that marine bacteria may use solar power to facilitate transport of substrates for growth into the cell, an effective strategy for existence in nutrient-poor habitats.

Recent studies have also identified high diversity among eukaryotic microbes. As found for prokaryotes, protistan communities have different phylogenetic compositions in the ocean surface compared to those at subsurface depths. A common group of marine algae, the haptophytes, has an extreme degree of biodiversity; their evolutionary success may be due, in part, to mixotrophic ingestion of bacteria. Picobiliphytes have been identified as a major new taxonomic class of marine phytoplankton. Diverse ribotypes related to the Syndiniales, parasitic dinoflagellates, are ubiquitous in the sea, but whether these open ocean strains are indeed parasitic is unknown. Microscopic fungi such as yeasts and chytrids are found in both freshwater and marine aquatic ecosystems and may be significant as decomposers of organic matter as well as parasites of algae and zooplankton. Understanding how the rich biodiversity of prokaryotic and eukaryotic organisms is partitioned among aquatic habitats and how this diversity affects biogeochemical processes and food web interactions continues to be a central theme for microbial ecologists.

FURTHER READING

Azam, F., Fenchel, T., Field, J. G., Meyer-Reil, R. A., & Thingstad, F. (1983). The ecological role of water-column microbes in the sea. *Marine Ecology Progress Series*, *10*, 257–263.

Caron, D. A. (2009). Past president's address: Protistan biogeography: Why all the fuss? *Journal of Eukaryotic Microbiology*, *56*, 105–112.

Countway, P. D., Gast, R. J., Dennett, M. R., Savai, P., Rose, J. M., & Caron, D. A. (2007). Distinct protistan assemblages characterize the euphotic zone and deep sea (2500 m) of the western North Atlantic (Sargasso Sea and Gulf Stream). *Environmental Microbiology*, *9*, 1219–1232.

Ducklow, H. W. (1983). Production and fate of bacteria in the oceans. *BioScience*, *33*, 494–501.

Fenchel, T., & Finlay, B. J. (1995). *Ecology and evolution in anoxic worlds*. Oxford University Press: New York.

Fuhrman, J. A., & Steele, J. A. (2008). Community structure of marine bacterioplankton: Patterns, networks, and relationships to function. *Aquatic Microbial Ecology*, *53*, 69–81.

Fuhrman, J. A., Steele, J. A., Hewson, I., Schwalbach, M. S., Brown, M., Green, J. L., *et al.* (2008). A latitudinal diversity gradient in planktonic marine bacteria. *Proceedings of the National Academy of Sciences of the United States of America*, *105*, 7774–7778.

Gasol, J., Pedros-Alio, C., & Vaque, D. (2002). Regulation of bacterial assemblages in oligotrophic plankton systems: Results from experimental and empirical approaches. *Antonie van Leeuwenhoek*, *81*, 435–452.

Guillou, L., Viprey, M., Chambouvet, A., Welsh, R. M., Kirkham, A. R., Massana, R., *et al.* (2008). Widespread occurrence and genetic diversity of marine parasitoids belonging to Syndiniales (Alveolata). *Environmental Microbiology*, *10*, 3349–3365.

Jobard, M., Rasconi, S., & Sime-Ngando, T. (2010). Diversity and functions of microscopic fungi: A missing component in pelagic food webs. *Aquatic Sciences*, doi:10.1007/s00027-010-0133-z.

Kirchman, D. (Ed.), (2000). *Microbial ecology of the oceans*. Wiley-Liss: New York.

Legendre, L., & Le Fevre, J. (1995). Microbial food webs and the export of biogenic carbon in oceans. *Aquatic Microbial Ecology*, *9*, 69–77.

Liu, H., Probert, I., Uitz, J., Claustre, H., Aris-Brosou, S., Frada, M., *et al.* (2009). Extreme diversity in noncalcifying haptophytes explains a major pigment paradox in open oceans. *Proceedings of the National Academy of Sciences of the United States of America*, *106*, 12803–12808.

Matz, C., Deines, P., Boenigk, J., *et al.* (2004). Impact of violacein-producing bacteria on survival and feeding of bacterivorous nanoflagellates. *Applied and Environmental Microbiology*, *70*, 1593–1599.

Morris, R. M., Nunn, B. L., Frazar, C., Goodlett, D. R., Ting, Y. S., & Rocap, G. (2010). Comparative metaproteomics reveals ocean-scale shifts in microbial nutrient utilization and energy transduction. *The ISME Journal*, *4*, 673–685.

Not, F., Valentin, K., Romari, K., Lovejoy, C., Massana, R., Töbe, K., *et al.* (2007). Picobiliphytes, a new marine picoplanktonic algal group with unknown affinities to other eukaryotes. *Science*, *315*, 252–254.

Pohnert, G., Steinke, M., & Tollrian, R. (2007). Chemical cues, defence metabolites, and the shaping of pelagic interspecific interactions. *Trends in Ecology and Evolution*, *22*, 198–204.

Pomeroy, L. R. (1974). The ocean's food web, a changing paradigm. *BioScience*, *24*, 499–504.

Rivkin, R. B., Legendre, L., Deibel, D., *et al.* (1996). Vertical flux of biogenic carbon in the ocean: Is there food web control? *Science*, *272*, 1163–1166.

Sakka, A., Legendre, L., Gosselin, M., & Delesalle, B. (2000). Structure of the oligotrophic planktonic food web under low grazing of heterotrophic bacteria: Takapoto Atoll, French Polynesia. *Marine Ecology Progress Series*, *197*, 1–17.

Samuelsson, K., & Andersson, A. (2003). Predation limitation in the pelagic microbial food web in an oligotrophic aquatic system. *Aquatic Microbial Ecology*, *30*, 239–250.

Sherr, E. B., & Sherr, B. F. (2008). Understanding roles of microbes in marine pelagic food webs: A brief history. In D. Kirchman (Ed.), *Advances in microbial ecology of the oceans* (pp. 27–44). Wiley-Blackwell: Hoboken, NJ.

Sieburth, J. Mc.N., Smetacek, V., & Lenz, J. (1978). Pelagic ecosystem structure: Heterotrophic compartments of the plankton and their relationship to plankton size fractions. *Limnology and Oceanography*, *23*, 1256–1263.

Walker, C. B., de la Torre, J. R., Klotz, M. G., Urakawa, H., Pinel, N., Arp, D. J., *et al.* (2010). Nitrosopumilus maritimus genome reveals unique mechanisms for nitrification and autotrophy in globally distributed marine crenarchaea. *Proceedings of the National Academy of Sciences*, doi:10.1073/pnas.0913533107.

Wolfe, G. V. (2000). The chemical defense ecology of marine unicellular plankton: Constraints, mechanisms, and impacts. *Biological Bulletin*, *198*, 225–244.

Index

Note: Page numbers followed by 'f' indicate figures and 't' indicate tables.

A

Abacus mutants (AbaA), 37
Abnormal growth, in plants, 111
ABPA. *See* Allergic bronchopulmonary aspergillosis
Abscission, 111
Acanthamoebidae family, 202, 206t
Acharius, Erik, 90
Acid rain, 245
Acidifying pollutants, 80
Acquired immunodeficiency syndrome (AIDS), 143, 152
- assessing, 153
- *Candida albicans* and, 16
- intestinal protozoa and, 324t
- microsporidia and, 65
- *Toxoplasma* and, 387

Acrasiomycetes, 280
AFLP. *See* Amplified fragment length polymorphism
AFM. *See* Atomic force microscopy
Agriculture. *See also* Food
- Clavicipitaceae impact on, 47
- emerging diseases and changes in, 98–99
- fungi control in, 124–125
- fungi impacting, 105–106
- mycorrhizae and, 79
- oomycetes impact on, 347–348
- *Phytophthora infestans* impact of, 347–348
- wheat stem rust and, 122
- yeasts and, 16

Agrobacterium tumefaciens, 149
AIDS. *See* Acquired immunodeficiency syndrome
ALA. *See* Amoebic liver abscess
Albuginaceae, 114t
ALC. *See* Anterior-like cell
Algal blooms, 435–450
- autumn, 445
- cyanobacteria and, 438–440, 439f
- diatoms and, 439–440, 439f
  - spring, 444
- dinoflagellates and, 441–442
- environment and
  - chemical, 437–438
  - physical, 436–437
- future research avenues for, 448–450
  - abandoning Sverdup's critical depth hypothesis, 449
  - marine genomics, 448–449
  - role of grazing in suppression and recycling iron, 449–450
- haptophytes and, 440–441, 441f
- harmful, 435–436, 446–447
- iron-fertilized, 447–448
- life cycle of, 436, 436f
- major contributors to, 438–442
- miscellaneous, 446
- pathogens and grazers of, 442–443, 443f
- peak of, 435
- phytoplankton, 435–436
- recurrent and unusual, 443–446
- spring, 443–445
- in upwelling regions of low latitudes, 445–446

Algirosphaera robusta, 236
Allen, R. D., 208
Allergic bronchopulmonary aspergillosis (ABPA), 143, 162–163
Allogromiida order, foraminifera, 301t
Alternaria solani, 123
Alveolates, 264
- groups of, 215–216

Alzheimer's disease, *Pin1* and, 35
Amitochondriate protists, 177–189. *See also* Diplomonads; Oxymonads; Parabasalids
- cell organization of, 182–185
- evolution of, 178–179, 179f
- example, 180t
- genetics and genomics of, 187–188
- habitats of, 182
- hydrogenosomes, 185–187, 186f
- important pathogenic species of, 188–189
- mitosomes, 185–187, 186f
- systematics of, 179–182
- types of, 177–178

Ammonia, 286–287
Amoebas
- intestinal protozoa, 324t, 329–331
- naked lobose, 191–211

Amoebic liver abscess (ALA), 323
- presentation of, 330

Amoebidae family, 201, 201f, 204t
Amoeboid movement, 207–209, 209f
Amplified fragment length polymorphism (AFLP), 3
Amylase, 108t
Anamorph state, 109
Anaphase-promoting complex (APC), 19, 35
Ancient asexuals, 228, 231
Anterior-like cell (ALC), 279
Anthracnose, 112
APC. *See* Anaphase-promoting complex
Apicomplexa, 324–326, 324t, 329t
Apicoplast, 390–391, 390f
Apple scab, 118–119
Arbuscular mycorrhizae, 74–75
Arbutoid mycorrhizae, 76
Archiascomycetes, 118t
ARISA. *See* Automated ribosomal interspace analysis
Armophorea class, ciliates, 221f, 222
Arthropods
- entomogenous fungi and, 127
- immune systems of, 129–130

Ascomycota, 117–119, 118t
Aspartate (asp), 279
Aspergillosis, 162f
- causative organisms of, 161
- clinical features of, 162–163
- diagnosis and treatment of, 163
- epidemiology of, 161–162

Aspergillus
- definition and classification of, 19–21, 20f
- enzymes produced by, 25–26
- food contamination by, 21
- in genomic era, 38–39
- as human pathogens, 21–24, 22f
  - barriers preventing, 23–24
  - increase in, 22
- organic acid production and, 26
- Oriental food uses of, 25
- as secondary metabolites, 26
- uses of, 25–26
- in veterinary medicine, 24

Aspergillus flavus, 21, 33, 161–162
- genomic data on, 38–39

Aspergillus fumigatus, 21, 33, 161–162, 162f
- genomic data on, 38–39
- prevalence of infections with, 22–23
- sex of, 229
- in veterinary medicine, 24

Aspergillus nidulans, 19, 23
- conidiophore of, 36, 36f
  - development of, 36–37, 37f
- developmental pathways of, 35–38, 36f
  - characteristics of, 35–36
- gene expression control of, 29–33
  - by external pH, 30–32, 32f

Aspergillus nidulans (*Continued*)
genetic system of, 26–27
genomic data on, 38–39
life cycles of, 27, 28f
mitochondrial DNA of, 27–28
as model for cell biology, 33–35, 33f
as model for genetic metabolic diseases, 28–38, 29f
as model organism, 26–28
nitrogen and carbon utilization in, 29–30, 30–31f
secondary metabolites regulated by, 32–33
transporter regulation with, 32, 32f
Aspergillus niger, 25–26
genomic data on, 38–39
Aspergillus oryzae, 25
genomic data on, 38–39
Aspergillus parasiticus, 21
Aspergillus sojae, 25
Aspergillus sydowii, 24–25, 25f
Astrorhizida order, foraminifera, 301t
Athlete's foot, 170
Atomic force microscopy (AFM), 3
of yeast, 6, 6f
Autogamy, 294
Automated ribosomal interspace analysis (ARISA), 355, 365
Autotrophic eukaryotes, in microbial food webs, 455
Autotrophic prokaryotes, in microbial food webs, 454–455
Autumn blooms, 445
Avr. *See* Pathogen avirulence

B

BAC. *See* Bacterial artificial chromosomes
Bacillaria paradoxa, 256
Bacterial artificial chromosomes (BAC), 355, 368
Bacterial growth efficiency (BGE), 459
Balansia, 44–45
Basidiomycota, 119–123, 120–121t
Bassi, Agostino, 138
Bathycoccus prasinos, 359
Beauveria bassiana, 42, 134–135, 137–140
Behrenfeld, M. J., 449
benA33, 34
Benthic foraminifera, 297
Benzer, Seymour, 27
Berenil, 408
BGE. *See* Bacterial growth efficiency
Bigelowiella natans, 433
Bim mutants, 35
Bioactive compounds, endophytic fungi for, 63
Bioethanol, 17
Biofilms, 464
Biogeochemical cycling, microbial food webs in, 458–460, 459f
Biostratigraphy, coccolithophores used in, 247
Biotrophs
Clavicipitaceae in plant, 43–45, 44–45f
nutrition of, 106–107
powdery mildews, 118t, 119
Blastodinids, 270
Blastodinium contortum, 275f
Blastodinium crassum, 274f
Blastodinium pruvon, 274f
Blastomyces dermatitidis, 150–151, 150f
Blastomycosis, 150f
causative organism of, 150
clinical features of, 151
diagnosis and treatment of, 151
epidemiology of, 150–151
Blooms. *See* Algal blooms
Bolidomonas pacifica, 360
Botanophila, 45
Bovee, E., 193
Braarudosphaera bigelowii, 237f
Bristle mutants (BrlA), 37–38
BrlA. *See* Bristle mutants
Brown patch, 121
Brown rot, 119
Bruce, David, Sir, 414
Budding yeast, 10–11, 10–11f
multilateral, 11, 11f
Buliminida order, foraminifera, 303–304t

C

Calkinsia, 319, 320f
Calyptrolithina multipora, 237f
cAMP. *See* Cyclic AMP
cAMP receptor (cAR), 279, 282, 284, 287
Candida albicans, 4, 145, 157–158, 157f, 167, 169–170
AIDS and, 16
dimorphism in, 12f
microarrays for study on, 147
Candida pintolopesii, 8t
Candida spp., 157–159, 157f
Candida utilis, 8t
Candidiasis, 157f
causative organisms of, 157
clinical features of, 158
cutaneous, 169–170
diagnosis and treatment of, 158–159
epidemiology of, 158
Candidosis, 16
Cankers, 112, 119
cAR. *See* cAMP receptor
Carbon
Aspergillus nidulans utilization of, 29–30, 30–31f
coccolithophore bloom sinking and release of, 245
metabolism, by yeasts, 8–9
Carbon catabolic repression (CreA), 29–30
Carbon concentrating mechanisms (CCMs), 449
Carbonate compensation depth (CCD), 291, 295
Card agglutination test for trypanosomiasis (CATT), 411, 417
Carpediemonas, 181
Carterinida order, foraminifera, 303t
Castanea spp., 99
CATT. *See* Card agglutination test for trypanosomiasis
CCD. *See* Carbonate compensation depth
CCMs. *See* Carbon concentrating mechanisms
CDI. *See* Cyclin-dependent kinase inhibitor
Cell envelope, 7t
Cellular reaction (CER), 127, 129f
Cellulase, 108t
Central nervous system (CNS), 143, 156, 397, 404
CER. *See* Cellular reaction; Chloroplast endoplasmic reticulum
Ceratium spp., 445
Cerebrospinal fluid (CSF), 143, 153, 156
CF. *See* Complement fixation
CFB. *See Cytophaga-Flavobacteria-Bacteriodes*
Chaetoceros spp., 439f, 440
Chagas' disease, 400–402, 401f
Chatton, Edouard, 263, 270
Chaunopycnis alba, 41–42
Chemosensory response, 463
Chemostats, 7–8
Chemotaxis, 285, 285f
Chlamydospores, 108
Chloroplast endoplasmic reticulum (CER), 373
Chloroplasts, euglenids, 315–316
Chromalveolata, 264, 265f
Chromoblastomycosis, 172
Chrysochromulina hirta, 235–236
Chrysochromulina polylepis, 446
Chrysophyceae, 239
Chytridiomycota, 116
Ciliates, 213–226
Armophorea class of, 221f, 222
cilium and rootlets of, 214, 214f
classification of, 217–218t
diversity of, 219, 221–224, 221f
Colpodea class of, 221f, 223
morphological diversity in, 227, 228f
sex in, 227, 229
conjugation of, 214
cortex of
evolution in, 216–217, 216f
ultrastructure and structural conservatism of, 215f, 219
11 classes of, 219, 221–224, 221f
Heterotricha class of, 220f, 221–222
history of, 213
Karyorelictea class of, 221
Litostomatea class of, 221f, 222
major, 218–219
morphological features of, 215, 216f
Nassophorea class of, 220–221f, 223
nuclear dimorphism evolution in, 217–218, 217–218t
nuclei of, 213–214, 217
Oligohymenophorea class of, 220–221f, 223–224
origin of, 215–218
Phyllopharyngea class of, 220–221f, 222–223
Plagiopylea class of, 220–221f, 223
Prostomatea class of, 221f, 223
sex distribution in, 230f
somatic kinetics of, 214, 215f
species level diversity of
molecular techniques in, 224–225f, 226
morphological diversity, 224

Spirotrichea class of, 220–221f, 222
SSU rDNA of, 219, 220f
subphyla of, 219
ciPCR. *See* Culture-independent PCR
Claviceps purpurea, 138
Clavicipitaceae, 41–49
agriculture impact of, 47
chemical diversity in, 45–46
defensive mutualism concept for, 47–48
evolution of, 48
function debate over, 48
family tree of, 41, 42f
grass endophytes and, 47–48
phylogenetic segregation of, 140
in plant biotrophs, 43–45, 44–45f
reclassification of, 137
saprotrophs and insect parasites of, 41–43
secondary metabolites and
biological activities of, 46–47
sources of, 45–46
in soft-bodied scale insects, 43–44f
soil inhabitants in, 43
Clay, Keith, 47
Cleistothecium, 27
Cleve, Per Teodor, 250
Climate change, 102
mycorrhizae and, 81
Clubroot, 111
of crucifers, 113
Clutterbuck, John, 27, 36–37
CMF. *See* Conditioned medium factor
CNS. *See* Central nervous system
Coastal Zone Color Scanner (CZCS), 235, 243
Coccidioides immitis, 153–155
Coccidioides posadasii, 153–155, 154f
Coccidioidomycosis, 154f
causative organisms of, 153–154
clinical features of, 154
diagnosis and treatment of, 155
epidemiology of, 154
Coccolithophores, 235–247
biogeography and ecology of, 241–242, 242f
by zone, 243t
biostratigraphic use of, 247
blooms of, 242–245
carbon release and sinking of, 245
detection of, 242–244
past, 246–247, 246f
cell features of, 235–237, 236f
cell structure of, 237
Chrysophyceae, 239
collection methods for, 240–241
environment impact of, 244–245
evolution of, 245–247
diversity and extinctions in, 246, 246f
function of, 238
Haptophytes, 238–239
morphology of, 237–238, 237f
production of, 237f, 238
seasonality and depth preferences of,
241–242, 243f, 244t
taxonomy of, 238–240
based on molecular genetics, 240
based on morphology, 239–240
vertical distribution of, 244t
viruses, 244
Coccolithus pelagicus, 237f
Cochliopodiidae family, 203, 203f, 207t
Cochliopodium spp., 210
Colpodea class, ciliates, 221f, 223
morphological diversity in, 227, 228f
sex in, 227, 229
Complement fixation (CF), 143, 153, 156
Complement receptor (CR), 335
Compustat, diatom selective breeding with,
258
Conditioned medium factor (CMF), 279, 286
Conductivity, temperature, depth (CTD), 241
Conidiophore, 108
of *Aspergillus nidulans*, 36, 36f
development of, 36–37, 37f
Cordyceps spp., 132, 137
biogeographic diversity of, 135
Corn smut, 122–123
Couch, John, 139
Cove, David, 27
COWP. *See* Cysteine-rich oocyst wall protein
COX1. *See* Cytochrome oxidase subunit 1
CR. *See* Complement receptor
Crabtree effect, 9t
CRAC. *See* Cystolic regulator of adenylate
cyclase
CreA. *See* Carbon catabolic repression
Cristamonadida, 180, 180t, 181f
Cronartium ribicola, 99, 100t
Crown rot, 112
Cryphonectria parasitica, 98, 100t
Crypthecodinium cohnii Biecheler, 265–266,
265–266f
Cryptococcosis, 160f
causative organisms of, 159
clinical features of, 159–160
diagnosis and treatment of, 160
epidemiology of, 159
Cryptococcus gattii, 159–160
Cryptococcus neoformans, 147–148, 160f
epidemiology of, 159
treatment for, 160
Cryptosporidium spp., 324–326, 324t
recent developments with, 331–332
CSF. *See* Cerebrospinal fluid
CTD. *See* Conductivity, temperature, depth
CTL. *See* Cytotoxic T lymphocytes
Culture-independent PCR (ciPCR), 355
on marine picoeukaryotes, 363–364, 364f
Custers effect, 9t
Cutaneous candidiasis, 169–170
Cutaneous fungal infections
chromoblastomycosis, 172
cutaneous candidiasis, 169–170
deep cutaneous and subcutaneous, 171–172,
172t
dermatophytosis, 170–171
host defenses and
antifungal substances for, 167–168
cutaneous immune system, 168–169
inflammatory response in, 168
innate immune system and, 168
keratinization and epidermal proliferation
in, 167
skin structure for, 167
malassezia folliculitis, 171
mycotic mycetoma, 172
recent developments with, 173
sporotrichosis, 171–172
superficial types of, 169t
tinea versicolor, 171
Cutaneous immune system, 168–169
Cuticle, 199
Cutinase, 108t
Cutting rot, 112
Cyanobacteria, algal blooms, 438–440, 439f
Cyclic AMP (cAMP), 279
Dictyostelium discoideum production of,
283–285, 284f
evolutionary history of signaling by,
288–289
PdsA degrading, 284
signals regulating gene expression of,
286–287, 286–287f
Cyclin-dependent kinase inhibitor (CDI), 3
Cysteine-rich oocyst wall protein (COWP), 323,
325
Cystolic regulator of adenylate cyclase (CRAC),
279, 284
Cystospora canker of spruce, 112
Cysts, 209
Cytochrome oxidase subunit 1 (COX1), 311,
314
Cytophaga-Flavobacteria-Bacteriodes (CFB),
451, 454
Cytoplasmic MTOC, 203, 208
Cytotoxic T lymphocytes (CTL), 68, 69f
CZCS. *See* Coastal Zone Color Scanner

D

DALY. *See* Disability-adjusted life year
Damping-off, 112
Dark septate endophytes, 76
Darnel ryegrass, *Neotyphodium occultans*
association with, 56
Darwin, Charles, 90
Dasyscypha willkommii, 100t, 102
DCL. *See* Diffuse cutaneous leishmaniasis
DCs. *See* Dendritic cells
de Bary, Heinrich Anton, 51
DEAE. *See* Diethylaminoethyl
Deep cutaneous mycoses, 171–172, 172t
Defensive mutualism
for Clavicipitaceae, 47–48
evolution of, 48
function debate over, 48
endophytes stress tolerance and, 63
Delayed-type hypersensitivity (DTH), 335,
341–342
Denaturing gradient gel electrophoresis
(DGGE), 355, 365
Dendritic cells (DCs), 68
Dense granules, 390f, 391
Dermatophytosis, 170–171
Deuteromycota, 123, 123f
DFMO. *See* Difluoromethylornithine

DGGE. *See* Denaturing gradient gel electrophoresis
DHFR. *See* Dihydrofolate reductase
Diatoms, 249–258
algal blooms and, 439–440, 439f
spring, 444
biophotonics of, 255–256, 255f
classifying, 382–383
compustat and selective breeding of, 258
for drug delivery, 257–258, 257f
genera of, 249, 250f
genomics of, 251–255
habitat of, 383
microfluidics within, 256
morphogenesis of, 252–253f, 252–254
models for, 254
for nanotechnology, 249
computing with, 258
drug delivery in, 257–258, 257f
opals compared to, 255
photoluminescence and, 256
professionals studying, 250
research of, 258
SEM of, 251f, 257f
silica of
biomineralization of, 251–252
intracellular pathway of, 255
structure of, 251, 252f
Dictyostelids, 279–289
culture of, 282–283
ecology of, 279–280
evolution of, 279–281
morphology and, 282f
genomics of, 281–283
modification and, 283
phylogeny wide genome sequencing of, 288
recent developments in, 288–289
taxonomy of, 279–281, 281f
tractability of, 281–283
experimental, 282–283
Dictyostelium discoideum, 279
culture of, 282–283
developmental program of, 283–288
cAMP production in, 283–285, 284f
chemotaxis in, 285, 285f
gene expression during, 286–287, 286–287f
morphogenesis in, 283–286
processing developmental signals in, 287–288
slug and fruiting body formation in, 285–286
experimentation of, 283
gene modification of, 283
genome of, 281–282
life cycle of, 284f
recent developments in, 288–289
Dieback, 112
Diethylaminoethyl (DEAE), 3, 12
Diffuse cutaneous leishmaniasis (DCL), 335, 341, 343
Diffusion limited aggregation (DLA), 254
Difluoromethylornithine (DFMO), 397, 407, 409
Dihydrofolate reductase (DHFR), 385, 387
Dimethyl sulfide (DMS), 235, 440, 463
in acid rain, 245
Dimethylsulfoniopropionate (DMSP), 235, 440
by-products of, 245
Diminazene aceturate (Berenil), 408
Dinoflagellates, 263–277
algal blooms and, 441–442
Crypthecodinium cohnii Biecheler, 265–266, 265–266f
diversity of, 265–271
evolution of, 263–265
eyespot of, 266, 268
feeding evolution of, 442
FeSODs and SODs in, 266, 267f
free-living bioluminescent, 264f
mitotic apparatus evolution and, 271, 274, 276–277f
mixotrophic, 270–271, 274–275f
Noctiluca scintillans McCartney, 268–269, 268f
nucleus of, 263, 271, 274
Prorocentrum micans, 269–270, 269–271f
sex in, 269–270, 271f, 272–273t
Dinophysis acuminata, 271
Dinophysis norvegica, 271
Diplomonads, 178, 178f
cell organization of, 182–183, 182f
evolution of, 178–179, 179f
example, 180t
genomics of, 187–188
habitats of, 182
mitosis of, 183
subgroups of, 179, 181f
Diplonemids, 314, 319, 319f
Disability-adjusted life year (DALY), 411, 418–419
Discomycetes, 118t
Discula destructiva, 100t
Diseases. *See* Emerging diseases
Dissolved organic matter (DOM), 355, 362–363
DLA. *See* Diffusion limited aggregation
DMS. *See* Dimethyl sulfide
DMSP. *See* Dimethylsulfoniopropionate
DNA
Aspergillus nidulans mitochondrial, 27–28
fungal infections and, transformation systems for, 148–150, 149t
DOM. *See* Dissolved organic matter
Downy mildews, 116
Dronkgras, *Neotyphodium melicicola* association with, 56
Drosophila melanogaster, 27
Drunken horse grass, *Neotyphodium gasuense* association with, 56
Dry rot, 112
DTH. *See* Delayed-type hypersensitivity
Dussiella spp., 43, 44f
Dysnectes, 181

E

The Ecological Theater and the Evolutionary Play (Hutchinson), 451
Ectomycorrhizae, 75
Effectors, oomycetes as, 352
Eflornithine, 417
EIA. *See* Enzyme immunoassay
18s rDNA libraries, marine picoeukaryotes, 364–365f, 365–368, 366t
Eisler, Klaus, 216–217
ELISA. *See* Enzyme-linked immunosorbent assay
Emerging diseases
agricultural changes and, 98–99
community-wide impact of, 99
environmental change causing, 102
examples of past, present, potential future, 100t
impact of, 99
plant pathogen changes causing, 99, 101
plants and, 97–98
predicting future, 102–103
preinvasion and postinvasion strategies for, 103
sexual recombination in, 101
Emericella nidulans, 20
Emiliania huxleyi, 237f, 439f, 440
blooms of, 242–244
genome sequencing of, 240
Encephalitozoon cuniculi, 66, 67t
Encephalitozoon intestinalis, 327
Endomycorrhizae
arbuscular mycorrhizae, 74–75
ericoid mycorrhizae, 75
Endophytes
classes of, 62
colonization of, 52
dark septate, 76
defensive mutualism and stress tolerance of, 63
developments in, 61–62
ecological impacts of, 57–58
fungal, 52–55, 53–54f
as bioactive compounds, 63
classification of, 62
of grasses, 53–54
groups of, 52
hyphal growth of, 63
marine, 55
pest control and, 59
grass
Clavicipitaceae and, 47–48
common associations of, 55–57
ergot alkaloids and, 59, 60f
indole diterpenoids and, 59, 60f
loline alkaloids and, 60–61, 61f
new associations with, 62–63
nontoxic, 58
peramine and, 60, 61f
secondary metabolite sources in, 58–61
technological advances with, 63–64
terminology of, 51
Endoplasmic reticulum (ER), 3, 6, 7t, 373, 431
of *Leishmania*, 339
Endopolygalacturonases (endoPGs), 347
Endosymbiosis
primary, 427–430, 428f, 429t
secondary, 429t, 430–434

Endosymbiotic gene transfer, 429–430
Entamoeba dispar, 329–330
Entamoeba histolytica, 329f, 329t
  clinical syndromes of, 330
  Entamoeba dispar distinguished from, 329–330
  global distribution of, 330
  immune response to, 331
  recent developments with, 332
Enterocytozoon bieneusi, 66, 67t
Entomogenous fungi, 127–140
  arthropods and, 127
  as biological control agents, 134–136
    host and geographical ranges of, 134–135
    as pathogen or saprobe, 136
    safety of, 136–137
    underutilization of, 140
    uses as, 135–136
  infection process and pathobiology of, 128–133
    development in hemocoel, 132–133, 133f
    germination in, 130–131
    host reactions in, 128–132, 129f
    hyphal development of, 131, 131f
  modernization of systematics and taxonomy for, 137
  nontraditional and nonorganismal uses of, 137–138
  secondary metabolites in, 132
Environment. *See also* Pollution
  algal blooms and
    chemical, 437–438
    physical, 436–437
  coccolithophore impact on, 244
  emerging diseases from changes in, 102
  fungi and, 110–111
  lichens and, 94
  oomycetes impact on, 347–348
  stramenopiles and, 377
  yeasts and, 16
Enzyme immunoassay (EIA), 143
Enzyme-linked immunosorbent assay (ELISA), 105, 124, 397, 403
Epibiosis, 52
Epichloë spp., 45–47, 45f
  transmission of, 52
Epidermal proliferation, 167
ER. *See* Endoplasmic reticulum
Ergots, 44
  alkaloids, 46
    grass endophytes and, 59, 60f
    host defense feature of, 48
    origin and evolution of, 47
    proposal of functionality for, 47
  sclerotium of, 46
Ericoid mycorrhizae, 75
Erysiphe alphitoides, 100t, 102
Erysiphe flexuosa, 100t
ES. *See* Expression site
ES-associated gene (ESAG), 397, 405
Escherichia coli, 339–340, 463
EST. *See* Expressed sequence tag
Etiolation, 111
Euglena gracilis, 315
Euglena spp., 313–317
  rRNA and, 316
Euglenids, 314–317
  chloroplasts of, 315–316
  evolution of, 316
  feeding apparatus in, 316–317
  flagella of, 315
  habitat of, 317, 317f
  pellicle of, 314–315, 315f
Euglenozoa, 311–321
  cell structure of, 312–314
  cytoskeleton of, 314
  diplonemids, 314, 319, 319f
  euglenids, 314–317, 315f, 317f
  evolutionary relationships of, 320–321
  flagellar structure of, 312–314, 313f
  kinetoplastids, 317–318f, 317–319
  mitochondria of, 314
  nucleus of, 314
  subgroups of, 312t
  taxonomy of, 311–312
Eukaryote, 228. *See also* Marine picoeukaryotes; Microbial eukaryotes; Phototrophic picoeukaryotes; Putative asexual microbial eukaryotes
  autotrophic, 455
Eukaryotic photosynthesis, 427–430
Eumycota, 116–123
  Ascomycota, 117–119, 118t
  Basidiomycota, 119–123, 120–121t
  Chytridiomycota, 116
  Deuteromycota, 123, 123f
  Zygomycota, 116–117
Euplotes, 224f
Eutreptia spp., 313
Eutreptiella spp., 313
Excavata, 179
Expressed sequence tag (EST), 427, 432
Expression site (ES), 397, 405

F

FACS. *See* Fluorescence-activated cell sorting
Facultative parasites, 107
Faeth, Stanley, 48
"Fescue foot," 46
FeSODs. *See* Iron SODs
Filamoebidae family, 203, 207t
FIND. *See* Foundation for Innovative New Diagnostics
Fingerprinting techniques, marine picoeukaryotes, 365
FISH. *See* Fluorescent in situ hybridization
Fission yeasts, 11–12f
Flabellulidae family, 202–203, 203f, 206t
Flagella
  of euglenids, 315
  euglenozoa structure of, 312–314, 313f
  of stramenopiles, 379–380, 380t
Flagellates, intestinal protozoa, 324t, 327–329
Flucytosine, 23
Fluffy mutants, 38
Fluorescence-activated cell sorting (FACS), 3, 6
Fluorescent in situ hybridization (FISH), 355, 358
  for marine picoeukaryotes, 364
    developments using, 369–370
  for *Micromonas pusilla*, 365
Food. *See also* Agriculture
  Aspergillus contaminating, 21
  fungi relationships with, 106–108
  mycorrhizae as, 79
  Oriental, *Aspergillus* uses in, 25
  Toxoplasma and, 388
  yeasts in, 4–5
    production and spoilage of, 15t
Food webs. *See* Microbial food webs
Foraminifera, 291–309
  abundance of, 291
    distribution and, 295
  Allogromiida order, 301t
  Astrorhizida order, 301t
  benthic, 297
  Buliminida order, 303–304t
  Carterinida order, 303t
  collection and maintenance of, 297
  cytological characteristics of, 292
  ecology of, 295
  evolutionary history and geological importance of, 308–309
  feeding strategies of, 295
  Globigerinida order, 305–306t
  granuloreticulpodia of, 292, 292f
  Involutinida order, 306t
  Lagenida order, 303t
  Lituolida order, 301–302t
  macroevolutionary relationships of, 300, 306–307, 307f
  Miliolida order, 302–303t
  molecular diversity of, 308
  molecular phylogeny and diversity of, 298
  morphological characteristics of, 292
  morphology-based classification of, 297–298, 298t, 299–300f
    family-level, 301–306t
  nucleus of, 292–293
  phylogenetic position of, 298–300
  Robertinida order, 306t
  Rotaliida order, 304–305t
  sex and life cycle of, 294–295, 294f
  Silicoloculinida order, 306t
  Spirillinida order, 303t
  symbiosis of, 295–297, 296f, 296t
    hosts and, 296t
  test morphology of, 293–294
  Textulariida order, 302t
  Trochamminida order, 302t
Ford, John, 416
Forest hedgehog grass, 57
Forestry, mycorrhizae and, 79–80
Foundation for Innovative New Diagnostics (FIND), 411, 422–423
Friz, C. T., 193
Fundamentals of Ecology (Odum, Eugene, & Odum, Howard), 452
Fungal infections
  aspergillosis, 161–163, 162f
  blastomycosis, 150–151, 150f
  candidiasis, 157–159, 157f
  classification of, 143–145
    by morphology, 145t
    by pathogenic potential, 145t

Fungal infections (*Continued*)
by phylogeny, 144t
coccidioidomycosis, 153–155, 154f
cryptococcosis, 159–160, 160f
cutaneous, 167–173
DNA transformation systems for, 148–150, 149t
genome organization of, 148t
histoplasmosis, 151–153, 152f
history of, 143
host defenses against, 145–147, 146t
iron defense against, 146
molecular approaches for studying, 147–150, 148–149t
paracocciodomycosis, 155–156, 155f
penicilliosis, 160–161
pneumocystis infections, 164–165
sporotrichosis, 156–157
zygomycosis, 163–164, 164f
Fungi
agriculture and
control of, 124–125
impact of, 105–106
Aspergillus, 19–39
cells, 106
characteristics of, 106–111, 107t
Clavicipitaceae, 41–49
disease diagnosis for, 123–124
dispersal of, 109–110
endophytes, 52–55, 53–54f
as bioactive compounds, 63
classification of, 62
of grasses, 53–54
groups of, 52
hyphal growth of, 63
marine, 55
pest control and, 59
entomogenous, 127–140
environment and, 110–111
food relationships with, 106–108
history of, 105–106
lichens, 85–94
microsporidia, 65–70
mycorrhizae, 73–82
mycorrhizae and conservation of, 81
interest in, 82
nutrition for, 106–108, 108t
pathogen groups of, 113–123
eumycota, 116–123
protozoa, 113
stramenopila, 113–116
as pathogen or saprobe, 136
plant disease symptoms caused by, 111–113
abnormal growth, 111
abscission, 111
hot tissue replacement, 111
necrosis, 111–112
permanent wilting, 112–113
plant pathogenic, 105–125
primary role of, 105
reproduction of
anamorph-teleomorph relationships in, 109
asexual, 108–109
sexual, 109
rust, 97
spore stages of, 121, 121t
survival of, 109–110
yeasts, 3–17

G

GAG. *See* Glycosaminoglycan
Galactose/*N*-acetyl-D-galactosamine (Gal/GalNAc), 323, 326, 330
Galls, 111
Gambian sleeping sickness. *See T. b. gambiense*
Gametogony, 294
Gamontogamy, 294
GAP. *See* General amino acid permease
GAPDH. *See* Glyceraldehyde–3-phosphate dehydrogenase
General amino acid permease (GAP), 3, 10
Genetic engineering, of yeasts, 13–14f
Genus Flamella, 203f, 207, 207t
Genus Gocevia, 203, 207t
Genus Paragocevia, 203, 207, 207t
Genus Stygamoeba, 203, 203f, 207t
Gephyramoeba spp., 209
Gephyramoebidae family, 203, 206t
Gephyrocapsa oceanica, 237f
blooms of, 242–244
GFP. *See* Green fluorescent protein
GGE. *See* Gross growth efficiency
Giardia intestinalis, 178f, 187, 328f
cell structure of, 183
clinical manifestations of, 327–328
genome of, 187
host immune response to, 328–329
humans and animals affected by, 188
MLCK and, 328
recent developments with, 332
Giardia lamblia, 329t
Giardiasis, 188
Giardiinae, 179, 180t, 181f
cell structure of, 183
GIPs. *See* Glucanase inhibitor proteins
Global positioning systems (GPS), 411, 416
Globigerinida order, foraminifera, 305–306t
Glomeribacter gigasporarum, 62
Glossina spp., 414–415
Glucanase inhibitor proteins (GIPs), 347, 351–352
Glyceraldehyde–3-phosphate dehydrogenase (GAPDH), 427, 432
Glycocalyx, 199
Glycogen, 130
Glycoprotein 63 (gp63), 338
Glycosaminoglycan (GAG), 385, 392
Glycosylphosphatidylinositol (GPI), 335
Goldacre, R. J., 208
Golgi apparatus, 7t
of *Leishmania*, 339
Gorgonian corals, *Aspergillus sydowii* and, 24–25, 25f
Gp63. *See* Glycoprotein 63
GPI. *See* Glycosylphosphatidylinositol
GPS. *See* Global positioning systems
Granuloreticulpodia, of foraminifera, 292, 292f
Grass
darnel ryegrass, *Neotyphodium occultans* association with, 56
dronkgras, *Neotyphodium melicicola* association with, 56
drunken horse grass, *Neotyphodium gasuense* association with, 56
endophytes
Clavicipitaceae and, 47–48
common associations of, 55–57
ergot alkaloids and, 59, 60f
fungal, 53–54
forest hedgehog grass, *Neotyphodium* association with, 57
huecu grass, *Neotyphodium tembladerae* association with, 57
perennial ryegrass, *Neotyphodium lolii* association with, 55–56
sleepygrass, *Neotyphodium chisosum* association with, 56
tall fescue, *Neotyphodium coenophialum* association with, 55
Grebecki, A., 198, 208
Green fluorescent protein (GFP), 6, 31f, 352
Gremmeniella abietina, 99, 100t
gRNA. *See* Guide RNA
Gross growth efficiency (GGE), 451, 458
Grunow, Albert, 250
Guide RNA (gRNA), 311
kinetoplastids and, 318
Guillardia theta, 433
Gymnamoebae. *See* Naked lobose amoebas

H

HABs. *See* Harmful algal blooms
Haptonema, 235–236
Haptophytes, 238–239
algal blooms and, 440–441, 441f
Harmful algal blooms (HABs), 435–436, 446–447
Hartmannellidae family, 201, 201f, 204t
Hartwell, Leland, 10
HAT. *See* Human African trypanosomiasis
Heat shock protein 70 (Hsp70), 397, 399
heavy metal pollutants
lichens and, 91
mycorrhizae and, 80–81
Hemibiotrophs, 108
Hemicellulase, 108t
Hemiselmis andersenii, 433
Heterakis gallinarum, 189
Heterobasidiomycetes, 120t
Heterokaryons, 27
Heterotricha class, ciliates, 220f, 221–222
Heterotrophic eukaryotes, in microbial food webs, 455–456
Heterotrophic flagellates (HF), 369
Heterotrophic nanoflagellates (HNF), 355, 361
Heterotrophic picoeukaryotes (HP), 355, 357
abundance and distribution of, 361
ecological role of, 362–363
Heterotrophic prokaryotes, in microbial food webs, 454

Hexamitidae, 179, 180t, 181f
cell structure of, 183
HF. *See* Heterotrophic flagellates
HGT. *See* Horizontal gene transfer
Hieracium pilosella, 230
High-nutrient, low-chlorophyll (HNLC), 435, 447
High-performance liquid chromatography (HPLC), 355, 358
for marine picoeukaryotes, 363
of prymnesiophytes, 366
Histidine (his), 279
Histomonas meleagridis, 189
Histoplasma capsulatum, 151–153, 152f
Histoplasmosis, 152f
causative organism of, 151
clinical features of, 152–153
diagnosis and treatment of, 153
epidemiology of, 151
HIV. *See* Human immunodeficiency virus
HMG-CoA. *See* 3-Hydroxy-3-methylglutaryl-coenzyme A
HNF. *See* Heterotrophic nanoflagellates
HNLC. *See* High-nutrient, low-chlorophyll
Holobasidiomycetes, 120t
Hoover, Richard B., 250
Horizontal gene transfer (HGT), 347
in oomycetes, 349
Host translocation (HT), 347, 349
hot tissue replacement, 111
HP. *See* Heterotrophic picoeukaryotes
HPLC. *See* High-performance liquid chromatography
HR. *See* Hypersensitive response
Hsp70. *See* Heat shock protein 70
HT. *See* Host translocation
Huecu grass, 57
Human African trypanosomiasis (HAT), 412
Human immunodeficiency virus (HIV), 143, 161
Human pathogens, *Aspergillus* as, 21–24, 22f
barriers preventing, 23–24
increase in, 22
Hutchinson, G. Evelyn, 451
HXGPRTase. *See* Hypoxanthine/xanthine/guanine phosphoribosyl transferase
Hyalolithus neolepis, 239–240
Hyaloplasm, 193
Hydrogenosomes, 185–187, 186f
3-Hydroxy-3-methylglutaryl-coenzyme A (HMG-CoA), 19
Hyperdermium spp., 43, 44f
Hypersensitive response (HR), 347
Hyphal cells
growth of, 63
structure of, 106
Hyphomycetes, 128
Hypocrella spp., 43–44f
Hypoxanthine/xanthine/guanine phosphoribosyl transferase (HXGPRTase), 385, 390
Hyrrokkin sarcophaga, 295

I

IAA. *See* 3-indoleacetic acid
ID. *See* Immunodiffusion
IELs. *See* Intraepithelial lymphocytes
IFN-γ. See *Interferon-γ*
IL. *See* Interleukin
IL12. *See* Interleukin 12
Immunodiffusion (ID), 143, 153
Indole diterpenoids, 46
endophytes and, 59, 60f
3-indoleacetic acid (IAA), 61, 61f
Innate immune system, 168
Inner-membrane complex, 390f, 391
Insect mycology, integration of multidisciplinary inputs to, 138–139
Interferon-γ (IFN-γ), 323, 326, 341
Interleukin (IL), 335
Interleukin 12 (IL12), 385, 394
Internally transcribed spacer (ITS), 213, 226, 291
Intestinal protozoa, 323–332
AIDS and, 324t
amoebae, 324t, 329–331
Apicomplexa, 324–326, 324t, 329t
flagellates, 324t, 327–329
Microsporidia, 324t, 326–327
morbidity and mortality from, 323
recent developments in, 331–332
taxonomic classification of, 324t
Intraepithelial lymphocytes (IELs), 68
Intraparasitophorous vacuolar network (IPN), 385, 392–393
Involutinida order, foraminifera, 306t
IPN. *See* Intraparasitophorous vacuolar network
Iron SODs (FeSODs), 263
in dinoflagellates, 266, 267f
Isozymogram, 224f
ITS. *See* Internally transcribed spacer

J

Jahn, T., 193
Jamot, Eugene, 419

K

Karyogamy, 12
Karyorelictea class, ciliates, 221
kDNA. *See* Kinetoplast DNA
Keller, Nancy, 33
Keratinization, 167
Kinetoplast DNA (kDNA), 338
Kinetoplastids, 317–319, 317f
glycosome in, 318–319
gRNAs and, 318
habitat of, 319
kinetoplast DNA structure in, 317–318, 318f
RNA editing in, 318, 339
Kluyver effect, 9t
Koch, Robert, 136
Krebs cycle, 9, 26, 318

L

LaeA, 23, 33
Lagenida order, foraminifera, 303t
Large subunit rDNA (LSU rDNA), 291, 308
LCL. *See* Localized cutaneous leishmaniasis
Leaf curl, 111
Leaf spot or bloch, 112
Lecanicillium, 43
Leishmania, 317f, 319, 321, 335–344
cellular biology of, 337–339
classification and morphology of, 335–336, 336f
control of, 344
diagnosis of, 343
epidemiology and disease of, 341–343
ER of, 339
flagellum of, 338
gene expression control in, 340
gene targeting in, 340
genomic organization of, 339
geographic distribution of, 342t
Golgi apparatus of, 339
gp63, 338
life cycle and ecology of, 336–337, 336f
LPG of, 337–338
major species of, 342t
mitochondria in, 338–339
mRNA processing of, 339–340
neutrophils and, 344
nucleus of, 338
pathogenesis and host response to, 340–341, 341f
plasma membrane of, 337
recent developments with, 344
trans-splicing of, 339–340
treatment of, 343–344
Leishmania enrietti, 339–340
Leishmania major, 337–338
Leotiomycetes, 118t
Leptocylindrus mediterraneus, 242
Leptomyxa spp., 209
Leptomyxidae family, 203, 206t
Letharia vulpina, 89
Lichens, 85–94
environment and, 94
diversity of, 85, 86f
environmental diversity of, 85, 86f
evolution of, 90–91
heavy metal pollutants and, 91
history of, 89–90
life span of, 90–91
moisture fluctuation in, 92–93
photobiont of, 86, 87f
nutrient exchange in, 92, 92f
pollution and, 91
plants compared to, 86
secondary metabolites of, 87, 88f, 93–94
symbiosis of, 92–94
terminology of, 89–90
thallus of, 85, 86f
appearance of, 89, 89f
organization of, 87–88f
Ligninase, 108t
Linneaus, 90
Lipase, 108t
Lipophosphoglycan (LPG), 335
of *Leishmania*, 337–338
Litostomatea class, ciliates, 221f, 222
Lituolida order, foraminifera, 301–302t
Livingstone, David, 414

Lobopodia, 198
Lobose amoebas. *See* Naked lobose amoebas
Localized cutaneous leishmaniasis (LCL), 335, 341, 343
Loculoascomycetes, 118t
LOHAFEX experiment, 449–450
Loline alkaloids, 46, 48
endophytes and, 60–61, 61f
Lorch, I. J., 208
Lower photic zone (LPZ), 235–236, 242, 243f, 244t
LPG. *See* Lipophosphoglycan
LPZ. *See* Lower photic zone
LSD. *See* Lysergic acid diethylamide
LSU rDNA. *See* Large subunit rDNA
Lynn, Denis, 219
Lysergic acid diethylamide (LSD), 105

M

MAC. *See* Membrane attack complex
Macrophages, 168
Malassezia folliculitis, 171
Malassezia furfur, 145
MALV. *See* Marine alveolates
Mannose binding lectin (MBL), 331
Marine alveolates (MALV), 355, 365, 365f
groups of, 367
Marine fungal endophytes, 55
Marine picoeukaryotes, 355–370
abundance and distribution of, 361–362
biogeography of, 367–368
biology of, 358–361
cell size of, 360–361
cell structure of, 356, 359–360
ciPCR on, 363–364, 364f
cloning and sequencing genes of, 363–364, 364f
cultured strains of, 358, 359t, 363
genome projects on, 368
18s rDNA libraries of, 364–365f, 365–368, 366t
environmental genomics of, 368
examples of, 357f
fingerprinting techniques for, 365
FISH for, 364
developments using, 369–370
HPLC for, 363
metagenomics for, 368
method-driven history of, 356–358, 357f
molecular tools for ecology study of, 363–365
physiological parameters of, 360, 361f
recent developments in, 368–370
SSU rDNA for, 363
TEM of, 356
Marine stramenopiles (MAST), 355, 365, 365f
groups of, 367, 367f
Marx, Don, 80
Massospora spp., 133
narrow host range of, 134
MAST. *See* Marine stramenopiles
Mast, S. O., 208
MAT-type-like loci (MTL), 143, 147, 157
MBL. *See* Mannose binding lectin
MedA. *See* Medusa mutants
Medusa mutants (MedA), 37
Melarsoprol, 408, 417
Membrane attack complex (MAC), 335, 337
MEMS. *See* Microelectromechanical systems
Mesodinium rubrum, 442
Messenger RNA (mRNA), 314, 318
Leishmania and, 339–340
mature, 340
Metabolites. *See* Secondary metabolites
Metagenomics, 464
for marine picoeukaryotes, 368
Metazooplankton, 442–443
Metschnikoff, Elie, 138–139
MHB. *See* Mycorrhizal helper bacteria
Michaelis-Menten equation, 360
Micheli, Pietro Antonio, 19, 89
Microbial eukaryotes
ancient asexuality and, 228, 231
autotrophic, 455
definition of, 228
heterotrophic, 455–456
putative asexual, 227–231
Microbial food webs, 452–465
autotrophic eukaryotes in, 455
autotrophic prokaryotes in, 454–455
in benthic habitats, 457–458, 457f
in biogeochemical cycling, 458–460, 459f
chemical interactions in, 462–463
components and pathways of, 452–454, 453f
heterotrophic eukaryotes in, 455–456
heterotrophic prokaryotes in, 454
in marine pelagic habitats, 456–457, 456f
modeling, 460–461f, 460–462
organization of, 451–452
recent developments in, 464–465
spatial structure of, 463–464
trophic level in, 453–454
understanding, 452
Microelectromechanical systems (MEMS), 249
Micromonas pusilla, 359
FISH for, 365
Micronemes, 390f, 391
Micropore, 390f, 391
Microsporidia, 65–70, 178, 324t, 326–327
AIDS and, 65
ecosystem functioning and, 65–66
genomes, 66, 67t, 68
in host cell invasion
host immune response in, 68–69, 69f
polar tube of, 66, 67f
spore structure of, 66, 67f
Microtubule-organizing center (MTOC), 191, 255
cytoplasmic, 203, 208
Middle photic zone (MPZ), 235, 242–243, 243f, 244t
MIF. *See* Migration inhibitory factor
Migration inhibitory factor (MIF), 335
Mildews
downy, 116
powdery, 118t, 119
Miliolida order, foraminifera, 302–303t
Mitochondria, 7t. *See also* Amitochondriate protists
Aspergillus nidulans DNA and, 27–28
of euglenozoa, 314
evolution of, 178
in *Leishmania*, 338–339
Mitochondrial DNA (mtDNA), 213, 225f, 226
Mitosomes, 185–187, 186f
Mixotricha paradoxa, 185
Mixotrophic dinoflagellates, 270–271, 274–275f
ML. *See* Mucosal leishmaniasis
MLCK. *See* Myosin light chain kinase
Molecular clock, 228
Monothalamids, 306
diversity in, 308
monotropoid mycorrhizae, 76
Motility organization vesicles (MOV), 291
MOV. *See* Motility organization vesicles
MPZ. *See* Middle photic zone
mRNA. *See* Messenger RNA
mtDNA. *See* Mitochondrial DNA
MTL. *See* *MAT*-type-like loci
MTOC. *See* Microtubule-organizing center
Mucosal leishmaniasis (ML), 335, 341, 343
Mycology, 105
insect, integration of multidisciplinary inputs to, 138–139
Mycorrhizae, 73–82
agriculture and, 79
arbutoid, 76
climate change and, 81
dark septate endophytes and, 76
developments in research of, 82
ecology of, 78–79
ectomycorrhizae, 75
endomycorrhizae
arbuscular mycorrhizae, 74–75
ericoid mycorrhizae, 75
as food source, 79
forestry and, 79–80
function of, 73–74, 76–77
nutrient acquisition in, 76–77
plant defense as, 77
water acquisition in, 77
fungal conservation and, 81
interest in, 82
global distribution of, 78
monotropoid, 76
orchid mycorrhizae, 75–76
plant communities influence of, 78
restoration from, 80
pollution and, 80–81
acidifying pollutants and, 80
heavy metal pollutants, 80–81
organic pollutants, 81
radionuclide pollutants, 81
soil nutrients and, 78
terminology of, 73
types of, 74–76
Mycorrhizal helper bacteria (MHB), 73–74
Mycotic mycetoma, 172
Myosin light chain kinase (MLCK), 323
Giardia intestinalis and, 328
Myriogenospora atramentosa, 44

N

NAD. *See* Nicotinamide adenine dinucleotide
Naegleria gruberi, 208–209
Naked lobose amoebas (gymnamoebae), 191–211
  Acanthamoebidae family, 202, 206t
  Amoebidae family, 201, 201f, 204t
  biogeography of, 210
  cell surface structure of, 199, 199f
  Cochliopodiidae family, 203, 203f, 207t
  combined morphological and molecular system of, 195–197t
  diversity of, 200–203, 201f, 203f, 204–207t, 207
  Filamoebidae family, 203, 207t
  Flabellulidae family, 202–203, 203f, 206t
  Genus Flamella, 203f, 207, 207t
  Genus Gocevia, 203, 207t
  Genus Paragocevia, 203, 207, 207t
  Genus Stygamoeba, 203, 203f, 207t
  Gephyramoebidae family, 203, 206t
  Hartmannellidae family, 201, 201f, 204t
  history of research in, 192
  importance of, 210–211
  Leptomyxidae family, 203, 206t
  life cycle of, 209–210
  lobopodia produced by, 198
  local-scale distribution of, 210
  morphological system of, 194t
  morphology of, 193, 198–199, 198–199f
  morphotypes of, 200, 200f
  movement of, 207–209, 209f
  nuclei in, 198, 199f
  Paramoebidae family, 201f, 202, 205t
  Pellitidae family, 202, 203f
  Rhizopoda incertae sedis, 203, 207t
  Stereomyxidae family, 203, 206t
  structure of, 191
  systematics and phylogeny of, 192–193, 194–197t
  Thecamoebidae family, 201–202, 201f, 204t
  uroidal structures in, 198, 199f
  Vannellidae family, 201f, 202, 204–205t
  Vexilliferidae family, 201f, 202, 205–206t
Nanotechnology, diatoms for, 249
  computing with, 258
  drug delivery in, 257–258, 257f
Nassophorea class, ciliates, 220–221f, 223
NCR. *See* Noncellular reaction
Necrosis, 111–112
Necrotrophs, 107–108
Needle cast, 112
Neotyphodium, 45
  animal reaction to, 47
  forest hedgehog grass association with, 57
  growth of, 54, 54f
  lox-toxicity, 58
  new associations and discoveries in, 62–63
Neotyphodium chisosum, 56
Neotyphodium coenophialum, 55
Neotyphodium gansuense, 56
Neotyphodium lolii, 55–56
Neotyphodium melicicola, 56
Neotyphodium occultans, 56
Neotyphodium tembladerae, 57
Nep1-like proteins (NLPs), 347, 351
Neutrophils, 168
  Leishmania and, 344
Nicotinamide adenine dinucleotide (NAD), 3
  regeneration of, 9
NirA, 29–30, 31f
Nitrogen
  Aspergillus nidulans utilization of, 29–30, 30–31f
  metabolism, by yeasts, 9–10
NLPs. *See* Nep1-like proteins
Noctiluca scintillans McCartney, 268–269, 268f
Nomuraea rileyi, 134
Noncellular reaction (NCR), 127, 129f
Nosema ceranae, 66, 67t
NPZ. *See* Nutrient-phytoplankton-zooplankton
Nuclear lamina, 199
Nucleomorphs, 433–434
Nucleus, 7t
  of ciliates, 213–214, 217
  of dinoflagellates, 263, 271, 274
  of euglenozoa, 314
  of foraminifera, 292–293
  of *Leishmania*, 338
  in naked lobose amoebas, 198, 199f
  Syndinium spp. division of, 276–277f
Nurse, Paul, 10
Nutrient-phytoplankton-zooplankton (NPZ), 451, 460
Nylander, William, 90

O

Octosporea bayeri, 66, 67t
ODC. *See* Ornithine decarboxylase
Odum, Eugene, 452
Odum, Howard, 452
Oleic acid, 129
Oligohymenophorea class, ciliates, 220–221f, 223–224
Olpidium brassicae, 116
Oomycetes (water mold), 113–114, 114t, 347–353
  adhesion, penetration, colonization of, 350–351
  agriculture impact of, 347–348
  biological features of
    general, 348–349
    unique, 349
  defense responses induced by, 351
  as effectors, 352
  environment impact of, 347–348
  evolutionary history of, 348–349
  genera of, 347, 348t
  genome structure of, 349
  HGT in, 349
  host enzyme inhibition and, 351–352
  infection cycle of, 349
  pathology of, 349–352
  sex and, 349–350
  taxonomic classes of, 349
Oospores, 110
Opals, diatoms compared to, 255
Open reading frame (ORF), 19, 28
Ophiostoma novo-ulmi, 100t, 101
Orchid mycorrhizae, 75–76
ORF. *See* Open reading frame
Organic acid, *Aspergillus* and production of, 26
Organic pollutants, 81
Oriental food, *Aspergillus* uses in, 25
Ornithine decarboxylase (ODC), 397, 409
Ostreococcus tauri, 356, 360
Oxymonads, 178
  cell organization of, 185, 185f
  example, 180t
  habitats of, 182
  subgroups of, 181, 181f

P

PacC, 30–32
Paclitaxel (Taxol), 59, 59f
Paecilomyces, 43
Page, F. C., 193, 200
PAMPs. *See* Pathogen-associated molecular patterns
Pan African Tsetse and Trypanosomosis Eradication Campaign (PATTEC), 411, 423
Pandora neoaphidis, 134
Pantin, C. F., 208
Parabasalids, 178
  cell organization of, 183–185, 184f
  evolution of, 178–179, 179f
  example, 180t
  habitats of, 182
  mitosis of, 184–185
  subgroups of, 179–180, 181f
Paracoccidtoides brasiliensis, 155–156, 155f
Paracocciodomycosis, 155f
  causative organism of, 155
  clinical features of, 156
  diagnosis and treatment of, 156
  epidemiology of, 155–156
Paraflagellar rod (PFR), 338
Paragymnodinium shiwhaense, 270
Paralytic shellfish poisoning (PSP), 435, 446–447
Paramoebidae family, 201f, 202, 205t
Parasitophorous vacuole membrane (PVM), 385, 391
  major features of, 392–393
Paraxial rod protein gene (PFR1), 313
PARP. *See* Pimaricin + ampicillin + rifampicin + pentachloronitrobenzene; Procyclic acid-rich protein
Pasteur, Louis, 138
Pasteur effect, 9t
Pateman, John, 27
Pathogen avirulence (*Avr*), 347, 352
Pathogen-associated molecular patterns (PAMPs), 143, 146
  TLRs identifying, 168
Pathogens. *See* Plant pathogens
PATTEC. *See* Pan African Tsetse and Trypanosomosis Eradication Campaign
Pavlova, 239
Pavlovophyceae, 239

PCD. *See* Programmed cell death
PCR. *See* Polymerase chain reaction
PdsA. *See* Phosphodiesterase
Pectinase, 108t
PEG. *See* Polyethylene glycol
Pelagomonas calceolata, 360
Pelagophyceae, 382
Pellicle, euglenids, 314–315, 315f
Pellitidae family, 202, 203f
Penard, E., 192
Penicilliosis
  causative organism of, 160
  clinical features of, 161
  diagnosis and treatment of, 161
  epidemiology of, 161
Penicillium marneffei, 160–161
Pentamidine, 407–408
PER. *See* Proplastidal endoplasmic reticulum
Peragallo, Hippolyte, 250
Peragallo, Maurice, 250
Peramine, endophytes and, 60, 61f
Perennial ryegrass, 55–56
Permanent wilting, 112–113
Peronosporaceae, 114t
Peronosporales family, 114t
Peroxisome, 7t
Pest control, 59, 135–136
Petalomonas cantuscygni, 319–320, 320f
Petch, Tom, 139
PFO. *See* Pyruvate: ferredoxin oxidoreductase
PFR. *See* Paraflagellar rod
PFR1. *See* Paraxial rod protein gene
PH. *See* Pleckstrin homology
pH, *Aspergillus nidulans* expression regulated by external, 30–32, 32f
Phaeocystis spp., 439f, 440–441
Phaeophyceae, 382–383
Phaeothamniophyceae, 382
Phosphatidyl inositolbisphosphate 2 (PIP2), 279, 285
Phosphatidyl inositolbisphosphate 3 (PIP3), 279, 285
Phosphodiesterase A (PdsA), 279
  cAMP degraded by, 284
Photoluminescence, 256
Photonic crystals, 256
Photosynthetic stramenopiles, 381–383
Phototrophic nanoflagellates (PNF), 355, 361
Phototrophic picoeukaryotes (PP), 355, 357
  abundance and distribution of, 361–362
  ecological role of, 362
Phyllopharyngea class, ciliates, 220–221f, 222–223
Phytophthora alni, 100t
Phytophthora cinnamomi, 99, 100t, 101
Phytophthora infestans, 115–116
  adhesion, penetration, colonization of, 350–351
  agriculture impact of, 347–348
  defense responses induced by, 351
  infection cycle of, 350
Phytophthora ramorum, 99, 100t
Picobiliphytes, 367
Picoeukaryotes. *See* Marine picoeukaryotes
Picophagus flagellatus, 360
Pimaricin + ampicillin + rifampicin + pentachloronitrobenzene (PARP), 105, 124
Pin1 gene, Alzheimer's disease and, 35
Pinguiochrysis pyriformis, 360
PIP2. *See* Phosphatidyl inositolbisphosphate 2
PIP3. *See* Phosphatidyl inositolbisphosphate 3
PKA-C. *See* Single catalytic subunit
PKA-R. *See* Single regulatory subunit
Plagiopylea class, ciliates, 220–221f, 223
Plankton nets, 240–241
Plant disease resistance (*R*), 347, 352
Plant pathogens
  anthropogenically generated changes in host distribution and, 98
  associations and evolutions of, 97
  countering invasive, 103
  emerging diseases from genetic change in, 99, 101
  fungi, 105–125
  host jumps and, 98–99
  hybridization between, 101
Plants
  biotrophs, Clavicipitaceae in, 43–45, 44–45f
  emerging diseases and, 97–98
  fungi causing symptoms of disease in, 111–113
    abnormal growth, 111
    abscission, 111
    hot tissue replacement, 111
    necrosis, 111–112
    permanent wilting, 112–113
  lichens compared to, 86
  mycorrhizae and defense of, 77
  mycorrhizae influence on, 78
    restoration from, 80
Plastids
  primary
    distribution of, 429t
    membranes of, 431
  protein import of, 429–430
  secondary
    distribution of, 429t
    diversity of, 430–431
    gene transfer and protein import in, 431
    membranes of, 431
    number of events with, 431–433
  stramenopiles, 377–379, 378t
Pleckstrin homology (PH), 279
Pneumocystis infections
  causative organism of, 164
  clinical features of, 165
  diagnosis and treatment of, 165
  epidemiology of, 164–165
Pneumocystis jiroveci, 164–165
PNF. *See* Phototrophic nanoflagellates
Pollution
  lichen photobionts and, 91
  mycorrhizae and, 80–81
    acidifying pollutants and, 80
    heavy metal pollutants, 80–81
    organic pollutants, 81
    radionuclide pollutants, 81
Polyethylene glycol (PEG), 143, 149
Polymerase chain reaction (PCR), 397, 403, 411, 417
Pontecorvo, Guido, 26
Postgaardi, 319, 319f
Powdery mildews, 118t, 119
PP. *See* Phototrophic picoeukaryotes
PPG. *See* Proteophosphoglycan
Prasinophytes, 365–366, 365f
Prestarvation factor (PSF), 279, 286
Primary endosymbiosis, 427–430, 428f, 429t
Pritchard, Bob, 27
Prochlorococcus spp., 454
Procyclic acid-rich protein (PARP), 397, 405
Programmed cell death (PCD), 435, 448
Prokaryotes
  autotrophic, 454–455
  heterotrophic, 454
Proplastidal endoplasmic reticulum (PER), 373
Prorocentrum micans, 269–270, 269–271f
Prostomatea class, ciliates, 221f, 223
Proteasome, 7t
Proteinase, 108t
Proteomics, 465
Proteophosphoglycan (PPG), 335, 337
Protistan grazers, 442
Protists
  algal blooms, 435–450
  amitochondriate protists, 177–189
  ciliates, 213–226
  coccolithophores, 235–247
  diatoms, 249–258, 382–383, 439–440, 439f, 444
  dictyostelids, 279–289
  dinoflagellates, 263–277, 441–442
  euglenozoa, 311–321
  foraminifera, 291–309
  intestinal protozoa, 323–332
  Leishmania, 317f, 319, 321, 335–344
  marine picoeukaryotes, 355–370
  microbial food webs, 452–465
  naked lobose amoebas, 191–211
  oomycetes, 113–114, 114t, 347–353
  putative asexual microbial eukaryotes, 227–231
  sleeping sickness, 411–425
  stramenopiles, 113–116, 373–383
  terminology of, 264
  Toxoplasma, 385–395
  trophic transfers involving, 460
  Trypanosoma spp., 397–409
Protocruzia, 218
Protozoa, 113
  intestinal, 323–332
Prymnesiophytes, 366
Prymnesium, 239–240
Pseudonitzschia spp., 439f, 440
PSF. *See* Prestarvation factor
PSP. *See* Paralytic shellfish poisoning
PTX3, 23
Puccinia komarovii, 100t
Puccinia lagenophorae, 99, 100t
Puccinia malvacearum, 99, 100t
Puccinia psidii, 100t

Putative asexual microbial eukaryotes, 227–231
ancient asexuals, 228, 231
roots of, 229–231
PVM. *See* Parasitophorous vacuole membrane
Pyrenomycetes, 118t
Pyruvate: ferredoxin oxidoreductase (PFO), 177, 186, 186f
Pythiaceae, 114t
Pythium spp., 114–115, 115f, 348

R

R. See Plant disease resistance
Radionuclide pollutants, 81
Random amplified polymorphic DNA (RAPD), 3, 105, 213
of *Euplotes*, 224f
RBBB. *See* Right bundle branch block
Reactive oxygen species (ROS), 19, 23
Redfield, Alfred C., 459
Redfield ratio, 459
REMI. *See* Restriction enzyme-mediated integration
Reproduction, 229. *See also* Sex
Reservoirs of disease, 413
Restriction enzyme-mediated integration (REMI), 279
Restriction-fragment-length-polymorphism (RFLP), 105, 385, 388
Reticulofenestra sessilis, 242
Reticulomyxa filosa, 292f
RFLP. *See* Restriction-fragment-length-polymorphism
Rhizaria, 299, 374
Rhizoctonia spp., 120–121
Rhizomorphs, 110
Rhizonin, 62
Rhizopoda incertae sedis, 203, 207t
Rhizopus, 145, 163–164, 164f
Rhizosolenia spp., 439f, 440
Rhizoxin, 62
Rhodesian sleeping sickness. *See T. b. rhodesiense*
Rhodotorula rubra, 8t
Rhoptries, 390f, 391
Ribosomal RNA (rRNA), 311
Euglena spp. and, 316
Right bundle branch block (RBBB), 397
RNA editing, in kinetoplastids, 318, 339
Robertinida order, foraminifera, 306t
Root rot, 112
Roper, Alan, 27
ROS. *See* Reactive oxygen species
Rotaliella elatiana, 295
Rotaliida order, foraminifera, 304–305t
rRNA. *See* Ribosomal RNA
Rust fungi, 97
spore stages of, 121, 121t
RXLR effectors, 352

S

Saccharomyces cerevisiae, 3, 149. *See also* Yeasts
in agriculture, 16
alcoholic fermentation of, 9
cellular age in, 11
genetic manipulation of, 12–13
genome sequencing of, 13
growth response to oxygen availability of, 8t
industrial commodities produced by, 14t
life cycle of, 12, 12f
shape of, 5, 5t
SALT. *See* Skin-associated lymphoid tissue
Saprolegniales, 114t
Saprophytes, nutrition of, 107
Saprotrophs, 41–43
Scabs, 112
Scale insects, Clavicipitaceae on soft-bodied, 43–44f
Scales, 379
Scanning electron microscope (SEM), 235, 237, 237f, 239
of diatoms, 251f, 257f
Schaeffer, A. A., 192–193
Schizosaccharomyces pombe, 10. *See also* Yeasts
genome sequencing of, 13
growth of, 11
life cycle of, 12
Schmidt, Adolf, 250
Schwendener, Simon, 90
Sclerotium, 43–44
of ergots, 46
SDF–2. *See* Spore differentiation factor 2
SDV. *See* Silica deposition vesicle
Sea-viewing Wide Field of view Sensor (SeaWiFS), 235, 244
Secondary endosymbiosis, 429t, 430–434
Secondary ion mass spectrometry (SIMS), 254
Secondary metabolites
Aspergillus as, 26
Aspergillus nidulans regulating, 32–33
Clavicipitaceae and
biological activities of, 46–47
sources of, 45–46
endophyte sources of, 58–61
in entomogenous fungi, 132
of lichens, 87, 88f, 93–94
SEM. *See* Scanning electron microscope
Seravin, L. N., 208
Serum resistance associated gene (SRA), 411, 420
Sex. *See also* Putative asexual microbial eukaryotes
ancient asexuals, 228, 231
Aspergillus fumigatus and, 229
ciliates distribution of, 230f
in Colpodea class, ciliates, 227, 229
definitions of, 228–229
in dinoflagellates, 269–270, 271f, 272–273t
foraminifera and, 294–295, 294f
oomycetes and, 349–350
reversing loss of, 229–231
Toxoplasma stages of, 386–387
Signal transducers and activators of transcription (STAT), 385
Silica deposition vesicle (SDV), 254
Silica transport vehicles (STVs), 252, 254–255
Silica transporter (SIT), 255
Silicoloculinida order, foraminifera, 306t
Simpicillium, 43
SIMS. *See* Secondary ion mass spectrometry
Single catalytic subunit (PKA-C), 279
activation of, 288
Single regulatory subunit (PKA-R), 279
activation of, 288
SIT. *See* Silica transporter
Skeletonema spp., 439f, 440, 445
Skin-associated lymphoid tissue (SALT), 168–169
Sleeping sickness, 411–425
background of, 411–412, 412f
burden of, 418, 418f
economic, 418–419
causes of, 411–412
clinical signs of, 416
control of, 419–420, 421f
diagnostics for, 422–423
future prospects for, 420, 422–423, 425
Stamp Out Sleeping Sickness campaign, 423, 424f, 425
diagnosis of, 416–417
distribution of, 412–413, 413f
epidemics of, 415–416, 415f
epidemiology of, 412–415
reservoirs of disease and, 413
transmission cycles of, 414–415, 414f
treatment of, 417
future of, 422
Sleepygrass, 56
Small, Eugene, 219
Small subunit rDNA (SSU rDNA), 291, 307–308
for marine picoeukaryotes, 363
Small subunit ribosomal RNA (SSU rDNA), 213
of ciliates, 219, 220f
Smith, William, 250
SML. *See* Surface mixed layer
SOD. *See* Superoxide dismutase
Soft rot, 112
Soil nutrients, mycorrhizae and, 78
Solenicola setigera, 242
SP. *See* Synthetic pyrethroids
Sphaerotheca mors-uvae, 100t, 101
Spirillinida order, foraminifera, 303t
Spironucleus salmonicida, 187
animals affected by, 188
Spirotrichea class, ciliates, 220–221f, 222
Spirotrichonymphida, 180, 180t, 181f
Spore differentiation factor 2 (SDF–2), 279, 282
Sporothrix schenckii, 156–157
Sporotrichosis, 171–172
causative organism of, 156
clinical features of, 156–157
diagnosis and treatment of, 157
epidemiology of, 156
Sporulation cascade, 288
Spring blooms, 443–445
SRA. *See* Serum resistance associated gene
SSU rDNA. *See* Small subunit rDNA; Small subunit ribosomal RNA
Stamp Out Sleeping Sickness campaign, 423, 424f, 425

Starvation. *See also* Prestarvation factor
surviving, 280f
STAT. *See* Signal transducers and activators of transcription
Stem rust of wheat, 122
Stereomyxidae family, 203, 206t
Stramenopiles, 113–116, 373–383
cell coverings of, 378t, 379
classification of, 380–383
colorless groups of, 379–381, 381f
diversity of, 376–380
environment and, 377
evolutionary history of, 373–376, 374f
examples of, 377f
flagella of, 379–380, 380t
fossil record and divergence times for, 374–376, 376f
origin of, 374, 375f
Phaeophyceae, 383
photosynthetic, 381–383
plastids, 377–379, 378t
Strobilidium spp., 213
Strombidium spp., 213
Strongwellsea spp., 131, 131f, 133
Stunted mutants (StuA), 37
STVs. *See* Silica transport vehicles
Subcutaneous mycoses, 171–172, 172t
Sugar
yeast metabolism of, 9, 9t
yeast transport of, 8–9
Superoxide dismutase (SOD), 263
in dinoflagellates, 266, 267f
Suramin, 408
Surface mixed layer (SML), 435
depth of, 436
nutrients mixed into, 437
seasons and, 437
Sverdup's critical depth hypothesis, 449
Symbiomonas scintillans, 360
Synchococcus spp., 454
Syndinium spp., 270, 275f
nuclear division in, 276–277f
Synthetic pyrethroids (SP), 411, 420
Syracolithus schilleri, 237f

T

T. b. gambiense, 411
clinical signs of, 416
control of, 419
future of, 423
distribution of, 412
reservoirs of, 413
transmission cycles of, 414–415, 414f
treatment of, 417
T. b. rhodesiense, 411
clinical signs of, 416
control of, 419–420, 422f
future of, 423, 425
distribution of, 412
reservoirs of, 413
transmission cycles of, 414–415, 414f
treatment of, 417
Tall fescue, 55
Taxol, 59, 59f
TCA. *See* Tricarboxylic acid
Tectum, 199
Teleomorph state, 109
TEM. *See* Transmission electron microscope
TEP. *See* Transparent expolymer particles
Terminal-restriction fragment length polymorphism (T-RFLP), 355, 365
Textulariida order, foraminifera, 302t
Thalassiosira spp., 242
Thaxter, Roland, 138
Thecamoebidae family, 201–202, 201f, 204t
Thrombospondin-related anonymous protein (TRAP), 385, 392
TIC. *See* Translocons of the inner chloroplast membrane
Timberlake, Bill, 27
Tinea nigra, 171
Tinea pedis (athlete's foot), 170
Tinea unguium, 170
Tinea versicolor, 171
TLRs. *See* Toll-like receptors
TNF-α. *See* Tumor necrosis factor-α
TOC. *See* Translocons of the outer chloroplast membrane
Toll-like receptors (TLRs), 143, 146
PAMPs identified by, 168
Torrubiella spp., 137
Toxoplasma, 385–395
AIDS and, 387
behavioral effects of, 394
chemotherapy and, 395
diagnosis of, 387
food and, 388
genetics of, 389–390
genome and gene expression of, 389
host response to, 393–394
genetics of, 394
in humans, 386
immunization studies on, 394
lytic cycle of, 391–393, 392f
attachment, 392
egress, 393
replication, 392–393, 393f
major genotypes of, 388–389
molecular genetic tools for study of, 390
organelles of, 390–391, 390f
public health and, 388
improvements for, 394
sexual stages of, 386–387
strain-specific virulence of, 389
symptoms of, 387
treatment of, 387–388
vaccination and, 394–395
Toxoplasma gondii, 325
classification of, 386
host family of, 385
life cycle of, 386–387, 386f
Toxoplasmosis, 385–395
in animals, 386
Translocons of the inner chloroplast membrane (TIC), 427, 430
Translocons of the outer chloroplast membrane (TOC), 427, 430
Transmission electron microscope (TEM), 177, 235, 237, 239, 355
of marine picoeukaryotes, 356
Transparent expolymer particles (TEP), 451, 464
Transporters, *Aspergillus nidulans* regulating, 32, 32f
TRAP. *See* Thrombospondin-related anonymous protein
T-RFLP. *See* Terminal-restriction fragment length polymorphism
Tricarboxylic acid (TCA), 177, 186
Trichomonadida, 180, 180t, 181f
cell structure of, 183–184
Trichomonas gallinae, 189
Trichomonas suis, 188–189
Trichomonas vaginalis, 178, 187
humans affected by, 188
Trichonymphida, 180, 180t, 181f
Trimastix, 180t, 181, 181f
Triparma spp., 369
Trochamminida order, foraminifera, 302t
Trypanosoma brucei, 339, 411
clinical manifestations of, 407
diagnosis of, 407
epidemiology of, 405–406, 406t
life cycle of, 404–405, 404f
pathology and pathogenesis of, 406–407
prevention of, 409
subspecies of, 404
treatment of, 407–409
Trypanosoma cruzi, 397–398, 412
Chagas' disease, 400–402, 401f
clinical manifestations of, 401–402
diagnosis of, 403
epizootiology and epidemiology of, 400–401
life cycle of, 398–399f, 398–400
pathogenesis of, 402–403
prevention of, 404
transfusion-associated infection of, 402
transmission of, 398
treatment of, 403–404
Trypanosoma spp., 397–409
Trypanosomes, 412
Tumor necrosis factor-α (TNF-α), 335

U

Uncinula necator, 100t
Upper photic zone (UPZ), 235, 242, 243f, 244t
UPRTase. *See* Uracil phosphoribosyl transferase
UPZ. *See* Upper photic zone
Uracil phosphoribosyl transferase (UPRTase), 385, 390
Urediniomycetes, 120t
Ustilaginomycetes, 120t

V

Vacuole, 7t
van Heurck, Henri, 250
van Leeuwenhoek, Anthony, 314
Vanella spp., 210
Vannellidae family, 201f, 202, 204–205t
Variable series of antigenic types (VATs), 412

Variant surface glycoprotein (VSG), 311, 319, 397, 405
Variant surface protein (VSP), 323, 329
VATs. *See* Variable series of antigenic types
VeA1 mutation, 37–38
Veterinary medicine, *Aspergillus* in, 24
Vexilliferidae family, 201f, 202, 205–206t
Visceral leishmaniasis (VL), 335, 341, 343
Volatile organic compound (VOC), 51, 59
VSG. *See* Variant surface glycoprotein
VSP. *See* Variant surface protein

W

Wallich, G. C., 192
Wart, 111
Vater mold. *See* Oomycetes
t mutants (WetA), 37
es' broom, 111

extract peptone glucose (YEPG), 3, 7
t nitrogen base (YNB), 3, 7
sts
AFM of, 6, 6f
biodiversity of, 4
budding, 10–11, 10–11f
multilateral, 11, 11f
carbon metabolism by, 8–9
sources for growth, 8
cell structure of, 5–6
functional components of, 7t
cellular characteristics and shapes of, 5, 5t
culture media of, 7–8
cytology methods for, 5–6
definition and characterization of, 3
ecology of, 4–5
environmental and agricultural significance of, 16
fission, 11–12f
in food chain, 4–5
in food production and spoilage, 15t
genetic engineering of, 13–14f
genetic manipulation of, 12–13, 13–14f, 14t
genome and proteome projects with, 13
growth factors, 7
growth of, 10–12
habitats of, 4, 4t
industrial significance of, 14, 14t, 15f
life cycle of, 12–13f
medical research and, 16, 16t
microbial ecology of, 5
nitrogen metabolism by, 9–10
nutritional requirements of, 6–7
physical growth requirements for, 8, 8t
population growth of, 11–12
science and technology progress in, 16–17
subcellular architecture and function of, 6, 7f
sugar metabolism in, 9, 9t
sugar transport in, 8–9
taxonomy of, 3, 4t
vegetative reproduction in, 10–11, 11–12f
YEPG. *See* Yeast extract peptone glucose
YNB. *See* Yeast nitrogen base

Z

Zoospores, 108–109
Zygomycetes, 128, 146
Zygomycosis, 164f
causative organisms of, 163
clinical features of, 163–164
diagnosis of, 164
Zygospores, 110